THE FUNGAL SPORE

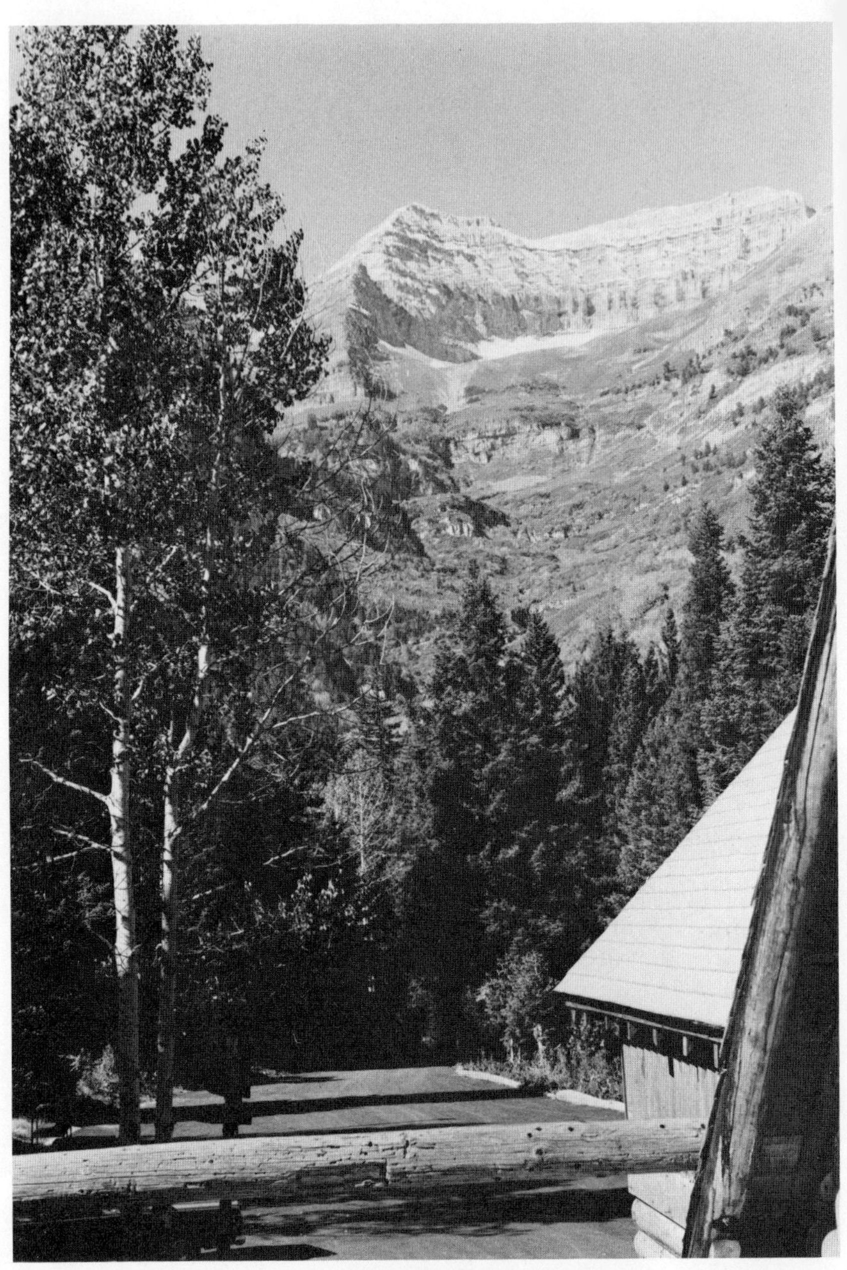

Mt. Timpanogos from Timp Lodge

THE FUNGAL SPORE

Form and Function

Edited by

DARRELL J. WEBER

WILFORD M. HESS

Department of Botany and Range Science
Brigham Young University
Provo, Utah

A WILEY-INTERSCIENCE PUBLICATION

JOHN WILEY AND SONS

New York • London • Sydney • Toronto

Library of Congress Cataloging in Publication Data:

International Fungal Spore Symposium, 2d, Brigham
Young University, 1974.
The fungal spore.

"A Wiley-Interscience publication."
Includes index.

1. Fungi—Spores—Congresses. I. Weber, Darrell, J., 1933- II. Hess, Wilford, M., 1934-

III. Title. [DNLM: 1. Spores, Fungal—Congresses.
W3 IN697 1974f/QE180 156 1974f]
QK601.I55 1974 589'.2'04334 75-38889
ISBN 0-371-92332-X

Printed in the United States of America

10 9 8 7 6 5 4 3 2 1

As we reflected on the research developments relating to ultrastructural and biochemical aspects of fungal spores, two scientists stood out as pioneers in these areas of research. Therefore it is only natural that this symposium on the form and function of fungal spores be dedicated to Professors Lilian Hawker and David Gottlieb.

Over a period of many years Professor Hawker has written significant papers, review articles, and books on the physiology and ultrastructure of fungal spores, and the First International Fungal Spore Symposium was organized under her direction. However, those who know her personally also consider her pleasant and enthusiastic personality as an outstanding trait.

Professor Gottlieb was among the first investigators to use biochemical techniques to study fungal spores. His enzymatic studies with metabolic pathways in fungal spores are now classic in the field. He has continued to produce a long list of significant contributions on the biochemistry of fungal spores.

It is with great pleasure that we dedicate this volume to Professors Lilian Hawker and David Gottlieb.

D.W.
W.M.H.

PREFACE

Almost a decade has passed since the First International Fungal Spore Symposium was held in Bristol, England. During this time there has been a tremendous increase in ultrastructural and biochemical research on fungal spores. This research has been conducted by scientists from many different disciplines who publish their papers in a wide variety of journals. These scientists attend different scientific meetings and seldom have an opportunity to interact with other scientists who are investigating fungal spores. One of the goals of the Second International Fungal Spore Symposium was to bring together scientists who are actively involved in form and function investigations of fungal spores. Another goal of the symposium was to have the participants review the literature in their respective areas of specialization so the proceedings of the symposium would be a useful reference and would also contribute findings about recent investigations.

There are many individuals and organizations that have contributed to the success of the Second International Fungal Spore Symposium. Although it is not possible to list all who assisted, we would like to give special recognition to Brigham Young University for providing financial support and for the use of Timp Lodge for the symposium. Thanks is extended to Gary Bascom and his associates from Brigham Young University Special Courses and Conferences for attending to many of the administrative details, and the Brigham Young University Food Services for providing meals. We are grateful to Kimball Harper, Chairman of the Department of Botany and Range Science at Brigham Young University for providing resources to assist with the conference and for his assistance in the preparation of the symposium proceedings for publication. We particularly appreciate the financial grant from the Research Corporation which made it possible to bring outstanding scientists from many parts of the world.

We also acknowledge the assistance of the four graduate students, John Gardner, George Lawler, Mark Richtsteig, and Dennis Madsen, who were the chauffeurs for the many trips from Provo to the lodge, and the technical assistance of Connie Swensen for the many details to which she attended. A dosen typists worked on the camera-ready copies of the manuscripts; we wish to especially thank Nina Lewis, Nancy Young, and Tracy Lundquist for their efforts.

We realize that many individuals were involved in the final publication of the proceedings, but we want to particularly thank Suzanne Ingrao for her help, and Wiley-Interscience for publishing the proceedings. Finally we wish to recognize the symposium participants for their presentations and for the stimulating discussions.

DARRELL WEBER
W. M. HESS

Provo, Utah
October 1975

CONTENTS

A DORMANCY AND ACTIVATION OF FUNGAL SPORES

B GERMINATION OF FUNGAL SPORES

C FORM AND FUNCTION OF FUNGAL SPORES

THE FUNGAL SPORE

A

Dormancy and Activation of Fungal Spores

CHAPTER 1 THE DORMANT SPORE

Lillian E. Hawker and M. F. Madelin
University of Bristol

INTRODUCTION

When the first symposium on the fungus spore was held at the University of Bristol, England, with the generous support of the Colston Research Society of the City of Bristol (Madelin, 1966a) the contributors were able to survey the whole field from the initiation of the spore, through its development and maturation to discharge, dispersal, and ultimate germination and to give some consideration to the significance of spores for man, his crop plants, and domestic animals. Since then, active research has continued on all the topics we discussed in 1966, and so great is the volume of recent literature that the organizers of this second symposium have wisely limited its scope to the mature spore and its germination.

A dormant spore may be considered in relation to its (a) external shape and dimensions, morphology of the surface, and physical properties; (b) internal structure and organization; (c) chemical composition (both of the whole spore and of its component parts); and (d) biological function. We shall describe it in these several ways and then seek to interrelate and interpret these various facets, but first we must define a dormant spore.

Sussman (1965) defined dormancy as "any rest period or reversible interruption of the phenotypic development of the organism." He went on to distinguish between constitutional dormancy of a propagule, a condition in which some innate property such as "a barrier to the penetration of nutrients, a metabolic block or the production of a self-inhibitor" delays development, and exogenous dormancy where delay in development is due to unfavorable external conditions. It is not always possible to make a clear-cut distinction between these two types of dormancy.

Sussman defined also maturation as "the complex of changes associated with the resting stage of dormant organisms or of the germinable stage in those without a dormant period," and after-ripening as "that part of the dormant period during which the changes occur which lead to germination."

Gregory (1966), in his introductory paper to the Colston Symposium on the fungus spore, considered spores to be of two main types: those which become detached and are dispersed from their place of origin and which usually germinate readily in suitable conditions (xenospores), and those remaining in their place of origin (usually still attached to the parent hypha), the majority of which need a resting period before germination or germinate only after the application of some specific stimulus or counter inhibitor (memnospores). These two categories correspond closely, but not exactly, with spores showing exogenous and constitutional dormancy, respectively. Spores can be divided more sharply into two classes by Gregory's classification than by Sussman's definitions.

Physiologically, a dormant spore is a quiescent phase situated in the life cycle of a fungus between two phases of active growth. It arises from cells or hyphae which were sites of intense metabolic activity and synthesis while they underwent special differentiation, and it is able to generate germ tubes or even zoospores when it germinates, but is itself in a state of almost suspended animation.

The principal role of the spore in the life cycle of the fungus is to provide for the survival, and usually also the dispersal, of the species. Spores may have to survive heat, cold, soaking, desiccation, radiation, exogenous chemical influences, physical pressure, abrasion, and attack by other microorganisms. Dispersal spores need to have properties that are related to those of the dispersal agency, which may be a fluid (air or water) or another organism. Size and shape, density, and "wettability" are thus important.

Though spores may display features that equip them to cope with the stresses of survival and dispersal, they must not be so insulated from the environment that they cannot detect and respond to circumstances likely to afford an opportunity to resume vegetative growth. It is not surprising that evolution has led to the appearance of mechanisms that protect the spore against responding to possibly false alarms. Such mechanisms principally involve the dependence of germination on two sequential "signals" from the environment, the first, the dormancy-breaking one, priming the spore for receipt of a second signal, which is the presence of congenial physiological conditions that evoke the actual events of germination.

As Gregory (1952, 1966) makes clear, fungus spores are not necessarily analogous structures; moreover, their basic biological functions may be different. Further, they very obviously are not homologous structures. The conidia (sporangia) and oospores of Oomycetes; the sporangiospores, conidia, and zygospores of Zygomycetes; the conidia and ascospores of Ascomycetes; the basidiospores, conidia, aecidiospores, uredospores, and teleutospores of Basidiomycetes; the conidia of Deuteromycetes; and the chlamydospores which appear in very diverse fungi, although all dormant spores in one or other of the senses defined by Sussman (1966), are very obviously different in origin. Even the conidia of higher fungi are of many ontogenetically distinguishable types (Hughes, 1953; Hawker et al., 1970; Kendrick, 1971; Subramanian, 1962). Despite such a diversity of processes by which discrete resting cells can be generated, fungus spores are a readily recognized class of biological objects, a situation which surely points to the powerful influence that adaptation has had in shaping them.

THE DORMANT SPORE

EXTERNAL FEATURES

As is pointed out by Gregory (1973), developing a view expressed by Savile (1954):

> "colour, size, shape, and texture of fungal spores must be looked upon as probably functional adaptations resulting from interplay of tensions in growth, modified in evolution by requirements of the liberation process, of flotation in air, and ultimately of deposition and survival."

Size, Shape, and Color

Nearly all spores, with a few exceptions such as the sexually produced zygospores of the Mucorales, are initially single spherical or ellipsoidal cells which are hyaline and smooth walled. Many remain so, but many become septate, assume a complex shape, develop surface ornamentation (spines, warts, ridges, etc.), and/or become pigmented.

Mature spores are of a wide range of size. Some, such as the conidia of *Beauveria* spp., are only ca. 3 μm in diameter, but some, such as the ascospores of some species of *Tuber* or of certain lichens, the zygospores of the Mucorales, or the chlamydospores of some species of *Endogone* may reach a diameter of 200 μm or more. Most are around 10 μm in diameter.

The taxonomic literature abounds with data on the linear dimensions of spores, but few workers have described their size in terms of volume. McCallan (1958) listed spore volumes ranging from 5.06 μm^3 in *Cephalosporium acremonium* up to 15,600 μm^3 in *Uromyces caryophyllinus*, a 3000-fold range. The proportion of the total volume of a spore that is occupied by the cytoplasm is easily seen to differ in different species, but even in an individual spore the volume of the lumen sometimes diminishes as it ages. Campbell (1970a) illustrated the centripetal thickening of the wall of ageing conidia of *Alternaria brassicicola*.

Most mature spores retain their original globose or ellipsoidal shape, but others during development assume more complex shapes. Some become spindle-shaped (e.g., conidia and ascospores of the apple scab fungus, *Venturia inaequalis*), crescent-shaped (e.g., conidia of *Fusarium* spp.), kidney-shaped (e.g., species of

Melanographium and *Virgaria*), star-shaped (e.g., many aquatic Hyphomycetes; Ingold, 1959), filiform or needle-shaped (e.g., conidia of *Septoria*, ascospores of *Cordyceps* spp.), or even coiled like a spring (e.g., species of *Helicoon* and *Helicodendron*).

Many species are septate, giving two-celled spores (e.g., the teleutospores of *Puccinia*), a single row of cells (e.g., conidia of *Helminthosporium* spp. or *Fusarium* spp.; ascospores of some species of *Geoglossum*), or they may be septate in more than one plane (e.g., conidia of *Alternaria* spp.). Occasionally the spore is a cluster of cells (e.g., the teleutospore of *Triphragmium ulmariae*). In some multicellular spores one or more of the cells may be infertile (e.g., conidia of *Pestalotia* spp.) or the spores may be aggregated in balls (e.g., smut spores of *Urocystis*).

The most common type of pigmentation is the development of the black pigment melanin, but many spores are brightly colored (e.g., the blue-green conidia of *Penicillium* spp., the pink ones of *Fusarium culmorum*, the orange or rust-red uredospores of many rusts, and the pink or yellow-brown basidiospores of some genera of agarics). The pigment may be situated in the spore wall or in the interior. The nature of these pigments is imperfectly known and will be discussed in the section dealing with the chemical composition of spores.

Surface Configuration

The surface ornamentation of spores occurs at different orders of magnitude. Ornamentation of the greatest magnitude comprises appendages (Fig. 1), as, for example, on the ascospores of *Podospora* and numerous marine fungi, on the conidia of many Deuteromycetes, such as *Pleiochaeta* and *Pestalotia*, and also those of Trichomycetes of the order Harpellales, and on the sporangiospores of various Zygomycetes, including *Gilbertella*, *Choanephora*, and *Blakeslea*. Ornamentation of a smaller magnitude, still discernible under the light

---

anserina to show primary appendage cut off by secondary wall layer laid down inside spore (after Beckett et al., 1968). (I) *Halosphaeria torquata* and (J) *Herpotrichiella ciliomaris*, ascospores of aquatic species, bearing hairlike appendages. After Ingold (1971).

(K--N) Spores of Trichomycetes (after Manier , 1969). (K) *Harpella melusinae*; (L) *Smittium avernense*; (M) *Stachylina* sp; (N) *Genistella ramosa*.

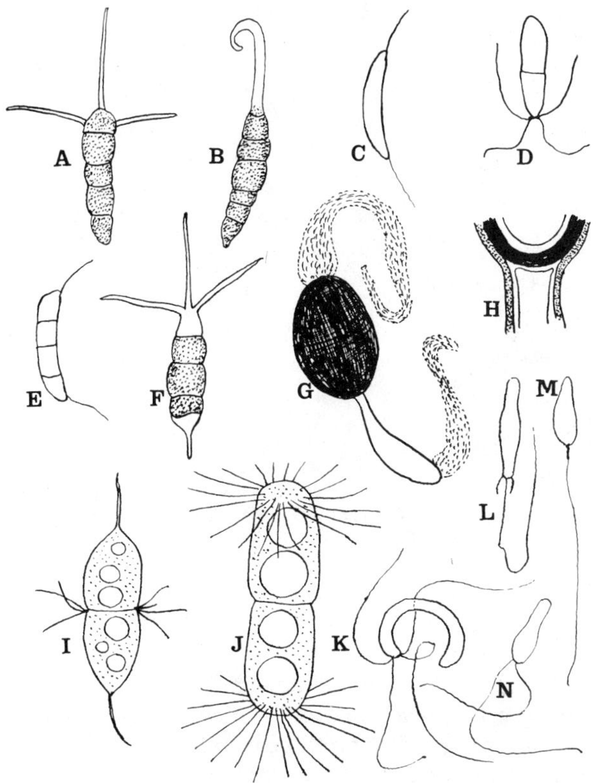

Fig. 1 Examples of spore appendages (not drawn to scale).

(A--B) Conidia of Hyphomycetes: (A) *Pleiochaeta setosa*; (B) *Ceratophorum uncinatum*.

(C--E) Conidia of Sphaeropsidales: (C) *Dinemasporium* sp; (D) *Robillardia muhlenbergiae*; (E) *Discosia maculicola*.

(F) Conidium of *Pestalotia macrotricha* (Melanconiales). (A--F) after Barnett (1955).

(G--J) Ascospores of ascomycetes; (G) *Podospora fimicola* with primary basal appendage and secondary apical and basal gelatinous appendages (after Ingold, 1971). (H) *Podospora*

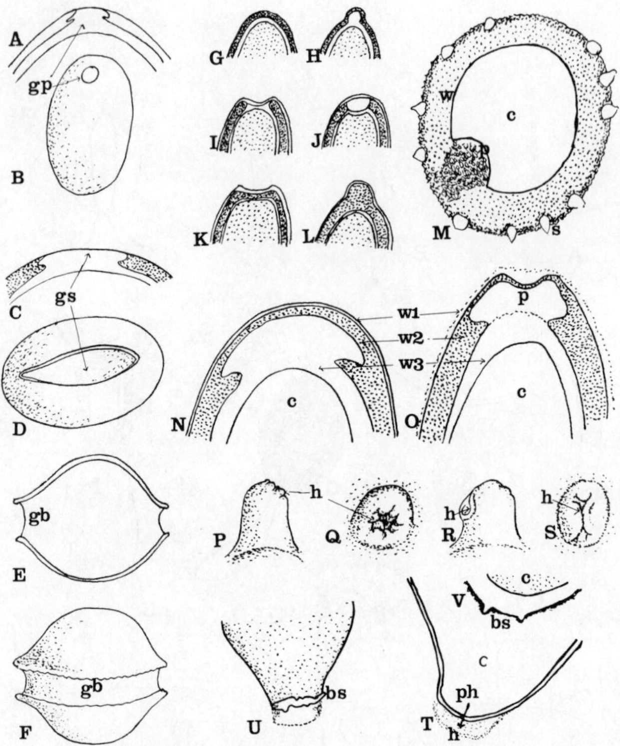

Fig. 2 Examples of germ "pores," abscission scars and hilums (not drawn to scale). gp = germ pore, gb = circular germination band, gs = germ slit, p = plug, bs = birth scar (abscission scar), h = hilum, W_1, W_2 and W_3 = outer, middle and inner layers of spore wall respectively, s = surface spines, c = cytoplasmic contents of spore, ph = parahilar electron transparent region.

(A--F) Germ "pores" of Ascomycetes, after Melendez-Howell (1969). (A) section through simple apical germ "pore" of ascospore of *Chaetomium* sp. (B) ascospore of same species tilted to show surface view of "germ pore." (C) section through germ "slit" of ascospore of *Xylosphaera polymorpha*. (D) ascospore of same species to show surface view of "slit." (E) section through ascospore of *Eurotium* sp. showing thin wall over germination "ring." (F) ascospore of same species to show circular germination ring or band.

microscope, comprises the ridges, spines, warts, reticulations, and pits that are characteristic features of many spores. The scanning electron microscope reveals fine ornamentation on some spores which under the light microscope appear to be smooth, for example, the basidiospores of *Hymenogaster luteus* (Hawker, 1975). Electron microscopy has also revealed even finer ornamentation comprising fine-structural patterns or texturings (Plate 3, Fig. 17).

Many spores also have surface features such as the scars, pegs, or tubes which mark the sites of attachment of many exogenous spores, for example, conidia, rust spores, and basidiospores; and the germ pores (or papillae or plugs) whose detailed structure within the spore wall is elucidated best by thin sections and electron microscopy (Fig 2; Plate 1, Figs. 1 and 4; Plate 2, Figs. 10, 12, 13, 14; Plate 3, Fig. 15).

(G--O) Germ "pores" of Basidiomycetes.

(G--L) Longitudinal sections of apices of basidiospores of genera of agarics; G with no definite predetermined pore, H--L with various types of "pore," after Pegler & Young (1971).

(M) Section through aecidiospore of *Cronartium mucronatum* showing plugged germ "pore"; (after electron micrograph by Von Hofsten & Holm, 1968).

(N) L.S. apex of basidiospore of *Coprinus micaceus* showing thinning of middle wall layer at apex (after electron micrograph by Melendez-Howell, 1966).

(O) L.S. apex of basidiospore of *Psilocybe* sp. showing plugged type of "pore" (after electron micrograph by Stock & Hess, 1970).

(P--T) Hilums of basidiospores of agarics. (P, Q) side and surface views of nodulose type of hilum. (R, S) similar views of open-pore type (P--S after Pegler & Young, 1971). (T) L.S. hilar region of recently discharged basidiospore of *Schizophyllum commune* showing hilum and parahilar electron-transparent region (ph) (after electron micrograph by Wells, 1965).

(U) Birth (abscission) scar of *Kickxella alabastrina* showing irregular scar and probable remains of cap of slime (after electron micrograph by Young, 1968).

(V) L.S. through birth (abscission) scar of *Botrytis cinerea* (after electron micrograph by Buckley et al., 1966).

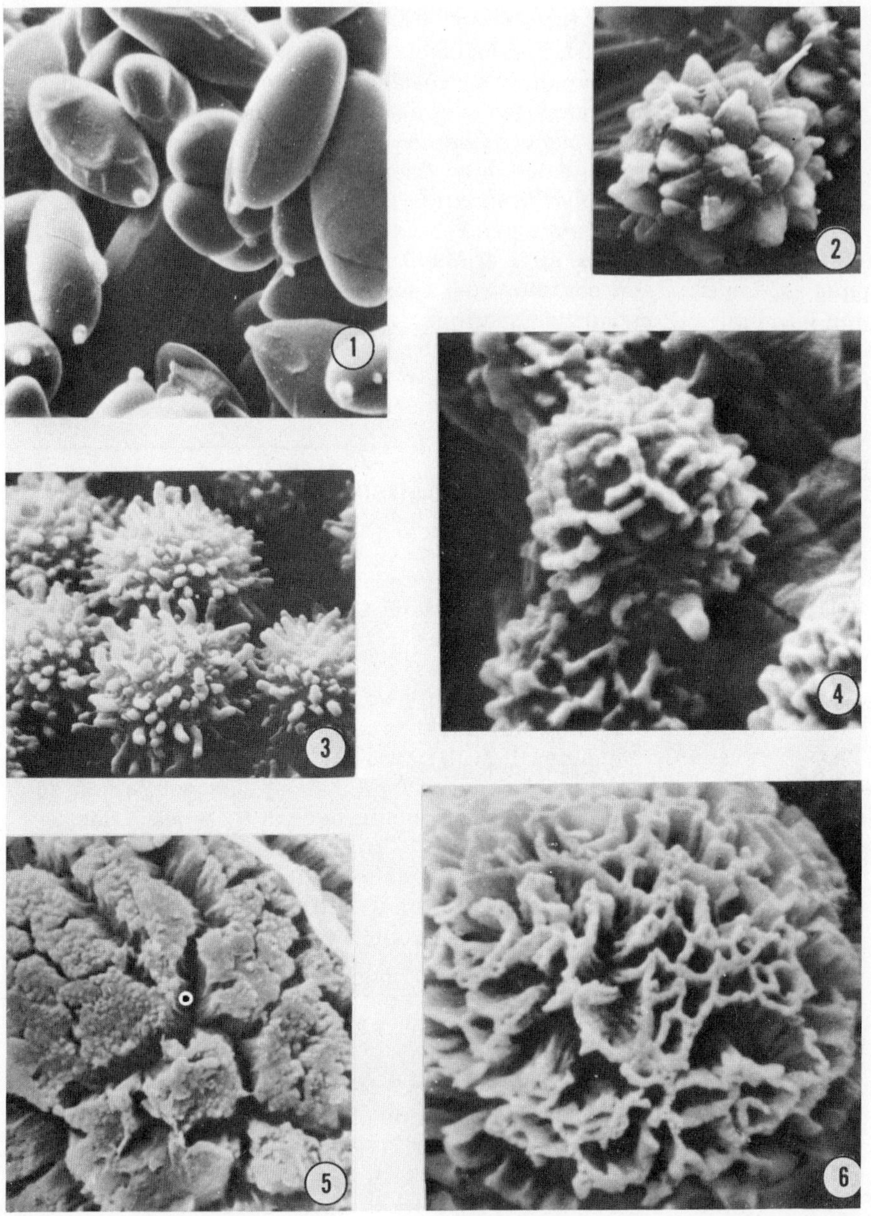
1
2
3
4
5
6

Appendages

Appendages may be hollow outgrowths of spores and at least initially have cytoplasmic contents, for example, in *Pleiochaeta setosa* (Harvey, 1974), or may be mucilaginous extensions or peeled-back portions of the spore wall (Madelin, 1966b). Those of ascospores of marine fungi are described in detail by Kirk (1966) who made a comprehensive cytochemical study. The often very long appendages of conidia (trichospores) of many Trichomycetes of the order Harpellales are initially in a highly coiled form within the stalk cell of the conidium (Fig. 1 K--N). The structure of the appendage differs from genus to genus of these fungi. It is amorphous in *Stachylina grandispora* (Moss, 1972), consists of concentric membranous lamellae in *Smittium mucronatum* (Manier & Coste-Mathiez, 1968), has a periodic structure in *Harpella melusinae* (Reichle & Lichtwardt, 1972) and consists of a core of finely granular compact material surrounded by a fibrillar zone and a thin dense sheath in *Genistella ramosa* (Manier, 1973). The modes of formation of these structures are also different.

Plate 1 Spore surfaces (scanning electron micrographs: approximate magnification given).

Fig. 1 *Gastroboletus turbinatus* (basidiospores): smooth spores with eccentrically placed hilar tubes (x 1650).
Fig. 2 *Octavianina asterospora* (basidiospore): surface covered with pyramidal blunt spines (x 2450).
Fig. 3 *Gymnomyces xanthoderma* (basidiospore): surface ornamented with curved blunt-ended processes (spines) of uneven lengths (x 2050).
Fig. 4 *Macowanites americanus* (basidiospore): spore covered with spines, ridges, and branched flanges, suggesting anastomosis of spines (x 3250).
Fig. 5 *Elaphomyces granulatus* (ascospore): spines aggregating in tufts to give appearance of irregular warts (x 3100). From Hawker & Gooday (1968).
Fig. 6 *Elaphomyces muricatus* (ascospore): spines aggregating to give irregular reticulate appearance (x 3600). From Hawker & Gooday (1968).

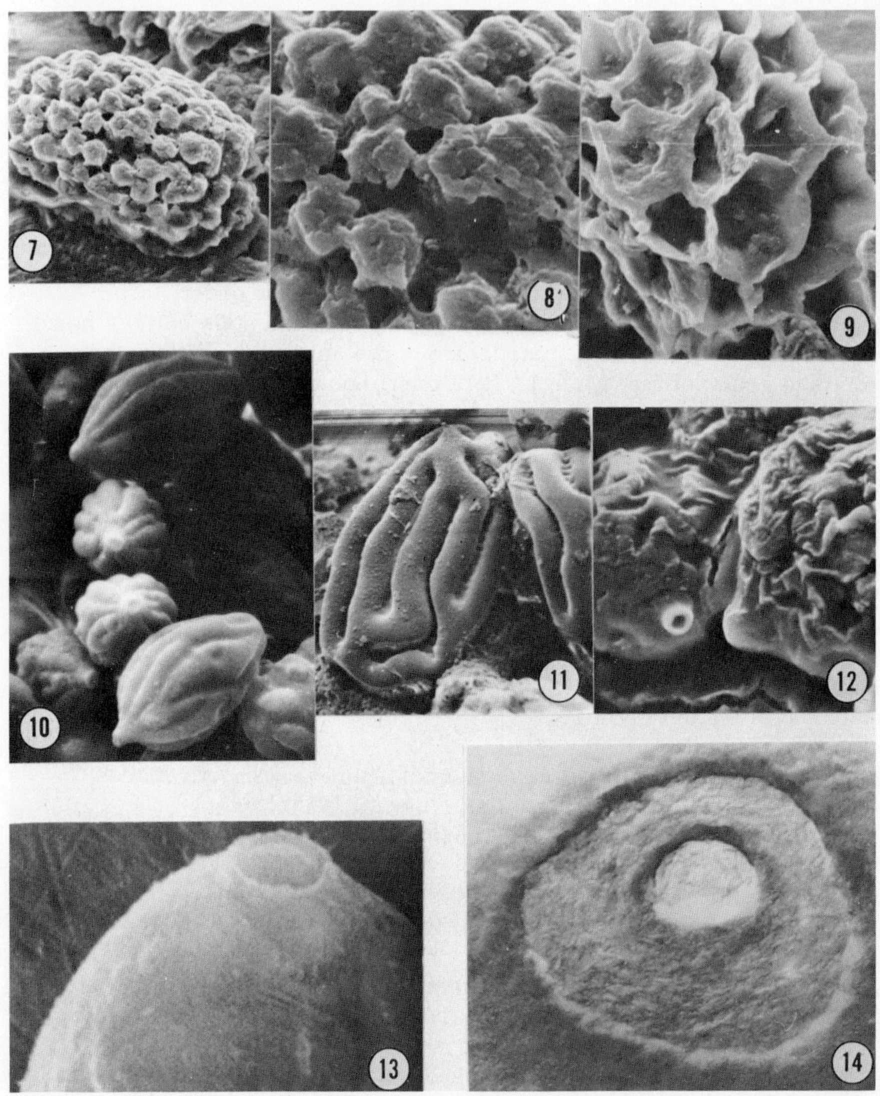

Plate 2 Spore surface features (scanning electron micrographs: except Fig. 14 which is a shadowed preparation: approximate magnification given).

Fig. 7 *Genea hispidula* (ascospores): spore ornamented with irregular warts (x 1070). From L. E. Hawker & A. H. Linton (ed.)

Ornamentation visible under the light microscope

Spores of many types and formed by particular species of nearly all the major groups of fungi have characteristic ornamentations which are valuable criteria in classification and identification of species. Many of these surface features attain their final form only after the spore has reached maximum size. Although of dimensions up to several microns and thus readily visible under the light microscope, these ornamentations often show further detail when examined by electron microscopy. The scanning electron microscope (Hawker 1971b; Heim & Perreau, 1971) and surface-replica techniques (Young, 1968, 1969a,b) are particularly useful for the examination of surface detail (Plates 1 and 2). Several such studies have initiated or confirmed ideas relating to the interrelationship of families, genera, or species, for example, the delimitation of species of *Elaphomyces* (Hawker, 1968); the relationship between groups of hypogeous Gasteromycetes and families of agarics or epigeous Gasteromycetes (Hawker, 1971b, 1975); and the grouping of agarics (Pegler & Young, 1971; Heim & Perreau, 1971).

Although spore ornamentation is a useful criterion in the identification of fungi and similarities often imply relationship, this is not always so. Thus the ribbed ascospores of species of *Neurospora* (Austin, Frederick, & Roth, 1974) superficially resemble the

---

Microorganisms: Form, Function and Environment (1971), Arnold's, London.

Fig. 8 Warts enlarged to show details. (x 3250).

Fig. 9 *Tuber puberulum* (ascospore) reticulate ornamentation (x 1400).

Fig. 10 *Gautieria morchellaeformis* (basidiospores): longitudinally ribbed spores, with short hilar tubes (x 1400).

Fig. 11 *Coelomomyces indicus* (resting sporangium): ribbed pattern with fine trabeculae (top right hand corner) linking ribs (x 900).

Fig. 12 *Hymenogaster tener* (basidiospore) wrinkled more or less ribbed surface with short hilar tubes (x 4100).

Fig. 13 & 14 *Botrytis fabae* (conidia): side and surface views of "birth" or abscission scar. Note fibrous zone surrounding scar. (x 9000 and 37,000 respectively). From Richmond & Pring (1971a).

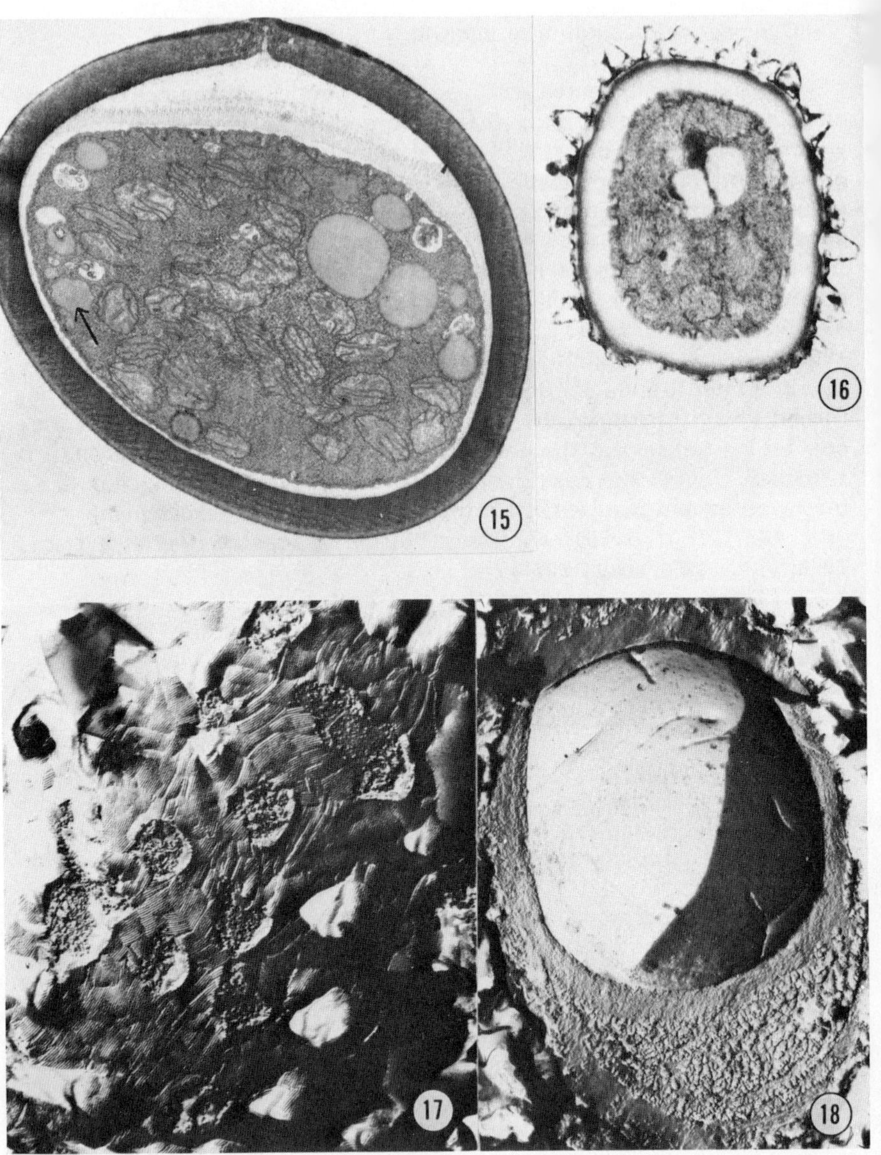

Plate 3 Surface and internal details.

basidiospores of *Gautieria* spp. (Plate 2, Fig. 10), the resting spores of *Coelomomyces* (Plate 2, Fig. 11), and the sporangiospores of *Rhizopus* spp., but these species are unrelated and the spores are all of different types. Neither do differences in external form necessarily imply taxonomic differences. Different types of spore produced by the same species may differ widely in surface features, as with uredospores and teleutospores of most rust fungi or the conidia and ascospores of species of *Aspergillus* (Locci, 1972).

Electron microscopy has also shown that ornamentations of similar final form are produced in very different ways. Although spore development does not strictly fall within the subject of this symposium, mature form of spores can often be understood only by reference to their mode of formation. Spines on spores may be formed by outgrowths of the wall (e.g., ascospores of *Tuber rufum*); they may be formed as solid structures in a gelatinous wall layer which afterwards dries and contracts (e.g., ascospores of *Elaphomyces granulatus* (Hawker, 1968)) or through which the spines gradually protrude as in uredospores of many rusts (Savile, 1954; Thomas & Isaac, 1966; Littlefield & Bracker, 1971); or they may be formed by

---

Fig. 15 *Daldinia concentrica* (ascospore): Transmission electron micrograph of transverse section through mature exogenously dormant spore stored $2\frac{1}{2}$ years at room temperature, showing germ slit as specially differentiated discontinuity in the melanized wall layer. Cytoplasm with a few small vacuoles with inclusions, irregular mitochondria, microbodies (arrowed), and endoplasmic reticulum principally as a single layer just beneath the plasmalemma. Glutaraldehyde/formaldehyde fixation (x 15,600). (Unpublished micrograph by courtesy of Dr. A. Beckett, University of Bristol.)

Figs. 16--18 *Wallemia sebi* (conidia). Fig. 16. Transmission electron micrograph of conidium, showing relatively thick wall with warts arising from surface layers, and plasmalemma with numerous deep invaginations (x 16,400). Fig. 17. Freeze-etch fracture of conidium showing outer surface with warts, sites where warts have been broken off, and rodlet pattern (x 31,000). Fig. 18. Freeze-etch fracture of conidium showing interior surface of thick wall with ribs corresponding to invaginations of plasmalemma (x 16,400). (Unpublished micrographs by M. F. Madelin and Miss Mary Syrop).

other methods. Similarly wartlike outgrowths may be superficially alike but of very different origin. They may be due to irregular thickness of the spore wall as in *Wakefieldia macrosporus* (Hawker, in press), or to discontinuous deposition of secondary inner wall material (e.g., the warts on the zygospores of some members of the Mucorales (Hawker & Gooday, 1968; Hawker & Beckett, 1971)) with consequent sloughing off of the original outer layer of the wall, or to the aggregation of spines owing to rapid drying of a gelatinous matrix (e.g., ascospores of *Elaphomyces granulatus* (Hawker, 1968)).

In contrast, ornamentations that appear to differ radically under the light microscope may be shown by scanning electron microscopy to be essentially similar. This is particularly striking when comparing spiny and reticulate spores of species of the same genus. Thus Locci (1972) showed that among spiny conidia of *Aspergillus* spp. there is a trend towards fusion and branching of papillae leading to the reticulate ornamentation seen in some other species (e.g., *A. ornatus*). Similarly, scanning electron microscope examination of spiny and reticulate ascospores of species of *Tuber* (Hawker, unpublished) shows that the spines on spores of *Tuber rufum* are connected by low ridges, not detectable by light microscopy, and forming a shallow reticulum, whereas the corners of the "boxes" of the reticulum of spores of such species as *Tuber excavatum* are surmounted by spines of greater or lesser prominence according to the species.

Particular attention has been paid to groups of economically important fungi such as the Aspergilli and the rusts and smuts.

The scanning electron microscope study by Locci (1972) of the surfaces of conidia and ascospores of Aspergilli showed that most were patterned. Conidia of various species were smooth, minutely or strongly warty, or spiny. The ascospores which, except for *A. spinulosus*, had walls composed of two symmetrical valves, had surfaces that were ribbed, warty, or spiny. Although spines tended to be of such lengths as to give the spore a more or less spherical or ellipsoidal contour, the tendency in some species for the development of flanges around the peripheries of the valves led to somewhat disc-shaped spores. Locci's data do not reveal any correlation between the ornamentation of conidia and that of ascospores.

Teleutospores of smuts (Swinburne & Matthews 1963; Schwinn, 1969; Zogg & Schwinn, 1971; Khanna et al., 1966; & Payak, 1968) and basidiospores of Homobasidiomycetes (Perreau, 1969, 1971; Bigelow & Rowley, 1968; Pegler & Young, 1971; Eckblad, 1971) show a range of ornamentation rather similar to that of Aspergilli. They may be smooth, echinulate, ridged, verrucose, or reticulate. The aecidiospores and uredospores of rust fungi (Groves, 1913; Cummings, 1971; Wilson & Henderson, 1966; Payak, 1962, 1964) are nearly all patterned with ornamentation that ranges from echinulate to verrucose and spiny, but rarely if ever are patterned with ribs or reticulations. On the other hand, teleutospores of rusts are commonly smooth. Amongst those that are patterned, verrucose and echinulate ornamentation predominates, but a few have striate or reticulate surfaces (Henderson et al., 1972; Durrieu, 1974). Savile (1973) in a study of aecidiospores of certain species of *Puccinia* and *Uromyces* recognized five classes of ornamentation. He observed that aecidiospores have a background pattern of small warts that is sometimes associated with discrete zones of the surface which bear coarse warts; occasionally small to large, definite plugs of material, usually surrounded by narrow smooth zones, are superimposed on this background. Savile suggests that the existence of areas of different intensity of ornamentation on the "bizonate" spores might be a manifestation of the uneven patterning that occurs in the cells of the aecidial peridium, which are homologous with aecidiospores. The outward-facing surface of peridial cells is typically minutely verrucose; that facing inwards toward the aecidiospore chains is typically warted. A slight breakdown in epigenetic control in the aecidiospores could thus result in bizonate ornamentation. The unequal thickness of the walls of some aecidiospores could have a similar basis since typical peridial cells are unevenly thickened.

Germ "pores"

The germ tubes of probably the majority of fungus spores may emerge through any part of the spore wall. In others they emerge through the apex or at other definite positions. Frequently with such spores the area(s) at which emergence will later take place are distinguished in the dormant spore by being structurally different from the rest of the spore wall. These are termed "germ pores." However during the dormant stage this term is a misnomer since the eventual pore is covered by at least one continuous layer of the spore wall, or is plugged with material, physically and perhaps also chemically different from the rest of the wall; or the wall at the site of future germination may be actually thickened. These so-called pores can usually be seen with the light microscope, but electron microscopy has revealed greater detail (Fig. 2, A--O). The form, number, and position of germ pores, is sometimes of diagnostic value (as in the rusts) or of evolutionary significance.

Apical germ papillae

The sporangia of *Phytophthora* spp. often behave as conidia and germinate by a germ tube. These, and the conidia of *Peronospora* and of some other downy mildews, possess a distinctive apical papilla through which germination later occurs.

In *Phytophthora erythroseptica* the structure of this "germ papilla" has been studied in detail (Chapman & Vujičić, 1965) during the maturation and germination of the sporangium/conidium. Under the light microscope the papilla of a young sporangium appears as a mucilaginous thickening of the apical wall (Blackwell, 1949). The electron microscope confirms this greater thickness and reveals a vesicular layer within the thickened wall situated near to and parallel with the plasmalemma. In older sporangia that are showing no signs of germination the apical wall is still much thickened but appears to be of similar structure to the rest of the wall.

Germ pores of higher fungi

The spores of many higher fungi have clearly defined germ pores. They are commonly present in one or more of the types of rust spores (aecidiospore (Fig. 2, M), uredospore, and teleutospore), in the teleutospores (chlamydospores) of certain smuts and in the basidiospores of many agarics (Fig. 2, G--L,N,O), but are less common among Ascomycetes (Fig. 2, A,B).

Savile (1973) distinguishes two types of germ pore in rust aecidiospores. The first type, as with the pores of uredospores, is differentiated during wall formation, and from its inception comprises material different from that forming the rest of the wall. These "differentiated pores" are thus not truly pores until germination begins. Because the material constituting the pore generally has a refractive index similar to that of the rest of the wall such pores are not easily detected by light microscopy. Von Hofsten & Holm (1968), in an electron microscope study, showed these pores to be electron-transparent regions of the wall (Fig. 2, M). Savile's second type of aecidiospore germ pore are the "impressed pores" in which an outer plug of wall material eventually falls out, leaving a very thin wall-covering over the base of the pore. The mature aecidiospores of numerous species of *Puccinia* and a few of *Gymnosporangium* have similar large bodies embedded in the wall. Klebahn (1912--1914) termed these bodies Plättchen, and Dodge (1924) called them pore plugs. Savile believes pore plugs to have evolved from "normal aecidiospore warts," presumably the type formed as discrete bodies embedded in the spore wall but not those formed by the shedding of fragments of wall from between the prospective warts (Moore & McAlear, 1961; Walkinshaw et al., 1967; Von Hofsten & Holm, 1968).

Payak and his co-workers (Payak, 1962; Payak et al., 1967) described the ultrastructure of germ pores of the uredospores of several species of rusts, and showed that the inner secondary wall of the spore failed to form at the sites of the germ pores but thinned out towards them. They demonstrated umbonate coverings over the germ pores in the equatorial region of uredospores of *Puccinia graminis* var. *tritici*. The germ pores of teleutospores are often very conspicuous, as in species of *Puccinia* and *Phragmidium* and are readily seen under the light microscope. The thick walls of these spores make staining, fixing, and ultrathin sectioning difficult, and it is, therefore, not surprising that information on the fine structure of germ pores of teleutospores is lacking.

Several studies have been made of the germ pores of the basidiospores of agarics. Pegler & Young (1971) described a number of different types, none of which are true openings since at least one of the component layers of the spore wall remains continuous. Meléndez-Howell (1967a,b) recognized five types of germ pore in the basidiospores of agarics: (a) the apex of the spore appears as a clear spot under the light microscope, and in section is seen to be due to a thinning of the "episporium" (Pegler & Young, 1971) and a thickening of the "exosporium"; (b) the episporium is interrupted; (c) similar to (b) but the gap in the episporium is plugged with additional material; (d) both episporium and exosporium are depressed inwards, but there is no break in either layer and no plug is formed; and (e) the episporium is thickened towards the interior of the spore. Some species do not have such definite pores. Pegler & Young (1971) consider that these pore types are of considerable diagnostic value as many families of agarics have a characteristic pore form (e.g., in Cortinariaceae most species show pores of type (a)). Meléndez-Howell (1966) gave a detailed description of the germ pores of *Coprinus* (Fig. 2, N). Heintz & Niederpruem (1970) published an electron micrograph of the pore of *Coprinus lagopus*.

Stocks & Hess (1970) studied the fine structure of the germ pores of basidiospores of a new species of *Psilocybe* (Fig. 2, O). Here a concave plug of electron-transparent material largely replaces the thick middle layer of the spore wall.

Meléndez-Howell (1969) studied the germ pores and the emergence of germ tubes of ascospores of representative species of Ascomycetes. She distinguished three types of pore. The germ tubes emerged through a simple pore (*Chaetomium*), a longitudinal slit along a previously differentiated line (*Poronia* and *Xylosphaera*), and a circular equatorial opening in *Aspergillus* spp. (Fig. 2, A--F). Beckett (personal communication) has examined structurally differentiated germination lines in dormant ascospores of members of the Xylariaceae (Plate 3, Fig. 15).

Abscission scars, the hilum, and hilar appendix

The light microscope reveals abscission scars on the conidia of some species, notably in some, but by no means all, of those where the spores are formed in chains. The spores in some chains have special wall structures leading to separation, as in certain species of *Oidium* and *Monilinia*. Often a minute pore remains open between the spores of a chain until just before release as in *Monilinia fructicola* (Willetts & Calonge, 1969).

Conidia borne singly on a conidiophore may also show definite basal scars as with some species of *Botrytis* (e.g., *B. allii*, Scurfield & da Costa, 1969; *B. fabae*, Richmond & Pring, 1971a). In *B. fabae* the basal scar has a circular zone of microfibrils surrounding a plug closing the pore through which, presumably, contact was maintained during development (Plate 2, Figs. 13 and 14).

The most striking basal abscission structures are to be found in the higher Basidiomycetes. The scanning electron microscope and carbon replica techniques have elucidated the structure of the hilum which is often difficult to determine by light microscopy alone. In the agarics the hilum is usually a very small region and is situated eccentrically at the base of the spore. In some species there is a subhilar depression just above the hilum with an area of surface different from that of the spore. Pegler & Young (1971) recognize two types of hilum among agarics: the nodulose type characteristic of many species with hyaline spores, and consisting of a circular area bearing small protuberances but with no indication of a pore; and the open type, consisting of a small depression usually with a pore t the center, found only in species with pigmented spores (Fig. 2, --S). However, some of their plates, for example, those illustra-.ng *Rozites caperata* and *Lacrymaria velutina*, show the hilar appa-atus as a short hollow tube. Basidiospores of many Gasteromycetes nd some species of *Russula* (Perreau & Heim, 1969; Heim & Perreau, 971; Perreau, 1971; Hawker, 1975) bear prominent hilar tubes the orm of which can best be seen in scanning electron micrographs Plates 1 and 2). In immature spores of hypogeous Gasteromycetes still attached to the sterigmata these tubes can be seen to fit over the apex of the sterigmata, and in detached spores they can be seen as hollow tubes of length varying with the species. The similarity in orm of the hilar tube in species of *Russula* and in hypogeous mem-bers of the Asterosporales is further evidence of a relationship be-tween these groups.

Ultrafine ornamentation revealed only by electron microscopy

Although some spores are virtually smooth even at the level of scrutiny of the electron microscope (e.g., conidia of *Botrytis fabae*, Richmond & Pring 1971a), many have fine ornamentation that is revealed only by electron microscopy. Examples of such ornamentation are the background roughness (approximately 0.1 μm high and wide) of the verrucose aecidiospores of *Phragmidium mucronatum* (Von Hofsten & Holm, 1968) and the highly ordered "rodlet" patterning first described by Hess et al. (1966) on the surfaces of conidia of ten species of *Penicillium*. This pattern of ordered arrays of minute linear series of particles having a diameter of approximately 5 nm has been described more fully in *Penicillium megasporum* (Sassen et al., 1967); and subsequently on conidia of *Aspergillus* species (Hess & Stocks, 1969; Sleytr et al., 1969; Ghiorse & Edwards, 1973), of *Neurospora crassa* (F. Schwegler, in Hess et al., 1972), of *Scopulariopsis* (Cole & Aldrich, 1971), of *Epicoccum nigrum* (Griffiths, 1973b), of *Gonatobotryum apiculatum, Oidiodendron truncatum,* and *Geotrichum candidum* (Cole, 1973b, 1973c); and on basidiospores of *Lycoperdon* (Hess et al., 1972; Bronchart & Demoulin, 1971), and *Scleroderma, Tulostoma brumale, Geastrum fimbriatum, Podaxis pistillaris,* and *Lacrymaria velutina* (Bronchart & Demoulin, 1971). The same pattern is to be seen also on conidia of the hyphomycete *Wallemia sebi* (Madelin & Syrop, unpublished; Plate 3, Fig. 17). Rodlets are thus present in representatives of the classes Ascomycetes, Deuteromycetes, and Basidiomycetes. However, despite the use of techniques that should have revealed rodlets had they been present, they have not been seen on chlamydospores of smuts, basidiospores of *Psilocybe* sp., ascospores of *Aspergillus nidulans* var. *echinulatus* and *Chaetomium gelasinosporum* (references in Hess et al., 1972), or conidia of *Botrytis fabae* (Richmond & Pring, 1971) and *Drechslera sorokiniana* (Cole, 1973a). They have been seen also on prokaryote spores. Wildemuth et al. (1971) and Moor (in Hess et al., 1972) reported them on the surface of conidia of *Streptomyces* spp., and Gould et al. (1970) on the surface of spores of *Bacillus coagulans*. The rodlet pattern has also been seen on the surfaces of phialides and conidiophores of *Aspergillus fumigatus* (Ghiorse & Edwards, 1973), phialides of species of *Aspergillus*, *Penicillium*, and *Phialophora*, and vegetative (evidently aerial) hyphal tips of *Oidiodendron truncatum* (Cole, 1973c). It is thus not peculiar to spores.

The composition of the rodlet material has not been unequivocally established in fungi, but evidence now points strongly to its being proteinaceous. Sassen et al. (1967) and Hess et al. (1968) originally suggested that the rodlets consisted of cutin, sporopollenin, or some similar substance. Since then, sporopollenin has been demonstrated to occur in the walls of certain zygospores (Gooday et al., 1973), but there is as yet no evidence associating its occurrence with rodlet patterning. Fisher & Richmond (1970) were unable to detect fatty acids characteristic of cutin in extracts of walls of conidia of *Penicillium expansum*. Sassen & Heinen (in Hess et al., 1968) found that rodlets survived boiling chloroform or methanol, but were removed by ethanolic or aqueous potassium hydroxide. Gould et al. (1970) identified the rodlet material on certain bacterial spores as a protein which could be removed by cold dilute alkali after the spores had been exposed to treatments (e.g., with mercaptoethanol) which ruptured disulphide bonds. Although Fisher & Richmond (1970) failed to remove rodlets from conidia of *Penicillium expansum* by incubation with mercaptoethanol in urea followed by 0.1 M sodium hydroxide, the rodlets became less distinct. These authors found that 11.2% of the nondialyzable, alkali-soluble material extracted by washing the spores was protein, the rest probably being polysaccharides.

Fisher & Richmond (1969, 1970) concluded from microelectrophoretic studies that a thin polyphosphate layer sometimes covered the amino-carboxyl surface of *Penicillium expansum* conidia. It is significant that Hess et al. (1968) remarked on the occasional obscurity of rodlet patterns in *Penicillium* conidia and suggested that a ery thin layer sometimes covered the rodlets. Ghiorse & Edwards 1973) also observed a layer which occasionally covered the rodlet pattern in *Aspergillus fumigatus*, and suggested that this might cor- respond to a superficial polyphosphate layer.

A specialized type of ornamentation at a fine-structural level oc- curs on conidia of the Hyphomycete *Pithomyces chartarum* whose surface layer consists of more or less perpendicular arrays of crystalline spicules, 800 nm long and 30--50 nm wide, composed of sporides- molides together with some lipids. Benzene or diethyl ether remove these spicules completely. The surface sporidesmolides represent a high proportion (3--4%) of the spores' weight, and may well contrib- ute to their remarkable ability to repel water (Bertaud et al., 1963).

INTERNAL STRUCTURE

The study of the internal structure of dormant spores is compli- cated by the structure and nature of the spore wall. This usually consists of more than one layer and may have as many as five or more distinct layers. One or other of these, particularly in resting spores (Gregory's memnospores), may be largely impermeable to the fixa- tives and stains used for study with the light or electron microscopes and/or may provide a mechanical obstacle to satisfactory preparation of ultrathin sections. The use of freeze-etch techniques has partially overcome these difficulties but again some spores with particularly thick multilayered walls (e.g., the zygospores of *Rhizopus sexualis* and other members of the Mucorales) are difficult to freeze satisfac- torily owing to their low permeability. The examination, by means of the scanning electron microscope, of fragments of these spores pre- pared by gentle grinding in liquid nitrogen yielded information on wall structure (Plate 4, Figs. 19--22), but the cytoplasm was lost (Hawker & Gooday, 1968; Hawker & Beckett 1971). Good fixation of exogenously dormant xenospores is usually more easily attained, and hence a number of reports of the structure of the cytoplasm of these spores is available.

The Spore Wall

The surface features of the spore wall have already been consid- ered. The transmission electron microscope reveals details of the stratification and "architecture" of the wall as seen in section. The

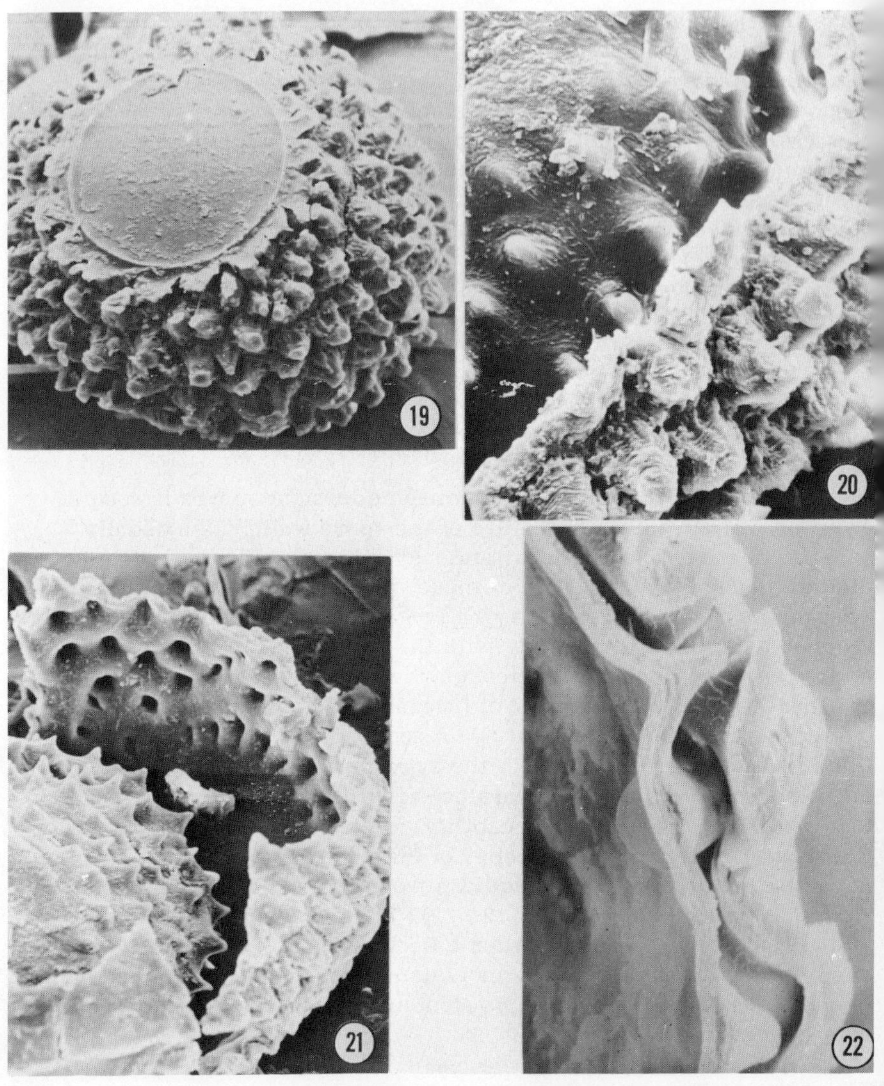

Plate 4 Multilayered wall of zygospore of *Rhizopus sexualis*. (scanning electron micrographs of broken spores, treated with liquid nitrogen).

Fig. 19 Spore with suspensor broken off leaving flat plate; the

walls of typical memnospores are all exceedingly complex, consisting of several distinct layers. The old terminology, based on light microscopy, for epispore, exospore, perispore, mesospore, endospore, and so on, used to denote the various layers, is not readily applied to some of these multilayered walls, and its use is liable to cause confusion by including layers of widely different origin under the same term or by different authors using different terms to describe layers of similar origin. These terms will therefore be used in the present essay only when quoting from the work of other authors. Even the relatively thin wall of many exogenously dormant xenospores may consist of more than one layer. Electron micrographs of unsoaked dormant spores demonstrate that the architecture of the walls, or of some of their constituent layers as seen in section, is usually radically different from that of the vegetative hyphae. The microfibrils of the latter are arranged more or less tangentially, whereas the elements of the spore walls may be arranged differently in at least one of the wall layers or may even lack obvious fibrillar structure; for example, Fig. 3 (A and B) shows the contrast between the wall of the unsoaked sporangiospore of *Rhizopus* spp. and that of the vegetative hyphae (Hawker & Abbott, 1963a,b). A particularly complex spore wall is that of the zygospore of *Rhizopus sexualis* (Hawker & Gooday, 1968; Hawker & Beckett, 1971). This consists of five or possibly more distinct layers laid down at different stages in spore development and maturation. Figure 3C and Plate 4, Figs. 19--22 illustrate this complex wall.

During maturation of the zygospore a number of wall layers is laid down in succession within the original gametangial wall until finally the wall of the dormant spore consists of the following layers (Fig. 3C): (i) a few tattered fragments of the original outer layer of the primary gametangial wall adhering to the blunt tops of the

---

remainder of spore covered with flower-pot shaped warts with fragments of original primary or gametangial walls adhering to some (see Fig. 3C) (x 490).

Figs. 20--21 Fragments of zygospores showing warts on outer wall and spiny inner wall which fitted into outer layer. Fig. 21 shows the depressions on the inner side of this layer into which the spines fitted (See Fig. 3C). (x 820 and 490 respectively).

Fig. 22 Broken edge of wall showing the various layers partially separated (x 1640).

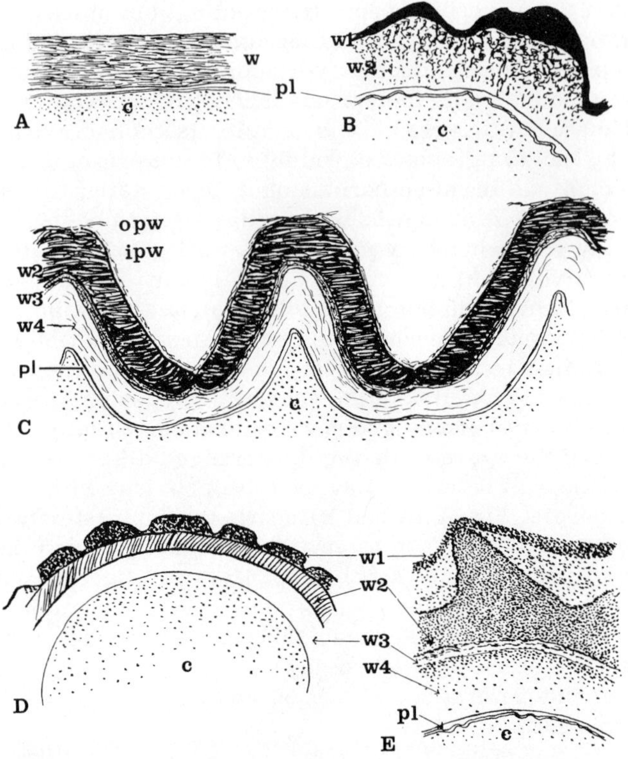

Fig. 3 Examples of multilayered spore walls (diagrammatic, drawn from electron-micrographs, not to same scale).

(A) L.S. wall of vegetative hypha of *Rhizopus homothallicus* (for comparison) showing wall (w) with microfibrils running parallel with axis of hypha. (After Hawker & Abbott, 1963a).

(B) Section through longitudinally ridged wall of sporangiospore of *Rhizopus stolonifer* (an endogenously dormant xenospore) showing outer electron-dense layer (w1), thicker inner layer with irregularly arranged granular elements (w2) and invaginated plasmalemma (pl); (after Hawker & Abbott, 1963b).

(C) Section through wall of zygospore of *Rhizopus sexualis* (an endogenously dormant memnospore), showing torn outer layer of primary (gametangial) wall (opw); inner layer of primary wall which was gelatinous and finally dried down to haracteristic black wartlike projections; (ii) a thin horny layer rep-esenting the formerly gelatinous inner layer of the primary wall, ow collapsed and forming an almost continuous sheath closely ad-ressed to the warts; (iii) the black wartlike projections, shaped like nverted flower pots, the bases of which are closely adpressed, and hus not circular, consisting of layers of hard black material laid own tangentially to the spore surface and progressively from the flat ops of the warts towards their bases, with each more or less circu-ar layer being of greater diameter than the last thus giving the final "flower pot" form; (iv) a thin "smoothing" layer lining the warts and aid down after their completion, obliterating the "joins" at their ases; and (v) a thick layer or layers laid down inside the smoothing ayer. Material similar to that forming the warts is laid down evenly ver the inner sides of the original gametangial septa simultaneously with the formation of the warts on the lateral walls. Layers (iv) and (v) are also continuous over these, the plasmodesmata which allowed passage of solutes from the suspensors to the developing spore

---

thin horny layer (ipw); warts consisting of stratified electron-dense secondary material (w2); "smoothing" tertiary layer (w3) laid down parallel to inner surface of warts; quaternary layer(s) of striate material (w4); diagram compiled from scanning and transmission electron micrographs after Hawker & Beckett (1971); details of plasmalemma (pl) and cytoplasm (c) not readily seen in dormant spores owing to impermeable wall but in nearly mature spores plasmalemma was invaginated.

(D) Section through ascospore of *Neurospora tetrasperma* (an endogenously dormant xenospore), showing outer ribs consisting of electron-dense granular material (w1); layer of electron-dense material (w2); and inner layer of homogeneous electron-transparent material (w3). After electron micrograph by Sussman (1966).

(E) Section through teleutospore of *Tilletia caries* (an endogenously dormant memnospore) showing outer granular layer (w1), a more homogeneous uneven layer which imparts the reticulate pattern to the surface (w2); a "partition layer" (w3) and a thick innermost electron-transparent layer (w4). The invaginated plasmalemma (pl) is deduced from an electron micrograph of freeze-etch material. After electron micrographs by Allen et al. (1971).

(Hawker & Beckett, 1971) becoming occluded.

When the smoothing layer (iv) has been completed the spore becomes almost impermeable to water and dissolved substances and probably to gases also. Fixation of undamaged spores for examination by transmission electron microscopy and preparation for freeze etching becomes difficult. Fragmented spores were examined in the scanning electron microscope revealing the separate layers of the spore (Plate 4, Figs. 19--22).

The zygospore walls of other genera of the Mucorales are also multilayered (Hawker, unpublished) differing from those of *Rhizopus* in the shape of the warts (e.g., *Mucor* and *Zygorhynchus*, where they are fluted) or in having no ornamentation (e.g., *Phycomyces* and *Absidia*). Multilayered walls have been described or figured for memnospores of a number of unrelated fungi. Allen et al. (1971), using both freeze-etch and thin section techniques, demonstrated four distinct layers in the walls of dormant teleutospores of the smut, *Tilletia caries*. The outermost and innermost layers appear to be of a fibrous nature. The second layer, which gives the reticulate surface pattern, is not fibrous. The thin layer separating the latter from the inner fibrous one appears as a thin undulating electron-dense line in transmission electron micrographs (Fig. 3E).

The resistant sporangia of *Allomyces neo-moniliformis* may not be considered as dormant spores by all mycologists, but they are homologous with those of some other fungi with which their origin, structure, and function are comparable. They are included here as another example of a complex wall (Skucas, 1967). The outer wall is thick, pitted, and five-layered; next come two layers of "cementing substances" (which later dissolve during germination), and finally an inner wall of homogeneous material. Architecturally this complex wall differs widely from that of the hyphae, but apart from containing melanin and lipids its chemical composition is qualitatively similar.

Other examples of memnospores are the chlamydospores formed in the vegetative hyphae of a number of unrelated fungi. Griffiths (1973a) showed that chlamydospores of *Fusarium oxysporum* have a two-layered wall, the outer layer of which represents the original hyphal wall and the inner one having been laid down during maturation. He observed striations in the wall due to layers of osmiophilic material which he considered to be concerned with the survival of the spores in dry soil.

Ascospores, with the exception of those of yeasts and genera such as *Aspergillus* and *Penicillium* (which are not explosively discharged or extruded in slime), cannot be classed as memnospores. Nevertheless, ascospores of many species possess complex walls which may be multilayered, sculptured, and/or pigmented. These spores are capable of retaining their viability for long periods of desiccation, or can withstand periods of high temperature or exposure to chemical action, as do those of coprophilous species which pass uninjured through the gut of herbivores. Many germinate only after shock treatment with heat or chemicals.

Examples of complex ascospore walls are those of the coprophilous species *Podospora anserina* (Beckett et al., 1968), which has a three-layered wall, of *Neurospora tetrasperma* (Sussman, 1966), with at least five layers, one of which is melanized (Fig. 3D), of the lichen *Pertusaria pertusa* (Pyatt, 1969), with five layers of very different thickness and appearance, of *Ascobolus stercorarius* (Wells, 1972), and of species of yeasts which Besson (1966) showed to have two, three, or four layers according to the species. Many more examples could be cited.

The walls of typical xenospores are usually thinner and less complex than those of memnospores. Many are single layered, but may show a graded appearance from an outer electron-dense zone to a more electron-transparent inner one without any clearly marked division into distinct layers, as with the conidia of *Botrytis* spp. (Hawker & Hendy, 1963; Buckley et al., 1966; Richmond & Pring, 1971) or sporangiospores of *Rhizopus* spp. (Hawker & Abbott, 1963a, b; Ekundayo, 1966; Buckley et al., 1968; Hess & Weber, 1973). The walls of others may be two-layered, as in the case of conidia of *Alternaria brassicicola* (Campbell, 1968, 1969), or three-layered, as in basidiospores of *Psilocybe* sp. (Stocks & Hess, 1970).

The Cytoplasm and Cytoplasmic Organelles

Since spores are of many functional types, formed in varied ways, and with walls varying in permeability to water and gases, it is not surprising that published accounts of the cytoplasmic spore contents present some differences. Nevertheless, a number of general characteristics of spore structure can be distinguished (Hawker, 1971a).

Most spores possess the same range of organelles as seen in the vegetative hyphae of the same species with the exception of vacuoles which are usually absent, although they have been shown to be present in some xenospores such as the uredospores of *Puccinia graminis* (Thomas & Isaac, 1966) and the conidia of *Aspergillus nidulans*

(Florance et al., 1972), which contain small vacuoles, and of *Penicillium megasporum* (Sassen et al., 1967), which contain large ones. Other organelles may show form or structure different from those of the parent hyphae or they may show a distinct pattern of distribution within the spore. A striking feature of most spores, particularly memnospores, is a visible abundance of reserve food material.

The plasmalemma

The cytoplast is enveloped in a distinct membrane, the plasmalemma, as in vegetative cells, but this membrane is commonly invaginated in spores. In section the shallow invaginations give the plasmalemma an undulating appearance (Plate 3, Fig. 16), and in fractured freeze-etched material they can be seen as randomly arranged more or less straight channels (e.g., in conidia of *Penicillium megasporum*, Sassen et al., 1967; of *Aspergillus niger*, Sleytr et al., 1969; of *Alternaria brassicicola,* and *Stachybotrys atra*, Campbell, 1969, 1972; and of *Wallemia sebi*, Madelin & Syrop, unpublished (Plate 3, Fig. 18); in basidiospores of *Psilocybe* sp., Stocks and Hess, 1970; and many others). Campbell (1970b) found that conidia of an albino strain of *Alternaria brassicicola,* unlike those of the wild type, lacked these channels unless induced to develop brown pigment by appropriate supplementation of the culture medium with tyrosine. He suggested that they may be connected with melanin synthesis. Allen et al. (1971) in an investigation of dormant teleutospores of *Tilletia caries* distinguished numerous particles associated with the sparsely invaginated plasmalemma. They suggested that these particles were "multienzyme complexes."

Endoplasmic reticulum (e.r.)

In mature dormant spores the e.r. is usually sparse, although it may be locally very abundant in immature developing spores (e.g., zygospores of *Rhizopus sexualis*, Hawker & Beckett, 1971). Weisberg & Turian (1971) also observed a reduction in the amount of e.r. in resting conidia of *Aspergillus nidulans* compared with that in young chains of spores. This poor development of e.r. in dormant spores has been reported in a number of investigations for a variety of spore types as in the case of sporangiospores of *Rhizopus* spp. (Hawker & Abbott, 1963b), conidia of *Botrytis* spp. (Hawker & Hendy, 1963; Richmond & Pring, 1971a) and of *Fusarium culmorum* (Marchant,

1966a,b), basidiospores of *Schizophyllum commune* (Voelz & Niederpruem, 1964), and ascospores of *Neurospora tetrasperma* (Sussman & Halvorson, 1966). Some authors (Richmond & Pring, 1971a; Beckett, personal communication) report that the e.r. tends to concentrate in the periphery of the spore and to be arranged parallel to the plasmalemma (Plate 5, Fig. 24).

Ribosomes

Ribosomes have been demonstrated in most spores when suitable methods of fixation have been employed (e.g., in sporangiospores of *Rhizopus* spp., Buckley et al., 1968; in conidia of *Aspergillus* spp., Florance et al., 1972, Ghiorse & Edwards, 1973).

Vesicles

Vesicles of unknown function have been reported in a few mature spores, as in the conidia of *Botrytis fabae* (Richmond & Pring, 1971b) where they are said to be arranged in the periphery near the plasmalemma. Vesicles are, however, numerous in many spores during development, and in view of their delicate nature they could readily be destroyed during the difficult process of fixing and sectioning thick-walled dormant spores.

Mitochondria

In many thick-walled spores, including examples of both memnospores and xenospores, the mitochondria are large and irregularly lobed, resembling those in sporangiospores of *Rhizopus arrhizus* subjected to anaerobic conditions (Ekundayo, 1966). This finding suggests that their unusual shape may be the result of partial anaerobic conditions obtaining in spores with largely impermeable walls. Such lobed mitochondria have been seen in sporangiospores of *Rhizopus* spp. (Fig. 4 A--H; Hawker & Abbott, 1963b); in ascospores of *Neurospora crassa* (Weiss, 1965) and of *Sordaria fimicola* (Hawker, unpublished); and in aecidiospores of *Cronartium fusiforme* (Walkinshaw et al., 1967). Voelz & Niederpruem (1964) reported that the mitochondria of the basidiospores of *Schizophyllum commune* were poorly defined and had a few cristae. In mature zygospores of *Rhizopus sexualis* the mitochondria show signs of disintegration (Plate 5, Fig. 23; Fig. 4, I).

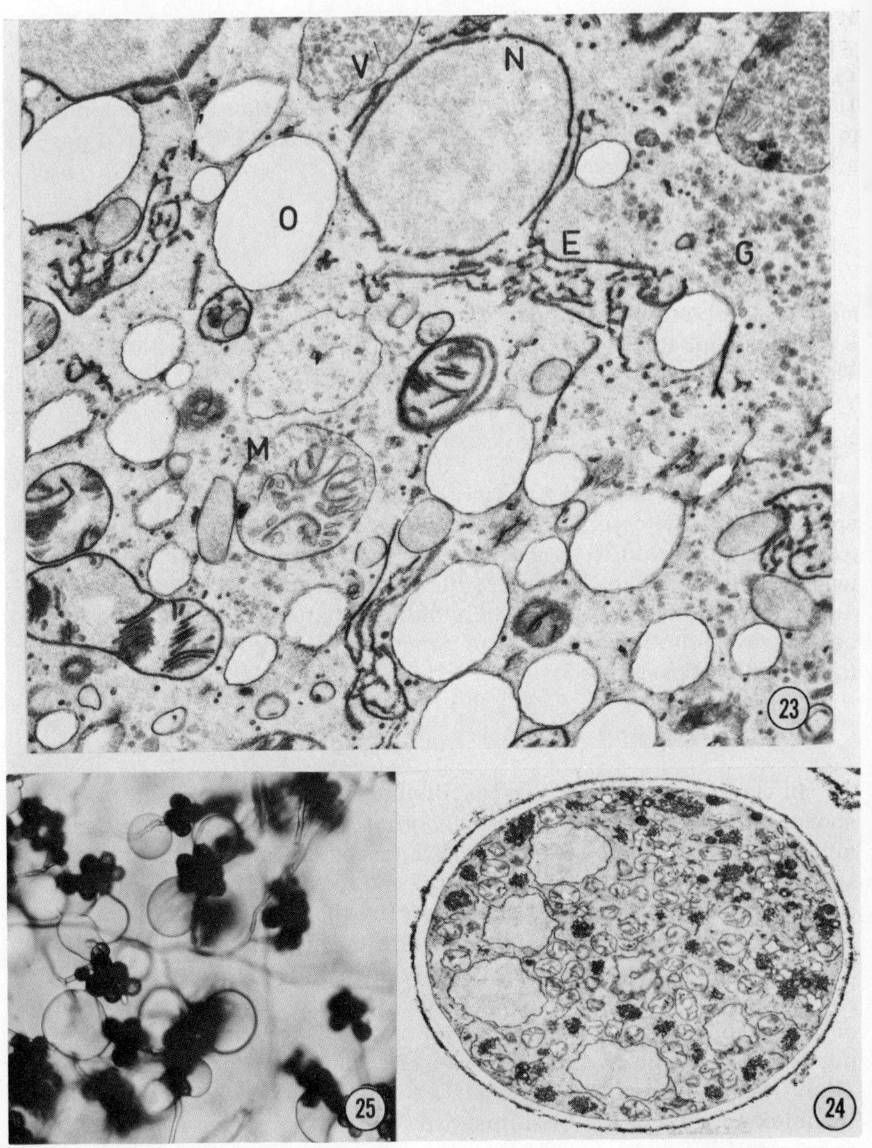

Plate 5 Figs. 23--25.

In contrast many relatively thin-walled xenospores contain numerous small, more or less spherical or ellipsoidal mitochondria with clearly defined cristae, which resemble the mitochondria of the parent mycelium and the resulting germ tubes (e.g., *Botrytis* spp.; Richmond & Pring, 1971a,b; Buckley et al., 1966; Plate 5, Fig. 24; Fig. 4J--L).

Nuclei

The nuclei do not normally differ greatly in appearance from those of the hyphae. An exception is the report of lobed nuclei in sporangiospores of *Mucor rouxii* (Bartnicki-Garcia et al., 1968). Campbell (1970a) saw no nuclear pores in the nuclei of 17-week-old conidia of *Alternaria brassicicola*, but his electron micrographs show them in younger mature spores (Campbell, 1968).

Storage bodies

Lipid bodies, patches of glycogen granules, and membrane-bounded storage vacuoles containing phospholipids, polysaccharides, or undetermined material are frequent in dormant spores. However, some spores lack one or the other of these types of storage materials, and the actual quantities present also vary with the species, type,

---

Fig. 23 *Rhizopus sexualis*. Transmission electron micrograph of section through cytoplasm of mature zygospore showing nuclei of normal appearance (N); distorted mitochondria (M); oil drops (O); storage vacuoles (V); glycogen in cytoplasm (G); and sparse e.r. (E); to show internal structure of typical memnospore (x 15,600).

Fig. 24 *Botrytis fabae*. Transmission electron micrograph of section of mature conidium, showing numerous small mitochondria of normal appearance, though with dense spherical inclusions; presence of aggregates of glycogen; and peripheral endoplasmic reticulum. To show internal structure of typical xenospore. From Richmond & Pring (1971a). (x6150).

Fig. 25 Droplets of watery exudate on conidia of *Ulocladium* sp. developing in a thin layer of hydrophobic liquid (perfluorotributylamine: FC43, 3M Co., St. Paul, Minnesota) in a slide culture (method modified from Madelin, 1969). (x 180).

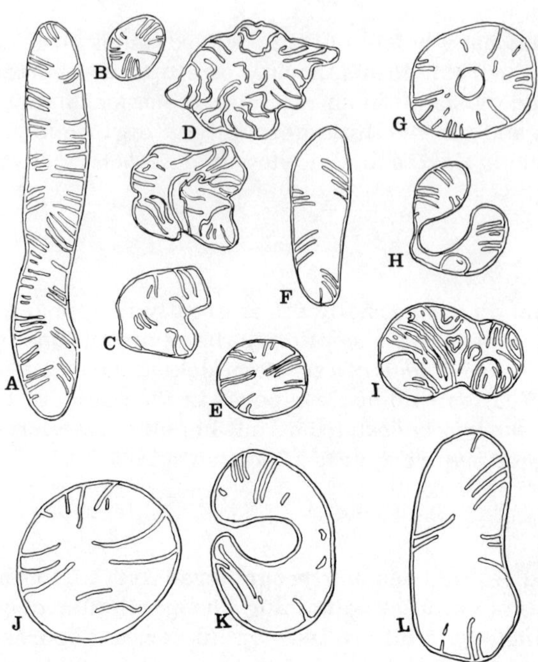

Fig. 4 Mitochondria from spores compared with those from hyphae and germ tubes. (Drawn from electron micrographs of material fixed with permanganate.)

(A--G) *Rhizopus* spp. drawn to approximately same scale. (A, B) *R. homothallicus*, L.S. and T.S., respectively, of mitochondria from young hypha (for comparison with those of spores). (C) Mitochondrion from nearly mature sporangiospore of *R. stolonifer*, showing incipient lobing and curving of cristae. (D) Section through parts of two lobed mitochondria with distorted cristae, from dormant sporangiospore of *R. stolonifer* (an endogenously dormant xenospore). (E, F) *R. sexualis*, mitochondria from germinated sporangiospore and germ tube (in which cytoplasmic streaming had been taking place), respectively, showing restoration of entire outline and parallel cristae. (G, H) *R. sexualis*, T.S. and L.S., respectively, of basin-shaped mitochondrion from nearly mature zygospore. (I) Section through mitochondrion from nearly dormant zygospore of *R. sexualis* (an endogenous

and age of spore. Lipid bodies may be concentrated in the periphery (e.g., in conidia of *Fusarium culmorum*, Marchant, 1966a; or of *Botrytis fabae*, Richmond & Pring, 1971a), or may be distributed throughout the cytoplasm (e.g., in zygospores of *Rhizopus sexualis*, Hawker & Beckett, 1971) where they are so numerous that they crush other organelles into the surrounding remnants of the cytoplasm (Plate 5, Fig. 23).

Unidentified bodies

Unidentified bodies of varied nature have been seen in many spores. These may be free in the cytoplasm (e.g., in teleutospores of *Tilletia caries*, Allen et al., 1971, where they bear numerous particles on their surfaces, and in basidiospores of *Lycoperdon perlatum*, Hess et al., 1972), or they may be contained inside other organelles (e.g., crystals seen in the disorganized mitochondria of mature zygospores; Hawker & Beckett, 1971). Buckley et al. (1966) describe membrane-bounded dense inclusions or "storage bodies" in conidia of *Botrytis cinerea*. These bodies contained whorled structures, spherules, granules or membrane loops and it was thought that they might contain phospholipids. Bodies such as these were not seen in *B. cinerea* by Hawker & Hendy (1963) or in *B. fabae* by Richmond & Pring (1971a) but are reported by Campbell (1968) in conidia of *Alternaria brassicicola*.

Chemical Composition and Water Content

Comparisons of the gross chemical composition of dormant spores are hampered by differences in the various analytical approaches that have been adopted. However the information available indicates the

---

memnospore) showing disorganized cristae.

(J--L) *Botrytis cinerea* drawn to slightly larger scale than (A--G) after electron-micrographs by Buckley et al. (1966). (J) Section of mitochondrion from "dormant" conidium (an exogenously dormant xenospore; note absence of the contortion seen in an endogenously dormant xenospore, Fig. 4D). (K) L.S. basin-shaped mitochondrion from base of a newly germinated conidium. (L) L.S. mitochondrion from germ tube.

sort of composition that is likely to prove widespread, and provides a basis for understanding some of the properties of spores. Some of this information is reviewed below.

Gross composition

Owens et al. (1958) reported that the composition of conidia of *Neurospora sitophila* (expressed as percentage of dry weight) was: spore wall, 27.2; protein, 25.8; lipids, 17.7; carbohydrates, 9.3; RNA, 9.0; ash, 6.8; DNA, 1.4; free amino acids and amides, 1.3; adenosine polyphosphates, 0.24; and organic acids, 0.2. Air-dried conidia contained water equivalent to 9.1% of the dry weight, but if the spores were dried over calcium chloride the water content fell to 5.1% of the dry weight which was probably roughly equivalent to bound water. This contrasts with the wet weight / dry weight ratio of mycelium of *Neurospora crassa* of 4.27 (Slayman & Tatum, 1964) which is equivalent to a water content of 77% of the fresh weight or 327% of the dry weight. Analyses of conidia of other species are less comprehensive. Conidia of *Aspergillus niger* contain (also as a percentage of dry weight) only 4.1% of total lipid, of which 54.3% was polar and 45.7% neutral, but this was more saturated than was the mycelial lipid (Gunasekaram et al. 1972a,b). Ninety-five percent of the neutral lipid of conidia of *Glomerella cingulata* was triglyceride, less than 5% being accounted for by sterol ester, free fatty acid, unesterified sterol (of which ergosterol was tentatively identified as the main component), and diglyceride. The phospholipid fraction consisted of phosphatidylethanolamine, phosphatidylserine, phosphatidylcholine, and an unidentified phospholipid (Jack, 1964).

Shaw (1964) assembled published data relating to uredospores of *Puccinia graminis tritici*. His figures are percentages based on fresh weight. Since the water contents of the uredospores were 10--15% the following figures become directly comparable with the foregoing ones if multiplied by 1.11--1.18. The adjusted figures are given in parentheses. Shaw reports mannan and polyglucose carbohydrate as 21.9% (24.3--25.8); chitin as 1.4% (1.56--1.66); crude protein as 25.9% (28.8--30.5); lipids as 19.7% (21.9--23.2) with fatty acids comprising 16.4% of the fresh weight of the uredospores, glycerol 2% and unsaponifiable lipid 1.6%; organic acids were 0.8% (0.89--0.94), ATP 0.029% (0.032--0.034), and ADP 0.085% (0.094--0.1). Between 15.2 and 20.2% of the fresh weight was not accounted for in these analyses.

In broad respects there are similarities between the analyses of conidia of *Neurospora* and uredospores of *Puccinia*. Both are xenospores. Roughly a quarter of the dry weight of the spore represents the wall, or materials which very likely comprise the wall. Roughly another quarter is protein and another fifth lipid. The balance (i.e., about a third) includes nucleic acids, low molecular weight intermediates and cofactors, and minerals.

By contrast the sporangiospores of Mucorales, like the conidia of *Aspergillus niger*, lack large lipid reserves (Sumner & Morgan, 1969; Weete et al., 1970). Sporangiospores of *Mucor mucedo, M. ramannianus, M. racemosus* and *M. hiemalis* contained (as percentages of their dry weight) only 3.7, 7.6, 4.1, and 8.4% of lipid, respectively (Sumner & Morgan, 1969). The lipid content of the spores was only about a third to a half of the amount in vegetative mycelium of the same species. Gunasekeram et al. (1972a,b) reported that the total lipid in ungerminated sporangiospores of *Rhizopus arrhizus* was only 2.9% of the fresh weight, and of this about two-thirds was neutral lipid and one third polar. The proportion of polar lipids increased during subsequent germination.

Thus in a few xenospores considered above, some emerge as lipid-rich and some as lipid-poor. The bulk of the difference presumably lies in lipids other than membrane components, since there is no evidence of correspondingly large differences in amounts of membrane in the different types. The fatty acid composition of the lipids in some fungal spores has however been found to be distinctive and unusual. The oil in uredospores contained up to 40% of 10-epoxyoctadecanoic acid; that of conidia of *Sphaerotheca humuli* 42% behenic acid, which otherwise is rare in fungal lipids; and that of conidia of *Erysiphe graminis* 45% of an unidentified fatty acid thought to have a branched or cyclic structure (Tulloch et al., 1959; Tulloch & Ledingham, 1960, 1962). Sixty-five percent of the fatty acid in conidia of *Penicillium atrovenetum* and 63% in chlamydospores of *Tilletia foetens* was linoleic acid (Van Etten & Gottlieb, 1965; Tulloch & Ledingham, 1960). By contrast, the fatty acid composition of spores of *Mucor* species quantitatively resembled that of the mycelium and contained no unusual components (Sumner & Morgan, 1969).

Free sugars constitute a rather variable component of spores. The diglucoside trehalose comprises up to 15% of the dry weight of ascospores of *Neurospora* and 5--7% of the dry weight of macroconidia, in both of which it is used during germination (Sussman et al., 1971). However Andrews (1964) found that conidia of *Pithomyces chartarum* contained only 0.5--1.4% (w/w) of glucose, 0.05--0.2% ribose, 0.07%

mannose, 0.03% fructose, 0.01--0.15% galactose, traces (<0.01%) of arabinose and xylose, and, as the only oligosaccharide present in measurable quantities, trehalose at 0.025%. Thus in these spores free sugars constituted only about 1% of the dry weight and trehalose was only a minor constituent.

Polyols have been reported in the basidiospores of *Schizophyllum commune* (Niederpruem & Hunt, 1967), the uredospores of *Puccinia graminis tritici* (Prentice & Guendet, 1954; Prentice et al. (1959), conidia of *Aspergillus* species (Sumi, 1928; Horikoshi et al., 1965) and of *Penicillium chrysogenum* (identified as erythritol and glycerol; Ballio et al., 1964). In uredospores of *Puccinia graminis tritici*, Reisener et al. (1962) found almost 12% (w/w) of polyols, viz. 4.1% mannitol, 5.6% arabitol, and 1.9% glyceritol. Polyols in fungi are however not confined to the spores, but occur also in many mycelia and fruit bodies, mannitol and arabitol being frequently reported. Holligan & Lewis (1973) report that, to their knowledge, ribitol as a product of glucose catabolism has never been found in extracts of molds.

Volutin has been demonstrated by cytochemical means in ascospores of marine fungi (Kirk, 1966) and, on the basis of the detection of polyphosphate in other spores (Gottlieb, 1966), is probably a widely occurring storage material. The volutin granules in spores of marine Ascomycetes are sites where polyphosphates and proteins accumulate, apparently in a loose complex with ribosomal RNA; their four principal components in apparent order of increasing abundance are lipid, RNA, polyphosphate and proteins (Kirk, 1966).

The enzymes of dormant spores are more the concern of other contributors to this symposium, but their presence, albeit in a largely nonfunctioning state, is important because it implies the existence of conditions in which they can remain stable for long periods. Though most of the enzyme protein of spores lies in the cytoplasm, some has been shown to occur in the wall, for example, some of the trehalase in conidia of *Neurospora crassa* (Sussman et al., 1971) and in *Neurospora* ascospores (Hecker, cited in Sussman & Douthit, 1973) and β-glucosidase, invertase and trehalase in *Aspergillus oryzae* (Horikoshi & Ikeda, 1965).

Walls

The walls of fungus spores often have a complex structure, though they are rarely so complex as the eight-layered wall of the resistant sporangium (functionally a memnospore) of *Allomyces neomoniliformis* which has six structural components, glucose, glucosamine, chitin, melanin, protein, and lipids (Skucas, 1967). Except for melanin and lipids, these are also the constituents of the walls of vegetative hyphae. More usually spore walls have from one to three layers discernible by transmission electron microscopy, though cytochemical techniques may reveal more. For example Graham (1960) detected five in the teleutospore of *Tilletia contraversa* (viz. two inner layers, a lipid-containing layer, a reticulate layer, and an outer sheath) and identified the mixtures of major components in each layer (hemicelluloses, chitin, pectic materials, proteins, lipid and melanin), and Kirk (1966), in a cytochemical study of ascospores of marine Pyrenomycetes, recognized that in several species the walls consisted of at least four layers.

Because analyses of spore walls have not often been accompanied by analyses of the walls of vegetative mycelia or of associated reproductive structures (sporangiophores, conidiophores, asci, etc.) of the same species, it is seldom clear what special chemical features the spore walls possess. Among xenospores, the conidia of *Aspergillus oryzae* have walls which contain polysaccharide (based on mannose, glucose, galactose, and glucosamine), ash, phosphate, protein, and nucleic acid. These are the constituents of the hyphal walls also, from which the conidial walls differ in containing considerably more protein and considerably less glucosamine (Horikoshi & Iida, 1964). Some of the spore wall protein was enzyme (Horikoshi & Ikeda, 1965). The walls of conidia of *Pithomyces chartarum* contained 45% carbohydrate, 3.3% nitrogen, 3.6% hexosamine, and 17.5% of protein, while the hyphal walls contained 40, 4.5, 10, and 25% of these constituents, respectively. It is chiefly in respect of hexosamine that the difference is large. Otherwise the composition is similar; glucose, galactose, mannose, and amino acids are present in acid hydrolysates of both, and two glycoproteins from each contain the same sugar and amino-acid components although those from the spore walls contain less carbohydrate and more protein (Sturgeon, 1966). The specialized surface layer of spicules which occurs on the conidia of *P. chartarum* has been referred to above. A distinctive feature of the asexual spores of Zygomycetes, is that not only the architectural structure but also the chemical composition of their walls (Bartnicki-Garcia & Reyes, 1964; Bartnicki-Garcia, 1970; Jones et al., 1968) is vastly different from that of the hyphal walls, which contain chitin, chitosan and polyuronides but no glucans. The walls of the asexual spores consist largely of glucan, polyglucosamine (tentatively identified as chitin) and melanin. Bartnicki-Garcia (1968) suggested that these differences in composition reflected an ontogenetic recapitulation of the phylogenetic history of the cell walls of Mucorales, the sporangiospore wall indicating a Chytridiomycete ancestry for the class.

Among memnospores, the oospores of *Phytophthora megasperma*, in association with the oogonia and antheridia with which they form the entire sexual apparatus, have recently been analyzed by Lippman et al. (1974). The walls of the sexual apparatus have a composition which is qualitatively similar but quantitatively different from that of the vegetative cells of related species. Glucans comprise almost 80% of the wall, mostly being highly branched noncellulosic β-glucan(s) linked mainly through the 3-position. Less than 10% of the wall is cellulose. Minor amounts of mannose and glucosamine are also present, these and glucose being in amounts comparable to those in mycelial walls of *Phytophthora* and *Pythium*. There is more protein, lipid, and phosphorus and less glucan than in vegetative walls. Although the analyses pertain to a mixture of walls, the rather massive oospore wall is likely to be the major component. It represents about 47% of the total dry weight of the oogonium - oospore - antheridium complex, one of the highest proportions known for wall material in fungal cells.

A recent study by Gooday et al. (1973) on another Phycomycetous memnospore, the zygospore of *Mucor mucedo*, revealed for the first time in fungi the presence of 1--3% of sporopollenin, formerly known only as a higher-plant product. It probably occurs as a component of the outer warty layer of the zygospore wall. The *Mucor mucedo* sporopollenin (empirical formula $C_{90}H_{130}O_{33}$) is very similar to those from many different pollens. Sporopollenin formation is probably a role for the increased production of carotenoids during sexual reproduction of many of the Mucorales. Histochemical methods indicate the presence of chitin and chitosan in the warty thickenings of the zygospore wall and chitin in the inner impermeable elastic layer. The zygospores of *Blakeslea trispora* contain a material which might be sporopollenin, but those of *Rhizopus sexualis*, like the vegetative mycelium of these fungi, contain no detectable amounts. This feature may be related to the absence of detectable carotene in this fungus. Because the structure and the formation of zygospores in Mucorales appear to be indifferent to the presence or absence of sporopollenin the latter presumably plays no essential structural role. Its extreme resistance to both chemical and biological degradation (Shaw, 1971) suggests a protective role. However the zygospores of *R. sexualis* which lack sporopollenin are highly resistant to unfavorable external factors. These spores germinate only with difficulty and it may be suggested that sporopollenin could be associated with ease of germination.

Although quantitative differences between spores and hyphae in respect of their composition are generally not large, it is evident from work on dimorphism of *Mucor* (Bartnicki-Garcia, 1963) that comparatively small changes in certain components (e.g., mannans) are correlated with profound changes of texture and shape of walls. So far there appears to be no clear correlation of wall composition with morphology of spores. Such differences between hyphal walls and spore walls as have been discovered are more readily interpreted as adaptive differences related to the need for protection against radiation damage (e.g., melanin in the walls of sporangiospores of *Mucor rouxii*) or microbial and chemical attack (e.g., in the previous example again and that of sporopollenin in the walls of zygospores of *Mucor mucedo*).

Surfaces

The surface chemistry of spores can be studied by particle-electrophoresis techniques which are particularly informative if combined with chemical and enzymic treatments of the spores. Fisher & Richmond (1969) by such means obtained results that were consistent with the view that the surface of sporangia of *Phytophthora infestans* was composed of cellulose devoid of ionizable groups; that the surface of basidiospores of *Stereum purpureum* bore carboxyl groups, was protein-free, and probably consisted wholly of carbohydrate; and that the surfaces of conidia of *Alternaria tenuis* and *Botrytis fabae* bore amino groups and carboxyl groups and were probably proteinaceous. Those of *Penicillium expansum* also bore amino and carboxyl groups, but these were overlain with easily removable phosphate. Those of *Eysiphe graminis*, and *Podosphaera leucotricha* showed electrophoretic mobility consistent with the presence of superficial amino and carboxyl groups. Fisher & Richmond (1970) later showed that the pH-electrophoretic mobility curves of conidia of five species of *Penicillium* were different and characteristic, but concluded that the differences were probably the result of the presence of a polyphosphate layer which was not an integral part of the conidium surface. Removal of the phosphate by enzyme action or prolonged washing exposed an amino-carboxyl surface. Fisher et al. (1972) showed by similar techniques that superficial lipid was also present on the asexual spores of *Alternaria tenuis, Botrytis fabae, Neurospora crassa,* and *Rhizopus stolonifer*, which are all airborne spores and difficult to wet, but that there was none on those of *Erysiphe cichoracearum, E. graminis, Penicillium expansum, Mucor rouxii, Nectria galligena,* and *Verticillium albo-atrum*, of which the last two species have water-dispersed conidia. The hydrophobicity of conidia of the first three species must be sought in other mechanisms than the presence of lipid layers: Fisher & Richmond (1970) suggest that in some spores the physical conformation of the surface may itself be sufficient to prevent wetting.

Pigments

Spores of fungi vary greatly in color, and much use is, or has been, made of this character in classification, especially in the Agaricales and Deuteromycetes. There have been recent attempts to quantify spore color objectively by means of a wavelength-scanning spectrophotometer to measure reflectance (Kulik & Johnson, 1969). Color may reside in either the wall or the cytoplasm or both. The commonest wall pigment appears to be melanin, though only rarely has this identification been based on specific tests. The fact that the dark conidia of *Aspergillus niger* contain a black pigment, aspergilline, which is not melanin argues for some caution. Aspergilline is a high molecular weight substance that contains a perylenequinone and several different amino acids (Barbetta et al., 1967). The distribution of presumed melanin in the walls of spores is revealed by many electron-microscopic studies. In different species, Durrell (1964) observed distribution of melanin as an outer layer, as scattered granules, as an impregnated outer layer, as granules at the surface which could be sloughed off or embedded in slime, or as radial bars through the wall.

The distinctive color of rust uredospores is partly due to color in the spore walls, which are generally brownish, and partly to the cytoplasmic carotenes which may figure among the "unsaponifiable lipids" of various spore analyses. Varying proportions of lycopene, β-carotene and γ-carotene occur in different species (Irvine et al., 1954). In uredospores of *Puccinia graminis tritici* Hougen et al. (1958) found minor amounts of carotenoids, together representing 0.12% of the freshly harvested spores, which contained 10 ppm of phytoene, 7 ppm of *cis*-β-carotene, 240 ppm of β-carotene, 45 ppm of *cis*-γ-carotene, 860 ppm of γ-carotene and 8 ppm of lycopene, but none of the colorless carotenoid, phytofluene (a common constituent of fungal carotenoids), nor of the oxygenated carotenoids, the xanthophylls.

Though spores of Agaricales vary greatly in color in the mass as a result of pigments in the individual spores, virtually nothing is known of the coloring materials responsible. It is possible that some of these are similar to those that color the fruit bodies of Basidiomycetes. These include carotenoid and quinonoid pigments and others of hydrocarbon and phenoxazine type (Arpin & Fiasson, 1971). One would expect melanin also to be present in brown, purple, and black spores of Agarics and many Gasteromycetes, but this seems to be unproven.

Water content

Published data on the proportion of water in fungus spores are surprisingly variable. This variability may sometimes reflect different experimental procedures. Somers & Horsfall (1966) found that estimations based on heating spores to constant weight suggested a higher water content than did freeze-drying and drying *in vacuo* over phosphorus pentoxide, presumably because heating drove off volatile constituents in addition to water. However it appears that some of the variation is real and reflects the intrinsic characteristics of different species and the conditions under which spores have been formed and harvested. The reported water contents of the spores of a number of species are given in Table 1. The conidia of powdery mildews (Erysiphaceae) should be considered separately owing to their possession of special features. Their conidia have often been considered to be unique in needing neither free water nor a high atmospheric humidity for germination and in even being hampered by wetting. However it is now apparent that the conidia of only a few powdery mildews (including *Erysiphe betae*, *E. polygoni*, and *Uncinula necator*) are capable of appreciable germination at very low humidities, and that although with other species a usually small proportion of spores will germinate under dry conditions, the highest germination is almost invariably recorded at relative humidities approaching 100% (Somers & Horsfall, 1966). Brodie & Neufeld (1942) thought that there was probably very little water in conidia of *Erysiphe polygoni* and that the protoplast consisted of a gellike material. However Yarwood (1950) found that water comprised 72% of the fresh weight of conidia of this species, 75% in *Erysiphe graminis* and 52% in *Erysiphe cichoracearum.* These figures were much higher than those which Yarwood found (Table 1) for spores of other sorts of fungi and led him not only to doubt Brodie & Neufeld's view but also to suggest that powdery mildew conidia carried with them the water that they need for germination. In the light of the evidence that other sorts of spores had water contents of the same order, Somers & Horsfall (1966) concluded that there appeared to be no justification for regarding the water content of powdery mildew conidia as being consistently greater than in spores of other fungi. They added that what may distinguish powdery mildew conidia from other fungi is the manner in which the water is bound. They nevertheless found by nuclear magnetic resonance spectrophotometry that the intracellular water in *Erysiphe graminis* conidia could exchange readily with external water, which was found also in ungerminated sporangiospores of *Rhizopus arrhizus* by Ekundayo (1966), using a heavy-water exchange technique. This is similar to the finding of Murrell & Scott (1958: cited in Sussman & Halvorson, 1966) who found that heavy water exchanged freely with almost all of the water in bacterial spores. From dielectric measurements, Carstensen et al. (1971) concluded that the water in bacillus spores was in normal states and not as tightly bound as in clathrate hydrates or hexagonal ice. Whether this is also true of fungus spores remains to be determined.

The water content of spores may even be different according to conditions prevailing while they form. Somers & Horsfall (1966) found that uredospores of *Uromyces fabae* contained 18% water when produced on plants in the open greenhouse and 51% when plants had been placed in damp chambers for four days. Under different conditions of cultivation conidia of *Erysiphe cichoracearum* contained 35, 45, and 66% water, and those of *Erysiphe graminis* 31, 44, and 69%.

The foregoing results suggest that although spores generally have water contents less than those of mycelium (in which early workers report that 85--90% of the fresh weight is water (Cochrane, 1958) and the data of Slayman & Tatum (1964) for *Neurospora crassa* indicate 77%), the amount of water which spores contain is very variable, partly as a result of certain intrinsic differences between species, and partly because of differences in environmental moisture conditions during their formation as well as subsequently during the period before they are actually analysed. The surface/volume ratio of spores is high and despite the presence of barriers which impede water loss (McKeen et al., 1967) their eventual equilibration with the

Table 1. Water Contents, as Percentages of the Fresh Weight, of the Asexual Spores of Some Species of Fungi

Water Content (%)	Species	Spore type
6	*Penicillium digitatum* 1	Conidia
10--15	*Cronartium fusiforme* 2	Aecidiospores
12	*Uromyces phaseoli* 1	Uredospores
13	*Aspergillus niger* 1	Conidia
17	*Peronospora destructor* 1	Conidia
17	*Botrytis cinerea* 1	Conidia
17.4	*Aspergillus oryzae* 3	Conidia
25	*Monilinia fructicola* 1	Conidia
35	*Penicillium glaucum* 4	Conidia
42.5	*Aspergillus oryzae* 5	Conidia
42.5	*Mucor* sp. 4	Sporangiospores
44	*Stemphylium sarcinaeforme* 6	Conidia
50	*Alternaria tenuis* 6	Conidia
55	*Aspergillus niger*[a] 7	Conidia
65	*Penicillium expansum*[a] 7	Conidia
65	*Penicillium italicum*[a] 7	Conidia
74	*Neurospora crassa*[a] 7	Conidia
88	*Botrytis fabae*[a] 7	Conidia

[a] Calculated on basis of fresh weight of spores, their measured volume, and assumption of a density factor of 1.1.

Authority

1. Yarwood, 1950
2. Walkinshaw et al., 1967
3. Sumi, 1928
4. Cramer, 1891
5. Aso, 1900
6. Somers & Horsfall, 1966
7. Richmond & Somers, 1963

water activity of the environment is to be expected. It is highly desirable to have information on the water content of individual spores when just mature and as they subsequently age. The data of Barer & Joseph (1958) are relevant. These workers employed refractometry to determine the total solid concentration of fungus spores. They found that the mature spores of many Penicillia and Phycomycetes (including *Piptocephalis* and *Zygorhynchus*) had concentrations of solids that were among the highest known in any living cell. These were mostly 45--55% (wt/vol) but some even exceeded 55%. These values correspond to water contents of 60% or less of the fresh weight. A particularly interesting finding was that spores from fresh cultures gave consistently smaller values for solid-content than did similar samples from old cultures. For example, conidia of *Penicillium roquefortii* gave a modal value of 44.8% solids when taken from a fresh culture, but after the culture had been stored in a stoppered bottle in which the atmosphere appeared to be water-saturated, the modal value had increased to 54.3% solids. The authors concluded that the water loss which accompanied this rise in proportion of solids was probably the result of a natural process of dehydration or syneresis (the spontaneous expulsion of liquid from a gel) rather than simple drying. In the light of evidence of expulsion of water from spores, a further observation by Barer & Joseph assumes special significance. In the long basipetal chains of conidia of some Penicillia the youngest (i.e., proximal) conidia contained the same concentration of solids as the phialide (about 27%) but the concentration increased progressively along the chain till it exceeded 37% in the more mature distal conidia.

The available evidence therefore suggests that, in general, mature fungus spores have a water content which is appreciably below that of the parent mycelium as a result of internal processes; that the final water content is determined in part by direct physical effects of the environment during the spore's formation, maturation, and ageing; and that viability is retained over a range of water contents that may fall to very low values.

SOME PHYSICAL PROPERTIES

The density of fungal spores is commonly of the order of 1.1--1.2, but lower densities have been recorded which are probably attributable to buoyancy adaptations such as internal gas bubbles or empty cell fragments (Gregory, 1973). As yet no general correlation between density and type of spore has emerged. Diverse spore-types are represented in the above density range which includes even the physiologically distinctive conidia of powdery mildews. Thus Jhooty & McKeen (1965) report that conidia of *Sphaerotheca macularis* have densities of 1.10--1.11 g/ml.

Wettability of the surface is a very variable physical property of spores. Many that are formed in slime are intrinsically highly wettable and rely on water for their dispersal (e.g., most Coelomycetes). The mucilaginous layer around conidia of *Fusarium culmorum* which abuts on a layer of chitin microfilbrils is probably mainly composed of xylan (Marchant, 1966b). However not all slimes in which spores are produced are water miscible, and, related to this, one can expect differences in the surface properties of the spores themselves. Whitney & Blauel (1972) reported that although the slimy spore masses of three species of *Ceratocystis* dispersed in water, those of seven other species and one of *Europhium* did not. In contrast, all but one dispersed in resin. Cirri of *Ceratocystis montia* dispersed in pinene, α-pinene, β-pinene, Δ^3 carene, undecane, toluene, benzene, xylene, ligroine, oleic acid, linoleic acid, or corn oil, but not in water, ethanol, tertiary butanol, methyl cellosolve, methyl benzoate, aniline, acetone, or various detergents. Such data suggest that slimes associated with water dispersal of spores may have very different physical properties from those of spores dispersed by arthropods.

Fisher et al. (1972) found that, although certain fungus spores with surface lipid were difficult to wet (conidia of *Alternaria tenuis*, *Botrytis fabae*, and *Neurospora crassa*; and sporangiospores of *Rhizopus stolonifer*), and certain others which were water-dispersed (conidia of *Verticillium albo-atrum* and *Nectria galligena*) were without lipid on their surfaces, there were others which were hydrophobic yet lacked surface lipids. These included the conidia of *Penicillium expansum* and of the powdery mildews, *Erysiphe cichoracearum* and *E. graminis*. Powdery mildew conidia are resistant to desiccation as well as being water repellant. McKeen et al. (1967) considered that a gelatinous, almost transparent, approximately 1 μm thick, outer layer of the conidium of *Erysiphe cichoracearum* was responsible for the impervious properties of the spores. Fisher & Richmond (1969), from a study of the electro-kinetic properties of the conidia of *Podosphaera leucotricha* and *Erysiphe graminis*, concluded that unionized compounds were possibly responsible for their water repellancy. They further concluded that the water repellant properties of *Penicillium* conidia were not due to lipids or cyclic peptides. Neither was the polyphosphate layer on the conidia responsible, since it could be lacking yet spores still be hydrophobic (Fisher & Richmond, 1970). They consequently suggested that the physical conformation of the surface might itself be sufficient to prevent wetting. Fogg (1948) pointed out that the extent of spreading of a droplet over a surface is determined by the nature of the chemical groups at the surface, its degree of roughness, and the presence or absence of an air film below the droplet. Fisher et al. (1972) suggested that the trapping of air in surface invaginations in the rodlet pattern of certain conidia might prevent wetting. Because rodlet patterning occurs not only on spores but also on other aerial parts, Morton's (1961) discovery of an abrupt change in the physical condition of the surface of *Penicillium* mycelium soon after it emerges from a liquid medium into air is of particular interest. He found that surface-active material, which tests suggested might have been protein, had formed at the mycelial surface even within 30 seconds of emergence. Whether this rapidly formed surface-active material comprised a rodlet layer is unknown.

The spores of *Pithomyces chartarum* in the mass are particularly difficult to wet, and will even travel up a continuous water film by a surface tension effect, thereby ascending at least 2--5 mm up grass leaves (Crawley et al., 1962). This remarkable water repellancy, as well as the control of movement of water in and out of the spore, may be in part attributable to the sporidesmolides and lipids which form the spicular surface layer of the spore (Bertaud et al., 1963).

INTERPRETATION OF SPORE FEATURES

The fungus spore may be considered as a device which performs certain essential functions in the life cycle. It is possible to recognize in many of the features of spores characteristics which appear to adapt spores to perform particular functions better. In man-made devices a penalty is almost always incurred whenever there is any special elaboration of a device, and compromises become necessary. The same is likely to be true of spores. There is conflict between some of the theoretical requirements of efficient harnessing of dispersal agencies and those of adequate provision for longevity, or of adequate protection against environmental hazards and of capacity to germinate when conditions are favorable. Natural selection has probably led to compromise. Consequently one is rarely, if ever, able to establish single simple correlations between particular features of spores and particular functions, because the presence of one feature may have compelled, directly or indirectly, the modification of another, or one feature may function in several different capacities (as for example does melanization). Also some features may be intrinsically neutral in that their presence confers no selective advantages or disadvantages; they are simply neutral manifestations of important underlying events. Despite these difficulties one can seek at least the more obvious correlations between form and composition on the one hand, and function on the other, and, if unsuccessful, consider whether some aspects of form are merely evolutionarily neutral consequences of the spore-producing process.

Mature dormant spores have many different shapes, yet most tend towards globose or ellipsoidal form. Adaptation to function appears to be a powerful influence which has determined the shapes of spores though this evolutionary influence does not, of course, account for the way in which shape is imparted to an individual developing spore. Ontogeny is not, however, the concern of this symposium.

The relationships between features of spores and their liberation, dispersal and deposition have been considered by many workers, but good recent accounts have been given by Ingold (1971) and Gregory (1973). Ingold gives special consideration to the relationship between size and form of spores and mechanisms for their liberation or discharge. In the Ascomycetes in particular, in which forcible spore discharge is highly developed, correlations emerge. Ascomycetes which can shoot their spores to relatively great distances have large spores or spores bound together in a single mass, or both these features. Also, ascospores in general tend to have a bipolar symmetry, with the distal end being blunter. This shape can be interpreted as heightening the efficiency of the discharge mechanism (Ingold, 1954). Gregory points out that after a spore has been discharged or released into the air, its size, density, surface roughness, and possibly electrostatic charge combine to control the speed at which it falls in still air (i.e., its terminal velocity of sedimentation) and consequently its prospect of long distance dispersal in turbulent air. Size is important in determining impaction efficiency. Although it is desirable that a spore should eventually alight on a substratum, excessively high impaction efficiency would in some situations, such as in dense vegetation, prevent efficient dispersal. Gregory remarks that 10-μm diameter spores may represent a working compromise between the conflicting requirements of dispersal and deposition, evolved under the normal range of winds encountered among vegetation. He points out that the spores of dry-spored airborne leaf pathogens, such as uredospores, aecidiospores, and spores of *Phytophthora* and *Helminthosporium*, are usually comparatively large. On the other hand dry-spored soil inhabitants (e.g., *Penicillium* and *Aspergillus*) are characterized by small spores unsuitable for impaction.

One would expect size also to have some bearing on longevity insofar as a larger spore has a greater capacity for storage of reserves, a potentially greater thickness of wall, and proportionately less surface through which exchanges with the environment can occur. Longevity is however much influenced by environmental conditions, and correlations between size and survivability are not readily made. Even small, thin-walled conidia can survive for long periods under appropriate conditions. The tiny spores (ca. μm^3) of *Beauveria bassiana* showed 90% germinability after 635 days in darkness at 0% relative humidity at 8°C. However within 28 days in light at 75% relative humidity at 18°C none were germinable (Clerk & Madelin, 1965). Possibly largeness in spores provides a resilience which widens the variety of conditions under which survival is long. Gregory (1966) notes that memnospores which often are very durable also tend to be large. Xenospores, in which survival tends to be short, vary much in size, but are often minute.

During the period in which a nonflagellate spore remains quiescent, it slowly metabolises and very slowly consumes oxygen (e.g., at ca. 0.5 μl/mg dry wt/h in uredospores of *Puccinia graminis* and ascospores of *Neurospora*, Allen, 1965). Insofar as there is respiration, substrates are being utilized. Energy, and hence respiratory substrate, is needed also in large amounts during subsequent germination, though in the transition from dormancy to incipient germination there may be a change in the particular reserves which are being used, as for example, from lipid to carbohydrate in *Neurospora* (Lingappa & Sussman, 1959). Lipids, carbohydrates, and, as more quickly available sources of energy, polyphosphates and phospholipids (Gottlieb, 1966) have all been reported in dormant spores, and are sometimes conspicuous in spores studied cytochemically or electronmicroscopically. The wall, which in most spores represents a relatively large quantity of chiefly polysaccharide material, might possibly also act as a storage material. The fact that the internal portion of the oospore wall of *Phytophthora* is markedly eroded on germination suggested to Lippman et al. (1974) that a main role for the massive spore wall is carbohydrate storage, the surrounding thinner oogonium wall probably being protective, for example against microbial lysis.

The polyols that have been detected in a number of dormant spores, and which are in large amount (ca. 12% of the dry weight) in uredospores of *Puccinia graminis tritici* (Reisener et al., 1962), probably also serve as respiratory substrates. However another role may be envisaged. Spores are exposed for much of the time to conditions of low environmental water activity. Though bounded by more or less impervious walls, they will tend towards equilibrium with their environments, and metabolic functions within the spore will be performed under conditions likely to be of physiologically low water activity. Brown & Simpson (1972) found that polyols were a major intracellular component of yeasts able to tolerate high sugar concentrations (i.e., environments of low water activity), and functioned in them as "benign solutes," that is, ones which at high concentration allow enzymes to function effectively. Similar adaptations might exist in spores and so allow at least limited functioning of enzymes under these physiologically dry conditions. Further Hecker & Sussman (1973) have recently reported that the component in extracts of ascospores of *Neurospora* which protects trehalase from heat inactivation is trehalose, itself a polyol, and that the heat stability of trehalose *in vitro* is increased by the presence of polyols, including mannitol and inositol near the limits of their solubility. Polyols thus probably serve more than one function in spores.

The dark pigment (usually, but not always, melanin) in the walls of many spores renders spores less vulnerable to radiation damage. Sussman & Halvorson (1966) reviewed literature which demonstrates that fungal spores with darker pigmentation have greater resistance to ultraviolet radiation than have light-colored ones. Even within a single species spores may become less vulnerable to radiation damage as they mature and darken (Uduebo & Madelin, 1974). Similarly, when the black-brown color of conidia of *Aspergillus carbonarius* was reduced by cultivation of the fungus on media with various additives, the ultraviolet resistance was also reduced (Curtis, 1970). However, among dark pigments, melanin also has structural and protective roles. Campbell et al. (1968) found that the multicellular conidia of an albino mutant of *Alternaria brassicicola* were fragile and commonly disrupted in aqueous mountants so that the component cells, each surrounded by a wall, escaped from the envelope of the conidium. Also, unlike the wild-type conidia, albino conidia dissolved completely if heated in zinc chlor-iodide solution. Thus roles for melanin both in contributing to the structural integrity of the spore and in protecting its walls against lysis were indicated. There are numerous reports of pigmented propagules being shorter lived in soil than unpigmented ones. For example, Old & Robertson (1970) report that conidia of hyaline isolates of *Cochliobolus sativus* had a very brief existence on soil compared with those of wild-type isolates; they also were very readily lysed by snail-gut enzyme or chitinase.

It is clear that a number of biological functions and physiological activities are profoundly affected by the actual surface layers of spores and that there is much need for further study. The optical properties of spore surfaces (as opposed to spore walls) have received little direct study. It is nevertheless possible that the reduction of the amount of radiation penetrating a spore by reflection at the surface is important. Many fungal spores are bright in the mass. This is particularly evident in the spore prints of agarics and in cultures of many molds. They could not appear bright if they were not reflective. Possibly reflectivity is an alternative protective device to screening by pigments. The recent demonstration that reflection in glass lenses can be reduced by simulating the finely papillate surface texture of the eye of certain moths (Clapham & Hutley, 1973) shows that the physical form of fine ornamentation at the surface of spores might affect their optical properties also, and thereby become subject to the workings of natural selection.

The absence of a correlation between the ornamentation of conidia and ascospores in individual species of Aspergilli, which emerges from such work as Locci's (1972), is not easily interpreted. If the type of spore surface confers no competitive advantage, then this feature is largely freed from the pressures of natural selection, in which case any pattern that is not disadvantageous might fortuitously arise. Since rather similar ranges of ornamentation occur in both conidia and ascospores of Aspergilli there is no evidence that the very different ways in which these spores are generated intrinsically favor particular sorts of ornamentation. In the absence of selective pressures and ontogenetic influences one might expect a natural economy to lead to similar patterns appearing in the one species on both its conidia and ascospores, genes controlling both processes being shared. However the surfaces of conidia and ascospores in single *Aspergillus* species often are not similar suggesting that some sort of selection pressure does in fact operate. The absence of any mechanism for ascospores of Aspergilli to be discharged forcibly through pores in ascus tips has freed their evolution from the constraints which such discharge imposes, and some have evolved elaborately flanged shapes. Ingold

(1954) has pointed out the advantages for forcible discharge of the ovoidal shape of many ascospores. If the selection pressure that shapes the surfaces of conidia and ascospores is not one that operates during production or release of the spores, and as argued above, this seems unlikely, it must function later, in which case the difference between conidium and ascospore ornamentation in individual species may be related to their different roles as xenospores and memnospores. The ornamentation of ascospores of Aspergilli may thus be related to their role of survival. A particular example can be selected to illustrate this possibility. The ascospores of *A. striatus* have many simple or anastomosing bar or bandlike ridges arranged in more or less concentric rings or in an irregular fashion (Raper & Fennell, 1965; Locci, 1972). Their external form very closely resembles that of the resistant sporangia of certain species of the phylogenetically distant Blastocladialean fungus, *Coelomomyces* (e.g., *C. indicus*; Plate 2, Fig. 11), although their size is much smaller. This resemblance may be more than coincidental if this unusual sort of structure is advantageous in memnospores that need to survive in mud, for it was from mangrove mud that Rai et al. (1964) isolated this *Aspergillus*, and it is in mud that many species of *Coelomomyces* (obligate parasites of chiefly mosquitoes) have to survive. Mud almost certainly presents spores with special problems of survival owing to physical stresses as it dries and moistens. Study of the ornamentation of memnospores of fungi in relation to their ecology might prove illuminating, as also might studies of mutational changes in spore ornamentation (Littlefield, 1971) with the purpose of establishing whether the surface characteristics of different spore types of the one species are genetically linked. The broad structural similarities between some conidia and ascospores is suggestive of considerable sharing of the genome (cf. conidia and ascospores of *Pleospora herbarum*, and of *Venturia inaequalis*). However as Ingold (1954) remarks, in speculations on the shapes of ascospores two types of explanation can be advanced, viz. a morphogenic one and a teleological one. These are not mutually exclusive though a simple morphogenic explanation may suggest that it is superfluous to look for a teleological one.

While some species show similarity in form between asexual and "perfect" stage spores, a very large number show striking differences in form, for example, the various spore types of certain rusts or the sporangiospores and zygospores of the Mucorales.

As noted above, the water content of spores is below that of vegetative mycelium by a greater or lesser amount, probably owing to both internal and external factors. It is especially significant that Barer & Joseph (1958) found that spores diminished in water content as they aged even within the parent culture under apparently saturation conditions. The application of their refractometric technique to other spores, especially those formed in aqueous surroundings (e.g., pycnidiospores, ascospores, and spores of aquatic higher fungi), is desirable. A basic need is to understand how spores can become of lower water content than the cells which support or surround them (e.g., conidiophores and asci). Provided that there is a conduction path, water will move by diffusion or mass flow from a region of higher water potential to one of more negative potential. Therefore if spores are to lose water by physical means as they mature they must maintain a water potential which is positive with respect to adjacent systems with which they are in connection, a seemingly difficult task when these are likely to be of relatively high water potential themselves during the process of sporulation. There are nevertheless possible mechanisms which would allow this. One would involve reducing the number of molecules in solution in the aqueous phase of the spore through polymerization or condensation reactions. The appearance of polysaccharides, lipids, polyphosphate, peptides, proteins, and nucleic acids within the osmotic volume of the spore (i.e., within the confines of its plasmalemma) is consistent with the operation of this mechanism. A second mechanism would be for the pressure potential within the cytoplasm to be increased. This could be effected in a spore no longer in continuity with the parent cell if the lumen were diminished as a result of centripetal wall thickening. The water potential of the spore would thereby become more positive. However for this to lead to a loss of water by the spore as a whole the newly formed wall material, if produced by the cytoplasm of the spore, would have to have a lower density than the cytoplasm it displaces. Bartnicki-Garcia (1963) reports that in the mycelium and yeast-phase of *Mucor rouxii* the bulk of the cell wall can in fact vary greatly with little change in the amount of solid material present, presumably as a result of textural changes. Wall material of unusually low solid content is at least possible. The centripetal thickening of walls observable by light or electron microscopy in some spores as they mature and age (e.g., Campbell, 1970a) could therefore represent a mechanism by which at least the cytoplasm of spores becomes more concentrated. It might even be a cause of folds in the plasmalemmas of mature spores. Lewis et al. (1960) suggested that the low water content of bacterial spores was maintained by mechanical pressure exerted by a slow contraction of the cortex during maturation, and Carstensen et al. (1971) refer to the possibility that polymer contraction, especially in the cortex, could be the cause of their high density. Critical attention to the dimensions, fine-structural organization and physical characteristics of the walls of maturing fungal spores might reveal whether comparable mechanisms operate in them. Because of the range of water contents involved it seems unlikely that alteration of the matric water potential of the solids in and around spores could significantly raise the water potential of the whole spore, though the introduction of lipids into wall layers might lesson the affinity of other wall constituents for water.

There is however the possibility that an active biological process might augment the physical processes that promote elimination of water. Sporulation is commonly accompanied by the formation of guttation droplets on or in the vicinity of the newly formed spores (Madelin, 1969; Plate 5, Fig. 25). A mechanism for the active elimination of water has not been demonstrated in spores, but a possible one involves lomasomes and other vesicles which fuse with the plasmalemma. Lomasomes or similar structures or vesicles fusing with the plasmalemma have been reported in many maturing spores (including conidia of *Aspergillus nidulans*, Weisberg & Turian, 1971; of *Fusarium culmorum*, Marchant, 1966; of *Alternaria brassicicola*, Campbell, 1968; and of *Botrytis fabae*, Richmond & Pring, 1971a; ascospores of *Podospora anserina*, Beckett et al., 1968; and of *Penicillium vermiculatum*, Wilsenach & Kessel, 1965). While the export of materials involved in wall synthesis may prove to be their major role, the contents of the discharged vesicles are presumably in aqueous solution or suspension. Extensive fusion of vesicles with the plasmalemma of a spore during its maturation would expel appreciable amounts of water.

We know of no published determination of the equilibrium water activity of the spores of any fungus. In bacterial spores rehydrated to various moisture contents after initial drying, this activity is almost the same as that of pure water except when spores are of very low water content (i.e., less than about a quarter of their fresh weight, Waldham & Halvorson, 1954). A similar situation is to be expected in fungal spores during the period over which their water content is falling, to account for the fact of their water loss. Their capacity to imbibe water when about to germinate implies that there is then a reduction of their equilibrium water activity (i.e., a corresponding shift to a more negative water potential), so that water diffuses or flows into them. Ekundayo (1966) suggests that swelling of *Rhizopus arrhizus* sporangiospores in early germination is a result of increased plasticity of the wall. Reduction of wall pressure by plasticizing the wall would make the water potential of the spore more negative and so lead to water uptake. Similar mechanisms, but working in opposite directions, might therefore account for dehydration during maturation and for hydration during germination.

One should question whether the generally low water content of spores is an adaptive feature in itself, or is perhaps incidental to other features which happen to lead to dehydration, such as the accumulation of reserves stored as insoluble macromolecules, or the secretion of materials for wall synthesis, or the building up of a thick wall by addition of interior layers. There is a certain amount of indirect evidence that a state of dehydration is advantageous in spores. James & Werner (1965) report that in yeast, substances including glycerol, glucose, and ethanol, which control free water and either dehydrate the cells or tie up water in them, serve to protect against radiation damage. Further, whatever may be the details of the underlying physiological basis, for many spores, though not for all (Cochrane, 1958; Clerk & Madelin, 1965), survival is longest under conditions of environmental dryness, which presumably soon become reflected in internal dryness of the spores. Bearing in mind the roles of spores, it thus seems that dehydration is advantageous, but its occurrence might partly be a more or less beneficial corollary to other more important processes involved in the formation of mature spores. As stated above, simple correlations between the composition of spores and their function may not exist.

From this summary of much of the extensive literature relating to dormant fungus spores the following conclusions emerge:

1. Spores of different species and of different phases of the same species show a wide range of size, shape, and surface characters. Some of these characters are seen to be advantageous to some spores, but many may well have arisen fortuitously.

2. The internal structure of memnospores and endogenously dormant xenospores differs in many features from that of exogenously dormant xenospores. Differences in wall structure and in the form of the mitochondria are particularly striking.

3. The chemical composition of spores is varied in detail but shows some common factors, such as low water content compared with that of vegetative hyphae and the presence of abundant reserve food material.

An attempt has been made to interpret some of the features noted. The later contributions in this symposium will we hope make such interpretation easier and thus increase our understanding of the biology of fungi and their spores.

ACKNOWLEDGMENTS

We would like to thank the organizers and Brigham Young University for arranging this symposium and inviting us to contribute to it; Miss C. Y. Fleming, Clerk to the Dean of the Faculty of Science of the University of Bristol, her colleagues, and Miss Rosemary Hamilton for typing the manuscript; and Mrs. Mary Dowden, Miss Mary Syrop and Dr. A. Beckett for photographic assistance.

REFERENCES

Allen, P. J. (1965). Metabolic aspects of spore germination in fungi. Annu. Rev. Phytopathol. 3, 313--342.

Allen, J. V., Hess, W. M., & Weber, D. J. (1971). Ultrastructural investigations of dormant *Tilletia caries* teliospores. Mycol. 43, 144--156.

Andrews, I. G. (1964). Free sugars in spores of *Pithomyces chartarum*. Nat. Lond. 203, 204--205.

Arpin, N. & Fiasson, J.-L. (1971). The pigments of Basidiomycetes: their chemotaxonomic interest. In Evolution in the Higher Basidiomycetes (ed. R. H. Petersen) pp. 63--96. Knoxville: University of Tennessee Press.

Aso, K. (1900). The chemical composition of the spores of *Aspergillus oryzae*. Bull. Coll. Agric. Tokyo Imp. Univ. 4, 81--96.

Austin, W. L., Frederick, F., & Roth, I. L. (1974). Scanning electron microscope studies on ascospores of homothallic species of *Neurospora*. Mycol. 66, 130--138.

Ballio, A., DeVittorio, V., & Russi, S. (1964). The isolation of trehalose and polyols from the conidia of *Penicillium chrysogenum* Thom. Arch. Biochem. Biophys. 107, 177--183.

Barbetta, M., Casnati, G., & Ricca, A. (1967). Aspergillina. Rend. Ist. Lombardo Sci. Lett. A, 101, 75--99.

Barer, R. & Joseph, S. (1958). Concentration and mass measurements

in microbiology. J. Appl. Bacteriol. 21, 146--159.
Barnett, H. L. (1955). Illustrated genera of imperfect fungi. Minneapolis: Burgess Publishing Co.
Bartnicki-Garcia, S. (1963). Mold-yeast dimorphism of *Mucor*. Bacteriol. Rev. 27, 293--304.
Bartnicki-Garcia, S. (1968). Cell wall chemistry, morphogenesis, and taxonomy of fungi. Annu. Rev. Microbiol. 22, 87--108.
Bartnicki-Garcia, S. (1970). Cell wall composition and other biochemical markers in fungal phylogeny. In Phytochemical Phylogeny (ed. J. B. Harborne). pp. 81--103. London and New York: Academic Press.
Bartnicki-Garcia, S., Nelson, N., & Cota-Robles, E. (1968). Electron microscopy of spore germination and cell wall formation in *Mucor rouxii*. Arch. Mikrobiol. 63, 242--255.
Bartnicki-Garcia, S. & Reyes, E. (1964). Chemistry of spore wall differentiation in *Mucor rouxii*. Arch. Biochem. Biophys. 108, 125--133.
Beckett, A., Barton, R., & Wilson, I. M. (1968). Fine structure of the wall and appendage formation in ascospores of *Podospora anserina*. J. Gen. Microbiol. 53, 89--94.
Bertaud, W. S., Morice, I. M., Russell, D. W., & Taylor, A. (1963). The spore surface in *Pithomyces chartarum*. J. Gen. Microbiol. 32, 385--395.
Besson, M. (1966). Les membranes des ascospores de levures au microscope électronique. Bull. Soc. Mycol. Fr. 82, 489--503.
Bigelow, H. E. & Rowley, J. R. (1968). Surface replicas of the spores of fleshy fungi. Mycol. 60, 869--887.
Blackwell, E. M. (1949). Terminology in *Phytophthora*. Mycol. Pap. no. 30 Kew, Surrey: C.M.I.
Brandenburger, W. & Schwinn, F. J. (1971). Über Oberflächenfeinstrukturen von Rostsporen. Arch. Mikrobiol. 78, 158--165.
Brodie, H. J. & Neufeld, C. C. (1942). The development and structure of the conidia of *Erysiphe polygoni* D.C. and their germination at low humidity. Can. J. Res. 20, Sect. C, 41--61.
Bronchart, R. & Demoulin, V. (1971). Ultrastructure de la paroi des basidiospores de *Lycoperdon* et de *Scleroderma* (Gastéromycètes) comparée à celle de quelques autres spores de champignons. Protoplasma 22, 179--189.
Brown, A. D. & Simpson, J. R. (1972). Water relations of sugar-tolerant yeasts: the role in intracellular polyols. J. Gen. Microbiol. 72, 589--591.

Buckley, P. M., Sjaholm,V. E., & Sommer, N. F. (1966). Electron microscopy of *Botrytis cinerea* conidia. J. Bacteriol. 91, 2037--2044.

Buckley, P. M., Sommer, N. F., & Matsumoto, T. T. (1968). Ultrastructural details in germinating sporangiospores of *Rhizopus stolonifer* and *Rhizopus arrhizus*. J. Bacteriol. 95, 2365--2375.

Campbell, R. (1968). An electron microscope study of spore structure and development in *Alternaria brassicicola*. J. Gen. Microbiol. 54, 381--392.

Campbell, R. (1969). Further electron microscope studies of the conidium of *Alternaria brassicicola*. Arch. Mikrobiol. 69, 60--68.

Campbell, R. (1970a). An electron microscope study of exogenously dormant spores, spore germination, hyphae and conidiophores of *Alternaria brassicicola*. New Phytol. 69, 287--293.

Campbell, R. (1970b). Ultrastructure of an albino strain of *Alternaria brassicicola*. Trans. Br. Mycol. Soc. 54, 309--312.

Campbell, R. (1972). Ultrastructure of conidium ontogeny in the deuteromycete fungus *Stachybotrys atra* Corda. New Phytol. 71, 1143--1149.

Campbell, R., Larner, R. W., & Madelin, M. F. (1968). Notes on an albino mutant of *Alternaria brassicicola*. Mycol. 60, 1122--1125.

Carstensen, E. L., Marquis, R. E., & Gerhardt, P. (1971). Dielectric study of the physical state of electrolytes and water within *Bacillus cereus* spores. J. Bacteriol. 107, 106--113.

Chapman, J. A. & Vujičić, R. (1965). The fine structure of sporangia of *Phytophthora erythroseptica* Pethybr. J. Gen. Microbiol. 41, 275--282.

Clapham, P. B. & Hutley, M. C. (1973). Reduction of lens reflection by the 'moth eye' principle. Nat. Lond. 244, 281.

Clerk, G. C. & Madelin, M. F. (1965). The longevity of conidia of three insect-parasitising Hyphomycetes. Trans. Br. Mycol. Soc. 48, 193--209.

Cochrane, V. W. (1958). Physiology of fungi. New York: Wiley.

Cole, G. T. (1973a). Ultrastructure of conidiogenesis in *Drechslera sorokiniana*. Can. J. Bot. 51, 629--638.

Cole, G. T. (1973b). Ultrastructural aspects of conidiogenesis in *Gonatobotryum apiculatum*. Can. J. Bot. 51, 1677--1684.

Cole, G. T. (1973c). A correlation between rodlet orientation and conidiogenesis in Hyphomycetes. Can. J. Bot. 51, 2413--2422.

Cole, G. T. & Aldrich, H. C. (1971). Ultrastructure of conidiogenesis in *Scopulariopsis brevicaulis*. Can. J. Bot. 49, 745--755.

Cramer, E. (1891). Die Ursache der Resistenz der sporen gegen trockne Hitze. Arch. Hyg. 13, 71--112.

Crawley, W. E., Campbell, A. G., & Smith, J. D. (1962). Movement of spores of *Pithomyces chartarum* on leaves of rye grass. Nat. Lond. 193, 295.

Cummings, G. B. (1971). The rust fungi of cereals, grasses and bamboos. Berlin, Heidelberg, and New York: Springer-Verlag.

Curtis, C. R. (1970). Comparison of UV-induced delay in germination in pigmented and pigment-inhibited conidia of *Aspergillus carbonarius*. Radiat. Bot. 10, 125--130.

Dodge, B. O. (1924). Aecidiospore discharge as related to the characters of the spore wall. J. Agric. Res. 27, 749--756.

Durrell, L. W. (1964). The composition and structure of walls of dark fungus spores. Mycopathol. Mycol. Appl. 23, 339--345.

Durrieu, G. (1974). Teliospore ornamentation in *Puccinia* parasitic on Umbelliferae. Trans. Brit. Mycol. Soc. 62, 406--410.

Eckblad, F.-E. (1971). Spores of Gasteromycetes studied in the scanning electron microscope (S.E.M.) I. Norw. J. Bot. 18, 145--151.

Ekundayo, J. A. (1966). Further studies on germination of sporangiospores of *Rhizopus arrhizus*. J. Gen. Microbiol. 42, 283--290.

Fisher, D. J., Holloway, P. J., & Richmond, D. V. (1972). Fatty acid and hydrocarbon constituents of the surface and wall lipids of some fungal spores. J. Gen. Microbiol. 72, 71--78.

Fisher, D. J. & Richmond, D. V. (1969). The electrokinetic properties of some fungal spores. J. Gen. Microbiol. 57, 51--60.

Fisher, D. J. & Richmond, D. V. (1970). The electrophoretic properties and some surface components of *Penicillium* conidia. J. Gen. Microbiol. 64, 205--214.

Florance, E. R., Denison, W. C., & Allen, T. C. (1972). Ultrastructure of dormant and germinating conidia of *Aspergillus nidulans*. Mycol. 49, 115--123.

Fogg, G. E. (1948). Adhesion of water to the external surfaces of leaves. Discuss. Faraday Soc. 3, 162--166.

Ghiorse, W. C. & Edwards, M. R. (1973). Ultrastructure of *Aspergillus fumigatus* conidia development and maturation. Protoplasma 76, 49--59.

Gooday, G. W., Fawcett, P., Green, D. & Shaw, G. (1973). The formation of fungal sporopollenin in the zygospore wall: a role for the sexual carotenogenesis in the Mucorales. J. Gen. Microbiol. 74, 233--239.

Gottlieb, D. (1966). Biosynthetic processes in germinating spores.

In The Fungus Spore (ed. M. F. Madelin), 18th Symposium of the Colston Research Society, pp. 217--233. London: Butterworths.

Gould, G. W., Stubbs, J. M., & King, W. L. (1970). Structure and composition of resistant layers in bacterial spore coats. J. Gen. Microbiol. 60, 347--355.

Graham, S. O. (1960). The morphology and chemical analysis of the teliospore of the dwarf bunt fungus *Tilletia contraversa*. Mycol. 52, 97--118.

Gregory, P. H. (1952). Presidential address: Fungus spores. Trans. Br. Mycol. Soc. 35, 1--18.

Gregory, P. H. (1966). The fungus spore: what it is and what it does. In The Fungus Spore (ed. M. F. Madelin), 18th Symposium of the Colston Research Society, pp. 1--13. London: Butterworths.

Gregory, P. H. (1973). The microbiology of the atmosphere. 2nd Ed. Aylesbury, Bucks: Leonard Hill.

Griffiths, D. A. (1973a). Fine structure of the chlamydospore wall in *Fusarium oxysporum*. Trans. Br. Mycol. Soc. 61, 1--6.

Griffiths, D. A. (1973b). The fine structure of conidial development in *Epicoccum nigrum*. J. Microsc. 17, 55--64.

Groves, W. B. (1913). The British rust fungi (Uredinales): their biology and classification. Cambridge: University Press.

Gunasekaran, M., Hess, W. M., & Weber, D. J. (1972). The fatty acid composition of conidia of *Aspergillus niger* V. Tiegh. Can. J. Microbiol. 18, 1575--1576.

Gunasekaran, M., Weber, D. J., & Hess, W. M. (1972). Changes in lipid composition during spore germination of *Rhizopus arrhizus*. Trans. Br. Mycol. Soc. 59, 241--248.

Harvey, I. C. (1974). Light and electron microscope observations of conidiogenesis in *Pleiochaeta setosa* (Kirchn.) Hughes. Protoplasma 82, 203--221.

Hawker, L. E. (1968). Wall ornamentation of ascospores of *Elaphomyces* as shown by the scanning electron microscope. Trans. Br. Mycol. Soc. 51, 493--498.

Hawker, L. E. (1971a). Ultrastructure of fungus spores with special reference to changes taking place during germination. J. Indian Bot. Soc. 50A, 12--25.

Hawker, L. E. (1971b). Scanning electron microscopy of fungi and its bearing on classification. In Scanning Electron Microscopy (ed. V. H. Heywood), pp. 237--249. London and New York: Academic Press.

Hawker, L. E. (1975) (in press). Scanning electron microscopy of basidiospores as an indication of relationships among hypogeous gasteromycetes. Nova Hedwigia.

Hawker, L. E. & Abbott, P. McV. (1963a). Fine structure of vegetative hyphae of *Rhizopus*. J. Gen. Microbiol. 30, 401--408.

Hawker, L. E. & Abbott, P. McV. (1963b). An electron microscope study of maturation and germination of sporangiospores of two species of *Rhizopus*. J. Gen. Microbiol. 32, 295--298.

Hawker, L. E. & Beckett, A. (1971). Fine structure and development of the zygospore of *Rhizopus sexualis* (Smith) Callen. Phil. Trans. R. Soc. Ser. B. 263, 71--100.

Hawker, L. E. & Gooday, M. A. (1968). Development of the zygospore wall in *Rhizopus sexualis* (Smith) Callen. J. Gen. Microbiol. 54, 13--20.

Hawker, L. E. & Hendy, R. J. (1963). An electron microscope study of germination of conidia of *Botrytis cinerea*. J. Gen. Microbiol. 33, 44--46.

Hawker, L. E., Thomas, B., & Beckett, A. (1970). An electron microscope study of structure and germination of conidia of *Cunninghamella elegans* Lendner. J. Gen. Microbiol. 60, 181--189.

Hecker, L. I. & Sussman, A. S. (1973). Activity and heat stability of trehalase from the mycelium and ascospores of *Neurospora*. J. Bacteriol. 115, 582--591.

Heim, R. & Perreau, J. (1971). Étude ornementale de basidiospores au microscope électronique à balayage. In Scanning Electron Microscopy (ed. V. H. Heywood), pp. 251--284. London and New York: Academic Press.

Heintz, C. E. & Niederpruem, D. J. (1970). Ultrastructure and respiration of oidia and basidiospores of *Coprinus lagopus* (sensu Buller). Can. J. Microbiol. 16, 481--484.

Henderson, D. M., Eudall, R., & Prentice, H. T. (1972). Morphology of the reticulate teleutospores of *Puccinia chaerophylli*. Trans. Brit. Mycol. Soc. 59, 229--232.

Hess, W. M., Bushnell, J. L., & Weber, D. J. (1972). Surface structures and unidentified organelles of *Lycoperdon perlatum* Pers. basidiospores. Can. J. Microbiol. 18, 270--271.

Hess, W. M., Sassen, M. M. A., & Remsen, C. C. (1966). Surface structures of frozen etched *Penicillium* conidiospores. Naturwiss. 24, 708--709.

Hess, W. M., Sassen, M. M. A., & Remsen, C. C. (1968). Surface characteristics of *Penicillium* conidia. Mycol. 60, 290--303.

Hess, W. M. & Stocks, D. L. (1969). Surface characteristics of *Aspergillus* conidia. Mycol. 61, 560--571.

Hess, W. M. & Weber, D. J. (1973). Ultrastructure of dormant and germinated sporangiospores of *Rhizopus arrhizus*. Protoplasma 77, 15--33.

Holligan, P. M. & Lewis, D. H. (1973). The soluble carbohydrates of *Aspergillus clavatus*. J. Gen. Microbiol. 75, 155--159.

Horikoshi, K. & Iida, S. (1964). Studies of the spore coats of fungi. Isolation and composition of the spore coats of *Aspergillus oryzae*. Biochim. Biophys. Acta 83, 197--203.

Horikoshi, K., Iida, S., & Ikeda, Y. (1965). Mannitol and mannitol dehydrogenases in conidia of *Aspergillus oryzae*. J. Bacteriol 89, 326--330.

Horikoshi, K. & Ikeda, Y. (1965). Studies on the spore coats of *Aspergillus oryzae*. II. Conidia coat-bound β-glucosidase. Biochim. Biophys. Acta 101, 352--357.

Hougen, F. W., Craig, B. M., & Ledingham, G. A. (1958). The oil of wheat stem rust uredospores. I. The sterol and carotenes of the unsaponifiable matter. Can. J. Microbiol. 4, 521--529.

Hughes, S. J. (1953). Conidiophores, conidia, and classification. Can. J. Bot. 31, 577--659.

Ingold, C. T. (1954). Ascospore form. Trans. Br. Mycol. Soc. 37, 19--21.

Ingold, C. T. (1959). Submerged aquatic Hyphomycetes. J. Quekett microscop. Club, Series 4, 5, 115--130.

Ingold, C. T. (1966). Aspects of spore liberation: violent discharge. In <u>The Fungus Spore</u> (ed. M. F. Madelin), 18th Symposium of the Colston Research Society, pp. 113--132. London: Butterworths.

Ingold, C. T. (1971). <u>Fungal Spores: Their Liberation and Dispersal.</u> Oxford: University Press.

Irvine, G. N., Golubchuk, M., & Anderson, J. A. (1954). The carotenoid pigments of the uredospores of rust fungi. Can. J. Agric. Sci. 34, 234--239.

Jack, R. C. M. (1964). Lipid metabolism in fungi. I. Lipids of the conidia of *Glomerella cingulata*. Contrib. Boyce Thompson Inst. Plant Res. 22, 311--333.

James, A. P. & Werner, M. M. (1965). The radiobiology of yeast. Radiat. Bot. 5, 359--382.

Jhooty, J. S. & McKeen, W. E. (1965). Water relations of asexual spores of *Sphaerotheca macularis* (Wallr. ex Fr.) Cooke and

Erysiphe polygoni D.C. Can. J. Microbiol. 11, 531--538.

Jones, D., Bacon, J. S. D., Farmer, V. C., & Webley, D. M. (1968). Lysis of cell walls of *Mucor ramannianus* Möller by a *Streptomyces* sp. Antonie van Leeuwenhoek 34, 173--182.

Kendrick, B. (ed.) (1971). Taxonomy of Fungi Imperfecti. 309 pp. Toronto and Buffalo: University of Toronto Press.

Khanna, A. & Payak, M. M. (1968). Teliospore morphology of some smut fungi. II. Light microscopy. Mycol. 60, 655--662.

Khanna, A., Payak, M. M. & Mehta, S. C. (1966). Teliospore morphology of some smut fungi. I. Electron microscopy. Mycol. 58, 562--569.

Kirk, P. W., Jr. (1966). Morphogenesis and microscopic cytochemistry of marine pyrenomycete ascospores. Nova Hedwigia 22 (Suppl.) 128 pp. and 19 Plates.

Klebahn, H. (1912--1914). Uredineen. In Kryptogamenflora der Mark Brandenburg 5a. Leipzig: Borntraeger.

Kulik, M. M. & Johnson, R. M. (1969). Feasibility of characterising spore colour in Aspergilli by reflectance spectrophotometry. Mycol. 61, 1142--1148.

Lewis, J. C., Snell, N. S., & Burr, H. K. (1960). Water permeability of bacterial spores and the concept of a contractile cortex. Science 132, 544--545.

Lingappa, B. T. & Sussman, A. S. (1959). Endogenous substrates of dormant, activated and germinating ascospores of *Neurospora tetrasperma*. Plant Physiol., Lancaster 34, 466--472.

Lippman, E., Erwin, D. C., & Bartnicki-Garcia, S. (1974). Isolation and chemical composition of oospore-oogonium walls of *Phytophthora megasperma* var. *sojae*. J. Gen. Microbiol. 80, 131--141.

Littlefield, L. J. (1971). Scanning electron microscopy of urediospores of *Melampsora lini* J. Microsc. 10, 225--228.

Littlefield, L. J. & Bracker, C. E. (1971). Ultrastructure and development of urediospores of *Melampsora lini*. Can. J. Bot. 49, 2067--2073.

Locci, R. (1972). Scanning electron microscopy of ascosporic aspergilli. Riv. Patol. Veg., Padova, 8 Serie, 4 (Suppl.).

Madelin, M. F. (ed.) (1966a). The Fungus Spore. 18th Symposium of the Colston Research Society, 338 pp. London: Butterworths.

Madelin, M. F. (1966b). The genesis of spores of higher fungi. In The Fungus Spore (ed. M. F. Madelin), 18th Symposium of the Colston Research Society, pp. 15--36. London: Butterworths.

Madelin, M. F. (1969). Conidium production by higher fungi within thin layers of liquid paraffin: A slide-culture technique. J. Gen. Microbiol. 55, 319--324.

Manier, J.-F. (1969). Trichomycètes de France. Ann. Sci. Nat. (Bot.), 12th Series, 10, 565--672.

Manier, J.-F. (1973). L'ultrastructure de la trichospore de *Genistella ramosa* Léger et Gauthier, Trichomycète Harpellale parasite du rectum des larves de *Baetis rhodani* Pict. C. R. Hebd. Séance Acad. Sci., Paris 276, Série D. 2159--2162.

Manier, J.-F. & Coste-Mathiez, F. (1968). L'ultrastructure du filament de la spore de *Smittium mucronatum* Manier, Mathier 1965 (Trichomycète, Harpellale). C. R. Hebd. Séance Acad. Sci. Paris 266, Série D, 341--342.

Marchant, R. (1966a). Fine structure and spore germination in *Fusarium culmorum*. Ann. Bot. 30, 441--445.

Marchant, R. (1966b). Wall structure and spore germination in *Fusarium culmorum*. Ann. Bot. 30, 821--830.

McCallan, S. E. A. (1958). Determination of individual fungus spore volumes and their size distribution. Contrib. Boyce Thompson Inst. Plant Res. 19, 303--320.

McKeen, W. E., Mitchell, N., & Smith, R. (1967). The *Erysiphe cichoracearum* conidium. Can. J. Bot. 45, 1489--1496.

Meléndez-Howell, L.-M. (1966). Ultrastructure du pore germinatif sporale dans le genre *Coprinus* Link. C. R. Hebd. Séance Acad. Sci. Paris 263, 717--720.

Meléndez-Howell, L.-M. (1967a). Les rapports entre le pore germinatif sporal et la germination chez les Basidiomycètes en microscopie électronique. C. R. Hebd. Séance Acad. Sci. Paris 264, 1266--1269.

Meléndez-Howell, L.-M. (1967b). Recherches sur le pore germinatif des basidiospores. Annls. Sci. Nat. (Bot.) Paris. 12th series, 8, 487--638.

Meléndez-Howell, L.-M. (1969). Rapports entre le pore germinatif sporal et les fentes de germination longitudinales ou circulaires chez les ascospores. C. R. Hebd. Séance Acad. Sci. Paris 268, 1273--1274.

Moore, R. T. & McAlear, J. H. (1961). Fine structure of Mycota. 8. On the aecidial stage of *Uromyces caladii*. Phytopathol. Z. 42, 297--304.

Morton, A. G. (1961). The induction of sporulation in mould fungi. Proc. R. Soc. Series B, 153, 548--569.

Moss, S. T. (1972). Ph.D. Thesis, University of Reading, England. 340 pp. (cited by Manier, 1973).

Niederpruem, D. J. & Hunt, S. (1967). Polyols in *Schizophyllum commune*. Am. J. Bot. 54, 241--245.

Old, K. M. & Robertson, W. M. (1970). Effects of lytic enzymes and natural soil on the fine structure of conidia of *Cochliobolus sativus*. Trans. Br. Mycol. Soc. 54, 343--350.

Owens, R. G., Novotny, H. M., & Michels, M. (1958). Composition of conidia of *Neurospora sitophila*. Contrib. Boyce Thompson Inst. Plant Res. 19, 355--374.

Payak, M. M. (1962). Spore morphology and "Lo-analysis" of spore surface ornamentation in some rust funti. Mycopathol. Mycol. Appl. 41, 70--82.

Payak, M. M. (1964). Palynology of fungi. In Recent Advances in Palynology (ed. P. K. K. Nair), pp. 27--55. Lucknow: National Botanical Garden.

Payak, M. M., Joshi, L. M., & Mehta, S. C. (1967). Electron microscopy of urediospores of *Puccinia graminis* var. *tritici*. Phytopath. Z. 60, 196--199.

Pegler, D. N. & Young, T. W. K. (1971). Basidiospore morphology in the Agaricales. Nova Hedwigia 35, (Suppl.) 210 pp. and 53 Plates.

Perreau, J. (1969). L'ornementation des basidiospores au microscope électronique à balayage. Rev. Mycol. 33, 331--340.

Perreau, J. (1971). L'ornementation sporale chez les Lycoperdons. Ann. Sci. Not. (Bot.) 12th Série, 12, 9--145.

Prentice, N. & Guendet, L. S. (1954). Chemical composition of uredospores of wheat stem rust (*Puccinia graminis tritici*). Nat. Lond. 174, 1151.

Prentice, N., Guendet, L. S., Geddes, W. F., & Smith, F. (1959). The constitution of a glucomannan from wheat stem rust (*Puccinia graminis tritici*). J. Am. Chem. Soc. 81, 684--688.

Pyatt, F. B. (1969). The ultrastructure of the ascospore wall of the lichen *Pertusaria pertusa*. Trans. Br. Mycol. Soc. 52, 167--169.

Rai, J. N., Tewari, J. P., & Mukerji, K. G. (1964). A new *Aspergillus* from Indian soils. *A. striatus* spec. nov. Can. J. Bot. 42, 1521--1524.

Raper, K. B. & Fennell, D. I. (1965). The genus *Aspergillus*. Baltimore: Williams & Wilkins.

Reichle, R. E. & Lichtwardt, R. W. (1972). Fine structure of the trichomycete, *Harpella melusinae*, from black-fly guts. Arch.

Mikrobiol. 81, 103--125.

Reisener, H. J., Goldschmid, H. R., Ledingham, G. A., & Perlin, A. S. (1962). Formation of trehalose and polyols by wheat stem rust (*Puccinia graminis tritici*) uredospores. Can. J. Biochem. Physiol. 40, 1248--1250.

Richmond, D. V. & Pring, R. J. (1971a). Fine structure of *Botrytis fabae* Sardiña. Ann. Bot. 35, 175--182.

Richmond, D. V. & Pring, R. J. (1971b). Fine structure of germinating *Botrytis fabae* Sardiña conidia. Ann. Bot. 35, 493--500.

Richmond, D. V. & Somers, E. (1963). Studies on the fungitoxicity of captan. III. Relation between the sulphydryl content of fungal spores and their uptake of captan. Ann. Appl. Biol. 52, 327--336.

Sassen, M. M. A., Remsen, C. C., & Hess, W. M. (1967). Fine structure of *Penicillium megasporum* conidiospores. Protoplasma 64, 75--88.

Savile, D. B. O. (1954). Cellular mechanics, taxonomy and evolution in the Uredinales and Ustilaginales. Mycol. 46, 736--761.

Savile, D. B. O. (1973). Aeciospore types in *Puccinia* and *Uromyces* attacking Cyperaceae, Juncaceae and Poaceae. Rep. Tottori Mycol. Inst. 10, 225--241.

Schwinn, F. J. (1969). Die Darstellung von Pilzsporen im Raster-Electronemikroscop. Phytopathol. Z. 64, 376--379.

Scurfield, G. & Da Costa, E. W. B. (1969). Fine structure of the walls of the conidia of various fungi and of the conidiophores of *Fomes annosus*. Aust. J. Bot. 17, 13--24.

Shaw, G. (1971). The chemistry of sporopollenin. In Sporopollenin (ed. J. Brooks, P. R. Grant, M. D. Muir, P. van Gijzel, & G. Shaw), pp. 305--350. London: Academic Press.

Shaw, M. (1964). The physiology of rust uredospores. Phytopathol. Z. 50, 159--180.

Skucas, G. P. (1967). Structure and composition of the resistant sporangial wall in the fungus *Allomyces*. Am. J. Bot. 54, 1152--1158.

Slayman, C. W. & Tatum, E. L. (1964). Potassium transport in Neurospora: I. Intracellular sodium and potassium concentration and cation requirements for growth. Biochim. Biophys. Acta 88, 578--592.

Sleytr, v. U., Adam, H., & Klaushofer, H. (1969). Die Feinstruktur der Konidien von *Aspergillus niger* V. Tiegh., dargestellt mit Hilfe der Gefrieratztechnik. Mikroskopie 25, 320--331.

Somers, E. & Horsfall, J. G. (1966). The water content of powdery mildew conidia. Phytopathol. 56, 1031--1035.

Stocks, D. L. & Hess, W. M. (1970). Ultrastructure of dormant and germinated basidiospores of a species of *Psilocybe*. Mycol. 62, 176--191.

Sturgeon, R. J. (1966). Components of the spore wall of *Pithomyces chartarum*. Nat. Lond. 209, 204.

Subramanian, C. V. (1962). The classification of Hyphomycetes. Bull. Bot. Surv. India 4, 249--259.

Sumi, M. (1928). Uber die chemischen Bestandteile der Sporen von *Aspergillus oryzae*. Biochem. Z. 195, 161--174.

Sumner, J. L. & Morgan, E. D. (1969). The fatty acid composition of sporangiospores and vegetative mycelium of temperature-adapted fungi in the Order Mucorales. J. Gen. Microbiol. 50, 215--221.

Sussman, A. S. (1965). Physiology of dormancy and germination in the propagules of cryptogamic plants. In Encyclopedia of Plant Physiology, (ed. A. Lang), Vol. 15, pp. 933--1025. Berlin: Springer.

Sussman, A. S. (1966). Types of dormancy as represented by conidia and ascospores of *Neurospora*. In The Fungus Spore (ed. M. F. Madelin), 18th Symposium of the Colston Research Society, 235--256. London: Butterworths.

Sussman, A. S. & Douthit, H. A. (1973). Dormancy in microbial spores. Annu. Rev. Plant Physiol. 24, 311--352.

Sussman, A. S. & Halvorson, H. O. (1966). Spores, their dormancy and germination, 354 pp. New York, and London. Harper and Row.

Sussman, A. S., Yu, S.-A., & Wooley, S. (1971). Localization of trehalase in *Neurospora*. J. Indian Bot. Soc. 50A, 60-69.

Swinburne, T. K. & Matthews, H. I. (1963). The surface structures of chlamydospores of *Tilletia caries*. Trans. Br. Mycol. Soc. 46, 245--248.

Thomas, P. L. & Isaac, P. K. (1966). The development of echinulation in urediospores of wheat stem rust. Can. J. Bot. 45, 287--289.

Tulloch, A. P., Craig, B. M., & Ledingham, G. A. (1959). The oil of wheat stem rust uredospores II. The isolation of *cis*-9,10-epoxyoctadecanoic acid, and the fatty acid composition of the oil. Can. J. Microbiol. 5, 485--491.

Tulloch, A. P. & Ledingham, G. A. (1960). The component fatty acids of oils found in spores of plant rusts and other fungi. Can. J. Microbiol. 6, 425--434.

Tulloch, A. P. & Ledingham, G. A. (1962). The component fatty acids of oils in spores of plant rusts and other fungi. II. Can. J. Microbiol. 8, 379--387.

Uduebo, A. E. & Madelin, M. F. (1974). Germination of conidia of *Botryodiplodia theobromae* in relation to age and environment. Trans. Br. Mycol. Soc. 63, 33--44.

Van Etten, J. L. & Gottlieb, D. (1965). Biochemical changes during the growth of fungi. II. Ergosterol and fatty acids in *Penicillium atrovenetum*. J. Bacteriol. 89, 409--414.

Voelz, H. & Niederpruem, D. J. (1964). Fine structure of basidiospores of *Schizophyllum commune*. J. Bacteriol. 88, 1497--1502.

Von Hofsten, A. & Holm, L. (1968). Studies on the fine structure of aeciospores, I. Grana Palynol. 8, 237, 251.

Waldham, D. G. & Halvorson, H. O. (1954). Studies on the relationship between equilibrium vapor pressure and moisture content of bacterial endospores. Appl. Microbiol. 2, 333--338.

Walkinshaw, C. H., Hyde, J. M., & van Zandt, J. (1967). Fine structure of quiescent and germinating aeciospores of *Cronartium fusiforme*. J. Bacteriol. 94, 245--254.

Weete, J. D., Weber, D. J., & Laseter, J. L. (1970). Lipids of *Rhizopus arrhizus* Fischer. J. Bacteriol. 103, 536--540.

Weisberg, S. H. & Turian, G. (1971). Ultrastructure of *Aspergillus nidulans* conidia and conidial lomasomes. Protoplasma 72, 55--67.

Weiss, B. (1965). An electron microscope and biochemical study of *Neurospora crassa* during development. J. Gen. Microbiol. 39, 85--94.

Wells, K. (1965). Ultrastructural features of developing and mature basidiospores of *Schizophyllum commune*. Mycol. 57, 236--261.

Wells, K. (1972). Light and electron microscopic studies of *Ascobolus stercorarius*. II. Ascus and ascospore ontogeny. Univ. Calif. Publ. Bot. 62, 33.

Whitney, H. S. & Blauel, R. A. (1972). Ascospore dispersion in *Ceratocystis* spp. and *Europhium clavigerum* in conifer resin. Mycol. 64, 410--414.

Wildemuth, H., Wehrli, E., &, Horne, W. R. (1971). The surface structure of spores and aerial mycelium in *Streptomyces coelicolor*. J. Ultrastruct. Res. 35, 168--180.

Willetts, H. J. & Calonge, F. de D. (1969). Spore development in the brown rot fungi (*Sclerotinia* spp.). New Phytol. 68, 123--131.

Wilsenach, R. & Kessel, M. (1965). The role of lomasomes in wall formation in *Penicillium vermiculatum*. J. Gen. Microbiol. 40,

401--404.

Wilson, M. & Henderson, D. M. (1966). British Rust Fungi, 384 pp. Cambridge, England: University Press.

Yarwood, C. E. (1950). Water content of fungus spores. Am J. Bot. 37, 636--639.

Young, T. W. K. (1968). Electron microscopic study of the asexual structures in Mucorales. Proc. Linn. Soc. Lond. 179, 1--8.

Young, T. W. K. (1969a). Asexual spore morphology in species of *Piptocephalis*. New Phytol. 68, 359--362.

Young, T. W. K. (1969b). Electron and phase-contrast microscopy of spores in two species of the genus *Mycotypha* (Mucorales). J. Gen. Microbiol. 55, 243--249.

Zogg, H. & Schwinn, F. J. (1971). Surface structure of spores of the *Ustilaginales*. Trans. Br. Mycol. Soc. 57, 403--410.

DISCUSSION

The Dormant Spore

Chairman: R. W. Lichtwardt
University of Kansas
Lawrence, Kansas

The question was asked whether there is any standardization in the terms used for sculpturing on spores. Professor Hawker replied that she knows of no standardization. Terms like spines, ridges, and reticulations are in general use, and most people understand what they are, but it would be useful for somebody to tabulate them.

Dr. G. M. Olah stated that he and co-workers believe the term "dormant spore" has too broad a definition and reflects but one part of the physiological development that can come after liberation of spores. This development is particularly evident in the Basidiomycetes wherein, according to their cytochemical and electronmicroscopic studies (Reisinger & Olah, 1973, 1974), the basidiospore wall can undergo transformations after liberation; that is, the number of layers of the wall can be notably changed. As a consequence, the use of the number of spore wall layers in systematics appears to them to be premature in the present state of knowledge.

Professor Hawker agreed that unless care is exercised one can easily be looking at spores of different physiological ages. She stressed the need to study development of the spore, as the Bristol group does, and to look at spores of all ages until one has what is obviously the oldest, and one that can be shown to be physiologically a dormant spore. However, in this paper she avoided talking about various developmental stages since predormancy development was excluded from these discussions.

Dr. Lafayette Frederick suggested that if one attempts to standardize terminology some significance should be placed on the manner in which sculpturing forms. In *Neurospora* ascospores he found an internal membrane that prevents rib formation along certain parts of the wall; thus there is a precise mechanism that determines where ribs can or cannot form. More work along these lines might lead to standardization of terminology based on ontogeny of the spore wall.

Professor Hawker again stated that she did not intend to go into developmental aspects of spores, but cited some of the examples in her paper (uredospores, hypogeous Gasteromycetes, *Elaphomyces)* that demonstrate that similar ornamentations may be produced in different ways. She agreed that further studies of modes of development of apparently similar types of ornamentation should be done.

In answer to a question about the viability of spores fractured by liquid nitrogen treatment, Professor Hawker answered that they lose their viability.

She was not able to supply information on the source of nutrients involved in sculpturing.

CHAPTER 2

SELF-INHIBITORS OF FUNGAL SPORE GERMINATION

V. Macko, R. C. Staples, Z. Yaniv, and R. R. Granados

Boyce Thompson Institute

The research of the authors reported here was supported in part by a grant (GB30698X) from the U. S. National Science Foundation.

INTRODUCTION

It is increasingly recognized that fungi produce substances during growth that influence their development. For example, among the first to be isolated and characterized were hormones that regulate sexual reproduction of several fungi (Barksdale, 1969; Gooday, 1974). Culture filtrates of fungi contain many metabolites that induce morphogenetic changes when applied to actively growing hyphal-tip cells (Park & Robinson, 1967; Robinson, 1972) and sporostatic factors that inhibit spore germination (Robinson & Garrett, 1969; Robinson et al., 1968). In addition, sporogenetic substances have been isolated from several species of fungi (Trione & Leach, 1969).

In this article we review research on endogenous effector molecules that regulate spore germination, especially the recent work on rust uredospores which has progressed from a characterization of these substances to attempts to elucidate their mode of action. Newer literature on self-inhibition in other types of spores will also be reviewed, but such studies have usually not progressed far enough for a truly comparative account.

POPULATION EFFECT

Spores of many fungi germinate poorly or not at all in dense suspensions or when crowded upon a surface. A spore usually remains ungerminated in the culture in which it was produced. In some instances this is due to spore dormancy, but more commonly it is due to the presence of inhibitory substances formed at the time of sporulation.

The crowding effect alone is not sufficient evidence for the presence of self-inhibitors, although it suggests that inhibitory materials from the spores might be present. In many fungi self- inhibition has been traced to inhibitory substances given off by the spores themselves, but in others competition for key substances such as oxygen or nutrients may be a factor. The numerous reports on self-inhibition in fungi have been summarized in reviews by Cochrane (1958), Allen (1965), Sussman (1965), Sussman & Halvorson (1966), Allen & Dunkle (1971), and Macko & Staples (1973). Emphasis has been placed on more recent investigations in this presentation.

The selective advantages of self-inhibition seem obvious. In the first place, it prevents germination in the sorus. Secondly, self-inhibitors prevent rapid germination of all spores, which ensures against the effects of sudden adverse changes in environmental conditions. Self-inhibition of germination is probably the means by which spore germination is delayed until conditions are appropriate for germination and growth, especially within the fructification structure where survival would be poor.

Generally, spore survival would be maximal at low population density, and a requirement for high cell density for germination of resting cells is unusual. Myxospores of the prokaryotic *Myxococcus xanthus*, which require a high cell density for germination in distilled water, are exceptional. Their requirement for a high cell density for germination under conditions of limited nutritional availability is interpreted as providing a guarantee that under conditions where cells must swarm to maximize nutrient collections, large numbers must germinate simultaneously (Dworkin, 1973).

FUNGI FOR WHICH SELF-INHIBITION WAS REPORTED

A few recent reports are included here to illustrate the breadth of the self-inhibition phenomenon in fungi. A population effect has been demonstrated for germination of zoospores of *Blastocladiella emersonii* in which the encystment was inhibited at high population densities. The self-inhibition was apparent above 6 x 10^5 spores/ml (Truesdell & Cantino, 1971; Soll & Sonneborn, 1969) and it was postulated that an encystment inhibitor was released from the spores.

An inhibitor, which has many of the properties of this factor, was recently shown to be associated with ribosomes of *B. emersonii* zoospores (Adelman & Lovett, 1974). The inhibitor apparently reduces the activity of ribosomes by blocking translation of RNA in cryptic zoospore until zoospore encystment begins when it is released and protein synthesis resumes. Adelman & Lovett (1974) proposed that the inhibitor is produced in response to the same environmental triggers which induce differentiation of the zoosporangium.

An inhibition was apparent in conidiospores of *Aspergillus niger* under conditions of spore crowding that affected only the initiation of spore growth. Neither the growth process itself nor the formation of the germ tubes was involved (Anderson & Smith, 1972). The inhibitor appeared to be present in the spores at the time of spore formation and could be removed by washing (Krishnan et al. 1954).

In experiments with conidia of *Sclerotinia fructicola*, the crowding effect was also demonstrated. When spore density was high (500 to 3000 spores/mm^3), germination was drastically reduced (Van den Ende & Cools, 1967). These authors demonstrated that the crowding effect was caused by a volatile substance which could be absorbed by charcoal. In similar experiments, a low concentration of spores (3--50 spores/mm^3) stimulated the rate of germ tube growth. This was shown to be due to the presence of a stimulator, but the possibility that both effects were caused by the same compound at different concentrations could not be excluded.

In teliospores of the common bunt fungus (*Tilletia caries*), trimethylamine was reported to be a natural inhibitor of germination (Ettel & Halbsguth, 1963). Treating these spores with octanal, octanol, nonanal, and nonanol was without effect on their germination. Since these alcohols and aldehydes effectively counteracted the inhibition in rust uredospores, it was concluded that germination of bunt teliospores was regulated by different mechanisms than those occurring in rust uredospores (Trione, 1973). Water-soluble inhibitors of germination were detected in teliospores of dwarf bunt (*Tilletia controversa*) and were thought to be the cause of the extreme dormancy of these spores (Macko et al., 1964).

Domsch (1954) clearly demonstrated the inverse relationship between the number of conidia of *Erysiphe graminis* in suspension and their germination. Self-inhibition was removed by washing the conidia with water and these extracts of the spores contained an inhibitory principle. The inhibitory substance was apparently a small molecule since distillates of spore washings were inhibitory and the inhibitors could be removed by dialysis (Janitor, 1967).

Information about self-inhibition in *Glomerella* has come from studies by Lingappa and his group who used conidia grown in liquid-shake cultures (Lingappa & Lingappa, 1965, 1966). Most spores germinated when suspended in nutrient solution at a concentration of 10^5 spores/ml, but less than 1% germinated when the titre was raised to 10^9 spores/ml. The self-inhibition was not due to limiting concentrations of nutrients or oxygen, and the Lingappas eventually showed that an inhibitory fraction could be extracted from conidia (Lingappa et al., 1973). The chemical identity of the fraction was not determined.

In a study of the effect of the self-inhibitor on spore germination, Lingappa et al. (1973) found that spore germination occurred in two steps. The first step which lasted 5 hours was insensitive to the self-inhibitor, while the second step of 2 hours was inhibitor-sensitive. The authors found that the self-inhibitor reduced incorporation of amino acids into protein and concluded that the self-inhibitor blocked spore germination by retarding protein synthesis. This conclusion is in agreement with their work which showed that washing the spores with water eliminated the crowding effect as well as inhibition of protein synthesis.

Spores are formed in *Geotrichum candidum* when the hyphae fragment into separate cells called arthrospores. At high spore concentrations the arthrospores required an exogenous source of nutrients to germinate, although they germinated on distilled water when the spore concentration was low (Steele, 1973). This nutrient requirement induced by crowding was traced to the presence of a self-inhibitor, but the inhibitor was too unstable to isolate for identification. Working with self-inhibited concentrations of arthrospores, Steele (1973) reported that as the concentration of the spores was increased, oxygen consumption per spore declined. This suggested that oxygen consumption may be sensitive to the self-inhibitor. Recently, Robinson (1973) has suggested that oxygen shortage at high spore titres limits germination of these spores. The possible involvement of a self-inhibitor in these spores will require additional information.

Other observations of crowding effects on spore germination promise to extend the list of fungal species that produce self-inhibitors.

SELF-INHIBITORS OF GERMINATION

The germination of *Penicillium griseofulvum* conidia declined as the number of spores per ml rose above 1.32 x 10^6 (Fletcher & Morton, 1970). These authors believe that the inhibition was due to the presence of an inhibitory substance and not to a shortage of oxygen due to overcrowding. Germination of conidia of *Botryodiplodia theobromae* was decreased if more than 1 mg of spores was used per ml of germinaton medium. This suggested to Van Etten (1968) that self-inhibitors may have been present. Finally, Trinci & Whittaker (1968) found that conidial germination of *Aspergillus nidulans* was reduced at spore densities above 10^6 conidia/ml. The authors attributed the inhibition to an insufficient supply of carbon dioxide during early stages of germination.

FUNGI IN WHICH SELF-INHIBITORS HAVE BEEN IDENTIFIED

Dictyostelium discoideum.

Spores from this slime mold germinate with the emergence of an amoeba. The process of germination can be triggered by amino acids or heat shock, but germination was poor above a titre of 10^5 spores/ml (Cotter & Raper, 1966). Eventually, Ceccarini & Cohen (1967) found that the spores contained a self-inhibitor which was responsible for the crowding effect. A crowding effect was reported earlier for the related species, *D. mucoroides* (Russell & Bonner, 1960).

A self-inhibitor was isolated from spores of *D. discoideum* and chemically characterized as 2-dimethylamino-6-oxypurineriboside (N,N-dimethylguanosine) (Bacon et al. 1973). Fifty micrograms per milliliter of either the synthetic compound or of the native inhibitor completely prevented germination of activated spores. A gram of spores contained about 300 μg of this inhibitor.

Dimethylguanosine is not the only inhibitory compound in slime-mold spores. A second compound, as yet unidentified, has been isolated by Obata et al. (1973) and shown to separate chromatographically from dimethylguanosine (Tanaka et al. 1974). This compound had a significant absorption maximum at 288 nm in water compared to 265 nm in phosphate buffer for dimethylguanosine, and was so potent that 0.14 μg/ml completely inhibited spore germination. The relationship between these two compounds and their relative roles as self-inhibitors of spore germination must await final identification of the latter compound.

Bacon & Sussman (1973) postulated that dimethylguanosine inhibited spore germination by preventing the synthesis of protein necessary for subsequent development. When added to swollen spores, the self-inhibitor delayed outgrowth, but inhibition was complete if the compound was added before the spores were swollen. In fact, based on respiratory response, the authors concluded that the inhibitor acted within the first 1.5--2 hours after activation by heat.

To pinpoint the action of the self-inhibitor, Bacon & Sussman (1973) studied glucose incorporation and found that dimethylguanosine principally inhibited incorporation of glucose carbon into protein while incorporation into organic acids, amino acids, and lipids was not inhibited. In addition, incorporation of leucine into protein was reduced by the presence of dimethylguanosine in the first hour after activation, but leucine was incorporated into protein if exposure of the spores to the inhibitor after activation was delayed until the second hour. Thus dimethylguanosine acted on the very earliest stage of spore germination, apparently by restraining the protein synthesis required for emergence of the amoeba.

Peronospora tabacina.

This phycomycete is an obligate parasite on tobacco. The conidiophores emerge through stomata and bear conidia which germinate by producing a stout germ tube. A self-inhibitor of conidial germination was first implicated in studies by Shepherd & Mandryk (1962). They showed by chromatography on filter paper that extracts from conidia and infected tobacco leaves contained two active self-inhibitors. No activity was found in uninfected leaves.

In a further study, Leppik et al. (1972) also isolated inhibitory compounds by ethanol extraction of infected leaves. In addition, a small quantity of one of these inhibitors was obtained from conidia although the availability of conidia was limited. The leaf extracts were resolved into two zones by chromatography on silica gel. The first of these compounds was isolated and identified as 5-isobutyroxy-β-ionone (quiesone) (Fig. 1). About 650 μg of quiesone was obtained from 250 kg of infected leaves.

Quiesone was extremely potent---the ED_{50} is 0.1 ng/ml. It inhibited germination only within the first 30 min of incubation even though the germ tubes do not ordinarily appear for 60 min. Addition after 30 min was without effect on either germination or subsequent germ tube elongation. The inhibitor was active only against conidia of *P. tabacina* and failed to prevent the germination of spores of five unrelated fungi including the wheat stem rust fungus.

Quiesone prevented the activation of protein synthesis which normally occurred during germination of the conidia (Leppik et al., 1972). However, the inhibitor did not alter the capacity of conidial ribosomes to incorporate amino acids *in vitro*. This indicates that the effect of the compound on protein synthesis in viable spores must be an indirect one.

The synthesis of quiesone was attempted by Mori (1973) who prepared a racemate of the suspected compound.* Although most of the spectral properties of synthetic *dl*-5-isobutyroxy-β-ionone were closely similar to natural quiesone, small differences were apparent especially in the NMR spectra. Furthermore, quiesone was 100 times more active as a germination inhibitor than the synthetic compound. Since the synthetic inhibitor was a racemate, it should have been at least half as active as the natural product. Mori concluded that the identification of quiesone as 5-isobutyroxy-β-ionone was not established rigorously. Either quiesone contains a small amount of active material or, as seems more likely, small structural details of the inhibitor have not yet been elucidated. For example, the effect of *cis-trans* isomerism in the side chain of 5-isobutyroxy-β-ionone was not tested, yet only the *cis*-isomers of the self-inhibitors from rust spores are active.

Microsporum gypseum.

An unusual regulation of germination was described for *M. gypseum* by Leighton & Stock (1970). Spore coat proteolysis was an essential event in the germination of the conidia. Spore germination was accompanied by the release of alkaline protease, calcium ions, and inorganic phosphate into the germination medium. Germination at high densities was regulated by the ratio of released phosphate to protease protein, which resulted in a low percentage of germination at high spore densities. The germination of macroconidia thus appeared to be regulated by the release of phosphate ions, which

---

*Mori used a different numbering system for the ring in compounds related to quiesone based on its analogy with the carotenoids. Consequently Mori used *dl*-3-isobutyroxy-β-ionone (Mori, 1973) as a synonym for *dl*-5-isobutyroxy-β-ionone (Leppik et al., 1972).

inhibited the alkaline protease (Page & Stock, 1971).

In subsequent work, Page & Stock (1972) have shown that the alkaline protease was contained in one of three types of lysosomes and that an acid protease and a phosphodiesterase were contained in separate lysosomal types as resolved on Ficoll gradients. The authors concluded from this and subsequent data (Page & Stock, 1974) that spore germination was initiated by insertion of the alkaline protease-containing lysosomes into the spore coat prior to germination. This was followed by release of protease which lysed the spore coat.

Rust Fungi.

Compounds identified

Work on physiology of germination of rust spores has been carried out almost entirely with less than a dozen species of *Melampsora, Puccinia,* and *Uromyces*, particularly *U. phaseoli* and *P. graminis tritici,* the bean rust and wheat stem rust fungi, respectively. Moreover, virtually all the work on rust fungi has dealt with uredospores so that information about other spore forms of these fungi is rare (LeRoux & Dickson, 1957).

Uredospores constitute the repeating stage of the rusts and are regarded as conidia of these fungi. On moist surfaces of leaves, uredospores imbibe water, swell, and form a germ tube. This germ tube elongates, and upon contact with a stoma forms an appressorium. The appressorium is the first of several infection structures, and invasion of the host is completed by sequential formation of an infection peg, a substomatal vesicle, vesicular hyphae, and haustoria. Intercellular mycelia develop in a matter of a few days, and new uredia

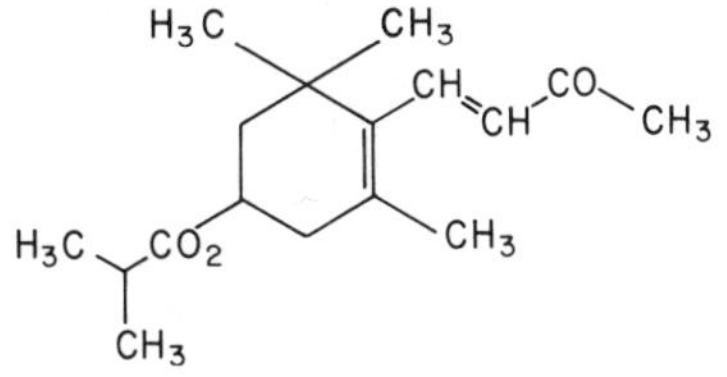

Fig. 1 5-Isobutyroxy-β-ionone (quiesone).

and uredospores are formed subepidermally. As the uredospores form, they press against the host epidermis from inside and rupture it to liberate the uredospores.

Germination of rust uredospores is reversibly inhibited by endogenous inhibitors carried by the spores and released along with many other soluble constituents when the spores are floated on water. In 1955, Allen showed that the water extracts of the uredospores of *P. graminis tritici* contained an inhibitory principle responsible for the population effect. He also showed that the inhibition of germination was reversible and coined the term "self-inhibitor." More than a decade of experimentation led to the generally accepted view that uredospores of many species of rust fungi contained potent germination inhibitors and that germination depended on their removal from the spores (Yarwood, 1954, 1956; Wilson, 1958; Hoyer, 1962; Schröder & Hassebrauk, 1964; Roy & Prasada, 1966; Pritchard & Bell, 1967; Musumeci et al. 1974).

The main problem in identification of the inhibitors has been their very low concentration in the spores. But techniques were gradually developed, and it was learned that the inhibitors can be partitioned from water into ether and separated into two active zones by paper chromatography (Bell & Daly, 1962). Eventually, self-inhibitors from uredospores of six species were identified by using chromatographic and spectrometric techniques (Macko et al. 1970, 1971a, 1971b, 1972; Foudin & Macko, 1974). The self-inhibitor of bean, sunflower, snapdragon, corn, and peanut rust uredospores was found to be methyl *cis*-3,4-dimethoxycinnamate (MDC), while that from wheat stem rust uredospores was methyl *cis*-ferulate (MF) (Fig. 2).

Purification and Bioassay.

The self-inhibitors can be routinely extracted from uredospores by stirring them in water. Purification involves partitioning the water extracts into ether followed by chromatography on silica gel thin-layer plates. The inhibitory zone from the plates can be purified by gas chromatography to yield a compound sufficiently pure for spectrometric analyses (Macko et al., 1972). The purification must rely on bioassay during the successive stages to avoid loss of the compound (Allen & Dunkle, 1971; Allen et al., 1971).

Synthesis of Inhibitory Compounds.

MDC and MF were prepared by methylation of the *trans*-3,

4-dimethoxycinnamic acid and the *trans*-ferulic acid. The resulting methyl esters were irradiated in solution to produce a *cis-trans* mixture of the isomers. The active *cis*-isomer was then separated from the mixture either by chromatography on cellulose or by preparative gas chromatography.

Parameters of Effective Dose.

The cinnamate inhibitors were active in a concentration range of 1--10 n*M* (ED_{50}) when tested with uredospores of the parent species (Macko et al. 1972). Spores of *P. arachidis* were sensitive to a dose as low as 36 p*M* (Foudin & Macko, 1974). This activity is comparable to the minimum effective dose of sexual hormones of fungi (Gooday, 1974). A range of values for the ED_{50} can be expected since a seasonal variability in sensitivity of uredospores of *P. graminis tritici* has been reported (Allen, 1972). The uredospores produced in the winter months appear to be less sensitive toward MF, and this in turn may be related to a higher content of germination stimulants in winter spores.

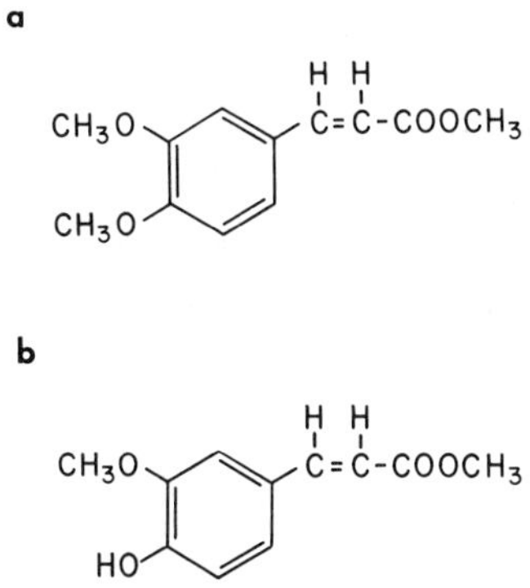

a

CH_3O, CH_3O, H H, $C=C-COOCH_3$

b

CH_3O, HO, H H, $C=C-COOCH_3$

Fig 2. Cinnamate esters identified as self-inhibitors in uredospores of rust fungi. (a) Methyl *cis*-3,4-dimethoxycinnamate. (b) Methyl *cis*-ferulate.

Structural Specificity of the Self-Inhibitors.

While the natural isomer of the rust inhibitors was the *cis*-isomer (Fig. 3a), the double bond in the side chain of the cinnamate compounds allows for geometrical *cis--trans* isomerism. Irradiation of the isomers with ultraviolet light converted one form to the other to produce an equilibrated mixture of the *cis*- and *trans*-isomers in solution (Fig. 3b). The *cis*-isomer also possessed all of the inhibitory activity, whereas, the *trans*-isomer was not active (Allen, 1972; Macko et al. 1972). The activity of the *trans* forms reported initially (Macko et al. 1970, 1971a) can now be accounted for as a result of the interconversion of the isomers handled in a lighted room. Growth responses in higher plants were produced by a *cis*-cinnamic acid, whereas the *trans*-isomer was inactive (cf. Fredga & Åberg, 1965). Thus the *cis*-isomeric specificity of the self-inhibitors is not exceptional.

The high potency of the natural germination inhibitors made it desirable to assay the close structural analogues of these cinnamate compounds. For example, dimethoxycinnamic acid can exist in six isomeric forms depending on the positions of the two methoxy groups on the ring. Each of these isomers is also possible in a *cis* and *trans* form. These and other cinnamates and their methyl esters were tested for their ability to inhibit germination of bean and wheat stem rust uredospores. Of these synthetic compounds, only MDC and MF were inhibitory in the low nanomolar concentration range (Allen, 1972; Macko et al. 1970, 1972; Macko, unpublished information).

Identification of the structure of germination self-inhibitors from rust uredospores makes it possible to investigate the phenomenon of self-inhibition in more depth. A plethora of questions can be asked, but the answers are obscure. Where is the inhibitor located in the spore? At what stage of development does it appear in the organism? How is the inhibitor biosynthesized? Does it have a role in the development of the fungus other than just inhibition of spore germination? How do stimulators counteract self-inhibitors? But the most interesting question probably is about its mode of action. How does it prevent germination, and what is the molecular basis for the inhibition?

Stimulators of Germination.

Several germination stimulants have been isolated from wheat stem rust uredospores including coumarin derivatives and phenolic compounds (Van Sumere et al. 1957), *n*- nonanal (French & Weintraub,

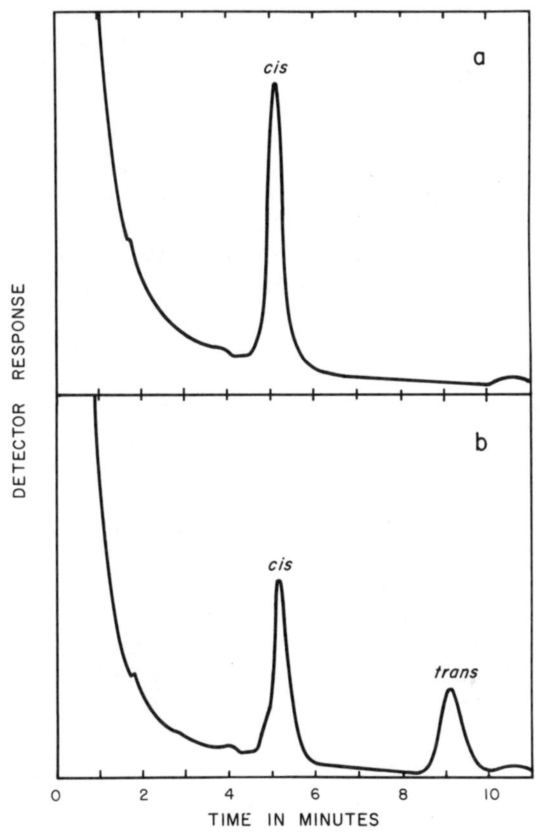

Fig. 3 Gas chromatograms of self-inhibitor from bean rust uredospores. Spores extracted with water, the water filtrate partitioned with ether and the ether extract injected into a 1% FFAP column at 200°C. (a) Extraction and assay under red safelight. (b) Same extract as (a) chromatographed after UV irradiation.

1957) and 6-methyl-5-hepten-2-one (Rines et al. 1974). Many other chemicals have been shown to stimulate germination of these spores

(French, 1961; French & Gallimore, 1971, 1972), but *n*-nonanol appeared to be the most effective (Allen & Dunkle, 1971).

Stimulators overcome the self-inhibition which prevails in dense populations of wheat stem rust uredospores and they prevent the action of added self-inhibitor. Higher concentrations of *n*-nonanol were required to counteract inhibition by high concentrations of methyl *cis*-ferulate than were required for lower concentrations (Allen, 1972).

Self-inhibited wheat stem rust uredospores were also reported to be stimulated by dilute solutions of peroxidase and hydrogen peroxide (Macko et al. 1968). It has now been found that methyl *cis*-ferulate was dimerized by the action of peroxidase and hydrogen peroxide, and the dimer was inactive as an inhibitor (Macko & Renwick, 1974).

Uredospores of the bean rust fungus were stimulated to germinate by β-ionone (Macko, unpublished information) which had the stimulatory properties of nonanol as they were described for wheat stem rust uredospores (Allen, 1972). Dense suspensions of bean rust uredospores were stimulated to germinate in 10^{-4} *M* β-ionone. From the filtrates of these suspensions, all the endogenous MDC was recovered which indicates that the molecule of the self-inhibitor was not changed chemically and thus rendered inactive as with peroxidase in the former example. It appeared as if β-ionone facilitated the release of the inhibitor from the spores. Future knowledge of the site and mode of action of the self-inhibitors may suggest the mode of action of the low molecular weight stimulators.

The reader should keep in mind that while *n*-nonanal is a native constituent of wheat stem rust uredospores, a native stimulator has not yet been detected in bean rust uredospores.

Origin of Self-Inhibitors.

The inhibitors appear to be present in the mature spores in a preformed, bound form; and no metabolic events are needed for their production or release. The inhibitor was extracted directly from the spores with water but not with ether. Since the inhibitor was more soluble in ether, water was apparently necessary for the release of the inhibitor (Bell & Daly, 1962; Allen & Dunkle, 1971). Heating spores in an oven at 120°C did not change this extraction pattern, and the same amount of inhibitor was recovered with water from heated and unheated spores (Macko, unpublished). The quick release of the inhibitor from spores by water suggested that the inhibitor is located on the surface or close to the surface of the spore. If so, the high temperature treatment of the spores should have removed any free inhibitor which can be sublimed at this temperature.

Additional studies have shown that nonpolar solvents did not extract the inhibitor; polar solvents such as ethanol or methanol did so partially, while water was the best solvent. Yet, the self-inhibitors are cinnamate esters and should have been readily soluble in nonpolar solvents. How the inhibitor is held in the spore is not known, and the location of it is uncertain.

The inhibitor was shown to be present in infected host plants as well as the spores when the fungus was fully sporulating, and it was not detected in uninoculated wheat plants (Allen & Dunkle, 1971). Gas chromatographic analyses of extracts made at daily intervals of bean plants inoculated with bean rust fungus (Macko, unpublished information) showed that neither *cis*- nor *trans*-methyl 3,4-dimethoxycinnamate was present in these plants in the first four days after infection, a time when the mycelium of the fungus proliferates. However, the inhibitor was detected by the fifth day when white flecks (caused as the developing uredospores pushed the epidermis upwards) appeared on the leaves. Only the *cis*-isomer of MDC was detected at this time. Thus MDC was synthesized at the same time as the spores were formed. The fact that the synthesis of the inhibitor coincided with spore formation suggested that at least the final steps of its synthesis occur in the fungus when it sporulates. It will be most interesting to discover the enzymatic process by which only the *cis*-isomer is synthesized. It will also be interesting to find out if there is a dependent relationship between the formation of the self-inhibitor and sporulation.

Mode of Action.

Germination of uredospores begins with hydration and swelling followed by growth of the germ tube. Germ tube elongation begins by extension of the inner spore wall at a place predetermined by the position of the germ pore. The definitive germ tube that emerges from the spore wall elongates further and can be induced to form infection structures. Neither swelling of the spore nor growth of the germ tube outside the spore wall was inhibited by self-inhibitors (Jones & Haburn, 1969; Allen & Dunkle, 1971). The inhibitory action of the chemicals was restricted to the initiation and growth of the germ tube which occurred prior to emergence from the spore.

Electron micrographs of thin sections of bean rust uredospores showed the germ pore region at stages of early germ tube growth (Fig. 4). The growth of the germ tube started immediately after the self-inhibitor was removed or the spore was placed on a stimulator solution, and this stage was characterized by the appearance of cytoplasmic vesicles at the tip of the tube (Fig. 4c) as was described for other fungi by Grove & Bracker (1970). Tube elongation was accompanied by loss of germ pore plug material next to the tip. At a later stage, it appeared as if this plug material had been partially pushed out of place by the advancing germ tube (Fig. 4d). Finally, the germ tube protruded beyond the periphery of the spore wall (Fig. 4e). In the presence of the self-inhibitor, elongation of the germ tube was not initiated and the appearance of the germ pore region was similar to that of the dormant nonhydrated spore (Fig. 4a,b).

Germination experiments in which inhibitor was added to the bean rust spores at variable times after the growth of the germ tube was initiated indicated that while this growth can be stopped reversibly during the growth of the early germ tube, the inhibitor was not active once the germ tube protruded beyond the outer periphery of the spore wall (Macko, unpublished). More direct evidence on this point is presented by Hess et al. (1975) with wheat stem rust uredospores.

After the germ tube protruded beyond the outer spore wall, the inhibitor was no longer effective and the cell was committed to germinate. The term commitment is generally applied to differentiating biological systems and is defined as the point of no return, the time when the biochemical and morphological events associated with the developmental system are irreversibly channeled toward the differentiated form and cannot be reversed (Szulmajster, 1973).

The self-inhibitor did not block leucine incorporation into protein of bean rust uredospores. The rate of incorporation was similar to that for the spores which were floated in stimulator and germinated normally (Table 1). This indicates that the self-inhibitor did not interfere with protein metabolism of the spores. In addition to the above, it has also been shown that germination of bean rust uredospores was not blocked by inhibitors of protein and RNA synthesis (unpublished information).

In a study of the mode of action of methyl *cis*-ferulate in wheat stem rust uredospores, Hess et al. (1975) found that germination began by digestion of wall material in the region of the germ pore plug. Germination of these spores progressed in a manner similar to that shown in Fig. 4 for bean rust uredospores. Methyl *cis*-ferulate prevented digestion of the pore plug at concentrations similar to those

Table 1. Incorporation of (^{3}H) Leucine into Protein by Uredospores of *Uromyces phaseoli* during 2 h of Germination[a]

Treatment	Leucine incorporation into protein (cpm/mg spore)	Germination (%)
β-Ionone (100 μ*M*)	1130	92
MDC (2 μ*M*)	1150	0

[a]Radioactivity was assayed in hot trichloroacetic acid precipitates. MDC refers to methyl *cis*-3,4-dimethoxycinnamate. Spores were floated on the chemical solutions, and leucine incorporation was assayed at 30-min intervals over 2 h.

required to inhibit germination (10--100 n*M*). Removal of the inhibitor allowed digestion to proceed, but it was interrupted if inhibitor was reintroduced before digestion progressed to the outer periphery of the spore wall. Digestion was insensitive to inhibitors of RNA and protein synthesis.

The authors proposed that the self-inhibitor blocks dissolution of the germ pore plug and that the inhibitor can no longer be active when the material has largely disappeared. Removal of the inhibitor at any time prior to disappearance of the plug material leads to reversal of the inhibition of the postulated enzyme and to an ensuing rapid digestion of the plug. Testing this hypothesis should provide an understanding of the mode of action of the self-inhibitor at the molecular level.

Physiological Correlates of Self-Inhibitors.

Germination characteristics have been intensively studied in uredospores of economically important species belonging to *Puccinia* and *Uromyces*. Differences were often noted between species in the same genus, and similarities have been found among species belonging to the two different genera. Thus it was learned that exposing uredospores to cold induced a germination dormancy which could be relieved only by exposure to elevated temperatures. Only a few species responded in this manner while others were unaffected by cold storage (Bromfield, 1964). In some species, the germ tubes grow away

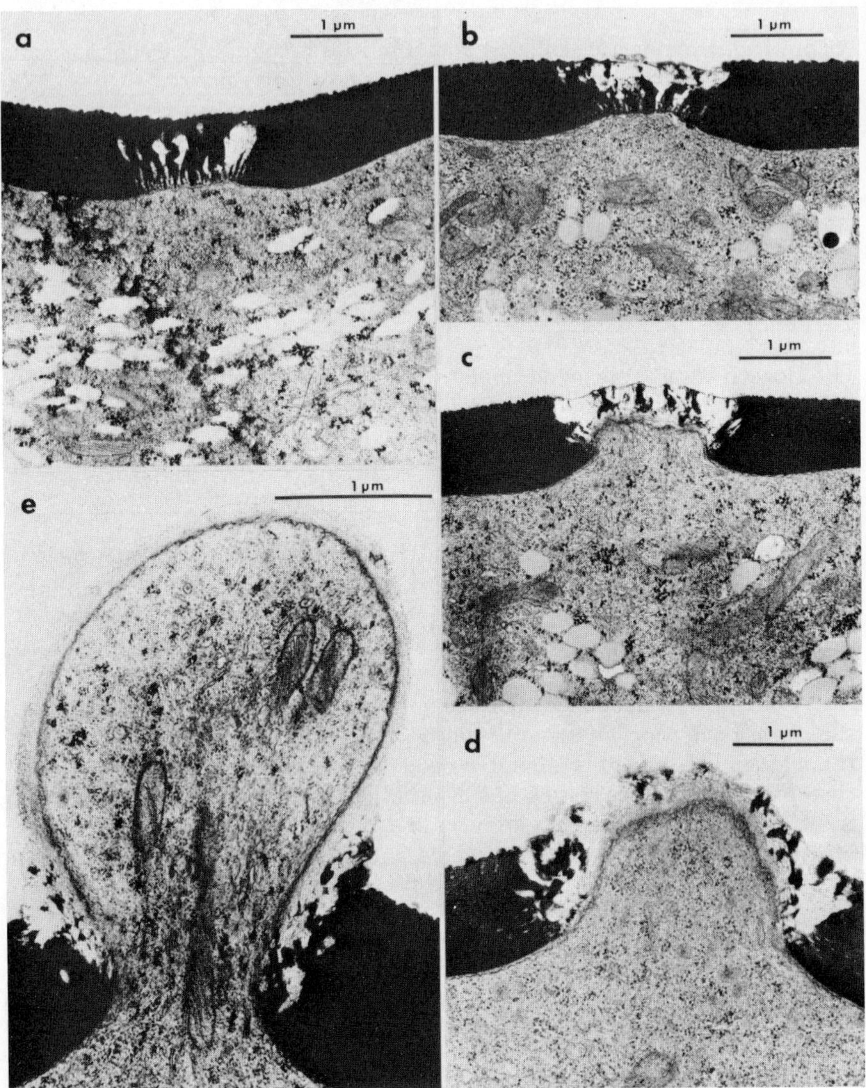

Fig. 4 Electron micrographs of thin sections of bean rust uredospores showing the germ pore region. (a) Dormant spore. (b) Spore floated for 120 min on 2 μ*M* MDC. (c) Spore floated 120 min on 2 μ*M* MDC and then 15 min on 10^{-4} *M* β-ionone. (d) Spore floated 120 min on MDC and then for 45 min on β-ionone. (e) Spore floated for 120 min on MDC and

from light while in others, growth is indifferent to light (Gettkandt, 1954). Differences among the species were also found in the response to differentiation stimuli. Thus wheat stem rust uredospores formed infection structures after a heat shock treatment or treatment with a volatile fraction extracted from the spores (Allen, 1957; Allen & Dunkle, 1971; Maheshwari et al. 1967). Other species such as bean rust required an oil-containing collodion membrane and did not respond to a heat shock. In addition, an active chemical has not yet been found to induce differentiation (Maheshwari et al. 1967). Using sensitivity of germination to water extracts of uredospores, Hoyer (1962) grouped *P. graminis tritici, P. triticina,* and *P. dispersa* into one class and *P. antirrhini, P. sorghi,* and *Uromyces fabae* into another class. Thus there appear to be at least two subgroups of species within the *Pucciniaceae*.

A summary of these germination characteristics is shown in Table 2. The data suggest that all of the uredospores examined segregate

Table 2. Some Germination Characteristics of the Uredospores of Several Species of Rusts[a]

Rusts	Natural germination self-inhibitor	Dormancy induced by cold	Negative phototropism of the germ tubes	Infection structures induced by		
				Heat shock	Volatile fraction	Collodion oil membrane
U. phaseoli	MDC	N	N	N	N	Y
P. helianthi	MDC	N	N	N	N	Y
P. antirrhini	MDC	N	N	N	N	Y
P. arachidis	MDC	N		N	N	Y
P. sorghi	MDC	N	Y	N	N	Y
P. graminis tritici	MF	Y	Y	Y	Y	N
P. recondita tritici		Y	Y			N

[a]Compiled from Allen, 1957; Bromfield, 1964; Foudin & Macko, 1974; Gettkandt, 1954; Macko et al. 1972; Maheshwari et al. 1967; and unpublished observations. MDC, methyl *cis*-3,4-dimethoxycinnamate; MF, methyl *cis*-ferulate; Blank spaces, not assayed.

then for 60 min on β-ionone.

into two groups. One is represented by the species in which the MDC was the self-inhibitor, and these species had similar characteristics of germination. The wheat stem rust uredospores, in which MF was the native self-inhibitor, had germination characteristics different from the former group. Although only a limited number of species have been examined, it appears probable that the rusts having MDC as the natural self-inhibitor are more closely related to each other and more distantly related to rusts which contain other self-inhibitors. If confirmed, this grouping may have taxonomical implications. From a physiological point of view, it would also be interesting to learn if there is a physiological link in these groupings.

REFERENCES

Adelman, T. G. & Lovett, J. S. (1974). Evidence for a ribosome-associated translation inhibitor during differentiation of *Blastocladiella emersonii*. Biochim. Biophys. Acta 335, 236--245.

Allen, P. J. (1955). The role of a self-inhibitor in the germination of rust uredospores. Phytopathology 45, 259--266.

Allen, P. J. (1957). Properties of a volatile fraction from uredospores of *Puccinia graminis* var. *tritici* affecting their germination and development. 1. Biological activity. Plant Physiol. 32, 385--389.

Allen, P. J. (1965). Metabolic aspects of spore germination in fungi. Annu. Rev. Phytopathol. 3, 313--342.

Allen, P. J. (1972). Specificity of the *cis*-isomers of inhibitors of uredospore germination in the rust fungi. Proc. Nat. Acad. Sci. U.S.A. 69, 3497--3500.

Allen, P. J. & Dunkle, L. D. (1971). Natural activators and inhibitors of spore germination. In Morphological and Biochemical Events in Plant-Parasite Interaction (ed. S. Akai & S. Ouchi), pp. 23--58. Tokyo: The Phytopathological Society of Japan.

Allen, P. J., Strange, R. N., & Elnaghy, M. A. (1971). Properties of germination inhibitors from stem rust uredospores. Phytopathology 61, 1382--1389.

Anderson, J. G. & Smith, J. E. (1972). The effects of elevated temperatures on spore swelling and germination in *Aspergillus niger*. Can. J. Microbiol. 18, 289--297.

Bacon, C. W. & Sussman, A. S. (1973). Effects of the self-inhibitor of *Dictyostelium discoideum* on spore metabolism. J. Gen. Microbiol. 76, 331--344.

Bacon, C. W., Sussman, A. S., & Paul, A. G. (1973). Identification of a self-inhibitor from spores of *Dictyostelium discoideum*. J. Bacteriol. 113, 1061--1063.

Barksdale, A. W. (1969). Sexual hormones of Achlya and other fungi. Science 166, 831--837.

Bell, A. A. & Daly, J. M. (1962). Assay and partial purification of self-inhibitors of germination from uredospores of the bean rust fungus. Phytopathology 52, 261--266.

Bromfield, K. R. (1964). Cold-induced dormancy and its reversal in uredospores of *Puccinia graminis* var. *tritici*. Phytopathology 54, 68--74.

Ceccarini, C. & Cohen, A. (1967). Germination inhibitor from the cellular slime mould *Dictyostelium discoideum*. Nat. Lond. 214,

1345--1346.

Cochrane, V. W. (1958). Physiology of fungi. 524 pp. New York: Wiley.

Cotter, D. A. & Raper, K. B. (1966). Spore germination in *Dictyostelium discoideum*. Proc. Nat. Acad. Sci. U.S.A. 56, 880--887.

Domsch, K. H. (1954). Keimungsphysiologische Untersuchungen mit Sporen von *Erysiphe graminis*. Arch. Mikrobiol. 20, 163--175.

Dworkin, M. (1973). Cell-cell interactions in the myxobacteria. In Microbial Differentiation (ed. J. M. Ashworth & J. E. Smith), pp. 125--142. Cambridge, England: Cambridge U. P.

Ettel, G. E. & Halbsguth, W. (1963). Über die Wirkung von Trimethylamin, Calciumnitrat und Licht bei der Keimung der Brandsporen von *Tilletia tritici* (Bjerk.) Winter. Beitr. Biol. Pflanz. 39, 451--488.

Fletcher, J. & Morton, A. G. (1970). Physiology of germination of *Penicillium griseofulvum* conidia. Trans. Br. Mycol. Soc. 54, 65--81.

Foudin, A. S. & Macko, V. (1974). Identification of the self-inhibitor and some germination characteristics of peanut rust uredospore germination. Phytopathology 64, 990-993.

Fredga, A. & Åberg, B. (1965). Stereoisomerism in plant growth regulators of the auxin type. Annu. Rev. Plant Physiol. 16, 53--72.

French, R. C. (1961). Stimulation of uredospore germination in wheat stem rust by terpenes and related compounds. Bot. Gaz. 122, 194--198.

French, R. C. & Gallimore, M. D. (1971). Effect of some nonylderivatives and related compounds on germination of uredospores. J. Agric. Food Chem. 19, 912--915.

French, R. C. & Gallimore, M. D. (1972). Effect of pretreatment with stimulator and water vapors on subsequent germination and infectivity of uredospores of *Puccinia graminis* var. *tritici*. Phytopathology 62, 116--119.

French, R. C. & Weintraub, R. L. (1957). Pelargonaldehyde as an endogenous germination stimulator of wheat rust spores. Arch. Biochem. Biophys. 72, 235--237.

Gettkandt, G. (1954). Zur Kenntnis des Phototropismus der Keimmyzelien einiger parasitischer Pilze. Wiss. Z. Univ. Halle-Wittenberg, Math.-Naturwiss. 3, 691--710.

Gooday, G. W. (1974). Fungal sex hormones. Annu. Rev. Biochem. 43, 35--49.

Grove, S. N. & Bracker, C. E. (1970). Protoplasmic organization of hyphal tips among fungi: Vesicles and Spitzenkörper. J. Bacteriol. 104, 989--1009.

Hess, S. L., Allen, P. J., Nelson, D., & Lester, H. (1975). Mode of action of methyl *cis*-ferulate, the self-inhibitor of stem rust uredospore germination. Physiol. Plant Pathol. 5, 107-112.

Hoyer, H. (1962). Ein Beitrag zur Kenntnis der Selbsthemmung der Uredosporenkeimung bei *Puccinia triticina* Erikss. und anderen Rostarten. 1. Die Hemmstoffbildung. Zentbl. Bakt. Parasitkde Abstr. II, 115, 266--296.

Janitor, A. (1967). Inhibičný vplyv extraktov a hustých suspenzií huby *Erysiphe graminis* f. sp. hordei Marchal na vlaste klíčenie. Česká Mykol. 21, 185--191.

Jones, J. P. & Haburn, M. A. (1969). Swelling phenomena of germinating crown rust uredospores. Phytopathology 59, 1034.

Krishnan, P. S., Bajaj, V., & Damle, S. P. (1954). Some observations on the growth of *Aspergillus niger* from spore inoculum. Appl. Microbiol. 2, 303--308.

Leighton, T. J. & Stock, J. J. (1970). Biochemical changes during fungal sporulation and spore germination. 1. Phenyl methyl sulfonyl fluoride inhibition of macroconidial germination in *Microsporum gypseum*. J. Bacteriol. 101, 931--940.

Leppik, R. A., Hollomon, D. W., & Bottomley, W. (1972). Quiesone: An inhibitor of the germination of *Peronospora tabacina conidia*. Phytochemistry 11, 2055--2063.

LeRoux, P. M. & Dickson, J. G. (1957). Physiology, specialization, and genetics of *Puccinia sorghi* on corn and of *Puccinia purpurea* on sorghum. Phytopathology 47, 101--107.

Lingappa, B. T. & Lingappa, Y. (1965). Effects of nutrients on self-inhibition of germination of conidia of *Glomerella cingulata*. J. Gen Microbiol. 41, 67--75.

Lingappa, B. T. & Lingappa, Y. (1966). The nature of self-inhibition of germination of conidia of *Glomerella cingulata*. J. Gen. Microbiol. 43, 91--100.

Lingappa, B. T., Lingappa, Y., & Bell, E. (1973). A self-inhibitor of protein synthesis in the conidia of *Glomerella cingulata*. Arch. Mikrobiol. 94, 97--107.

Macko, V., Novacký, & Škrobal, M. (1964). Inhibition in "slide germination test" caused by *Tilletia controversa* spores extract. Biologia 19, 869--870.

Macko, V. & Renwick, J. A. A. (1974). Reversal of self-inhibition by peroxidase and hydrogen peroxide in wheat stem rust uredospores: Mode of action. Phytopathology 64, 901--902.

Macko, V. & Staples, R. C. (1973). Regulation of uredospore germination and germ tube development. Bull. Torrey Bot. Club 100, 223--229.

Macko, V., Staples, R. C., Allen, P. J., & Renwick, J. A. A. (1971a). Identification of the germination self-inhibitor from wheat stem rust uredospores. Science 173, 835--836.

Macko, V., Staples, R. C., Gershon, H., & Renwick, J. A. A. (1970). Self-inhibitor of bean rust uredospores: Methyl 3,4-dimethoxycinnamate. Science 170, 539--540.

Macko, V., Staples, R. C., & Renwick, J. A. A. (1971b). Germination self-inhibitor of sunflower and snapdragon rust uredospores. Phytopathology 61, 902.

Macko, V., Staples, R. C., Renwick, J. A. A., & Pirone, J. (1972). Germination of self-inhibitors of rust uredospores. Physiol. Plant Pathol. 2,347--355.

Macko, V., Woodbury, W., & Stahmann, M. A. (1968). The effect of peroxidase on the germination and growth of mycelium of *Puccinia graminis* f. sp. *tritici*. Phytopathology 58, 1250--1254.

Maheshwari, R., Allen, P. J., & Hildebrandt, A. C. (1967). Physical and chemical factors controlling the development of infection structures from urediospore germ tubes of rust fungi. Phytopathology 57, 855--862.

Mori, K. (1973). Synthesis of *dl*-3-isobutyroxy-β-ionone and *dl*-dehydrovomifoliol. Agric. Biol. Chem. 37, 2899--2905.

Musumeci, M. R., Moraes, W. B. C., & Staples, R. C. (1974). A self-inhibitor in uredospores of the coffee rust fungus. Phytopathology 64, 71--73.

Obata, Y., Abe, H., Tanaka, Y., Yanagisawa, K., & Uchiyama, M. (1973). Isolation of a spore germination inhibitor from a cellular slime mold *Dictyostelium discoideum*. Agric. Biol. Chem. 37, 1989--1990.

Page, W. J., & Stock, J. J. (1971). Regulation and self-inhibition of *Microsporum gypseum* macroconidia germination. J. Bacteriol. 108, 276--281.

Page, W. J., & Stock, J. J. (1972). Isolation and characterization of *Microsporum gypseum* lysosomes: Role of lysosomes in macroconidia germination. J. Bacteriol. 110, 354--362.

Page, W. J., & Stock, J. J. (1974). Sequential action of cell wall hydrolases in the germination and outgrowth of *Microsporum gypseum* macroconidia. Can. J. Microbiol. 20, 483--489.

Park, D., & Robinson, P. M. (1971). A fungal hormone controlling internal water distribution normally associated with cell aging in fungi. Symp. Soc. Exp. Biol. 21, 323--336.

Pritchard, N. J. & Bell, A. A. (1967). Relative activity of germination inhibitors from spores of rust and smut fungi. Phytopathology 57, 932--934.

Rines, H. W., French, R. C., & Daasch, L. W. (1974). Nonanal and 6-methyl-5-hepten-2-one: Endogenous germination stimulators of uredospores of *Puccinia graminis* var. *tritici* and other rusts. J. Agric. Food Chem. 22, 96--100.

Robinson, P. M. (1972). Isolation and characterization of a branching factor produced by *Fusarium oxysporum*. Trans. Br. Mycol. Soc. 59, 320--322.

Robinson, P. M. (1973). Oxygen--positive chemotropic factor for fungi? New Phytol. 72, 1349--1356.

Robinson, P. M. & Garrett, M. K. (1969). Identification of volatile sporostatic factors from cultures of *Fusarium oxysporum*. Trans. Br. Mycol. Soc. 52, 293--299.

Robinson, P. M., Park, D., & Garrett, M. K. (1968). Sporostatic products of fungi. Trans. Br. Mycol. Soc. 51, 113--124.

Roy, M. K. & Prasada, R. (1966). Studies on the identification of a self-inhibitor of uredospore germination in maize rust (*Puccinia sorghi* Schw.). Phytopathol. Z. 57, 216--220.

Russell, G. K. & Bonner, J. T. (1960). A note on spore germination in the cellular slime mold *Dictyostelium mucoroides*. Bull. Torrey Bot. Club 87, 187--191.

Schröder, J. & Hassebrauk, K. (1964). Untersuchungen über die Keimung der Uredosporen des Gelbrostes *(Puccinia striiformis* West). Zentbl. Bakt. Parasitkde Hyg. Abstr. II, 118, 622--657.

Shepherd, C. J. & Mandryk, M. (1962). Auto-inhibitors of germination and sporulation in *Peronospora tabacina* Adam. Trans. Br. Mycol. Soc, 45, 233--244.

Soll, D. R. & Sonneborn, D. R. (1969). Zoospore germination in the water mold, *Blastocladiella emersonii*. II. Influence of cellular and environmental variables on germination. Dev. Biol. 20, 218-235.

Steele, S. D. (1973). Self-inhibition of arthrospore germination in *Geotrichum candidum*. Can. J. Microbiol. 19, 943--947.

Sussman, A. S. (1965). Physiology of dormancy and germination in the propagules of cryptogamic plants. Handbuch der Pflanzenphysiologie (ed. W. Ruhland), Vol. 15/2, pp. 933--1025. Berlin and New York: Springer.

Sussman, A. S. & Halvorson, H. O. (1966). Spores: Their dormancy and germination. 354 pp. New York and London: Harper & Row.

Szulmajster, J. (1973). Initiation of bacterial sporogenesis. In Microbial Differentiation (ed. J. M. Ashworth & J. E. Smith), pp. 45--83. Cambridge, England: Cambridge U. P.

Tanaka, Y., Yanagisawa, K., Hashimoto, Y., & Yamaguchi, M. (1974). True spore germination inhibitor of a cellular slime mold *Dictyostelium discoideum*. Agric. Biol. Chem. 38, 689--690.

Trinci, A. P. J. & Whittaker, C. (1968). Self-inhibition of spore germination in *Aspergillus nidulans*. Trans. Br. Mycol. Soc. 51, 594--596.

Trione, E. J. (1973). The physiology of germination of *Tilletia teliospores*. Phytopathology 63, 643--648.

Trione, E. J. & Leach, C. M. (1969). Light-induced sporulation and sporogenic substances in fungi. Phytopathology 59, 1077--1083.

Truesdell, L. C. & Cantino, E. C. (1971). The induction and early events of germination in the zoospore of *Blastocladiella emersonii*. Current Topics in Developmental Biology 6, 1--44.

Van den Ende, G. & Cools, P. J. M. (1967). Selbsthemmung and Selbststimulation der Konidiosporen von *Sclerotinia fructicola* bei der Keimung und Keimschlauchbildung. Phytopathol. Z. 59, 122--128.

Van Etten, J. L. (1968). Protein synthesis during fungal spore germination. I. Characteristics of an *in vitro* phenylalanine incorporating system prepared from germinated spores of *Botryodiplodia theobromae*. Arch. Biochem. Biophys. 125, 13--21.

Van Sumere, C. F., Van Sumere-De Preter, C., Vining, L. C., & Ledingham, G. A. (1957). Coumarins and phenolic acids in the uredospores of wheat stem rust. Can. J. Microbiol. 3, 847--862.

Wilson, E. M. (1958). Aspartic and glutamic acid as self-inhibitors of uredospore germination. Phytopathology 48, 595--600.

Yarwood, C. E. (1954). Mechanism of acquired immunity to a plant rust. Proc. Nat. Acad. Sci. U.S.A. 40, 374--377.

Yarwood, C. E. (1956). Simultaneous self-stimulation and self-inhibition of uredospore germination. Mycologi 48, 20--24.

DISCUSSION

Self-Inhibitors of Fungal Spore Germination

Chairman: Edward Trione
Oregon State University
Corvallis, Oregon

Professor Allen offered a few comments about the recent research in his laborabory on the wheat stem rust inhibitor, methyl ferulate. He noted that the effect of cycloheximide does not prevent the dissolution of the germ pore plugs, even at a concentration which completely inhibits protein synthesis. During the first 10 minutes of floating on water, the plugs dissolve completely in the presence of cycloheximide. Professor Allen and co-workers are examining the mode of action of the inhibitor and trying to find out how the dissolution of the germ pore plug is controlled. They are examining the products of hydrolysis that are released when the plug is dissolved and are also looking for the hydrolytic enzymes involved. Some enzyme preparations like zymolyase will digest the plug, but the specific enzymes in zymolyase that are responsible for this action are unknown. Current research is focused on a carbohydrase and a protease. If the self-inhibitor is found to be a specific inhibitor of carbohydrase it will be very interesting, because inhibitors of carbohydrases are very rare, and the levels of the inhibitors effective against these spores are so small that it would be perhaps the best example of a specific carbohydrase inhibitor.

A question was asked about the chemistry of the germ pore plug in relation to the rest of the cell wall. Professor Allen stated that the mature spore wall is over 90% mannose, and the germ pore plug has very little mannose in it. Professor Allen agreed with a suggestion that it should be possible to assist digestion of the germ pore plug in the rust uredospore wall by applying hydrolytic enzymes from the outside.

A question was asked concerning the effect of the rust self inhibitors on oxygen uptake. Dr. Macko cautioned that in the inhibitor work it is very easy to confuse the condition necessary for the phenomenon to occur with the site of action or mode of action. If the uredospore germinates there is an increase in the oxygen uptake. If uredospores are floated on the inhibitor this increase in oxygen does not occur, but this does not necessarily mean that the mode of action of the inhibitor is the prevention of oxygen uptake.

In response to a question, Dr. Staples noted that MDC is not found in the healthy bean plant. It is present only in the fungus, and shows up only during early fleck stage when the rust begins to sporulate. Drs. Staples and Macko did not know if the stimulator, β-ionone, enhanced the germination of bean rust uredospores after the endogenous inhibitor had been removed by washing.

Professor Allen thought there was an argument against dimethylguanosine being the self-inhibitor in *Dictyostelium*. The natural self-inhibitor interferes with swelling which occurs during the first 1.5 hours of germination, but dimethylguanosine is an inhibitor of protein synthesis and it is not active during the swelling stage, but rather at a later stage in germination, after the swelling has occurred. The question of dimethylguanosine being the only inhibitor in *Dictyostelium* may also be in doubt from some work done by Dr. Cotter. He purchased dimethylguanosine commercially and tested it on five different strains of *Dictyostelium discoideum*, two strains of *D. purpureum*, and a few Mucorales. Under no conditions did it inhibit germination at twice the concentration recommended. He used lyophilized spores and three to four different methods of activation. It is possible that different laboratories are working with different strains of *D. discoideum* and that some of the strains do not react with that particular inhibitor. Dr. Cotter agreed that there may be different types of inhibitors that have different modes of action, however, in the case of the seven strains of *Dictyostelium* tested, none of them were inhibited by dimethylguanosine.

CHAPTER 3

ACTIVATORS OF FUNGAL SPORE GERMINATION

A. S. Sussman
University of Michigan

DEDICATION

This paper is dedicated to Professor Alexander H. Smith upon the occasion of his retirement as Professor of Botany at the University of

Michigan, and in honor of his distinguished contributions to mycology.

THE PHENOMENOLOGY OF ACTIVATION

I would like to promulgate the suggestion by Herrman & Tootle (1964) that, ". . . the absence of a 'general theory' of differentiation may be of less importance (to biology) and the complete analysis of the molecular mechanisms of specific developmental processes a worthwhile end in itself." Although I subscribe to the notion that fungal spore germination is a potentially useful paradigm of differentiation, given the need to explicate mechanisms, my goal will be a modest one: to raise some questions about activation and activators and to point directions to the future, rather than to advance general theories of differentiation via activation. To the extent that activation is a type of control, it represents the broad class of means through which arrests of development are relieved, my assumption being that the duration of the dormant state is less general than the array of mechanisms used in imposing and releasing such arrests. Therefore, while eschewing the presentation of a "general theory" I will, nevertheless, suggest that such generalities must derive from the analysis of mechanisms of activation.

Definitions

The definitions I have used before (Sussman, 1966) will be accepted in what follows, including:

1. Dormancy. Any rest period or reversible interruption of the phenotypic development of an organism.
2. Activation. A treatment that leads to an increase in germination.
3. Germination. The first irreversible stage which is recognizably different from the dormant organism, as judged by morphological, cytological, physiological, or biochemical criteria.

In the definition of activation given above, the word "treatment" may be misleading. Thus, some dormant organisms may germinate as a result of storage in nature or the laboratory, without what might ordinarily be called a treatment. In such cases I would argue that the changes which transform such a spore from the dormant to the active mode should be subsumed under the heading of activation. In fact, a further question might be raised: Should at least two general types of activation be distinguished, including that which does not require deliberate treatment ("gratuitous activation") and that which does ("applied activation")? Perhaps such a dichotomy has similar advantages to those that accrue from the distinction between exogenous and constitutive dormancy (Sussman, 1966), and its difficulties as well (Cochrane, 1974).

Types of Activation Treatments

Possibly one of the most important functions of dormancy in microbes is the provision of a means for their dispersal in time (Sussman & Halvorson, 1966). That this is true as well for animals (Lees, 1961) and higher plants (Black, 1972) is likely so spores, seeds, and diapausing animals can be viewed as timing devices that ensure that active growth ensues when conditions are most propitious. Consequently, this type of adaptation confers selective advantage upon organisms whose environments undergo fluctuations in moisture, nutrients, temperature, and other factors. Therefore, it is to be expected that the treatments that effect activation would include temperature extremes, drying and wetting, various light regimes, and chemical addenda and, indeed, this is the case. These treatments have been reviewed before (Cochrane, 1958, 1960; Gottlieb, 1959; Sussman, 1965, 1966; Allen, 1965) so they will be discussed herein only in the context of the mechanisms of activation.

CAUSAL EXPLANATIONS AND ACTIVATION

The assignment of causality to developmental phenomena like activation is exceptionally difficult because of the distance of morphological events from primary sites of action such as genes, enzymes, etc., and because of the possibility of pleiotropic effects stemming from a single causal event or of simultaneous causal events converging on a single morphological effect. Some of these difficulties are illustrated in Fig. 1 and chasten we who might tread unwarily in this area of research.

Another difficulty is: At which level of organization should the causal event be sought? In order to set a perspective for this discussion, I present a hierarchy of organization in cells through Fig. 2 in

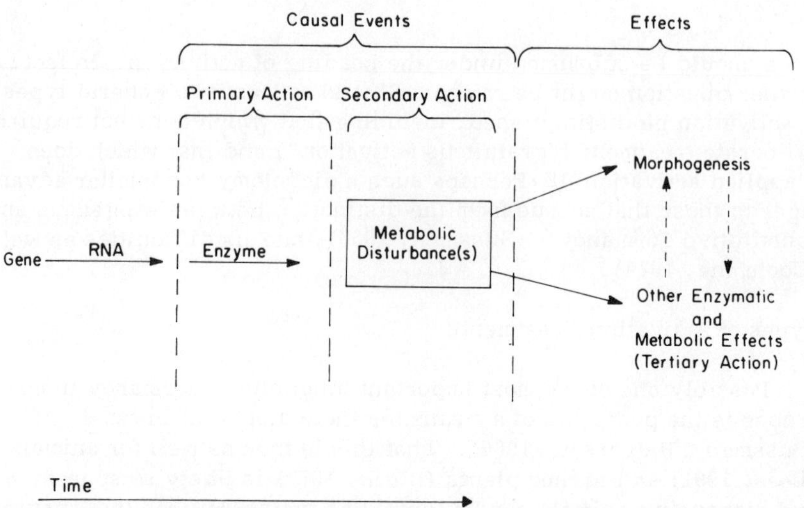

Fig. 1 The complexity of causal relations in development beginning with gene action. (The original ideas for this figure derive from Dr. David Sternberg.)

---

which seven levels are illustrated, ranging from environmental precursors of low molecular weight and size to cells themselves. A theorem that I believe is crucial might be presented as follows: The cause of an effect observed at any level must be sought at a lower level. For example, because activation results in an immediate effect upon respiration, and because respiratory metabolism is centered in mitochondria, the explanation must be sought at the level of supramolecular assemblies, or at lower ones (Fig. 2). Thus, an upper limit is set for the level at which experimentation should proceed in the search for a causal explanation of this effect. Can a lower level be set as well?

This question might be answerable through two approaches, the first through an examination of the properties of the entities at each of the levels, and the second through an empirical examination of the changes in energetic parameters as a result of activation.

Platt (1961) has taken the first approach by examining the ways in which the properties of molecules depend upon their size (Table 1). In fact, he suggests that these further properties of macromolecules might not have been anticipated from a knowledge of the properties of their polyatomic chemical subgroups or those of smaller molecules. The properties in the 50--500 atom range that are discussed include

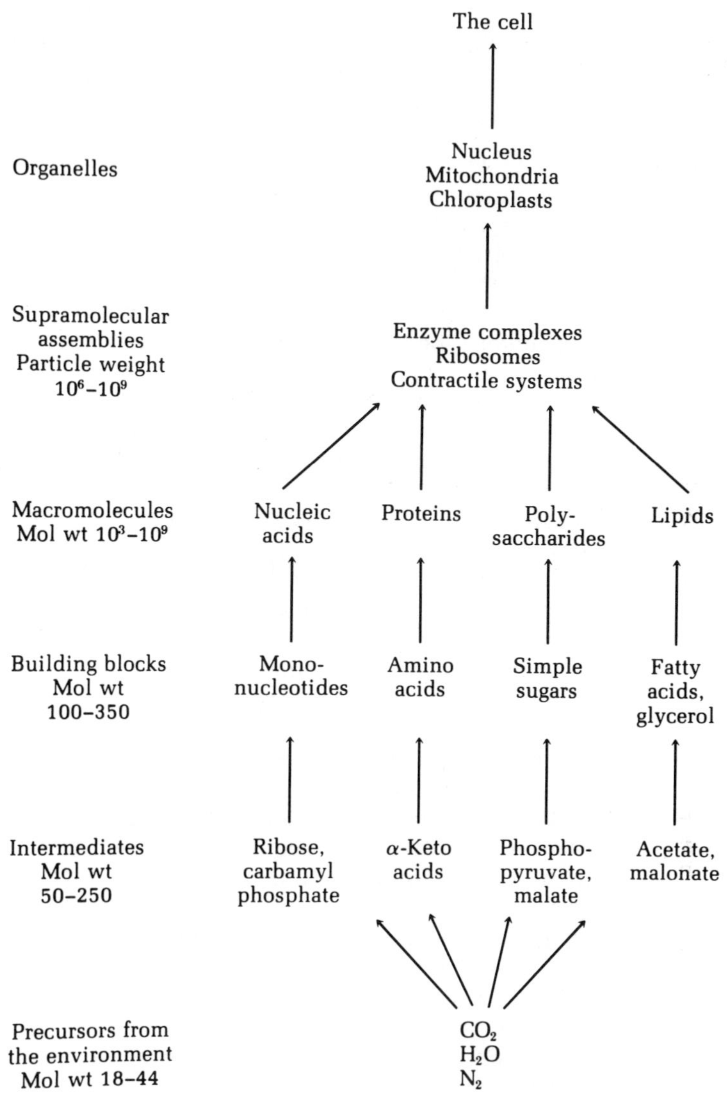

Fig. 2 The hierarchy of molecular organization in cells (taken from Lehninger, 1970).

---

contractility, interactions between groups attached to the chain,

Table 1. Properties of Molecules in the 50--500 Atom Range and Greater That Go Beyond the Properties of Their Chemical Subgroups (Taken from Platt, 1961).

Long-chain properties

Properties arising from connectedness

1. Possible contractility (from multiple internal rotations)
2. Variable viscosity in solution (from 1)
3. Mechano-chemical contractive forces, solvent-dependent or reaction-dependent, between points on the chain (from 1)
4. Long-range specific and directed interactions between groups attached to the chain (from 3)
5. Stability and reproducibility in successive amplification processes such as catalysis and response

Properties arising from continuity

6. Ease of repetitive construction and repetitive operation in reactions
7. Reduction of entropy factors in hand-to-hand reactions along the chain

Properties arising from sequential complexity

8. Specificity of sequence, multiplying with chain length
9. Self-replication (from 8 for chemically reacting chain with a sequence longer than some unknown minimum size)

Other cooperative properties

10. Long-range order (secondary; tertiary; and long-range "crystallization")
11. Polyelectrolyte behavior
12. Possible "wrap-around" cooperative effects on binding energies and activation barriers
13. Possible selective transfer of electrons through large ions and radicals

stability and reproducibility in catalysis and response, reduction of entropy factors, self-replication, possible "wrap-around" cooperative effects on binding energies and activation barriers and transfer of electrons through large ions and barriers. In short, many of the properties we recognize in enzymes, nucleic acids, and other macromolecules are included and these are the very ones that we might predict to be important in activation. Thus, as Platt puts it:

> Selective response and "enzyme action" or "catalysis" are names for biological amplification processes which convert a small selective input stimulus into an amplified output response. In order for such responses or reactions to be reproducible time after time, it is necessary to have a stable amplifer, that is, an amplification system that comes back to its original configuration no matter what intermediate forms it may assume.

Clearly, the statistical variation in the arrangement of a collection of small molecules cannot meet this requirement but arguments have been advanced to suggest that a chain is a particularly favorable way to construct a reliable amplifier. This is just another way of saying that the stability and reproducibility of processes in living organisms are furthered by the tight bonding of large molecules, with their elevated energetic barriers to thermal disruption. Thus, whereas random and small thermal fluctuations would disrupt the weakly bonded arrangements of small molecules, they would not affect the arrangement of large ones. Furthermore, chain processes, wherein stepwise reactions proceed down a chain, or in which directed contractions of, for example, microfilaments, bring together reactants, also are advantageous to the extent that they reduce entropy by the elimination of diffusion-controlled steps. In addition, there is a reduction in diffusional delays and in the amounts of reactants needed, and a gain in the accuracy by which intermediates or reactants are brought to the site of the activity. Other advantages could be adduced but the main point is that these arguments lead to the conclusion that it is at the level of macromolecules (Fig. 2) where considerable emphasis probably should be placed if we are to arrive at causal explanations of the mechanism of activation.

Another way of setting a lower limit to the level at which our search for mechanisms of activation might begin is to analyze the thermodynamic parameters involved, as was done by Cotter (1973) in an ingenious and enlightening way. First of all, he argued that the sharpness and large slope of the usual heat-activation curves for spores (Fig. 3) suggested a process that required a high degree of cooperativity and that the thermal energy effected a conformational change in some molecules or aggregates thereof. Furthermore, like Platt (1961), he assumed that the sharpness of the transition points in the curves, and their large slope and large activation energy ruled out either small molecules or those of varied molecular weight. Consequently, he concluded that the key substances that should be considered as the prime candidates for heat-mediated transitions of the kind he studied are the macromolecules, RNA, DNA, lipids, and proteins. To buttress this argument further he calculated, by the Van't Hoff equation, the change in entropy (ΔS^o) for the heat-activation of spores of *Dictyostelium discoideum* and other microorganisms and compared these with the values for protein denaturation by high temperature, as in Table 2. Values such as these are obtained by the Arrhenius method whereby the logarithm of a parameter is plotted against the reciprocal of the temperature. When straight lines are obtained the slope yields the ΔH, or enthalpy of activation. These data reveal surprisingly high changes in entropy for the heat-activation of fungal spores, as compared with protein denaturation, but this discrepancy was reconciled by invoking a role for water in helix$\rightarrow$coil transitions such that very high ΔS^o values are obtained (Bresler, 1971). Therefore, Cotter favors proteins as the source of the heat-mediated transitions in the activation of spores of *Dictyostelium* and, indeed, the denaturation curve transitions are sharp enough to merit such consideration (Fig. 4). It should also be observed that the hydrophobicity of the poly-L-alanine helix makes it much more stable than that of poly-L-glycine.

Nevertheless, it would be dangerous to conclude that the level of macromolecules alone is the one at which the basic mechanism of all activations should be sought for supramolecular assemblies like ribosomes and enzyme complexes, or organelles like mitochondria, are entities to which causal effects might be assigned. Yet, to be really causal such effects would have to be exerted in ways that go beyond the properties, as we know them, of macromolecules themselves. If, for example, mitochondria were to be considered the locus of an effect, it would have to be established that the observed changes were not due simply to recognized properties of macromolecular constituents of these organelles. Going in the other direction, to the level of small molecules (Fig. 2), it might be imagined that an ordinary substrate might activate if by so doing it removed, or rendered innocuous, a small toxic molecule. Just such a mechanism is used to explain the

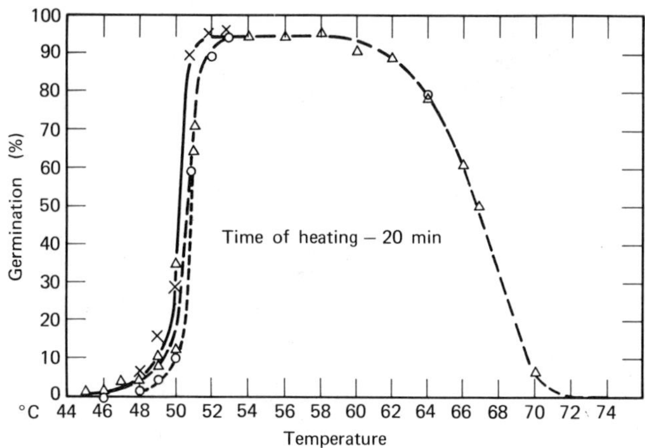

Fig. 3 The effect of the temperature of activation on the germination of ascospores of *Neurospora tetrasperma* (taken from Goddard, 1935).

Table 2. The Change in Standard Entropy with Protein Denaturation or Spore Activation (Taken from Cotter, 1973).

Protein or spore	Temperature used for ΔS^o Calculation	ΔS^o
Trypsin (pH 2) [1]	44°C	213
Ribonuclease (pH 2.5) [2]	50°C	340
Chymotrypsinogen (pH 2.5) [2]	50°C	360
Chymotrypsinogen (pH 2.5) [2]	75°C	680
Bacillus subtilis [3]	5--95°C	5--8
Bacillus megaterium [4]	50--75°C	147
Neurospora tetrasperma [5]	51°C	1500
Dictyostelium discoideum [6]	44°C	1700

References

1. Anson & Mirsky (1934)
2. Tanford (1970)
3. Busta & Ordal (1964)
4. Levinson & Hyatt (1969)
5. Calculated from Goddard (1935)
6. Calculated

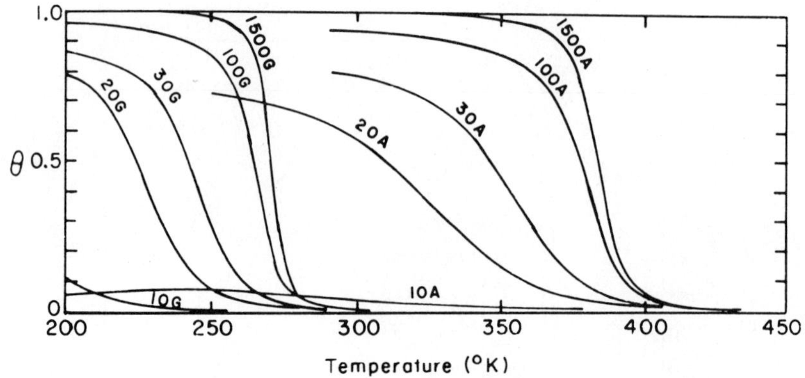

Fig. 4 Computed curves for helix→random-coil transition for polyglycine (G) and poly-L-alanine (A) for the various degrees of polymerization indicated on the curves (taken from Scheraga, 1966).

---

activation of spores of *Agaricus bisporus* by isovalerate or isoamyl alcohol, which are believed to lead to the removal of CO_2 as a result of the activity of carboxylating enzymes in the pathway whereby these activators are metabolized (Rast & Stäuble, 1970). However, it could be argued that the causal event in this case is the initiation of the carboxylation reaction which, of course, involves a protein. Even so, I would not wish to rule out the possibility that causal explanations for activation should be sought at other levels but I suggest that macromolecules offer more clear-cut approaches at this time so I shall limit my discussion to this level of organization in what follows.

HEAT ACTIVATION

Before beginning the discussion of the involvement of macromolecules in activation it is well to point out that organisms require a diversity of temperatures and durations of treatment for heat-activation (Table 3). Furthermore, it has been established in at least one case, that of *Coprinus radiatus* (Mills & Eilers, 1973), that combined treatment with both heat and chemicals is required for maximal germination, so the mechanism in such cases probably is quite complex.

Table 3. Optimal temperatures and duration of treatment required for the heat-activation of selected fungal spores.

Organism	Temperature	Duration
Dictyostelium discoideum	$40^{o}C$	5 min
Puccinia graminis	$40^{o}C$	5 min
Coprinus radiatus	44--$46^{o}C$	4 h
Mucor miehi	$45^{o}C$	5 h
Phycomyces blakesleeanus	$50^{o}C$	3 min
Neurospora tetrasperma	50--$65^{o}C$	10--20 min
Byssochlamys fulva	$75^{o}C$	5--10 min

Involvement of Proteins

Cotter's conclusion (1973) that proteins mediate the changes that result in the heat-activation of spores of *D. discoideum* was based not only upon comparisons of the ΔS^{o} but upon the rejection of such a role for nucleic acids and lipids. Thus, he argued that the melting curves of RNA are not as sharp as those for heat-activation of spores, nor is the slope for the transition from the native to the denatured state large enough (Cox & Littauer, 1960). By contrast, as Fig. 5 discloses, the melting curve for DNA can be very sharp as in the case of amoebae of *D. discoideum* (Leach & Ashworth, 1972). But the nuclear DNA of this organism has a melting temperature (T_m) of $78.8^{o}C$ (Fig. 5) and its mitochondrial equivalent melts at $80.0^{o}C$ (Sussman & Rayner, 1971), whereas the spores are activated between 40--$45^{o}C$. Nor can high internal ionic strength be expected to change the T_m sufficiently *in vivo* (Gruenwedel et al., 1971). A similar argument can be applied to *Neurospora* ascospores which are activated optimally between 50 and $60^{o}C$. In this case the T_m for DNA varies between 90.5 and $85.2^{o}C$, depending upon the species (Dutta & Chakrabartty, 1971). Other evidence against the participation of DNA in the heat-activation of spores is that those killed by ultraviolet light can still be heat-activated, even though the DNA is functionally inactivated (Cotter, 1973). And, finally, other data have revealed that the informational content of DNA is not used for RNA or protein synthesis in spores of *D. discoideum* until just before the emergence of the myxamoebae (Cotter & Raper, 1970; Feit et al., 1971).

According to Cotter, lipid transitions seem not to be sharp enough to account for heat-activation curves in *Dictyostelium* and *Neurospora*. Moreover, at least in the case of spores of the former, lipid transitions are alleged not to take place at the same temperature as activation

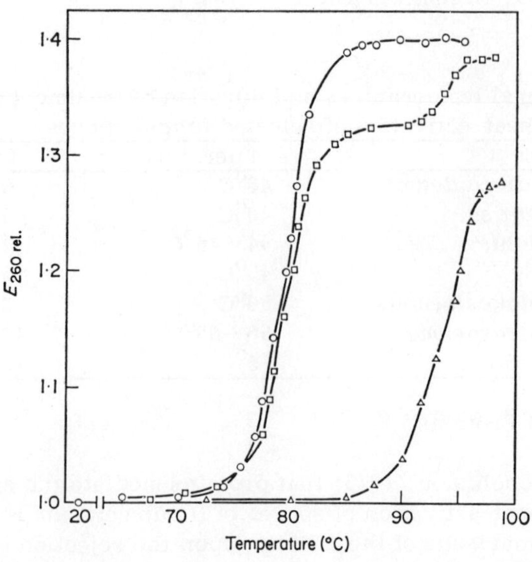

Fig. 5 Melting profiles of purified DNA from amoebae of *Dictyostelium discoideum*. Results are plotted as the O.D.$_{260\ nm}$ at room temperature to yield the $E_{260\ rel}$. -o-o-, thermal transition profile of DNA prepared from axenically grown amoebae; -□-□-, profile of DNA prepared from amoebae grown on bacteria; -Δ-Δ-, thermal transition profile of DNA isolated from *Aerobacter aerogenes* (taken from Leach & Ashworth, 1972).

(Raison & Lyons, 1971).

So it is concluded by Cotter and others (Morowitz, 1968; Sadoff, 1969) that proteins seem to be the most likely among the macromolecules to mediate thermal effects, and, in particular, denaturation has been suggested to be the process responsible for the activation of bacterial and fungal spores (Sussman & Halvorson, 1966; Gould & Hurst, 1969). However, as was pointed out above, the data in Table 2 show that the ΔS^{o} for fungal spore germination is much higher than that for protein denaturation. But, accompanying the order→disorder transition due to the unfolding of proteins during denaturation is one of disorder→order as hydrophobic amino acid side-chains are exposed to water (Tanford, 1968). Consequently, Cotter argues that if denaturation is partial and involves only the hydrophilic residues then a relatively high and positive ΔS^{o} would result during a helix→random coil transition. In fact, Bresler (1971) has calculated that such a transition for a protein with a molecular weight of 50,000 has a ΔS^o of 2000 e.u. per mole. This analysis led Cotter to conclude that a protein in the mitochondrial membrane might be the mediator of the activation of spores of *Dictyostelium* on the basis that respiration commonly increases upon activation, spores produced on azide cannot be activated and germination can be stopped at any point in the process by inhibitors of mitochondrial function (Cotter & Raper, 1968). Furthermore, Bogin et al. (1970) have shown that oxidative phosphorylation in the electron transport particles of *Mycobacterium phlei* is stimulated by exposure to 50°C for 15 min. To summarize Cotter's (1973) hypothesis: ". . . heat-activation partially denatures a regulatory or control protein in the phosphorylation complex of dormant procaryotic and eucaryotic spores; the relaxed state of the complex in the activated spores then results in full phosphorylation capacity and subsequent germination." The large array of protein inhibitors that are known to exist, including trypsin-inhibitors and others, might extend the hypothesis were such controls to exist in spores. Thus, denaturation could lead directly to activation by the removal of inhibition. The inhibitor of the trehalase of *Neurospora* is an example of such a possibility (Garrett et al., 1972) which has not yet been adequately tested.

But, as cogent as these arguments are there is the need for more data, including those bearing upon the analysis of mitochondrial functions and proteins in the spores of *D. discoideum* and other organisms. Thus, the data on *Mycobacterium phlei* are not from spores, nor has the phenomenon been demonstrated elsewhere. Other evidence that can be cited, such as that showing that the cytochrome chain is in the reduced state in bacterial spores (Sussman & Douthit, 1973) does not deal with the locus of the blockage. Moreover, other data contradict the generalization, at least as far as procaryotes are concerned, for it has been shown that the activation of bacterial spores is unaffected by cyanide and azide whereas later stages in germination are inhibited (Sussman & Halvorson, 1966). As for heat-activated enzymes themselves, the endonuclease of *Bacillus subtilis* (McCarthy & Nestor, 1969) remains unchanged after activation so even just partial denaturation seems not to have occurred. In the case of eucaryotes, data on *Neurospora* ascospores exist which show that there is, if anything, a decrease in ATP upon heat-activation due probably to the formation of large amounts of phosphorylated sugars which accumulate immediately after activation (Fig. 6; Eilers et al., 1970b). Of course it might be argued that increased oxidative phosphorylation could be

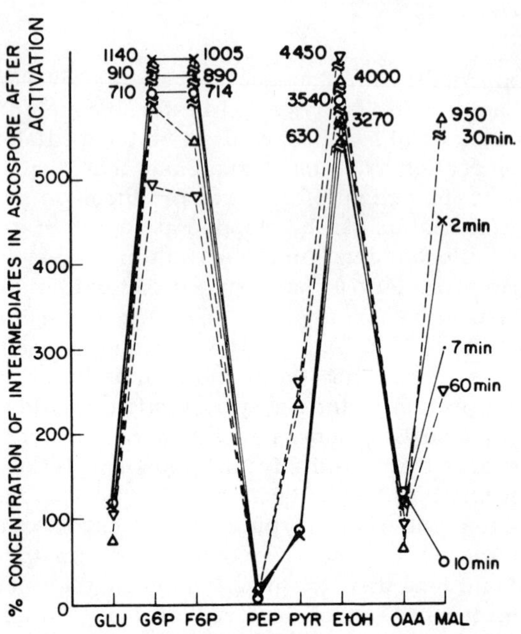

Fig. 6 Phase-plot of some metabolic intermediates from heat-activated ascospores of *Neurospora tetrasperma* during the first 60 min after activation, using the concentrations in dormant spores as the reference point (100%). The following symbols designate the times after activation: X, 2 min; ●, 7 min; Δ, 30 min; ∇, 60 min. GLU=glucose; PYR= pyruvate; EtOH=ethanol; OAA=oxaloacetic acid; MAL-malate (taken from Eilers, 1968).

---

responsible for the accumulation of these materials but the fact that the latter accumulate within 3 min after activation and that there is a sizable pool of ATP in dormant spores suggests otherwise. Therefore, I would suggest that the data are incomplete and, in addition, alternative possibilities exist, some of which will be discussed below.

Alternatives to partial denaturation of proteins as an activation mechanism include conformational and/or allosteric changes. For example, the activation of spores of *Phycomyces blakesleeanus* involves exposure to 50°C for a few minutes and the subsequent breakdown of trehalose (Van Assche et al., 1972). If trehalase is extracted and subjected to this heating regime its activity increases 10- to 15-fold immediately (Fig. 7) because of what probably is an allosteric change.

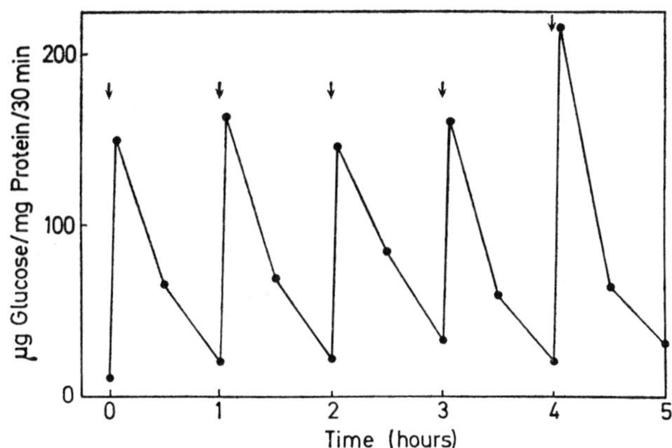

Fig. 7 Increase and decay in trehalase activity in spores of *Phycomyces blakesleeanus* following repeated heat treatments at 50°C for 3 min (arrows), interrupted by periods of 1 h at 18°C (taken from Van Assche et al., 1972).

Table 4. Possible Points of Control of Enzyme Activation (Taken from Varner, 1971).

Oxidation of sulphydryl groups
Reduction of disulphide bonds
Hydroxylation of proline and lysine
Glycosylation of hydroxyl groups of amino acid side-chains
Peptide bond cleavage
Phosphorylation
Acetylation
Adenylation
Removal of an inhibitor
Addition of an activator

The enhanced activity decays soon afterward and the increase and decline can be repeated several times. This remarkable cycle of enhanced and declining activity is not inhibited by cycloheximide so protein synthesis and its arrest cannot explain the data. Other examples of heat-activated enzymes are known (Swartz et al., 1956), including a variety of nucleases and carbohydrases, along with a temperature-induced conformational change in pyruvate kinase (Kayne & Suelter, 1968).

Varner (1971) has summarized possible points of control of enzyme activation as in Table 4 but, except for the removal of an inhibitor, he has listed reactions which probably follow the mediating reactions discussed above. In fact, some of the control reactions he has listed can be very complex. For example, an activating factor (AF) for chitin synthetase in *Saccharomyces cerevisiae* has been discovered to be a protease which is itself inhibited by another protein in the cell (Cabib & Ulane, 1973). However, it is likely that the enzyme and its inhibitor are compartmentalized in the cell so that another element of control is introduced. Proteolytic activation of enzymes is common in a variety of organisms and is reviewed by Ottesen (1967) who, along with Walther et al. (1974), provides insights into the events that lead to an activated protein through a "cascade" of reactions. However, as useful as these examples are as models of mechanisms of activation, they are only models because just a few systems have been developed in enough detail to permit tests of the possibilities discussed above.

Involvement of Lipids

Although Cotter (1973) discards the possibility that lipids are involved causally in heat-activation I believe there are good reasons for invoking a key role for these substances and I shall use mainly ascospores of *Neurospora* to make this point. But it should be observed at the start that clear distinctions between lipids and proteins sometimes may be hard to make, as when membrane complexes and lipoproteins are considered. A striking illustration of this difficulty is the case of "retinal isomerase" whose status as an enzyme has not been questioned since its discovery almost 20 years ago. Yet, this year it has been suggested that this "enzyme" is a lipid, possibly phosphatidylethanolamine (Shichi & Somers, 1974).

In order to set a basis for some of this discussion, without reviving past controversies over membrane models, I should like to assume the following about the "fluid-mosaic" model because they seem adequately confirmed by experiment: ". . . that in most membranes the majority of the lipid is in the form of a fluid bilayer matrix, and that the membrane proteins are both loosely bound and in some cases deeply or transversely embedded in this matrix" (Oseroff et al., 1973).

First of all, there are thermodynamic parameters involving lipids whose inflection points are very sharp, as in the case of the swelling rate in glycerol of liposomes prepared from lecithins (Fig. 8). In this case, the increased thermal mobility of the chains of the lipid greatly

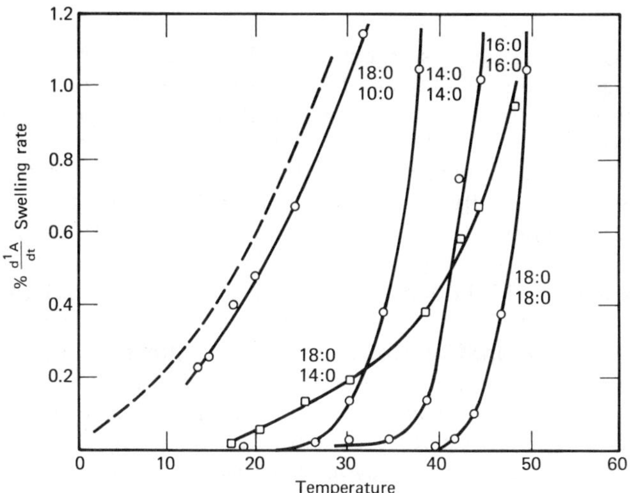

Fig. 8 Initial swelling rate in isotonic glycerol of liposomes of various lecithins as a function of temperature. Curves of (distearoyl)-, (dipalmitoyl)-, (dimyristoyl)-, (1-stearoyl-2-myristoyl)-, and (1-stearoyl-2-decanoyl) lecithin (taken from DeGier et al., 1968).

---

increases the entry rates of nonelectrolytes, as a possible model for permeability effects of other kinds as well. In addition, several abrupt phase transitions are known to occur in lipids, such as those outlined in Table 5. As in the case of the liposomes whose permeability was studied in Fig. 8, the thermal data for phospholipids reveal that a wide range of temperatures are encompassed by these phase transitions, depending upon the particular compounds involved, so that the diversity of activation temperatures observed for microbial spores could be explained by this means.

The large amount of evidence which establishes the "fluid-bilayer" model of membrane structure (Oseroff et al., 1973; Machtiger & Fox, 1973) suggests that phospholipids diffuse rapidly in the plane of the membrane. If this property of lipids is combined with their ability to activate certain enzymes (Table 6), then it is possible that high temperatures cause the movement of essential lipids to the site of enzymes requiring them for activity, thereby initiating reactions that could lead to germination. In fact, it has been shown that the stimulation of the membrane-bound sialyl transferase complex from *Escherichia coli* by lipids is markedly temperature-dependent (Vijay et al., 1974). Thus, it is suggested by these workers that the failure of lipids to activate this enzyme at lower temperatures is because they are relatively immobile under these conditions. However, at higher temperatures the increased fluidity of the membrane components results in activation. Another case is that of lecithin: cholesterol acyl transferase which is activated by peptides only above 24° C, which is the transition temperature of the phospholipid substrate (Soutar et al., 1974). The work of Chuang & Crane (1973) and of others listed by Rothfield & Romeo (1971) demonstrates that cytochrome oxidase and other enzymes of the Krebs cycle require lipids so that energy metabolism could be triggered by this means, as an alternative to the mechanism proposed by Cotter. Moreover, since membranes are organized as ordered and continuous arrays of interacting molecules, sequential perturbation of adjacent molecules could be caused by a change in a single molecule such that a cascade effect would result (Rothfield & Romeo, 1971). Consequently, changes in permeability and/or enzyme activity might be caused by heating and could explain activation, in some cases at least. Are there data on spores which support this conjecture?

That trehalose is almost ubiquitous in fungi has been established along with its possible role as one of the chief sources of energy for the germination of some spores (Sussman, 1961a; Sussman, 1969;

Table 5. Thermal Data for Phospholipids (1,2-Diacylphosphatidycholines (Taken from Ladbrooke & Chapman, 1969).

	Solid (β-crystals) liquid crystalline		Solid (α-form) liquid crystalline		Gel→liquid crystalline	
	T°C	ΔH Kcal/mole	T°C	ΔH Kcal/mole	T°C	ΔH Kcal/mole
C22 (behenol)	120	16.7	105	7.1	75	14.9
C18 (stearoyl)	115	13.8	97	6.2	58	10.7
C16 (palmitoyl)			93	4.3	41	8.7
C14 (myristol)			89	3.4	23	6.7
C12 (lauroyl)					20	---
C18-2 (oleyl)					-22	7.6

Rousseau & Halvorson, 1973). In ascospores of *Neurospora* this sugar disappears only after activation and may reside, unused, for years in dormant cells. Therefore, we studied the accumulation of intermediates as a function of activation, as mentioned earlier, and obtained the phase-plot shown in Fig. 6. There is a striking increase in glucose-6-phosphate and fructose-6-phosphate within two minutes after activation, suggesting a cross-over in the region of 6-phosphofructokinase which may be inhibited or present in limiting amounts. That this may also be the case for spores of *Phycomyces blakesleeanus* is suggested by the finding that glycolysis in extracts of these spores is "activated" by the addition of fructose 1,6-diphosphate and ATP (Rudolph, 1973). It is interesting to note that the lack of this enzyme may be responsible for the failure of early mouse embryos to grow on glucose alone, although glycogen phosphorylase and hexokinase may be involved as well (Barbehenn et al., 1974). In this case, citrate, which accumulates in large amounts, may inhibit 6-phosphofructokinase and such an effect also should be sought in spores. In any event, the immediate metabolic effect is to trigger glycolysis and the secretion of large amounts of ethanol (Eilers et al., 1970b), an effect parallel to that described for the heat-activation of spores of *Phycomyces blakesleeanus* (Rudolph & Furch, 1970; Furch, 1972). Consequently, mitochondrial function probably is reduced because of the accumulation of malate (Fig. 6). It has been our contention that activation occurs because of the flood of glucose that is released by the utilization of trehalose so that we have postulated a role for trehalase in the process. However, it should be noted that this is not the mechanism favored by Rudolph and his co-workers with spores of *P. blakesleeanus*: rather, they suggest that some step(s) in glycolysis are activated.

Returning to *Neurospora*, we have found no large increase in trehalase activity upon activation, it does not exist in alternative forms (isozymes) in ascospores, nor have we found an inhibitor of the enzyme in the spores. Therefore, Dr. Hecker, in my laboratory, investigated the location of the enzyme by immunofluorescent techniques and discovered that trehalase is associated with the innermost ascospore wall, or the endosporium (Hecker & Sussman, 1973). Moreover, the fact that trehalase is much more resistant to temperature extremes in intact cells that when isolated (Yu et al., 1967) led Hecker to examine whether the enzyme could be protected in the latter state. As Fig. 9 discloses, he discovered that ground ascospore walls and purified mycelial walls are almost equally effective in protecting trehalase during lyophilization and heating to 65°C. The only one of the four fractions of the wall that protected as well as the wall itself was that which contains peptides, glucan and polygalactosamine (fraction I), and further experiments with Pronase digestion revealed that the active fraction probably is a protein or proteins. These experiments led to the conclusion that trehalase is separated, within its mural compartment, from the cytoplasm which probably contains its substrate. Heat-activation could be accomplished through a change in the permeability of the membrane separating the cytoplasm from the mural compartment containing trehalase, thereby allowing access of the substrate to the enzyme. This conclusion is fortified by the finding that trehalase appears to remain in its compartment for the first four hours after activation (Hecker & Sussman, 1973).

Further support for this hypothesis comes from some unpublished work we have been doing which reveals that washed mycelial walls from *Neurospora crassa* will yield solubilized trehalase after exposure to $60^{o}C$. We have worked with mycelium up to now because of the difficulty in getting large numbers of ascospores and because the grinding of these releases most of the enzyme so a direct test of the hypothesis remains to be performed.

Work on other spores lends some credence to this hypothesis for reversible changes in membrane structure and permeability are caused in rust urediospores by cycles of extreme cold and subsequent

Table 6. Membrane Enzymes Requiring Phospholipid or Detergent for Activity (Taken from Machtiger & Fox, 1973).

Enzyme	Lipid activators[a]
α-Methyl glucoside phosphotransferase (Enzyme II A + B)	Activated most effectively by PG, markedly less effectively by PS and CL, and not at all by PE. Requires Mg^{2+} or Ca^{2+} for activity. Order of addition of protein components, lipid and divalent cation important in reconstituting active enzyme complex.
β-Methyl glucoside phosphotransferase (Enzyme II)	Activated by anionic phospholipid or nonionic detergent (Tween 40, Triton X-100). Not activated by PE. Detergent activation maximal at critical micelle concentration. Excess detergent is inhibitory.
Pyruvate oxidase	Activated by LPE, PE, LPC, and PC;

Table 6. (continued)

	less effectively by PS. Phospholipids serve as allosteric effectors, markedly reducing the K_Ms for pyruvate and TPP.
Isoprenoid alcohol phosphokinase	Solubilized enzymes requires phospholipid for activity. Activation greatest with PG or CL, lower with LPG or LCL, and much lower with PC or PE. Activity maximal in presence of both PG and DOC.
Phospho-N-acetylmuramic acid: pentapeptide translocase	Enzyme solubilized by Triton X-100; this results in significant loss of activity. Activity of solubilized enzyme preparation has been assayed only in the presence of Triton X-100, and under this condition, PG, PS, PE, and PI serve as "activators."
UDP-galactose: LPS α,3-galactosyltransferase	Requires phospholipid substituted with unsaturated fatty acid. Dioleyl PE most effective, PG less effective, PS and methylated PEs not effective. Order of addition of protein, substrate, and lipid components important in reconstituting active enzyme complex.
UDP-glucose: LPS glucosyltransferase I	Phospholipid required for activity. PE most effective; PA and PG much less effective. Divalent cations are also required. Order of addition of protein, substrate, and lipid components important in reconstituting active enzyme complex.
CDP-diglyceride hydrolase	Activated by Triton X-100.

[a]Abbreviations used: CL=cardiolipin or diphosphatidylglycerol; LCL=lysocardiolipin; LPC=lysophosphatidylcholine; LPE=lysophosphatidylethanolamine; LPG=lysophosphatidylglycerol; PA=phosphatidic acid; PC=phosphatidyl choline or lecithin; PE=phosphatidylethanolamine; PG=phosphatidylglycerol; PI=phosphatidylinositol; PS=phosphatidylserine; DOC=deoxycholate; TPP=thiamine pyrophosphate.

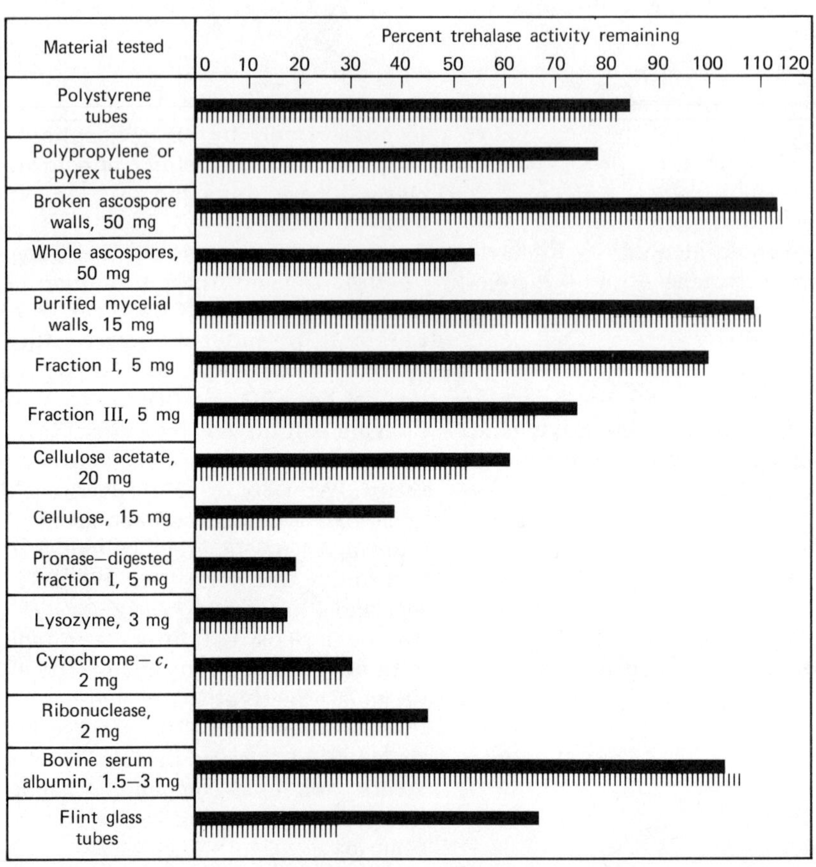

Fig. 9 Effect of lyophilization and subsequent heating at 65°C on mixtures of trehalase and various substances. The solid lines represent the percentage of trehalase remaining after lyophilization and the striped lines the percentage of trehalase activity after the freeze-dried mixture had been heated at 65°C for 30 min. One hundred percent activity is that of trehalase in 0.05 M acetate buffer, pH 5.6. All assays were done in 0.0005 M acetate buffer, pH 5.4. Chitin obtained from crab shells or mycelial walls of *Neurospora* completely abolished trehalase activity during lyophilization so the results are not given in this figure (taken from Hecker & Sussman, 1973).

---

heating to 40°C (Maheshwari & Sussman, 1971). The tantalizing coincidence in the kinetics of the respiration of these urediospores

(Maheshwari & Sussman, 1970) and the catabolism of phospholipids (Langenbach & Knoche, 1971) is as yet unexplained but could relate to changes in the membrane structure of mitochondria, for example, thereby affecting respiratory metabolism. Another instance where membranal elements may be concerned in heat-activation is that of spores of *Phycomyces blakesleeanus* in which particles in the membrane have been found to be altered upon activation (Malhotra & Tewari, 1973). Permeability changes which release endogenous substrates from confinement are proposed as a common initial cause of the stimulation of the endogenous respiration of spores of *Myrothecium verrucaria* by heat and a variety of other physical and chemical treatments (Mandels & Maguire, 1972). However, it should be noted that germination was stimulated only slightly by these treatments. Whether the mechanism recently proposed (Najjar & Constantopoulos, 1973) for the activation of adenyl cyclase by fluoride obtains in the case of *M. verrucaria* is not known, but the key role of c-AMP and the location of the enzyme on membranes make it an attractive possibility. In addition, temperature-induced changes in the E_a (energy of activation) of some mitochondrial enzymes and correlations of these effects with plant respiration have been reviewed by Raison (1973) who concludes that they are due to phase transitions induced in lipids.

To summarize this section on heat-activation, it is clear that there is a dramatic change in many of the biophysical properties of membranes at the temperature of the phase transition of lipids, and these changes may influence the rate of many physiological processes strongly. Thus, "in addition to the direct effects of temperature on the velocity of reactions, it is important to consider the consequences of temperature-induced changes in the physical properties of membranes in regard to their role as physical boundaries for compartments within cells, as supporting structures for enzymes involved in the active transport of metabolites and as a surface for subcellular particles such as ribosomes" (Raison, 1973).

CHEMICAL ACTIVATION

How do these considerations bear upon the mechanism of chemical activation? But before beginning, a selected outline of spores that are activated by detergents, organic solvents and lipids is presented in Table 7. I suggest that there is at least a *prima facie* case for the involvement of lipids in the activation process of these spores. Of course, many other compounds serve as activators and lists of these may be found elsewhere (Sussman & Halvorson, 1966).

In the case of *Neurospora* ascospores, there are some data that support the possibility that lipids are mediators of chemical activation. Some of the chemicals that activate these spores in concentrations lower than 10^{-2} M include furans, thiophenes, pyrroles (Emerson, 1948; Sussman, 1953; Sussman et al., 1959; Mills & Eilers, 1973) phenethyl alcohol, phenyl acetaldehyde, phenyl acetate and benzyl alcohol (Lingappa et al., 1970). In addition, organic solvents like aliphatic alcohols, ketones and esters are effective in concentrations greater than 1M (Sussman et al., 1959). Among these substances, furfural seems to be one of the most active so most studies have been

Table 7. Selected List of Fungal Spores That Are Activated by Detergents, Organic Solvents or Lipids

Organism	Activator	Reference
More than 50 slime molds	Detergents	Elliott, 1949
Colletotrichum trifolii	Tween-20	Miehle & Lukezic, 1972
Phycomyces blakesleeanus	Acetic, propionic and butyric acids	Sommer & Halbsguth, 1957; Borchert, 1962; Delvaux, 1973
Penicillium frequentans (conidia)	Ethanol, acetone	Dobbs & Hinson, 1953
Neurospora spp. (ascospores)	In low concentrations: furans, thiophenes, pyrrole and aromatic alcohols In high concentrations: ethanol, acetone and other organic solvents	Emerson, 1948; Sussman, 1953; Sussman et al., 1959; Lingappa et al., 1970
Coprinus radiatus (basidiospores)	Furfural and other heterocyclics[a]	Mills & Eilers, 1973
Anthracoidea spp.	Quaternary detergents	Kukkonen & Hyytiainen, 1969

[a]Maximal germination requires the simultaneous provision of heat.

performed with it although, as Lingappa et al. (1970) point out, phenethyl alcohol offers some advantages for study.

Furfural uptake has been studied with a view to determine its locus of action. With this in mind, ^{14}C-furfural was prepared and added to dormant ascospores, which removed it from solution very rapidly for the first half-hour, after which uptake leveled off (Fig. 10; Eilers et al., 1970). The discrepancy between uptake measured by the decrease in the absorption of ultraviolet light and radioactivity in Fig. 10 is purely a digression and is due to the conversion of furfural to furoic acid and furfuryl alcohol (Eilers & Sussman, 1970). Inasmuch as conversion is not necessary for activation by furfural I do not consider it further. Previous data reveal that at least a third of the spores are activated within a few minutes after exposure to furfural and the maximum number absorb enough of the activator in about an hour (Sussman, 1961b), or at about the time when the threshold in uptake is reached. Furthermore, the kinetics of furfural uptake are similar to an adsorption isotherm, and uptake occurs at 4°C (Sussman, 1961b), suggesting that furfural, like many other substances, may bind to the cell surface (Sussman & Lowry, 1955). This suggestion is corroborated by the fact that the uptake of furfural by heat-killed ascospores is almost identical to that of living spores (Fig. 11) and by the finding that about 75% of the ^{14}C-furfural taken up can be eluted by unlabeled furfural (Eilers et al., 1970b). Therefore, it may be concluded that furfural uptake appears to involve a large absorptive (reversible) component and a smaller irreversible one.

Where is the furfural located when it is taken up by ascospores? It was reasoned that if the reversible component of uptake is due to adsorption to the spore surface then removal of wall layers by soaking in Clorox solution (Lowry & Sussman, 1958) might reduce uptake. Although uptake was reduced by about half, suggesting that most of the reversible component disappeared, considerable germination still occurred, so that the site of binding is internal to the wall layers removed by Clorox (Eilers & Sussman, 1970). The site of furfural uptake also was investigated by exposing fractions of ascospores, prepared by grinding and centrifugation, to furfural. Figure 12 discloses that no furfural was found in the supernatant fraction and almost all was found in the wall fraction, which also contains membranes. Although this experiment indicates that the spore surface is a major site of furfural absorption, it does not distinguish between the reversible and irreversible phases of the process. Accordingly, intact dormant spores were exposed to ^{14}C-furfural for 10 min and the

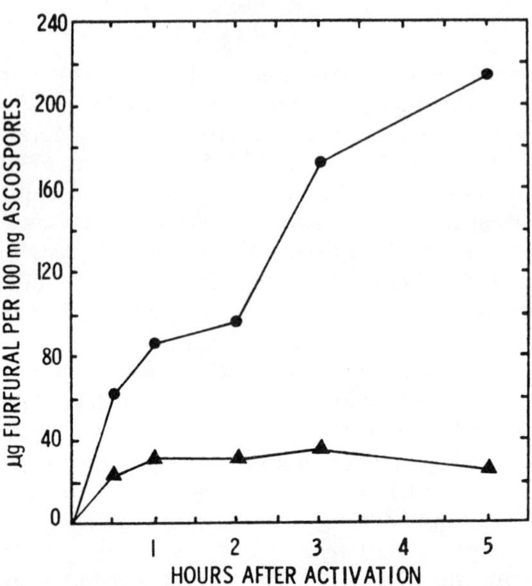

Fig. 10 Furfural uptake and conversion by ascospores of *Neurospora tetrasperma* as determined by radioactivity measurements (▲) and ultraviolet light absorption (●) per 100 mg untreated ascospores at room temperature. The spores were incubated in 2×10^{-3} M furfural at a ratio of 100 mg spores per 1 ml of furfural solution (taken from Eilers et al., 1970b).

---

exchangeable label removed by repeated washing. The washed spores, containing only the nonexchangeable furfural, were homogenized and fractionated (Table 8). Only about 10% of the nonexchangeable activator was found in the supernatant fraction in contrast to the wall/membrane complex which, upon combustion, yielded the equivalent of almost 80% of the furfural. These data reveal that most of nonexchangeable furfural is bound to the surface of the spore and a small amount may enter its interior although it is difficult to exclude the possibility that the supernatant contains mainly furfural removed from other fractions by grinding and other operations. This possibility is fortified by previous data which show that compounds with more than two carbon atoms do not penetrate dormant ascospores (Sussman et al., 1958). Granted that most, if not all, of the nonexchangeable furfural is on the cell surface, is it bound to the same sites as those which adsorb cations? The answer is no, because treatments which

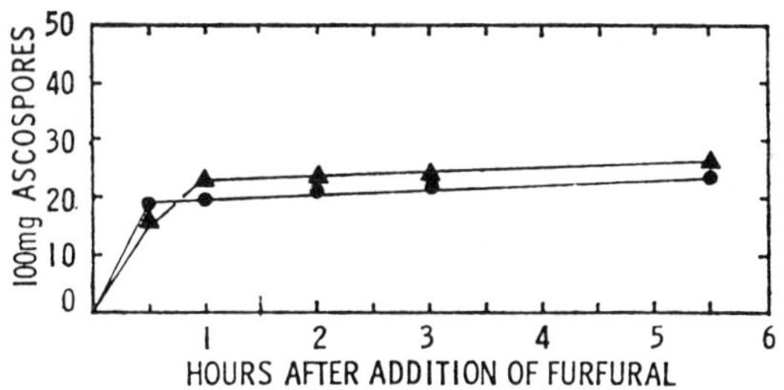

Fig. 11 Furfural uptake at room temperature by ascospores of *Neurospora tetrasperma* as determined by radioactivity measurements (▲) and ultraviolet light absorption (●) per 100 mg "heat-killed" ascospores (taken from Eilers et al., 1970b).

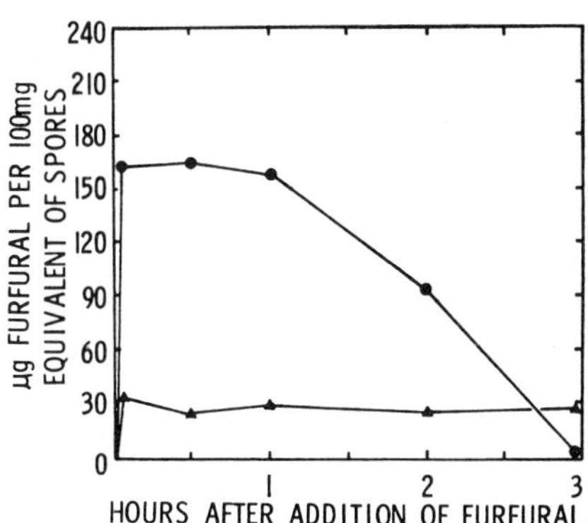

Fig. 12 Furfural conversion and uptake in cell-free extracts of ascospores of *Neurospora tetrasperma*. The spores were homogenized, centrifuged at 1,000 xg and separated into a wall fraction (●), an intermediate fraction (15,000 xg) and a supernatant fraction (▲), (over 15,000 xg) that did not show any activity. The spores were incubated in 2 x 10^{-3} M furfural per 100 mg of spores. All concentrations were measured by ultraviolet light absorption (taken from Eilers et al., 1970b).

block charged sites on the cell do not prevent activation by furfural, nor does the latter compete with cations for adsorption (Sussman & Lowry, 1955) so they must occupy different sites. These data lead to the conclusion that furfural is bound by, or incorporated into, a component of the cell surface which is interior to the wall layers removed by Clorox. And, given the solubility of furfural in lipids, the fact that it is taken up by, and activates spores at 4°C, and is bound by dead cells, it is likely that this activator is bound physically to lipids in the cell membrane, possibly in combination with elements of the wall. If so, then its locus of action is at the membrane, as in the case of heat-activation, whose effects are mimicked by chemical activation to a large extent (Eilers et al., 1970).

Although there are other fungal spores that are activated by compounds similar to those that work with *Neurospora* ascospores (Table 7), there is very little known about the mechanism of the process. Nevertheless, in the case of spores of *Phycomyces blakesleeanus* at least, there are so many similarities to the way chemical activation occurs in *Neurospora* that it is likely that the mechanisms are not too disparate. These similarities include the magnitude of the respiratory increase, stimulation of glycolysis and the secretion of fermentation products, as well as the lack of reversibility of activation by chemicals

Table 8. The Distribution of Furfural in Ascospores of *Neurospora tetrasperma*[a]

^{14}C-Furfural concentration expressed as	g	cpm	% of total uptake
^{14}C-Furfural taken up by the ascospores	55	240,264	100
^{14}C-Furfural removed from the ascospore in three washes	43	187,290	77.9
^{14}C-Furfural remaining in the ascospore	12	52,974	22.1 (100)
^{14}C-Furfural found in the supernatant fraction	1.2	5,213	9.8
^{14}C-Furfural found in the wall fraction upon combustion of the wall to $^{14}CO_2$	9.4	36,489	78.5
Per cent recovery			88.3

[a]Two-hundred milligram ascospores were activated in ^{14}C-furfural for 10 min before the excess furfural was removed. The spores were then washed in three 5-ml portions of water to remove the non-absorbed furfural. After washing, the spores were ground and separated by centrifugation at 10,000g into a wall and a supernatant fraction. The supernatant fraction was assayed directly, while the wall fraction was oxidized to $^{14}CO_2$ by concentrated sulfuric acid before its activity was assayed (taken from Eilers et al., 1970).

although heat-activation is fully reversible (Rudolph & Furch, 1970; Delvaux, 1973).

Still, even were the effect of the chemicals listed in Table 7 to be upon lipids and membranes the mechanism through which they act in spores is largely unknown. If the mechanism proposed to explain heat-activation in *Neurospora* ascospores were extended to chemical activation, a change in the cell membrane which allows access of trehalose to its hydrolase would be invoked. That this is at least reasonable is suggested by data which show that N-alkanols influence the permeability of liposomes markedly (Jain et al., 1973), as well as modifying their structure (Jain & Cordes, 1973a). Or the effect might be more indirect whereby lecithinase, or other enzymes affecting membrane activity, are activated by solvents and/or heat (Jain & Cordes, 1973b). On the other hand, alcohols, including ethanol, propanol (Rambeck & Simon, 1972) and phenethyl alcohol (Yamashiro et al., 1970) initiate the metabolism of endogenous substrates like glycogen and trehalose in yeast, and phenethyl alcohol also is known to increase the viscosity of the interior of these fungi (Burns, 1971). Other effects of solvents include the fostering of the association of subunits of β-galactosidase (Shifrin & Hunn, 1969), stimulation of phenylalanine hydroxylase (Brase & Westfall, 1973), and the induction of cellulase (Stranks, 1973), phenoloxidase (Preston & Taylor, 1970) and glutamic dehydrogenase (Sinohara, 1973), the latter occurring in dormant spores of *Aspergillus oryzae*. The well-known effect of detergents in activating "latent" enzymes is illustrated in the case of a lipase from *Candida paralipolytica* (Ota et al., 1972), several hydrolases and catalase (Vaes, 1965), arabinofuranosidase and acid phosphatase (Hislop et al., 1974), plastid phosphatidase C (Kates, 1957), glyceraldehyde 3-phosphate dehydrogenase (Shin & Carraway, 1973) and other enzymes.

Consequently, there are a variety of ways through which activation by this class of chemicals might occur and although I favor lipids as the site of action at this time I must concede that the exact mechanism remains to be discovered.

CONCLUSION

This exercise has not been to establish with certainty that one or another class of macromolecules is involved in activation, to the exclusion of all others. Rather, I have meant to point out that, as attractive as partial protein denaturation may be as a mechanism (Cotter, 1973), it is not the only plausible hypothesis. In fact, as I have tried to indicate there are reasons for believing that the lipid components of membranes may be the immediate locus of activation in some cases although other mechanisms are not yet ruled out. Moreover, the large bulk of activable systems remain to be explored in detail and it is likely that varied mechanisms will be found to exist among the fungi, whose diversity is legendary.

ACKNOWLEDGMENTS

Valuable discussions with my colleagues, Dr. Harry Douthit, Charles Yocum, and William Lands are gratefully acknowledged, as well as the help of Miss Meredy Gockel and Mr. Louis P. Martonyi in the typing and preparation of photographs, respectively.

REFERENCES

Allen, P. J. (1965). Metabolic aspects of spore germination in fungi. Annu. Rev. Phytopathol. 3, 313--342.

Anson, M. L. & Mirsky, A. E. (1934). The equilibrium between active native trypsin and inactive denatured trypsin. J. Gen. Physiol. 17, 393--398.

Barbehenn, E. K., Wales, R. G., & Lowry, O. H. (1974). The explanation for the blockade of glycolysis in early mouse embryos. Proc. Nat. Acad. Sci. U.S.A. 71, 1056--1060.

Black, M. (1972). Control Processes in Germination and Dormancy. Oxford Biology Readers. Oxford: Oxford University Press.

Bogin, E., Higashi, T., & Brodie, A. F. (1970). Effects of heat treatment of electron-transport particles on bacterial oxidative phosphorylation. Proc. Nat. Acad. Sci. U.S.A. 67, 1--6.

Borchert, R. (1962). Über die Azetat-Aktivierung der Sporangiosporen von *Phycomyces blakesleeanus*. Beitr. Biol. Pflanzen 38, 31--61.

Brase, D. A. & Westfall, T. C. (1973). Studies on the stimulation of phenylalanine hydroxylase activity by short-chain alcohols. Biochim. Biophys. Acta. 309, 271--279.

Bresler, S. E. (1971). Introduction to Molecular Biology. New York and London: Academic Press.

Burns, V. W. (1971). Microviscosity and calcium exchange in yeast cells and effects of phenethyl alcohol. Exp. Cell Res. 64, 35--40.

Busta, F. E. & Ordal, Z. J. (1964). Heat-activation kinetics of endospores of *Bacillus subtilis*. J. Food Sci. 29, 345--353.

Cabib, E. & Ulane, R. (1973). Chitin synthetase activating factor from yeast, a protease. Biochem. Biophys. Res. Commun. 50, 186--191.

Chuang, T. F. & Crane, F. L. (1973). Phospholipids in the cytochrome oxidase reaction. Bioenerg. 4, 563--578.

Cochrane, V. W. (1958). Physiology of Fungi. New York: Wiley.

Cochrane, V. W. (1960). Spore germination. In Plant Pathology: An Advanced Treatise (ed. J. Horsfall and A. Diamond), Vol. 2, pp. 167--202. New York and London: Academic Press.

Cochrane, V. W. (1974). Dormancy in spores of fungi. Trans. Amer. Microsc. Soc. (in press).

Cotter, D. A. (1973). Spore germination in *Dictyostelium discoideum* I. The thermodynamics of reversible activation. J. Theor. Biol. 41, 41--51.

Cotter, D. A. & Raper, K. B. (1968). Properties of germinating spores of *Dictyostelium discoideum*. J. Bacteriol. 96, 1680--1689.

Cotter, D. A. & Raper, K. B. (1970). Spore germination in *Dictyostelium discoideum:* trehalase and the requirement for protein synthesis. Devel. Biol. 22, 112--128.

Cox, R. A. & Littauer, U. Z. (1960). Properties of low molecular weight RNA from *E. coli*. J. Mol. Biol. 2, 166--168.

DeGier, J., Mandersloot, J. G., & Van Deene, L. L. M. (1968). Lipid composition and permeability of liposomes. Biochim. Biophys. Acta. 150, 666--675.

Delvaux, E. (1973). Some aspects of germination induction in *Phycomyces blakesleeanus* by an ammonium-acetate pretreatment. Arch. Mikrobiol. 88, 273--284.

Dobbs, C. G. & Hinson, W. H. (1953). A widespread fungistasis in soils. Nat. (Lond.) 172, 197--199.

Dutta, S. K. & Chakrabartty, P. K. (1971). Characterization and measurements of nucleotide sequence similarities of DNA's from several *Neurospora* species. *Neurospora* Newsl. 18, 6--7.

Eilers, F. I. (1968). The effect of furfural upon ascospores of *Neurospora*. Ph.D. Thesis, University of Michigan, Ann Arbor.

Eilers, F. I., Ikuma, H., & Sussman, A. S. (1970a). Changes in metabolic intermediates during activation of *Neurospora* ascospores. Can. J. Microbiol. 12, 1351--1356.

Eilers, F. I., Ikuma, H., & Sussman, A. S. (1970b). Furfural uptake by *Neurospora* ascospores. Planta 94, 265--272.

Eilers, F. I. & Sussman, A. S. (1970). Conversion of furfural to furoic acid and furfuryl alcohol by *Neurospora* ascospores. Planta 94, 253--264.

Elliott, E. W. (1949). The swarm-cells of myxomycetes. Mycol. 41, 141--170.

Emerson, M. R. (1948). Chemical activation of ascospore germination in *Neurospora crassa*. J. Bacteriol. 55, 327--330.

Feit, I. N., Chu, L. K., & Iverson, R. M. (1971). Appearance of polyribosomes during germination of spores in the cellular slime mole *Dictyostelium purpureum*. Exp. Cell Res. 65, 439--444.

Furch, B. (1972). Zur Wärmeaktivierung der Sporen von *Phycomyces blakesleeanus*. Das Auftreten von Gärungen unter aeroben Bedingungen. Protoplasma 75, 371--379.

Garrett, M. K., Sussman, A. S., & Yu, S. A. (1972). Properties of an inhibitor of trehalase in trehalaseless mutants of *Neurospora*. Nat. New Biol. 235, 119--121.

Goddard, D. R. (1935). The reversible heat activation inducing germination and increased respiration in the ascospores of *Neurospora tetrasperma*. J. Gen. Physiol. 19, 45--60.

Gottlieb, D. (1959). The physiology of spore germination in fungi. Bot. Rev. 16, 229--257.

Gould, G. W. & Hurst, A. (1969). The Bacterial Spore. New York and London: Academic Press.

Gruenwedel, D. W., Hsu, C. H., & Lu, D. (1971). The effects of aqueous neutral-salt solutions on the melting temperatures of deoxyribonucleic acids. Biopolymers 10, 47--68.

Hecker, L. I. & Sussman, A. S. (1973). Localization of trehalase in the ascospores of *Neurospora*. J. Bacteriol. 115, 592--599.

Herrmann, H. & Tootle, M. L. (1964). Specific and general aspects of the development of enzymes and metabolic pathways. Physiol. Rev. 44, 289--371.

Hislop, E. C., Barnaby, V. M., Shellis, C., & Laborda, F. (1974). Localization of α-L-arabinofuranosidase and acid phosphatase in mycelium of *Sclerotinia fructigena*. J. Gen. Microbiol. 81, 79--99.

Jain, M. K. & Cordes, E. H. (1973a). Phospholipases I. Effect of n-alkanols on the rate of enzymatic hydrolysis of egg phosphatidylcholine. J. Membrane Biol. 14, 101--118.

Jain, M. K. & Cordes, E. H. (1973b). Phospholipases II. Enzymatic hydrolysis of lecithin: effects of structure, cholesterol content,

and sonication. J. Membrane Biol. 14, 119--134.
Jain, M. K., Toussaint, D. G., & Cordes, E. H. (1973). Kinetics of water penetration into unsonicated liposomes. Effects of n-alkanols and cholesterol. J. Membrane Biol. 14, 1--16.
Kates, M. (1957). Effects of solvents and surface-active agents on plastid phosphatidase C activity. Can. J. Biochem. Physiol. 35, 128--142.
Kayne, F. J. & Suelter, C. H. (1968). The temperature-dependent conformational transitions of pyruvate kinase. Biochem. 7, 1678--1684.
Kukkonen, I. & Hyytiäinen, R. (1969). A method for sterilizing spores of *Anthracoidea* (Ustilaginales) and the effect of sterilization on spore germination. Ann. Bot. Fenn. 6, 243--254.
Ladbrooke, B. D. & Chapman, D. (1969). Thermal analysis of lipids, proteins and biological membranes. A review and summary of some recent studies. Chem. Phys. Lipids 3, 304--367.
Langenbach, R. J. & Knoche, H. W. (1971). Phospholipids in the uredospores of *Uromyces phaseoli* II. Metabolism during germination. Plant Physiol. 48, 735--739.
Leach, C. K. & Ashworth, J. M. (1972). Characterization of DNA from the cellular slime mould *Dictyostelium discoideum* after growth of the amoebae in different media. J. Mol. Biol. 68, 35--48.
Lehninger, A. L. (1970). Biochemistry. New York: Worth Publishers.
Lees, A. D. (1961). Discussion. In Cryptobiotic Stages in Biological Systems, 5th Biology Conference "Oholo", (ed. N. Grossowicz, S. Hestrin and A. Keynan), p. 224. Amsterdam: Elsevier.
Levinson, H. & Hyatt, M. T. (1969). Heat activation kinetics of *Bacillus megaterium* spores. Biochem. Biophys. Res. Commun. 37, 909--916.
Lingappa, B. T., Lingappa, Y., & Turian, G. (1970). Phenethyl alcohol induced germination of ascospores of *Neurospora*. Arch. Mikrobiol. 72, 97--105.
Lowry, R. J. & Sussman, A. S. (1958). Wall structure of ascospores of *Neurospora tetrasperma*. Amer. J. Bot. 45, 397--403.
Machtiger, N. A. & Fox, C. F. (1973). Biochemistry of bacterial membranes. Annu. Rev. Biochem. 42, 575--600.
Maheshwari, R. & Sussman, A. S. (1970). Respiratory changes during germination of urediospores of *Puccinia graminis* f. sp. *tritici*. Phytopathology 60, 1357--1364.
Maheshwari, R. & Sussman, A. S. (1971). The nature of cold-induced

dormancy in urediospores of *Puccinia graminis tritici*. Plant Physiol. 47, 389--395.

Malhotra, S. K. & Tewari, J. P. (1973). Molecular alterations in the plasma membrane of sporangiospores of *Phycomyces* related to germination. Proc. Royal Soc. Lond. B. 184, 207--216.

Mandels, G. R. & Maguire, A. (1972). Endogenous metabolism of fungal spores. Stimulation by physical and chemical means. Plant Physiol. 50, 425--431.

McCarthy, C. & Nestor, E. W. (1969). Heat-activated endonuclease in *Bacillus subtilis*. J. Bacteriol. 97, 1426--1430.

Miehle, B. R. & Lukezic, F. L. (1972). Studies on conidial germination and appressorium formation by *Colletotrichum trifolii* Bain & Essary. Can. J. Microbiol. 18, 1263--1269.

Mills, G. L. & Eilers, F. I. (1973). Factors influencing the germination of basidiospores of *Coprinus radiatus*. J. Gen. Microbiol. 77, 393--401.

Morowitz, H. J. (1968). Energy Flow in Biology. New York: Academic Press.

Najjar, V. A. & Constantopoulos, A. (1973). The activation of adenylate cyclase I. A postulated mechanism for fluoride and hormone activation of adenylate cyclase. Mol. Cell. Biochem. 2, 87--93.

Oseroff, A. R., Robbins, P. W., & Burger, M. M. (1973). The cell surface membrane: biochemical aspects and biophysical probes. Annu. Rev. Biochem. 42, 647--682.

Ota, Y., Yoshioka, K., Minoda, Y., & Yamada, K. (1972). Demonstration of the lipid lipase activators produced by *Candida paralipolytica*. Agric. Biol. Chem. 37, 2879--2883.

Ottesen, M. (1967). Induction of biological activity by limited proteolysis. Annu. Rev. Biochem. 36, 55--76.

Platt, J. R. (1961). Properties of large molecules that go beyond the properties of their chemical sub-groups. J. Theor. Biol. 1, 342--358.

Preston, J. W. & Taylor, R. L. (1970). Observations on the phenoloxidase system in the haemolymph of the cockroach *Leucophaea maderae*. J. Insect Physiol. 16, 1729--1744.

Raison, J. K. (1973). Temperature-induced phase changes in membrane lipids and their influence on metabolic regulation. In Rate Control of Biological Processes. Soc. Exper. Biol. Symp. 27, 485--512.

Raison, J. K. & Lyons, J. M. (1971). Hibernation: alteration of mitochondrial membranes as a requisite for metabolism at low temperature. Proc. Nat. Acad. Sci. U.S.A. 68, 2092--2094.

Rambeck, W. & Simon, H. (1972). Decrease of gycogen and trehalose in yeast during starvation and during ethanol formation under the influence of propanol or ethanol. Hoppe-Seyler's Z. Physiol. Chem. 353, 1107--1110.

Rast, D. & Stäuble, E. J. (1970). On the mode of action of isovaleric acid in stimulating the germination of *Agaricus bisporus* spores. New Phytol. 69, 557--566.

Rothfield, L. I. & Romeo, D. (1971). Enzyme reactions in biological membranes. In Structure and Function of Biological Membranes, (ed. L. I. Rothfield), pp. 251--284. London and New York: Academic Press.

Rousseau, P. & Halvorson, H. O. (1973). Physiological changes following the breaking of dormancy of *Saccharomyces cerevisiae* ascospores. Can. J. Microbiol. 19, 547--555.

Rudolph, H. (1973). Uber die Wirkung von D-glycerat-3-phosphat und D-Glycerat-2-phosphat-Zusätzen auf Sporenhomogenate von *Phycomyces blakesleeanus*. Z. Pflanzenphysiol. 67, 212--222.

Rudolph, H. & Furch, B. (1970). Untersuchungen zur Aktivierung von Sporenhomogenaten durch Wärmebehandlung. Arch. Mikrobiol. 72, 175--181.

Sadoff, H. L. (1969). Spore Enzymes. In The Bacterial Spore (ed. G. W. Gould and A. Hurst), pp. 275--299. London and New York: Academic Press.

Scheraga, H. A. (1966). Principles of protein structure. In Molecular Architecture in Cell Physiology, (ed. T. Hayashi and A. G. Szent-Gyorgyi), pp. 39--61. Englewood Cliffs: Prentice-Hall.

Shichi, H. & Somers, R. L. (1974). Is retinal isomerase an enzyme? Fed. Proc. 33, 1526.

Shifrin, S. & Hunn, G. (1969). Effect of alcohols on the enzymatic activity and subunit association of β-galactosidase. Arch. Biochem. Biophys. 130, 530--535.

Shin, B. C. & Carraway, K. L. (1973). Association of glyceraldehyde 3-phosphate dehydrogenase with the human erythrocyte membrane. Effect of detergents, trypsin and adenosine triphosphate. J. Biol. Chem. 248, 1436--1444.

Sinohara, H. (1973). Stimulation of induction of glucose dehydrogenase by homologous series of aliphatic alcohols in dormant spores of *Aspergillus niger*. Experientia 29, 32--33.

Sommer, L. & Halbsguth, W. (1957). Grundlegende Versuche zur

Keimungsphysiologie von Pilzsporen. Fortschr. Ber. Wirtsch. und Verkehrsminist. Nordrh. Westf., Nr. 411.

Soutar, A. K., Pownall, H. J., & Smith, L. C. (1974). Correlation of apolipoprotein activation of LCAT with phase transition of phospholipid substrate. Fed. Proc. 33, 1526.

Stranks, D. W. (1973). Influence of phenethyl alcohol and other organic solvents on cellulase production. Can. J. Microbiol.

Sussman, A. S. (1953). The effect of heterocyclic and other compounds upon the germination of ascospores of *Neurospora tetrasperma*. J. Gen. Microbiol. 8, 211--216.

Sussman, A. S. (1961a). The role of trehalose in the activation of dormant ascospores of *Neurospora*. Q. Rev. Biol. 36, 109--116.

Sussman, A. S. (1961b). The role of endogenous substrates in the dormancy of *Neurospora* ascospores. In Spores II., Proc. 2d Allerton Park Spore Conference, (ed. H. O. Halvorson), pp. 198--213. Minneapolis: Burgess Publishing Co.

Sussman, A. S. (1965). Physiology of dormancy and germination in the propagules of cryptogamic plants. In Encyclopedia of Plant Physiology, (ed. A. Lang), Vol. XV, 933--1025. Berlin, Heidelberg and New York: Springer.

Sussman, A. S. (1966). Dormancy and spore germination. In The Fungi: An Advanced Treatise, (ed. G. C. Ainsworth and A. S. Sussman), Vol. II, 733--764. London and New York, Academic Press.

Sussman, A. S. (1969). The dormancy and germination of fungus spores. Symp. Soc. Exp. Biol. 23, 99--121.

Sussman, A. S. & Douthit, H. A. (1973). Dormancy in microbial spores. Annu. Rev. Plant Physiol. 24, 311--352.

Sussman, A. S. & Halvorson, H. O. (1966). Spores: Their Dormancy and Germination. New York and London: Harper & Row.

Sussman, A. S., Holton, R., & von Böventer-Heidenhain, B. (1958). Physiology of the cell surface of *Neurospora* ascospores. Entrance of anions and non-polar compounds. Arch. Mikrobiol. 29, 38--50.

Sussman, A. S. & Lowry, R. J. (1955). Physiology of the cell surface of *Neurospora* ascospores I. Cation binding properties of the cell surface. J. Bacteriol. 70, 675--685.

Sussman, A. S., Lowry, R. J., & Tyrrell, E. (1959). Activation of *Neurospora* ascospores by organic solvents and furans. Mycol. 51, 237--247.

Sussman, R. S. & Rayner, E. P. (1971). Physical characterization of deoxyribonucleic acids in *Dictyostelium discoideum* Arch.

Biochem. Biophys. 144, 127--137.

Swartz, M. N., Kaplan, N. O., & Frech, M. E. (1956). Significance of "heat-activated" enzymes. Science 123, 50--53.

Tanford, C. (1968). Protein denaturation. Adv. Protein Chem. 23, 121--282.

Tanford, C. (1970). Protein denaturation. Part C. Theoretical models for the mechanism of denaturation. Adv. Protein Chem. 24, 1--95.

Vaes, G. (1965). Studies on bone enzyme. The activation and release of latent acid hydrolases and catalase in bone-tissue homogenates. Biochem. J. 97, 393--402.

Van Assche, J. A., Carlier, A. R., & Dekeersmaeker, H. I. (1972). Trehalase activity in dormant and activated spores of *Phycomyces blakesleeanus*. Planta 103, 327--333.

Varner, J. (1971). The control of enzyme formation in plants. Symp. Soc. Exp. Biol. 25, 197--205.

Vijay, I., Tesche, N., & Troy, F. A. (1974). Fluidity of membrane lipids associated with function of membrane-bound sialyltransferase complex in *E. coli*. Fed. Proc. 33, 1240.

Walther, P. J., Steinman, H. M., Hill, R. L., & McKee, P. A. (1974). Activation of human plasminogen by urokinase. Partial characterization of a pre-activation peptide. J. Biol. Chem. 249, 1173--1181.

Yamashiro, K., Nishihara, H., Tani, Y., & Fukui, S. (1970). Studies on the metabolism of *Saccharomyces saki* II. Effect of ethanol and phenethyl alcohol on respiration of *Saccharomyces saki*. J. Ferment. Technol. 48, 318--388.

Yu, S. A., Sussman, A. S., & Wooley, S. (1967). Mechanisms of protection of trehalase against heat-inactivation in *Neurospora*. J. Bacteriol. 94, 1306--1312.

DISCUSSION

Activators of Fungal Spore Germination

Chairman: V. Macko
Boyce Thompson Institute
Yonkers, New York

Dr. Reisener stated that coumarin stimulates germ tube wall synthesis in uredospores of the wheat stem rust fungus. Ten to twelve minutes after coumarin was added to the spores, label from glucose was found incorporated in chitin. The enzymes and substrates for chitin synthesis appeared to be present in nongerminated spores. There are many stimulators that act similarly, mostly lipophilic substances, which tend to support Dr. Sussman's reasoning about the action of stimulators.

Dr. Sussman agreed that in general there is a possibility of other spores that use the same mechanism of stimulation. In the case of coumarin, one has also to consider Dr. Allen's model, where there might be a plug which could be removed by the action of an enzyme.

Dr. Cotter showed electron micrographs of *Dictyostelium* spores. In the spores that have been normally activated at 45°C for 30 min, the mitochondria were dense. Overstimulation by heat caused the destruction of mitochondrial membranes. In some cases the outer membrane remained but the inner membrane disappeared. Oxygen uptake stopped after destruction of the inner membrane. Also Dr. Cotter's model predicted that protein denaturants such as urea, DMSO, and analogs of urea would activate spores and these predictions were verified experimentally. There may also be other ways to activate dormant spores.

Dr. Sussman suggested that DMSO has mixed effects. It has solubility properties that are close to those of furfural, and in fact DMSO has some activating effects. Also urea may have effects upon lipids as well as proteins. One of the problems in distinguishing between lipids and proteins as the mediators of these effects is the overlap in the effectors of changes in these materials.

To the question, "Why does glucose when added exogenously not activate if trehalose is the key to the activation," Dr. Sussman replied that glucose does not penetrate in amounts large enough to induce the effects on metabolism that he postulates. Consequently, a test of the hypothesis is not possible by this means.

Dr. Sussman was asked about the mechanism for changing membranes to allow trehalose to get to trehalase and the reversibility of the system. Is there a flood of trehalose into the trehalase compartment and, if so, how is the sugar so readily hydrolyzed? Is trehalose entering slowly through the membrane so that if membrane permeability is changed a little bit the trehalose that gets through gets metabolized? Or is trehalose suddenly leaked out?

Dr. Sussman answered that it is likely that trehalose penetrates slowly, possibly by active means. Moreover, there are metabolic changes that follow the physical cues of activation.

B

Germination of Fungal Spores

CHAPTER 4

CARBOHYDRATE METABOLISM AND SPORE GERMINATION

David Gottlieb
University of Illinois

CARBOHYDRATE METABOLISM AND SPORE GERMINATION

INTRODUCTION

A discussion of the relationship between carbohydrate metabolism and spore germination appears as a rather prosaic undertaking in an era when most advances in cellular biology are being made in the area of the complex processes such as protein and nucleic acid synthesis and of the cell's regulatory mechanisms. Surely, carbohydrate metabolism must have been intensively studied and completely reviewed. However, a search of the literature reveals that this impression is misleading and there is no plethora of good information in this field. Very few fungi have been intensively studied and, except for the data from these few, most of the information is "spot data" in which the relationship of one phenomenon is not placed in perspective to other factors concerned with germination. Proof of the presence of a respiratory pathway most often rests on the demonstration of only a few steps in a pathway or perhaps its inhibition by one or two respiratory inhibitors. Probably, we confuse the information that is available concerning the growth phase of the fungi by unconsciously extrapolating such information as valid also for the spore. Although much of the data from the hyphal studies does apply to spore germination, other data require confirmation. Furthermore, many biochemical relationships are special to germination and are not concerned in vegetative growth --- such as the role of endogenous reserves and the synthesis of enzymes that might be deficient in spores. Special studies in such areas are vital to understanding the activities of spores during their germination phase.

In most spores the biochemical materials and the mechanism for germination are present and the system is ready to begin functioning. Sometimes there is a deficiency of a nutritive substrate, in which case the spores must obtain it from exogenous sources. At other times a particular enzyme may be deficient in amount and is only synthesized slowly during the incubation of the spores; then germ tube formation must wait until the enzyme is present in high enough concentration to allow the reaction to proceed. Moreover, one can postulate an overall regulatory device in the spores, an autoinhibitor, which prevents the metabolic systems from functioning when the spores are present in high concentration. Under such conditions the spores cannot successfully establish themselves in order to eventually give rise to the vegetative phase.

Carbon metabolism in spore germination can be best seen in the framework of the entire process of germination. The spore of a fungus is a differentiated structure of one to many cells which functions in sexual or asexual propagation, resulting in the multiplication of a species, or which allows the species to survive under adverse environmental conditions. Metabolically, the process is a change from a resting, quiescent state to an active state approximating that of a growing hypha. These changes are often associated with the production of a germ tube, the protrusion of the contents of the spore cell that eventually forms a filament. The term germination can also be applied to the formation of yeastlike cells or even plasmodia. The term has also been applied to the formation of zoospores from the polynucleate protoplasm of a sporangium, but this usage should be abandoned since the formation of zoospores is sporogenesis and not germination. However it can of course still be used for the germination of the entire sporangium to form a germ tube. The *sina qua non* of germination is the change from the low metabolic state of the quiescent spore to the relatively high metabolic state of the vegetative phase of the organism.

Spore germination is a vital process that requires energy producing mechanisms and synthetic processes. The energy forming reactions are dependent on the oxidation of carbon compounds, usually sugars or their polyalcohols. These materials are stored in the spore during its formation, and they are utilized even when the spore is in a quiescent, nongerminating state. Germination involves, among other features, the more rapid metabolism of such carbon sources. A supply of carbon in a form which is convertible to glucose or acetyl CoA is usually needed for spore germination. This supply sometimes consists of endogenous sources which, if present in sufficient concentration and if all other requirements are satisfied, allows the spore to germinate.

In contrast are those spores that require exogenous carbon sources such as glucose for germination; compounds readily convertible to glucose often serve this purpose. Some fungal species do not have sufficient amounts of such reserves to allow even the initial swelling phase of germination to proceed (Mandels & Darby, 1953). *Penicillium atrovenetum* conidia, for example, failed to germinate if a suitable carbon source was not supplied exogenously. Adding glucose and phosphate to the incubation medium allowed the greatest percentage of swelling and subsequent germination. Moreover, both swelling and germ tube formation increased with increasing substrate concentration (Gottlieb & Tripathi, 1968). The swelling of spores of *Rhizopus arrhizus* similarly required a carbohydrate source as well as other compounds (Ekundayo & Carlile, 1964). Germ tube formation followed swelling and seemed dependent on it. In both species, the carbon served for respiration at least, for when respiratory poisons such as azide or cyanide were added, both respiration and swelling were inhibited (Ekundayo, 1966; Gottlieb & Tripathi, 1968).

EMBDEN--MEYERHOF--PARNAS PATHWAY

The Embden--Meyerhof--Parnas (EMP) system with alcohol or lactic acid as the end product is an anaerobic system functioning in a number of species. It furnishes the cell with only a limited amount of energy and with a relatively small number of intermediates useful for synthetic activities. Usually, however, part of the EMP is linked with aerobic systems such as the citric acid cycle and the terminal electron transport system so that much more energy is obtained and more intermediates become available for synthesis.

EMP enzymes have been shown in the spores of all fungi that have been studied. Very early, Shu & Ledingham (1956) demonstrated the presence of two enzymes of this pathway in uredospores of *Puccinia graminis tritici*, triose isomerase and aldolase, as well as α-glycerol phosphate dehydrogenase. Later, Caltrider & Gottlieb (1963) revealed the presence of most enzymes of the EMP pathway in the spores of this species. All the EMP enzymes cannot be demonstrated in spores of some species. *Penicillium oxalicum* has a complete set of these enzymes including alcohol and lactic acid dehydrogenases. These two enzymes are, however, absent in *P. graminis tritici, Uromyces phaseoli*, and *Ustilago maydis*, indicating that the corresponding alcohols are probably not produced by these fungi. Under aerobic conditions, the EMP pathway in most spores is probably not utilized for the production primarily of ethanol or lactic acid but to furnish acetyl CoA that will be oxidized in the citric acid cycle. Though pyruvic dehydrogenase could not be demonstrated in any of the four species, this lack can probably be explained by the difficulty of assay. Fungal spores usually have all the enzymes of the EMP that are needed to form acetyl CoA from glucose. One example is the spores of *Tilletia tritici* which have all the enzymes that are used for glycolysis (Newburgh et al., 1955).

The presence of a functional EMP system under aerobic conditions has been indicated by a variety of data. Among them are the inhibitory effects of such reagents as sodium fluoride and sodium arsenate, which inhibit the enzyme enolase and prevent the formation of ATP in the glyceraldehyde-3-phosphate dehydrogenase reaction, respectively. Iodoacetate also inhibits the action of this dehydrogenase. When glucose is used as the exogenous carbon source, the cessation of oxygen consumption in the presence of such inhibitors reflects the role of these two enzymes of the EMP in aerobic respiration. This type of evidence is sometimes not reliable, for such inhibitors might not even penetrate the spore. In *Curvularia geniculata*, for example, oxygen consumption was inhibited by flouride but not by arsenate (Townsley & Bell, 1965). In *Penicillium oxalicum*, fluoride inhibited oxygen consumption only 12--25% (Caltrider & Gottlieb, 1963). Iodoacetate was a poor inhibitor and fluoride was entirely inactive as an inhibitor of endogenous respiration in *Fusarium solani* f. *phaseoli* (Cochrane et al., 1963c). Both endogenous and exogenous types of aerobic respiration in *Verticillium albo-atrum* spores were essentially refractile to inhibition by fluoride, iodoacetate and arsenate (Bechtol & Throneberry, 1966).

Even if the presence of all enzymes of the EMP has been demonstrated, as in spores of *Fusarium solani* f. sp. *phaseoli*, the complete system might not be operative. In this species ungerminated spores did not produce carbon dioxide anaerobically despite the presence of all EMP enzymes (Cochrane et al., 1963b; Cochrane & Cochrane, 1966). A similar situation was present in *Penicillium oxalicum* spores where, despite the presence of the enzymes of the EMP, the Q_{CO_2} of endogenous respiration was too low to be detected and even after 12 h of incubation had increased only to Q_{CO_2} 2.3. Exogenous respiration of glucose was absent at first and then increased to Q_{CO_2} 4.5 at the end of 12 h. That the EMP was functional after the incubation period was shown by the inhibition of carbon dioxide production following the addition of fluoride. *Ustilago maydis* spores may be another example of nonfunctional systems; all the EMP enzymes are present but carbon dioxide was not produced under anaerobic conditions (Caltrider & Gottlieb, 1963). Of course the system may be operative, with the cells accumulating pyruvate and lactate. Apparently, the system is under an inhibitory control which loses its effect during incubation.

The activity of the EMP can also be inferred from the fermentation of variously labeled glucose. Germinating *Fusarium solani* spores produced more radioactive carbon dioxide from glucose labeled with ^{14}C in C_3 or C_4 than from C_1-labeled glucose, a result consistent with the operation of this pathway. The C_1:C_6 ratio decreased with time indicating an acceleration of the EMP with increasing incubation time (Cochrane et al., 1963b). Germination was accompanied by a 2.5-fold increase in the evolution of labeled carbon dioxide from all positions except C_1, from which the increase was sixfold. Adding a source of nitrogen with the sugar also stimulated the formation of germ tubes. No carbon dioxide was evolved anaerobically from ungerminated spores, but this gas was evolved under aerobic conditions presumably by the decarboxylation of the pyruvate.

HEXOSE MONOPHOSPHATE SHUNT

The hexose monophosphate shunt (HMP) carries out the oxidation of glucose, yielding five-carbon compounds of which the most important probably is ribose-5-phosphate. Subsequent steps degrade the pentoses to erythrose and triose. Since the system can work in the reverse direction, it is also a pathway for the formation of glucose from some three-carbon compounds.

D-Glucose-6-phosphate is oxidized by a NADP-linked dehydrogenase first to 6-phosphogluconate and then to D-ribulose-5-phosphate. The ribulose-5-phosphate can be isomerized to D-ribose-5-phosphate and xylulose-5-phosphate. A transketolase reaction between these compounds gives D-sedoheptulose-7-phosphate and D-glyceraldehyde-3-phosphate. A transaldolase step follows, giving D-Fructose-6-phosphate and erythrose-4-phosphate. Various other reactions between the components and intermediates of this system are also possible which link it with the EMP.

The presence of NADP-linked dehydrogenases of the HMP has been demonstrated in *Puccinia graminis tritici* uredospores by Shu & Ledingham (1956). Cellfree extracts of ungerminated spores converted glucose-6-phosphate and other sugar phosphates to pentose phosphates, sedoheptulose-7-phosphate and glyceraldehyde-3-phosphate. Caltrider & Gottlieb (1963) showed the presence of all the HMP enzymes in spores of this species, and the entire enzyme complement was also present in *Uromyces phaseoli* and *Penicillium oxalicum*. Sedoheptulose and ribose-5-phosphate were also detected. *Tilletia caries* spores also have all the enzymes of the HMP, and the carbon dioxide from glucose-1-^{14}C appeared more rapidly than from the glucose labeled in the $C_3 + C_4$ carbons (Newburgh & Cheldelin, 1958). Ungerminated teliospores of *Ustilago maydis* did not have all the HMP enzymes, as will be discussed later.

Both glucose degrading systems can be present and operative at the same time. In *Verticillium albo-atrum* spores, the EMP and HMP, measured in the presence of an air-sweep, were 40 and 60%, respectively, but with an oxygen sweep the proportions were 40 and 52%. Under pure nitrogen, operation of the EMP was very slight (Brandt & Wang, 1960). *Tilletia caries* spores had a more active HMP than EMP (Newburgh & Cheldelin, 1958). Such results are approximate since they are derived from respiratory measurements; the calculations must be based on all the systems that operate simultaneously, and these are not always known.

Studies of the various systems that take part in the glucose metabolism of *Schizophyllum commune* basidiospores showed changes in these systems with the development of germination. Ungerminated spores appeared to have equally active EMP and HMP systems operating. Carbon dioxide from C_1-labeled glucose was produced at about the same rate as from C_3 + C_4-labeled glucose, and was greater than from the C_6 or C_2-labeled sugar. At 4 h of incubation the HMP was predominant, but from 8 to 12 h of incubation, the EMP seemed to play a greater role (Aitken & Neiderpruem, 1973).

ENTNER--DOUDOROFF PATHWAY

Another system for the degradation of glucose, the Entner--Doudoroff Pathway, may be present in some fungi. In this scheme, glucose is oxidized to 6-phosphogluconate, and carbons 2 and 3 are subsequently dehydrated to from an enol; the compound is then converted to 2-keto-3-deoxy-6-phosphogluconate. An aldol cleavage splits the compound into pyruvate and glyceraldehyde-3-phosphate. There is some evidence that such a system is present in teliospores of *Tilletia caries* (Newburgh & Cheldelin, 1958). Both three-carbon products can then be incorporated into the EMP and citric acid pathways to be oxidized to carbon dioxide.

CITRIC ACID CYCLE

The citiric acid cycle (CAC) is the main channel for oxidizing the acetate carbons which are derived from the EMP as acetyl CoA. It also serves to oxidize the acetate carbons obtained from other sources such as fatty acids or benzene derivatives. At each turn of the cycle, one acetyl CoA combines with oxaloacetate to form citric acid. This in turn is oxidized to the final end products, 2 moles of carbon dioxide and 1 mole of the four-carbon oxaloacetate. The oxaloacetate then serves to carry another acetyl moiety into the CAC to be oxidized. The components of this cycle are, in order, citrate, *cis*-aconitate, isocitrate, oxalosuccinate, α-ketoglutarate, succinyl CoA, succinate, fumarate, malate, and oxaloacetate.

The component enzymes of the citric acid cycle (CAC) are also present in spores of many fungi. The studies of Caltrider & Gottlieb (1963) showed the presence of aconitase, isocitric dehydrogenase, succinic dehydrogenase, fumarase, and malic dehydrogenase in *Penicillium oxalicum, Puccinia graminis tritici*, and *Uromyces phaseoli*. α-Ketoglutaric dehydrogenase could not be found but this was attributed to difficulties with the assay. The succinic dehydrogenase was malonate-sensitive. Conidia of *Neurospora sitophila* probably contain a functional citric acid cycle since they could oxidize the various acids that are part of the cycle (Owens, 1955). The component acids of this system have been shown to be present in *Puccinia graminis tritici* and *P. rubigo-vera* spores. The operation of the CAC has also been shown in the breakdown of carbohydrate by spores of *Schizophyllum commune* (Wessels, 1959).

Bechtol & Throneberry (1966) reported that the intermediates of the CAC were not utilized as respiratory substrates by *Verticillium albo-atrum* spores and that malonate and arsenite did not inhibit respiration. They doubted the importance of this respiratory pathway in such spores. However, Brandt & Wang (1960) showed that acetate was oxidized through the CAC. Niederpruem (1964, 1965) demonstrated the presence of isocitric dehydrogenase and fumarase in basidiospores of *Schizophyllum commune*.

Puccinia graminis tritici uredospores were shown to fully oxidize most of the CAC intermediates (White & Ledingham, 1961). Staples (1957) demonstrated the presence of the non-keto acids, except isocitrate, in uredospores of this organism. Malonate was also present, but during germination, as the citrate increased, the malonate disappeared.

The operation of the CAC in *Puccinia graminis tritici* is also suggested by the labeling studies with acetate, propionate and valerate. In each case the labeling of the intermediates was consistent with the operation of this cycle (Reisener et al., 1961; 1963; 1964; Reisener & Jäger, 1966).

TERMINAL ELECTRON TRANSPORT SYSTEM

In aerobic eukaryotic organisms, the terminal electron transport system (TET) is the mechanism for sequentially moving electrons from NADH, flavoproteins or other intermediates to oxygen. Another way of visualizing the sequence is as a means of removing hydrogens from such substrates and combining them with oxygen to form water. The components of the system are usually flavoproteins, various cytochromes, and coenzyme Q. The entire system is contained in the mitochondria which have been found in all fungi. The electrons from substrates pass through the flavoproteins and successively through CoQ, cytochromes B, C_1, C, and A + A_3, and then combine with oxygen to form water. Energy is lost to the electron at three specific steps in the sequence and is stored as high energy phosphate in adenosine triphosphate for use in various cellular syntheses and other energy requiring reactions.

The terminal electron transport system appears in spores of all species that have been examined and is usually sensitive to cyanide and azide. NADH-cytochrome C reductase, NADH-cytochrome C oxidase and cytochrome C oxidase were found in spores of *Penicillium oxalicum*, *Puccinia graminis tritici*, *Uromyces phaseoli*, and *Ustilago maydis*; all of these were sensitive to the two cytochrome C oxidase inhibitors (Caltrider & Gottlieb, 1963). Succinic oxidase has also been found in uredospores of *P. rubigo-vera* (Staples, 1957). Respiration of *Schizophyllum commune* basidiospores with glucose as the substrate was inhibited by cyanide, azide, antimycin A, and atabrine, thus implicating the normal terminal electron transport system in basidiospore respiration (Niederpruem, 1964; Ratts et al. 1964). Similar results with *Puccinia graminis tritici* uredospores have been reported by White & Ledingham (1961). The following systems at least were present: cytochrome oxidase (which was inhibited by azide and cyanide), NADH-cytochrome C and malic-cytochrome C reductases (which were inhibited by antimycin A), succinic-cytochrome C reductase, and an NADH oxidase. Antimycin A blocked respiration almost completely. In this species the cytochrome C oxidase, succinic-cytochrome C reductase, and NADH-cytochrome C reductase were more active after the appearance of germ tubes than before.

Conidia of *Verticillium albo-atrum* generally are reported to have a weak TET. Only NADH dehydrogenase was very active. NADH oxidase could not be detected and succinic-cytochrome C reductase was so low as to be negligible. NADPH dehydrogenase, NADH-cytochrome

C reductase and cytochrome oxidase activities were also low. Only the activity of NADH-cytochrome C reductase increased with germination (Throneberry, 1967). One might conclude that the low activities of the mitochondria from these spores might be due to their fragility, for the 105,000 g pellet had only cytochrome C reductase and small amounts of succinic dehydrogenase; all other component activities were in the supernatant fraction. Amytol and antimycin A inhibited particulate NADH- and NADPH-cytochrome C reductases, but had no such effect on the activities in the supernatant. Another anomaly was the lack of effect of azide on respiration whereas cyanide and dinitrophenol inhibited respiration. Cyanide inhibited the exogenous glucose respiration more than it affected endogenous respiration (Bechtol & Throneberry, 1966).

Another odd effect of inhibitors is the increased endogenous respiration of *Fusarium solani* f. *phaseoli* when the spores were treated with cyanide or azide (Cochrane et al. 1963c).

ENZYME DEVELOPMENT

The respiratory enzymes of spores may vary in kind and activity as the spore develops from its quiescent condition to its active germ tube stage. In ungerminated conidia of *Verticillium albo-atrum*, NADH-cytochrome C reductase and succinic dehydrogenase had only slight activity, but this increased during germination (Throneberry, 1968). Conidia of *Aspergillus niger* at first had no detectable phosphoglucomutase, phosphohexose isomerase, and aldolase of the EMP, but acquired them as incubation proceeded (Bhatnagar & Krishnan, 1960c). Thus, there was no EMP at this stage of development. Except for hexokinase, the remainder of the missing enzymes appeared between 12 and 18 hours of incubation; the EMP was then functional and germ tubes appeared (Bhatnagar & Krishnan, 1960c). No acid phosphatase was present in the resting spore, but it was synthesized during germination (Bhatnagar & Krishnan, 1960b). In contrast, catalase was present in the resting spore, but decreased during germination (Bhatnagar & Krishnan, 1960a).

In teliospores of *Ustilago maydis* many of the respiratory enzymes were absent in the nongerminated spore: for example, α-glycerophosphate dehydrogenase; glucose-6-phosphate dehydrogenase, transketolase, and transaldolase; aconitase, fumarase, and succinic dehydrogenase (Gottlieb & Caltrider, 1963). These enzymes developed slowly and were not detected until after 6 h, but before 12 hours of incubation. This portion of the incubation period was the time when there also was a rise in oxygen consumption and protein synthesis was detected. Even those respiratory enzymes that are present in spores prior to incubation increased during the germination process.

In intensive studies on conidia of four fungi, Ohmori & Gottlieb (1965) examined eleven enzymes from the beginning of incubation through the first appearance of the germ tube and its later development. During this period the dry weight of the spores and germ tubes increased four to five times and protein about four times. There is reason to believe that the increased protein was primarily enzyme protein. The increases of enzyme were both in total activity and in specific activity, and usually were evident before the formation of the germ tube was first observed. The total and specific activities increased further during the growth of the germ tube.

Except for succinic dehydrogenase in one species, the conidia of all four fungi, *Trichoderma viride, Aspergillus niger, Penicillium atrovenetum*, and *P. oxalicum*, had increased total respiratory enzyme activity of between 2- and 210-fold. The increases differed depending on the species and the enzyme that was studied. Of this relatively small sampling of fungi, those fungi with the highest initial enzyme activity had the smallest increase. Increases in specific activity were not found in *P. oxalicum* conidia for glucose-6-phosphate dehydrogenase and 6-phosphogluconic acid dehydrogenase. The remainder of the enzymes in the conidia of this fungus had relatively small increases during germination. For the most part, the greatest changes were in NADH oxidase and NADH-cytochrome C reductase.

Another area in which the enzyme complement of the germinating spores changes is the formation of adaptive enzymes or, in more contemporary terms, derepression. For trehalose to be respired, it must first be transported into the spores of *Myrothecium verrucaria*. These spores have both a constitutive and an induced mechanism for transporting this disaccharide into the cells where it is hydrolyzed to glucose before being respired (Mandels & Vitols, 1967).

OXYGEN CONSUMPTION

The oxidation of carbon substrates by spores is often measured by the amount of oxygen consumed as microliters per mg dry weight, the Q_{O_2}. When conidia of *Penicillium oxalicum* were incubated in the absence of carbohydrate, the reserves of the spores were gradually depleted as shown by the decreased Q_{O_2}. For the first 12 h of incubation the Q_{O_2} was 3.5, but after 18 h it dropped to 0.57. When the spores were incubated in the presence of glucose, however, the Q_{O_2} remained 17.5 even after 24 h (Caltrider & Gottlieb, 1963).

In contrast to that of *Penicillium oxalicum* conidia, the oxygen consumption of *Schizophyllum commune* basidiospores was relatively high, Q_O 10--20 (Niederpruem, 1964). The endogenous oxygen consumption of *Ustilago maydis* spores increased with the period of incubation from 1.2 to 9.0 in the absence of glucose, and from 1.3 to 23.5 in its presence (Caltrider & Gottlieb, 1963). The Q_{O_2} values of *Puccinia graminis* and *Uromyces phaseoli* spores were very low and not materially increased by exogenous glucose or succinate. *Puccinia coronata* spores also had decreased oxygen consumption as the time of incubation increased; at 0, 3, and 6 h, the Q_{O_2} values were 9.0, 6.0, and 5.25, respectively, and the addition of exogenous substrates had little if any effect on the Q_{O_2} (Naito & Tani, 1967).

Fusarium solani f. *phaseoli* macroconidia consumed more oxygen after germination than before, the respective Q_{O_2} values were 20.4--29.8 and 2.6--4.4. The Q_{O_2} of the germinated spores was even higher than that of young 48-h mycelium, Q_{O_2} 19.4. Ungerminated spores had a Q_{O_2} about the same as 6-day-old mycelium (Cochrane et al., 1963a). The endogenous Q_{O_2} of ungerminated spores did not vary with spore concentration, but the exogenous glucose oxidation was sensitive to spore density (Cochrane et al. 1963c). The ability to form germ tubes was not directly correlated with the ability of exogenous carbohydrates to support respiration. The respiration of these spores was unusual in that inhibitors, such as azide, that commonly inhibit respiration stimulated endogenous respiration and thus resembled other agents that uncouple oxidative phosphorylation such as 2,4-dinitrophenol. The rate limiting effect was thought to be the supply of phosphate acceptors.

Oxygen consumption varied with the carbon source available to the spore. Conidia of *Curvularia geniculata*, for example, had Q_{O_2} values in salt solution to which various carbon sources had been added as follows: maltose, raffinose, fructose, and sucrose, about 14.0; glucose and sugar alcohols, 12.6; trehalose, mannose, galactose, and xylose, between 12.6 and 8.0; acetate, 8.0; ribose, 4.0; and citrate, 1.0 (Townsley & Bell, 1965).

Odd effects sometimes also occurred with inhibitors. Conidia of *Myrothecium verrucaria* in a salt solution that did not allow germ tube formation unless a carbohydrate was included had a Q_{O_2} 3--4. However, when azide was added it caused a 10-fold increase (Mandels et al., 1965). The Q_{O_2} also can vary with the metabolic state of the spore. Dormant ascospores of *Neurospora tetrasperma* had a Q_{O_2} 0.6, whereas activated spores had a Q_{O_2} 1.0 (Budd et al., 1966).

As one expects, there is usually increasing oxygen consumption as the germination process proceeds (Mandels & Norton, 1948), but this does not always occur. The Q_{O_2} of ungerminated spores of *Fusarium solani* f. *phaseoli* with glucose as the carbon source was no greater when germ tubes had been produced than before germination (Cochrane et al., 1963b). With glucose or sucrose as exogenous carbon source for *Myrothecium verrucaria* spores, oxygen consumption began immediately after placing them in the incubation medium and continued linearly for at least 3 h (Mandels, 1954).

The intensity of respiration can vary with the concentration of the substrate. Thus, oxygen consumption by spores of *Curvularia geniculata*, which require glucose to germinate, increased as the glucose concentration increased from 0.002 to 0.125 M (Townsley & Bell, 1965).

Among the unusual relationships of exogenous substrates and respiration is that reported for basidiospores of *Schizophyllum commune* in which oxygen consumption was stimulated by sucrose, some hexoses, and xylose, but not by glucose (Niederpruem & Hafiz, (1964).

Exogenous substrate will sometimes spare the endogenous source. In *Neurospora tetrasperma* ascospores, 55% of the endogenous carbon dioxide production was quenched when exogenous glucose was available. However, the total amount of $^{14}CO_2$ released from glucose-^{14}C was unchanged (Budd et al., 1966).

ANAEROBIC RESPIRATION

Anaerobic respiration is not common in fungus spores, and the spores usually do not germinate in the absence of oxygen. *Verticillium albo-atrum* under anaerobic conditions evolved only 10% as much CO_2 as under aerobic conditions. This occurred whether or not exogenous glucose was supplied (Bechtol & Throneberry, 1966; Brandt & Wang, 1960). The use of inhibitors of cytochrome C oxidase effectively duplicated anaerobic respiration and thus prevented germination.

ASSIMILATION AND RESPIRATION

Even ungerminated spores may have a strong assimilative capacity. Ungerminated macrospores of *Fusarium solani* f. *phaseoli* oxidatively assimilated 85% of the glucose that was utilized. The major product was mannitol (Cochrane et al., 1963a). Spores of *F. roseum* (Sisler & Cox, 1954) and *F. oxysporum* have increased exogenous oxygen consumption when a nitrogen source such as the ammonium ion is added to the incubation medium. A similar phenomenon was observed with conidia of *Verticillium albo-atrum*. Ammonium and nitrite ions stimulated endogenous respiration but nitrate failed to do so (Bechtol & Throneberry, 1966). In ungerminated spores of *F. solani* endogenous oxygen consumption was greatly increased by the addition of ammonium ions (Cochrane et al., 1963c), but exogenous oxygen consumption was not affected (Cochrane et al., 1963d).

RESPIRATORY QUOTIENTS

The respiratory quotient (RQ) is a rough guide to the general type of endogenous substrate that is being oxidized. It is the ratio of carbon dioxide evolved to the oxygen consumed. For *Puccinia graminis tritici* ungerminated uredospores, the RQ was 0.6 (Farkas & Ledingham, 1959) and for *P. rubigo-vera*, 0.4 (Staples, 1957). Conidia of *Neurospora sitophila* had an RQ of 0.74. For ungerminated macrospores of *Fusarium solani* f. *phaseoli* the RQ was 0.55--0.68 (Cochrane et al., 1963c). All these values were in the range that would be expected if lipids were being oxidized, though the *P. rubigo-vera* value is even low for such substrates. In contrast, the RQ for *Schizophyllum commune* spores was near 1.0, suggesting the oxidation of a carbohydrate; these spores are known to contain glycogen which might be the endogenous respiration substrate for germination (Niederpruem, 1964). *Fusarium solani* is interesting because the substrate seems to shift from lipid to carbohydrate during germination, as shown by the increase in RQ from 0.68 to 0.89 (Cochrane et al., 1963c). Adding sugars greatly enhances oxygen consumption of spores of most fungi. The rusts are usually an exception and show no significant change when glucose is added to the incubation medium (Caltrider & Gottlieb, 1963). *Verticillium albo-atrum* conidia required no exogenous substrates for respiration, but when 16 sugars and sugar phosphates were separately added, RQ values from 0.84 to 1.35 were obtained (Bechtol & Throneberry, 1966).

ENDOGENOUS CARBON RESERVES

Spores of many fungi contain sufficient carbon reserves to allow respiration to go on, but in some species this is insufficient to allow germination to proceed to the formation of germ tubes. Under these restrictions, such spores require the addition of exogenous carbon substrates to complete the process. The endogenous reserves that are most frequently associated with germination are carbohydrates and their analogues, such as polyols. Lipid reserves have also been implicated as respiratory reserves, primarily in uredospores of rusts. The amount of reserve that is present in the spores may depend on the medium in which the parent culture was grown. For example, the lipid content of macrospores of *Fusarium solani* f. *phaseoli* increased as the concentration of glucose in the medium was increased (Cochrane et al., 1963b).

The metabolism of lipid for respiration seems common only in the rusts, such as *Puccinia graminis tritici* and *Uromyces phaseoli* in which the lipid content of the uredospores decreased with increasing time of incubation (Shu et al., 1954; Caltrider & Gottlieb, 1963). However, more recent studies have shown that at least in *P. graminis triciti* spores there is also a rapid utilization of some sugars and sugar alcohols (Daly et al., 1967). In other fungi such as *F. solani* f. *phaseoli*, endogenous respiration of spores does not seem to be dependent on their lipid content (Cochrane et al., 1963a).

Trehalose, mannitol, and glucose are the most common endogenous respiratory carbon reserves that are found in fungus spores. A hot water extract of *Myrothecium verrucaria* spores contained trehalose as the major reserve, 18.6% of the spore dry weight; about 30% of this disaccharide was at the spore surface. Mannitol, which is also widely found in spores, was 2% of the dry weight while glucose was only 0.3% (Mandels et al., 1965). Trehalase and mannitol as well as arabitol, erythritol and glycerol occurred in conidia of *Aspergillus oryzae* (Horikoshi & Ikeda, 1966). The trehalose decreased with incubation time. Trehalase which hydrolyzed this disaccharide was in bound and unbound forms. Mannitol competes with trehalose for this enzyme, and acts as a competitive inhibitor. Probably the mannitol is used by the spore before the trehalose.

Trehalose exists in two cell pools in *Neurospora tetrasperma* (Budd et al., 1966), and its utilization is related to spore respiration. In *Myrothecium verrucaria* spores, when the respiratory systems were activated by azide or heat treatment, the consumption of oxygen decreased the the endogenous trehalose content of the conidia also was reduced. In contrast, exogenous trehalose respiration was unaffected by such activation. As trehalose decreased, the concentration of mannitol increased (Mandels et al., 1965). In spores of this fungus, endogenous trehalose was respired slowly, Q_{O_2} 3--4 and the loss of the sugar was about 5 μg per ml of spores per hour. These data indicate that the reserve trehalose could have been completely oxidized to carbon dioxide and water. Activating the spores resulted in an eightfold increase in trehalose utilization in the early stages before the germ tubes were formed; later in germination trehalose was synthesized.

Mannitol and arabitol have been identified in basidiospores of *Schizophyllum commune*. Arabitol apparently is the more favorable carbon source since it disappeared first during germination (Niederpruem & Hunt, 1967).

EXOGENOUS CARBON RESERVES

As previously indicated, if the concentrations of endogenous carbon reserves are too low or of inadequate types to promote the formation of germ tubes, exogenous sources of such compounds are necessary. In nature such exogenous compounds are widely spread on plant parts and in the general milieu of soil. Germination on chemically defined media, however, often requires their addition. The most favorable exogenous substrate varies with different species (Lin, 1940, 1945). *Fusarium solani* f. *phaseoli* macrospores in the presence of glucose, maltose, or trehalose as a carbon source readily germinated (Cochrane et al., 1963a). Similarily, *Penicillium atrovenetum* could use a wide variety of carbohydrates (Gottlieb & Tripathi, 1968). *Ustilago maydis* teliospores were different; they germinated in the presence of neither glucose nor fructose, but 85% of them germinated when exogenous sucrose was present. A mixture of glucose and fructose could not replace the disaccharide. The trisaccharides raffinose and melezitose, which contain α-1,2-D-glucosyl-β-D-fructosidic linkages similar to sucrose, promoted germination in 50% of the teliospores. Molasses which contains about 50% sucrose also stimulated germ tube formation (Caltrider & Gottlieb, 1966).

Although exogenous glucose usually stimulates the germination of fungi which require a carbohydrate for germination, its addition might not greatly stimulate total respiration because of a high endogenous respiration. In *Fusarium solani* f. *phaseoli* spores, exogenous glucose sometimes stimulated oxygen consumption but at other times endogenous respiration was so strong that it suppressed the exogenous component (Cochrane et al., 1963c).

The sugars or polyalcohols that support germination vary with the species of fungi. For most fungi, D-glucose and compounds which are readily converted to glucose are effective in promoting germination of those spores that require exogenous sources. Glucose or galactose supported germination of up to 90% of *Penicillium atrovenetum* spores whereas there was no germination in the absence of either sugar. Melibiose, fructose and sucrose allowed 50, 30, and 10% of the spores to germinate, respectively. Arabinose, lactose, melibiose, sorbose, and soluble starch were entirely ineffective (Gottlieb & Tripathi, 1968). In *Fusarium solani* f. *phaseoli* spores, glucose and trehalose, which is hydrolyzed to glucose, were effective in promoting germination, but mannose and fructose were not. The ability of spores to oxidize a carbohydrate is not always indicative of their ability to allow germ tube formation since with this fungus all four carbohydrates had stimulated respiration; however, mannose and fructose were not as effective (Cochrane et al., 1963a).

A wide variety of exogenous carbon sources stimulated basidiospore germination in *Schizophyllum commune*. Among these substrates were glucose, fructose, sucrose, glycogen, and some sugar alcohols (Niederpruem & Dennen, 1966).

An interesting relationship exists between the germination of basidiospores of *Fomes igniarius* var. *populinus* and the glucose available to the spore in saplings. Germination was better on the outer than inner sapwood. The outer sapwood also contained the most free carbohydrates. Paper chromatography showed the presence of glucose, sucrose, and fructose among other unknown sugars. Only the spot indicating glucose supported germination of the spores when agar plugs were placed on the paper. Spotting various sugars, as standards, on filter paper allowed germination only with glucose, galactose, and mannose; glucose was the best, inducing germination 57% within 7 days, whereas mannose and galactose were much less effective (Manion & French, 1969).

CONVERSION OF SUBSTRATES

Fungus spores have mechanisms that serve to convert various

compounds to more readily metabolizable forms of carbon. *Penicillium chrysogenum* conidia have a xylose reductase in germinated spores but not in ungerminated ones. The enzyme activity increased with time of incubation, and the specific activity increased with increasing sugar concentration up to 10^{-3} M. NADPH was the hydrogen donor to form xylitol. Probably its function under natural conditions would be the conversion of xylitol to xylose (Chiang & Knight, 1959).

Mannitol and mannitol dehydrogenase have been reported in spores of *Aspergillus oryzae*. Extracts of the spores have two types of enzyme, one linked to NAD and the other to NADP. Both converted the alcohol to fructose and were more active on mannitol than on other sugar alcohols. The mannitol content of the spores decreased during early germination and then returned to its original value after 60 min, that is, after mannitol conversion had ceased. Thus mannitol may function to provide fructose for energy for germination before the mechanism for transporting glucose into the spore has developed (Horikoshi et al., 1965).

Basidiospores of *Schizophyllum commune* contain mannitol dehydrogenase in cell-free extracts. This enzyme was not active in ungerminated spores but the spores acquired this activity during germination, as shown by the increased oxygen consumption when mannitol was supplied; the enzyme is NAD- and NADP-linked. Arabitol can also be oxidized by this fungus (Niederpruem & Hafiz, 1964).

Myrothecium verrucaria spores readily hydrolyzed sucrose to glucose and fructose with almost no lag period, and the hydrolysis was proportional to spore concentration. The enzyme was located near the external surface of the cell membrane but was not on the cell wall (Mandels, 1951). Sucrose and melezitose were metabolized by a non-hydrolytic system such as sucrose phosphorylase, and raffinose by invertase (Mandels, 1954). In conidia of *Neurospora crassa*, trehalase activity was highest in the ungerminated spores, where as invertase was highest in the mycelium. The lowest activity was in the ungerminated ascospores but this increased threefold during spore germination. In conidia of this fungus, both enzymes were present in the soluble as well as the cell wall fraction (Hill & Sussman, 1964). The addition of exogenous glucose prevented the decrease of mannitol in the spores. In *Fusarium solani* f. *phaseoli*, endogenous mannitol disappeared during respiration of the spores but it probably was not the main energy source. Furthermore, endogenous respiration was not suppressed by the addition of exogenous substrate. The endogenous respiration might have contributed some intermediates for synthetic activities but not enough to support germ tube formation, and therefore exogenous carbon was necessary (Cochrane et al., 1963c).

RESPIRATION AND SYNTHESES

Although this discussion has been confined mainly to the degradation and respiratory aspects of spore germination, these pathways are intimately linked to the assimilative, synthetic properties of the germinating spore. For example, one can point to the number of enzymes in the glyoxylate pathway that are also part of the citric acid cycle. Furthermore, there are many intermediates of the degradative pathways that function directly as components of vital synthetic systems. Among them are pyruvate and phosphoenolpyruvate from the Embden--Meyerhof--Parnas pathway and oxaloacetate and α-ketoglutarate from the citric acid cycle. Acetyl CoA is used not only in respiration but also in syntheses of various moieties and compounds such as phenyl groups, sterols and fatty acids. Similarly, erythrose and ribose from the hexose monophosphate shunt are utilized in the synthetic activities of the germinating spore. Thus, it is readily apparent from even these few examples that the respiration of carbon compounds can be considered not only as an energy producing function but also as a vital source for intermediates in the synthetic processes of the spores.

REFERENCES

Aitken, W. B. & Niederpruem, D. J. (1973). Isotopic studies of carbohydrate metabolism during basidiospore germination in *Schizophyllum commune*. Arch. Mikrobiol. 88, 331--344.

Bechtol, M. K. & Throneberry, G. D. (1966). Characteristics of respiration by conidia of *Verticillium albo-atrum*. Phytopathol. 56, 963--966.

Bhatnagar, G. M. & Krishnan, P. S. (1960a). Enzymatic studies on the spores of *Aspergillus niger*. Part I. Catalase. Arch. Mikrobiol. 36, 131--138.

Bhatnagar, G. M. & Krishnan, P. S. (1960b). Enzymatic studies on the spores of *Aspergillus niger*. Part II. Phosphatases. Arch. Mikrobiol. 36, 169--174.

Bhatnagar, G. M. & Krishnan, P. S. (1960c). Enzymatic studies on the spores of *Aspergillus niger*. Part III. Enzymes of Embden--Meyerhof--Parnas pathway in germinating spores of *Aspergillus niger*. Arch. Mikrobiol. 37, 211--214.

Brandt, W. H. & Wang, C. H. (1960). Catabolism of C^{14}-labeled glucose, gluconate and acetate in *Verticillium albo-atrum*. Am. J. Bot. 47, 50--53.

Budd, K., Sussman, A. S., & Eilers, F. I. (1966). Glucose-C^{14} metabolism of dormant and activated ascospores of *Neurospora*. J. Bacteriol. 91, 551--561.

Caltrider, P. G. & Gottlieb, D. (1963). Respiratory activity and enzymes for glucose catabolism in fungus spores. Phytopathol. 53, 1021--1030.

Caltrider, P. G. & Gottlieb, D. (1966). Effect of sugars on germination and metabolism of teliospores of *Ustilago maydis*. Phytopathol. 56, 479--484.

Chiang, C. & Knight, S. G. (1959). D-Xylose metabolism by cell free extracts of *Penicillium chrysogenum*. Biochim. Biophys. Acta 35, 454--463.

Cochrane, J. C., Cochrane, V. W., Simon, F. G., & Spaeth, J. (1963a). Spore germination and carbon metabolism in *Fusarium solani*. I. Requirements for spore germination. Phytopathol. 53, 1155--1160.

Cochrane, V. W., Berry, S. J., Simon, F. G., Cochrane, J. C., Collins, C. B., Levy, J. A., & Holmes, P. K. (1963b). Spore germination and carbon metabolism in *Fusarium solani*. III. Carbohydrate respiration in relation to germination. Plant Physiol. 38, 533--541.

Cochrane, V. W., Cochrane, J. C., Collins, C. B., & Serafin, F. G. (1963c). Spore germination and carbon metabolism in *Fusarium solani*. II. Endogenous respiration in relation to germination. Am. J. Bot. 50, 806--814.

Cochrane, V. W., Cochrane, J. C., Vogel, J. M., & Coles, R. S., Jr. (1963d). Spore germination and carbon metabolism in *Fusarium solani*. IV. Metabolism of ethanol and acetate. J. Bacteriol. 86, 312--319.

Cochrane, V. W. & Cochrane, J. C. (1966). Spore germination and carbon metabolism in *Fusarium solani*. V. Changes in anaerobic metabolism and related enzyme activities during development. Plant Physiol. 41, 810--814.

Daly, J. M., Knoche, H. W., & Wiese, M. V. (1967). Carbohydrate and lipid metabolism during germination of uredospores of

Puccinia graminis tritici. Plant Physiol. 42, 1633--1642.

Ekundayo, J. A. (1966). Further studies on germination of sporangiospores of *Rhizopus arrhizus*. J. Gen. Microbio. 42, 283--291.

Ekundayo, J. A. & Carlile, M. J. (1964). The germination of sporangiospores of *Rhizopus arrhizus*; spore swelling and germ tube emergence. J. Gen. Microbiol. 35, 261--269.

Farkas, G. L. & Ledingham, G. A. (1959). The relation of self-inhibition of germination to the oxidative metabolism of stem rust uredospores. Can. J. Microbiol. 5, 141--151.

Gottlieb, D. & Caltrider, P. G. (1963). Synthesis of enzymes during the germination of fungus spores. Nat. & Lond. 197, 916--917.

Gottlieb, D. & Tripathi, R. K. (1968). The physiology of swelling phase of spore germination in *Penicillium atrovenetum*. Mycol. 60, 571--590.

Hill, E. P. & Sussman, A. S. (1964). Development of trehalase and invertase activity in *Neurospora*. J. Bacteriol. 88, 1556--1566.

Horikoshi, K., Iida, S., & Ikeda, Y. (1965). Mannitol and mannitol dehydrogenases in conidia of *Aspergillus oryzae*. J. Bacteriol. 89, 326--330.

Horikoshi, K. & Ikeda, Y. (1966). Trehalase in conidia of *Aspergillus oryzae*. J. Bacteriol. 91, 1883--1887.

Lin, C. K. (1940). Germination of the conidia of *Sclerotina fructicola*, with special reference to the toxicity of copper. Cornell Univ. Memoir 233.

Lin, C. K. (1945). Nutrient requirements in the germination of the conidia of *Glomerella cingulata*. Am. J. Bot. 32, 296--298.

Mandels, G. R. (1951). The invertase of *Myrothecium verrucaria* spores. Am. J. Bot. 38, 213--221.

Mandels, G. R. (1954). Metabolism of sucrose and related oligosaccharides by spores of the fungus *Myrothecium verrucaria*. Plant Physiol. 29, 18--26.

Mandels, G. R. & Darby, R. T. (1953). A rapid cell volume assay for fungitoxicity using fungus spores. J. Bacteriol. 65, 16--26.

Mandels, G. R. & Norton, A. B. (1948). Studies on the physiology of spores of the cellulolytic fungus *Myrothecium verrucaria*. Research Report Quartermaster General Laboratories, U.S.A., Microbiol. Ser. 11.

Mandels, G. R. & Vitols, R. (1967). Constitutive and induced trehalose transport mechanisms in spores of the fungus *Myrothecium verrucaria*. J. Bacteriol. 93, 159--167.

Mandels, G. R., Vitols, R., & Parrish, F. W. (1965). Trehalose as an endogenous reserve in spores of the fungus *Myrothecium verrucaria*. J. Bacteriol. 90, 1589--1598.

Manion, P. D. & French, D. W. (1969). The role of glucose in stimulating germination of *Fomes igniarius* var. *populinus* basidiospores. Phytopathol. 59, 293--296.

Naito, N. & Tani, T. (1967). Respiration during germination in uredospores of *Puccinia coronata*. Ann. Phytopath. Soc. Japan 33, 17--22.

Newburgh, R. W. & Cheldelin, V. H. (1958). Glucose oxidation in mycelia and spores of the wheat smut fungus *Tilletia caries*. J. Bacteriol. 76, 308--311.

Newburgh, R. W., Claridge, C. A., & Cheldelin, V. H. (1955). Carbohydrate oxidation by the wheat smut fungus, *Tilletia caries*. J. Biol. Chem. 214, 27--35.

Niederpruem, D. J. (1964). Respiration of basidiospores of *Schizophyllum commune*. J. Bacteriol. 88, 210--215.

Niederpruem, D. J. & Dennen, D. W. (1966). Kinetics, nutrition and inhibitor properties of basidiospore germination in *Schizophyllum commune*. Arch. Mikrobiol. 54, 91--105.

Niederpruem, D. J. & Hafiz, A. (1964). Polyol metabolism during basidiospore germination in *Schizophyllum commune*. Plant Physiol. Proc. Suppl. 39 P VI.

Niederpruem, D. J., Hafiz, A., & Henry, L. (1965). Polyol metabolism in the basidiomycete *Schizophyllum commune*. J. Bacteriol. 89, 954--959.

Niederpruem, D. J. & Hunt, S. (1967). Polyols in *Schizophyllum commune*. Am. J. Bot. 54, 241--245.

Ohmori, K. & Gottlieb, D. (1965). Development of respiratory enzyme activities during spore germination. Phytopathol. 55, 1328--1336.

Owens, R. G. (1955). Metabolism of fungus spores. I. Oxidation and accumulation of organic acids by conidia of *Neurospora sitophila*. Contrib. Boyce Thompson Inst. 18, 125--144.

Ratts, T., Hafiz, A., Niederpruem, D. J., & Egbert, P. (1964). Physiological studies on basidiospore germination in *Schizophyllum commune*. Bacteriol. Proc. p. 111.

Reisener, H., Finlayson, A. J., & McConnell, W. B. (1964). The metabolism of valerate-3-C^{14} and -5-C^{14} by wheat stem rust uredospores. Can. J. Biochem. 42, 327--332.

Reisener, H., Finlayson, A. J., McConnell, W. B., & Ledingham, G. A. (1963). The metabolism of propionate by wheat stem rust uredospores. Can. J. Biochem. Physiol. 41, 737--743.

Reisener, H. & Jäger, K. (1966). The quantitative importance of some metabolic pathways in uredospores of *Puccinia graminis* var. *tritici*. Planta 72, 265--283.

Reisener, H., McConnell, W. B., & Ledingham, G. A. (1961). Studies on the metabolism of valerate-1-C^{14} by uredospores of wheat stem rust. Can, J. Biochem. Physiol. 39, 1559--1566.

Shu, P. & Ledingham, G. A. (1956). Enzymes related to carbohydrate metabolism in uredospores of wheat stem rust. Can. J. Microbiol. 2, 489--495.

Shu, P., Tanner, K. G., & Ledingham, G. A. (1954). Studies on the respiration of resting and germinating uredospores of wheat stem rust. Can. J. Bot. 32, 16--23.

Sisler, H. D. & Cox, C. E. (1954). Effects of tetramethylthiuram disulfide on metabolism of *Fusarium roseum*. Am. J. Bot. 41, 338--345.

Staples, R. C. (1957). The organic acid composition and succinoxidase activity of the uredospore of the leaf and stem rust fungi. Contrib. Boyce Thompson Inst. 19, 19--31.

Throneberry, G. O. (1967). Relative activities and characteristics of some oxidative respiratory enzymes from conidia of *Verticillium albo-atrum*. Plant Physiol. 42, 1472--1478.

Townsley, W. W. & Bell, A. A. (1965). Respiratory metabolism of germinating conidia of *Curvularia geniculata*. Univ. Maryland Agric. Esp. Sta. Bull. A-140, 1--20.

Wessels, J. G. H. (1959). The breakdown of carbohydrate by *Schizophyllum commune* Fr. The operation of the TCA cycle. Acta Botan. Neerl. 8, 497--505.

White, G. A. & Ledingham, G. A. (1961). Studies on the cytochrome oxidase and oxidation pathway in uredospores of wheat stem rust. Can. J. Bot. 39, 1131--1148.

CHAPTER 5

LIPID METABOLISM OF FUNGAL SPORES DURING SPOROGENESIS AND GERMINATION

H. J. Reisener

Technische Hochschule

LIPID METABOLISM OF FUNGAL SPORES

INTRODUCTION

The accumulation of fatty or lipid materials can be observed in almost any fungus. In some filamentous fungi the lipid content is as low as 1%, in other species the accumulation amounts to one-third and almost as much as one half the dry mycelial weight (Steiner, 1957). A similar fluctuation with respect to lipid content is found in fungal spores. The literature on fungal lipids has recently been organized by Weete (1974) in his book Fungal Lipid Biochemistry and by a review article by Weete and Laseter (1974).

The degree to which lipid synthesis proceeds is a function of two factors:

1. genetic capacity for synthesis;
2. cultural conditions.

LIPID CONTENT OF DORMANT SPORES

While spores of *Aspergillus oryzae* contain only 2.2% fat calculated on a dry weight basis (Sumi, 1928; 1929) the uredospores of *Puccinia graminis tritici* contain 20% lipid (Shu, et al., 1954), and dormant ascospores of *Neurospora tetrasperma* contain as much as 26.6% lipid (Lingappa & Sussman, 1959). A lack of uniformity in analytical approaches makes comparative evaluation difficult, but these differences are substantial and are reflected in the physiological behavior of the spores. For example, there are physiological differences between conidiospores of *Aspergillus*, uredospores of *Puccinia*, and ascospores of *Neurospora*. While conidiospores of *Aspergillus* completely rely on a supply of exogenous substrates for germination, the spores of *Neurospora* and all rust fungi are independent of an external supply and use endogenous material during germination.

The chemical nature of different lipid fractions are similar only in their solubility characteristics, for example, solubility in lipid solvents such as chloroform and their near insolubility in water. Lipid compounds can be classified into the following groups:

1. hydrocarbons;
2. fatty acids;
3. glycerides;

4. phospholipids and glycolipids;
5. sterols;
6. carotenoids.

The bulk of the lipid fraction in spores consists of triglycerides. In stem rust spores the triglyceride fraction is greater than the nonsaponifiable material, which comprises about 8% of total lipids (Hougen et al., 1958). The lipid fractions are similar in dormant ascospores of *Neurospora tetrasperma* (Lingappa & Sussman, 1959). In conidiospores of *Neurospora sitophila*, however, the phospholipid fraction predominates (Owens, et al., 1958) (Fig. 1).

species	sporeform	total lipids	nonsaponif. lipids	phospho lipids
Puccinia graminis	uredospores	19,7 (16-23)	1,6	-
Melampsora lini	uredospores	22,0	6,0	0,23
Neurospora sitophila	conidia	17,7	3,8	13,8
Neurospora **tetrasperma**	dormant ascospores	26,6	6,2	1,1

Figure 1. Composition of lipids in different spores

Using gas chromatography, Tulloch and co-workers (Tulloch, et al., 1959; Tulloch,1960; 1964; Tulloch & Ledingham, 1960; 1962; and 1964) analyzed the fatty acid composition of spores from different rusts, mildews, and other fungi. The survey included different spore forms of rust fungi. Some fatty acids were common to all spores analyzed, for instance the saturated myristic, palmitic, and stearic acid and the unsaturated oleic, linoleic, and linolenic acid.

Cis-9, 10-epoxyoctadecanoic acid, a very unusual fatty acid, was found to be a characteristic component of the oil of most rust species. However, it is completely absent from corn rust spores and from two species of Gymnosporangium. In some cases, the epoxy acid is partially replaced by the corresponding (+) *threo*-9, 10-dihydroxy-octadecanoic acid. The existence of the epoxy acid seems to be confined to rusts only, as it could not be detected in lipids of conidia of mildews or the chlamydospores of cereal smuts.

In addition to the fatty acids already mentioned, smaller and higher homologues of saturated and unsaturated fatty acids such as lauric, palmitoleic, heptadecanoic, arachidic, eicosenoic, and behenic acid are present in different spore oils. The quantitative contribution of the different acids to the total fatty acid content varies from species to species. In most species palmitic (6--52% of the total fatty acids), oleic (2--62%), linoleic (2--63%), linolenic (6--72%), and *cis*-9, 10-epoxyoctadecanoic acid (0--78%) were major components.

The composition of the oils from different rusts, mildews, and other fungi suggests that each is characteristic of the fungal species and is not greatly influenced by the host (Shaw, 1964). The oils from different uredospores of eight races of *Puccinia graminis tritici* were virtually identical, but oils from different genera were less similar. Significant differences were found between higher taxa.

Phospholipids are recognized as important structural components of cellular membranes. In studies carried out with uredospores of *Uromyces phaseoli* the identity of its phospholipids was established (Langenbach & Knoche, 1971a). Phosphatidylcholine and phosphatidylethanolamine were detected as major phospholipids of resting spores. Diphosphatidylglycerol, phosphatidylinisitol and other phosphoinositides were present only as minor components.

HYDROCARBONS AND RELATED SUBSTANCES IN SPORES

Aliphatic hydrocarbons have been found in almost every plant, animal, and microorganism examined (Weete, 1972). Baker and Strobel (1965) were the first to report the presence of aliphatic hydrocarbons as constituents of fungi. From uredospores of *Puccinia striiformis* a series of n-alkanes was extracted. Similar investigations with chlamydospores of several smut fungi (Oro et al., 1966) and spores of selected phytopathogenic fungi were in accord with the results on rust spores (Laseter et al., 1968; Weete et al., 1969; Weete, 1974; Jackson et al., 1973). Generally, fungal spore alkane carbon chain lengths range from C_{18} to C_{35}. The principal hydrocarbons are the C_{27}, C_{29}, and C_{31} carbon compounds.

Molecules with an odd number of carbon atoms predominate. Smaller amounts of branched hydrocarbons and chains containing an even number of carbon atoms were detected. The exact origin and location of fungal spore hydrocarbons are not yet certain.

Together with hydrocarbons, surface lipid extracts of uredospores of wheat stripe rust contained β-diketones, long chain alcohols, free fatty acids, triglycerides, wax esters, and sterol esters (Jackson et al., 1973).

In *Ustilago maydis* in addition to hydrocarbons, free fatty and methyl esters of free fatty acids were detected. The free fatty acid fraction contained fatty acids ranging in carbon number from C_{12} through C_{20}. The free fatty acids and natural methyl esters of the fatty acids were similar in chain length and the predominant compounds had even carbon number chains (Laseter et al., 1967).

STEROLS IN FUNGI

Interest in sterols in fungal spores has been primarily centered around the stimulation of germination, particularly in the phycomycetes (Weete, 1973; Hendrix, 1965). Sterols have also attracted interest indirectly in the sense that stimulation or control of fungal sporulation has been attributed to sterols in a number of fungi (Hendrix, 1965). Sterols have also been of interest in fungi because of their possible role as hormones in the reproduction process (Raper, 1950). The subject of fungal sterols has recently been reviewed by Weete (1973) and Weete & Laseter (1974). The most common sterol in fungi has normally been considered to be ergosterol (24-methyl-delta-5, 7, 22-cholestatrien-3 beta-01, however, more recent analysis indicates a number of sterols commonly found in higher plants and animals are also present in fungi (Weete, 1974).

In the phycomycetes, few comparisons have been made between the composition of sterols in the mycelium and sterols in the spores (Weete, 1973). However, in those cases where comparisons have been made, the sterol composition in the spores is qualitatively very similar or often even identical to the sterol composition in the mycelium of the organism, such as in *Rhizopus arrhizus* (Weete et al., 1973). The conidial sterols of another mucorales fungus, *Linderina pennispora* were atypical since fungisterol was the predominant sterol and ergosterol could not be detected (Weete & Laseter, 1974). Evidence was also obtained for the C-28 and C-29 sterols in *L.*

pennispora. The spores of *Plasmodiophora brassicae* were analyzed and no sterols were detected (Knights, 1970). While the sterol composition of oomycetes has not been investigated to determine sterol content, the inference is that sterols are necessary for sporulation in a number of the groups.

Ergosterol appears to be the predominant sterol in the ascomycetes and deuteromycetes. However, a wide variety of other sterols have also been reported, particularly in yeasts. Ergosterol is the most abundant sterol in the conidia of *Aspergillus niger* (Weete, 1973), although smaller concentrations of a di and tetra unsaturated C-28 sterol (ergostra-delta-5, 7, 14, 22-tetraenol were detected. Conidia from *Pennicillium claviforme* was different in that the predominant sterol was fungisterol, although there were several other sterols in small concentrations, such as ergosterol.

In basidiomycetes, only a limited number spores have been analyzed for sterol content such as *P. graminis tritici*, *P. striiformis*, *Melampsora lini*, *Ustilago maydis*, and *U. nuda*. The principal sterol component in *U. maydis* and *U. nuda* was ergosterol with fungisterol being a minor component (Weete, 1974). However, in *U. maydis*, cholesterol was also present in minor concentrations. The sterol contents of *P. graminis* var. *tritici*, and *P. striiformis* were remarkably similar with the most abundant sterol, stigmast-ϕ^7-enol (Nowak et al., 1972). *Melampsora lini* also contained the same C_{29}-sterol as its major component (Jackson & Frear, 1968). The lipid changes during germination were followed by Bushnell (1972) in spores of *U. maydis*. He found that the sterols increase slightly during the germination process, but still represented a low percentage of the lipid fraction in the spore. Gunasekaran et al., (1972) investigated the changes in lipids during the germination of spores of *Rhizopus arrhizus* and observed that the sterols and sterol esters varied during the germination process. There was a general decrease occurring during germination. Many of the changes in lipids that occur during the germination process in spores were recently reviewed by Weber & Hess (1974).

FATTY ACID SYNTHESIS DURING SPOROGENESIS

Certain spores contain considerable amounts of neutral fats. These have to be synthesized by the mycelium or by the developing spores during sporogenesis. If a fungus is grown on a medium containing glucose as the sole carbon source, it is evident that glucose is used as starting material for fatty acid synthesis. The problem is more complicated if we include parasitic fungi that use living host plants as substrates.

The question may be asked, what kinds of compounds are used under these circumstances. What does the parasite use as a substrate for fungal fatty acid synthesis? Several possibilities may be considered:

a. Short chain acids (for instance, acetate, acetyl CoA or malonyl CoA) are taken up by the mycelium from the host and are directly incorporated into fungal fatty acids.

b. Long chain fatty acids (for instance palmitic acid or stearic acid) are absorbed from the host. The long chain fatty acids could be oxidized to two-carbon compounds prior to fungal synthesis. As an alternative, the possibility exists that the fungus uses the carbon skeleton of the host fatty acids directly. The specific spore fatty acids are formed by additional chemical reactions, for example chain-elongation by condensation of two carbon units, desaturation, hydroxylation, etc.

c. Fungal mycelium could use other carbon sources for starting material for fatty acid synthesis, for instance carbohydrates. Hexoses have been demonstrated to be withdrawn from host plants by the mycelium (Pfeiffer et al., 1969).

Knoche (1968) incubated sliced host tissue, infected with *P. graminis* with 1-^{14}C-acetate, 1-^{14}C-stearate and 1-^{14}C-oleate. The 1-^{14}C-acetate was incorporated into the spore fatty acids, including *cis*-9,10-epoxyoctadecanoic acid. Chemical degradation of this compound revealed that it is formed by condensation of acetate units. Labeled stearic and oleic acid are also incorporated directly into the epoxy acid without first undergoing β-oxidation, which suggests that chain elongation and modification are taking place. However, it is highly probable that in these experiments the fungal mycelium or parts of it came into direct contact with the labeled compounds of the incubation medium. Therefore, though these experiments confirm that fatty acids in the host parasite complex are synthesized by normal metabolic routes, they offer no information about the natural starting materials for fatty acid synthesis that are taken up by the fungus from the host tissue.

In order to avoid the difficulties that arise from direct contact of the medium and the mycelium, Ziegler (1971) incubated rust infected wheat leaves *in situ* with possible precursors. A comparison of 1-^{14}C-acetate and 2-^{14}C glucose revealed that glucose carbon was a much better substrate for fatty acid synthesis than acetate carbon. These findings could be interpreted as evidence that though both compounds are taken up by the mycelium from the host, glucose is the favoured substrate for fatty acid synthesis. However, a comparison of specific activities (dpm/mole) of glucose isolated from the spore wall with that of a two carbon unit of spore palmitic acid showed that this explanation cannot be correct (Fig. 2). If 2-^{14}C-glucose is applied to rust infected leaves, the glucose is incorporated into spore wall material. At the same time, it seems to be metabolized via acetate units to fatty acids. On a molar basis the specific activity of the spore wall glucose should be two to three times as high as that of an acetate unit of palmitic acid. The experimental results agree with this expectation.

If acetate is withdrawn directly from the host, the 1-^{14}C-acetate should label a C_2 unit of palmitate to a higher degree than a spore wall hexose unit. However, the acetate experiment furnished the same results as the glucose experiment. Therefore, it appears that ^{14}C enters the mycelium in both experiments in the same form, most likely as a hexose unit. Partial degradation of spore palmitic acid isolated after applying 1-^{14}C-acetate to rust infected wheat leaves was contrary to the experiments carried out by Knoche (1968) for nearly complete randomization of ^{14}C activity was obtained. This result is to be expected if prior to uptake of ^{14}C by the mycelium, acetate carbon is oxidized to $^{14}CO_2$ in the host tissue and if this CO_2 is refixed by photosynthesis and then incorporated into hexoses. If 2-^{14}C-glucose is applied to rust infected leaves, the specific labeling of the 2-^{14}C-glucose should become evident in the ^{14}C distribution of the fatty acids.

Partial degradation of spore palmitic acid revealed an uneven distribution of ^{14}C in the molecule. The odd numbered carbon atoms contained 36--38%, the even numbered carbon atoms 62--64% of

tracer applied to rusted leaves	specific activities [dpm/μmol] of compounds isolated from uredospores		ratio of specific activities
	glucose	palmitic acid (1 acetate-unit)	
Acetate-1-^{14}C	479	104	4.6
Glucose-2-^{14}C	991	277	3.7

Figure 2. Specific activities and ratios of specific activities of spore-wall-glucose and spore-palmitic acid isolated from uredospores of *Puccinia graminis* after application of acetate-1-^{14}C and glucose-2-^{14}C, respectively to rusted wheat leaves.

---

the total radioactivity in palmitic acid. Obviously randomization is considerable if 2-^{14}C-glucose is applied. Nevertheless, this experiment is in accord with the view that glucose or another hexose is the starting material for fatty acid synthesis in the rust mycelium. Surprisingly, the even numbered carbon atoms have more ^{14}C than the odd numbered carbon atoms. If glucose is degraded via the Embden--Meyerhof--Parnas (EMP) sequence of reactions or via the hexose monophosphate pathway (HMP), which is known to be active in rust fungi, the opposite labeling pattern should be found. Carbon 2 of glucose should become the carboxyl group and carbon 1 of glucose should become the methyl group of acetyl-CoA. Therefore, 2-^{14}C-glucose should label the odd numbered carbon atoms of palmitic acid preferentially if it is metabolized via the indicated pathways. The result of the degradation can only be rationalized by reactions that in addition to EMP and HMP, allow carbon 2 of glucose to be transformed to the methyl group of acetate. Such a metabolic sequence is known from certain bacteria, for instance *Lactobacillus plantarum* (Heath et al., 1958), that contain an active phosphoketolase. This enzyme splits pentoses to acetyl phosphate and triosephosphate. Carbon 1 of the pentose is converted to the methyl carbon and carbon 2 to the carboxyl carbon of acetyl phosphate. *Lactobacillus plantarum* also ferments glucose by the oxidative pentose phosphate pathway. In this instance, the carbon 1 of glucose is quantitatively removed as CO_2 by the dehydrogenases of the oxidative pathway prior to the C_3--C_2 cleavage of the pentose.

Bacteria are able to transfer the high energy phosphate group of acetyl phosphate to ADP. It is conceivable that this two-carbon compound is used in the rust mycelium as a precursor for acetyl-CoA and consequently for fatty acid synthesis. This idea was substantiated by comparing ^{14}C distribution of spore wall glucose with that of palmitic acid. The labeling pattern results can be accounted for by a phospho-ketolase-type reaction on a pentosephosphate arising from 2-^{14}C-glucose. Apparently, carbon 2 and carbon 3 and carbon 5 and carbon 6 and in addition carbon 1 of glucose contribute to the acetate units that build up the fatty acids. Carbon 2, carbon 3 and carbon 5 are converted to carboxyl carbons, carbon 1, carbon 2 and carbon 6 to methyl carbons of acetate. This scheme also explains the considerable randomization of ^{14}C in the fatty acids. It is evident that the reasoning is entirely based on tracer work and the results still have to be established by enzymatic evidence. Hence, these suggestions can only be regarded as tentative.

Most of our knowledge concerning fatty acid synthesis in fungi stems from work that has been done on yeast. Lynen has demonstrated that fatty acid synthetase is a highly organized multienzyme complex (Lynen, 1966; Lynen et al., 1967). Purified preparations of the fatty acid synthetase from yeast behave as a single particulate component, of molecular weight 2,300,000. This complex appears to be composed of seven enzymes. The yeast enzyme is two to five times larger than that of the enzyme of the fatty acid synthetase complexes of other organisms. It could be that the fatty acid synthetases of other fungi are of similar complexity. But the evidence from *Penicillium patutum* indicates that they are not similar (Holtermüller et al., 1970). The synthesis of unsaturated fatty acids has been investigated mainly on yeast as far as fungi are concerned (O'Leary, 1970). Further attempts will be necessary to clarify the pathway in other fungi.

PATHWAYS FOR FATTY ACID OXIDATION DURING GERMINATION

The important role of lipids in germinating wheat stem rust uredospores was demonstrated by Shu and co-workers (Shu et al., 1954). These observations have led to the concept that lipid components are the chief storage metabolites and are the substrates in spores of many fungi primarily used during the germination process. The idea is supported by low RQ of respiration during germination. But even in spores that have a high fat content, it is obvious that the germination is not based solely on the utilization of lipids. Other material, for instance, carbohydrates, contribute to respiration as well as to synthetic processes (Lingappa & Sussman, 1959; Daly et al., 1967).

The mechanism of fatty acid oxidation during spore germination was elucidated in investigations done on uredospores of *P. graminis tritici* (Reisener et al., 1961; 1963a). Short-chain fatty acids added to a suspension of wheat stem rust uredospores are readily incorporated into the spores. The uptake depends (among other things) on the chain length of the fatty acids. A considerable percentage of the applied ^{14}C was released as carbon dioxide. The rest of the radioactivity was incorporated into different fractions of the spore material such as soluble amino acids, soluble and insoluble carbohydrates, and proteins.

The high specific activity in glutamic acid allowed a chemical degradation of this amino acid. In experiments with 1-^{14}C-valerate as the substrate, the two carboxyl groups of glutamic acid were predominantly labeled. This result is consistent with the view that glutamic acid became labeled as a result of β-oxidation of the 1-^{14}C-valerate to yield 1-^{14}C-acetate which in turn was metabolized by the tricarboxylic acid cycle. This reasoning was substantiated by incorporation experiments with 2-^{14}C-valerate where the predominant incorporation of ^{14}C was into the internal carbons of glutamic acid with the maximum activity in carbon 4. This result shows that carbon 2 of valerate is metabolized in the same way as the methyl group of acetate.

According to normal β-oxidation schemes of fatty acids, carbon 3, carbon 4 and carbon 5 of valerate should be converted to propionic acid or propionyl CoA. Therefore, the metabolism of propionate was investigated with 1-^{14}C-propionate as the substrate (Reisener et al., 1963b). Most of the ^{14}C taken up was released as carbon dioxide. As in the experiments with 1-^{14}C-valerate and

2-^{14}C-valerate, respectively, carbon 2 and carbon 3 of propionate labels glutamic acid preferentially either in the carboxyl groups or in the internal carbons. These results can be rationalized by the assumption that carbon 2 and carbon 3 of propionate are utilized by the rust uredospores as acetate. Carbon 2 serves as the carboxyl group and carbon 3 as the methyl group of acetate, whereas, carbon 1 of propionate is released as carbon dioxide.

The mode of propionate metabolism indicated by these data is not in accord with most commonly accepted schemes of propionate utilization in which this compound is either converted by a carboxylation reaction to a dicarboxylic acid or metabolized by a β-oxidation to β-hydroxypropionyl-CoA (Mahler, 1964). Instead, in rust spores propionate is metabolized by a reaction that includes an oxidation of the α-carbon atom. Intermediates in this pathway are acrylyl CoA, lactate and pyruvate. Such a metabolic sequence has been found in certain bacteria. That this pathway is used by rust spores could be substantiated by the finding that two intermediates, lactate and pyruvate, label glutamic acid in exactly the same way as propionate, if they contain the ^{14}C in the appropriate positions (Kottke, 1969). Furthermore, it could be demonstrated that carbon 5 and carbon 3 of valerate are metabolized as carbon 3 and carbon 1 of propionate (Reisener et al., 1964). These investigations account for oxidation of both even and odd numbered fatty acids. The main distinction is in the end products. Oxidation of even numbered fatty acids leads to acetylCoA only, oxidation of odd numbered fatty acids leads to a terminal propionylCoA residue. That this scheme can be extended to longer chain fatty acids has been demonstrated by incubating rust uredospores with 1-^{14}C-nonanoate (Suryanarayanan & McConnell, 1965).

QUANTITATIVE CONTRIBUTION OF FATTY ACID OXIDATION TO RESPIRATION

Determinations of C_6/C_1 ratios of $^{14}CO_2$ after application of 1-^{14}C-glucose and 6-^{14}C-glucose respectively, demonstrated the existence of the Embden--Meyerhof--Parnas (EMP) sequence of reactions and of the hexose monophosphate (HMP) pathway in rust and other spores (Shu & Ledingham, 1956; Shaw & Samborski, 1957). In addition, most of the enzymes active in these metabolic sequences have been shown to be present (Shu & Ledingham, 1956; Caltrider & Gottlieb, 1963). Obviously, both fatty acid oxidation and glucose oxidation contribute to total carbon dioxide evolution. There is a problem in evaluating the relationship of the two pathways operating simultaneously.

Chemical analysis of germinating rust spores revealed that soluble carbohydrates and fatty acids are used up during germination (Daly et al., 1967). As the decrease of either substance accounts for the total carbon dioxide liberation, no conclusion can be drawn from these findings.

By incubating spores in separate experiments with 1-^{14}C-glucose 2-^{14}C-glucose and 6-^{14}C-glucose, respectively, the contribution of EMP, HMP and nontriosepathways (NTP) to total glucose-6-phosphate turnover can be estimated. Experiments of this kind revealed that in uredospores of *P. graminis tritici* only a small amount of glucose-6-phosphate is metabolized via EMP and HMP as compared to NTP (Jäger, 1968). This implies that most of the exogenously supplied glucose is used in synthetic processes. This is in line with results of a chemical analysis using uniformly labeled glucose (Burger et al., 1972). On the other hand, experiments with 1-^{14}C-valerate showed that the specific activity of carbon dioxide and the carboxyl groups of certain TCA-cycle acids that are direct precursors of CO_2 are nearly equal (Ziegler, 1969). These data can be taken as evidence that carbon dioxide arising by reactions of the HMP and EMP cannot dilute the carbon dioxide set free by reaction of the TCA cycle with 1-^{14}C-valerate as substrate to any larger extent. The same conclusion can be drawn from an experiment in which the specific activities of glucose-6-phosphate, carbon dioxide and the carboxyl groups of malate were compared after application of 1-^{14}C-glucose. Malate is known to be a pool that is not compartmentalized in rust spores. The two carboxyl groups of malate are therefore direct precursors of the CO_2 set free in the TCA-cycle

(Böcker, 1971). If the HMP contributes extensively to the total carbon dioxide evolution, the specific activity of glucose-6-phosphate and CO_2 should be similar. If, on the other hand, the contribution of the HMP and EMP is low, the specific activity of glucose-6-phosphate should be much higher as compared to the specific activities of the carbon dioxide and the malate, due to dilution of the acetylCoA arising from fatty acids. The results are fully in accord with the latter suggestion.

The data from rust uredospores cannot be extended to other spores. Even spores of the same species can behave quite differently, if they have been formed under different external conditions. Moreover, it seems that certain spores change their metabolic activities. During the course of germination a shift from one endogenous substrate to another endogenous substrate may take place. So generalizations have to be avoided.

Uredospores of rust and spores of other fungi do not germinate or germinate poorly if they are floating on a solution in high spore loads. This inhibition is due to the presence of natural inhibitors, which have recently been characterized (Macko et al., 1970; 1971). The germination inhibition can be overcome by many different compounds. Among these coumarin breaks the self inhibition very effectively. Coumarin, besides starting germ tube wall synthesis in the spores (Burger, 1970) has a marked effect on respiration. Only 3 min. after addition of coumarin to a dense population of rust spores, CO_2 evolution increases considerably.

In order to determine whether coumarin affects the oxidation of carbohydrate material or of fatty acids, labeled glucose and labeled valerate were added to the medium in separate experiments. Besides total CO_2 evolution the amount of $^{14}CO_2$ was registered. While coumarin has no effect on glucose uptake, $^{14}CO_2$ evolution is markedly reduced in the experiment with 1-^{14}C-glucose and slightly increased in an experiment with 6-^{14}C-glucose. This can be interpreted tentatively as a reduction of the HMP and a slight increase of the EMP. Certainly, this cannot explain the large increase of respiration caused by coumarin. Using 2-^{14}C-valerate as an exogenous substrate not only a better uptake of valerate is visible in the presence of coumarin, but a much higher turnover of the TCA-cycle with a higher rate of fatty acid oxidation is indicated. Most probably, there exists an interrelationship between the onset of the germ tube wall synthesis and the increased fatty acid consumption. But this has still to be demonstrated by experimental evidence.

SYNTHETIC PROCESSES DURING GERMINATION

During spore germination fatty acids are not only used as respiratory material in catabolic pathways, but also for anabolic processes. Experiments with labeled acetate (Staples, 1962) and valerate (Reisener et al., 1961) indicated that in rust uredospores fatty acid carbon is incorporated into ethanol- and water-soluble compounds such as amino acids, organic acids, and carbohydrates, as well as into insoluble spore material. Some labeled activity was detected in proteins during spore germination. However, the bulk of the ^{14}C incorporated into the insoluble fraction is found in the polymeric carbohydrates that build up the germ tube wall. The glyoxalate pathway would be the obvious method whereby acetylCoA derived from β-oxidation of fatty acids could give rise to labeled sugars. The key enzymes of this reaction sequence have been demonstrated in uredospores of *Melampsora* (Frear & Johnson, 1961), *Uromyces* and *Puccinia* (Gottlieb & Caltrider, 1962; Caltrider et al., 1963) and in many other species. This doesn't imply that all the germ tube wall material originates from fatty acids. A considerable part must be derived from other sources, especially soluble sugars or reserve polysaccharides.

Some of the fatty acid components which disappear during germination may be accounted by increases in the nonsaponifiable lipid (Caltrider et al., 1963) and the phospholipid fraction (Jackson & Frear, 1967). Phospholipids, after extensive degradation in the very early stages of germination, are resynthesized at later stages in uredospores of *Uromyces phaseoli* (Langenbach & Knoche, 1971b). Several reactions involved in biosynthesis of different phospholipids were demonstrated. The changes in phospholipid level may reflect a degradation and synthesis of membranes.

Acknowledgments

I am grateful to Professor D. J. Weber for reading and revising the manuscript of this contribution.

References

Baker, K. & Strobel, G.A. (1965). Lipids of the cell wall of *Puccinia striiformis* uredospores. Proc. Mont. Acad. Sci. 25, 83--86.

Böcker, J.F. (1971). Untersuchungen über die Beteiligung des Pentosephosphatcyclus zur Gesamt-CO_2-Entwicklung keimender Uredosporen von *Puccinia graminis* var. *tritici*. Diplomarbeit, Universität Freiburg i. Br., Freiburg i. Br., Germany.

Burger, A. (1970). Zum Mechanismus der Keimungsenthemmung der Uredosporen von *Puccinia graminis* var. *tritici* durch Cumarin. Dissertation Universität Freiburg i. Br., Freiburg, i. Br., Germany.

Burger, A., Prinzing, A., & Reisener, H.J. (1972). Untersuchungen über den glucosestoffwechsel keimender Uredosporen von *Puccinia graminis* var. *tritici*. Arch. Mikrobiol. 83, 1--16.

Bushnell, J.L. (1972). Lipids of fungal spores: Identification and metabolism. Thesis, Brigham Young University.

Caltrider, P.G. & Gottlieb, D.,(1963). Respiratory activity and enzymes for glucose catabolism in fungus spores. Phytopathology 53, 1021--1030.

Caltrider, P.G., Ramanchandran, S. & Gottlieb, D. (1963). Metabolism during germination and function of glyoxalate enzymes in uredospores of rust fungi. Phytopathology 53, 86--92.

Daly, J.M., Knoche, H.W., & Wiese, M.V. (1967). Carbohydrate and lipid metabolism during germination of uredospores of *Puccinia graminis tritici*. Plant Physiol. 42, 1633--1642.

Frear, D.S. & Johnson, M.A. (1961). Enzymes of the glyoxalate cycle in germinating uredospores of *Melampsora lini* (Pers.) Lev. Biochim. Biophys. Acta. 47, 419--421.

Gottlieb, D., & Caltrider, P.G. (1962). Synthesis during germination of uredospores of rust fungi. Phytopathology 52, 11.

Gunasekaran, M., Weber, D.J. & Hess, W.M. (1972). Changes in lipid composition during spore germination of *Rhizopus arrhizus*. Trans. Br. Mycol. Soc. 59, 241--248.

Holtermüller, K.H., Ringelmann, E. & Lynen, F. (1970). Reinigung und Charakterisierung der Fettsaure-Synthetase aus *Penicillium patulum*. Hoppe-Seyler Z. Physiol-Chem. 351, 1411.

Heath, E.C., Hurwitz, J., Horecker, B.L., & Ginsburg, A. (1958). Pentose fermentation by *Lactobacillus plantarum*. I. The cleavage of xylulose-5-phosphate by phosphoketolase. J. Biol. Chem 231, 1009--1029.

Hendrix, J. W. (1965). Influence of sterols on growth and reproduction of *Pythium* and *Phytophthora* spp. Phytopathology 55, 790--797.

Hendrix, J.W. (1970). Sterols in growth and reproduction of fungi. Ann. Rev. Phytopathology 8, 111--130.

Hougen, F. W., Craig, B.M. & Ledingham, G.A. (1958). The oil of wheat stem rust uredospores I. The sterol and carotenes of the unsaponifiable matter. Can. J. Microbiol. 4, 521--529.

Jackson, L.L. & Frear, D.S. (1967). Lipids of rust fungi. I. Lipid metabolism of germinating flax rust uredospores. Can. J. Biochem. 45, 1309--1315.

Jackson, L.L. & Frear, D.S. (1968). Lipids of rust fungi. II. Stigmast-7-enol and stigmasta-7, 24 (28)-dienol in flax rust uredospores. Phytochemistry 7, 651--654.

Jackson, L. L., Dobbs, L., Hildebrand, A. & Yokiel, R.A., (1973). Surface lipids of wheat stripe rust uredospores, *Puccinia striiformis*, compared to those of the host. Phytochemistry 12, 2233--2237.

Jäger, K. (1968). Untersuchungen zum Glukosestoffwechsel der keimenden Uredosporen von *Puccinia graminis* (Erikss. and Henn.) und zum Stoffwechsel des Wirt-Parasit-Komplexes. Dissertation, Universität Freiburg i. Br., Freiburg i. Br., Germany.

Knights, B. A. (1970). Lipids of *Plasmodiophora brassicae*. Phytochemistry 9, 70.

Kottke, H.J. (1969). Untersuchungen über den Propionsäureabbau in den Uredosporen des Weizenrostes (*Puccinia graminis* var. *tritici*). Thesis, University of Freiburg i. Br., Freiburg i. Br., Germany.

Knoche, H.W. (1968). A study on the biosynthesis of *cis*-9,10-epoxyoctadecanoic acid. Lipids 3, 163--169.

Langenbach, R.J. & Knoche, H.W. (1971a). Phospholipids in the uredospores of *Uromyces phaseoli*. I. Identification and localization. Plant. Physiol. 48, 728--734.

Langenbach, R.J. & Knoche, H.W. (1971b). Phospholipids in uredospores of *Uromyces phaseoli*. II. Metabolism during germination Plant. Physiol. 48, 735--739.

Laseter, J.L., Weete, J.D. & Weber, D.J. (1967). Alkanes, fatty acids methylesters, and free fatty acids in surface wax of *Ustilago maydis*. Phytochemistry 7, 1177--1181.

Laseter, J.L., Hess, W.M. Weete, J.D., Stocks, D.L. & Weber, D.J. (1968). Chemotaxonomic and ultrastrucutral studies on three species of *Tilletia* occurring on wheat. Can. J. Microbiol. 14, 1149--1154.

Lingappa, B.T., Sussman, A.S. (1959). Endogenous substrates of dormant, activated and germinating ascospores of *Neurospora tetrasperma*. Plant. Physiol. 34, 466--472.

Lynen, F. (1961). Biosynthesis of saturated fatty acids. Proc. Fed. Am. Socs. Exp. Biol. 20, 941--951.

Lynen, F., Hagen, A., Hopper-Kessel, I., Oesterhelt, D., Reeves, H.C., Schweizer, E., Willecke, K. & Yalpani, M. (1967). Multi-enzyme complex of fatty acid synthetase. In Organizational Biosynthesis (ed. H.J. Vogel, J.O. Lampen & V. Brysen), pp. 243--266. New York: Academic Press.

Macko, V., Staples, R.C., Gershon, H., & Renwick, J.A.A. (1970). Self-inhibitor of bean rust uredospores: Methyl 3,4-dimethoxycinnamate. Science 170, 539--540.

Macko, V., Staples, R.C., Allen, P.J. & Renwick, J.A.A. (1971). Identification of the germination self-inhibitor from wheat stem rust uredospores. Science 173, 835--836.

Mahler, H.R. (1964). Oxidation of odd-chain fatty acids. In Fatty Acids (ed. K.S. Markeley) Vol. 3, pp. 1514--1525. New York: Wiley-Interscience.

Nowak, R., Kim, W. K., & Rohringer, R. (1972). Sterols of healthy and rust-infected primary leaves of wheat and non-germinated and germinated uredospores of wheat stem rust. Can. J. Bot. 50, 185--190.

O'Leary, W.M. (1970). Bacterial lipid metabolism. In Comprehensive Biochemistry (ed. M. Florkin and E.H. Stotz), Vol. 18, pp. 229--264. Amsterdam: Elsevier.

Oro, J., Laseter, J.L. & Weber, D.J. (1966). Alkanes in fungal spores. Science 154, 339--400.

Owens, R.G., Novotny, H.M. & Michels, M, (1958). Composition of conidia of *Neurspora sitophila*. Contrib. Boyce Thompson Inst. 19, 355--374.

Pfeiffer, E., Jäger, K., & Reisener, H.J. (1969). Untersuchungen über Stoffwechselbeziehungen zwischen Parasit und Wirt am Beispiel von *Puccinia graminis* var. *tritici* auf Weizen. Planta (Berl.) 85, 194--201.

Raper, J.R. (1950). Sexual hormones in Achlya. VII. The hormonal mechanism in homothallic species. Botan. Gaz. 112, 1--24.

Reisener, H.J., McConnell, W.B. & Ledingham, G.A. (1961). Studies on the metabolism of valerate-1-C^{14} by uredospores of wheat stem rust. Can. J. Biochem. Physiol. 39, 1559--1566.

Reisener, H.J., Finlayson, A.J. & McConnell, W.B. (1963a). The metabolism of valerate-2-C^{14} by uredospores of wheat stem rust. Can. J. Biochem. Physiol. 41, 1--7.

Reisener, H.J., Finlayson, A.J. McConnell, W.B. & Ledingham, G.A. (1963b). The metabolism of propionate by wheat stem rust uredospores. Can. J. Biochem. Physiol. 41, 737--743.

Reisener, H.J., Finlayson, A.J. & McConnell, W.B. (1964). The metabolism of valerate-3-C^{14} and -5-C^{14} by wheat stem rust uredospores. Can. J. Biochem. 42, 327--332.

Shaw, M. (1964). The physiology of rust uredospores. Phytopath. Z. 50, 159--180.

Shaw, M. & Samborski, D.J. (1957). The physiology of host-parasite relations III. The pattern of respiration in rusted and mildewed cereal leaves. Can. J. Botan. 35, 389--394.

Shu, P. Ledingham, G.A. (1956). Enzymes related to carbohydrate metabolism in uredospores of wheat stem rust. Can. J. Microbiol. 2, 489--495.

Shu, P. Tanner, K.G., Ledingham, G.A. (1954). Studies on the respiration of resting and germinating uredospores of wheat stem rust. Can. J. Bot. 32, 16--23.

Staples, R.C. (1962). Initial products of acetate utilization by bean rust uredospores. Contrib. Boyce Thompson Inst. 21, 487--497.

Steiner, M. (1957). Die Fette der Pilze. In Encyclopedia of plant physiology (ed. W. Ruhland) Vol. 7, pp. 59-89. Berlin: Springer.

Sumi, M. (1928). Über die chemischen Bestandteile der Sporen von *Aspergillus oryzae*. Biochem. Z. 195, 161--174.

Sumi, M. (1929). Nachtrag zur Arbeit "Über die chemischen Bestandteile der Sporen von *Aspergillus oryzae*." Biochem. Z. 204, 412.

Suryanarayanan, S. McConnell, W.B. (1965). The metabolism of pelargonate-1-C^{14} by wheat stem rust uredospores. Can. J. Biochem. 43, 91--95.

Tulloch, A.P. (1960). The oxygenated fatty acids of the oil of wheat stem rust uredospores. Can. J. Chem. 38, 204--207.

Tulloch, A.P. (1964). The component fatty acids of oils found in spores of plant rusts and other fungi. Part IV. Can. J. Microbiol. 10, 359--364.

Tulloch, A.P. & Ledingham, G.A. (1960). The component fatty acids of oils found in spores of plant rusts and other fungi. Can. J. Microbiol. 6, 425--434.

Tulloch, A.P. & Ledingham, G.A. (1962). The component fatty acids of oils found in spores of plant rusts and other fungi. Part II. Can J. Microbiol. 8, 379--387.

Tulloch, A.P. & Ledingham, G.A. (1964). The component fatty acids of oils found in spores of plant rusts and other fungi. Part III. Can J. Microbiol. 10, 351--358.

Tulloch, A.P., Craig, B.M. & Ledingham, G.A. (1959). The oil of wheat stem rust uredospores. II. The isolation of *cis*-9,10-epoxyoctadecanoic acid, and the fatty acid composition of the oil. Can. J. Microbiol. 5, 485--491.

Weber, D.J. & Hess, W.M. (1974). Lipid metabolism and ultrastructure during spore germination in Fungal Lipid Biochemistry (ed. J. Weete). New York: Plenum Press.

Weete, J.D. (1972). Aliphatic hydrocarbons of the fungi. Phytochem. 11, 1201--1205.

Weete, J.D. (1973). Sterols in fungi. Phytochem. 12, 1843.

Weete, J.D. (1974). Fungal Lipid Biochemistry. Plenum Press, New York.

Weete, J.D. & Laseter, J.L. (1974). Distribution of sterols in fungi. I. Fungal spores. Lipids 9, 575--581.

Weete, J.D. & Laseter, J.L., (1974). Sterols of *Linderina pennispora*. Lipids (in press).

Weete, J.D., Laseter, J.L. & Lawler, G.C., (1973). Total lipid and sterol components of *Rhizopus arrhizus*: identification and metabolism. Arch. Biochem. Biophys. 155, 411.

Weete, J.D.,Laseter, J.L. Weber, D.J., Hess, W.M. & Stocks, D.L. (1969). Hydrocarbons, fatty acids, and ultrastructure of smut spores. Phytopathology 59, 545--548.

Ziegler, E. (1969). Untersuchungen über die Kompartimentierung von Äpfelsäure und Citronensäure in keimenden Uredosporen des Weizenrostpilzes (*Puccinia graminis* var. *tritici*). Diplomarbeit, Universität Freiburg i. Br., Freiburg i. Br., Germany.

Ziegler, E. (1971). Untersuchungen über den Aminosäure- und Fettstoffwechsel des auf der Wirtspflanze parasitierenden Mycels des Weizenrostpilzes (*Puccinia graminis* var. *tritici*). Dissertation Ruhr-Universität-Bochum, Bochum, Germany.

DISCUSSION

Chairman: John D. Weete
Auburn University
Auburn, Alabama

The principal topics of discussion following Dr. Reisener's presentation focused on the importance of lipid as carbon substrate during spore germination and the influence of light on lipid formation, particularly sterols.

Dr. D. Maxwell asked if there is any indication of the relative importance of fatty acids versus other storage materials as building blocks in, for example, cell wall formation during uredospore germination and is there any indication of carbohydrate reserves going into the cell wall. Dr. Reisener feels that both fatty acids and carbohydrate reserves are used for building the germ tube, but the quantitative importance of each has not been shown. He is planning experiments to determine this. Dr. A. Sussman commented that whether or not an organism requires exogenous substrates for germination, it is clear that most spores require the simultaneous degradation of lipid and carbohydrate reserves. Dr. Sussman asked why lipids cannot be broken down in the absence of carbohydrates or vice versa. Dr. Reisener said he had no immediate answer to the question.

Dr. E. Cantino asked whether or not visible light has any effect on the incorporation of [^{14}C]-acetate into lipids during sporogenesis or spore germination. Dr. J. Weete commented that in his studies on the biosynthesis of sterols by a cell-free homogenate of *Rhizopus arrhizus*, a 40% reduction in the incorporation of [^{14}C]-mevalonate into the hexane-soluble products was obtained when samples were incubated in the light. He has not looked at the incorporation of [^{14}C]-acetate in the *Rhizopus* system.

Dr. Cantino continued to say that light has profound effects on *Blastocladiella emersonii*, one of which concerns the incorporation of [^{14}C]-acetate into sterols and their esters. In the dark both the free and esterified sterols are formed, but in the light inhibition of ester formation occurs. Dr. Weete added that during the initial growth stages of *R. arrhizus* the incorporation of [^{14}C]-acetate into total lipids and sterols occurs at about the same rate, but during sporulation there is a decrease in radioactivity in sterols. However, toward the end of the sporulation period the mycelial lipid content drops very rapidly while radioactivity in the sterols increases steadily relative to the total lipids. When this fungus is grown in the presence of $[^{14}C]$-mevalonate, radioactivity in the sterols increases steadily throughout the growth period with no decrease during sporulation. This suggests that sterols are not readily produced during the sporulation process by this fungus because acetate is utilized in other processes and is not available as a substrate for sterol biosynthesis. We are just getting to the point in our studies where we are comparing sterol metabolism and sporulation in light and dark.

Dr. Cantino further commented that since so much has been said about lipid utilization during spore germination, it might be of interest to the group to know that while *Blastocladiella* spores are swimming and for up to 20 h. they continue to use lipid at a fairly linear rate. The minute encystment occurs the spores stop using lipid, at least triglycerides, but begins using them again upon germination.

Dr. T. Hashimoto asked why coumarin was selected for his studies on spore germination. Dr. Reisener noted that coumarin had been used by others before and continued to use it when it was shown that coumarin possessed structural similarities to natural spore germination inhibitors. He added that coumarins are not as active as the natural inhibitors.

CHAPTER 6

REGULATION OF PROTEIN METABOLISM DURING SPORE GERMINATION

James S. Lovett
Purdue University

INTRODUCTION

The status of the system for protein synthesis in fungal spores, and its activation during spore germination have both been reviewed extensively in the past few years (Allen, 1965; Gottlieb, 1966; Van Etten, 1969; Sussman & Douthit, 1973). The discussion to follow will of necessity cover some of the same ground, though in part from a somewhat different point of view; the reason for this is that any critical analysis of regulation during germination will require a clear understanding of the beginning state, the dormant spore itself. By using both old and new information I should like to attempt to identify the probable mechanisms through which all fungal spores may regulate protein synthesis. Even if tentative, or wrong in detail, it is my hope that such an approach will help to direct more attention and thought toward testing some as yet unexamined mechanisms of regulation.

THE REQUIREMENT FOR PROTEIN SYNTHESIS DURING SPORE GERMINATION

In considering regulation it is worth asking at the start whether protein synthesis occurs during the germination of most or all fungus spores, and if it does whether it is essential? The results of a reasonably comprehensive survey of the literature, given in Table 1, indicate that protein synthesis is indeed a normal part of spore germination. Furthermore, the stage at which it is detected depends upon the method used; chemical measurements of total protein frequently fail to show an increase in protein content before one or two hours, while more sensitive amino acid incorporation experiments have detected synthesis within the first 15 min. The actual necessity for the production of new proteins has only been examined with a few of the organisms in Table 1, but again the results seem quite consistent. Appropriate concentrations of the antibiotic cycloheximide (Actidione) were found to inhibit spore germination in each species tested. Though it is not indicated in the table, cycloheximide was also shown to effectively inhibit amino acid incorporation in most of the same experiments. I

Table 1 Protein Synthesis During Fungal Spore Germination

Organism	Earliest Measured Change[a]		Start of Germination[d]	Evidence for Required Synthesis	Reference
	Protein[b]	RNA[b]			
Rhizophlyctis rosea	30 min (I)	5 min (I)	4 h	NM[c]	LeJohn & Lovett, 1966
Allomyces arbuscula	20 min (I)	20 min (I)	30 min	Cyclohex.	Burke et al., 1972
Blastocladiella emersonii	80 min (C) 15 min (I)	40 min (C) 15 min (I)	25 min 25 min	NM Cyclohex.	Lovett, 1968
Peronospora tabacina	10 min (I)	15 min (I)	1 h	Cyclohex.	Holloman, 1971
Phycomyces blakesleanus	2 h (C) 10 min (I)	2 h (C) 15 min (I)	6 h 6 h	NM NM	Van Assche & Carlier, 1973
Rhizopus stolonifer	2 h (C) 30 min (I)	1 h (C) 30 min (I)	3 h 3 h	NM NM	Van Etten et al., 1974
Saccharomyces cerevisiae	1 h (C) 15 min (I)	1 h (C) 15 min (I)	--- ---	Cyclohex. NM	Rousseau & Halvorson 1973a,b

Organism	Earliest Measured Change		Start of Germination	Evidence for Required Synthesis	Reference
	Protein	RNA			
Aspergillus niger	6 h (C)	3 h (C)	6 h	NM	Yanagita, 1957
Aspergillus niger	3 h (C)	NM	5 h	NM	Ohmori & Gottlieb, 1965
Aspergillus oryzae	1 h (C)	20 min (C)	3--4 h	NM	Kimura et al., 1965
	15 min (I)	NM	---	NM	Horikoshi & Ikeda, 1968
Aspergillus nidulans	2.5 h (C)	30 min (C)	3.5 h	NM	Bainbridge, 1971
Penicillium atrovenetrum	3 h (C)	NM	6 h	NM	Ohmori & Gottlieb, 1965
Penicillium oxalicum	2 h (C)	NM	5.5 h	NM	Ohmori & Gottlieb, 1965
Trichoderma viride	4 h (C)	NM	6 h	NM	Ohmori & Gottlieb, 1965

Organism	Earliest Measured Change		Start of Germination	Evidence for Required Synthesis	Reference
	Protein	RNA			
Botryodiplodia theobromae	20 min (I)	2 h (I)	2.5 h	Cyclohex.	Brambl & Van Etten 1970
Fusarium solani	2 h (C) 10 min (I)	4 h (C) 10 min (I)	2 h 2 h	NM ---	Cochrane et al., 1971
Glomerella cingulata	15 min (I)	NM	6 h	Cyclohex.	Lingappa et al., 1973
Geotrichum candidum	1 h (C)	1 h (C)	2 h	NM	Barash, 1968
Microsporum gypseum	30 min (C)	30 min (C)	1 h	NM	Barash, et al., 1967
Microsporum gypseum	10 min (I)	10 min (I)	3 h	Cyclohex.	Dill et al., 1972
Neurospora crassa	15 min (I)	10 min (I)	2 h	Cyclohex.	Mirkes, 1974; Inoue & Ishikawa, 1970

Organism	Earliest Measured Change		Start of Germination	Evidence for Required Synthesis	Reference
	Protein	RNA			
Puccinia graminis	NM	NM	1.5 h	Cyclohex.	Dunkle et al., 1969
	decrease (C)	NM	---	NM	Reisner, 1967
Uromyces phaseoli	1.5 h (C)	3 h (C)	1.5 h	NM	Trocha & Daly, 1970
	1 h (I)	---	2 h	Puromycin	Staples et al., 1962
Schizophyllum commune	4 h (C)	4 h (C)	4--8 h	Cyclohex.	Aitkin & Niederpruem 1970
Lenzites saepiaria	15 min (I)	15 min (I)	3 h	NM	Scheld & Perry, 1970

[a]The times given represent the earliest data point; actual synthesis could have started earlier.

[b]The letters in brackets refer to the method of measurement used: (C), chemical determination; (I), isotopic precursor incorporation by cells.

[c]NM means the quantity or effect was not measured.

[d]Refers to germ-tube emergence.

am not aware of any examples where cycloheximide has failed to block the completion of spore germination as defined by germ tube emergence. On the other hand, some very early events such as "activation" of spores, or encystment of aquatic zoospores (Lovett, 1968; Soll & Sonneborn, 1971b) may take place in its presence. Although inhibitor experiments have to be interpreted with caution, the ribosomal site of action (Siegel & Sisler, 1965) and the failure of cycloheximide to prevent amino acid uptake by spores (Cochrane et al. 1971; Tisdale & DeBusk, 1970), suggest that it is a reliable test for a synthetic requirement when positive results are obtained.

If we accept the evidence that protein synthesis is required as an early event in spore germination, a second and perhaps even more critical question arises: Can the essential protein(s) be identified? The answer at present is no, and on the basis of what is presently known it may be difficult to identify the specific proteins that are required. There is abundant evidence for increases in the specific activities of various enzymes during spore germination. Some of these, such as glutamate dehydrogenase (Tuveson et al. 1967), and a post-conidial amino acid transport system (Tisdale & DeBusk, 1970) in *Neurospora crassa* fail to increase in cells exposed to cycloheximide. Ohmori & Gottlieb (1965) have also shown that a number of metabolic enzymes increase in both specific activity and total amount during germination in species of *Trichoderma, Aspergillus,* and *Pencillium.* However, none of these increased enzyme activities have been shown necessary for the success of germination. Furthermore, in a few cases where the sudden appearance of a specific enzyme has been implicated in spore germination, the enzyme has been shown to be preformed and either activated or released at germination. Notable examples are trehalase activation in *Neurospora* ascopores (Sussman, 1969) and *Phycomyces blakesleanus* sporangiospores (Van Assche et al. 1972), and the release of an alkaline protease from vesicles in *Microsporum gypseum* (Page & Stock, 1972).

The complexity of the problem is even more apparent when the numbers of newly synthesized proteins are examined by polyacrylamide gel electrophoresis. When the spectrum of proteins synthesized in the first hour of *Botryodiplodia* germination was compared to that produced between 4 and 5 h by Van Etten et al. (1972), only one band out of a large number seemed clearly different at early germination. Silverman et al. (1974) have shown that an equally impressive array of new proteins are synthesized by *Blastocladiella* during germination, even when *Actinomycin D* is added to block new messenger RNA (mRNA) synthesis (Fig 1). We have also found that a significant number of ribosomal proteins are produced during early stages in *Blastocladiella* (Fig 2; Abe & Lovett, unpublished). The large variety of proteins assembled at very early times in spore germination suggest that the probability of finding a single essential protein may be low.

Much of the work on early events up to now has been concentrated on pathways of energy metabolism. It seems probable, however, that critical levels of these enzymes would be preformed in spores since new enzyme proteins for these functions could not be synthesized unless precursors and energy were first made available for protein biosynthesis. On the other hand, a new activity which could be expected to require new proteins is the production of new cell wall material. Growth by deposition of apical vesicles is a well established mechanism for hyphae (Grove et al. 1970) as well as, for example, the germ tubes of *Mucor* and *Gilbertella* sporangiospores (Bartnicki-Garcia & Lippman, 1969; Bracker, 1971), *Aspergillus* conidia (Grove, 1972) and the primary rhizoid of *Blastocladiella* (Barstow & Lovett, 1974). Any step in the organization of the golgi membranes and synthesis of the secretory vesicle contents could be a critical site. The identification of essential proteins for germination that may also participate in normal growth functions, as suggested above, could

---

Fig. 1 Autoradiograms of proteins synthesized by germinating zoospores of *Blastocladiella emersonii*. The cells were labeled with a ^{14}C-amino acid mixture in the presence (left series) and absence (right series) of 5 μg/ml Actinomycin D; the labeled cellular proteins were then isolated and separated by gel electrophoresis. The pulse intervals were: 15--25 min (lane 1), 25--35 min (lane 2), 35--45 min (lane 3), 45--55 min (lane 4), and 55--65 min (lane 5). Reproduced by permission of P. M. Silverman, M. M. Huh & L. Sun, Dev. Biol. 40, 59--70 (1974).

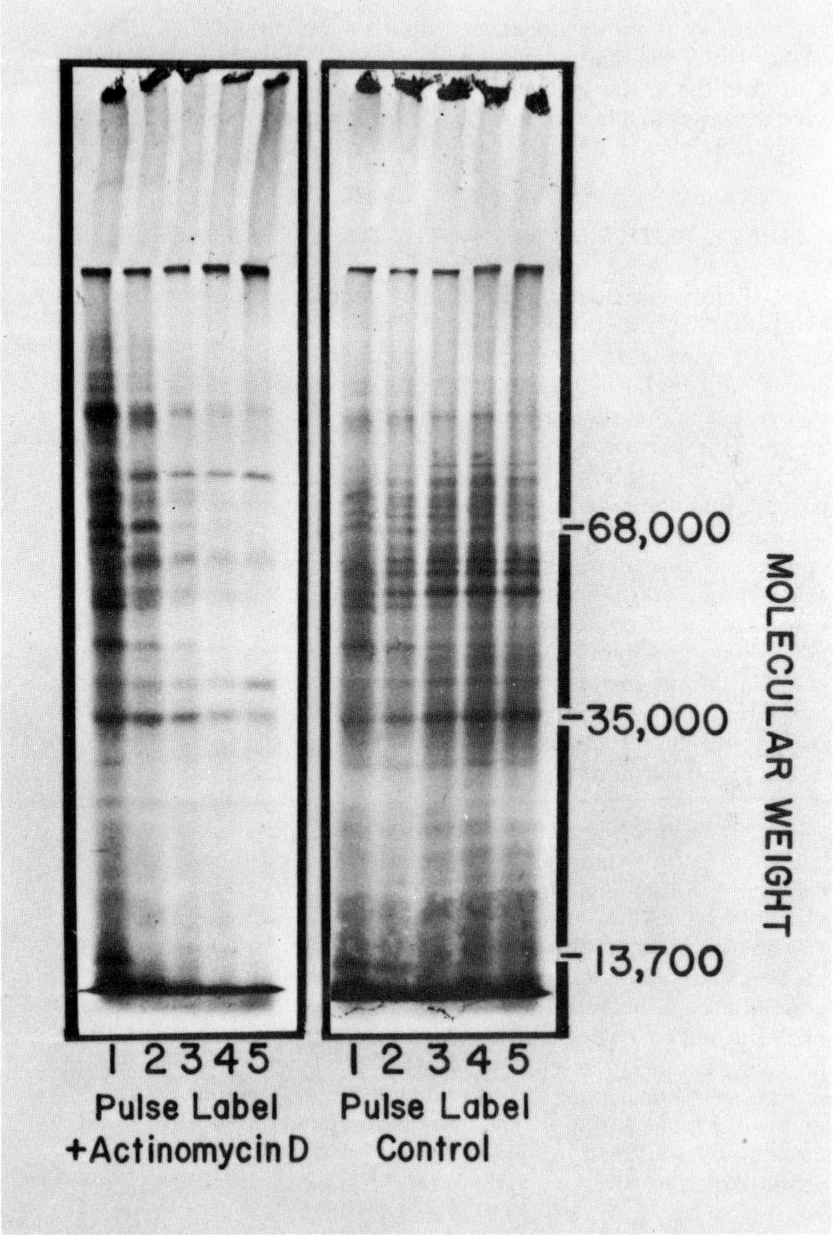

-68,000
-35,000
-13,700
MOLECULAR WEIGHT
1 2 3 4 5
Pulse Label
+Actinomycin D
1 2 3 4 5
Pulse Label
Control

be analyzed using temperature sensitive mutants. The utility of this technique has been elegantly demonstrated by Hartwell (1974) in the analysis of essential cell cycle functions of *Saccharomyces*, but has been used in few spore germination experiments (e.g., *Neurospora;* Inoue & Ishikawa, 1970).

CHARACTERISTICS OF THE SYSTEM FOR PROTEIN BIOSYNTHESIS

Before considering how the system for protein synthesis exists in dormant spores, and functions during germination, a brief review of the known components and the steps involved seems appropriate. A list of the known requirements is given in Table 2 and even the most casual survey should immediately suggest the complexities we may expect to encounter in attempting to identify regulatory mechanisms; the synthesis of a single polypeptide chain requires the integrated function of a system with more than 50 different small molecules, large molecules, and macromolecular aggregates. The principal steps in polypeptide synthesis are outlined diagramatically in Figure 3. Some details are still uncertain, but it is established that ribosomes participate in a cyclic fashion; this cycle starts with the formation of an initiation complex by the free subunits, and ends when the subunits are released at the 3'-end of the messenger RNA. Depending upon the conditions, the released subunits may reenter the synthetic cycle as "native" subunits or complex

---

Fig. 2 Ribosomal protein synthesis during germination of *Blastocladiella* zoospores. The spores were germinated in complete medium containing 0.5 μ Ci/ml of ^{3}H-leucine under standard conditions (Lovett, 1968). After 30 min (A), and 45 min (B) germination time the cells were collected and the ribosomes isolated and purified by centrifugation to remove non-ribosomal proteins (Adelman & Lovett, 1972); the ribosomal proteins were extracted with LiCl-urea and separated on 10% polyacrylamide-urea gels at pH 4.5, along with a small amount of marker ribosomal protein labeled with ^{14}C-leucine. The gels were frozen, and 0.1-mm slices were cut, solubilized and counted by standard procedures (S. S. Abe & J. S. Lovett, unpublished). ^{3}H cpm, solid line; ^{14}C cpm, dotted line.

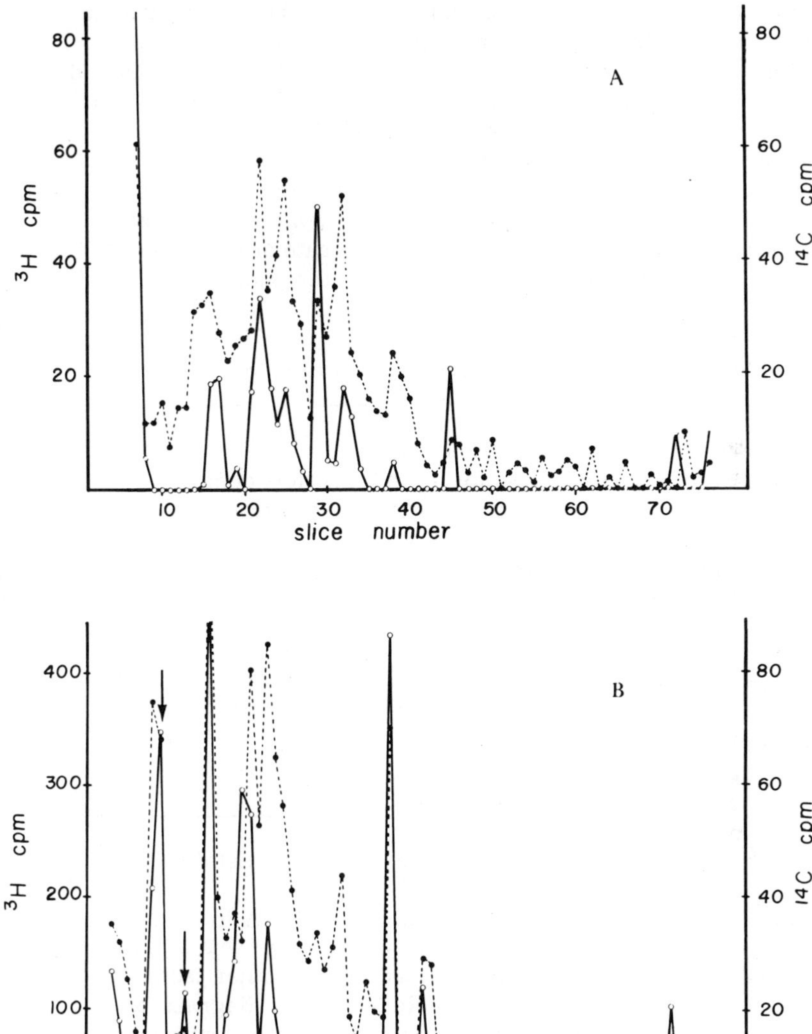

80
60
40
20
^{3}H cpm
A
80
60
40
20
14C cpm
10
20
30
40
50
60
70
slice number
400
300
200
100
^{3}H cpm
B
80
60
40
20
14C cpm
10
20
30
40
50
60
70
slice number

together to form inactive 80S monoribosomes. The concentration of free subunits in cells generally remains rather constant, but the ratio between the concentration of free 80S ribosomes and ribosomes in polysomes, varies with the rate of cellular protein synthesis (Hogan & Korner, 1968). Thus, if the rate of overall synthesis is slow, fewer subunits can enter initiation complexes per unit time, and more of the subunits enter the 80S ribosome pool as they are released from the mRNA templates. Increased synthetic rates on the other hand raise the rate of initiation and the needed subunits are released from the pool of inactive 80S ribosomes.

Table 2 Components Required for Protein Synthesis[a]

Components
40S ribosomal subunit (initiation sites)
60S ribosomal subunit (acceptor *a* and peptide donor *p* sites, peptidyltransferase)
Messenger RNA
Initiator aminoacyl-tRNA (methionyl-tRNA $^{met}_{F}$)
Amino acids (20)
Transfer RNAs (for 20 amino acids)
Aminoacyl-tRNA synthetases (for 20 amino acids)
Initiation factors (IF-1, IF-2, IF-3)
Elongation factors (Transferase I, Transferase II)
Release factors
ATP, GTP

[a]Refer to the legend of Figure 3 for additional details.

The normal initiation-specific aminoacyl-tRNA in eukaryotes is methionyl-tRNA $^{met}_{F}$ as it is in bacteria except that in eukaryotes it is used only in the unformylated form. Some other factors are still controversial but have been included in the diagram for completeness (e.g., the number of initiation factors and release factors required, and the existence of a specific 80S ribosome dissociation factor). The very complexity of the system suggests that many different sites of regulation could exist, and some of these will be discussed in later sections. From an experimental point of view it is equally obvious that such a multicomponent system will be difficult, if not impossible, to duplicate exactly outside the cell. Despite this, the analysis of individual steps does require the use of *in vitro* experiments that, however, need to be interpreted with considerable caution when extrapolated back to deduce the activities of living cells---in our case the fungal spore.

The details of eukaryotic protein synthesis have been derived largely from work with animal systems, but the conditions found necessary for *in vitro* protein synthesis with extracts from a variety of fungi have been found to be basically similar. Furthermore, most of the important components have been demonstrated to occur in fungi including: methionyl-tRNA $^{met}_{F}$ as the initiation-specific aminoacyl-tRNA in yeast (Takeishi et al. 1968) and *Neurospora* (Rho & DeBusk, 1971), transferases I and II in yeasts (Perani et al. 1971) and *Uromyces* (Yaniv & Staples, 1971), peptidyltransferase in yeast (Van der Zeijst et al. 1972), a ribosome dissociation factor in yeast (Petre, 1970), and messenger RNA with poly (A) regions in yeast (McLaughlin et al., 1973; Reed & Wintersberger, 1973). On this basis it will be assumed for this discussion that most of the components and mechanisms in fungi will differ in only minor detail from other eukaryotes.

THE BIOSYNTHETIC SYSTEM IN SPORES

The question of whether fungal spores can support protein synthesis has been of interest for some time and the spores of several species have been studied in this regard. In this section I should like to briefly review the available information and draw some specific conclusions concerning the status of the system and potential regulatory sites or mechanisms. One particular point I consider self-evident, and this is that all the proteins needed for polypeptide synthesis will be found in fungal spores in one form or another. The cells cannot produce a new biosynthetic system without at least a minimum number of preformed enzymes, ribosomes, and other protein factors.

If the assumption that all proteins are conserved is correct, the problem of regulation may be simplified to mechanisms for suppressing the activities of one or more protein functions, or controlling the availability of a non-protein component, such as the energy supply or aminoacyl-tRNA's. One might legitimately ask whether the evidence supports such an assumption, and

Fig. 3 The ribosome cycle. In eukaryotic cells the ribosome cycle involves the formation of an initiation complex between the "native" 40S ribosomal subunit, initiation factors (a complex of met-tRNA $_F^{met}$, initiation factor IF-2, and GTP, plus IF-1) and the AUG initiation codon at the 5'-end of the mRNA. The native 60S ribosomal subunit then joins the complex while the initiation factors are released to recycle (The IF-3 prior to the addition of the 60S subunit; IF-1 and IF-2 leave afterward with simultaneous hydrolysis of GTP and regeneration of IF-2-GTP). The steps in peptide bond formation require the sequential binding of the mRNA-specified aminoacyl-tRNAs at the acceptor (a) site via an aminoacyl-tRNA--Transferase I--GTP complex; peptide bond formation by the peptidyltransferase of the 60S subunit, and release of the deacylated tRNA from the peptidyl (p) site; transfer of the peptide-bearing tRNA (peptidyl-tRNA) from the (a) to the (p) site by the second elongation factor, translocase (transferase II), followed by the entry of a new aminoacyl-tRNA at the (a) site. These steps are repeated for each amino acid until a termination condon is reached (e.g., UAA). At this point one or more release factors enter and cause the release of the completed polypeptide, deacylated tRNA, and both of the ribosomal subunits. If an initiation factor IF-3 is available, it binds to the newly released 40S subunit converting it into a "native" 40S which can re-enter an initiation complex (a similar factor for the 60S has not been identified). In the absence of IF-3 the 40S subunit complexes with a free 60S subunit to form an inactive 80S monoribosome. The 60S subunit may also enter a new initiation complex if not diverted to form part of an inactive 80S ribosome. The 80S ribosomes may dissociate to yield "native" 40S and 60S subunits if the rate of protein synthesis increases. The exact mechanism for this conversion is unclear, but may involve either a specific dissociation factor and/or the availability of free IF-3 molecules to complex with transiently dissociated 40S subunits. This is a somewhat over-simplified description; for more details refer to Haselkorn & Rothman-Denes (1973).

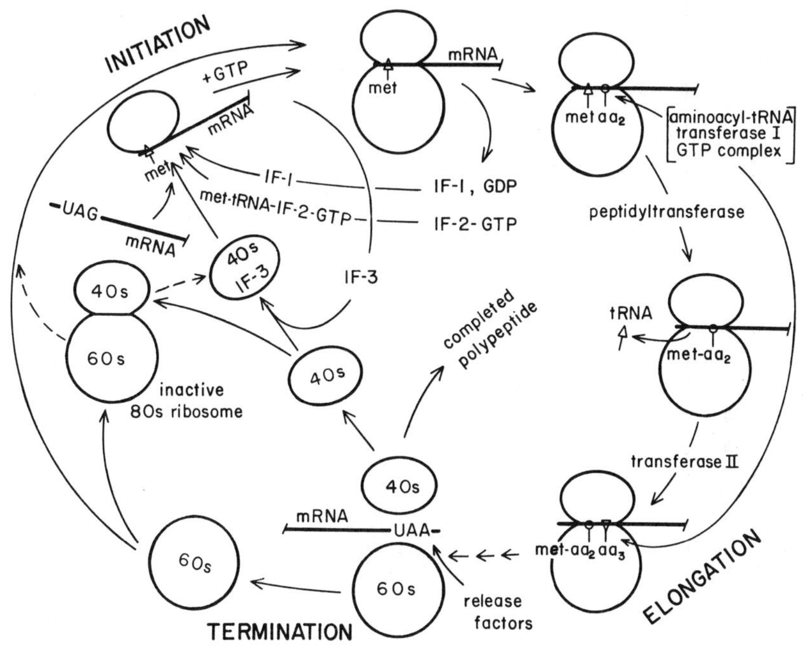

I think an examination of Table 3 suggests that it does. In each instance where a careful examination has been made, the appropriate protein factors have been found; only in the case of *Phytophthora* spores has a claim been made for the actual absence of intact 80S ribosomes (Leary et al., 1974) and this remains to be substantiated. In several fungi the ribosomes isolated from spores have been less active than those from germinated spores when tested for *in vitro* activity by amino acid incorporation assays. However, much if not all of this difference may be attributable to isolation artifacts. In most instances the integrity of the ribosomal RNA (rRNA) has not been assessed; in our experience identical extraction procedures can give rise to ribosomes with significantly different *in vitro* activities and degrees of rRNA degradation when cells of different stages or cell densities are used (Adelman & Lovett, 1974b). Examples of both cell density and culture stage effects on *in vitro* ribosome activity and the integrity of the rRNA are given in Fig 4. Since direct extraction

Table 3 Components for Protein Synthesis in Fungal Spores[a]

Organism	Ribosomes	Synthetases	Elongation factors	Amino acid pool	Transfer RNA	mRNA
Aspergillus oryzae (1,2,3)	+	+			+	
Botryodiplodia theobromae (4,5,6,7)	+	+ (17)	+		+ (20)	+
Blastocladiella emersonii (8,9,10)	+	+			+	+
Uromyces phaseoli (11,12,13)	+		+		+	+
Fusarium solani (14,15)	+					+
Peronospora tabacina (16)	+		+		+	+

Organism	Ribosomes	Synthetases	Elongation factors	Amino acid pool	Transfer RNA	mRNA
Neurospora crassa (17,18,19)	+				+	+
Rhizopus stolonifer (20,21)		+ (20)			+ (20)	
Allomyces arbuscula (22)						+
Erysiphe graminis (23)	+					
Schizophyllum commune (24)	+					
Phytophthora spp. (25)	(-)					
Saccharomyces cerevisiae (26)				+		

Organism	Ribosomes	Synthetases	Elongation factors	Amino acid pool	Transfer RNA	mRNA
Puccinia graminis (27)				+		
Rhizopus arrhizus (28)				+		

[a]The presence of (+) means the component(s) was found present in spores, a (-) means it was reported absent; numbers in parenthesis indicate the number of specific synthetases or tRNAs shown to occur in spores.

References

1. Horikoshi & Ikeda, 1968
2. Kimura et al., 1965
3. Horikoshi et al., 1969
4. Van Etten, 1968
5. Van Etten & Brambl, 1968, 1969
6. Van Etten et al., 1969
7. Brambl & Van Etten, 1970
8. Lovett, 1963, 1968
9. Schmoyer & Lovett, 1969

10. Silverman et al., 1974
11. Staples et al., 1966
12. Yaniv & Staples, 1969, 1971
13. Ramakrishnan & Staples, 1970
14. Rado & Cochrane, 1971
15. Cochrane et al., 1971
16. Holloman, 1971, 1973.
17. Henney & Storck, 1963, 1964
18. Mirkes, 1974
19. Bhagwat & Mahadevan, 1970
20. Van Etten et al., 1969
21. Merlo et al., 1972
22. Burke et al., 1972
23. Leary & Ellingboe 1971
24. Leary et al., 1969
25. Leary et al., 1974
26. Rousseau & Halvorson, 1973a
27. Reisner, 1967
28. Weber, 1966

of the cellular RNA from the same cells yields undegraded RNA, one can only conclude that the cells at some stages (particularly stationary phase and during spore differentiation) contain higher levels of nucleases (or more readily released ones) which make it more difficult to isolate intact ribosomes. A similar conclusion was drawn by Horikoshi & Ikeda (1969). The main point is that such artifacts need to be eliminated before one can conclude that spore ribosomes are defective.

Rothschild et al. (1967) found no differences in the number of ribosomal structural proteins in a comparison between spore and mycelial ribosomes of *Neurospora*. The evidence for other proteins such as transferases and aminoacyl-tRNA synthetases is not at all extensive, but both have been found (Table 3). Elongation factors (Transferases I and II) have been reported to occur in spore extracts of *Uromyces*, *Botryodiplodia*, and *Peronospora*; and of course the demonstration of cell-free amino acid incorporation in extracts from several spores also indicates their presence. The presence of aminoacyl-tRNA synthetases is also indicated since most cell-free assays utilize free radioactive amino acids and deacylated tRNA. The spores of only three species have been examined in detail for their complement of activating enzymes. Van Etten & Brambl (1968) found charging activity for the tRNAs of all but three amino acids in spores of *Botryodiplodia*, and Merlo et al. (1972) for all 20 amino acids in *Rhizopus*. Interestingly enough the spore enzymes and the tRNAs of *Rhizopus* were more active than those from germinated spores. Horikoshi et al. (1969) also found no differences in the charging of 13 tRNAs between conidial and mycelial extracts of *Aspergillus oryzae*. Aminoacyl-tRNA synthetase activity has been found in *Blastocladiella* zoospores as well, but individual tRNAs were not assayed (Schmoyer & Lovett, 1969).

Neither initiation nor termination factors have been studied in fungal spores, but the evidence just discussed suggests that they too may be found. In any case there is no published data that seriously conflicts with the assumption that all essential proteins are present in spores. If this is so, can we also draw a similar conclusion concerning the presence or absence of the non-protein components? Unfortunately, the amount of information available on the pools of amino acids, aminoacyl-tRNAs and nucleotide triphosphates of dormant spores is quite unsatisfactory. The amino acid pools have not been measured

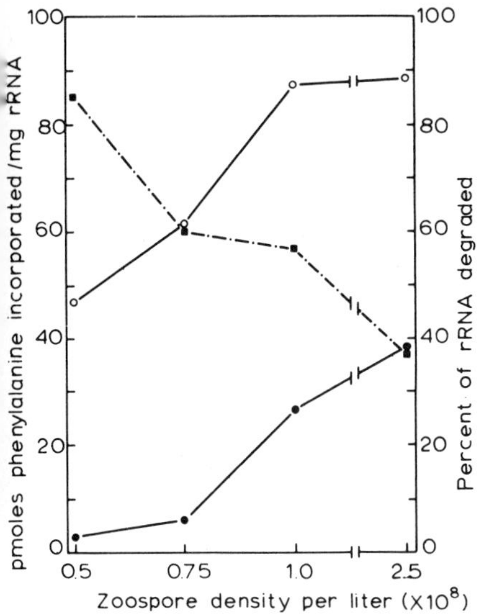

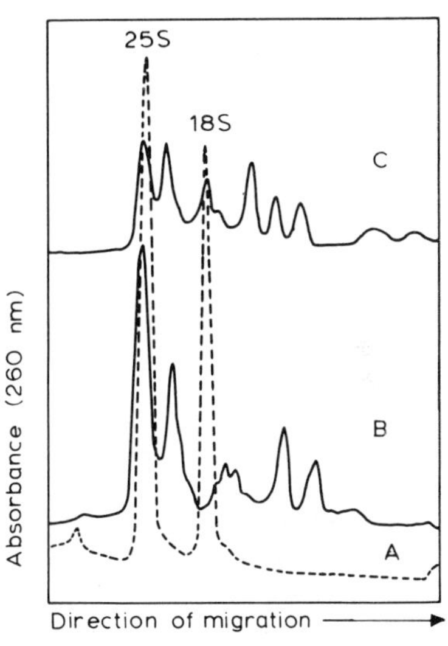

Fig. 4 The effects of culture density and stage on the structural and functional integrity of ribosomes isolated from *Blastocladiella*. (Left). Ribosomes were isolated at 15--15.5 h from cultures grown at different cell densities. Their functional capacities were tested by poly (U)-^{3}H-phenylalanine incorporation assays (■--■), and the extent of RNA degradation by gel electrophoresis (0——0, percent degradation of 18S RNA; ●—●, percent degradation of 25S RNA). (Right). Polyacrylamide gels were scanned at 260 nm after electrophoresis of 25 μg RNA obtained from: (A) zoospores by direct RNA extraction; (B) ribosomes isolated by bulk procedures from late log-phase cells at 15 h; (C) ribosomes isolated from cells during zoospore differentiation at 19 h. Reproduced by permission of T. G. Adelman & J. S. Lovett, Biochim. Biophys. Acta 349, 240--249 (1974).

in many spores, and then usually in fungi that have not been studied for other parts of the biosynthetic system (Table 3). In the case of *Rhizopus arrhizus*, Weber (1966) interpreted a very low proline pool, and the ability of added proline to induce germination, to mean that a deficiency of proline and a reduced capacity to synthesize it were the cause of dormancy. However, it was not shown that the proline was actually deficient for protein synthesis although present at one-tenth the level of other amino acids.

In the case of transfer RNA the spores of only three species have been examined in detail, i.e., *Botrodiplodia theobromae* (Van Etten. et al., 1969), *Rhizopus stolonifer* (Merlo et al., 1972), and *Aspergillus oryzae* (Horikoshi et al., 1969) (Table 3). No individual tRNAs were found to be absent, but some differences were observed in the iso-accepting tRNAs for a few amino acids when spore and vegetative tRNA preparations were compared. The significance of the change is not known, although a shift in the ratios of iso-accepting tRNA molecules has been proposed as a mechanism for the recognition of specific messenger RNA molecules during differentiation (Sueoka & Kano-Sueoka, 1970). A potential mechanism for tRNA alteration through methylation has been shown for *Neurospora crassa* (Wong et al., 1971) where some evidence for differences in methylation activity were observed between spore and mycelial extracts. However, shifts in iso-accepting tRNAs have also been shown to result from different culture conditions in *Neurospora* (Nazario, 1972), and the difference in spores and mycelia could represent simply different nutritional states.

The extent to which individual tRNA species are charged with amino acids *in vivo* has not been reported for any type of spore. Since the aminoacyl-tRNA synthetases seem to be present in spores, a deficiency of amino acids should be reflected in low levels of tRNA charging. In *Blastocladiella* just the reverse was found (Schmoyer & Lovett, 1969); the ratio of charged to uncharged tRNA in spores was much higher than found with tRNA from growth-phase plants. The result though interesting is not conclusive because individual tRNAs were not assayed in these experiments and a lack of charging for a single amino acid would not have been detected. A deficiency in the initiation-specific met-tRNA $_F^{met}$ in particular could be expected to cause an effective block in synthesis. Increased levels of uncharged tRNA have been proposed as inhibitors of initiation in animal cells (Kyner et al., 1973). However, experiments with a temperature sensitive isoleucyl-tRNA synthetase mutant of *Saccharomyces cerevisiae*, though not definitive, suggest this mechanism may not occur in fungi (McLaughlin et al., 1969). On the basis of this limited evidence a possible regulatory role for tRNA in spores cannot be ruled out with certainty, but, on the other hand, there is no direct experimental evidence which supports such a function.

With the exception of *Neurospora crassa* nothing is known about the levels of ATP or GTP in spores. Nevertheless, adequate pools or regeneration systems would be essential in order to maintain synthesis and these pools cannot be ignored as potentially limiting factors. The ATP concentration in *Neurospora crassa* conidia is about half that achieved in the exponential phase, and does not increase until after 6 h (Slayman, 1973). The spores of two other fungi have been shown to convert detectable amounts of radioactive nucleosides to intracellular triphosphates immediately after being placed in the germination medium, i.e., uridine to UTP in *Botryodiplodia* (Brambl & Van Etten, 1970) and adenosine to ATP in *Blastocladiella* (Lovett & Wilt, unpublished).

The last major requirement for protein synthesis is the mRNA to serve as templates for polypeptide assembly. There is some evidence for the presence of mRNA in spores of several fungi (Table 3); however, all of the evidence is largely circumstantial, as it is based on the failure of RNA inhibitors (actinomycin D or ethidium bromide) to block either germination or protein synthesis (*Allomyces*, *Blastocladiella*, *Peronospora*, *Botrydiplodia*, and *Neurospora*), competitive hybridization (*Neurospora*), or the stimulation of amino acid incorporation when RNA is added to a pre incubated cell-free system from *Escherichia coli* (*Uromyces*). All of these reports will require independent verification by more direct methods, but taken together are certainly suggestive that spores do contain mRNA.

If mRNA occurs in spores it could exist either in a stored and inactive form or as part of polyribosomal aggregates. The latter would seem unlikely to occur in spores which are not making protein; but, because polysomes are normally only found when polypeptides are being formed at a measurable rate, their existence in spores would indicate both the presence of mRNA and at least some protein synthesis. Unfortunately the results obtained by this straightforward approach are clouded by technical difficulties inherent in the isolation of polysomes (Marcus et al., 1967) and the usual pretreatment of the spores. The severe mechanical operations necessary to homogenize most types of spores can cause some shearing of polysomes, but the primary problem is exposure of the polysomes to the cellular nucleases that are also released when the cells are broken. Nevertheless at least some detectable material with the characteristics of polysomes has been found after density gradient centrifugation of spore extracts from *Botryodiplodia* (Brambl & Van Etten, 1970), *Fusarium* (Cochrane et al., 1971), *Uromyces* (Staples et al., 1968), *Schizophyllum* (Leary et al., 1969), and *Erysiphe* (Leary & Ellingboe, 1971). On the face of it, this might be considered good evidence for polysomes in spores, except that in each case the spores had been washed <u>before</u> the polysomes were extracted. Such washing could reasonably be argued to cause at least partial activation of the spore synthetic systems. When Cochrane et al. (1971) tested this possibility they found that unwashed spores of *Fusarium* incorporated ^{14}C-leucine into protein at ca. one-third the rate of washed but ungerminated spores. Even more dramatic evidence for the activation of protein synthesis as a result of washing spores has been reported by Mirkes (1974). He found that when *Neurospora* conidia were collected air dry they contained less than three percent of the ribosomes in polysomes, while conidia exposed to water for only 10--15 min contained an average of 30% of their ribosomes in polysomes (see Fig 6). While some spores may have a low level of synthesis at all times, as suggested by the spores of *Fusarium* which are produced in a sticky matrix, the results with *Neurospora* show that non-activated dry spores are not likely to contain significant levels of polysomes as a form of mRNA storage.

Polysomes do not occur in the swimming zoospores of *Blastocladiella emersonii* (Schmoyer & Lovett, 1969; Leaver & Lovett, 1974) and we have recently attempted to verify the existence of the postulated stored mRNA (Lovett, 1968), and to determine its location in the zoospore. In these spores the ribosomes occur in a distinct, extranuclear, cytoplasmic compartment called the nuclear cap, which can be readily isolated (Lovett, 1963).

The probable locations for the stored mRNA are the nucleus, the extra cap cytoplasm, or inside the nuclear cap where the mRNA molecules could exist either as free particles or in complexes with the 80S monoribosomes.

Our experimental approach was based on the general occurrence of polyadenylic acid sequences (polyA) in the polysomal mRNA and nuclear RNA of many eukaryotic organisms (McLaughlin et al., 1973; Blobel, 1973). Although some cytoplasmic mRNAs do not carry a poly(A) segment at their 3'-end (e.g., histone mRNA) the majority do and the poly(A) segment itself provides a convenient means for both the separation of the mRNA from the bulk cellular RNA (rRNA, tRNA) and for estimating the amount of messenger present. This work is not complete but the poly(A) and poly(A)-associated RNA content of total zoospore RNA, and the nuclear cap subfraction have been determined. The poly(A) segments themselves comprise ca. 0.05--0.1% of the total zoospore RNA, and the poly(A)-associated RNA approximately 3.25--4.0% of the total. When nuclear caps were isolated, it was further found that they contain 80% of the total zoospore poly(A)-RNA; the ramaining 20% has not been localized and could be in either the nuclear or extra-cap cytoplasmic compartment. To establish whether the poly(A)-RNA in the caps was associated with the ribosomes, or existed as free molecules, we lysed the caps by the procedure of Adelman & Lovett (1974a) and centrifuged the lysates through linear 10--30% sucrose gradients (Fig 5). When the RNA was isolated from different parts of the gradients and analyzed for poly(A)-RNA, it was apparent that the poly(A)-containing RNA was not specifically associated with the ribosomes, but occurred throughout the gradient. Aliquots from the gradients were also fixed with glutaraldehyde and centrifuged in cesium chloride and their sedimentation characteristics indicated a ribonuclear protein composition and a range of densities.

We are encouraged to consider the poly(A)-containing RNA as representing the mRNA in *Blastocladiella* because poly(A) has been reported to occur in the mRNA isolated from polysomes in yeast (McLaughlin et al., 1973; Reed & Wintersberger, 1973). Dr. Van Etten will, I am sure, have more to say about the significance of poly(A) in fungi. If the assumption that poly(A)-containing RNA is mRNA is correct, then these results provide one piece of direct evidence for the existence of stored mRNA in the zoospores. The lack of a direct association with the cap ribosomes is consistent with an earlier observation that salt-washed cap ribosomes support little amino acid incorporation without added synthetic message (Schmoyer & Lovett, 1969). It remains to be demonstrated that the nuclear cap poly(A)-RNA actually has an mRNA function, and provides templates for polypeptide synthesis at germination; our ability to separate this RNA from isolated nuclear caps suggests that experiments to assess this should be feasible.

With this additional evidence for stored mRNA I am encouraged to expand my earlier conclusions concerning the presence of all needed protein factors in spores, and propose that most fungal spores probably carry a virtually complete and potentially functional system for protein synthesis. Certainly the evidence now available cannot bolster a very strong case for a contrary position, and it is actually a rather conservative hypothesis. It does have the virtue of focusing our attention on some specific and testable predictions concerning regulation. If we eliminate controls that operate by a complete omission of any factors, we should expect to find positive mechanisms to keep synthesis either turned off or turned down in the dormant or semidormant spore. It is not difficult to propose a variety of positive controls, at least some of which have already been discussed by Van Etten (1969) and Sussman & Douthit (1973). The most obvious controls would appear to be: (1) separation of one or more factors from the rest of the system by physical or functional compartmentalization; (2) reversible inhibition of one or more of the many reactions or components; or (3) alteration of one or more components in a reversible way so that they cannot function. There certainly is no reason to assume that more than one type of control could not exist in the same spore.

With the exception of the ribosomal nuclear caps in the *Blastocladiales* (or ribosomal aggregates in some Chytrids) there is little evidence for physical compartmentalization in spores. Even in the case of *Blastocladiella* we have not found any of the required factors (e.g., ribosomes, aminoacyl-tRNA, aminoacyl-tRNA synthetases, mRNA) to be excluded from the cap, and the functional significance of the compartmentalization remains unknown.

Autoinhibition has already been discussed in detail by Staples and Macko (Chapter 2). Evidence for direct effects of self-inhibitors on spore protein synthesis has come from work with the cellular slime mold *Dictyostelium discoideum* (Bacon &

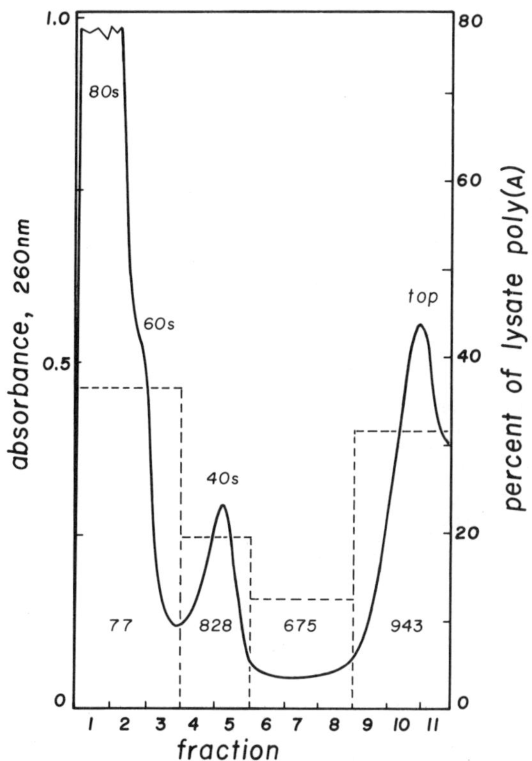

Fig. 5 The location of poly(A)-containing RNA in nuclear cap lysates from *Blastocladiella* zoospores. *Blastocladiella* zoospore nuclear caps were isolated, lysed (Adelman & Lovett, 1974a), and centrifuged in 30 ml, 10--35% sucrose gradients (with a 2 ml 60% cushion) for 19.5 h at 22,000 rpm (Spinco SW 25.1 rotor). The regions indicated by the histogram bars were pooled, and the RNA extracted and assayed for the amount of poly(A) by the procedure of Wilt (1973). The poly(A) content is expressed as the percent of the total on the gradient found in each fraction. The numbers under the bars indicate the relative specific activities of the fraction, expressed as the cpm ^{3}H-poly(U) hybridized with poly(A) per A_{260} of RNA. J. S. Lovett & F. Wilt, manuscript in preparation.

Sussman, 1973; Bacon et al., 1973), and the fungi *Glomerella cingulata* (Lingappa et al., 1973), and *Blastocladiella emersonii* (Schmoyer & Lovett, 1969; Adelman & Lovett, 1974a). With both *Dictyostelium* and *Glomerella* the original effect was observed as a cell density-dependent inhibition of spore germination. The inhibitor in *D. discoideum* has been indentified as $N^1 N^1$-dimethylguanosine (Bacon et al., 1973) and reported to have a primary effect on protein synthesis (Bacon & Sussman, 1973). In *Glomerella* the inhibitor also inhibits amino acid incorporation (Lingappa et al., 1973) but has not been identified. The inhibitor of *Blastocladiella* was originally detected by its effect on *in vitro* protein synthesis when spore extracts were mixed with incorporation systems from growing cells (Schmoyer & Lovett, 1969), while Truesdell & Cantino (1971) independently reported the presence of a germination inhibitor in media used to wash zoospores. Material with characteristics of the zoospore ribosome-inhibitor has also been isolated from the medium following zoospore encystment (Gong & Lovett, unpublished). This inhibitor has not been identified but binds reversibly to ribosomes and has fractionation characteristics which indicate a possible nucleoside or nucleotide, and it may thus be analogous to the *D. discoideum* inhibitor.

The specific mode of action is not known for any of the inhibitors; the *Dictyostelium* compound apparently inhibits peptide bond formation (See, Sussman & Douthit, 1973), while that of *Blastocladiella* inhibits both aminoacyl-tRNA binding and peptide bond formation; this suggests that both compounds could be translation (elongation) inhibitors.

The reversible modification of some component remains a possible mechanism of control, such as the iso-accepting tRNAs for specific amino acids discussed earlier. Extra proteins have also been reported to occur on the ribosomes of *Drosophila* larvae when compared with ribosomes from adult flies (Lambertsson et al., 1970) suggesting possible developmental modifications by addition or substitution of specific proteins. There is evidence for a turnover of ribosomal proteins during zoospore differentiation in *Blastocladiella* (Adelman & Lovett, 1972). But, despite this activity, no unique proteins have been detected in the zoospore ribosomes; the spore ribosomes can also be activated for synthesis *in vitro* by high salt washes to remove the inhibitor which would remove non-structural proteins but should not alter the basic ribosome protein composition. Differential phosphorylation of ribosomal proteins by protein kinases and cyclic AMP has also been proposed as a mechanism for altering ribosome specificity (Kabat, 1970). The *in vitro* experiments of Eil & Wool (1973) to test this hypothesis with rat liver ribosomes yielded negative results, but the authors could not exclude the possibility that they had either used the wrong conditions for specific phosphorylation or tested the wrong ribosome functions. The possibility thus remains attractive for reversible modifications of ribosomal or other proteins through phosphorylation and phosphate removal by phosphatases.

Our demonstration of apparently free, poly(A)-containing ribonuclear protein particles in the nuclear cap of *Blastocladiella* zoospores suggests another potential site for regulation. There is by now considerable evidence that most if not all eukaryotic mRNA molecules are complexed with protein both in the nucleus and in the cytoplasm; moreover, only a few specific proteins seem to be associated with mRNAs when isolated under high salt conditions (Bryan & Hayashi, 1973; Morel et al., 1973), and one of these may be specific for the poly(A) segment found on most polysomal mRNAs of eukaryotic cells (Blobel, 1973). Other pertinent observations with animal cells are that mRNA may exist in the cytoplasm free of ribosomes and rapidly enter polysomes when cultures are enriched to increase the rate of protein synthesis (Lee et al., 1971; Christman, 1973). And, finally, that the proteins associated with the free cytoplasmic mRNA are different from those attached to polysomal mRNA (Sander et al., 1973). These results point toward an important regulatory mechanism(s) which may entail the association of specific regulatory proteins with the 5'-end of the messenger molecules, thereby determining if and when they may enter initiation complexes and be translated. This subject has been discussed in some detail by Scherrer (1974).

The concept of mRNA regulation by specific proteins seems to me to present an extremely interesting and potentially useful approach to the control of protein synthesis in fungal spores. A specific regulatory mRNA--protein complex could provide an explanation for the way some if not all spores can store mRNA in an inactive, yet rapidly mobilizable form. Regulation at the point of initiation would seem to be a very effective mechanism, and such regulation has been proposed as the most efficient type on theoretical grounds (Vassart et al., 1971). In addition to the attractiveness of its simplicity, a regulation of the initiation events seems to fit the known facts concerning spores better than other mechanisms. In particular, it is consistent with the existence of most or all of the ribosomes as free 80S monomers in spores. This would not otherwise be expected unless no mRNA at all were available. If mRNAs were present and available for initiation, some polysomes would be expected to form even with low levels of ATP, GTP, and amino acids. However, if aminoacyl-tRNAs were severely restricted, and constituted the primary limitation on synthesis, the probable result would be polysome turnover by the premature release of ribosomes and incomplete polypeptides (Brunschede & Bremer, 1971) which would be an energetically wasteful process. Even in severe step-down conditions (e.g., the sudden removal of most or all medium amino acids) we have found no more than a 50% decrease in the level of polysomes in *Blastocladiella* (Lovett, unpublished). Alternatively, if synthesis in spores were restricted by the inhibition of translation or termination alone, polysomes would tend to accumulate with little or no net polypeptide formation, as occurs in cells treated with cycloheximide (Hartwell et al., 1970).

While none of these arguments alone is conclusive, taken together they may be interpreted to mean that a more direct and positive control should be expected if the spores are to achieve efficient regulation. To summarize my conclusions concerning the spore system, it seems logical to propose that dormant fungal spores carry a virtually complete system for protein synthesis, including 80S ribosomes (with or without inhibitors or a mechanism to prevent their turnover), aminoacyl-tRNA synthetases, the initiation, elongation and termination proteins, tRNA, and mRNA in an inactive form. Although ATP, GTP and amino acids may be present in limited amounts these seem unlikely to serve as primary controls.

In discussing spores I have stressed the regulation of mRNA availability for initiation as a probable control point, but it should be obvious that the initiation factor proteins themselves could be targets for regulation, as could the equally important initiation-specific tRNA $_F^{met}$. Indeed, effective repression or modulation of protein synthesis could involve all three since all are required for formation of the critical initiation complex. Whether this hypothesis will hold up as more spores are studied, and we learn more of the necessary details, remains to be seen.

Transient, exogenously dormant spores may be regulated less stringently than endogenously dormant spores, but should still be readily accommodated without difficulty by such a mechanism. The above suggestions will have served a useful purpose if they do no more than stimulate a careful examination of initiation events in spores, a sorely neglected area in all of the work published up to this time.

ACTIVATION OF PROTEIN SYNTHESIS DURING SPORE GERMINATION

The discussion of protein synthesis during spore germination will of necessity be both brief and somewhat speculative. There is some information available concerning the appearance of synthetic activity as spores germinate, but very little is known about the mechanisms which may regulate how much synthesis occurs or the types of proteins produced. The conclusions drawn in the preceding section, have, however, provided a framework for some additional speculations on regulation during germination.

It is not yet possible to contribute any very useful suggestions concerning the initial events that trigger the start of protein synthesis in germinating spores. In both *D. discoideum* (Bacon et al., 1973) and *Glomerella* (Lingappa et al., 1973) the beginning of protein synthesis is correlated with the release of a germination inhibitor, which in *Dictyostelium* can be induced by heat shock, but this does not mean that the inhibitor release is in itself the primary event. In *Blastocladiella* a similar release may be involved, but the start of synthesis is also correlated with a complex set of structural changes associated with zoospore encystment (Lovett, 1968; Truesdell & Cantino, 1971; Soll & Sonneborn, 1971b) which can be induced by manipulating the ionic conditions (Suberkropp & Cantino, 1972; Soll & Sonneborn, 1972). It does seem probable that protein synthesis *per se* is not a primary event in germination despite its essential role for completion of the process. Active protein synthesis would depend upon a *continued* supply of ATP, GTP and amino acids, which could in turn only be supplied by biosynthesis or turnover of preexisting materials. This type of requirement seems to be reflected by the early and rapid utilization of carbohydrates such as trehalose by *Neurospora* (Lingappa & Sussman, 1959) and *Phycomyces* (Rudolph & Ochsen, 1969), or glycogen by *Blastocladiella* (Suberkropp & Cantino,

1972), and the early appearance of metabolic enzyme activity (Cochrane, 1966; Gottlieb, (Chapter 4).

The process of protein synthesis activation during germination has been well documented for only a few fungi, e.g., *Uromyces* (Yaniv & Staples, 1969), *Botryodiplodia* (Brambl & Van Etten, 1970), *Fusarium* (Cochrane et al., 1971), *Blastocladiella* (Lovett, 1968; Soll & Sonneborn, 1971a,b; Leaver & Lovett, 1974; Silverman et al., 1974), *Peronospora* (Holloman, 1971, 1973), and *Neurospora* (Mirkes, 1974). Even in these organisms many of the important details which would be needed to understand the regulatory mechanisms are still lacking. To simplify the discussion I will concentrate on the simpler systems represented by the germination of conidia and zoospores produced by saprophytic fungi. Obligate parasites seem in many ways to represent a special case and will be taken up briefly after discussing these less complicated systems.

As illustrated in Table 1 the amount of protein per cell was found to increase fairly rapidly during spore germination in some species, but in many fungi little or no net change was detected for an hour or even longer. Yet, when the spores of many of the same fungi were exposed to labeled amino acids they actively incorporated radioactivity into hot trichloroacetic acid-insoluble material soon after exposure to the germination medium. It should be stressed, nevertheless, that whole-cell amino acid incorporation experiments, in the absence of data on internal pool specific activities, cannot measure the amount of protein being made. Actual physical pool compartmentalization has now been shown to occur in both *Neurospora* (Subramanian et al., 1973; Brooks & DeBusk, 1973) and *Candida* (Wiemkin & Nurse, 1973; Nurse & Wiemkin, 1974). This phenomenon will further complicate the interpretation of incorporation experiments because the average internal precursor specific activity may differ significantly from the actual pool used for protein synthesis (Subramanian et al., 1973). Our near total ignorance of the types of protein produced during these early germination stages has been alluded to earlier. Certainly until we know what is produced it will be very difficult to interpret the function of the new synthesis in the germination process. The inhibitor experiments which have indicated the requirements for synthesis, and provided evidence for pre formed mRNA, have generally not revealed either how much new protein is required, nor often with any precision the time during which it is needed.

A second approach used to evaluate the activation of protein biosynthesis has been to extract and measure the percentage of the cellular ribosomes in polysomes during spore germination. The difficulties inherent in this procedure have already been discussed, but in each organism so far examined at least some polysomes have been found present at the same time that the first amino acid incorporation was detectable. It is uncertain yet whether the generally low yields of polysomes obtained throughout the germination of *Botryodiplodia* (Brambl & Van Etten, 1970) and *Fusarium* (Cochrane et al., 1971) actually reflect low rates of protein synthesis in these fungi, or destruction of polysomes as a result of breakdown during the isolation procedures. The latter seems probable, however, since the polysome levels remained low even after several hours of germination when the rates of synthesis should have become significant.

The experiments of Mirkes (1974) using *Neurospora crassa* conidia, and ours with *Blastocladiella emersonii* zoospores (Leaver & Lovett, 1974) have provided a clear picture of the process in these rapidly growing fungi. Figure 6 illustrates the rapidity with which polysomes form in the macroconidia of *Neurospora* after they are placed in growth medium. The polysome profiles obtained from *Blastocladiella* before and during zoospore germination, and the kinetics of their formation, are also shown in Fig. 7 and 8. The rates of polysome formation are remarkably similar in these two organisms; in each case less than 20 min is required to reach a state where at least 50% of the ribosomes have been shifted into the polyribosome fraction. Since the results with these fungi are quite similar, I will confine my description of events primarily to *Blastocladiella* with which I am familiar, and for which more complete information is available. Cycloheximide at a concentration of 5 μg/ml effectively inhibits both protein synthesis and germ tube formation in *Blastocladiella* (Lovett, 1968; Soll & Sonneborn, 1971b), but it does not completely block the entry of some ribosomes into polysomes (Fig. 8). The small percentage that do appear presumably represent an accumulation of ribosomes on mRNA molecules with little or no polypeptide formation (Hartwell et al., 1970). In sharp contrast to the result with cycloheximide, the addition of 20 μg/ml of Actinomycin D at zero time has virtually no effect on the initial rate or yield of

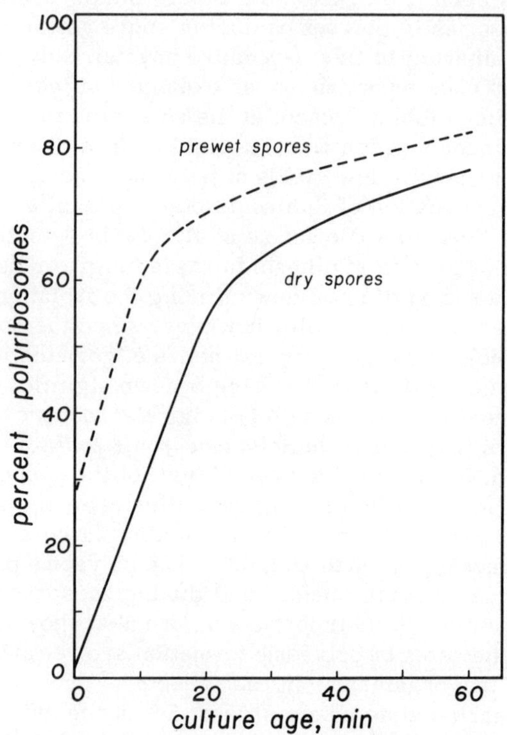

Fig. 6 The kinetics of polysome formation during germination of *Neurospora* conidia. The percent of the total cellular ribosomes in polysomes has been plotted as a function of germination time for conidia either harvested dry (dry spores) or by washing (wet spores). Redrawn by permission of P. E. Mirkes, J. Bacteriol. 117, 196--202 (1974).

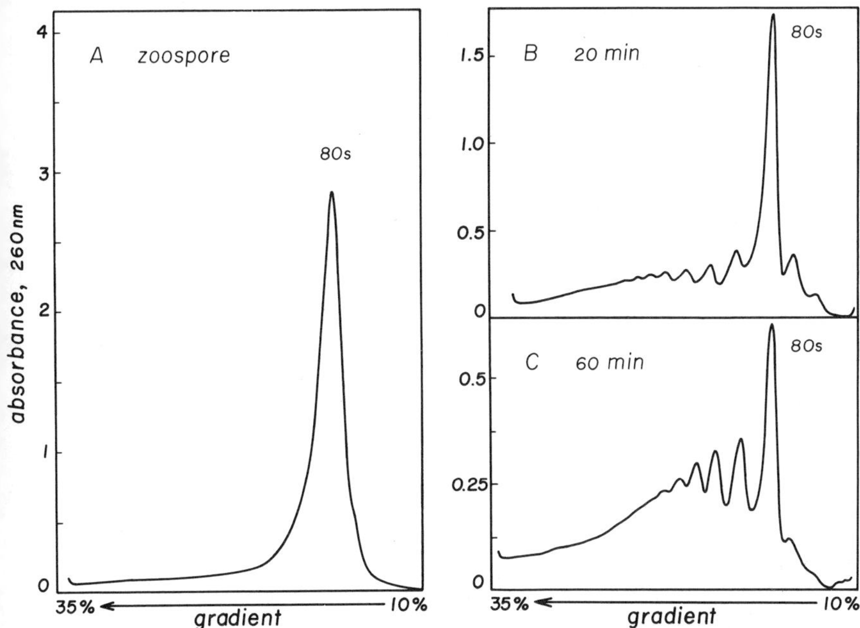

Fig. 7 Changes in ribosome distribution during encystment and germination of *Blastocladiella* zoospores. The polysome regions of the 10--35% exponential sucrose gradients contained (A) 0%, (B) 48.5%, and (C) 70.4% of the total ribosomes, respectively. Reproduced by permission of C. J. Leaver & J. S. Lovett, Cell Diff. 3, 165--192 (1974).

polysomes by 20 minutes. Actinomycin D had also been shown earlier to have no effect on the rate of amino acid incorporation through ca 40 minutes if added 5 minutes before each radioactive pulse (Lovett, 1968). Silverman et al (1974) have shown this antibiotic to cause a progressively severe inhibition of incorporation if present throughout germination (Fig. 1). The decrease in polysomes after 20 minute exposure to Actinomycin D suggests a deficiency of new mRNA and will be discussed below. Polysome assembly is considerably less rapid when spores are germinated in an

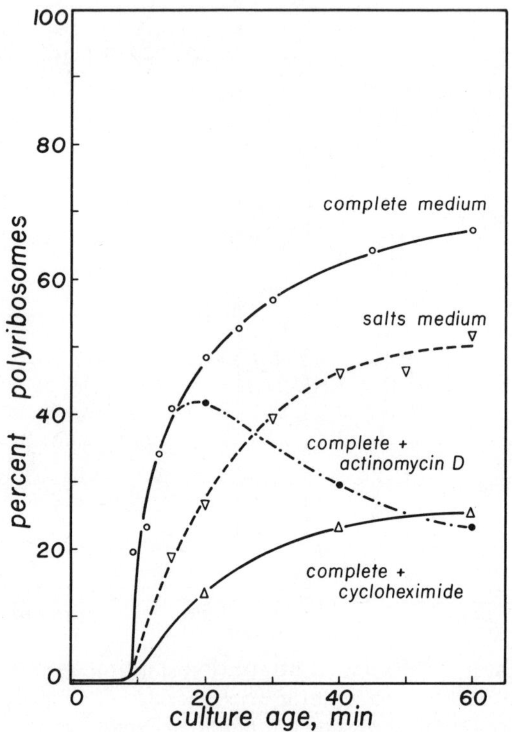

Fig. 8 The kinetics of polysome formation during germination of *Blastocladiella* zoospores; effects of medium composition and inhibitors. Polysome preparations were extracted and analyzed from cultures grown in complete medium, an inorganic salts medium (BSM; Lovett, 1968), complete medium plus 5 μg/ml cycloheximide, and complete medium plus 20 μg/ml Actinomycin D. Reproduced by permission of C. J. Leaver & J. S. Lovett, Cell Diff. 3, 165--192 (1974).

inorganic salts medium, and only reaches a level of 50% by 60 min of germination. These starved cells also fail to increase their RNA and protein content, but do develop a normal branching rhizoid by one hr. In this particular respect *Neurospora* conidia differ from *Blastocladiella* zoospores in that a carbon source is essential for a significant and continued increase in polysome content and germ tube production (Mirkes, 1974). *Fusarium* macroconidia incubated without ethanol failed to germinate but showed no difference in the polysome profiles when compared to germinating spores in medium with ethanol through the first four hours (Cochrane et al., 1971). Thus in both these cell types, unlike *Blastocladiella,* a utilizable carbon source is required for the spores to procede beyond the early activation of the system.

The rapidity with which polysomes form seems to provide a strong indication that initiation takes place using the stored mRNA and ribosomes of the zoospore nuclear cap in *Blastocladiella.* Several lines of evidence can be marshalled to support this argument. First, if new mRNA were required for the assembly of 50% of the ribosomes into polysomes in only 10 min, the synthesis of the mRNA, followed by its transport to the cytoplasm and entry into initiation events would have to be exceedingly rapid. Our measurements of ^{3}H-adenosine incorporation into the ATP pools and RNA have failed to detect new synthesis of any RNA until about 15 min at which time half of the initial polysome rise has already occurred (Lovett & Wilt, unpublished). Second, the failure of Actinomycin D to prevent polysome formation is also consistent with a mechanism that does not require new RNA synthesis. Thirdly, as shown in Fig. 9, the rapid increase in amino acid incorporation that accompanies polysome formation is delayed by 10 min and inhibited more than 88% by the specific initiation inhibitor D-MDMP (2-(4-methyl-2,6-dinitroanilino)-N-methylpropionamide; Weeks & Baxter, 1972) at 5×10^{-5} M (Leaver & Lovett, 1974). The same concentration of D-MDMP blocked germ tube formation in 65% of the cells and 1×10^{-4} M resulted in 92% inhibition. Finally, if one is to conclude that the new polysomes are organized without new mRNA synthesis there should be enough preformed mRNA to provide templates for the level of polysomes found in the cells. Now that the amount of putative mRNA in the zoospore nuclear cap is known the

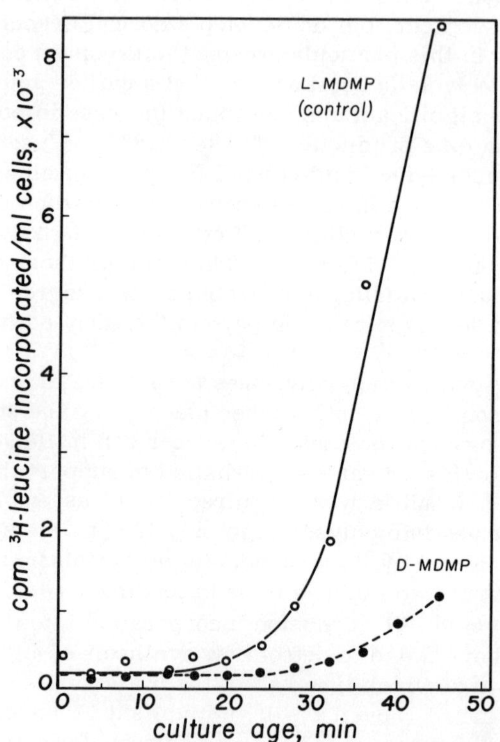

Fig. 9 The effect of a specific initiation inhibitor D-MDMP on ^{3}H-leucine incorporation during germination of *Blastocladiella* zoospores. Two identical cultures of complete medium containing 1 µ Ci/ml of ^{3}H-leucine and 5 x 10^{-5} M L-MDMP (control) or D-MDMP (experimental), respectively, were inoculated with 2 x 10^{6} zoospores/ml. At the times indicated, 1-ml aliquots were removed and the amount of radioactivity incorporated into hot TCA insoluble material determined. Reproduced by permission of C. J. Leaver & J. S. Lovett, Cell Diff. 3, 165--192 (1974).

sufficiency of the mRNA for polysome formation can be estimated if it is assumed that: (a) the average mRNA contains 750 nucleotides with a molecular weight of 250,000 (coding for a protein of molecular weight ca. 26,750), and (b) an average polyribosome consists of an mRNA and ten ribosomes. By using the poly(A)-RNA content of the nuclear cap (0.165 pg), the total ribosomal RNA molecular weight (2.11×10^6), the RNA/cell (8.25 pg), and the percent of polysomes present at 20 min (50%), the stored nuclear cap mRNA is calculated to equal 4.7% of the total polysome-associated RNA at 20 min. Based on the preceding assumptions, only one-fourth of this mRNA would be needed to yield a 50% polysome level; or, put in another way, there are enough mRNA molecules in the cap to form polysomes with an average of only 2.5 ribosomes per messenger <u>if all the mRNA were used</u>.

Even if the above calculations are in error by a factor of two, the results indicate that there is more than enough mRNA to combine with half of the cellular ribosomes to produce polysomes. If all of the nuclear cap poly(A)-associated RNA is actually stored mRNA, this calculation also leads to the conclusion that whatever regulates the initiation events must also discriminate among the mRNA molecules available. If this were not the case one would expect more than 50% polysomes to form in Actinomycin D-treated cells, yet the level decreases after 20 min. We interpret this loss of polysomes in the absence of new RNA synthesis to result from turnover (i.e., random breakdown) of the preformed mRNA in the polysomes as a result of their normal function in polypeptide synthesis because the decay is exponential with a half-life of ca. 45 min. Since untreated cells increase their polysome content after 20 min, we assume that as the preformed mRNA turns over it is both replaced and supplemented by new mRNA synthesis. The excess stored mRNA could either be utilized after a delay, or even be degraded without participation in polypeptide synthesis (Scherrer, 1974).

Evidence for the participation of newly synthesized mRNA in new protein synthesis during germination is at present only circumstantial. With both *Blastocladiella* zoospores (Leaver & Lovett, 1974) and *Neurospora* macroconidia (Mirkes, 1974) new polydisperse RNA appears in the cytoplasmic polysomes within 15--20 min after the cells are placed in germination medium with radioactive uridine or adenine. The relative contributions of the preformed and newly synthesized mRNA to the polysomes, and the actual rates of new protein formation still need to be evaluated. Because there is little or no change in total protein during these early states, attempts have been made to measure the *in vivo* rates by amino acid incorporation experiments. To avoid the uncertainties of variable specific activities in the precursor pools, we have estimated the "weight average transit time" of ribosomes (i.e., the time required for a ribosome to move across an "average" mRNA, which is equal to the time required to synthesize an "average" protein). The procedures have been described in detail elsewhere (Leaver & Lovett, 1974) and need not be described here. The results are of interest because they help explain the apparent anomaly in the case of *Blastocladiella* where significant polysome levels occur while there is little change in protein content for at least an hour (Lovett, 1968). When the rates were compared between cells at the 30 min stage of germination and after 120 min of growth, they were found to differ by twofold. At 30 min the average ribosome transit time was 66 s, while at 120 min it was 32 s. This reduced translation rate in *Blastocladiella* can be ascribed to the effect of the ribosome inhibitor which can still be found associated with the ribosomes at 30 min in the high density cultures normally used (Adelman & Lovett, 1974a).

When the translation rate per ribosome was further corrected for the lower polysome level at 30 min, the maximum rate of protein synthesis compared to that at 120 min was 80% less on a per cell basis. Even at this low rate some increase should have been detectable chemically and recent experiments of Lodi & Sonneborn (1974) have shown why it was not; they found that during germination in *Blastocladiella* proteolytic activity caused a turnover of preformed proteins at the rate of five percent per hour. Thus, the small amount of new synthesis (1.5--2 pg/cell/h) would be nearly matched by the decrease in old protein due to proteolysis (ca. 1.2 pg/cell/h). The important general conclusion to be derived from these experimental results with *Blastocladiella* is that the rate of protein accumulation in germinating spores represents the net effect of both new synthesis and degradation. In organisms whose spores can germinate in the absence of external nutrients such as *Blastocladiella, Uromyces, Peronospora,* and others, protein turnover may well be the source for all of the amino acids needed to produce the early essential proteins.

In spores which require external nutrients for germination this may not be an obligatory activity, but it seems likely to occur in the many spores that only increase their protein content slowly in complete medium. The significance of protein turnover in germination has not been the subject of much study although a direct role for proteolysis is well established in *Microsporum gypseum* (Leighton & Stock, 1970).

The extensive observations of Staples and co-workers on uredospore germination in the bean rust *Uromyces phaseoli* are basically similar to the other fungi we have discussed, except in one important way. The germinating spore produces a germ tube, an appressorium, and an infection hypha under appropriate conditions, but this developmental activity is not accompanied by any increase in dry weight, i.e., net growth. In this respect there is a good analogy between *Uromyces* spores and *Blastocladiella* zoospores germinated in an inorganic medium. Washed uredospores germinated on water produce only a germ tube, while those germinated on collodion membranes with paraffin oil produce in addition an appressorium and infection hypha (Staples et al., 1970). In the latter case polysomes appear after 4 h, reach a maximum level of ca. 64% by 8 h and then decline; spores germinated on water produce a maximum level of 55% polysomes at 4 h after which they decrease steadily (Yaniv & Staples, 1969). At least one major activity of these germinating spores is apparently the turnover of stored glucans for cell wall synthesis (Wynn & Gajdusek, 1968), and it seems probable that such spores represent a closed system with virtually all of the new protein synthesis occurring as a result of turnover (Staples et al., 1970). The condia of the obligate parasite *Peronospora tabacina* germinate more rapidly than *Uromyces phaseoli*, but in other ways the systems are very similar (Holloman, 1971; 1973).

To conclude, the start-up of new protein synthesis in all fungal spores can be proposed to involve most if not all of the following important steps: (1) an increased availability of ATP, GTP, and amino acids as a result of metabolic activation and variable proteolysis; (2) a change in the preformed "stored" messenger ribonuclearproteins to a state in which they can enter initiation complexes; (3) an alteration of the inactive 80S monoribosomes to permit dissociation and release of the required 40S and 60S subunits for initiation (for which a specific dissociation factor could be required); (4) the activation of initiation factors if these are stored in an inactive state; and (5) a release of translation inhibitors from the ribosomes and/or the cell cytoplasm in spores where these occur. The actual kinetics of the overall activation of protein synthesis may be expected to vary with the type of spore, the presence or absence of self-inhibitors, and the sufficiency of the medium in which it germinates. I have tried to outline a plausible series of interrelated events in an attempt to focus attention on what seem to be the most probable areas for control of the very complex system which protein synthesis represents. As was the case with the spores themselves, the regulation of the events surrounding initiation stands out as the area which most deserves more experimental study.

ACKNOWLEDGMENTS

The author would like to acknowledge the very important contributions of Irvin R. Schmoyer, Thomas G. Adelman, Christopher J. Leaver, Cheng-Shung Gong, and Fred H. Wilt to various portions of the research described for *Blastocladiella*. These investigations were also supported by Public Health Service Grant AI-04783 from the National Institutes of Allergy and Infectious Diseases.

REFERENCES

Adelman, T. G. & Lovett, J. S. (1972), Synthesis of ribosomal protein without *de novo* ribosome production during differentiation in *Blastocladiella emersonii*. Biochem. Biosphys. Res. Comm. 49, 1174--1182.

Adelman, T. G. & Lovett, J. S. (1974a). Evidence for a ribosome-associated translation inhibitor during differentiation of *Blastocladiella emersonii*. Biochim. Biophys. Acta 335, 236--245.

Adelman, T. G., & Lovett, J. S. (1974b). Ribosome function *in vitro* and *in vivo* during the life cycle of *Blastocladiella emersonii*. Biochim. Biophys. Acta 349, 240--249.

Aitkin, W. B. & Niederpruem, D. J. (1970). Ultrastructural changes and biochemical events in basidiospore germination of *Schizophyllum commune*. J. Bacteriol. 104, 981--988.

Allen, P. J. (1965). Metabolic aspects of spore germination in fungi. Annu. Rev. Plant Path. 3, 313--342.

Bacon, C. W. & Sussman, A. S. (1973). Effects of the self-inhibitor of *Dictyostelium discoideum* on spore metabolism. J. Gen. Microbiol. 76, 331--344.

Bacon, C. W., Sussman, A. S. & Paul, A. G. (1973). Identitication of a self-inhibitor from spores of *Dictyostelium discoideum*. J. Bacteriol. 113, 1061--1063.

Bainbridge, B. W. (1971). Macromolecular composition and nuclear division during spore germination in *Aspergillus nidulans*. J. Gen. Microbiol. 66, 319--325.

Barash, I. (1968). Liberation of polygalacturonase during spore germination by *Geotrichum candidum*. Phytopathology. 58, 1364--1371.

Barash, I., Conway, M. L., & Howard, D.H. (1967). Carbon catabolism and synthesis of macromolecules during spore germination of *Microsporum gypseum*. J. Bacteriol. 93, 656--662.

Barstow, W. E. & Lovett, J. S. (1974). Apical vesicles and microtubules in rhizoids of *Blastocladiella emersonii*: Effects of Actinonycin D and cycloheximide on development during germination. Protoplasma, 82, 103--117.

Bartnicki-Garcia, S. & Lippman, E. (1969). Fungal morphogenisis: Cell wall construction in *Mucor rouxii*. Science 165, 302--304.

Bhagwat, A. S. & Mahadevan, P.R. (1970). Conserved mRNA from the conidia of *Neurospora crassa*, Mol. Gen. Genetics 109, 142-151.

Blobel, G. (1973). A protein of molecular weight 78,000 bound to the polyadenylate region of eukaryotic messenger RNAs. Proc. Nat. Acad. Sci. U.S.A. 70, 924--928.

Bracker, C. E. (1971). Cytoplasmic vesicles in germinating spores of *Gillertella persicaria*. Protoplasma 72, 381--397.

Brambl, R. M. & Van Etten, J. L. (1970). Protein synthesis during fungal spore germination. V. Evidence that the ungerminated conidiospores of *Botryodiplodia theobromae* contain messenger ribonucleic acid. Arch. Biochem. Biophys. 137, 442--452.

Brooks. C. J. & DeBusk, A. G. (1973). Cellular compartmentalization of aromatic amino acids in *Neurospora crassa*. I. Occupation of a protein synthesis pool by phenylalanine in Tyr-1 mutants. Biochem. Genet. 10, 91--103.

Brunsched, H. & Bremer, H. (1971). Synthesis and breakdown of proteins in *Escherichia coli* during amino acid starvation. J. Mol. Biol. 57, 35--57.

Bryan, R. N. & Hayashi, M. (1973). Two proteins are bound to most species of polysomal mRNA. Nat. New Biol. 244, 271--274.

Burke, D. J., Seale, T. W. & McCarthy, B. J. (1972). Protein and ribonucleic acid synthesis during the diploid life cycle of *Allomyces arbuscula*. J. Bacteriol. 110, 1065--1072.

Christman, J. K. (1973). Effect of elevated potassium level and amino acid deprivation on polysome distribution and rate of protein synthesis in L. cells. Biochim. Biophys. Acta 294, 138--152.

Cochrane, V. W. (1966). Respiration and spore germination. In The Fungus Spore (ed. M.F. Madelin), pp. 201--213, Colston Papers, No. 18. London, Butterworths.

Cochrane, J. C., Rado, T. A., & Cochrane, V. W. (1971). Synthesis of macromolecules and polyribosome formation in early stages of spore germination in *Fusarium solani*. J. Gen. Microbiol. 65, 45--55.

Dill, B. C., Leighton T. J. & Stock, J. J. (1972). Physiological and biochemical changes associated with macroconidial germination in *Microsporum gypseum*. Appl. Microbiol. 24, 977--985.

Dunkle, L. D., Maheshwari, R. & Allen, P. J. (1969). Infection structures from rust urediospores: Effect of RNA and protein inhibitors. Science 163, 481--482.

Eil, C. & Wool, L. G. (1973). Function of phosphorylated ribosomes. J. Biol. Chem. 248, 5130--5136.

Gottlieb, D. (1966). Biosynthetic processes in germinating spores. In The Fungus Spore (ed. M. F. Madelin), pp. 217--230, Colston Papers, No. 18. London, Butterworths.

Grove, S. N. (1972). Apical vesicles in germinating conidia of *Aspergillus parasiticus*. Mycologia 64, 638--641.

Grove, S. N., Bracker, C. F. & Morre, D. J. (1970). An ultrastructural basis for hyphal tip growth in *Pythium ultimum*. Amer. J. Bot. 57, 245--266.

Hartwell, L. H. (1974). *Saccharomyces cerevisiae* cell cycle. Bacteriol. Rev. 38, 164--198.

Hartwell, L. H., Hutchison, H. T., Holland, T. M. & McLaughlin, C. S. (1970). The effect of cycloheximide upon polysome stability in two yeast mutants defective respectively in the initiation of polypeptide chains and in messenger RNA synthesis. Mol. Gen. Genetics 106, 347--361.

Haselkorn, R. & Rothman-Denes, L. B. (1973). Protein synthesis. Annu. Rev. Biochem. 42, 397--438.

Henney, H. R. & Storck, R. (1963). Ribosomes and ribonucleic acid in three morphological states of *Neurospora*, Science 142, 1675--1676.

Henney, H. R. & Storck, R. (1964). Polyribosomes and morphology in *Neurospora*. Proc. Nat. Acad. Sci. U.S.A. 51, 1050--1055.

Hogan, B. L. M. & Korner, A. (1968). Ribosomal subunits of Landschultz ascites cells during changes in polysome distribution. Biochim. Biophys. Acta 169, 129--138.

Hollomon, D. W. (1971). Protein synthesis during germination of *Peronospora tabacina* (Adam) conidia. Arch. Biochem. Biophys. 145, 643--649.

Hollomon, D. W. (1973). Protein synthesis during germination of *Peronospora tabacina* conidia: An examination of the events involved in the initiation of germination. J. Gen. Microbiol. 78, 1--13.

Horikoshi, K. & Ikeda, Y. (1968). Studies on the conidia of *Aspergillus oryzae*. IX. Protein synthesizing activity of dormant conidia. Biochim. Biophys. Acta 190, 187--192.

Horikoshi, K., Ohtaka, Y. & Ikeda, Y. (1969). Properties of ribosomes and transfer ribonucleic acid in dormant conidia of *Aspergillus oryzae*. In Spores IV, (ed. L. L. Campbell,) pp. 175--179, Bethesda, MD. American Society for Microbiology.

Inoue, H. & Ishikawa, T. (1970). Macromolecule synthesis and germination of conidia in temperature sensitive mutants of *Neurospora crassa*. Jap. J. Genet. 45, 357--369.

Kabat, D. (1970). Phosphorylation of ribosomal proteins in rabbit reticulocytes. Characterization and regulatory aspects. Biochemistry 9, 4160--4175.

Kimura, K., Ono, T. & Yanagita, T. (1965). Characterization of ribosomes from dormant and germinating conidia of *Aspergillus oryzae*. J. Biochem. 58, 569--576.

Kyner, D., Zabas, P. & Leven D. H. (1973). Inhibition of protein chain initiation in eukaryotes by de-acylated transfer RNA and its reversibility by spermine. Biochim. Biophys. Acta 324, 386--396.

Lambertsson, A. G., Rasmuson, S. B., & Bloom, G. D. (1970). The ribosome proteins of *Drosophila Melanogaster*. I. Characterization in polyacrylamide gels of proteins from larvel, adult, and ammonium chloride treated ribosomes. Mol. Gen. Genet. 108, 349--357.

Leary, J. V. & Ellingboe, A. H. (1971). Isolation and characterization of ribosomes from non-germinated conidia of *Erysiphe graminis* f. sp. *tritici*. Phytopathology 61, 1030--1031.

Leary, J. V., Morris, A. J., & Ellingboe, A. H. (1969). Isolation of functional ribosomes and polysomes from lyophilized fungi. Biochim. Biophys. Acta 182, 113--120.

Leary J. V., Roheim, J. R. & Zentmyer, G. A. (1974). Ribosome content of various spore forms of *Pytophthora* spp. Phytopathology 64, 404--408.

Leaver, C. J. & Lovett, J. S. (1974). An analysis of protein and RNA synthesis during encystment and outgrowth (germination) of *Blastocladiella* zoospores. Cell Diff. 3, 165--192.

Lee, S. L., Krsmanovic, V. & Brawerman, G. (1971). Initiation of polysome formation in mouse sarcoma 180 ascites cells: Utilization of cytoplasmic messenger RNA. Biochemistry 10, 895--900.

Leighton, T. J. & Stock, J. J. (1970). Biochemical changes during fungal sporulation and spore germination. I. Phenyl methyel sulfonyl flouride inhibition of macroconidial germination in *Microsporum gypseum*. J. Bacteriol. 101, 931--940.

LeJohn, H. B. & Lovett, J. S. (1966). Ribonucleic acid and protein synthesis in *Rhizophlyctis rosea* zoospores. J. Bacteriol. 91, 709--717.

Lingappa, B. T., Lingappa, Y. & Bell, E. (1973). A self-inhibitor of protein synthesis in the conidia of *Glomerilla cingulata*. Arch. Mikrobiol. 94, 97-107.

Lingappa, B. T. & Sussman A. S. (1959). Endogenous substrates of dormant, activated and germinating ascospores of *Neurospora tetrasperma*. Plant Physiol. 34, 466--472.

Lodi, W. R. & Sonneborn, D. R. (1974). Protein degradation and protease activity during the life cycle of *Blastocladiella emersonii*. J. Bacteriol. 117, 1035--1042.

Lovett, J. S. (1963). Chemical and physical characterization of "nuclear caps" isolated from *Blastocladiella* zoospores. J. Bacteriol. 85, 1235--1246.

Lovett, J. S. (1968). Reactivation of ribonucleic acid and protein synthesis during germination of *Blastocladiella* zoospores and the role of the ribosomal nuclear cap. J. Bacteriol. 96, 962--969.

Marcus, L., Ris, H., Halvorson, H. O., Bretthauer, R. K. & Bock, R. M. (1967). Occurrence, isolation, and characterization of polyribosomes in yeast. J. Cell Biol. 34, 505--512.

McLaughlin, C. S., Magee, P. T. & Hartwell, L. H. (1969). Role of isoleucyl-transfer ribonucleic acid synthetase in ribonucleic acid synthesis and enzyme repression in yeast. J. Bacteriol. 100, 579--584.

McLaughlin, C. S., Warner, J. R., Edmunds, M., Nakazato, H. & Vaughn, M. H. (1973). Polyadenylic acid sequences in yeast messenger ribonucleic acid. J. Biol. Chem. 248, 1466--1471.

Merlo, D. J., Roker, H. & Van Etten, J. L. (1972). Protein synthesis during fungal-spore germination. VI. Analysis of transfer ribonucleic acid from germinated and ungerminated spores of *Rhizopus stolonifer*. Can. J. Microbiol. 18, 949--956.

Mirkes, P.E. (1974). Polysomes, ribonucleic acid, and protein synthesis during germination of *Neurospora crassa* conidia. J. Bacteriol. 117, 196--202.

Morel, C. M., Gander E. S., Herzberg, M., Dubochet, J. & Scherrer, K. (1973). The duck-globin messenger ribonuclear protein complex. Resistance to high ionic strength, particle gel electrophorisis, composition and visualization by dark field electron microscopy. Eur. J. Biochem. 36, 455--464.

Nazario, M. (1972). Different arginine transfer ribonucleic acid species prevalent in shaken and unshaken cultures of *Neurospora*. J. Bacteriol. 112, 1076--1082.

Nurse, P. & Wiemkin, A. (1974). Amino acid pools and metabolism during the cell division cycle of arginine grown *Candida utilis*. J. Bacteriol. 117, 1108--1116.

Ohmori, K. & Gottlieb, D. (1965). Development of respiratory enzyme activities during spore germination. Phytopathol. 55, 1328--1336.

Page, W. J. & Stock, J. J. (1972). Isolation and characterization of *Microsporum gypseum* lysosomes: Role of lysosomes in macrocondia germination. J. Bacteriol. 110, 354--362.

Perani, A., Tiboni, O. & Ciferri, O. (1971). Absolute ribosome specificity of two sets of transfer factors isolated from *Saccharomyces fragilis*. J. Mol. Biol. 55, 107--112.

Petre, J. (1970). Specific dissociation of ribosomes from *Saccharomyces cerevisiae* by a protein factor. Eur. J. Biochem. 14, 399--405.

Rado, T. A. & Cochrane, V. W. (1971). Ribosomal competence and spore germination in *Fusarium solani*. J. Bacteriol. 106, 301--304.

Ramakrishnan, L. & Staples R. C. (1970). Evidence for a template RNA in resting urdospores of the bean rust fungus. Contrib. Boyce Thomp. Inst. 24, 197--202.

Reed, J. & Wintersberger, E. (1973). Adenylic acid-rich sequences in messenger RNA from yeast polysomes. Fed. Eur. Biochem. Soc. Lett. 32, 213--217.

Reisner, H. G. (1967). Untersuchungen uber den Aminosauer- und Proteinstoffwechsel der Uredosporen des Weizenrostes. Arch. Mikrobiol. 55, 382--397.

Rho, H. M. & DeBusk, A. G. (1971). Protein chain initiation by methionyl transfer ribonucleic acid in the cytoplasm of *Neurospora*. J. Biol. Chem. 246, 6566--6569.

Rothschild, H., Itikawa, H. & Suskind, S. R. (1967). Ribosomes and ribosomal proteins from *Neurospora crassa*. II. Ribosomal proteins in different wild-type strains and during various stages of development. J. Bacteriol. 94, 1800--1801.

Rousseau, P. & Halvorson, H. O. (1973a). Macromolecular synthesis during the germination of *Saccharomyces cerevisiae* spores. J. Bacteriol. 113, 1289--1295.

Rousseau, P. & Halvorson, H. A. (1973b). Effect of metabolic inhibitors on germination and outgrowth of *Saccharomyces cerevisiae* ascospores. Can. J. Microbiol. 19, 1311--1318.

Rudolph, H., & Ochsen, B. (1969). Trehalose-Umsatz warmeaktiveter sporen von *Phycomyces blakesleanus*. VI. Beitrag zur Kausalanalyse der Warmeaktivierung von Pilzsporen. Arch. Mikrobiol. 65, 163--171.

Sander, E. S., Stewart, A. G., Morel, C. M. & Scherrer, K. (1973). Isolation and characterization of ribosome-free cytoplasmic messenger ribonucleoprotein complexes from avian erythroblasts. Eur. J. Biochem. 38, 443--452.

Scheld, H. W. & Perry, J. J. (1970). Basidiospore germination in the wood destroying fungus *Lenzites saepiaria*. J. Gen. Microbiol. 60, 9--21.

Scherrer, K. (1974). Control of gene expression in animal cells: The cascade regulation hypothesis revisited. In Control of Gene Expression (ed. A. Kohn and A. Shatkay), pp. 169--219. New York: Plenum Press.

Schmoyer, I. R. & Lovett, J. S. (1969). Regulation of protein synthesis in zoospores of *Blastocladiella*. J. Bacteriol. 100, 854--864.

Siegel, M. R. & Sisler, H. D. (1965). Site of action of cyclohex-imide in cells of *Saccharomyces pastorianus*. III. Further studies on the mechanism of action and the mechanism of resistance in *Saccharomyces* species. Biochim. Biophys. Acta 103, 558--567.

Silverman, P. M., Huh, M. W. & Sun, L. (1974). Protein synthesis during zoospore germination in the aquatic phycomycete *Blastocladiella emersonii*. Dev. Biol. 40, 59--70.

Slayman, C. L. (1973). Adenine nucleotide levels in *Neurospora*, as influenced by conditions of growth and by metabolic inhibitors. J. Bacteriol. 114, 752--766.

Soll, D. R. & Sonneborn, D. R. (1971a). Zoospore germination in *Blastocladiella emersonii*: Cell differentiation without protein synthesis? Proc. Nat. Acad. Sci. U.S.A. 68, 459--463.

Soll, D. R. & Sonneborn, D. R. (1971b). Zoospore germination in *Blastocladiella emersonii*. III. Structural changes in relation to protein and RNA synthesis. J. Cell Sci. 9, 679--699.

Soll, D. R. & Sonneborn, D. R. (1972). Zoospore germination in *Blastocladiella emersonii*. IV. Ion control over cell differentiation. J. Cell Sci. 10, 315--333.

Staples, R. C., App, A. A., McCarthy, W. J. & Gerosa, M. M. (1966). Some properties of ribosomes from uredospores of the bean rust fungus. Contrib. Boyce Thomp. Inst. 23, 159--164.

Staples, R. C., Bedigian, D. & Williams, P. H. (1968). Evidence for polysomes in extracts of bean rust uredospores. Phytopathology 58, 151--154.

Staples, R. C., Ramakrishnan, L. & Yaniv, Z. (1970). Membrane-bound ribosomes in germinated uredospores. Phytopathology 60, 58--62.

Staples, R. C., Syamananda, R., Kao, V. & Block, R. J. (1962). Comparative biochemistry of obligately parasitic and saprophytic fungi. II. Assimilation of C^{14}-labeled substrates by germinating spores. Contrib. Boyce Thomp. Inst. 21, 345--362.

Suberkropp, K. F. & Cantino, E. C. (1972). Environmental control of motility and encystment in *Blastocladiella emersonii* zoospores at high population densities. Trans. Brit. Mycol. Soc. 59, 463--475.

Subramanian, K. N., Weiss, R. L. & Davis, R. H. (1973). Use of external, biosynthetic, and organellar arginine by *Neurospora*. J. Bacteriol. 115, 284--290.

Sueoka, N. & Kano-Sueoka, T. (1970). Transfer RNA and Cell differentiation. Prog. Nucleic Acid Res. Mol. Biol. 10, 23--55.

Sussman, A. S. (1969). The dormancy and germination of fungus spores. Symp. Soc. Exp. Biol. 23, 99--121.

Sussman, A. S. & Douthit, H. A. (1973). Dormancy in microbial spores. Annu. Rev. Plant. Physiol. 24, 311--352.

Takeishi, K., Ukita, T. & Nishimura, S. (1968). Characterization of two species of methionine transfer ribonucleic acid from bakers' yeast. J. Biol. Chem. 243, 5761--5769.

Tisdale, J. H. & DeBusk, A. G. (1970). Developmental regulation of amino acid transport in *Neurospora crassa*. J. Bacteriol 104, 689--697.

Trocha, P. & Daly, J. M. (1970). Protein and ribonucleic acid synthesis during germination of uredospores. Plant Physiol. 46, 520--526.

Truesdell, L. C. & Cantino, E. C. (1971). The induction and early events of germination in the zoospore of *Blastocladiella emersonii*. Curr. Top. Dev. Biol. 6, 1--44.

Tuveson, R. W., West, D. J. & Barrett, R. W. (1967). Glutamic acid dehydrogenases in quiescent and germinating conidia of *Neurospora crassa*. J. Gen. Microbiol. 48, 235--248.

Van Assche, J. A., Carlier, A. R. & Dekeersmaeker, H. I. (1972). Trehalase activity in dormant and activated spores of *Phycomyces blakesleanus*. Planta 103, 327--333.

Van Assche, J. A. & Carlier, A. R. (1973). The pattern of protein and nucleic acid synthesis in germinating spores of *Phycomyces blakesleanus*. Arch. Mikrobiol. 93, 129--136.

Van Etten, J. L. (1968). Protein synthesis during fungal spore germination. I. Characteristics of an *in vitro* phenylalanine incorporating system prepared from germinated spores of *Botryodiplodia theobromae*. Arch. Biochem. Biophys. 125, 13--21.

Van Etten, J. L. (1969). Protein synthesis during fungal spore germination. Phytopathology 59, 1060--1083.

Van Etten, J. L. & Brambl, R. M. (1968). Protein synthesis during fungal spore germination. II. Amino-acyl soluble ribonucleic acid synthetase activities during germination of *Botryodiplodia theobromae* spores. J. Bacteriol. 96, 1042--1048.

Van Etten, J. L. & Brambl, R. M. (1969). Protein synthesis during fungal spore germination. III. Transfer activity during germination of *Botryodiplodia theobromae* spores. Phytopathol. 59, 1894--1902.

Van Etten, J. L., Koski, R. K. & El-Olemy, M. M. (1969). Protein synthesis during fungal spore germination. IV. Transfer ribonucleic acid from germinated and ungerminated spores. J. Bacteriol. 100, 1182--1186.

Van Etten, J. L., Roker, H. R. & Davies, E. (1972). Protein synthesis during fungal spore germination: Differential protein synthesis during germination of *Botryodiplodia theobromae* spores. J. Bacteriol. 112, 1029--1031.

Van Etten, J. L., Bulla, L. A. & St. Julian, G. (1974). Physiological and morphological Correlation of *Rhizopus stolonifer* spore germination. J. Bacteriol. 117, 882--887.

Van der Zeijst, A. M., Kool, A., & Bloemers, H. P. J. (1972). Isolation of active ribosomal subunits from yeast. Eur. J. Biochem. 30, 15--25.

Vassart, G., Dumont, J. E., & Cantraine, F. R. L. (1971). Translational control of protein synthesis: A simulation study. Biochim. Biophys. Acta 247, 471--485.

Weber, D. J. (1966). Role of proline and glutamic acid metabolism in spore germination of *Rhizopus arrhizus*. Phytopathology 56, 118--123.

Weeks, D. P, & Baxter, R. (1972). Specific inhibition of peptide-chain initiation by 2-(4-methyl-2, 6-dinitroanalino)-N-methylpropionamide. Biochem. 11, 3060--3064.

Wiemkin, A. & Nurse, P. (1973). Isolation and characterization of the amino acid pools located within the cytoplasm and vacuoles of *Candida utilis*, Planta 109, 293--306.

Wilt, F. H. (1973). Polyadenylation of Maternal RNA of sea urchin eggs after fertilization. Proc. Nat. Acad. Sci. U.S.A. 70, 2345--2349.

Wong, R. S. L., Scarborough, G. & Borek, E. (1971). Transfer ribonucleic acid methylases during the germination of *Neurospora crassa*. J. Bacteriol. 108, 446--450.

Wynn, W. K. & Gajdusek, C. (1968). Metabolism of glucomannan-protein during germination of bean rust spores. Contrib. Boyce Thomp. Inst. 24, 123--138.

Yanagita, T. (1957). Biochemical Aspects on the germination of conidiospores of *Aspergillus niger*. Arch. Mikrobiol. 26, 329--344.

Yaniv, Z. & Staples, R. C. (1969). Transfer activity of ribosomes from germinating uredospores. Contrib. Boyce Thomp. Inst. 24, 157--163.

Yaniv, Z. & Staples, R. C. (1971). The purification and properties of the aminoacyl-tRNA binding enzyme from bean rust uredospores. Biochim. Biophys. Acta 232, 717--725.

REGULATION OF PROTEIN METABOLISM

DISCUSSION

Regulation of Protein Metabolism During Spore Germination

Chairman: Larry D. Dunkle
University of Nebraska
Lincoln, Nebraska

Three mechanisms were discussed as potential sites for regulating protein synthesis: (a) the presence of messenger RNA (or poly(A)-containing RNA in spores), (b) the role of protein turnover and the level of aminoacylated transfer RNA in relation to the requirement of amino acids for germination, and (c) the role of endogenous germination inhibitors. The following is a summary of the major points of discussion.

It was emphasized that not all fungal spores necessarily contain polyribosomes although mRNA (detected as poly(A)-containing RNA) may be present. Thus, extrapolation of data obtained with one fungal species to another or from one spore type (e.g., conidia) to another (e.g., ascospores) of the same species cannot be justified. The objection to using poly(A)-containing RNA to strictly quantitate the amount of mRNA in the spore was recognized as a valid objection, since some mRNAs (e.g., histone mRNA in other organisms) may not contain poly(A) segments. However, Dr. Lovett maintained that, by making some assumptions about the amount and size of poly(A)-containing RNA, a reasonable (though minimal) estimate can be made of the amount of mRNA in the spore.

The question was raised whether protein turnover could be the sole mechanism for providing amino acids to spores that apparently require an amino acid for germination. Lovett contended that the spores of some species may be programmed to degrade proteins during germination, whereas in other species protein turnover occurs only under certain conditions. He suggested that, although free amino acids have been quantitated in a few species, the more important and meaningful information would be the amount of charged tRNA for each amino acid, since in order to significantly affect protein synthesis the absence (or near absence) of free amino acid must result in the absence of that aminocyl-tRNA. He further supported this view by indicating that, in "step-down" cultures of *Blastocladiella emersonii* where amino acids are removed, polysomes decrease (ca. 50%) but never completely disappear.

The role of endogenous germination inhibitors in regulating protein synthesis was discussed. The point was made that leucine incorporation (into a TCA-insoluble fraction) by self-inhibited rust uredospores is not affected and progresses at the same rate as in germinating spores. Lovett agreed that it is not possible to propose a general role for the self-inhibitors. He suggested that, even in those cases where the inhibitors have been shown to have an effect on protein synthesis, they may not be the most important or the only factor controlling protein synthesis in the spore. Further, Lovett indicated that blocking translation or termination of protein synthesis in the spore would be a very inefficient mechanism for maintaining "dormancy" and one would expect to find arrested polysomes in the spore. In fact, polysomes are not detected in the zoospores of *B. emersonii* in which the protein synthesis inhibitor may be the same substance as the self-inhibitor of germination.

CHAPTER 7

NUCLEIC ACIDS AND FUNGAL SPORE GERMINATION

J. L. Van Etten, L. D. Dunkle, and R. H. Knight
University of Nebraska

Published with the approval of the Director as paper No. 3826, Journal Series, Nebraska Agricultural Experiment Station. The work was conducted under Nebraska Agricultural Experiment Station Project No. 27--17.

INTRODUCTION

Fungal spore germination provides a useful experimental system for investigating the concept that cellular development and resumption of growth from a dormant or quiescent state involve unique macromolecular biosynthetic activities. Among the more demonstrable activities during spore germination is the biosynthesis of nucleic acids. This review summarizes studies of nucleic acids in dormant and germinated spores and the biosynthetic activities that accompany fungal spore germination. The mechanisms and significance of these activities as they relate to and perhaps regulate the germination process are discussed. Although, in many cases, the experimental approaches and rationale for specific experiments were derived from studies on the molecular biology and biochemistry of prokaryotic and higher eukaryotic organisms, this discussion is restricted to the true fungi and extrapolation from these other systems is used only to establish a basis of approach or a conclusion. The extensive amount of research which has been conducted on cellular and acellular slime molds is not covered in the present review. Other recent relevant reviews of fungal spore germination include those of Sussman & Douthit (1973) and Van Etten (1969).

Certain technical problems are encountered in studies on nucleic acids in fungal spores and nucleic acid synthesis associated with spore germination. These difficulties include: (1) the apparent inability of many common inhibitors of nucleic acid synthesis such as actinomycin D, cordycepin, and α-amanitin to penetrate dormant spores; (2) the harsh conditions required to disrupt most fungal spores which may result in extensive degradation as well as aggregation of nucleic acids; (3) the difficulty of isolating intact nuclei, nucleoli, and mitochondria from spores; and (4) the common use of water as a suspending medium which may trigger the occurrence of certain key events before the first measurements are made. Water is frequently used to remove a self inhibitor of germination or to harvest the spores from the parent mycelium. In spite of these difficulties, however, much information on nucleic acids in fungal spores has accumulated during the last fifteen years.

In this review the terms spore (or dormant spore), germinating spores, and germinated spore refer, respectively, to native spores, spores incubated under conditions favorable for germination, and spores with visible germ tubes (or budding cells in the case of yeasts).

DEOXYRIBONUCLEIC ACIDS

Nature of the Fungal Genome

Knowledge of the composition and compartmentalization of cellular DNA is essential to understanding the mechanisms by which nucleic acids may regulate the spore germination process. Most properties of the fungal genome have been determined with the vegetative phase of the fungal life cycle. However, the general properties of DNA from vegetative cells are apparently very similar to those of DNA from spores.

The base compositions of DNA from more than 400 species of fungi have been determined and collated by Storck and co-workers (Storck et al., 1969, 1971; Storck & Alexopoulos, 1970; Villa & Storck, 1968). Although the use of DNA base composition for taxonomic and phylogenetic studies is limited, species that are related morphologically, physiologically, or genetically tend to have similar base compositions. The guanine plus cytosine content (moles % GC) of fungal DNAs ranges from 27.5 -- 70%. The GC content of a particular isolate is the same, whether determined by buoyant density in CsCl or by thermal denaturation, suggesting that fungal DNA does not contain a significant proportion of unusual bases.

Nuclear DNA (nuc-DNA) comprises approximately 80--99% of the total cellular DNA. In a few fungal species from 10--20% of this nuc-DNA occurs as a satellite species with a distinctive buoyant density in CsCl (Table 1). The nuclear satellite DNA contains the majority (60--70%) of the rRNA cistrons in *Saccharomyces cerevisiae* (Cramer et al., 1972; Kaback et al., 1973; Finkelstein et al., 1972), but the genetic function of satellite DNA in other fungi has not been investigated.

Estimates of the size of the haploid fungal genome (see review by Storck, 1974) range from about 0.5×10^{10} daltons (in *Mucor rouxii*) to 3×10^{10} daltons (in *Talaromyces vermiculatus*), approximately two to ten times more complex than the bacterial genome. A portion of the total cellular DNA is reiterated (10--20% in *Neurospora crassa* (Dutta, 1973; Brooks & Huang, 1972); 5--16% among several yeasts (Christiansen et al., 1971)). Some of this reiterated DNA can be accounted for as multiple copies of rRNA and tRNA cistrons, which together comprise about 1.3--2.5% of the total DNA (Table 2). Thus, the fungal genome contains 100--200 rRNA cistrons and considerably more tRNA cistrons (320--400 in yeast; 2600--3000 in *Neurospora crassa*). Higher estimates of these values have been reported for *Achlya bisexualis* in which the diploid nucleus in the spores contains 2.8×10^{11} daltons of DNA and in which there are 435 rRNA cistrons per haploid genome (Jaworski & Horgen, 1973).

Studies on yeast chromosomes (Petes & Fangman, 1972; Petes et al., 1973; Finkelstein et al., 1972) indicate that each chromosome contains a single DNA molecule ranging in size from $0.5\text{--}12 \times 10^8$ daltons with an average mass of $5\text{--}7 \times 10^8$ daltons (four times smaller than the *Escherichia coli* chromosome). If the same relationship exists for the seven chromosomes in the haploid nucleus of *N. crassa*, then the average chromosome would contain $2\text{--}3 \times 10^9$ daltons of DNA (each approximately the same size as the *E. coli* genome).

The structural organization and composition of macromolecules associated with nuc-DNA have been studied in a few fungal species. Preparations of fungal chromatin contain proteins and RNA in addition to DNA. The relative amounts of the constitutents vary between

Table 1. DNA Composition of Selected Fungi

Fungus	nuc-DNA		satellite-DNA		mit-DNA		other DNA	
	% of total	ρCsCl	% of total	ρCsCl	% of total	ρCsCl	% of total	ρCsCl
Saccharomyces	90	1.699	10[a]	1.705	10	1.683	---	---
cerevisiae	---	1.699	1--11[b]	1.706	3--20[b]	1.683	---	---
1,2,3,4,5,6,7	---	1.699	---	1.706	11--28[c]	1.683	---	---
	95	---	---	---	5	---	---	---
	---	1.701	---	---	---	1.684	1--4[a,d]	1.701
	---	1.700	---	---	---	---	2.9[d]	1.701
Neurospora	99	1.713	---	---	1	1.701	---	---
crassa 8,9	---	1.710	---	---	---	1.702	---	---
Blastocladiella emersonii 10	77	1.731	12.5	1.715	9	1.705	1.5[e]	1.687
Saprolegnia sp. 11	---	1.717	---	1.707	5	1.685	---	---
Saprolegnia parasitica 12	---	1.718	10--20	1.710	---	---	---	---
Botryodiplodia theobromae 13	95	1.713	---	---	5	1.699	---	---
Uromyces phaseoli 14	98	1.695	---	---	2	1.718	---	---

[a] Expressed as % of nuc-DNA.
[b] Range during vegetative growth.
[c] Range found in numerous strains of *S. cerevisiae*.
[d] oDNA.
[e] γ-particle DNA.

References

1. Schweizer et al., 1969
2. Tingle et al., 1974
3. Moustacchi & Williamson, 1966
4. Williamson, 1970
5. Petes et al., 1973
6. Hollenberg et al., 1970
7. Clark-Walker & Miklos, 1974
8. Luck & Reich, 1964
9. Brooks & Huang, 1972
10. Myers & Cantino, 1971
11. Clark-Walker & Gleason, 1973
12. Green & Dick, 1972
13. Dunkle & Van Etten, 1972
14. Staples, 1974

Table 2. Haploid Size and Reiterated Sequences in Fungal DNA

Fungus	Haploid genome size (daltons)	% of total DNA reiterated	rRNA cistrons		tRNA cistrons	
			% of total DNA	#	% of total DNA	#
Saccharomyces	1.25×10^{10}	---	2.4	140	0.064--0.08	320--400
cerevisiae	0.92×10^{10}	---	---	---	---	---
1,2,3	1.0×10^{10}	---	2.2	---	---	---
Several yeasts 4	$0.6\text{--}1.4 \times 10^{10}$	5--16	---	---	---	---
Neurospora	1.4×10^{10}	20	1	100-200	0.3	3000
crassa 5,6,7,8	2.3×10^{10}	10	0.95	100	0.3	2600--3000
Achlya bisexualis 9	1.4×10^{11}	---	0.65	435	---	---

References
1. Schweizer et al., 1969
2. Bicknell & Douglas, 1970
3. Finkelstein et al., 1972
4. Christiansen et al., 1971
5. Brooks & Huang, 1972
6. Dutta, 1973
7. Chattopadhyay et al., 1972
8. Dutta & Ray, 1973
9. Jaworski & Horgen, 1973

species, and more disconcerting, analyses of chromatin from the same species have yielded widely varying results. For example, in *Neurospora crassa* the ratio of RNA/DNA in chromatin varies from 0.63 (Dwivedi et al., 1969) to 0.14 (Hsiang & Cole, 1973) and for the ratio of protein/DNA from 1.54 (Dwivedi et al., 1969) to 0.84 (Hsiang & Cole, 1973). The corresponding values reported for *Microsporum gypseum* are 0.05 and 1.05, respectively (Leighton et al., 1971). The association of histones, major components of chromatin in higher eukaryotes, with fungal chromatin is presently controversial. Some studies indicate that typical histones are not associated with DNA in *N. crassa, N. tetrasperma, M. gypseum, Phycomyces blakesleeanus* (Leighton et al., 1971) and *Blastocladiella emersonii* (Horgen et al., 1973). Other studies indicate that histones are constituents of chromatin in *Saccharomyces cerevisiae* (Tonino & Rozijn, 1966; Bhargava & Halvorson, 1971; Wintersberger et al., 1973), *N. crassa* (Hsiang & Cole, 1973), and *Achlya bisexualis* (Horgen et al., 1973). The histones isolated from *N. crassa* occurred in a ratio of histone/DNA of 0.24 and were similar in several respects to histones from calf thymus but were present at only 25% of the level detected in higher eukaryotes (Hsiang & Cole, 1973). Chromatin from *Achlya bisexualis* (Horgen et al., 1973) and from *S. cerevisiae* (Wintersberger et al., 1973) contain histones, present in a ratio of histone/DNA of 0.77 and 0.83--1.25, respectively.

Whereas *in vitro* transcription of *M. gypseum* chromatin (using RNA polymerases from *E. coli*) was considered to be significantly higher than with other eukaryotic chromatins (Leighton et al., 1971), some evidence for template restriction by DNA-associated proteins in *S. cerevisiae* chromatin was presented by Wintersberger et al. (1973). The yeast chromatin was poorly transcribed by isolated yeast RNA polymerases, but transcription by endogenous RNA polymerases was enhanced by increasing concentrations of salts, which apparently removed DNA-associated proteins from the chromatin. Unfortunately, since there have been no studies on chromatin isolated from spores, the effect of associated proteins on DNA transcription in germinating spores is unknown.

Mitochondrial DNA (mit-DNA) constitutes 1--20% of the total cellular DNA (Table 1). The base compositions (based on buoyant density determinations) of mit-DNAs from a variety of different fungal species vary less than those of nuc-DNAs (Villa & Storck, 1968; Storck, 1974). In all species except the bean rust fungus *Uromyces phaseoli* (Staples, 1974) mit-DNA has a GC content (and buoyant density in CsCl) less than or nearly equal to that of nuc-DNA.

Evidence has been presented for covalently closed circular mit-DNA in three fungi, *N. crassa* (the wall-less slime mutant) (Clayton and Brambl, 1972), *S. cerevisiae* (Hollenberg et al., 1970), and *Saprolegnia* sp. (Clark-Walker & Gleason, 1973). It is probable that mit-DNA in most fungi occurs in a circular form during some stage of its molecular life cycle. The contour lengths of isolated circular mit-DNA from yeast (26 μm), *N. crassa* (20 μm), and *Saprolegnia* (14 μm) correlate closely with size estimates of the mitochondrial genome (4--6.6 X 10^7 daltons) determined by renaturation kinetics or length of isolated linear molecules (see Storck, 1974).

Studies by Wood & Luck (1969) indicated that nearly 9% of the mit-DNA in *N. crassa* hybridizes with mitochondrial rRNAs and that mit-DNA contains at least four copies of mitochondrial rRNA genes. More recent investigations (Schäfer & Küntzel, 1972), however, indicate that mitochondrial rRNAs hybridize with only 2% of the mit-DNA and that there is only a single cistron for mitochondrial rRNAs in *N. crassa*. Likewise, Hollenberg et al. (1970) found that the mitochondrial genome of *S. cerevisiae* consists of unique sequences.

It is not known whether fungal mitochondria contain multiple DNA molecules as in other eukaryotes and whether these molecules are identical, but it is probable that in yeast mitochondria there are at least two molecules per organelle (Nass, 1969; Williamson, 1970). Pertinent to this aspect is the ultrastructural evidence recently presented by Hoffman & Avers (1973) that there is but one large mitochondrion per cell in at least one species of yeast.

Additional DNA species have been found in the cytoplasm of some fungi. Zoospores of *Blastocladiella emersonii* have from 8 to 16 organelles, designated gamma (γ) particles, which contain DNA with a distinct buoyant density comprising about 1.5% of the total cellular DNA (Myers & Cantino, 1971). Although the γ-particles disappear during germination, the role and fate of γ-DNA has not been established.

A class of non-mitochondrial closed circular DNA, termed omicron DNA (oDNA), has been isolated from *S. cerevisiae* (Sinclair et al., 1967; Guerineau et al., 1971; Clark-Walker, 1972). The oDNA comprises 2.9% of the total cell DNA, has the same buoyant density as nuc-DNA ($\rho = 1.701$), a contour length of 1.9 μm, and a molecular weight of 4 x 10^6 (Clark-Walker & Miklos, 1974). Several functions and origins of oDNA have been suggested. Using a respiratory deficient mutant of *S. cerevisiae* with an altered (0.05--1.6 μm mini circular) mit-DNA which enabled them to monitor cytoplasmic contamination of nuclear DNA preparations, Clark-Walker & Miklos (1974) concluded that oDNA is not a nuclear gene amplification product nor a type of the "informosomal" DNA reported in HeLa cells (Smith & Vinograd, 1972). The suggestion (Clark-Walker, 1972) that oDNA is localized within peroxisomes has not been eliminated. However, no circular DNA molecules other than mit-DNA were isolated from a species of *Saprolegnia* which has a large percentage of cyanide-insensitive respiration (Clark-Walker & Gleason, 1973).

Enzymes and Mechanisms of DNA Synthesis

The isolation (DeLucia & Cairns, 1969) of *E. coli* mutants (e.g., *pol* A-1) which lack DNA polymerase I activity has made possible the detection and isolation of two additional enzymes (DNA polymerase II and DNA polymerase III) involved in DNA synthesis in *E. coli* (see review by Smith, 1973). Since pol A-1 mutants grow and replicate DNA normally and are very sensitive to radiation, it is believed that DNA polymerase I is a DNA repair enzyme. Mutants with a thermolabile DNA polymerase III fail to replicate DNA at the restrictive temperature, suggesting that DNA polymerase III is the enzyme involved in DNA replication.

With these studies as a basis, Jeggo et al. (1973) studied DNA polymerase in temperature-sensitive mutants of *Ustilago maydis*. At the restrictive temperature (32° C) the mutant (*pol* 1-1) grows as uninucleate filamentous cells instead of budding yeast like cells, and after 5 h the cells contain only 10--25% of the polymerase activity present in extracts from wild type cells. In genetic analyses the temperature-sensitive phenotype and the reduced enzyme activity segregated together in 1:1 ratio and a diploid heterozygote exhibited a normal level of activity. It was concluded that the *pol* gene is the structural gene for the DNA polymerase and that a single recessive mutation in that gene results in the temperature-sensitive phenotype. Preliminary evidence suggested that more than one DNA polymerase exists and that the enzyme activity reduced by the mutation is not a repair enzyme.

DNA polymerase has also been studied in *S. cerevisiae* (Wintersberger & Wintersberger, 1970a) and *Rhizopus stolonifer* (Gong et al., 1973), but only in the latter species has the enzyme been investigated in relation to spore germination. The enzymes from both fungi have properties and requirements similar to DNA polymerases from other eukaryotes; the enzyme has an alkaline pH optimum and activity is dependent upon Mg^{2+}, DNA, a reducing agent, and four deoxyribonucleoside triphosphates. Activated (nicked) DNA is a better template than native DNA and, with both enzymes, greater activity is obtained with activated salmon sperm DNA than with activated calf thymus or homologous DNAs. On the other hand, mitochondrial DNA polymerase from *S. cerevisiae* exhibits considerable activity with many native DNAs and is equally active with native or denatured yeast mit-DNA (homologous) and with activated salmon sperm DNA (Wintersberger & Wintersberger, 1970b).

Gong et al. (1973) compared DNA polymerases isolated from dormant and germinated spores of *R. stolonifer*. The purified enzymes from both spore states were physically and chromatographically identical, and had the same reaction kinetics and requirements for activity. However, during early stages of purification, activity of the enzyme from germinated spores was independent of exogenous template DNA and, in glycerol density gradients, the enzyme sedimented heterogeneously and more rapidly than the enzyme from the dormant spore. The sedimentation rate was reduced considerably and the DNA independence was completely eliminated by treating the enzyme preparation with deoxyribonuclease prior to centrifugation. It was suggested that the enzyme is more tightly bound to DNA in germinated spores than in dormant spores.

Recent studies on yeast chromosomes (Petes et al., 1973) have demonstrated that DNA replication probably involves initiation at multiple sites within a molecule; molecules containing one or more "eye" structures and Y-shaped molecules have been observed in electron micrographs of yeast DNA. The same structures have been observed in replicating DNA from bacteriophage T7 (Dressler et al., 1972; Wolfson & Dressler, 1972) and from *Drosophila* (Kriegstein & Hogness, 1974) in which bidirectional synthesis is initiated at multiple sites. Hence, there is evidence that DNA replication in the fungi proceeds by mechanisms similar to those described for other organisms.

Characteristics of Dormant and Germinated Spore DNA

Regardless of the generally accepted concept of the metabolic stability of DNA, there are reports of qualitative alterations in DNA composition during the life cycle of several prokaryotic and eukaryotic organisms. In several cases the compositional changes are characteristic of a specific stage of the life cycle and may reflect changes in the number of reiterated sequences.

For example, a satellite DNA is present during certain stages of the life cycles of the bacteria *Caulobacter crescentus* (Shapiro et al., 1971) and *Bacillus cereus* T (Douthit & Halvorson, 1966). In both cases the satellite DNA disappears during the progression of the life cycle without altering the base composition (buoyant density) of the major DNA species. A similar situation exists in *Drosophila melanogaster* and *D. virilis* (Blumenfeld & Forrest, 1972); the amount of AT-rich satellite DNAs in adult insects is approximately 50% of that present in embryos. Different satellites and repeated sequences are apparently differentially under-replicated during polytenization which accompanies development.

DNA from germinated wheat embryos was shown by Chen & Osborne (1970) to have a lower GC content and to hybridize less with rRNA than DNA from dormant embryos. They suggested that rRNA gene amplification occurs in the embryo before maturation and that germination involves a deletion of rRNA cistrons and GC-rich regions of the chromosomal DNA. In contrast, Ingle & Sinclair (1972) determined the number of rRNA genes in dormant and germinating embryos of wheat and maize and in growing and nongrowing artichoke tuber explants and found no gross amplification or deletion of rRNA genes during germination and development.

The first comparison of fungal DNA from various morphological states was the accurate chemical analyses by Minagawa et al. (1959). They found no significant difference in base composition of DNA from resting and germinated conidia or mycelium of *N. crassa*. Similarly, based on chemical analyses of DNA from dormant and germinating spores of *Aspergillus oryzae*, Yanagita and co-workers (Kogane & Yanagita, 1964; Tanaka et al., 1965) concluded that DNA is not chemically altered during the early stages of germination. The buoyant density, thermal denaturation profiles, and spectral properties of DNA isolated from dormant spores of *Botryodiplodia theobromae* and *Rhizopus stolonifer* were indistinguishable from those of DNA isolated from germinated spores (Dunkle & Van Etten, 1972). These results corroborate the observations of Storck & Alexopoulos (1970), who indicated that they detected no difference in the GC content of DNA extracted from spores or mycelium of varying ages. Thus, in the fungi so far examined, spore formation and germination do not involve detectable amplification of genes which are then deleted or fail to replicate.

DNA Synthesis During Spore Germination

Evidence for nuc- and mit-DNA synthesis during spore germination has been obtained or inferred from the results of a variety of experimental approaches: cytological studies with light or electron microscopes, cell fractionation and colorimetric analyses, and radioactive precursor incorporation. Conclusions based on microscopic evidence are limited by the possibility that the number of nuclei or mitochondria may increase without prior DNA synthesis. For example, Van Assche & Carlier (1973) observed in germinating spores of *Phycomyces blakesleeanus* an increase in nuclear number 3--4 h before a significant incorporation of radioactive precursor was detected; they suggested that the dormant spore was in the premitotic (G2) phase of the cell cycle. Colorimetric analyses for DNA are often affected by contaminating substances in cell extracts, are limited by the sensitivity of the assay, and detect only net increases in DNA. Consequently, the onset of synthetic activity may actually precede the apparent time of synthesis by a considerable period. Radioactive labeling of DNA is hampered by the failure of most fungi to incorporate a precursor directly and specifically into DNA. Therefore, the labeled DNA must be isolated and separated from other nucleic acids before analysis. Grivell & Jackson (1968) established that the inability of many microorganisms (including *N. crassa, S. cerevisiae*, and *A. nidulans*) to incorporate thymidine into DNA results from an absence of thymidine kinase. Their observations have been confirmed for several other fungal species. Although numerous investigators have used thymidine to label DNA, only one report (Barash et al., 1967) stated (without the benefit of supportive data) that spore extracts contained thymidine kinase activity. Evidence has been presented that isolated mitochondria of *S. cerevisiae* (Wintersberger, 1966) and *N. crassa* (Koke et al., 1970) can incorporate deoxythymidine triphosphate into an acid precipitable product *via* a reaction that requires Mg^{2+} and four deoxynucleoside triphosphates (i.e., a DNA polymerase-dependent reaction).

Cytological studies show that nuclei divide during germination of spores of *Allomyces neo-moniliformis* (Olson & Fuller, 1971), *A. macrogynus* (Olson, 1973), *Geotrichum candidum* (Steele & Fraser, 1973), *Penicillium notatum* (Martin et al., 1973), *Phycomyces blakesleeanus* (Van Assche & Carlier, 1973), *Aspergillus nidulans* (Bainbridge, 1971), and *Trichoderma viride* (Rosen et al., 1974). In uredospores of the bean rust fungus *Uromyces phaseoli*, nuclear division occurs only in germ tubes which are induced to differentiate infection structures consisting of an appressorium, vesicle, and infection hyphae; the nuclei do not divide in germ tubes that are growing linearly (Maheshwari et al., 1967).

Changes in shape and appearance of mitochondria without any apparent increase in number have been observed in germinating spores of *Penicillium chrysogenum* (McCoy et al., 1971), *P. griseofulvum* (Fletcher, 1971), *Rhizopus arrhizus* (Hess & Weber, 1973), *Puccinia graminis* f. sp. *tritici* (Williams & Ledingham, 1964; Sussman et al., 1969), and *Cronartium fusiforme* (Walkinshaw et al., 1967). In contrast, increases in the number of mitochondria were reported during spore germination in *Rhizopus* spp. (Hawker & Abbott, 1963), *Botrytis cinerea* (Hawker & Hendy, 1963), *Mucor rouxii* (Bartnicki-Garcia et al., 1968), *Allomyces macrogynus* (Olson, 1973), *Geotrichum candidum* (Steele & Fraser 1973), *Penicillium notatum* (Martin et al., 1973), *Aspergillus niger* (Tsukahara, 1968), *A. nidulans* (Florance et al., 1972), *Sphaerotheca macularis* (Mitchell & McKeen, 1970), and *Trichoderma viride* (Rosen et al., 1974). In addition to observing an increase in the number of mitochondria in germinated *M. rouxii* spores, Bartnicki-Garcia et al. (1968) determined the proportion of the cytoplasm occupied by mitochondria and concluded that mitochondrial growth paralleled the overall growth of the cell. Zoospores of the water mold *Blastocladiella emersonii* contain a single, large mitochondrion (Cantino et al., 1963). Multiple mitochondrial profiles are observed in electron micrographs of synchronously germinating zoospores. By serial sectioning completely through zoospores in different stages of germination, Bromberg (1974) demonstrated that the mitochondrion initially undergoes drastic shape changes and then fragments into several separate mitochondria prior to germ tube emergence.

An increase in the number of mitochondria does not, of course, establish the occurrence of mit-DNA synthesis since mitochondria may contain multiple copies of DNA which are distributed into the new organelles. More conclusive evidence for mit-DNA synthesis was obtained by pulse-labeling the spores of *Botryodiplodia theobromae* at different stages of germination and then separating mit-DNA from nuc-DNA on CsCl density gradients (Dunkle & Van Etten, 1972). Initial incorporation of guanine into mit-DNA and nuc-DNA was detected at about the same time (ca. 2--3 h of germination), and synthesis of both components continued throughout the remainder of the germination period.

Similar studies were conducted with uredospores of *Uromyces phaseoli* (Staples, 1974). During germination and undifferentiated germ tube growth, labeled adenosine was incorporated only into mit-DNA. When the germ tubes stopped elongating upon differentiation of appressoria, mit-DNA synthesis ceased and nuc-DNA synthesis began. Thus, in this fungus, mit-DNA synthesis and nuc-DNA synthesis are apparently mutually exclusive processes during germination and germ tube differentiation, respectively.

Approaches to ascertain the essentiality of mit-DNA synthesis for spore germination have involved (1) pulse-labeling spores incubated in the presence of ethidium bromide, which selectively inhibits mit-DNA synthesis (Dunkle et al., 1972), and (2) studying the germination process in ascospores of a petite strain of yeast which lacks mit-DNA (Tingle et al., 1974). In the presence of ethidium bromide (5--50 μg/ml), spores of *Botryodiplodia theobromae* germinated at the same rate and to the same extent as spores incubated in the absence of the drug, but continued vegetative development was inhibited. Ethidium bromide inhibited the synthesis of mit-DNA without having an immediate effect upon the rates of nuc-DNA, protein, and RNA synthesis or of aerobic respiration. Thus, it was concluded that nuc- and mit-DNA synthesis are independent processes and that the activity of the mitochondrial genetic system is not essential to the spores until the germ tubes have developed and have begun rapid vegetative growth (Dunkle et al., 1972).

In wild-type ascospores of *S. cerevisiae* mit-DNA synthesis and an increase in cytochrome oxidase activity occur early in the germination process and nuc-DNA synthesis begins later in outgrowth (Tingle et al., 1974). Ascospores of a mit-DNA-less petite strain germinated at a rate and extent only slightly less than wild-type spores; no mit-DNA synthesis was detected and cytochrome oxidase activity decreased and remained at a very low level. Therefore, germination and outgrowth of *S. cerevisiae* ascospores are not dependent on a functional mitochondrial genome (Tingle et al., 1974).

The relationship of total DNA synthesis to the synthesis of other macromolecules during spore germination is indicated in Table 3. In general, DNA synthesis begins after the initiation of RNA and protein syntheses. Hence, DNA replication is apparently not required for early transcription.

In the majority of the fungi cited in Table 3, DNA synthesis begins prior to or at about the time of germ tube emergence. Consequently, it is not possible to establish with these criteria alone whether or not

DNA synthesis is essential for germination and germ tube growth. The difficulty of labeling DNA with a specific radioactive precursor further complicates the study of DNA synthesis. Other than the studies with ethidium bromide (cited above), only limited information is available from studies with inhibitors of DNA synthesis, e.g., mitomycin C, naladixic acid, hydroxyurea, and the base analogues 6-fluorodeoxyuridine, 5-fluorouracil, adenine arabinoside, and cytosine arabinoside. The effects of some of these inhibitors on germination and germ tube growth of *R. stolonifer* and *B. theobromae* are summarized in Table 4. None of the inhibitors of DNA synthesis except 5-fluorouracil, and possibly naladixic acid, affects the rate and extent of spore germination, although some of these antibiotics considerably reduce germ tube growth. However, it is not known whether the growth response is the direct result of inhibition of DNA synthesis or a secondary effect, since precursor incorporation into DNA was not monitored. But the tentative conclusion from these results is that DNA synthesis is not essential for germination.

RIBONUCLEIC ACIDS

Characteristics of Dormant and Germinated Spore RNA

Fungi contain all of the major classes of RNA. Ribosomal (r)RNAs comprise approximately 80% of the total cellular RNA; transfer (t)RNA and 5S RNA usually make up 15--20% and messenger (m)RNA approximately 5% or less of the total RNA. There may be additional species of RNA associated with DNA in chromatin, but these species, if they exist, comprise a small percentage of the total RNA. In addition, the mitochondria contain unique species of rRNAs, tRNAs, and mRNAs which will be discussed in a separate section.

Ribosomal RNA

Four distinct RNA classes, not including tRNA, have been isolated from fungal ribosomes. Large rRNA and small rRNA are present in the large and small ribosomal subunits, respectively (Udem et al., 1971; Van Etten, 1971). A 5S RNA species is associated with the large ribosomal subunit and an RNA species designated 5.8S RNA is tightly associated with, but noncovalently linked to, the large rRNA in yeast and probably other fungi (Udem et al., 1971).

Table 3. The Times at Which Germ Tubes Are Observed and RNA, DNA, and Protein Synthesis Occur During Fungal Spore Germination[a]

Fungus	Spore	Time, min[b]			
		RNA	Protein	DNA	Germ Tube
Category I					
Peronospora tabacina 1	Conidia	15	15	---	60
Rhizopus stolonifer 2,3	Sporangiospores	15	15	60--120	240
Phycomyces blakesleeanus 4	Sporangiospores	10	10	360--480	360--420
Neurospora crassa 5	Conidia	5	5	---	90--120
Fusarium solani 6	Macrospores	10	10	---	180
Microsporum gypseum 7	Macroconidia	10	20	0--60	180--240
Microsporum gypseum 8	Macroconidia	5	5	5	60--120
Lenzites saepiaria 9	Basidiospores	60	60	180	180
Allomyces arbuscula 10	Mitospores	20	20	---	30[e]
Category II					
Allomyces neo-moniliformis 11	Zoospores	10	5	---	---
Botryodiplodia theobromae 3,12	Pycnidiospores	120	15	120--180	240
Saccharomyces cerevisiae 13	Ascospores	20 or 60[c]	45	150	30[d]
Category III					
Blastocladiella emersonii 14	Zoospores	5	25	120	15--30[e]
Rhizophlyctis rosea 15	Zoospores	5	30	---	---
Aspergillus nidulans 16	Conidia	30	150	180	---
Aspergillus niger 17	Conidia	240	420	180	540

[a] The germinating spores are classified into three categories: Category I includes fungi in which RNA and protein initiate synthesis concomitantly; Category II includes fungi in which protein synthesis precedes RNA synthesis; and Category III includes fungi in which RNA synthesis precedes protein synthesis.

Table 3. (continued)

[b] In many cases these values were the first time periods measured. The values may not be exactly correct since many were estimated from figures.

[c] Adenine incorporation occurred within 20 min and uracil and cytosine incorporation at about 60 min.

[d] Germination monitored by a change in $A_{600\ nm}$.

[e] Retraction of the zoospore flagellum occurs within the first 10--20 min. A germ tube first appears between 20 and 30 min.

References

1. Hollomon, 1969
2. Van Etten et al., 1974
3. Dunkle & Van Etten, 1972
4. Van Assche & Carlier, 1973
5. Bhagwat & Mahadevan, 1970
6. Cochrane et al., 1971
7. Dill et al., 1972
8. Barash et al., 1967
9. Scheld & Perry, 1970
10. Burke et al., 1972
11. Olson & Fuller, 1971
12. Brambl & Van Etten, 1970
13. Rousseau & Halvorson, 1973a
14. Lovett, 1968
15. LeJohn & Lovett, 1966
16. Bainbridge, 1971
17. Yanagita, 1957

Table 4. The Effect of Inhibitors of Nucleic Acid Synthesis on Spore Germination and Growth of *Botryodiplodia theobromae* and *Rhizopus stolonifer*[a,b]

Inhibitor	Concentration (μg/ml)	*R. stolonifer* Germination	*R. stolonifer* Growth	*B. theobromae* Germination	*B. theobromae* Growth
Hydroxyurea	760	No	Slight	No	Slight
Mitomycin C	10	No	No	No	Slight
Ethidium bromide	10	No	No	No	Yes
	100	No	Yes	No	Yes
Naladixic acid	100	Yes	Yes	No	Yes
Proflavin sulfate	10	No	No	No	N.D.
	100	Yes	Yes	Yes	Yes
Acridine orange	10	No	No	No	N.D.
	100	Yes	Yes	Yes	N.D.
Actinomycin D	25	No	No	No	Yes
Lomofungin	5	Yes	Yes	No	Yes
Daunorubicin	2.5	Yes	Yes	No	No
Cordycepin	12.5	No	No	No	No
α-Amanitin	20	No	N.D.	N.D.	N.D.
5-Fluorouracil	100	Yes	Yes	No	Yes
β-Exotoxin from *Bacillus thuringiensis*	5	No	No	No	No
Rifampin	200	No	No	No	No

a No signifies that the drug had no effect on germination or growth; Yes means that it inhibited germination or growth; and N.D. means not determined.

b In many cases the lack of inhibition may be due to the failure of the drug to enter the cells.

The properties of large and small rRNAs in fungi have been summarized by Lovett (Lovett & Leaver, 1969; Lovett & Haselby, 1971). A range of sedimentation coefficients have been reported for small and large rRNAs; S values of 16 and 25.6 (Dure et al., 1967) and 19 and 28 (Henney & Storck, 1963b) have been described for *Neurospora crassa*. This difference probably results from variations in centrifugation conditions (Taylor et al., 1967). Molecular weight determinations of rRNAs from several fungi by polyacrylamide gel electrophoresis led to the assignment of molecular weights of 0.71--0.74 x 10^6 and 1.30--1.40 x 10^6 for small and large rRNAs, respectively (Lovett & Haselby, 1971). The apparent molecular weights of the small rRNAs were similar for all the fungi examined, whereas the molecular weights of the large rRNAs decreased slightly in size from 1.4 x 10^6 in the Oomycetes to about 1.31 x 10^6 in the Basidiomycetes (Lovett & Haselby, 1971). Small and large rRNAs will be referred to as 18S and 25S rRNAs, respectively.

The A_{260nm} ratio of 25S/18S rRNA from purified ribosomes is usually about 2--2.5:1 (Van Etten, 1971). There is one unverified report that 18S rRNA from *Neurospora crassa* separates into two RNA species in a ratio of 1:1 during chromatography on a methylated albumin-coated Kieselguhr column (Michelson & Suyama, 1968).

Ribosomal RNAs are synthesized in the nucleolus from a large precursor RNA which undergoes specific cleavages yielding 18S and 25S rRNAs. However, there is disagreement on the initial size of the precursor rRNA in yeast (cf. Udem & Warner, 1972; and Retel & Planta, 1970). The following scheme was suggested by Udem and Warner (1972). The 35S (2.5 x 10^6) precursor rRNA is cleaved to 27S (1.6 x 10^6) RNA and 20S (0.8 x 10^6) RNA; the 27S RNA is cleaved to 25S (1.3 x 10^6) rRNA and the 20S RNA to 18S (0.7 x 10^6) rRNA. The 25S and 18S rRNAs of yeast (Taber & Vincent, 1969; Udem & Warner, 1972), *Phycomyces blakesleeanus* (Gamow & Prescott, 1972), *Neurospora crassa* (Sturani et al., 1973), and *Rhizopus stolonifer* (Knight et al., 1974; Roheim et al., 1974) contain methylated bases as do rRNAs from other eukaryotic organisms (Davidson, 1972); methylation of the rRNAs occurs at the 35S precursor rRNA stage in yeast (Taber & Vincent, 1969; Udem & Warner, 1972).

The degree of base sequence homology between 18S and 25S rRNA is not known. Indeed, Retel & Planta (1968; 1970), Lovett & Leaver (1969), and Lovett & Haselby (1971), have suggested that extensive base sequence homologies exist between the two rRNAs. However, 18S rRNA and 25S rRNA from *Saccharomyces cerevisiae* do not compete with each other in DNA: RNA hybridization experiments (Schweizer et al., 1969).

The 5S RNA and 5.8S RNA from *S. cerevisiae* contain few, if any, methylated bases (Udem et al., 1971). The origin of the 5.8S RNA is not known, although Udem & Warner (1972) suggest that it arises from the 27S precursor rRNA. In contrast, the 5S RNA is probably synthesized independently of the precursor rRNA (Udem & Warner, 1972).

A 20S RNA species accumulates during sporulation of *S. cerevisiae* (Kadowaki & Halvorson, 1971a), and ultimately comprises about 4% of the total RNA in these cells. The relation of this 20S RNA to the 18S rRNA is not known. Evidence that it is not a precursor of 18S RNA include the difference in base composition between the 20S RNA (57% GC) and 18S rRNA (45% GC) (Kadowaki & Halvorson, 1971b) and the paucity of methylated bases in the 20S RNA molecule (Sogin et al., 1972). However, the 20S RNA species is transcribed from the nuclear DNA and exhibits approximately 70% homology with 18S rRNA in DNA: RNA competitive hybridization experiments (Sogin et al., 1972). One interpretation of these results is that during sporulation methylation of 35S precursor rRNA becomes limiting; this undermethylated precursor rRNA is processed to 20S RNA which cannot be processed to 18S rRNA because of the low content of methylated bases and consequently accumulates in these cells (Sogin et al., 1972).

Qualitative differences have not been detected in 18S and 25S rRNAs isolated from dormant and germinated fungal spores. For example, rRNAs isolated from conidia, ascospores, and hyphae of *Neurospora crassa* have identical sedimentation properties and base compositions (Henney & Storck, 1963a,b). Likewise, rRNAs from dormant and germinated conidia of *Aspergillus oryzae* have nearly identical base compositions (Horikoshi et al., 1965).

Leary and his colleagues examined ribosomes and rRNA from different spore types of *Phytophthora* spp. Ribosomes were isolated from the mycelium, zoospores, cysts, and chlamydospores of the fungi. However, intact ribosomes were not detected in extracts from oospores (Leary et al., 1974a) although rRNAs were present in all morphological stages of the fungus (Leary, Roheim, & Grantham, personal communication). However, these interesting observations should be regarded with caution, since the ribosomal profiles from all spore stages of the *Phytophthora* spp. were severely contaminated with other A_{254nm} absorbing materials.

Transfer RNA

Transfer RNA has been isolated from dormant and germinated spores of *Neurospora crassa* (Henney & Storck, 1963a,b), *Aspergillus oryzae* (Tanaka et al., 1966a; Horikoshi et al., 1969), *Uromyces phaseoli* (Staples, 1968), *Ustilago maydis* (Lin et al., 1971) *Rhizopus stolonifer* and *Botryodiplodia theobromae* (Van Etten et al., 1969; Merlo et al., 1972). Base compositions of tRNA isolated from dormant ascospores and conidia of *Neurospora crassa* were indistinguishable from tRNA isolated from hyphae (Henney & Storck, 1963a,b). Likewise, the base composition, thermal denaturation properties, and sedimentation profiles of tRNA isolated from dormant conidia, germinated conidia and hyphae of *Aspergillus oryzae* were nearly identical (Tanaka et al., 1966a). Transfer RNAs which had amino acid acceptor activity for all 20 amino acids commonly found in protein were isolated from dormant and germinated spores of *Botryodiplodia theobromae* (Van Etten et al., 1969) and *Rhizopus stolonifer* (Van Etten et al., 1969; Merlo et al., 1972). Likewise, amino acid acceptor activity for 13 amino acids was observed with tRNA isolated from dormant conidia and vegetative cells of *Aspergillus oryzae* (Horikoshi et al., 1969).

However, differences in some of the isoaccepting species of tRNA exist between dormant and germinated spores. Horikoshi et al. (1969) analyzed aminoacyl-tRNAs prepared from vegetative cells and conidia of *Aspergillus oryzae* by methylated albumin-Kieselguhr column chromatography. Of the 13 amino acids examined, differences were observed only in the isoaccepting species of tRNAs for lysine and methionine. Two isoaccepting tRNA species for lysine were present in conidia and three in vegetative cells; a quantitative difference in the two isoaccepting tRNA species for methionine was also observed between the two cell stages. Aminoacyl-tRNAs were synthesized with tRNA and aminoacyl-tRNA synthetases isolated from dormant and germinated spores of *Rhizopus stolonifer* and analyzed by cochromatography on benzoylated diethyl-aminoethyl cellulose columns (Merlo et al., 1972). No significant differences were detected for seven of the ten aminoacyl-tRNAs examined from dormant and germinated spores. The two isoaccepting species of lysyl-tRNA and valyl-tRNA differed quantitatively and isoleucyl-tRNA changed qualitatively during germination. Therefore, both qualitative and quantitative changes in tRNA isoaccepting species are associated with fungal spore germination. These differences could result from increased synthesis of the particular isoaccepting species of tRNA during germination or by modification of preexisting isoaccepting species. However, the significance of these observations can not be assessed until more is known about the metabolic function of isoaccepting species of tRNA.

A comparison of tRNA methylases and the ability of tRNAs from hyphae and conidia of *Neurospora crassa* to accept methyl groups also indicate that tRNA changes during spore germination (Wong et al., 1971). Transfer RNA and tRNA methylases were isolated from conidia and 12-h-old vegetative cells. The enzyme fraction from the hyphae catalyzed significant methylation of conidial tRNA, whereas conidial enzymes were less active with tRNA isolated from vegetative cells. Thus, the tRNA methylase content and tRNA species differ in conidia and vegetative cells.

Analysis of N-terminal amino acids of proteins from conidia of *N. crassa* revealed a preponderance of phenylalanine when compared with those of proteins isolated from germinated spores (Rho and DeBusk, 1971). Analysis of the tRNAs isolated from conidia revealed a unique species of $tRNA_{Phe}$ which was undermethylated and more susceptible to nucleases than the other two $tRNA_{Phe}$ (Jervis & DeBusk, 1975). Jervis & DeBusk (1975) proposed that this unique $tRNA_{Phe}$ species might occupy a specific site on the ribosomes in dormant spores and thereby prevent initiation of protein synthesis. When the spores are placed under conditions favorable for germination, this $tRNA_{Phe}$ is degraded by nucleases and protein synthesis begins.

Messenger RNA

There is less information about fungal mRNA than rRNA and tRNA because mRNA is heterogeneous in size and only comprises about 5% of the total RNA in the cell. Evidence from animal cells indicates that mRNA is derived by post-transcriptional modification of large precursor RNAs (45 to >100S), termed heterogeneous nuclear RNA (HnRNA) (Darnell, 1968; Darnell et al., 1973). After HnRNA is synthesized in the nucleus, adenine residues are added to the 3'-hydroxy end by a stepwise addition resulting in polyadenylate segments of 150 to 200 nucleotides. In animal cells as much as 90% of the HnRNA stays in the nucleus; the remaining HnRNA, containing most of the polyadenylate segment, moves into the cytoplasm and serves as mRNA. The synthesis of mRNA in fungi may occur by a similar process, although the HnRNAs may be smaller in size. Firtel & Lodish (1973) reported that the mRNA precursor in the slime mold *Dictyostelium discoideum* is only 20% larger than mature mRNA. If, indeed, precursor mRNAs in the true fungi are smaller than in higher eukaryotes, it would explain the inability to detect large HnRNAs in *Phycomyces blakesleeanus* (Gamow & Prescott, 1972) and *Rhizopus stolonifer* (Roheim et al., 1974).

Recent techniques developed for the isolation of mRNAs utilize the fact that most mRNAs in eukaryotic organisms contain a polyadenylate segment (histone mRNA may be an exception, Schochetman & Perry, 1972). These techniques consist of trapping mRNAs on columns containing polyuridylic acid (dT)-cellulose (Aviv & Leder, 1972), polydeoxythymidylic acid-cellulose (Kates, 1970), or on nitrocellulose filters (Lee et al., 1971; Brawerman et al., 1972). Ribosomal RNA and tRNA, in contrast to mRNA, elute from these columns in high ionic strength buffers; mRNA is eluted with low ionic strength buffers. Thus, mRNAs containing polyadenylate segments have been shown in yeast cells (Edmonds & Kopp, 1970; McLaughlin et al., 1973; Reed & Wintersberger, 1973), although McLaughlin et al. (1973) have suggested that not all yeast mRNAs contain polyadenylate segments. The average length of the polyadenylate segment in yeast mRNAs is 40--60 nucleotides long (McLaughlin et al., 1973; Reed and Wintersberger, 1973) which is considerably shorter than the 150--200 nucleotide residues present in mammalian mRNAs (Darnell et al., 1973). Like mRNAs from mammalian cells, yeast mRNAs are monocistronic (Peterson & McLaughlin, 1973).

An RNA fraction isolated from germinating spores of *Rhizopus stolonifer* was eluted from a poly(dT)-cellulose column as if it contained a polyadenylate segment (Knight et al., 1974; Roheim et al., 1974). This RNA fraction composes approximately 4--5% of the total RNA, sediments heterogeneously in sucrose density gradient columns, contains few methylated bases, and does not compete with rRNA in DNA: RNA competitive hybridization experiments. Since all of these characteristics are consistent with those of mRNA, it has been termed mRNA. Likewise, Lovett (personal communication) has shown that about 3% of the total RNA from zoospores of *Blastocladiella emersonii* binds to a poly(dT)-cellulose column in the manner expected for mRNA.

Since the presence of polyribosomes is *a priori* evidence for the presence of mRNA, there has been considerable interest among investigators in the existence of polyribosomes in dormant fungal spores. Henney & Storck (1964) were unable to detect polyribosomes in conidia and ascospores of *Neurospora crassa*, although polyribosomes were easily demonstrated in germinated spores. Likewise Kimura et al.

(1965) and Horikoshi et al. (1965) were unable to detect polyribosomes in conidia of *Aspergillus oryzae*. Therefore, these early experiments suggested that synthesis of mRNA or ribosomal attachment to mRNA might regulate protein synthesis in the dormant spore.

In contrast, more recent investigations indicate that at least some fungal spores contain small quantities of polyribosomes and thus contain mRNA. Good evidence exists for polyribosomes in uredospores of *Uromyces phaseoli* (Staples et al., 1968), pycnidiospores of *Botryodiplodia theobromae* (Brambl & Van Etten, 1970), conidia of *Neurospora crassa* (Mirkes, 1974), and sporangiospores of *Rhizopus stolonifer* (Freer & Van Etten, unpublished data). Data supporting the existence of polyribosomes in basidiospores of *Schizophyllum commune* (Leary et al., 1969), conidia of *Erysiphe graminis* f. sp. *tritici* (Leary & Ellingboe, 1971), and macroconidia of *Fusarium solani* (Cochrane et al., 1971) are less convincing.

An RNA fraction, which has properties of mRNA, was isolated from uredospores of *Uromyces phaseoli* (Ramakrishnan & Staples, 1970a) and conidia of *Neurospora crassa* (Bhagwat & Mahadevan, 1970). The mRNA fraction from uredospores of *U. phaseoli* sedimented between 4S and 19S and stimulated amino acid incorporation into protein using a cell free protein synthesizing system from *E. coli*. Base analysis of this RNA fraction revealed that it was rich in AMP. The mRNA fraction from conidia of *N. crassa* hybridized to 2% of the cellular DNA and stimulated amino acid incorporation into protein using a cell free protein synthesizing system from mycelium of *N. crassa*.

That protein synthesis occurs in the absence of detectable RNA synthesis immediately after *B. theobromae* spores are placed under germinating conditions also suggests that the dormant spore contains mRNA (Brambl & Van Etten, 1970). Protein synthesis occurs in germinating zoospores of *Blastocladiella emersonii* in the presence of sufficient actinomycin D to inhibit RNA synthesis (Lovett, 1968). Although this observation suggests that mRNA is present in these zoospores, polysomes have not been observed (Schmoyer & Lovett, 1969). Likewise, conidia of *Peronospora tabacina* probably contain preformed mRNA since protein synthesis occurs during the germination of these spores in the presence of ethidium bromide or 5-fluorouracil (Holloman, 1971, 1973). An *in vitro* protein synthesizing system prepared from these conidia incorporated amino acids in the absence of added mRNA. Hollomon presented evidence that this mRNA is associated with a membranous fraction that sediments during low speed centrifugation. In conclusion, it appears that at least some fungal spores contain preformed mRNA.

However, some reservations must be noted in regard to the presence of polyribosomes in dormant spores. In many instances in which polyribosomes were identified in dormant spores, the spores were harvested in water. Even though dry- and water-harvested spores are morphologically similar, it is possible that water stimulates polyribosome formation. For example, water could cause polyribosome formation by stimulating the attachment of preexisting mature mRNA to the ribosomes, de novo synthesis of mRNA, or the attachment of adenylate residues to preexisting mRNA. A recent report (Mirkes, 1974) indicates that a short exposure of *Neurospora crassa* conidia to water increases the polyribosome content ten fold. Likewise, preliminary data obtained with *Rhizopus stolonifer* suggest that, whereas polyribosomes are present in water-harvested sporangiospores, they are not detected in dry-harvested spores (Freer & Van Etten, unpublished data). However, an RNA fraction can be isolated from these spores which elutes from a poly(dT)-cellulose column in a manner characteristic of mRNA.

Thus, the question of whether dry, dormant fungal spores contain mRNA specifically associated with ribosomes has not been satisfactorily resolved; it is possible that fungal spores differ in this regard. The best evidence suggests that some fungal spores contain latent mRNA(s); however, it is not known how this mRNA(s) is stored in the spore.

RNA Polymerases

Roeder & Rutter (1969) found multiple DNA-directed RNA polymerases in sea urchin embryos and rat liver nuclei and demonstrated that these enzymes transcribed specific classes of RNA. Since then multiple RNA polymerases have been reported in many eukaryotic organisms (cf. review by Jacob, 1973). Studies with isolated nuclei from mammalian cells have led to the concept that RNA polymerase I directs the synthesis of rRNA and RNA polymerase(s) II directs the synthesis of HnRNA (a precursor of mRNA).

Multiple RNA polymerases have been reported in fungi; two (Sebastian et al., 1973), three (Ponta et al., 1972), or four (Adman et al., 1972) have been isolated from yeast, three from germinated spores of *Rhizopus stolonifer* (Gong & Van Etten, 1972), three from vegetative cells of *Blastocladiella emersonii* (Horgen & Griffin, 1971a), *Achlya bisexualis* (Timberlake et al., 1972b), *Allomyces arbuscula*

(Cain & Nester, 1973), three from the yeast phase of *Histoplasma capsulatum* (Boguslawski et al., 1974), and four from the slime mutant of *Neurospora crassa* (Tellez de Inon et al., 1974). In contrast, only one RNA polymerase was detected in germinating uredospores of *Uromyces phaseoli* (Manocha, 1973). Usually RNA polymerases from fungi are relatively unstable and transcribe denatured DNA more efficiently than native DNA.

Indirect data support the concept that RNA polymerases I, II, and III from fungi serve similar functions to those of other eukaryotic organisms. The effects of increasing concentrations of $(NH_4)_2SO_4$ on enzyme activity of each of the three RNA polymerases from *R. stolonifer* and *A. arbuscula* were similar to those in mammalian systems. That α-amanitin only inhibits RNA polymerase II in *B. emersonii, A. arbuscula*, and *A. bisexualis*, and to a slight extent in *R. stolonifer* supports the concept that fungal RNA polymerase II corresponds to RNA polymerase II of other eukaryotic organisms (Jacob, 1973). Horgen & Griffin (1971a) reported that RNA polymerase I from *B. emersonii* was specifically inhibited by high concentrations of cycloheximide (> 100 μg/ml) and RNA polymerase III by high concentrations of rifampin (200 μg/ml). RNA polymerase I from *A. bisexualis* was also specifically inhibited by concentrations of cycloheximide as low as 0.2 μg/ml (Timberlake et al., 1972b). In contrast, very high concentrations of cycloheximide and rifampin do not have a selective effect on the activities of RNA polymerases I and III from *R. stolonifer* (Gong & Van Etten, 1972), *N. crassa* (Tellez de Inon et al., 1974), and *A. arbuscula* (Cain & Nester, 1973). A subsequent report from Griffin's laboratory (Timberlake et al., 1972a) indicated that cycloheximide inhibited RNA polymerase I only if the enzyme was purified by a specific procedure. Horgen & Griffin (1971b) reported that RNA polymerase III in *B. emersonii* is of mitochondrial origin. Mitochondria from *N. crassa* and yeast also contain a separate RNA polymerase(s), although there is disagreement on the number and properties of the mitochondrial enzyme(s) (see the section on Mitochondrial RNA).

It is difficult to evaluate the RNA polymerase isolated from germinating uredospores of *Uromyces phaseoli* (Manocha, 1973); enzyme activity is unaffected by cycloheximide and α-amanitin but is inhibited 58 and 80% by actinomycin D and rifampin, respectively. The rifampin sensitivity of the enzyme indicates that it might be of mitochondrial origin although Manocha suggested a nuclear origin.

RNA polymerases I and II from germinated spores of *Rhizopus stolonifer* (Gong & Van Etten, 1974) and RNA polymerases A(I) and B(II) from *Saccharomyces cerevisiae* (Ponta et al., 1972) have been purified several hundred fold. RNA polymerase I from *R. stolonifer* contains three major subunit proteins, with estimated molecular weights of 180,000, 104,000, and 32,000 and RNA polymerase II contains four major subunit proteins, with estimated molecular weights of 215,000, 114,000, 37,000, and 27,000. RNA polymerase A(I) from *S. cerevisiae* had two large polypeptides of molecular weight 190,000 and 135,000 and RNA polymerase B(II) contained two large polypeptides of molecular weight 175,000 and 140,000. Therefore, the two large subunit polypeptides in fungal RNA polymerases are similar in size to those of RNA polymerases I and II from other eukaryotic organisms. The RNA polymerases I and II from *S. cerevisiae* are related immunologically (Hildebrandt et al., 1973).

RNA polymerases isolated from vegetative cells and zoospores of *B. emersonii* (Horgen, 1971) or *A. arbuscula* (Cain & Nester, 1973) were indistinguishable. In contrast, both qualitative and quantitative differences were reported for RNA polymerases isolated from dormant and germinated sporangiospores of *R. stolonifer* (Gong & Van Etten, 1972). Germinated (6 h) spores contained three RNA polymerases (I, II, and III), whereas dormant and swollen (2 h) spores yielded only RNA polymerases I and III. RNA polymerase III was indistinguishable between the two spore states, whereas RNA polymerase I from the dormant spores responded to divalent cations differently and eluted from a DEAE-cellulose column earlier than the corresponding fraction from germinated spores. RNA polymerase II was first detected 3 h after the spores were placed on a germination medium and were initiating germ tube formation. At the same time the characteristics of RNA polymerase I were altered so that they resembled those of germinated spore enzyme. Perhaps the dormant spore is incapable of synthesizing mRNA until RNA polymerase II is synthesized 2--3 h into the germination process. Furthermore, the synthesis of RNA polymerase II would have to be translated from preexisting mRNA in the spores. However, during *R. stolonifer* spore germination the synthesis of tRNA, rRNA, and mRNA all began within the first 15 min of germination (Knight et al., 1974; Roheim et al., 1974; see the section, Pattern of RNA Synthesis).

Ribonucleases

Ribonucleases (RNases) have been isolated from many eukaryotic organisms including fungi (Barnard, 1969). Since the catalytic activity of many of these RNases depends on nucleotide sequences, they are used to determine RNA base sequences (Gilham, 1970). Some of the RNases from fungi which have been studied include RNase T_1 and RNase T_2 from *Aspergillus oryzae* (Egami et al., 1964), RNase U_2 from *Ustilago sphaerogena* (Arima et al., 1968 a,b), RNase Rh from *Rhizopus* sp. (Komiyama and Irie, 1971), RNase N from *Neurospora crassa* (Takai et al., 1966), and a nuclease from *N. crassa* which is specific for single stranded RNA and DNA (Rabin & Fraser, 1970). In addition, RNase activity is associated with ribosomes purified from *N. crassa* (Somberg & Davis, 1965; Ursino et al., 1969).

Horikoshi and his associates have investigated RNase activity during the germination of *Aspergillus oryzae* conidia. A cell free protein synthesizing system prepared from conidia of *A. oryzae* was less active than a similar preparation from germinated spores (Horikoshi & Ikeda, 1968; 1969), suggesting that conidia might contain more RNase activity than germinated spores. Investigation of RNase activity in a high speed (105,000 g) supernatant fraction from dormant conidia revealed three types of RNase activity; one RNase (designated as thermal labile RNase) was inactivated by heating at 55° C for 5 min, whereas a second one was stable to this treatment. In addition, a latent RNase was activated by several treatments, including heating at 85° C for 5 min, a pH of less than 5.0, proteolytic enzymes, or high (4--8 M) concentrations of urea (Horikoshi, 1971; Horikoshi & Ando, 1972). The latent RNase was not detected in germinated conidia, whereas thermal labile RNase activity was unchanged with germination.

In summary, fungi contain RNases with varying specificities. The involvement, if any, of these RNases in spore germination is not known. However, it is conceivable that spore specific RNAs (most likely mRNAs) are protected in some manner, e.g., bound to a membrane or to a subribosomal particle. During germination, this RNA could be cleaved in a specific fashion allowing it to interact with the ribosomes and synthesize a key protein(s).

Mitochondrial RNA

Mitochondria from vegetative cells of fungi contain rRNAs, tRNAs, and mRNAs which are transcribed from mitochondrial DNA. The majority of the mitochondrial RNA is associated with mitochondrial ribosomes and this rRNA differs distinctly from cytoplasmic rRNA (Table

5). Mitochondrial rRNAs like cytoplasmic rRNAs, however, contain methylated bases (Kuriyama & Luck, 1973). Molecular weights of 1.27×10^6 and 0.66×10^6, and 1.30×10^6 and 0.70×10^6 have been assigned to large and small mitochondrial rRNAs, respectively, from *Aspergillus nidulans* (Verma et al., 1970) and *Saccharomyces carlsbergensis* (Reijnders et al., 1973). However, difficulties arise in assigning molecular weights to mitochondrial rRNAs because their electrophoretic mobility in polyacrylamide gels is slower than that predicted from their sedimentation rates in sucrose density gradients. This phenomenon is exemplified by comparative studies of cytoplasmic and mitochondrial rRNAs from *A. nidulans* and rRNAs from *Escherichia coli* (Edelman et al., 1970; Verma et al., 1970; 1971). The mitochondrial rRNAs from *A. nidulans* have sedimentation values (15.5S and 23.5S) which are similar to rRNAs from *E. coli,* but lower than the cytoplasmic rRNAs of *A. nidulans* (17S and 26.5S). However, the mitochondrial rRNAs migrate at the same rate as the cytoplasmic rRNAs and slower than rRNA from *E. coli* on polyacrylamide gels. The base composition of the mitochondrial rRNAs (32% GC) differs from that of cytoplasmic rRNAs (51% GC). Also thermal denaturation studies and two dimensional electrophoresis of RNase T_1 digests of small and large rRNAs from the cytoplasm and from the mitochondria indicate major differences in all four species of rRNA.

Mitochondrial ribosomes of *Neurospora crassa* (Lizardi & Luck, 1971) and yeast (Grivell cited in Borst, 1972) have no detectable 5S RNA. However, if such an RNA were similar in size to tRNA (i.e., 4S), it would be difficult to detect.

Mitochondria also contain unique species of tRNA. Mitochondria of *Neurospora crassa* contain at least fifteen tRNAs which differ in their chromatographic properties from cytoplasmic tRNAs (Epler, 1969). Likewise, yeast mitochondria contain at least 14 unique tRNAs (Rabinowitz et al., 1974). Reijnders & Borst (1972) have shown that mitochondrial tRNA hybridizes with about 0.9% of the mitochondrial DNA. Assuming that yeast mitochondrial DNA weighs 55×10^6 daltons and that tRNAs weigh 25,000 daltons, then mitochondrial DNA would code for 20 tRNAs. Additional differences between mitochondrial and cytoplasmic tRNAs of *N. crassa* were revealed by studies with aminoacyl-tRNA synthetases. Mitochondria contained at least 15 aminoacyl-tRNA synthetases (Barnett et al., 1967). The mitochondrial leucyl-, phenylalanyl-, and aspartyl-tRNA synthetases were distinct from the corresponding cytoplasmic enzymes. The mitochondrial enzymes acylated only mitochondrial tRNAs and were inactive

Table 5. A Comparison of Properties of Ribosomal RNA Isolated From the Cytoplasm and Mitochondria of Three Fungi

Fungus	Source of RNA	rRNA	S Value	Base composition (mole %) C	A	G	U	G+C	G/U
Saccharomyces cerevisiae 1	Cytoplasm	Small	18	17.0	26.4	28.2	28.4	45.2	0.99
		Large	26						
	Mitochondria	Small	15	12.5	36.1	17.7	34.5	30.2	0.51
		Large	25	12.5	38.4	16.5	32.8	28.7	0.50
Neurospora crassa 2,3,4	Cytoplasm	Small	17	21.6	25.3	27.7	25.4	49.3	1.09
		Large	25						
	Mitochondria	Small	16	16.0	31.8	20.4	31.7	36.4	0.64
		Large	23	15.0	33.9	19.1	31.9	34.1	0.60
Aspergillus nidulans 5	Cytoplasm	Small	17	23	24	28	23	51	1.1
		Large	26.5						
	Mitochondria	Small	15.5	13	30	19	38	32	0.5
		Large	23.5						

1. Morimoto et al., 1970
2. Dure et al., 1967
3. Küntzel & Noll, 1967
4. Rifkin et al., 1967
5. Edelman et al., 1970

with a crude preparation of cytoplasmic tRNA. Conversely, the corresponding three cytoplasmic enzymes were more active with cytoplasmic tRNAs than with the mitochondrial tRNAs, although the pattern of specificity was not as absolute as it was with the mitochondrial enzymes. N-Formyl-methionyl-tRNA, which initiates protein synthesis in bacteria, is present in mitochondria from yeast (Smith & Marcker, 1968) and *N. crassa* (Epler et al., 1970). *Neurospora crassa* mitochondria contain two methionyl-tRNAs, only one of which is formylated by *E. coli* transformylase (Epler et al., 1970). The mitochondrial extracts contained transformylase activity and this enzyme formylated the same met-tRNA that was formylated by the *E. coli* enzyme. Attempts to identify transformylase activity in the cytoplasm of *N. crassa* were unsuccessful.

Cumulative evidence over the last few years indicates that mitochondria from *Neurospora crassa* and *Saccharomyces cerevisiae* synthesize mRNA which is translated on mitochondrial ribosomes. Most, if not all, of the resultant polypeptides are located in the inner mitochondrial membrane (e.g., Borst, 1972). Recently, polyribosomes have been isolated from yeast (Cooper & Avers, 1974) and *N. crassa* mitochondria (Agsteribbe et al., 1974). Preliminary evidence indicates that the mRNAs isolated from yeast mitochondria, like those in the cytoplasm, contain polyadenylic acid segments (Cooper & Avers, 1974).

Since isolated mitochondria from *N. crassa* (Luck & Reich, 1964) and yeast (Wintersberger & Tuppy, 1965; Wintersberger & Wintersberger, 1970c; South & Mahler, 1968) synthesize RNA, fungal mitochondria must contain an RNA polymerase(s). Studies on the properties of RNA polymerase from yeast and *N. crassa* mitochondria have resulted in disagreement on the number and molecular weight of the enzyme(s) and their sensitivity to inhibitors such as rifampin. An RNA polymerase isolated from *N. crassa* mitochondria (Küntzel & Schäfer, 1971) consisted of a single polypeptide chain with a molecular weight of about 64,000, utilized native mit-DNA as a template and was inhibited by rifampin. Likewise, Scragg (1974) reported the isolation of an RNA polymerase from *S. cerevisiae* mitochondria which had many of the properties of the *N. crassa* enzyme. The enzyme weighed 65,000--70,000 daltons and was inhibited by rifamycin. The enzyme preferred denatured DNA as a template but it also functioned with yeast mit-DNA. In contrast, Wintersberger (1972) isolated RNA polymerases from mitochondria of yeast and *N. crassa* which were insensitive to rifampicin. Eccleshall & Criddle (1974) found three mitochondrial RNA polymerases in yeast which weighed about 200,000 daltons, had multiple subunits, were insensitive to rifampicin and, in general, were similar to nuclear RNA polymerases.

In conclusion, mitochondria from vegetative cells of fungi and probably from fungal spores contain RNA and RNA polymerase(s) which are distinct from their cytoplasmic and nuclear counterparts. The role of these mitochondrial RNAs in dormant fungal spores and their involvement in fungal spore germination awaits investigation.

RNA Synthesis During Spore Germination

Among the earliest events of fungal spore germination are the initiation of RNA and protein syntheses. Protein synthesis is apparently essential for germ tube formation in all fungal spores although morphological changes, such as encystment of zoospores can occur in its absence (Lovett, 1968; Soll & Sonneborn, 1971a,b). Ribonucleic acid synthesis, however, may not be essential for germ tube formation for all fungal spores, even though RNA synthesis usually begins early in the germination process. In relating protein and RNA synthesis to spore germination, several considerations should be noted. (1) Protein and RNA synthesis can usually be detected early in the germination process by using radioactive precursors. Determinations of protein and RNA by less sensitive, conventional analytical procedures usually detect net increases in RNA and protein about the time of germ tube emergence. (2) Most studies that have relied on the incorporation of labeled precursors to detect RNA and protein synthesis have not considered the effect of endogenous precursor pools or phase-specific transport mechanisms. In addition, many such studies have not presented conclusive evidence that the precursor is actually incorporated into the macromolecule under investigation. This is especially important when long-term labeling experiments are conducted. For example, uracil can be converted to other bases which are incorporated into DNA as well as RNA in many fungi. (3) The age of the dormant spore may influence precursor incorporation. Ramakrishnan & Staples, (1971) observed that uredospores of *Uromyces phaseoli* stored for four days incorporated less uridine into RNA than freshly harvested spores although there were no differences in the viability of the spores.

Germinating fungal spores can be divided into three categories on the basis of the relative times at which RNA and protein syntheses begin during the germination process (Table 3). Category I includes spores in which RNA and protein syntheses are initiated concomitantly, usually shortly after the spores are placed in a germination medium. These spores must contain active RNA polymerases and preexisting mRNA. Fungi which fall into this class include *Allomyces arbuscula* (Burke et al., 1972), *Peronospora tabacina* (Holloman, 1969), *Rhizopus stolonifer* (Van Etten et al., 1974), *Phycomyces blakesleeanus* (Van Assche & Carlier, 1973), *Neurospora crassa* (Bhagwat & Mahadevan, 1970), *Fusarium solani* (Cochrane et al., 1971), *Microsporum gypseum* (Barash et al., 1967), and *Lenzites saepiaria* (Scheld & Perry, 1970). Category II consists of fungal spores in which protein synthesis precedes RNA synthesis. Spores in this group must also contain preexisting mRNA. The pattern of synthesis of macromolecules in these fungal spores is similar to that observed during development of higher eukaryotes, e.g., during activation of sea urchin eggs and *Xenopus* oocytes (Nemer, 1967; Gross, 1967). Fungi included in this class are *Allomyces neo-moniliformis* (Olson & Fuller, 1971), *Botryodiplodia theobromae* (Brambl & Van Etten, 1970), and possibly ascospores of *Saccharomyces cerevisiae* (Rousseau & Halvorson, 1973a). Fungi in Category III initiate RNA synthesis prior to protein synthesis during the germination process. These spores, therefore, must contain RNA polymerases. Fungi in this group include *Aspergillus niger* (Yanagita, 1957), *Rhizophlyctis rosea* (LeJohn & Lovett, 1966), *Blastocladiella emersonii* (Lovett, 1968), and *Aspergillus nidulans* (Bainbridge, 1971). However, it is likely that some fungi may be reassigned to other categories after they are studied under more precisely defined conditions.

The pioneering experiments by Yanagita and his colleagues (Yanagita, 1957; Nishi, 1961; Hoshino et al., 1962; Yanagita, 1963; Tsay et al., 1965) demonstrated the importance of RNA synthesis to spore germination. They first observed a net synthesis of RNA, followed by increases in DNA and protein during germination of *Aspergillus niger* conidia. In addition, ^{14}C-adenine and $^{14}CO_2$ were incorporated into RNA within 30 min after the spores were placed under germinating conditions. Since these original studies, many investigators have examined the association between RNA synthesis and spore germination.

Ribosomal RNA

Ribosomal RNA synthesis occurs during spore germination of *Aspergillus oryzae* (Ono et al., 1966; Tanaka et al., 1966b), *Rhizophlyctis rosea* (LéJohn & Lovett, 1966), *Uromyces phaseoli*

(Ramakrishnan & Staples, 1967; Stallknecht & Mirocha, 1971), *Ustilago maydis* (Lin et al., 1971), *Neurospora crassa* (Hollomon 1970; Mirkes, 1974), *Blastocladiella emersonii* (Leaver & Lovett, 1974), *Melampsora lini* (Chakravorty & Shaw, 1972), *Phycomyces blakesleeanus* (Van Assche & Carlier, 1973), *Rhizopus stolonifer* (Knight et al., 1974; Roheim et al., 1974), and *Allomyces macrogynus* (Smith & Burke, 1974). The synthesis of rRNA was detected almost immediately after *A. oryzae* conidia were placed on a germination medium (Ono et al., 1966; Tanaka et al., 1966b). When conidia were placed in a medium unfavorable for germination, maturation of the rRNA was delayed. Thus, these investigators suggested that synthesis of rRNA or its precursors might be one of the first events of germination. Subsequent studies of other fungi have generally supported this concept with one possible exception. Hollomon (1970) reported that rRNA was not synthesized during germination of *Peronospora tabacina* conidia although an earlier study (Hollomon, 1969) clearly indicated that rRNA synthesis occurred.

The synthesis of rRNA and precursor rRNAs is detected within 20--30 min of *Blastocladiella emersonii* zoospore germination (Leaver & Lovett, 1974). However, RNA synthesis is not required for germination since the zoospores germinate in the presence of sufficient actinomycin D to inhibit most RNA synthesis (Lovett, 1968). Likewise, the synthesis of 18S, 25S, and 5S rRNAs can be detected within the first 15 min of germination of *Allomyces macrogynus* mitospores (Smith & Burke, 1974). In these mitospores large and small rRNAs are synthesized in a one-to-one molecular ratio, but the 5S rRNA is synthesized at a significantly greater rate than either of the larger rRNAs. Analysis of RNA extracted from germinating sporangiospores of *Rhizopus stolonifer*, which had been pulsed with ^{3}H-adenine for 15-min periods, revealed incorporation of radioactivity into two precursor rRNAs, large and small rRNAs, and 5S rRNA during the first 15 min of germination (Knight et al., 1974; Roheim et al., 1974).

Total cellular RNA was analyzed in the studies cited above but not the rate of incorporation of the newly synthesized rRNA into ribosomes. That rRNA is processed and incorporated into ribosomes very rapidly is implied from experiments with germinating conidia of *Neurospora crassa* (Mirkes, 1974). Since labeled rRNA was isolated from polyribosomes 5 min after the initiation of germination, rRNAs were synthesized, assembled into ribosomes, and the ribosomes attached to mRNA within 5 min.

Transfer RNA

Synthesis of tRNA during fungal spore germination has not been examined as thoroughly as rRNA synthesis. As indicated (see section on Characteristics of tRNA from Dormant and Germinated Spores), several investigations have revealed differences in isoaccepting species of tRNAs between dormant and germinated spores. These differences could result from *de novo* synthesis of tRNA, molecular modification of specific nucleotides in preexisting tRNA, or by cytosine and adenosine addition to the 3'-hydroxy end of the tRNA.

Transfer RNA initiates synthesis between 3 and 15 min after conidia of *Aspergillus oryzae* are placed on a germination medium (Tanaka et al., 1966b; Ono et al., 1966). Hollomon (1970) reported that tRNA was synthesized during the first 20 min and 60 min of germination of *Peronospora tabacina* and *Neurospora crassa* conidia, respectively. Mirkes (1974) found labeled tRNA associated with polyribosomes within 15 min after placing *N. crassa* conidia in germinating conditions. Transfer RNA is also synthesized during the first 15 min of germination of mitospores of *Allomyces macrogynus* (Smith & Burke, 1974) and of sporangiospores of *Rhizopus stolonifer* (Knight et al., 1974; Roheim et al., 1974). In the latter report, the incorporation of ^{3}H-adenine into tRNA resulted from *de novo* synthesis of the tRNA and not simply from adenine addition to the 3'-hydroxy end. Germination of uredospores of *Uromyces phaseoli* may present an exception to the early synthesis of tRNA since tRNA synthesis is not observed until after germ tube formation (Ramakrishnan & Staples, 1967, 1970b; Stallknecht & Mirocha, 1971).

Since Smith and Burke (1974) reported that 5S rRNA was synthesized more rapidly than 4S RNA during the early stages of germination of *A. macrogynus* mitospores, it is necessary that 5S rRNA be separated from tRNA (4S) in studies of tRNA synthesis. If this differential rate of synthesis occurs in other spores, earlier studies which relied solely on sucrose density gradients for the analysis of RNA synthesis may be in error since centrifugation does not separate tRNAs from 5S RNA.

Messenger RNA

As indicated (see section on Characteristics of mRNA from Dormant and Germinated Spores), some dormant fungal spores contain preexisting mRNA which may or may not be associated with ribosomes.

Satisfactory methods for establishing unequivocably that mRNA is synthesized during germination have only recently been developed. However, several earlier reports strongly suggest that mRNA synthesis occurs during germination.

Ono et al. (1966) reported that an RNA fraction, which eluted from a methylated albumin Kieselguhr column in the region expected for mRNA, was synthesized about 60 min after the initiation of *Aspergillus oryzae* conidia germination. This observation is consistent with the appearance of polyribosomes 60 min into the germination process of these spores (Kimura et al., 1965). Likewise, a heterogeneous and metabolically unstable RNA is detected 30--60 min after the initiation of germination of *Peronospora tabacina* conidia (Hollomon, 1970). Uridine and $^{32}PO_4$, respectively, are incorporated into a heterogeneous RNA fraction as early as 15 min into germination of *Blastocladiella emersonii* zoospores (Leaver & Lovett, 1974) and *Allomyces macrogynus* mitospores (Smith & Burke, 1974).

Bhagwat & Mahadevan (1970) reported synthesis of mRNA during the first 30 min of germination of *Neurospora crassa* conidia. Subsequently, Mirkes (1974) reported incorporation of ^{3}H-adenine into RNA, which sedimented as mRNA, during the first 5--15 min of germination. Since this RNA fraction was isolated from polyribosomes, the synthesis of mRNA and its association with ribosomes must be rapid. Likewise, ^{3}H-adenine was incorporated into a polyadenylate containing RNA within the first 15 min of germination of *Rhizopus stolonifer* sporangiospores (Knight et al., 1974; Roheim et al., 1974).

In these latter two studies RNA synthesis was monitored with ^{3}H-adenine and it is possible that adenine is incorporated into preexisting spore precursor mRNA *via* polyadenylate addition rather than by *de novo* synthesis of mRNA. However, some polyadenylate containing RNA is synthesized *de novo* during the germination of *R. stolonifer* spores since uracil is also incorporated into this fraction during the first 15 min of germination (Knight et al., 1974; Roheim et al., 1974). Subsequent studies in our laboratory have revealed that the majority, if not all, of the adenine incorporated into the polyadenylate containing RNA in the first 15 min of germination of *R. stolonifer* is due to *de novo* synthesis (Van Etten & Mayama, unpublished data).

Ramakrishnan & Staples (1970b) reported that germinating uredospores of *Uromyces phaseoli* incorporated very little precursor into a mRNA fraction, whereas, if the germinating spores were induced to form appressoria, mRNA synthesis occurred. These data are consistent with the observation that actinomycin D has no effect on germination of *Puccinia graminis tritici* uredospores but does inhibit appressoria formation (Dunkle et al., 1969).

The synthesis of the enzymes α-amylase, invertase, and glucose dehydrogenase can be induced in *Aspergillus oryzae* conidia under nongerminating conditions (Sinohara, 1970). Therefore, mRNA synthesis can occur in ungerminated spores and, in this case, all three mRNAs had a half-life of about 20 min. As far as we are aware there have been no attempts to determine the half-lives of mRNAs normally synthesized during spore germination.

Pattern of RNA Synthesis

The various classes of RNA initiate synthesis sequentially during germination of some fungal spores. However, the sequence varies with the fungal species. Ono et al. (1966) and Tanaka et al. (1966b) found that rRNA synthesis in *Aspergillus oryzae* conidia began immediately after the initiation of germination and that tRNA synthesis began a few minutes later. A mRNA-like fraction was synthesized approximately 60 min into the germination process. In *Peronospora tabacina* tRNA synthesis began during the first 20 min of germination and an RNA fraction identified as mRNA was synthesized at 20--40 min (Hollomon, 1970). In *Uromyces phaseoli* uredospores rRNA synthesis began 90 min after initiation of germination, whereas tRNA and mRNA syntheses were detected only after germ tubes appeared (Ramakrishnan & Staples, 1967, 1970a, 1970b).

In contrast, recent studies with *Rhizopus stolonifer* sporangiospores (Knight et al., 1974; Roheim et al., 1974), *Neurospora crassa* conidia (Mirkes, 1974), *Allomyces macrogynus* mitospores (Smith & Burke, 1974), and *Blastocladiella emersonii* zoospores (Leaver & Lovett, 1974) suggest that all classes of RNA are synthesized early and probably concomitantly during germination. However, proportionately more mRNA may be synthesized during the first few min of germination (Knight et al., 1974; Roheim et al., 1974; Smith & Burke, 1974; Leaver & Lovett, 1974). Precursor incorporation into mRNA relative to tRNA and rRNA then decreases with germination time.

Essentiality of RNA Synthesis for Germination

As discussed in the previous section, fungal spores synthesize RNA during germination. However, unlike the absolute requirement for protein synthesis, RNA synthesis may not always be required for germ tube formation. Furthermore, the lack of precursor incorporation into RNA does not ensure the absence of RNA synthesis unless permeability and endogenous pools are considered. Obviously, RNA synthesis is necessary for germination if a dormant spore lacks mRNA, contains ribosomes deficient in rRNA, or is deficient in tRNA(s). In contrast, if a dormant spore contains all classes of RNA including mRNA, the spore might have the potential to germinate in the absence of RNA synthesis. Unfortunately, the question of whether RNA synthesis is required for spore germination cannot be satisfactorily answered for many fungi. Before this can be accomplished an inhibitor will have to be found which effectively penetrates dormant spores and specifically inhibits RNA synthesis.

The RNA synthesis which occurs during the germination of zoospores of *Blastocladiella emersonii* (Lovett, 1968; Soll & Sonneborn, 1971a) and *Allomyces arbuscula* (Burke et al., 1972) is inhibited by low (25--50 μg/ml) concentrations of actinomycin D. Since these spores germinate to form a primary rhizoid in the presence of actinomycin D, RNA synthesis may not be necessary for germination. There are several other reports that fungal spores form germ tubes in the presence of actinomycin D. However, in none of these reports has actinomycin D been shown to inhibit RNA synthesis in the spore prior to germ tube formation.

Hollomon (1970) used the inhibitor proflavin to examine the relationship of RNA synthesis and spore germination in four fungi. Germination of *Alternaria solani* and *Peronospora tabacina* conidia was unaffected by concentrations of proflavin which severely inhibited RNA synthesis. In contrast, the germination of conidia of *Aspergillus nidulans* and *Neurospora crassa* was inhibited by about the same concentration of proflavin required to inhibit RNA synthesis. However, Inoue & Ishikawa (1970) found that proflavin inhibited *N. crassa* germination by only 20% at concentrations of the drug which inhibited RNA synthesis by 80%. Tisdale & DeBusk (1972) found that proflavin prevented the uptake of uridine by *N. crassa* conidia and suggested that proflavin may inhibit precursor uptake rather than incorporation into RNA. Rousseau & Halvorson (1973b) obtained apparent total inhibition of RNA synthesis by high (800 μg/ml) concentrations of ethidium bromide during germination of *Saccharomyces cerevisiae* ascospores, and they tentatively concluded that RNA synthesis might not be required for germination.

We have investigated the requirement of RNA synthesis for

germination of *Botryodiplodia theobromae* and *Rhizopus stolonifer* spores. Whereas protein synthesis begins within 15 min after placing pycnidiospores of *B. theobromae* on germination medium, RNA synthesis is not detected until about 120 min into germination (Brambl & Van Etten, 1970). Thus, it was suggested that RNA synthesis might not be necessary for germ tube formation in these spores. In contrast, sporangiospores of *Rhizopus stolonifer* initiate RNA and protein synthesis immediately after they are placed on a germination medium (Van Etten et al., 1974). It is interesting that two reported RNA synthesis inhibitors, lomofungin and daunorubicin, inhibit germination in *R. stolonifer* but not *B. theobromae* (Table 4; Nickerson & Van Etten, unpublished data). Thus, RNA synthesis may be required for *R. stolonifer* germination but not for *B. theobromae* germination; however, additional experiments are required to establish that the inhibitors enter the spores and specifically inhibit RNA synthesis.

Thus, there is suggestive evidence that spores may be classified into two groups. Some fungi may have an absolute requirement for RNA synthesis to form germ tubes while others may not.

CONCLUDING REMARKS

A considerable amount of research has been conducted in the last fifteen years on the relationship of nucleic acid metabolism to fungal spore dormancy and germination. Information at present suggests that DNA remains unchanged during spore germination and that DNA synthesis is not required for germination. It is more difficult to generalize about the relationships between RNA synthesis and spore germination because they apparently vary among species. The question of whether fungal spores contain preexisting mRNA has not been adequately resolved, although the best information suggests that dormant spores of many fungi contain mRNA or its precursor. The manner in which this mRNA is conserved in the spore is not known. Although RNA synthesis generally accompanies spore germination, the question of whether dormant spores contain sufficient mRNA to form a germ tube cannot be answered for most fungi. Present data suggest that some spores require RNA synthesis for germination and others do not.

Many questions remain unanswered. In our opinion some of the most interesting areas of research remaining to be explored include: (1) the relationship, if any, of mitochondrial nucleic acid synthesis to spore germination, (2) the degree and mechanism of genome repression in fungal spores, (3) the regulation of transcription in the spores, (4) the quantification of endogenous nucleic acid precursors in the spore and their utilization during germination, (5) the degree of RNA degradation by specific RNases during the early stages of germination, and (6) the proteins encoded by the spore mRNA.

ACKNOWLEDGMENTS

We thank Ken Nickerson, Les Lane, Roger Storck, and Robert Brambl for critically reading the manuscript and other colleagues for providing reprints and preprints of articles on nucleic acids in fungal spores. Specificially we thank Gib DeBusk, Dick Staples, Jim Lovett, Jack Leary, and Marvin Edelman who allowed us to cite their results prior to publication. Research from our laboratory was supported by Public Health Service Grant AI 108057 from the National Institute of Allergy and Infectious Diseases.

REFERENCES

Adman, R., Schultz, L. D., & Hall, B. D. (1972). Transcription in yeast: separation and properties of multiple RNA polymerases. Proc. Nat. Acad. Sci. U.S.A. 69, 1702--1706.

Agsteribbe, E., Datema, R., & Kroon, A. M. (1974). Mitochondrial polysomes of *Neurospora crassa*. In The Biogenesis of Mitochondria, (ed. A.M. Kroon & C. Saccone), pp. 305--314. New York and London, Academic Press.

Arima, T., Uchida, T., & Egami, F. (1968a). Studies on extracellular ribonucleases of *Ustilago sphaerogena*. Purification and properties. Biochem. J. 106, 601--607.

Arima, T., Uchida, T., & Egami, F. (1968b). Studies on extracellular ribonucleases of *Ustilago sphaerogena*. Characterization of substrate specificity with special reference to purine-specific ribonucleases. Biochem. J. 106, 609--613.

Aviv, H. & Leder, P. (1972). Purification of biologically active globin messenger RNA by chromatography on oligo thymidylic acid-cellulose. Proc. Nat. Acad. Sci. U.S.A. 69, 1408--1412.

Bainbridge, B. W. (1971). Macromolecular composition and nuclear division during spore germination in *Aspergillus nidulans*. J. Gen. Microbiol. 66, 319--325.

Barash, I., Conway, M. L., & Howard, D. H. (1967). Carbon catabolism and synthesis of macromolecules during spore germination of *Microsporum gypseum*. J. Bacteriol. 93, 656--662.

Barnard, E. A. (1969). Ribonucleases. Annu. Rev. Riochem. 38, 677--732.

Barnett, W. E., Brown, D. H., & Epler, J. L. (1967). Mitochondrial-specific aminoacyl-RNA synthetases. Proc. Nat. Acad. Sci. U.S.A. 57, 1775--1781.

Bartnicki-Garcia, S., Nelson, N., & Cota-Robles, E. (1968). Electron microscopy of spore germination and cell wall formation in *Mucor rouxii*. Arch. Mikrobiol. 63, 242--255.

Bhagwat, A. S. & Mahadevan, P. R. (1970). Conserved mRNA from the conidia of *Neurospora crassa*. Mol. Gen. Genet. 109, 142--151.

Bhargava, M. M. & Halvorson, H. O. (1971). Isolation of nuclei from yeast. J. Cell Biol. 49, 423--429.

Bicknell, J. N. & Douglas, H. C. (1970). Nucleic acid homologies among species of *Saccharomyces*. J. Bacteriol. 101, 505--512.

Blumenfeld, M. & Forrest, H. S. (1972). Differential under-replication of satellite DNAs during *Drosophila* development. Nat. New Biol. 239, 170--172.

Boguslawski, G., Schlessinger, D., Medoff, G., & Kobayashi, G. (1974). Ribonucleic acid polymerases of the yeast phase of *Histoplasma capsulatum*. J. Bacteriol. 118, 480--485.

Borst, P. (1972). Mitochondrial nucleic acids. Annu. Rev. Biochem. 41, 333-376.

Brambl, R. M. & Van Etten, J. L. (1970). Protein synthesis during fungal spore germination V. Evidence that the ungerminated conidiospores of *Botryodiplodia theobromae* contain messenger ribonucleic acid. Arch. Biochem. Biophys. 137, 442--452.

Brawerman, G., Mendecki, J., & Lee, S. Y. (1972). A procedure for the isolation of mammalian messenger ribonucleic acid. Biochemistry 11, 637--641.

Bromberg, R. (1974). Mitochondrial fragmentation during germination in *Blastocladiella emersonii*. Dev. Biol. 36, 187--194.

Brooks, R. R. & Huang, P. C. (1972). Redundant DNA of *Neurospora crassa*. Biochem. Genet. 6, 41--49.

Burke, D. J., Seale, T. W., & McCarthy, B. J. (1972). Protein and ribonucleic acid synthesis during the diploid life cycle of *Allomyces arbuscula*. J. Bacteriol. 110, 1065--1072.

Cain, A. & Nester, E. W. (1973). Ribonucleic acid polymerase in

Allomyces arbuscula. J. Bacteriol. 115, 769--776.

Cantino, E. C., Lovett, J. S., Leak, L. V., & Lythgoe, J. (1963). The single mitochondrion, fine structure, and germination of the spore of *Blastocladiella emersonii*. J. Gen. Microbiol. 31, 393--404.

Chakravorty, A. K. & Shaw, M. (1972). Ribosomal RNA synthesis and ribosome formation during germination of flax rust uredospores. Physiol. Plant Pathol. 2, 1--6.

Chattopadhyay, S. K., Kohne, D. E., & Dutta, S. K. (1972). Ribosomal RNA genes of *Neurospora*: Isolation and characterization. Proc. Nat. Acad. Sci. U.S.A. 69, 3256--3259.

Chen, D. & Osborne, D. (1970). Ribosomal genes and DNA replication in germinating wheat embryos. Nat. Lond, 225, 336--340.

Christiansen, C., Bak, A. L., Stenderup, A., & Christiansen, G. (1971). Repetitive DNA in yeasts. Nat. New Biol. 231, 176--177.

Clark-Walker, G. D. (1972). Isolation of circular DNA from a mitochondrial fraction from yeast. Proc. Nat. Acad. Sci. U.S.A. 69, 388--392.

Clark-Walker, G. D. & Gleason, F. H. (1973). Circular DNA from the water mold *Saprolegnia*. Arch. Mikrobiol. 92, 209--216.

Clark-Walker, G. D. & Miklos, G. L. G. (1974). Localization and quantification of circular DNA in yeast. Eur. J. Biochem. 41, 359--365.

Clayton, D. A. & Brambl, R. M. (1972). Detection of circular DNA from mitochondria of *Neurospora crassa*. Biochem. Biophys. Res. Commun. 46, 1477--1482.

Cochrane, J. C., Rado, T. A., & Cochrane, V. W. (1971). Synthesis of macromolecules and polyribosome formation in early stages of spore germination in *Fusarium solani*. J. Gen. Microbiol. 65, 45--55.

Cooper, C. S. & Avers, C. J. (1974). Evidence of involvement of mitochondrial polysomes and messenger RNA in the synthesis of organelle proteins. In The Biogenesis of Mitochondria (ed. A. M. Kroon and C. Saccone), pp. 289--304. New York and London: Academic Press.

Cramer, J. H., Bhargava, M. M., & Halvorson, H. O. (1972). Isolation and characterization of γDNA of *Saccharomyces cerevisiae*. J. Mol. Biol. 71, 11--20.

Darnell, J. E. (1968). Ribonucleic acids from animal cells. Bacteriol. Rev. 32, 262--290.

Darnell, J. E., Jelinek, W. R., & Molloy, G. R. (1973). Biogenesis of mRNA: Genetic regulation in mammalian cells. Science 181, 1215--1221.

Davidson, J. N. (1972). The Biochemistry of the Nucleic Acids, 7th ed., London; Chapman and Hall.

DeLucia, P. & Cairns, J. (1969). Isolation of an *E. coli* strain with a mutation affecting DNA polymerase. Nat. Lond. 224, 1164--1166.

Dill, B. C., Leighton, T. J., & Stock, J. J. (1972). Physiological and biochemistry changes associated with macroconidial germination in *Microsporum gypseum*. Appl. Microbiol. 24, 977--985.

Douthit, H. A. & Halvorson, H. O. (1966). Satellite deoxyribonucleic acid from *Bacillus cereus* strain T. Science 153, 182--183.

Dressler, D., Wolfson, J., & Magazin, M. (1972). Initiation and re-initiation of DNA synthesis during replication of bacteriophage T7. Proc. Nat. Acad. Sci. U.S.A. 69, 998--1002.

Dunkle, L. D., Maheshwari, R., and Allen, P. J. (1969). Infection structures from rust urediospores: Effect of RNA and protein synthesis inhibitors. Science 158, 481--482.

Dunkle, L. D. & Van Etten, J. L. (1972). Characteristics and synthesis of deoxyribonucleic acid during fungal spore germination. In Spores V (ed. H. O. Halvorson, R. Hanson, and L. L. Campbell), pp. 283--289. Washington D.C.: American Society for Microbiology.

Dunkle, L. D., Van Etten, J. L., & Brambl, R. M. (1972). Mitochondria DNA synthesis during fungal spore germination. Arch. Mikrobiol. 85, 225--232.

Dure, L. S., Epler, J. L., & Barnett, W. E. (1967). Sedimentation properties of mitochondrial and cytoplasmic ribosomal RNA's from *Neurospora*. Proc. Nat. Acad. Sci. U.S.A. 58, 1883--1887.

Dutta, S. K. (1973). Transcription of non-repeated DNA in *Neurospora crassa*. Biochim. Biophys. Acta 324, 482--487.

Dutta, S. K. & Ray, R. (1973). Partial characterization of transfer RNA genes isolated from *Neurospora crassa*. Mol. Gen. Genet. 125, 295--300.

Dwivedi, R. S., Dutta, S. K., & Bloch, D. P., (1969). Isolation and characterization of chromatin from *Neurospora crassa*. J. Cell Biol. 43, 51--58.

Eccleshall, T. R. & Criddle, R. S. (1974). The DNA-dependent RNA polymerase from yeast mitochondria. In The Biogenesis of Mitochondria (ed. A. M. Kroon and C. Saccone), pp. 31--46. New York and London: Academic Press.

Edelman, M., Verma, I. M., & Littauer, U. Z. (1970). Mitochondrial ribosomal RNA from *Aspergillus nidulans*: characterization of a novel molecular species. J. Mol. Biol. 49, 67--83.

Edmonds, M. and Kopp, D. W. (1970). The occurrence of the poly-adenylate sequences in bacteria and yeast. Biochem. Biophys. Res. Commun. 41, 1531--1537.

Egami, F., Takahashi, & Uchida, T. (1964). Ribonuclease in Taka diastase: properties, chemical nature, and applications. In Progress in Nucleic Acid Research and Molecular Biology, Vol. 3 (ed. J. N. Davidson and W. E. Cohn), pp. 59--101. New York and London: Academic Press.

Epler, J. L. (1969). The mitochondrial and cytoplasmic transfer ribonucleic acids of *Neurospora crassa*. Biochemistry 8, 2285--2290.

Epler, J. L., Shugart, L. R., & Barnett, W. E. (1970). N-formyl-methionyl transfer ribonucleic acid in mitochondria from *Neurospora*. Biochemistry 9, 3575--3579.

Finkelstein, D. B., Blamire, J., & Marmur, J. (1972). Location of ribosomal RNA cistrons in yeast. Nat. New Biol. 240, 279--281.

Firtel, R. A. & Lodish, H. F. (1973). A small nuclear precursor of messenger RNA in the cellular slime mold *Dictyostelium discoideum*. J. Mol. Biol. 79, 295--314.

Fletcher, J. (1971). Fine-structural changes during germination of conidia of *Penicillium griseofulvum* Dierckx. Ann. Bot. 35, 441--449.

Florance, E. R., Denison, W. C., & Allen, T. C., Jr. (1972). Ultra-structure of dormant and germinating conidia of *Aspergillus nidulans*. Mycologia 64, 115--123.

Gamow, E. & Prescott, D. M. (1972). Characterization of the RNA synthesized by *Phycomyces blakesleeanus*. Biochim. Biophys. Acta. 259, 223--227.

Gilham, P. T. (1970). RNA sequence analysis. Annu. Rev. Biochem. 39, 227--250.

Gong, C., Dunkle, L. D., & Van Etten, J. L. (1973). Characteristics of deoxyribonucleic acid polymerase isolated from spores of *Rhizopus stolonifer*. J. Bacteriol. 115, 762--768.

Gong, C. & Van Etten, J. L. (1972). Changes in soluble ribonucleic acid polymerases associated with the germination of *Rhizopus stolonifer* spores. Biochim. Biophys. Acta 272, 44--52.

Gong, C. & Van Etten, J. L. (1974). Purification and properties of RNA polymerases I and II from germinated spores of *Rhizopus stolonifer*. Can. J. Microbiol. 20, 1267--1272.

Green, B. R. & Dick, M. W. (1972). DNA base composition and the taxonomy of the Oomycetes. Can. J. Microbiol. 18, 963--968.

Grivell, A. R. & Jackson, J. F. (1968). Thymidine kinase: Evidence for its absence from *Neurospora crassa* and some other microorganisms, and the relevance of this to the specific labeling of deoxyribonucleic acid. J. Gen. Microbiol. 54, 307--317.

Gross, P. R. (1967). The control of protein synthesis in embryonic development. In Current Topics in Developmental Biology, Vol. 2 (ed. A. A. Moscona and A. Monroy), pp. 1--46. New York and London: Academic Press.

Guerineau, M., Grandchamp, C., Paoletti, C., & Slonimski, P. (1971). Characterization of a new class of circular DNA molecules in yeast. Biochem. Biophys. Res. Commun. 42, 550--557.

Hawker, L. E. & Abbott, P. McV. (1963). An electron microscope study of maturation and germination of two species of *Rhizopus*. J. Gen. Microbiol. 32, 295--298.

Hawker, L. E. & Hendy, R. J. (1963). An electron microscope study of germination of conidia of *Botrytis cinerea*. J. Gen. Microbiol. 33, 43--46.

Henney, H. R. & Storck, R. (1963a). Nucleotide composition of ribonucleic acid from *Neurospora crassa*. J. Bacteriol. 85, 822--826.

Henney, H. R. & Storck, R. (1963b). Ribosomes and ribonucleic acids in three morphological states of *Neurospora*. Science 142, 1675--1676.

Henney, H. R. & Storck, R. (1964). Polyribosomes and morphology in *Neurospora crassa*. Proc. Nat. Acad. Sci. U.S.A. 51, 1050--1055.

Hess, W. M. & Weber, D. J. (1973). Ultrastructure of dormant and germinated sporangiospores of *Rhizopus arrhizus*. Protoplasma 77, 15--33.

Hildebrandt, A., Sebastian, J., & Halvorson, H. O. (1973). Yeast nuclear RNA polymerase I and II are immunologically related. Nat. New Biol. 246, 73--74.

Hoffman, H. P. & Avers, C. J. (1973). Mitochondrion of yeast: Ultrastructural evidence for one giant, branched organelle per cell. Science 181, 749--751.

Hollenberg, C. P., Borst, P., & Van Bruggen, E. F. J. (1970). Mitochondrial DNA. V. A 25-μ closed circular duplex DNA molecule in wild-type yeast mitochondria and genetic complexity. Biochim. Biophys. Acta 209, 1--15.

Holloman, D. W. (1969). Biochemistry of germination in *Peronospora tabacina* (Adam) conidia: Evidence for the existence of stable messenger RNA. J. Gen. Microbiol. 55, 267--274.

Hollomon, D. W. (1970). Ribonucleic acid synthesis during fungal spore germination. J. Gen. Microbiol. 62, 75--87.

Hollomon, D. W. (1971). Protein synthesis during germination of *Peronospora tabacina* conidia: (Adam) conidia. Arch. Biochem. Biophys. 145, 643--649.

Hollomon, D. W. (1973). Protein synthesis during germination of *Peronospora tabacina* conidia: An examination of the events involved in the initiation of germination. J. Gen. Microbiol. 78, 1--13.

Horgen, P. A. (1971). *In vitro* ribonucleic acid synthesis in the zoospores of the aquatic fungus *Blastocladiella emersonii*. J. Bacteriol. 106, 281--282.

Horgen, P. A. & Griffin, D. H. (1971a). Specific inhibitors of the three RNA polymerases from the aquatic fungus *Blastocladiella emersonii*. Proc. Nat. Acad. Sci. U.S.A. 68, 338--341.

Horgen, P. A. & Griffin, D. H. (1971b). RNA polymerase III of *Blastocladiella emersonii* is mitochondrial. Nat. New Biol. 234, 17--18.

Horgen, P. A., Nagao, R. T., Chia, L. S. Y., & Key, J. L. (1973). Basic nuclear proteins in the Oomycete fungus *Achlya bisexualis*. Arch. Mikrobiol. 94, 249--258.

Horikoshi, K. (1971). Studies on the conidia of *Aspergillus oryzae* XI. Latent ribonuclease in the conidia of *Aspergillus oryzae*. Biochim. Biophys. Acta 240, 532--540.

Horikoshi, K. & Ando, T. (1972). Conidia of *Aspergillus oryzae* XII. Ribonucleases in the conidia of *Aspergillus oryzae*. In Spores V. (ed. H. O. Halvorson, R. Hanson, and L. L. Campbell), pp. 456--460. Washington D.C.: American Society for Microbiology.

Horikoshi, K. & Ikeda, Y. (1968). Studies on the conidia of *Aspergillus oryzae* VII. Development of protein synthesizing activity during germination. Biochim. Biophys. Acta 166, 505--511.

Horikoshi, K. & Ikeda, Y. (1969). Studies on the conidia of *Aspergillus oryzae* IX. Protein synthesizing activity of dormant conidia. Biochim. Biophys. Acta 190, 187--192.

Horikoshi, K., Ohtaka Y., & Ikeda, Y. (1965). Ribosomes of dormant and germinating conidia of *Aspergillus oryzae*. Agr. Biol. Chem. 29, 724--727.

Horikoshi, K., Ohtaka, Y., & Ikeda, Y. (1969). Properties of ribosomes and transfer ribonucleic acid in dormant conidia of

Aspergillus oryzae. In Spores IV (ed. L. L. Campbell), pp. 175--179. Washington D.C.: American Society for Microbiology.

Hoshino, J., Nishi, A., & Yanagita, T. (1962). Alanine metabolism in conidiospores of *Aspergillus niger* in the early phase of germination. J. Gen. Appl. Microbiol. 8, 233-245.

Hsiang, M. W. & Cole, R. D. (1973). The isolation of histone from *Neurospora crassa*. J. Biol. Chem. 248, 2007--2013.

Ingle, J. & Sinclair, J. (1972). Ribosomal RNA genes and plant development. Nat. Lond. 236, 30--32.

Inoue, H. & Ishikawa, T. (1970). Macromolecule synthesis and germination of conidia in temperature-sensitive mutants of *Neurospora crassa*. (original reference not seen by the authors) Jap. J. Genet, 45, 357--369.

Jacob, S. T. (1973). Mammalian RNA polymerases. In Progress in Nucleic Acid Research and Molecular Biology, Vol. 13 (ed. J. N. Davidson and W. E. Cohn), pp. 93--126. New York and London: Academic Press.

Jaworski, A. J. & Horgen, P. A. (1973). The ribosomal cistrons of the water mold *Achlya bisexualis*. Arch. Biochem. Biophys. 157, 260--267.

Jeggo, P. A., Unrau, P., Banks, G. R., & Holliday, R. (1973). A temperature sensitive DNA polymerase mutant of *Ustilago maydis*. Nat. New Biol. 242, 14--16.

Jervis, H. H. & DeBusk, A. G. (1975). A rapid loss of a novel phenylalanyl-tRNA upon germination of *Neurospora crassa* conidia. Nat. Lond. (in press).

Kaback, D. B., Bhargava, M. M., & Halvorson, H. O. (1973). Location and arrangement of genes coding for ribosomal RNA in *Saccharomyces cerevisiae*. J. Mol. Biol. 79, 735--739.

Kadowaki, K. & Halvorson, H. O. (1971a). Appearance of a new species of ribonucleic acid during sporulation in *Saccharomyces cerevisiae*. J. Bacteriol. 105, 826--830.

Kadowaki, K. & Halvorson, H. O. (1971b). Isolation and properties of a new species of ribonucleic acid synthesized in sporulating cells of *Saccaromyces cerevisiae*. J. Bacteriol. 105, 831--836.

Kates, J. (1970). Transcription of the vaccinia virus genome and the occurence of polyadenylic acid sequences in messenger RNA. Cold Spring Harbor Symp. Quant. Biol. 35, 743--752.

Kimura, K., Ono., T., & Yanagita, T. (1965). Isolation and characterization of ribosomes from dormant and germinating conidia of *Aspergillus oryzae*. J. Biochem. (Tokyo) 58, 569--576.

Knight, R. H., Roheim, J. R., & Van Etten, J. L. (1974). Synthesis of ribonucleic acid during germination of *Rhizopus stolonifer* sporangiospores. Abstr. Annu. Meet. Amer. Soc. Microbiol. p. 144.

Kogane, F. & Yanagita, T. (1964). Isolation and purification of deoxyribonucleic acid from *Aspergillus oryzae* conidia. J. Gen. Appl. Microbiol. 10, 61--68.

Koke, J. R., Malhotra, S. K., & Bryan, L. E. (1970). Factors influencing incorporation of ^{3}H-TTP into an acid precipitable product by isolated mitochondria of *Neurospora crassa*. Cytobios 5, 19--27.

Komiyama, T. & Irie, M. (1972). Purification and properties of a ribonuclease from *Rhizopus* species. J. Biochem. (Tokyo) 70, 765--773.

Kriegstein, H. J. & Hogness, D. S. (1974). Mechanism of DNA replication in *Drosophila* chromosomes: Structure of replication forks and evidence for bidirectionality. Proc. Nat. Acad. Sci. U.S.A. 71, 135--139.

Küntzel, H. & Noll, H. (1967). Mitochondrial and cytoplasmic polysomes from *Neurospora crassa*. Nat. Lond. 215, 1340--1345.

Küntzel, H. & Schafer, K. P. (1971). Mitochondrial RNA polymerase from *Neurospora crassa*. Nat. New Biol. 231, 265--269.

Kuriyama, Y. & Luck, D. (1973). Ribosomal RNA synthesis in mitochondria of *Neurospora crassa*. J. Mol. Biol. 73, 425--437.

Leary, J. V. & Ellingboe, A. H. (1971). Isolation and characterization of ribosomes from nongerminated conidia of *Erysiphe graminis* f. sp. *tritici*. Phytopathology 61, 1030--1031.

Leary, J. V., Morris, A. J., & Ellingboe, A. H. (1969). Isolation of functional ribosomes and polysomes from lyophilized fungi. Biochim. Biophys. Acta. 182, 113--120.

Leary, J. V., Roheim, J. R., & Zentmyer, G. A. (1974a). Ribosome content of various spore forms of *Phytophthora* spp. Phytopathology 64, 404--408.

Leaver, C. J. & Lovett, J. S. (1974). An analysis of protein and RNA synthesis during encystment and outgrowth (germination) of *Blastocladiella* zoospores. Cell Differentiation. 3, 165--192.

Lee, S. Y., Mendecki, J., & Brawerman, G. (1971). A polynucleotide segment rich in adenylic acid in the rapidly-labeled polyribosomal RNA component of mouse sarcome 180 ascites cells. Proc. Nat. Acad. Sci., U.S.A. 68, 1331--1335.

Leighton, T. J., Dill, B. C., Stock, J. J., & Phillips, C. (1971).

Absence of histones from the chromosomal proteins of fungi. Proc. Nat. Acad. Sci. U.S.A. 68, 677--680.

Le'John, H. B. & Lovett, J. S. (1966). Ribonucleic acid and protein synthesis in *Rhizophlyctis rosea* zoospores. J. Bacteriol. 91, 709--717.

Lin, F. K., Davies, F. L., Tripathi, R. K., Raghu, K., & Gottlieb, D. (1971). Ribonucleic acids in spore germination of *Ustilago maydis*. Phytopathology 61, 645--648.

Lizardi, P. M. & Luck, D. J. (1971). Absence of 5S RNA component in the mitochondrial ribosomes of *Neurospora crassa*. Nat. New Biol. 229, 140--142.

Lovett, J. S. (1968). Reactivation of ribonucleic acid and protein synthesis during germination of *Blastocladiella* zoospores and the role of the ribosomal nuclear cap. J. Bacteriol. 96, 962--969.

Lovett, J. S. & Haselby, J. A. (1971). Molecular weights of the ribosomal ribonucleic acid of fungi. Arch. Mikrobiol. 80, 191--204.

Lovett, J. S. & Leaver, C. J. (1969). High-molecular weight artifacts in RNA extracted from *Blastocladiella* at elevated temperatures. Biochim. Biophys. Acta 195, 319--327.

Luck, D. J. L. & Reich, E. (1964). DNA in mitochondria of *Neurospora crassa*. Proc. Nat. Acad. Sci. U.S.A. 52, 931--938.

Maheshwari, R., Hildebrandt, A. C., & Allen, P. J. (1967). The cytology of infection structure development in urediospore germ tubes of *Uromyces phaseoli* var. *typica* (Pers.) Wint. Can. J. Bot. 45, 447--450.

Manocha, M. S. (1973). DNA-dependent RNA polymerase in germinating bean rust uredospores. Can. J. Microbiol. 19, 1175--1177.

Martin, J. F., Uruburu, F., & Villanueva, J. R. (1973). Ultrastructural changes in the conidia of *Penicillium notatum* during germination. Can. J. Microbiol. 19, 797--801.

McCoy, E. C., Girard, A. E., & Kornfeld, J. M. (1971). Fine structure of resting and germination *Penicillium chrysogenum* conidiospores. Protoplasma 73, 443--456.

McLaughlin, C. S., Warner, J. R., Edmonds, M., Nakazato, H., & Vaughan, M. H. (1973). Polyadenylic acid sequences in yeast messenger ribonucleic acid. J. Biol. Chem. 248, 1466--1471.

Merlo, D. J., Roker, H., & Van Etten, J. L. (1972). Protein synthesis during fungal spore germination VI. Analysis of transfer ribonucleic acid from germinated and ungerminated spores of

Rhizopus stolonifer. Can. J. Microbiol. 18, 949--956.

Michelson, E. L. & Suyama, Y. (1968). Evidence for two types of 18-S ribosomal RNA in *Neurospora crassa*. Biochim. Biophys. Acta 157, 200--203.

Minagawa, T., Wagner, B., & Strauss, B. (1959). The nucleic acid content of *Neurospora crassa*. Arch. Biochem. Biophys. 80, 442--445.

Mirkes, P. E. (1974). Polysomes, ribonucleic acid, and protein synthesis during germination of *Neurospora crassa* conidia. J. Bacteriol. 117, 196--202.

Mitchell, N. L. & McKeen, W. E. (1970). Light and electron microscope studies on the conidium and germ tube of *Sphaerotheca macularis*. Can. J. Microbiol. 16, 273--280.

Morimoto, H., Scragg, A. M., Nekhorocheff, J., Villa, V., & Halvorson, H. O. (1970). Comparison of the protein synthesizing systems from mitochondria and cytoplasm of yeast. In Autonomy and Biogenesis of Mitochondria and Chloroplasts, (ed. N. K. Boardman, A. W. Linnane, and R. M. Smillie), pp. 282--292. Amsterdam: North Holland.

Moustacchi, E. & Williamson, D. H. (1966). Physiological variations in satellite components of yeast DNA detected by density gradient centrifugation. Biochem. Biophys. Res. Commun. 23, 56--61.

Myers, R. B. & Cantino, E. C. (1971). DNA profile of the spore of *Blastocladiella emersonii*: Evidence for γ-particle DNA. Arch. Mikrobiol. 78, 252--267.

Nass, M. M. K. (1969). Mitochondrial DNA: Advances, problems, and goals. Science 165, 25--35.

Nemer, M. (1967). Transfer of genetic information during embryogenesis. In Progress in Nucleic Acid Research and Molecular Biology, Vol. 7 (ed. J. N. Davidson, and W. E. Cohn), pp. 243--301. New York and London: Academic Press.

Nishi, A. (1961). Role of polyphosphate and phospholipid in germinating spores of *Aspergillus niger*. J. Bacteriol. 81, 10--19.

Olson, L. W. (1973). Synchronized development of gametophytic germlings of the aquatic phycomycete *Allomyces macrogynus*. Protoplasma 78, 129--144.

Olson, L. W. & Fuller, M. S. (1971). Leucine-lysine synchronization of *Allomyces* germlings. Arch. Mikrobiol. 78, 76--91.

Ono, T., Kimura, K., & Yanagita, T. (1966). Sequential synthesis of various molecular species of ribonucleic acid in the early phase of conidia germination in *Aspergillus oryzae*. J. Gen.

Appl. Microbiol. 12, 13--26.

Petersen, N. S. & McLaughlin, C. S. (1973). Monocistronic messenger RNA in yeast. J. Mol. Biol. 81, 33--45.

Petes, T. D. & Fangman, W. L. (1972). Sedimentation properties of yeast chromosomal DNA. Proc. Nat. Acad. Sci. U.S.A. 69, 1188--1191.

Petes, T. D., Byers, B., & Fangman, W. L. (1973). Size and structure of yeast chromosomal DNA. Proc. Nat. Acad. Sci. U.S.A. 70, 3072--3076.

Ponta, H., Ponta, U., & Wintersberger, E. (1972). Purification and properties of DNA-dependent RNA polymerases from yeast. Eur. J. Biochem. 29, 110--118.

Rabin, E. Z. & Frazer, M. J. (1970). Isolation of *Neurospora crassa* endonuclease specific for single-stranded DNA. Can. J. Biochem. 48, 389--392.

Rabinowitz, M., Casey, J., Gordon, P., Locker, J., Hsu, H., & Getz, S. (1974). Characterization of yeast grande and petite mitochondrial DNA by hybridization and physical techniques. In The Biogenesis of Mitochondria, (ed. A. M. Kroon and C. Saccone), pp. 89--106. New York and London: Academic Press.

Ramakrishnan, L. & Staples, R. C. (1967). Some observations on ribonucleic acids and their synthesis in germinating bean rust uredospores. Phytopathology 57, 826.

Ramakrishnan, L. & Staples, R. C. (1970a). Evidence for a template RNA in resting uredospores of the bean rust fungus. Contrib. Boyce Thompson Inst. 24, 197--202.

Ramakrishnan, L. & Staples, R. C. (1970b). Changes in ribonucleic acids during uredospore differentiation. Phytopathology 60, 1087--1091.

Ramakrishnan, L. & Staples, R. C. (1971). Effect of storage on the rate of incorporation of uridine into ribonucleic acid by germinating uredospores. Phytopathology 61, 244-245.

Reed, J. & Wintersberger, E. (1973). Adenylic acid-rich sequences in messenger RNA from yeast polysomes. Fed. Eur. Biochem. Soc. Lett. 32, 213--217.

Reijnders, L. & Borst, P. (1972). The number of 4-S RNA genes on yeast mitochondrial DNA. Biochem. Biophys. Res. Commun. 47, 126--133.

Reijnders, L., Sloof, P., & Borst, P. (1973). The molecular weights of the mitochondrial-ribosomal RNAs of *Saccharomyces carlsbergensis*. Eur. J. Biochem, 35, 266--269.

Retel, J. & Planta, R. J. (1968). The investigation of the ribosomal RNA sites in yeast DNA by the hybridization technique. Biochim. Biophys. Acta 169, 416--429.

Retel, J. & Planta, R. J. (1970). On the mechanism of the biosynthesis of ribosomal RNA in yeast. Biochim. Biophys. Acta 224, 458--469.

Rho, H. M. & DeBusk, A. G. (1971). NH_2-terminal residues of *Neurospora crassa* proteins. J. Bacteriol. 107, 840--845.

Rifkin, M. R., Wood, D., & Luck, D. J. L. (1967). Ribosomal RNA and ribosomes from mitochondria of *Neurospora crassa*. Proc. Nat. Acad. Sci. U.S.A. 58, 1025--1032.

Roeder, R. G. & Rutter, W. J. (1969). Multiple forms of DNA-dependent RNA polymerase in eukaryotic organisms. Nat. Lond. 224, 234--237.

Roheim, J. R., Knight, R. H., & Van Etten, J. L. (1974). Synthesis of ribonucleic acids during the germination of *Rhizopus stolonifer* sporangiospores. Dev. Biol. 41, 137--145.

Rosen, D., Edelman, M., Galun, E., & Danon, D. (1974). Biogenesis of mitochondria in *Trichoderma viride*: Structural changes in mitochondria and other spore constituents during conidial maturation and germination. J. Gen Microbiol. 83, 31--49.

Rousseau, P. & Halvorson, H. O. (1973a). Macromolecular synthesis during the germination of *Saccharomyces cerevisiae* spores. J. Bacteriol. 113, 1289--1295.

Rousseau, P. & Halvorson, H. O. (1973b). Effect of metabolic inhibitors on germination and outgrowth of *Saccharomyces cerevisiae* ascospores. Can. J. Microbiol. 19, 1311--1318.

Schäfer, K. P. & Kuntzel, H. (1972). Mitochondrial genes in *Neurospora*: A single cistron for ribosomal RNA. Biochem. Biophys. Res. Commun. 46, 1312--1319.

Scheld, H. W. & Perry, J. J. (1970). Basidiospore germination in the wood-destroying fungus *Lenzites saepiaria*. J. Gen. Microbiol. 60, 9--21.

Schmoyer, I. R. & Lovett, J. S. (1969). Regulation of protein synthesis in zoospores of *Blastocladiella*. J. Bacteriol. 100, 854--864.

Schochetman, G. & Perry, R. P. (1972). Early appearance of histone messenger RNA in polyribosomes of cultured L cells. J. Mol. Biol. 63, 591--596.

Schweizer, E., MacKechnie, C., & Halvorson, H. O. (1969). The redundancy of ribosomal and transfer RNA genes in *Saccharomyces cerevisiae*. J. Mol. Biol. 40, 261--277.

Scragg, A. H. (1974). A mitochondrial DNA-directed RNA polymerase from yeast mitochondria. In The Biogenesis of Mitochondria, (ed. A. M. Kroon & C. Saccone), pp. 47--58. New York and London: Academic Press.

Sebastian, J., Bhargava, M. M., & Halvorson, H. O. (1973). Nuclear deoxyribonucleic acid dependent ribonucleic acid polymerases from *Saccharomyces cerevisiae*. J. Bacteriol. 114, 1--6.

Shapiro, L., Agabian-Keshishian, N., & Bendis, I. (1971). Bacterial differentiation. Science 173, 884--892.

Sinclair, J. H., Stevens, B. J., Sanghavi, P., & Rabinowitz, M. (1967). Mitochondrial-satellite and circular DNA filaments in yeast. Science 156, 1234--1237.

Sinohara, H. (1970). Induction of enzymes in dormant spores of *Aspergillus oryzae*. J. Bacteriol. 101, 1070--1072.

Smith, D. W. (1973). DNA synthesis in prokaryotes: replication. In Progress in Biophysics and Molecular Biology, Vol. 26 (ed. J. A. V. Butler & D. Noble), pp. 323--408. Oxford: New York, Toronto, Sydney, and Braunschweig: Pergamon Press.

Smith, B. A., & Burke, D. D. (1974). Pattern of RNA synthesis during germination of *Allomyces macrogynus* mitospores. Abstr. Annu. Meet., Amer. Soc. Microbiol. p. 175.

Smith, A. E. & Marcker, K. A. (1968). N-formylmethionyl transfer RNA in mitochondria from yeast and rat liver. J. Mol. Biol. 38, 241--243.

Smith, C. A. & Vinograd, J. (1972). Small polydisperse circular DNA of HeLa cells. J. Mol. Biol. 69, 163--178.

Sogin, S. J., Haber, J. E., & Halvorson, H. O. (1972). Relationship between sporulation-specific 20S ribonucleic acid and ribosomal ribonucleic acid processing in *Saccharomyces cerevisiae*. J. Bacteriol. 112, 806--814.

Soll, D. R. & Sonneborn, D. R. (1971a). Zoospore germination in *Blastocladiella emersonii*. III. Structural changes in relation to protein and RNA synthesis. J. Cell Sci. 9, 679--699.

Soll, D. R. & Sonneborn, D. R. (1971b). Zoospore germination in *Blastocladiella emersonii*: cell differentiation without protein synthesis. Proc. Nat. Acad. Sci. U.S.A. 68, 459--463.

Somberg, E. W. & Davis, F. F. (1965). Ribosomal nuclease activity of *Neurospora crassa*. Biochim. Biophys. Acta 108, 137--139.

South, D. J. & Mahler, H. R. (1968). RNA synthesis in yeast

mitochondria: a derepressible activity. Nat. Lond. 218, 1226--1232.

Stallknecht, G. F. & Mirocha, C. J. (1971). Fixation and incorporation of CO_2 into ribonucleic acid by germinating uredospores of *Uromyces phaseoli*. Phytopathology 61, 400--405.

Staples, R. C. (1968). Protein synthesis by uredospores of the bean rust fungus. Neth. J. Plant Pathol. 74 (Suppl. I) 25--36.

Staples, R. C. (1974). Synthesis of DNA during differentiation of bean rust uredospores. Physiol. Plant Pathol. 4, 415--424.

Staples, R. C., Bedigian, D., & Williams, P. H. (1968). Evidence for polysomes in extracts of bean rust uredospores. Phytopathology 58, 151--154.

Steele, S. D. & Fraser, T. W. (1973). Ultrastructural changes during germination of *Geotrichum candidum* arthrospores. Can. J. Microbiol. 19, 1031--1034.

Storck, R. (1974). Molecular Mycology. In Molecular Microbiology, (ed. J. B. G. Kwapinski), pp. 423--477. New York, London, and Sydney: Wiley.

Storck, R. & Alexopoulos, C. J. (1970). Deoxyribonucleic acid of fungi. Bacteriol. Rev. 34, 126--154.

Storck, R., Alexopoulos , C. J., & Phaff, H. J. (1969). Nucleotide composition of deoxyribonucleic acid of some species of *Cryptococcus, Rhodotorula*, and *Sporobolomyces*. J. Bacteriol. 98, 1069--1072.

Storck, R., Nobels, M. K., & Alexopoulos, C. J. (1971). The nucleotide composition of deoxyribonucleic acid of some species of *Hymenochaetaceae* and *Polyporaceae*. Mycologia 63, 38--49.

Sturani, E., Magnani, F., & Alberghina, F. A. M. (1973). Inhibition of ribosomal RNA synthesis during a shift-down transition of growth in *Neurospora crassa*. Biochim. Biophys. Acta. N58, 153--164.

Sussman, A. S. & Douthit, H. A. (1973). Dormancy in microbial spores. Annu. Rev. Plant Physiol. 24, 311--352.

Sussman, A. S., Lowry, R. J., Durkee, T. L., & Maheshwari, R. (1969). Ultrastructural studies of cold-dormant and germinating uredospores of *Puccinia graminis* var. *tritici*. Can. J. Bot. 47, 2073--2078.

Taber, R. L. & Vincent, W. S. (1969). The synthesis and processing of ribosomal RNA precursor molecules in yeast. Biochim. Biophys. Acta 186, 317--325.

Takai, N., Uchida, T., & Egami, F. (1966). Purification and

properties of ribonuclease N_1, an extracellular ribonuclease of *Neurospora crassa*. Biochim. Biophys. Acta. 128, 218--220.

Tanaka, K., Kogane, F., & Yanagita, T., (1965). Is deoxyribonucleic acid of *Aspergillus oryzae* conidia modified chemically in the early period of germination? J. Gen. Appl. Microbiol. 11, 85--90.

Tanaka, K., Motohashi, A., Miura, K., & Yanagita T., (1966a). Isolation and characterization of soluble RNA from dormant and germinated conidia of *Aspergillus oryzae*. J. Gen. Appl. Microbiol. 12, 277--292.

Tanaka, K., Ono, T., & Yanagita, T. (1966b). Further observations on the carbon dioxide incorporation into RNA in the early phase of conidia germination in *Aspergillus oryzae* with special reference to soluble RNA synthesis. J. Gen. Appl. Microbiol. 12, 329--336.

Taylor, M. M., Glasgow, J. E., & Storck, R. (1967). Sedimentation coefficients of RNA from 70S and 80S ribosomes. Proc. Nat. Acad. Sci. U.S.A. 57, 164--169.

Tellez de Inon, M. T., Leoni, P. D., & Torres, H. N. (1974). RNA polymerase activities in *Neurospora crassa*. Fed. Eur. Biochem. Soc. Lett. 39, 91--95.

Timberlake, W. D., Hogen, G., & Griffin, D. H. (1972b). Rat liver DNA-dependent RNA polymerase I is inhibited by cycloheximide. Biochem, Biophys. Res. Commun, 48, 823--827.

Timberlake, W. E., McDowell, L., and Griffin, D. H. (1972b). Cycloheximide inhibition of the DNA-dependent RNA polymerase I of *Achyla bisexualis*. Biochem. Biophys. Res. Commun, 46, 942--947.

Tingle, M. A., Kuenzi, M. T., & Halvorson, H. O., (1974). Germination of yeast spores lacking mitochondrial deoxyribonucleic acid. J. Bacteriol. 117, 89--93.

Tisdale, J. H. & DeBusk, A. G. (1972). Permeability problems encountered when treating conidia of *Neurospora crassa* with RNA synthesis inhibitors. Biochem. Biophys. Res. Commun. 48, 816--822.

Tonino, G. J. M. & Rozijn, T. H. (1966). On the occurrence of histones in yeast. Biochim. Biophys. Acta 124, 427--429.

Tsay, Y., Nishi, A., & Yanagita, T. (1965). Carbon dioxide fixation in *Aspergillus oryzae* conidia at the initial phase of germination. J. Biochem. (Tokyo) 58, 487--493.

Tsukahara, T. (1968). Electron microscopy of swelling and

germinating conidiospores of *Aspergillus niger*. Sabouraudia 6, 185--191.

Udem, S. A., Kaufman, K., Warner, J. R. (1971). Small ribosomal ribonucleic acid species of *Saccharomyces cerevisiae*. J. Bacteriol. 105, 101--106.

Udem, S. A. & Warner, J. R. (1972). Ribosomal RNA synthesis in *Saccharomyces cerevisiae*. J. Mol. Biol. 65, 227--242.

Ursino, D. J., Sturani, E., & Alberghina F. A. M. (1969). Ribonucleic acid-degrading enzymes associated with purified ribosomes from *Neurospora crassa* mycelia. Biochim. Biophys. Acta 179, 500-502.

Van Assche, J. A. & Carlier, A. R. (1973). The pattern of protein and nucleic acid synthesis in germinating spores of *Phycomyces blakesleeanus*. Arch. Mikrobiol. 93, 129--136.

Van Etten, J. L. (1969). Protein synthesis during fungal spore germination. Phytopathology 59, 1060--1064.

Van Etten, J. L. (1971). Preparation of biologically active ribosomal subunits from fungal spores. J. Bacteriol. 106, 704--706.

Van Etten, J. L., Bulla, L. A., & St Julian, G. (1974). Physiological and morphological correlation of *Rhizopus stolonifer* spore germination. J. Bacteriol. 117, 882--887.

Van Etten, J. L., Koski, R., El-Olemy, M. M. (1969). Protein synthesis during fungal spore germination. IV. Transfer ribonucleic acid from germinated and ungerminated spores. J. Bacteriol. 100, 1182--1186.

Verma, I. M., Edelman, M., Hertzberg, M., & Littauer, U. Z. (1970). Size determination of mitochondrial ribosomal RNA from *Aspergillus nidulans* by electron microscopy. J. Mol. Biol. 52, 137--140.

Verma, I. M., Edelman, M., & Littauer, U. Z. (1971). A comparison of nucleotide sequences from mitochondrial and cytoplasmic ribosomal RNA of *Aspergillus nidulans*. Eur. J. Biochem. 19, 124--129.

Villa, V. D. & Storck, R. (1968). Nucleotide composition of nuclear and mitochondrial deoxyribonucleic acid of fungi. J. Bacteriol. 96, 184--190.

Walkinshaw, C. H., Hyde, J. M., & Van Zandt, J. (1967). Fine structure of quiescent and germinating aeciospores of *Cronartium fusiforme*. J. Bacteriol. 94, 245--254.

Williams, P. G. & Ledingham. G. A. (1964). Fine structure of wheat stem rust uredospores. Can. J. Bot. 42, 1503--1508.

Williamson, D. H. (1970). The effect of environmental and genetic factors on the replication of mitochondrial DNA in yeast. In Control of Organelle Development, Vol. 24 (ed. P. L. Miller), pp. 247--276. New York: Academic Press.

Wintersberger, E. (1966). Occurrence of a DNA-polymerase in isolated yeast mitochondria. Biochem. Biophys. Res. Commun. 25, 1--7.

Wintersberger, E. (1972). Isolation of a distinct rifampicin-resistant RNA polymerase from mitochondria of yeast, *Neurospora* and liver. Biochem. Biophys. Res. Commun. 48, 1287--1294.

Wintersberger, U., Smith, P., & Letnansky, K. (1973). Yeast chromatin. Preparation from isolated nuclei, histone composition, and transcription capacity. Eur. J. Biochem. 33, 123--130.

Wintersberger, E. & Tuppy. H. (1965). DNA-abhängige RNA-synthese in isolierten Hefe-mitochondrien. Biochem. Z341, 399--408.

Wintersberger, U. & Wintersberger, E. (1970a). Studies on deoxyribonucleic acid polymerases from yeast. I. Partial purification and properties of two DNA polymerases from mitochondria-free extracts. Eur. J. Biochem. 13, 11--19.

Wintersberger, U. & Wintersberger E. (1970b). Studies on deoxyribonucleic acid polymerases from yeast. II. Partial purification and characterization of mitochondrial DNA polymerase from wild type and respiration deficient yeast cells. Eur. J. Biochem. 13, 20--27.

Wintersberger, E. & Wintersberger, U. (1970c). Rifamycin insensitivity of RNA synthesis in yeast. Fed. Eur. Biochem. Soc. Lett. 6, 58--60.

Wolfson, J. & Dressler, D. (1972). Regions of a single-stranded DNA in the growing points of replicating bacteriophage T7 chromosomes. Proc. Nat. Acad. Sci. U.S.A. 69, 2682--2686.

Wong, R. S. L., Scarborough, G. A., & Borek, E. (1971). Transfer ribonucleic acid methylases during the germination of *Neurospora crassa*. J. Bacteriol. 108, 446--450.

Wood, D. L. & Luck, D. J. L. (1969). Hybridization of mitochondrial ribosomal RNA. J. Mol. Biol. 41, 211--224.

Yanagita, T. (1957). Biochemical aspects on the germination of *Aspergillus niger*. Arch. Mikrobiol. 26, 329--344.

Yanagita, T. (1963). Carbon dioxide fixation in germinating conidiospores of *Aspergillus niger*. J. Gen. Appl. Microbiol. 9, 343--351.

DISCUSSION

Nucleic Acids and Fungal Spore Germination

Chairman: Roger Storck
Rice University
Houston, Texas

As a result of questions and comments from members of the audience and replies and comments by the Speaker (J. Van Etten), additional information was obtained which is summarized in the following paragraphs.

According to A. Sussman's definition of dormancy types, the conidiospores of *Botryodiplodia theobromae* and the sporangiospores of *Rhizopus stolonifer* are of the exogenous type. Thus the mere addition of given nutrients to the medium is sufficient for germination to occur. It has been demonstrated in the case of *R. stolonifer* that protein synthesis, when measured by amino acid incorporation into trichloroacetic insoluble material, is detectable during the first 15 min of germination. That incorporation, as well as germ tube formation, is inhibited by cycloheximide.

It is not yet known when nuclear division occurs in relationship to DNA synthesis during the germination of *R. stolonifer* and *B. theobromae* spores.

There is no evidence for the presence of virus like particles in these two organisms.

In *Blastocladiella emersonii*, as in *R. stolonifer*, adenine is incorporated into both ribosomal RNA and polyadenylic acid (poly (A)) containing RNA almost immediately after the initiation of germination. Furthermore, in both *R. stolonifer* and *B. emersonii* the length of the poly(A) segment associated with RNA appears to decrease with germination. It is not yet known, in the case of *R. stolonifer*, if the poly(A) containing RNA synthesized initially is associated with ribosomes or whether it is heterogenous nuclear RNA which eventually give rise to messenger RNA after cleavage of the poly(A) fragment.

CHAPTER 8 ORGANELLE CHANGES DURING FUNGAL SPORE GERMINATION

J. E. Smith, K. Gull, J. G. Anderson, and S. G. Deans
Stathclyde University

INTRODUCTION

All living organisms proceed through a series of programmed changes in which structures are built up and transformed by complex interactions between the organism and its external environment. In this way, the organism is the product of the interaction between the information which it receives by heredity and the resources and stimuli to which it is exposed from the environment. At a biochemical level the relationship between gene action and macromolecular synthesis is now unfolding and it is apparent that differentiation results from an interplay between an unchanging genome and a labile cytoplasm---the cytoplasm at any one period determining the sequence of gene expression while itself being modified as a consequence. A full understanding of how the environment of an organism can influence this process of cellular differentiation has not yet been realized.

Fungal spores offer a dual interest in developmental studies since they represent both the beginning and the end of the developmental cycle of the fungus. The genesis of spores comes normally at the end of the period of active growth of the fungus and results in the formation of a propagative unit characterized by a lack of cytoplasmic movement, reduced metabolic activity and low water content (Gregory, 1966). The biochemical and ultrastructural changes associated with this developmental process have been extensively reviewed (for references see Smith & Galbraith, 1971; Smith & Berry, 1974).

In general, most fungal spores display some degree of dormancy, the nature and extent of this dormancy differing according to the species. When all constitutional barriers of dormancy have been removed spores require certain favorable environmental parameters, such as pH, temperature, and humidity, to be present before the process of germination can commence. Aerobic respiration appears to be a normal requirement for most spores while many spores have a positive requirement for carbon dioxide. Some spores with adequate organic reserves appear to need only water for germination although the majority of spores that have been examined appear to have a requirement for organic carbon at some stage of the germination process.

Germination basically comprises the many processes and changes that occur during the resumption of development or growth of a resting structure and its subsequent transformation to a morphologically different structure. In most fungi, this will usually involve the transformation from a nonpolar to a polar germ tube which will continue growth by extension at the tip (Bartnicki-Garcia, 1973). While germination has been defined by Manners (1966) as the formation of a germ tube from the dormant spore, Sussman (1966) has considered that any measurable irreversible change can be the accepted criterion of germination. For this presentation germination will be considered in three distinct though relatively arbitrary phases viz.

I. The physiological and morphological changes which must occur within the spore before germ tube emergence can take place.

II. The act of protrusion of the germ tube from the spore wall. This constitutes the major visible criterion of germination although from a metabolic standpoint germination is well advanced by this time.

III. The elongation of the germ tube and the establishment of true polar growth giving rise to the characteristic growth form of the fungus.

Germination can be primarily considered as a process of regrowth from a more or less dormant condition and will involve an increase in the components of the cell viz. cytoplasm, cell organelles, plasmalemma, and cell wall. Disregarding the many differences that occur between spore types, the first stage of germination generally involves the formation of an enlarged (usually spherical) cell which results from wall changes which probably involve the uniform isotropic deposition of cell wall polymers. Such spherical cell formation may result in little increase in total volume (Bartnicki-Garcia, 1969; Truesdell & Cantino, 1971) or in considerable increase in volume due to growth (Bartnicki-Garcia, 1969; Anderson & Smith, 1972). In the second stage of germination cell wall formation is no longer formed isotropically but instead becomes restricted to a small area of the surface of the rounded cell where the germ tube will emerge. The continued polarized growth of the cell wall results in the characteristic shape of the fungus and in particular, in filamentous fungi, to the formation of the hyphal tube. The factors which control isotropic and polarized growth during spore germination are not yet clearly documented.

In recent years the ultrastructural events associated with growth in fungi have been extensively studied (for references see Bartnicki-Garcia, 1973; Smith & Berry, 1974). From these studies has emerged the concept of an endomembrane system in which the nuclear envelope, endoplasmic reticulum, golgi apparatus or its equivalent structure, and secretory vesicles form a structural and functional continuum. A vesicular hypothesis of growth has now become widely accepted and considers that the exocytotic vesicles produced by the endomembrane system of the cell accrue at the site of growth and contribute to the formation of new plasmalemma and cell wall.

Although there are many ultrastructural changes occurring during germination it is pertinent to note that perhaps the most critical events will be those leading to the net increase in surface area, i.e., plasmalemma and cell wall. Thus before dealing in detail with the ultrastructural changes accompanying germination it is of paramount importance to document the current appreciation of the role of cytoplasmic vesicles in plasmalemma and cell wall formation. The development of this concept has been derived almost entirely from investigations of vegetative cells but has now been confirmed in other conditions including germinating spores.

THE VESICULAR HYPOTHESIS OF CELL GROWTH

It has been shown that hyphal tip growth is the principal means by which filamentous fungi expand their somatic hyphae. The phenomenon of hyphal tip growth has been extensively described (for references see Bartnicki-Garcia, 1973) although the precise mechanism of incorporation and regulation of the materials for hyphal extension has yet to be fully explained. There is a considerable amount of ultrastructural and biochemical evidence to suggest that the apex differs both structurally and cytochemically from the main body of the hypha.

The Role of Cytoplasmic Vesicles

In a number of fungi it has been shown that actively growing hyphal apices contain accumulations of cytoplasmic vesicles (Moore & McAlear, 1961; Wilsenach & Kessel, 1965; Gay & Greenwood, 1966; Grove et al., 1970; Brenner & Carroll, 1968; McClure et al., 1968; Hemmes & Hohl, 1969; Grove & Bracker, 1970) and these vesicles have been suggested as having an important part to play in the process of hyphal extension. Similar vesicle conglomerations have been shown to be present in yeasts (Cortat et al., 1972; McCully & Bracker, 1972; Holley & Kidby, 1973), and in pollen tubes (Van der Woude et al., 1971). The occurrence of vesicles at the site of germ tube outgrowth in several fungal spores has also been noted (Bracker, 1971; Grove, 1972; Steele & Fraser, 1973).

Extensive investigations have demonstrated the apparently universal occurrence of large numbers of vesicles in the hyphal apex of actively growing fungi. Single membrane-bound vesicles are located in the apical dome, and appear in all longitudinal sections of apices undergoing extension growth at the time of fixation. Vesicles vary in size from 40 to 300 nm in diameter, electron-dense strands and particles being evident in some (McClure et al., 1968; Grove et al., 1970). Similar vesicles are occasionally present in regions of hyphae proximal to the apex, but in lesser numbers.

Under electron microscopy, the most vivid fine structural detail of the apical dome is an accumulation of cytoplasmic vesicles, generally to the exclusion of all other organelles. The vesicles can be grouped, according to size into two main categories: 40--120 nm diameter and 200--300 nm (Grove et al., 1970) corresponding to the small rapidly moving granules seen under the light microscope. There is a uniform distribution of both sizes of vesicles in the apex, evident from both longitudinal and transverse sections of the apical dome. Ribosomes are not usually present in the apical cytoplasm, endoplasmic reticulum also being of limited dispersion. This suggests that any ribosomal activity, i.e., genesis of polypeptides and proteins, must occur remote from the apex.

Three distinct patterns of vesicle aggregation have been demonstrated in apical domes of fungal hyphae (Fig. 1). In the Oömycetes there is a random distribution of the vesicles in the apical dome whereas in the Zygomycetes the vesicles occur predominently in a zone close to the surface of the dome. In septate fungi the apical dome contains numerous vesicles together with a denser central part containing numerous ribosomal tubules. This denser zone is now equated with the Spitzenkörper body readily seen in growing hyphal apices.

Immediately behind the apical dome there is the subapical zone which contains mitochondria, nuclei, ribosomes and an endomembrane system composed of rough-surfaced endoplasmic reticulum, smooth-surfaced cisternae and cytoplasmic vesicles. In the Oömycetes, the endomembrane system producing the vesicles consists of stacks of cisternae termed dictyosomes (Grove et al., 1970). Non-oömycetous fungi do not have dictyosomes but instead have an endomembrane system which includes cisternae or tubules and are associated with vesicle formation (Grove & Bracker, 1970).

Detailed electron micrographs reveal the nuclear envelope, endoplasmic reticulum, cisternae, and associated vesicles are structurally interassociated, appearing to form a functionally continuous

---

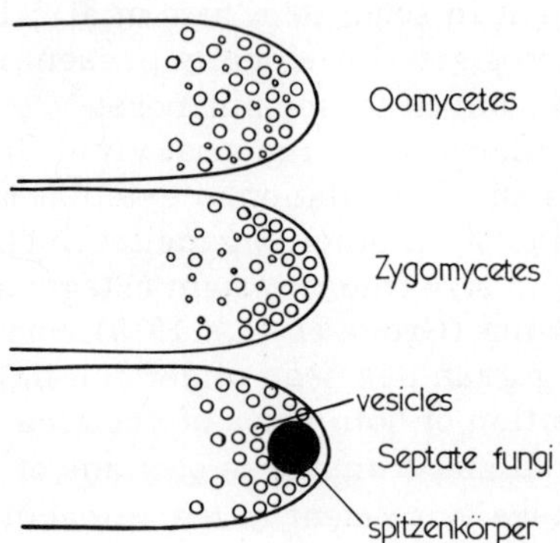

Fig. 1. Diagrammatic comparisons of the principal forms of apical organization in hyphae, based on representatives from major taxonomic groups. From Grove and Bracker, 1970.

endomembrane system. The membraned material is sometimes continuous with the nuclear membrane. The continuity between membranes is perhaps enabling protein/enzymes/or regulators to have a direct pathway between the nucleus and the vesicles. Smooth surfaces of endoplasmic reticulum or nuclear membrane are located in juxtaposition to the cisternae, this intimate contact again allowing for the transfer of "chemical messages" between the two structures. Endoplasmic reticulum profiles appear rough-surfaced, due to being lined on the outer membrane with ribosomes. The amount of "rough" endoplasmic reticulum decreases near the apex.

Dictyosomes occur as polarised (distal and proximal) stacks of ribosome-free cisternae, which are partially fenestrated, i.e., "pitted" or tubular at their peripheries. The polarity of dictyosomes is evident by electron micrographs showing a steady progressive transition in staining characteristics of the cisternal membranes and contents, as well as by the structural morphology of individual cisternae, i.e., curvature, continuity, size, and profile (Grove et al., 1970). Dictyosomes are situated in the hyphal cytoplasm such that the cisternae at the proximal pole of each dictyosome is adjacent to either endoplasmic reticulum or nuclear membrane. The distal pole of the dictyosomes has no such association. Profiles of endoplasmic reticulum or nuclear envelope adjacent to the proximal pole of a dictyosome are characterised by "blebs" (small droplets pinched off from the main body of the membrane) small vesicular profiles being aligned in the interspace between the endoplasmic reticulum and the dictyosome. When seen in 1% aqueous barium permanganate stained cells (fixed with OsO_4), these blebs and vesicular profiles lack ribosomes. Thus during the migration of the small section of constricted membrane across the space between the endoplasmic reticulum and the dictyosome, there has been a liberation of the surface-bound ribosomes into the cytoplasm. Numerous vesicles, termed secretory vesicles, occur at the periphery of the dictyosomes, especially near the distal pole. The profile of small vesicles may be continuous with dictyosome cisternae, but both large and small vesicles are found free in the apical and subapical cytoplasm. The granular contents of some of these vesicles appear identical to those of the distal cisternae of the dictyosome. Vesicles, regardless of their size, seem to have both the same type of membrane and to be smooth-surfaced, i.e., ribosome free.

Grove et al. (1968, 1970) established that the staining characteristics of the vesicle membranes are like both the plasmalemma and distal cisternae of dictyosomes, but unlike the nuclear membrane, endoplasmic reticulum and cisternae of the proximal pole of the dictyosomes. If a continuous endomembrane system is to be considered there must be membrane transition during membrane displacement from the nuclear envelope to the vesicles. In addition to there being vesicles of different sizes there are vesicles with joined membranes in configurations which suggest intervesicular fusion. Continuity has been observed between small vesicles, large vesicles, and between vesicles of different sizes. When membranes of two or more vesicles appear conjoined, two or more inclusions are evident in the common lumen. Vesicle accumulates have been noted at various developmental stages at periods when either apical growth or wall depositions were initiated.

Apparently double-membraned vesicles have been reported in *Gilbertella persicaria* (Bracker, 1966), *Mucor rouxii* (Bartnicki-Garcia et al., 1968), and *Phycomyces blakesleeanus* (Thorton, 1968) and several functions for such vesicles have been proposed. Bracker (1966) considers that they may be cisternal rings which act as a golgi apparatus and that possibly they are derived from endoplasmic reticulum. Thorton (1968) considers that they are stages in the formation of autophagic vesicles and are derived from the part of the endoplasmic cisternae which has surrounded regions of cytoplasm before its digestion. However, Steel (1973) stated that they may function as transporters of materials, corresponding possibly to the particles in the electron-dense layer of the membrane, either to or from the cell surface. Since the particles present on the vesicle and plasma membranes may be proteinaceous the vesicles may also be involved in secretion, possibly of enzymes. The polyribosomes abundant in regions where double-membraned vesicles are evident, may represent the sites of synthesis of such enzymes. The concept that the vesicles represent a stage in the formation of autophagic vacuoles cannot be disregarded, especially since cytochemical techniques have revealed lysosomelike particles in *Geotrichum candidum* (Reiss, 1971).

Generalized Schemes for Apical Growth

As a result of their extensive studies on vesicle occurrence and formation in fungi Grove et al. (1970) have presented an interesting hypothesis to explain the involvement of vesicles in apical growth of an Oömycete fungus and this has been further developed by Bartnicki-Garcia (1973) in an integrated model of cell wall growth (Fig. 2)

I. Cell growth necessitates a net increase in surface area which in turn requires the synthesis and assembly of both lipid and protein

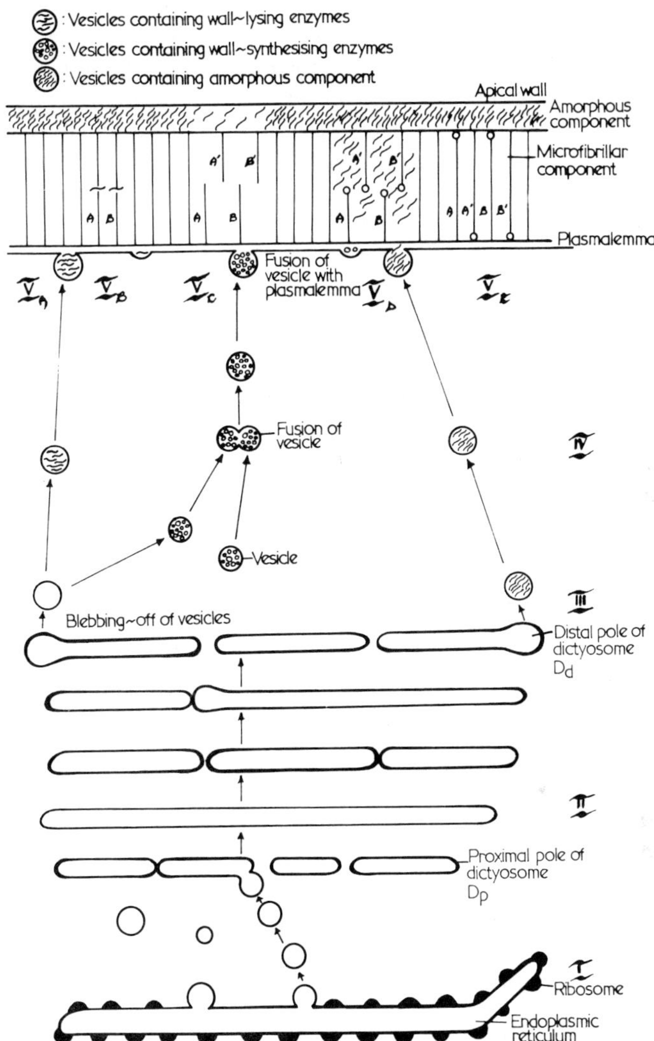

Fig. 2. Diagrammatic representation of the sequence of events believed to occur in a unit of cell wall growth. Adapted from Grove et al., 1970 and Bartnicki-Garcia, 1973.

constituents for the new plasmalemma. The suggestion that endoplasmic reticulum is a source of this new membrane presupposes the manufacture of membrane material de novo in the endoplasmic reticulum. This contention is supported by correlative biochemical and ultrastructural evidence from other systems, indicating that endoplasmic reticulum is a site of membrane biosynthesis (Dallner et al., 1966; Jones & Fawcett, 1964). The golgi apparatus or its equivalent structure is not believed to take part in protein synthesis and it appears unlikely that in growing hyphae the apical dome is involved with membrane protein synthesis, due to the infrequent presence of ribosomal units or rough endoplasmic reticulum. Ribosomes are found in greatest abundance in the subapical, nonvacuolated zone wherein there is constant association between endoplasmic reticulum and dictyosomes.

Thus this initiating stage is characterized by the transfer of material from endoplasmic reticulum to the dictyosomes by blebbing of the endoplasmic reticulum and refusion of vesicles to form a cisterna at the proximal pole of thedictyosome. (Fig. 2Dp).

II. As each cisterna is progressively displaced from the proximal pole of the dictyosome through continued production of new cisternae, the membrane is gradually altered until by the time it reaches the distal cisternae stage, and is beginning to bleb off as vesicles, it assumes characteristics of the plasmalemma. While this transition is taking place there are changes occurring within the lumen of the cisternae, resulting in the appearance of matrix materials contributing to the contents of the nascent secretory vesicles. Thus for endoplasmic reticulum to be the source of the new plasmalemma infers that a conversion from endoplasmic reticulumlike to plasmalemmalike membranes must occur at points on route to the apex. Evidence of polar membrane differentiation along stacks of cisternae (Grove et al., 1968), along with a dynamic concept of dictyosomal function makes the dictyosome a logical site for such a membrane transformation.

III. At the distal pole of the dictyosome, secretory vesicles are liberated into the cytoplasm with the eventual loss of the entire cisternae due to this blebbing mechanism. The view that vesicles are both formed and discharged by dictyosomes is in agreement with other examples of developmental continuity between cisternae and vesicles.

IV. In hyphal conditions as vesicles migrate forward towards the apex, some undergo an increase in size, often fusing to produce larger vesicles. Further accumulations and modifications of cell wall matrix materials may occur within the vesicles while they are in transit. Unless there is a similarity of membrane types there would be no reason to expect compatability for coalescence and/or growth of small vesicles to form larger ones, without evidence of progressive transformation from one type of membrane into another.

V. Vesicles accumulate at sites of active growth where the vesicle membranes fuse with the plasmalemma, the vesicle contents duly being deposited in the wall region. Grove et al. (1970) have calculated from measurements of vesicle diameters and plasmalemma surface areas that there are sufficient vesicles present in the apex of an actively growing hypha to sustain apical growth over a given period of time.

In addition to providing new membranes it seems reasonable that the vesicles might also transport some component for wall formation and also wall softening and other enzymes (Bartnicki-Garcia, 1973). The relationship between vesicle contents and extracellular material cannot be fully understood without the use of a specific marker. With some staining techniques vesicle contents may stain differently from the hyphal wall. Matile et al. (1971) have isolated exo- and endo-β-1, 3-glucanases from vesicles of budding *Saccharomyces cerevisiae,* while Grove & Bracker (1970) have observed vesicles which show a positive staining reaction for polysaccharides. Recently, Nolan & Bal (1974) have shown with *Achlya ambisexualis* that cellulase (β-1,4-glucan glucanohydrolase) appears to occur within the dictyosome and is then transported within the vesicles to the plasmalemma where it is released to the area between the plasmalemma and the cell wall.

The nature of the interaction between the wall synthesising and wall lytic enzymes in the overall control of wall growth has been elegantly considered by Bartnicki-Garcia (1973) and is outwith the scope of this review (Fig. 2).

Thus it can be seen that cytoplasmic vesicles play a major part in the growth of fungal cells by supplying essential new membranes for the plasmalemma while also transporting the essential components for new wall synthesis. The vesicular hypothesis for cell extension has been almost totally derived from studies with growing hyphae and it is only recently that attempts have been made to relate the hypothesis to the conditions of regrowth that occur during spore germination.

THE NONMOTILE SPORE

Morphology of Germination

Isotropic Growth Phase

ORGANELLE CHANGES DURING GERMINATION

Although the primary events responsible for the initiation of germination in fungal spores still remain elusive there is now, for some species, a considerable amount known about the physiological and structural changes which take place after the initiation stimulus.

An initial swelling phase during spore germination appears to be typical of most fungi. Exceptions have been noted with the spores of some powdery mildews (Yarwood, 1936), the blastospore of *Candida albicans* (Cassone et al., 1973) and the chlamydospore of *Fusarium oxysporum* (Griffiths, 1973). The degree of spore swelling may depend on the nutrient status of the spore: spores with a positive nutrient requirement for germination will normally undergo greater swelling than spores which are capable of germinating in water alone. The swelling phase has been shown to be biphasic in a number of fungi. The first phase is nonmetabolic and is probably due to rehydration or inhibition phenomena. This is then followed by a metabolic phase in which swelling is a growth process (Yanagita, 1957; Barnes & Parker, 1967). During spore swelling, spores become more readily stainable with cytoplasmic stains, suggesting an increase in permeability to solutes (Fletcher, 1969). The reduction in optical path length that was observed with *Penicillium griseofulvum* conidia (Fletcher, 1969) suggests that the volume increase could be due in part to water uptake with the concomitant dilution of protoplast solids.

The extent of spore swelling which occurs during germination varies in different fungi but normally occurs within the range of a factorial increase in diameter of between 1.25--3x (Baker, 1944, 1945; Mandels & Darby, 1953; Ekundayo & Carlile, 1964; Marchant & White, 1966; Barnes & Parker, 1966; Fletcher, 1969). The dynamics of swelling varies with the fungus (Ekundayo & Carlile, 1964, Fletcher, 1969). A peculiar reversible swelling of 20--30x has been observed in *Penicillium notatum* and *Trichoderma lignorum* conidia in response to low pH (Burguillo et al., 1972).

An abnormal and extensive swelling of about 6x has been observed in *Aspergillus niger* conidia in response to elevated temperatures (Anderson & Smith, 1972). Under normal conditions of germination *A. niger* conidia will undergo on average a 2x swelling before producing a germ tube. However, at temperatures from 38°--43°C the proportion of conidia which produce germ tubes progressively decreases and at 44°C germ tube formation is completely inhibited. At these temperatures swelling of the conidia continues resulting in the formation of large spherical cells (20--30 nm mean diameter). During this process an increase in spore surface area of 30x occurs (Fig. 3).

To begin to understand the organelle changes associated with swelling and germ tube formation in spores it is pertinent to examine briefly the nature of the pregermination spore wall and the subsequent changes which occur during the full germination process.

It is now clear that for most if not all fungal spores synthesis of new wall material occurs during the spore swelling phase of germination. The site of this synthesis has been the subject of considerable debate. The origin of the new germ tube wall has also been a subject of controversy. However, it is now clear that most early ultrastructural studies on spore germination were technically at fault since most methods permitted some degree of rehydration before the onset of fixation. In this way, the rapid formation of new wall layers was missed. Since most spores will undergo rehydration (endogenous swelling) to some extent when placed in water it is imperative to initiate the studies using dry harvested spores to obtain a valid representation of the ultrastructural changes which occur during germination. This was clearly shown by Florance et al. (1972) in a study of the ultrastructure of spore germination in *Aspergillus nidulans*. The mature dormant spore wall was observed to consist of three distinct layers, the outermost pigmented layer being composed of a network of rodlets **(Fig. 4)**. After a short period of immersion in water the spore was shown to possess two new rehydrated layers within the inner wall. It is significant that the time taken for the appearance of these two new layers is of such short duration that it is more likely that the new wall layers were derived from molecular reorientation within the innermost wall layer. The formation of one or more new wall layers at the innermost part of the spore wall during the swelling phase has been observed with a number of other fungi including *Rhizopus nigricans*, *R. sexualis* and *R. stolonifer* (Hawker & Abbot, 1963; Ekundayo, 1966; Bussel et al., 1969); *Fusarium culmorum* (Marchant, 1966b); *Gilbertella persicaria* (Bracker, 1966); *Mucor rouxii* (Bartnicki-Garcia et al., 1968a); *Cunninghamella elegans* (Hawker et al., 1970); *Botrytis cinerea* (Gull & Trinci, 1971; Hawker & Hendy, 1963); *Fusarium oxysporum* (Griffiths, 1973); *Botryodiplodia theobromae* (Wergin et al., 1973); and *Aspergillus nidulans* (Border & Trinci, 1970).

In the study by Border & Trinci (1970) on germinating spores of *Aspergillus nidulans* it was observed that two new wall layers were formed beneath the original spore wall. The outermost of these remained the same thickness during spore swelling whereas the

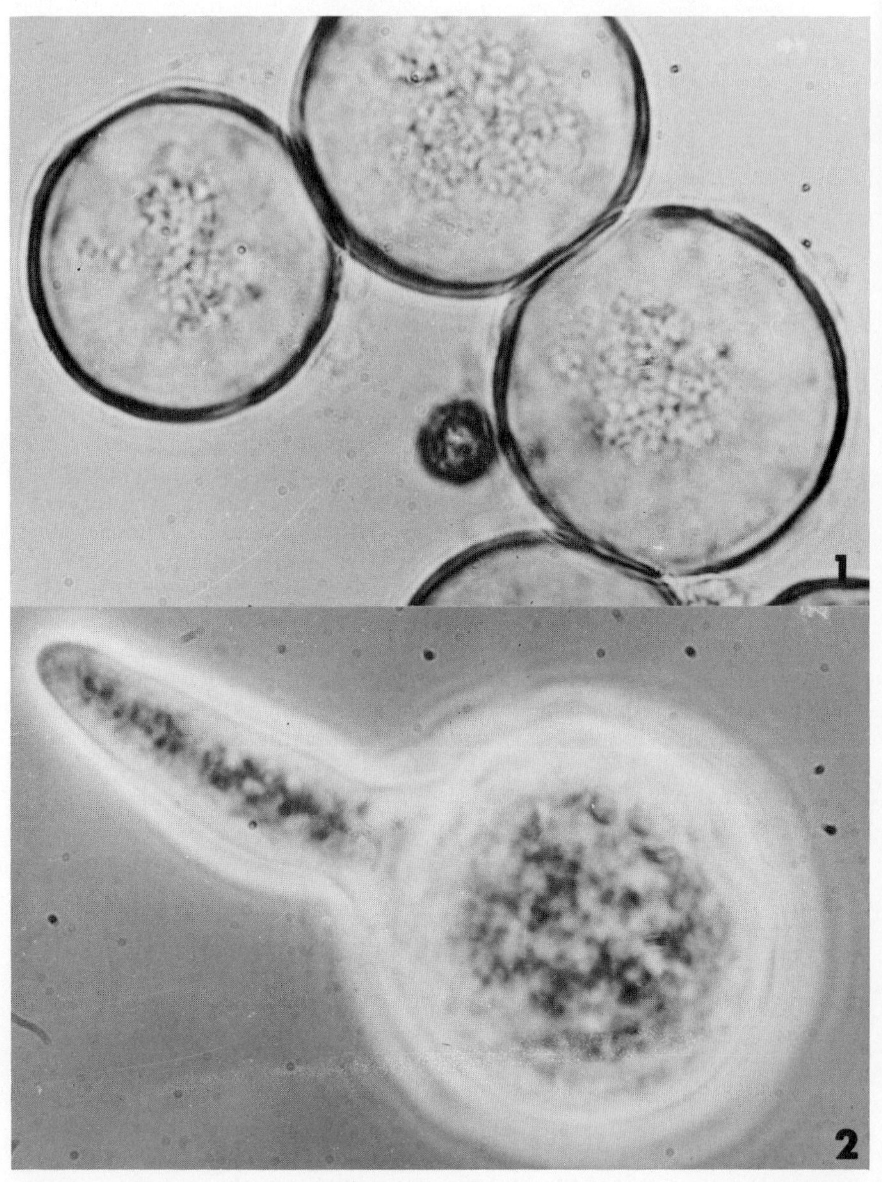

Fig. 3. Swelling and germination of giant cells of *Aspergillus niger*.

innermost layer increased in thickness implying active wall synthesis. That the synthesis of new wall material at the inner part of the spore wall does in fact occur during the swelling phase is clearly shown in the giant conidia of *Aspergillus niger* in Fig. 5 where the remains of the old cell wall can be seen. In these cells a progressive increase in wall thickness occurs throughout the swelling phase to produce a thick multilayered wall (Fig. 6).

Bartnicki-Garcia (1973) considers that spherical wall growth results from the unpolarized nature of the incorporation of wall polymers. Tritium labelled N-acetyl-D-glucosamine was pulsed into spores of *Mucor rouxii* and autoradiography was used to monitor the sites of deposition of new cell wall material (Bartnicki-Garcia & Lippman, 1969). The uniform deposition of chitin and chitosan (polymeric forms of N-acetyl glucosamine) enables the spore to perpetuate a sphere in a manner similar to yeast cell wall synthesis (Cortat et al., 1972), until a predetermined (presumably by a combination of genetical and environmental influences) diameter is achieved.

Polarized Growth Phase

Once the spore has achieved maximum or near maximum size the pattern of wall synthesis becomes polarized in that wall polymer deposition becomes limited to one or a few areas within the spore and these subsequently become the sites of incipient germ tube emergence. Polarized growth continues until apical dominance is established within the nascent hypha (Bartnicki-Garcia, 1973). In most genera, the wall of the germ tube is continuous with a new inner spore wall layer which either completely surrounds the spore protoplast or alternatively only occurs at the site of germ tube emergence. This has been observed in spores of *Botryodiplodia theobromae* (Wergin et al., 1973); *Candida albicans* (Jansons & Nickerson, 1970); *Fusarium oxysporum* and *F. culmorum* (Garcia-Acha et al.,

---

1. Conidia were incubated in the culture medium at 44 °C for 48 h. Conidia increased in size by spherical growth and did not produce germ tubes. Note presence of dormant spore. Phase contrast, x 565.80. 2. Production of a conidiophore stalk from an enlarged conidium 6 h after transfer of giant cells to 30 °C.

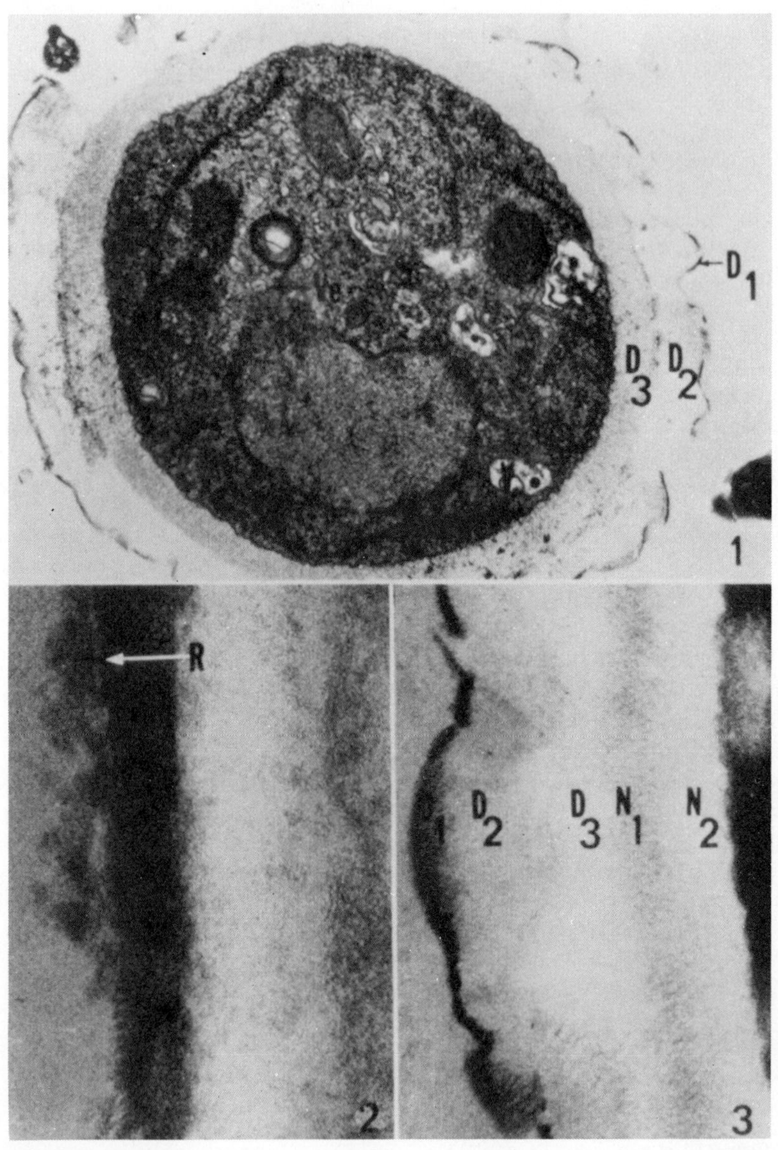

Fig. 4. Ultrastructure of *Aspergillus nidulans* conidium.
1. Thin section of dormant conidium, showing ER, nucleus,

1966; Marchant, 1966a, b; Marchant & White, 1966; Griffiths, 1973); *Geotrichum candidum* (Steele & Fraser, 1973); *Phytophthora parasitica* (Hemmes & Hohl, 1969; Bartnicki-Garcia, 1969); *Rhizopus nigricans* (Hawker & Abbot, 1963); *Rhizopus stolonifer, R. sexualis* and *Cunninghamella elegans* (Hawker, 1966; Bussel et al., 1969a,b; Hawker et al., 1970); *Aspergillus nidulans* (Border & Trinci, 1970; Florance et al., 1972); and *Mucor rouxii* (Bartnicki-Garcia et al., 1968a).

Further evidence that wall outgrowth from fungal spores involves an extension of the innermost wall layers of the spore is seen with the germination of the giant spores of *Aspergillus niger*. Following the period of spherical growth the giant cells will respond to a lowering of the temperature by either producing a germ tube or a conidiophore initial (Anderson & Smith, 1971). In each case the wall of the new outgrowth is continuous with the innermost layer of the spherical cell (Fig. 7) and emerges through the outer layers.

The physical act of germ tube emergence is, as yet, not fully understood. Attempts to clarify this problem have implied mechanical rupturing of the outer spore wall permitting the germ tube to break through. This suggests that a zone of weakness exists and germ pores have been identified (Hawker, 1966). However, it seems more probable that the germ tube passes through the spore wall by way of enzymic digestion. Under certain conditions the giant spores of *Aspergillus niger* can produce up to twenty germ tubes, occupying almost the entire surface area of the spore.

Thus the outer spore coat serves only in a protective capacity during the early stages of germination when the metabolism within the activated spore is in a state of high flux and is thus susceptible to adverse conditions. Furthermore, the emerging germ tube wall arises either from a new continuous wall layer innermost to the spore wall or is formed de novo at the site of the emergence but again deriving from the innermost site.

---

vacuoles (V), vesicles (Ve), mitochondria, ribosomes and three-layered wall (D_1, D_2 and D_3) (x 35,360). 2. Portion of conidium wall hydrated 10.5 h. Note rodlets (R) in outer (D_1) layer (x 108,160). 3. Portion of conidium wall hydrated 30 min. Note new layers (N_1 and N_2) (x 86,580). From Florance et al., 1972.

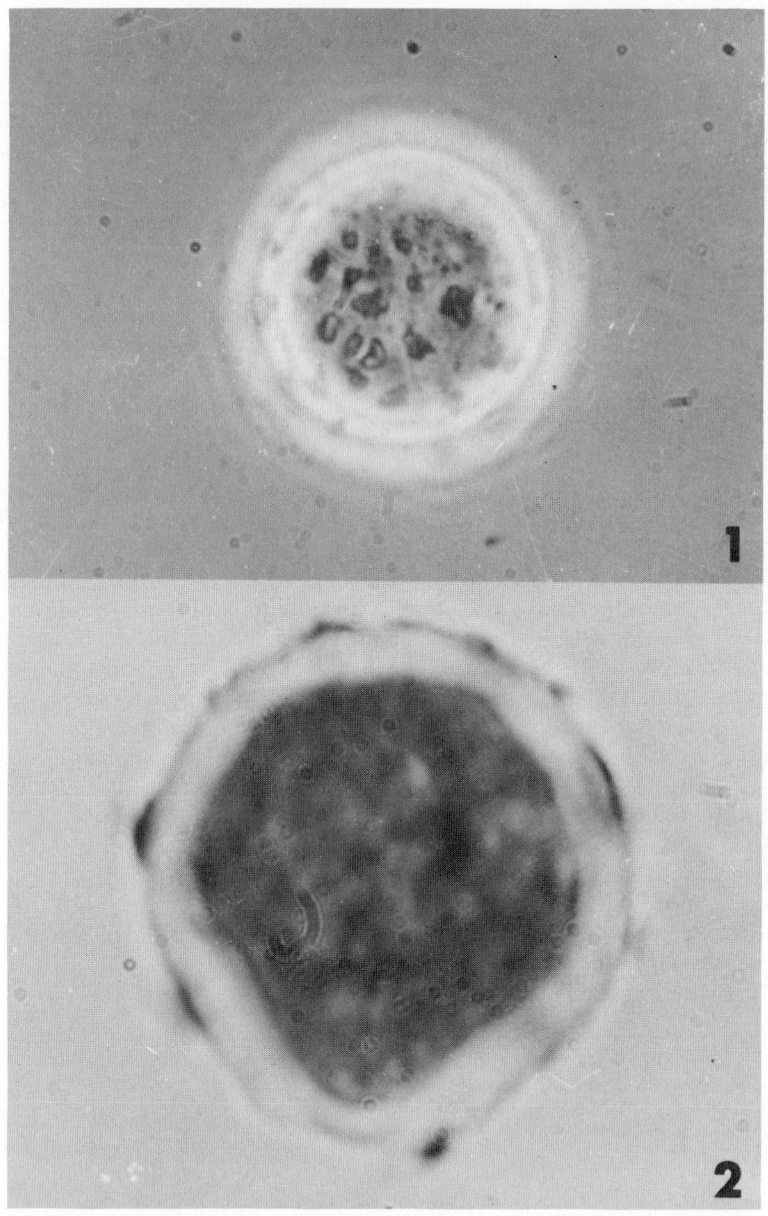

Fig. 5. Wall surface of giant cells of *Aspergillus niger*.

Organelle Changes Associated With Germination

The transition from the dormant state to the highly active metabolic condition of germination is frequently accompanied by major changes in the number and morphology of most subcellular organelles. Furthermore, many of the organelles show a tendency to move towards the side of the spore from which the germ tube will emerge and these may then continue into the germ tube during outgrowth.

The Nucleus and Endoplasmic Reticulum

In general spore nuclei are similar in structural detail to those described for parent vegetative cells (Tsukahara et al., 1966). The nucleus is enclosed by a double membrane, the nuclear membrane, which has discontinuities or nuclear pores to permit continuity between the nucleoplasm and cytoplasm. Nucleoli can be identified in most nuclei by their increased density relative to the nucleoplasm, lack of limiting membrane and their position adjacent to the nuclear membrane (Heale et al., 1968; Neiderpruem & Wessels, 1969; Dunkle et al., 1970).

An early event in spore germination is the increase in nuclear volume followed by nuclear divison (Lowry & Sussman, 1968; Fletcher, 1969; Hawker, 1966; Hawker et al., 1970). Nuclear division occurs either before or during germ tube emission in uninucleate spores and one of the daughter nuclei migrates into the germ tube (Baker, 1944, 1945; Fletcher, 1969). Multiple nuclear formation occurs during giant cell formation in *Aspergillus niger* (Fig. 8). The significance of nuclear division prior to germ tube formation in initially multinucleate spores has not been considered. The role of centriolelike bodies associated with dividing spore nuclei has not been fully interpreted (Fletcher, 1969).

Endoplasmic reticulum (ER) has been reported to be sparse in some spores (Lowry & Sussman, 1968; Hawker, 1966) and prominent in others (Buckley et al., 1968; Florance et al., 1972; Steele &

---

1. Giant cell focused to show fragments of the dark pigmented external coat adhering to the expanding spore wall. Phase contrast, x 2930. 2. Plasmolysed giant cell. Cytoplasm has collapsed while the spore wall remains extended. Bright field, x 2,200.

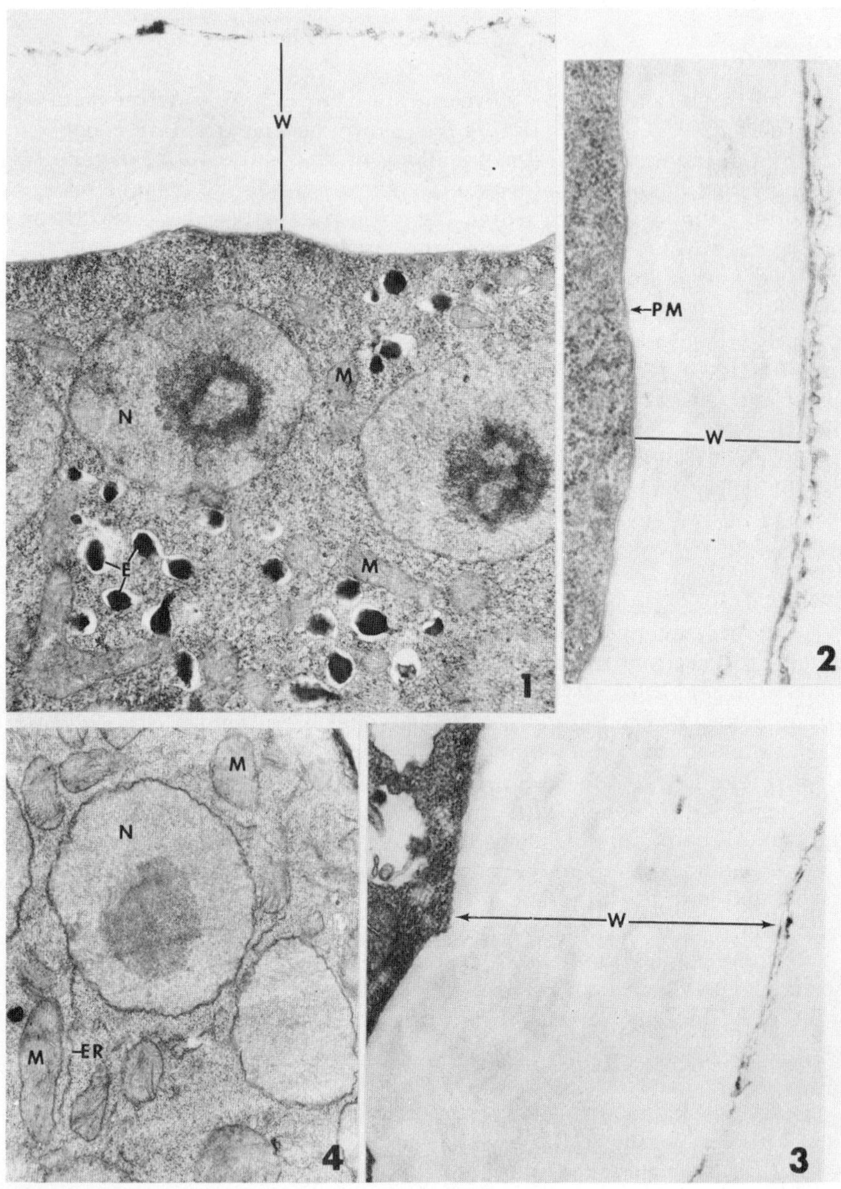

Fig. 6. Wall development during giant cell formation in *Aspergillus*

Fraser, 1973). In general, there is a marked increase in ER with germination and occasionally these structures are seen to be connected to the nuclear envelope (Campbell, 1971; Buckley et al., 1968; Richmond & Pring, 1971; Robb, 1972; Williams & Ledingham, 1964). Robb (1972) has suggested that the nuclear envelope may be the source of the abundant ER which forms during germination. The ER profile appears normally as paired parallel membranes which ramify throughout the cytoplasm and is often porous or fenestrated, although tubular or vesicular forms may also be found (Bracker, 1967). In resting spores ER normally occurs as short strands close to the plasmalemma while in germinating spores it can occur as long strands and frequently as multiple strands surrounding the nucleus. In *Ustilago hordei* Robb (1972) observed that one of the earliest ultrastructural changes which occurred during germination was the formation of stacks of ER. Although such stacks of ER in fungi have been referred to as simple golgi (Campbell, 1969), Robb (1972) considers that since these stacks are not always associated with the nucleus or with vesicle production they cannot justifiably be referred to as golgi.

Ribosomes are generally sparse in dormant spores but increase considerably as germination is initiated presumably as a result of the increase in the metabolic activity prior to and during germination. Most ribosomes occur free in the cytoplasm and as such will be involved in nonsecretory functions. Ribosomes associated with secretory functions are normally associated with ER giving the characteristic rough ER (Richmond & Pring, 1971).

---

niger. All material fixed in gluteraldehyde, postfixed in osmium tetroxide, and embedded in Spurr's epoxy resin (A. R. Spurr. J. Ultrastruct. Res. 26, 31--43 (1969). Sectioned on glass knives on an LKB ultramicrotome III; sections stained in lead citrate and viewed using an AEI 801A electron microscope.
1. 48 h incubation, showing cell wall (W); nuclei (N); mitochondria (M); and small electron dense bodies (E). x 16,560. 2. 36 h incubation showing wall (W) and plasma membrane (PM) (x 33,120). 3. 72 h incubation, showing wall (W) (x 24,840). 4. 24 h incubation showing nuclei (N); mitochondria (M); and some endoplasmic reticulum (ER) (x 13,800).

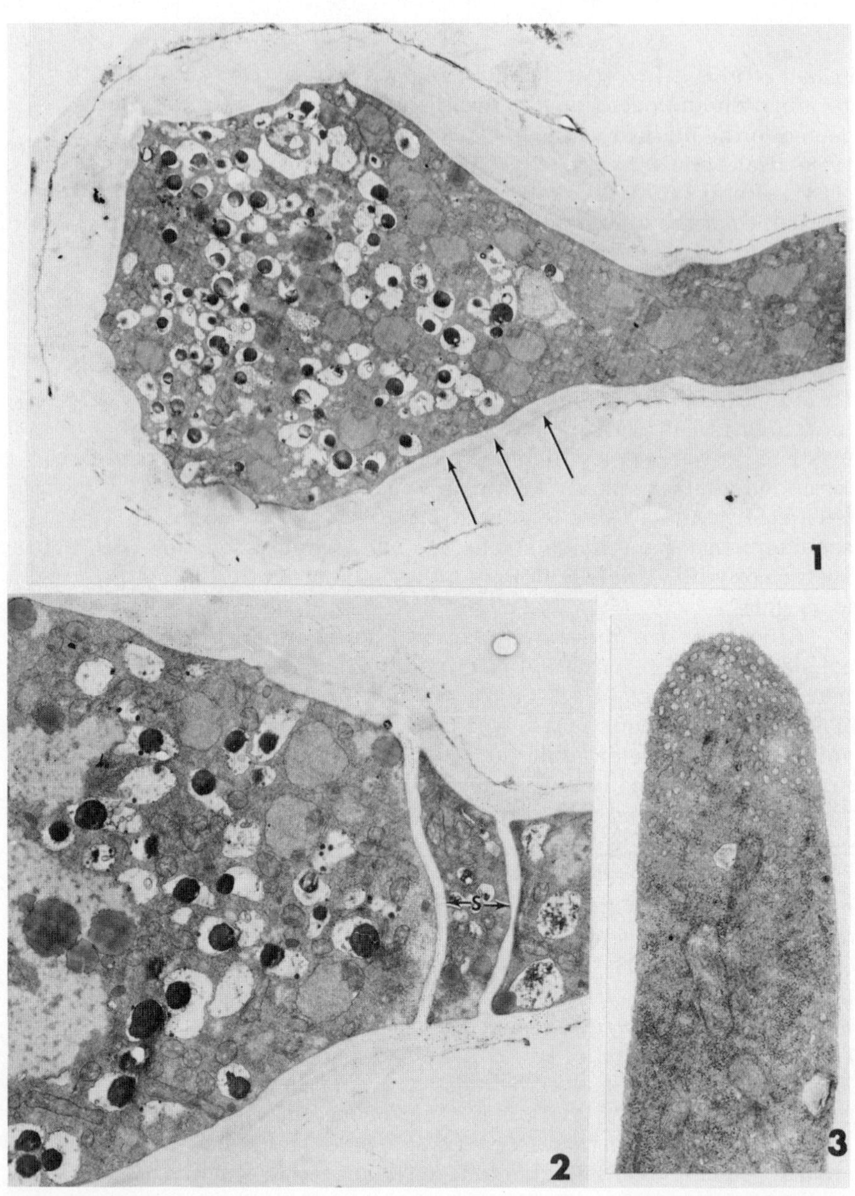

Fig. 7. Germination of giant cells of *Aspergillus niger*.
1. 72 h incubation. Note the cell wall of the germ tube is continuous

ORGANELLE CHANGES DURING GERMINATION

Cytoplasmic Vesicles

Since it has now become widely accepted that the cytoplasmic vesicles which occur at the expanding apices of hyphae serve to supply new plasmalemma and cell wall material it is indeed surprising that so many early ultrastructural studies of spore germination failed to observe vesicles at the sites of germ tube outgrowth (Bracker, 1971). However, several recent studies have clearly shown that cytoplasmic vesicles do in fact occur at the site of germ tube development and continue throughout to the establishment of apical dominance (Matile et al., 1969; Hemmes & Hohl, 1969; Sussman et al., 1969; Grove, 1972; Bracker, 1971; Florance et al., 1973; Steele & Fraser, 1973). In particular, Bracker (1971) has shown that the failure of previous investigations to observe vesicles was due almost entirely to the use of variable pretreatments, such as spore washing, before fixation.

In a series of carefully controlled studies Bracker (1971) showed that spores pretreated in any way before fixation failed to show vesicles in emerging germ tubes. Even particularly mild treatments lead to extensive lomasomelike deposits on the germ tube walls. Consequently, fixation should be as near to instantaneous as possible in order to prevent any momentary disturbance to the formation and flow of vesicles to sites of active growth. This was further evidence of the extreme lability of growing apices and demonstrates how any influence on the active growth process causes rapid changes in protoplasmic organization. Thus subapical organization can be rapidly distorted by environmental changes and by physical manipulation of the material. Thus in such studies extreme care must be exerted to ensure that there is no interruption from active growth prior to fixation. Undoubtedly, all ultrastructural studies which have involved some degree of pretreatment prior to fixation must be of questionable value.

---

with a new wall layer inside the spore wall (arrows). Spore contains lipid bodies, large electron dense bodies and nuclei. The germ tube cytoplasm is more uniformly organized (x 3,550). 2. 72 h incubation. After germ tube emergence two septa (S) are often laid down at the point of emergence. This is a slightly glancing section (x 5,680). 3. 72 h incubation. Germ tube tip, slightly off center, but showing apical vesicles (x 18,105).

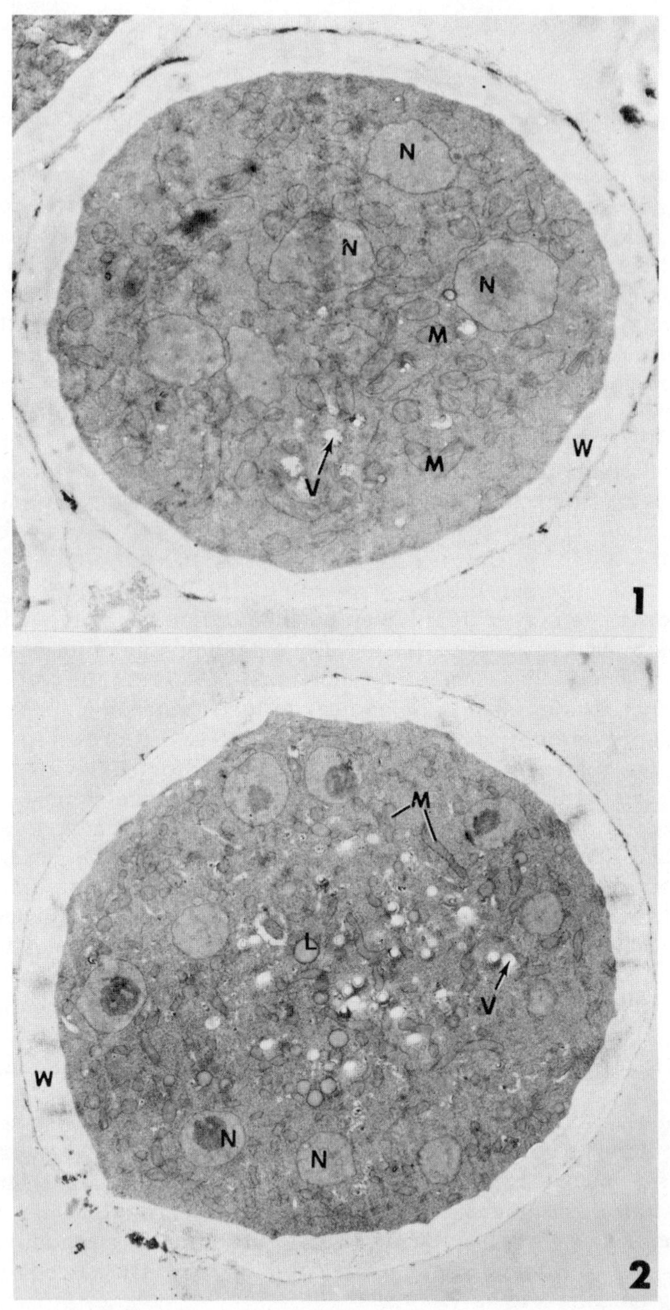
N
N
N
M
M
W
V
1
M
L
V
W
N
N
2

The number of vesicles at the sites of germination are less than are normally found at active hyphal apices. Such vesicles normally occur in a single layer, some of them possessing staining contents while others do not (Bracker, 1971). Vesicles of both types have membranes with dimensions and stainability that are similar to the plasma membrane. Single cisternae with densely staining contents have been seen in several germinating spores (Bracker, 1971; Steele & Fraser, 1973) and are probably the functional equivalent of the dictyosomes (Fig. 9). Few ribosomes are present in the peripheral cytoplasm where the vesicles are clustered. In some cases the membrane of the vesicles can be seen to be continuous with the plasmalemma (Fig. 10).

These results reinforce the concept that apical vesicles occur in tip-growing systems where they are associated with the growth of the expanding surfaces.

Whereas numerous studies have confirmed the involvement of cytoplasmic vesicles in wall changes at the hyphal and germ tube apices much less is known about vesicle involvement in the spore wall changes which occur during the swelling phase of germination. Vesicles have however been observed lying adjacent to the plasmalemma of swelling spores of *Penicillium notatum* (Martin et al., 1973) and *Cunninghamella elegans* (Hawker et al., 1970). In the latter study the vesicles were observed not only in the peripheral layers of the cytoplasm but also between the plasmalemma and the spore wall. With *Botrytis fabae* spores Richmond & Pring (1971) also observed that vesicles produced by the endoplasmic reticulum passed through the plasmalemma and into the spore wall. In *Fusarium oxysporum* chlamydospores, vesicles were abstricted from the plasmalemma into the paramural space and these then attached to the face of the newly developing spore wall (Griffiths, 1973).

Fig. 8 (1--2). Sections of giant cells of *Aspergillus niger* after incubation at various time intervals at 44°C.
1, 24 h incubation. The cell is multinucleate (N) (5 profiles in this section) and a few small vacuoles (V) and the spore wall (W) and mitochondria (M) can be seen. x 6,160. 2. 36 h incubation. Multinucleate (10 profiles), Nuclei (N) are arranged under the spore wall (W) with lipid bodies (L) and small vacuoles (V) towards the center. Mitochondria (M) can also be seen. (x 4,235.) 3. 48 h incubation. Note very thick wall (W). Multinucleate (22 nuclear profiles in this

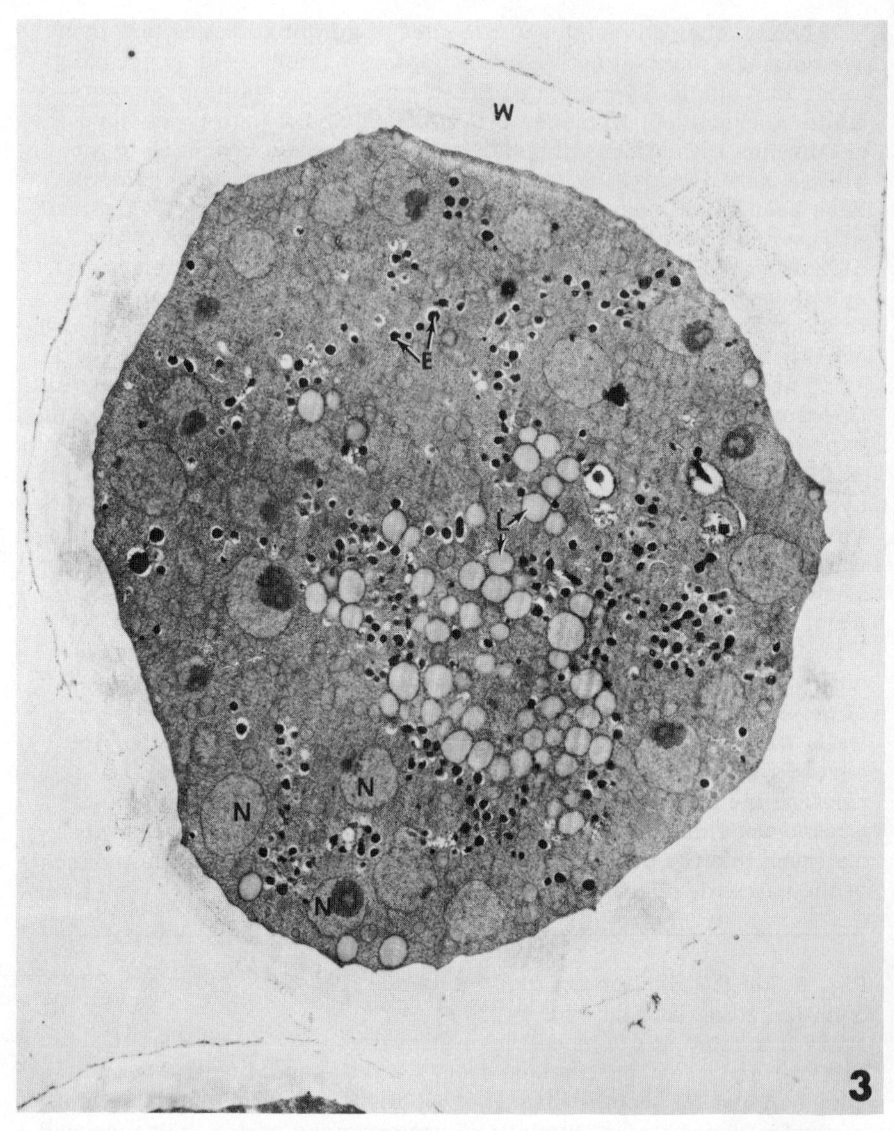

section). Nuclei (N) towards outside, under wall. Lipid bodies (L) have increased in numbers since 36 h. Also note the appearance of small electron dense bodies (E) in small vacuoles.

Apical Corpuscle

Bartnicki-Garcia et al. (1968b) reported the presence of an apical corpuscle in the germinating sporangiospore of *Mucor rouxii*. They described this corpuscular deposit on the apical wall as a single electron-dense structure 0.2 nm in size. It was seen to be closely associated with the apical wall of the incipient germ tube and it was suggested that it had a role in germ tube emission and for its continued apical growth. Subsequent investigations have revealed the presence of somewhat similar structures in germinating spores of *Botrytis fabae* (Richmond & Pring, 1971), *B. cinerea* (Gull & Trinci, 1971) and *Phytophthora parasitica* (Bartnicki-Garcia, 1969). Spores of *Aspergillus nidulans* (Weisberg & Turian, 1971) and *Microsporum gypseum* (Page & Stock, 1971) exhibit peripheral structures resembling the apical corpuscle which these authors categorize as lomasomes or lysosomes.

Such observations would perhaps tend to support the proposed role of the apical corpuscle in germ tube formation. However, where secretory vesicles can be found associated with germ tube formation no apical corpuscle can be observed (Bracker, 1971; Grove, 1972). Structures not unlike but not identical to apical corpuscles were observed by Bracker (1971) in *Gilbertella* when plasma membrane diverges from the germ tube wall in cells pretreated before fixation. Apical corpuscles have never been observed in the apices of growing hyphae and therefore its function in extension growth remains questionable. It may be that the apical corpuscle only has a function in the process of intial germ tube emergence and is not necessary for continued apical growth. Apical corpuscles may only occur in incipient germ tubes and they may have been overlooked in many studies because of the difficulty of obtaining medium longitudinal sections at a sufficiently early stage of germination. The organelle could well be involved not with germ tube growth but rather with the dissolution of the spore wall.

Lomasomes

The term lomasome was first used by Moore & McAlear (1961) who defined them as dome shaped spongelike intumescences, contiguous with the wall, containing granular or vesicular material internally bordered by the plasmalemma. Lomasomes have been more particularly defined by Heath & Greenwood (1970) as membranous vesicular

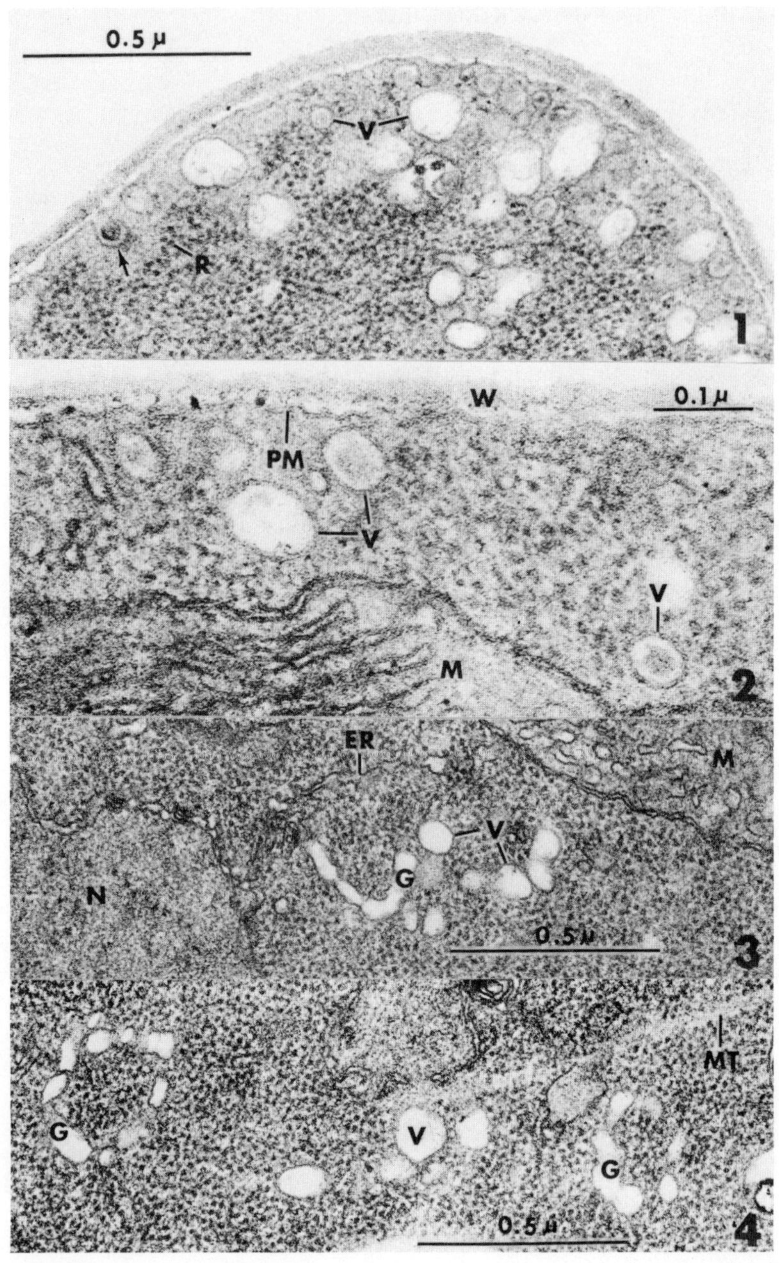
0.5 μ
V
R
1
W
0.1 μ
PM
V
V
M
2
ER
M
V
N
G
0.5 μ
3
MT
G
V
G
0.5 μ
4

material embedded in the wall external to the line of the plasmalemma and plasmalemmasomes as structures with various membrane configurations which are internal to the plasmalemma, often in a pocket projecting into the cytoplasm, and less obviously embedded in the wall material.

Lomasomes have been observed in many germinating spores and vegetative hyphae (Bracker, 1967; Bussel & Sommer, 1973). These vesicular structures have varying configurations being tubular, vesicular or dilated and would appear to differ in their modes of origin. A considerable degree of controversy has been generated on the nature and reality of these structures (Bracker, 1967; Bussel & Sommer, 1973). Lomasomes have not yet been identified in living cells nor have they been identified in cells prepared other than by chemical fixation.

The involvement of lomasomes with cell wall formation has been implied in many studies (Wilsenach & Kessel, 1965; Zachariah & Fitz-James, 1967; Bussel & Sommer, 1973). In particular, lomasome appearance seems to be related to stress conditions and this may account for their variable appearance. In controlled experiments Bussel & Sommer (1973) have shown that anoxia can be an important factor in the appearance of lomasomes during germination of spores of *Rhizopus stolonifer*. During germination in this fungus the inner wall of the sporangiospore develops to become continuous with the germ tube wall and under normal conditions does not become thick. However, under anoxia spore wall synthesis continues resulting in

---

Fig. 9. Vesicle formation in *Gilbertella persicaria*.
1. Part of germ tube apex showing vesicles (V) with staining and non-staining contents. The membrane of one vesicle (arrow) is continuous with the plasma membrane. Note the scarcity of ribosomes (R) in the peripheral cytoplasm where the vesicles are clustered. x 62, 480.
2. Several vesicles (V) near the tip of a germ tube. The dimensions and stainability of the vesicle membranes are similar to plasma membrane (PM). The hyphal walls (W) and a mitochondrion (M) are also shown (x 136,400). 3. Single cisternae golgi apparatus (G) and associated vesicles (V) in a germ tube. The nucleus (N), endoplasmic reticulum (ER), and a mitochondrion (M) are also shown (x 56,320). 4. Two ringlike golgi cisternae (G) and vesicles (V) in a germ tube. Microtubule (MT) is also shown (x 56,320). From Bracker, 1971.

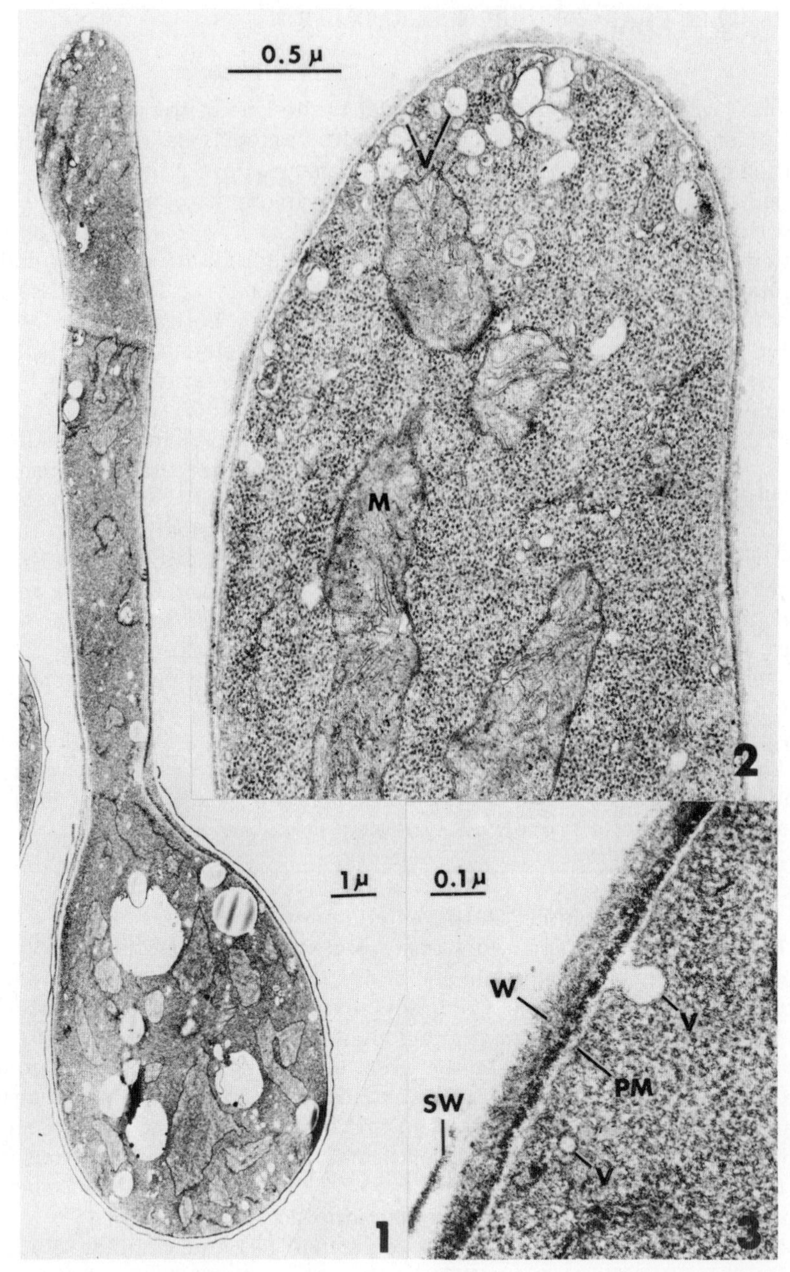
0.5 μ
V
M
2
1μ
0.1μ
W
V
PM
SW
V
1
3

obviously thick walls, together with an increase in the size and appearance of lomasomes. Thus these organelles are implicated in the formation of wall materials required for development and organization of the inner wall.

How the lomasomes are formed is still unclear. However, it would appear to involve the addition of vesicles, originating in the ER and dictyosome system, to the plasma membrane. Bussel & Sommer (1973) consider that the lomasomes arise by elaboration of the plasma membrane, the plasma membrane undergoing extensive invagination into the cytoplasm and thus greatly increasing the surface of the membrane. These extensively invaginated channels of the plasmalemma could provide a direct route from the interior of the cytoplasm to the inner wall. Heath & Greenwood (1970) have suggested that plasmalemmasomes are produced when plasmalemma production is not balanced by cell extension. When such plasmalemmasomes become sequestered in developing cell walls they may be termed lomasomes.

There is little doubt that lomasomes and plasmalemmasomes are real and not artifacts of the methods used for the preparation of samples. It is possible they only occur when environmental conditions impose stresses whch prevent normal vesicle movement and establishment of polarity. A complete understanding of lomasome function may not be found until the enigma of polarity in vesicle movement has been solved.

Mitochondria

The presence of mitochondria in dormant spores is now well documented. An increase in number and changes in the shape of mitochondria have frequently been observed during germination

---

Fig. 10. Germination of spore of *Gilbertella persicaria*.
1. Germinated spore with unbranched germ tube (x 6,160). 2. Enlargement of the germ tube apex from 1, but from a different section. Vesicles (V) with staining and nonstaining contents are clustered in ribosome-free cytoplasm at the tube apex. A mitochondrion (M) is also shown (x 30,800). 3. Part of the periphery of a germinating spore showing the germ tube wall (W) and the nonstaining contents is continuous with the plasma membrane (PM) (x 72,160).

(Tsukahara et al., 1966; Campbell, 1970; Hawker & Abbot, 1963; Buckley et al., 1968; Lowry & Sussman, 1968; Hawker et al., 1970; Park & Robinson, 1970; Florance et al., 1972; Steele & Fraser, 1973; Martin et al., 1973). It is generally considered that mitochondria are derived by the division of preexisting mitochondria. Constricted mitochondria have been seen which appear to be in the process of division (Campbell, 1970).

Bartnicki-Garcia et al. (1968a) have pointed out that whereas there is a net increase in the number of mitochondria in germinating sporangiospores of *Mucor rouxii*, this increase was commensurate with the overall increase in protoplasmic mass and not relative to other cell components. In a study on *Botrytis cinerea* Gull & Trinci (1971) calculated the cytoplasmic area occupied by mitochondria in dormant and germinating spores and concluded that although the number of mitochondria increased during spore swelling the areas occupied by mitochondria and cytoplasm remained relatively constant.

In most cases the morphology of spore mitochondria is relatively similar to the mitochondria present in vegetative mycelium. The mitochondria have a double membrane separated by an interspace of less density and the parallel internal lamellae project from the inner layer of membrane. During germination the cristae can be of variable appearance and can on occasion extend completely across the mitochondrion (Buckley et al., 1968). More cristae have been reported to be present in the mitochondria of the germ tubes of wheat stem rust than in the uredospores and their orientation is more regular (Williams & Ledingham, 1964). Changes in the size and shape of mitochondria have also been observed. In most cases the mitochondria in the dormant spores are larger than during germination (Hawker & Abbot, 1963; Marchant, 1966a; Buckley et al., 1968; Lowry & Sussman, 1968; Mitchell & McKeen, 1970). Exceptions have however been noted. During the early stages of germination of teliospores of *Ustilago hordei* the mitochondria increase in size with little or no increase in number (Robb, 1972). An increase in the size of mitochondria during germination has also been reported in *Sphaerotheca macularis* although in this case an increase in number also occurred (Mitchell & McKeen, 1970; McKeen et al., 1966). The large, convoluted mitochondria observed in spores of *Rhizopus* (Hawker & Abbot, 1963) may be related to the possible anaerobic internal environment of the spore. As germination proceeds with ensuing increase in permeability (Fletcher, 1969) the new mitochondria are smaller and more regular in shape and are probably formed by the rounding and splitting off of the lobes of the large mitochondria. The possibility that the large mitochondria could be the result of sudden osmotic changes during fixation has been considered by Hawker & Abbot (1963). There seems little doubt however that the large mitochonodria are real and not an artifact of the preparative technique.

Lowry & Sussman (1968) found that mitochondria in dormant ascospores of *Neurospora tetrasperma* were very large and swollen whereas in germinating spores they were smaller and of various shapes. During arthrospore germination in *Geotrichum candidum* (Park & Robinson, 1970) the mitochondria change in form from short and oval to filamentous and the change precedes an increase in oxygen uptake.

In general, the respiration level of all fungal spores increases substantially during germination (Sussman & Halvorson, 1966). The functional role of the respiratory system in cytodifferentiation is indicated by the observation that germination is inhibited by cyanide and DNP (Niederpruem & Dennen, 1966). The endogenous content of cytochrome c in dormant ascospores of *Neurospora sitophila* is much lower than in mitochondria of germinating spores (Holton, 1961). The role of the oxidative and phosphorylative capacity of the mitochondria during germination will be considered in a later section in connection with concepts of polarity.

Other Inclusions

An increase or decrease in the degree of vacuolation appears to be an important ultrastructural feature which is commonly observed during germination of fungal spores. Two different kinds of vacuoles were observed by Martin et al. (1973) in dormant spores of *Penicillium notatum*, and both types decreased in size during germination. In general, however, it is an increase in vacuolation which is usually noticed during germination. This was the case in *Lenzites saepiaria* (Hyde & Walkinshaw, 1966); *Aryria cinerea* (Mims, 1971); *Ustilago hordei* (Robb, 1972); *Cunninghamella elegans* (Hawker et al., 1970); *Mucor rouxii* (Bartnicki-Garcia et al., 1968a); *Schizophyllum commune* (Niederpruem & Dennen, 1966); *Botrytis cinerea* (Gull & Trinci, 1971); and *Coprinus lagopus* (Heintz & Niederpruem, 1971). In *Geotrichum candidum* Park & Robinson (1970) observed an initial decrease in vacuolar volume followed by an increase which occurred at the time of germ tube formation. Unusually high vacuolation appears to be a particular feature of the conidia of *Sphaerotheca macularis* (Mitchell & McKeen, 1970) since more than

50% of the volume of these conidia consists of vacuoles. Such vacuolation appears to be a necessary adjunct of hydration and germination in *Sphaerotheca macularis* since if the vacuoles are damaged or prevented from expanding the spore ceases to germinate (Mitchell & McKeen, 1970). In spite of the obvious importance of the development of these organelles in the spore little information is available on their origin during germination. Robb (1972) has however considered in depth the origin of vacuole formation in germinating teliospores of *Ustilago hordei*. It was suggested that the "primary hydration vacuoles" which were responsible for the initial swelling of the teliospore arise from dilation of the cisternae in whorls of ER. This idea was based on the observation that whorls of ER were produced just prior to vacuole formation and then disappeared. Additional evidence was the fact that the tonoplast membrane (average width, 96 Å) was the same width as the unit membranes of the ER (average width, 94 Å).

Lipid bodies are generally found in dormant fungal spores and these frequently disappear upon germination. This has been observed during the germination of spores of *Mucor rouxii* (Bartnicki-Garcia et al., 1968a); *Cuninghamella elegans* (Hawker et al., 1970); *Aryria cinerea* (Mims, 1971) and *Aspergillus fumigatus* (Campbell, 1971).

Foamy cytoplasm occurs in the spores of several fungi and various functions have been attributed to it including the storage of polysaccharide reserves (Hyde & Walkinshaw, 1966; Mims, 1971; Robb, 1972; Voelz & Niederpruem, 1964). It has been observed that such regions of the spore cytoplasm are often associated with clusters of large granules and this suggests that the foamy appearance results from the leaching out of some component (Hyde & Walkinshaw, 1966; Robb, 1972; Voelz & Niederpruem, 1964).

THE MOTILE SPORE

The Morphology of Germination

Studies on zoospore germination have presented intrinsic problems in that synchronous germination was until recently difficult to obtain. However, methods have now been developed which allow a much greater degree of synchrony to be achieved (Soll et al., 1969; Truesdell & Cantino, 1971). The fungi most studied have been *Blastocladiella emersonii* (see Truesdell & Cantino, 1971; Reichle &

Fuller, 1967); *Allomyces* sp (Fuller & Olson, 1971; Olson & Fuller, 1971; Olson, 1973a,b) and several other fungi to a lesser extent (see Chong & Barr, 1973).

A vast literature has accumulated on zoospore germination in *Blastocladiella emersonii* (Truesdell & Cantino, 1971) and consequently this organism will be used as the type example of zoospore germination. However, due recognition will be made of deviations in ultrastructural organization which occur in other types of zoospores germination.

The zoospore of *Blastocladiella emersonii* possesses a single polar whiplash flagellum by which it propels itself through its aquatic environment. Although the zoospore is capable of germinating immediately after its release from the sporangium there is normally a time (of variable duration) during which the zoospore swims around before settling on a firm surface and commencing the sequence of events culminating in the emergence of a germ tube and eventually a complex syncytial growth form.

The germination process can be conveniently resolved into a sequence of four distinct, morphological cell forms: zoospore, round cell I, round cell II, and germling (Lovett, 1968; Soll et al., 1969). From an external morphological point of view the transition from zoospore to round cell I is accompanied by a change in cell morphology from variably ovoid to round and retraction of the flagellar axoneme into the cell body proper. A new, rigid cell wall is rapidly formed and there is an overall decrease in cell volume. The transition from round cell I to round cell II is accompanied by absorption of the flagellar axoneme and an increase in cell size by spherical growth. Finally, germling appearance is accompanied by the formation of localized cellular outgrowth, the germ tube, which is the precursor of the eventual rhizoidal system of the growing chytrid (Table 1; Fig. 11).

The similarities between the germination processes of the nonmotile and motile spores are striking, each having to some extent a phase of round cell formation together with spherical growth, leading ultimately to polarized growth with germ tube formation. As with some nonmotile spores spherical growth of the germinating zoospores may continue after germ tube emission. Thus isotropic and polarized wall formation can exist at the same time in one cell.

Organelle Changes During Germination

The zoospore of *Blastocladiella emersonii* contains a remarkably ordered array of organelles which include: a single posteriorly situated flagellum; a single nucleus with nucleolus; a membrane-bound nuclear cap overlying the nucleus; a single giant mitochondrion positioned eccentrically around the nucleus; and gamma particles consisting of a unit membrane enclosing an ellipsoid, cresentric matrix, approximately 0.5 nm in length. In addition to these organelles the zoospore of *Allomyces* contains a set of 27 cytoplasmic microtubules which flare out from the functional kinetosome as well as several smaller mitochondria in the anterior region of the spore.

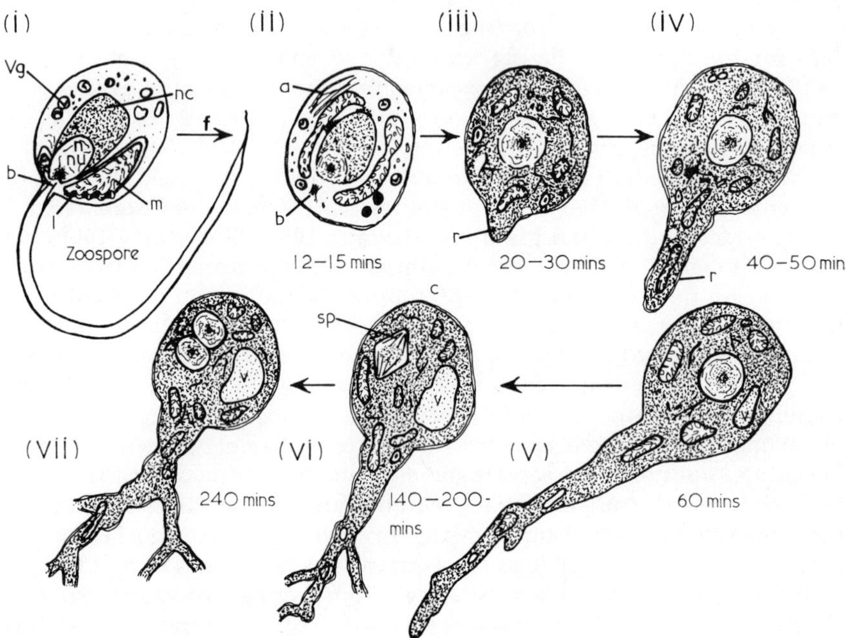

Fig. 11. Stages of zoospore germination and early growth of *Blastocladiella emersonii*. A series of diagrammatic sketches to illustrate the morphological and intracellular changes during the first 4 h of germination. Labels for this diagram are as follows: axoneme (a) of retracted flagellum; basal body (b); centriole (c); flagellum (f); mitochondrion (m); nuclear cap (nc); nucleolus (nu); nucleus (n); rhizoid (r); spindle (sp); vacuole (v); lipid (l); and vesicle-enclosed granule (vg). From Lovett, 1968.

Table 1. Summary of the Sequence of Events During Zoospore Germination of *Blastocladiella emersonii*. From Soll et al., 1969.

Feature	Zoospore	Round cell I	Round cell II	Germling
External flagellum	+	-	-	-
Internal flagellum	-	+	-	-
Cell wall	-	+	+	+
Cell adhesion (to dish bottom)	-	+	+	+
Resistance to lytic agents	-	+	+	+
Vesicle "opening"		+	+	
Intact nuclear cap	+	+	-	-
Single "unbranched" mitochondrion	+	+?	-	-
Germ tube	-	-	-	+

Allomyces zoospores in general appear to lack the gamma particles although they are present in the meiospores of *A. macrogynus*.

The change in pattern of organelles during germination of zoospores of *Blastocladiella emersonii* has been lucidly set out by Truesdell & Cantino (1971) in an extensive review. In the present context only three aspects of organelle change will be examined in detail viz: nucleus and nuclear cap; mitochondria; and gamma particles since in each of these areas there have been interesting recent developments.

The Nucleus and Nuclear Cap

A unique feature of the uninucleate zoospore of *Blastocladiella emersonii* is the formation of the nuclear cap organelle which contains the entire complement of zoospore ribosomes (Lovett, 1963; Lessie & Lovett, 1968). The nuclear cap is double-membraned and continuous with the nuclear membrane. The nucleus and cap share a common outer membrane but each possesses its own inner membrane (Soll et al., 1969). The nuclear cap remains intact in the discharged zoospore and the enclosed ribosomes are inactive until the later stages of germination. A single nucleolus is located in the posterior portion of the nucleus directly above the flagellar basal body.

The progression from round cell I to round cell II is marked by the fragmentation of the nuclear cap double membrane which continues until the membrane no longer confines the ribosomes of the cap and they become dispersed throughout the cytoplasm (Lovett, 1968; Soll et al., 1969). Polysome formation and protein synthesis begin immediately (Lovett, 1968; Adelman & Lovett, 1972). The role of the nuclear cap would appear to be for the conservation of inactive (but functional) ribosomes for use during germination. It has been considered that the nuclear cap ribosomes are inactive because of the presence of an unidentifiable ribosome-bound inhibitor (Schmoyer & Lovett, 1969). Ribosomal protein synthesis without de novo ribosome production has been shown (Adelman & Lovett, 1972).

The developmental aspects of *Blastocladiella* germination are remarkable since the structural changes associated with encystment can occur without apparent protein synthesis and the entire germination process including germ tube formation does not require concomitant synthesis of RNA. Thus many of the events of zoospore germination rather than requiring protein synthesis involve instead rearrangement of pre-existing structures (Soll et al., 1969; Soll & Sonneborn, 1971).

Distinct ribosome-filled nuclear caps have also been shown to be present in *Nowakowskiella profusa* (Chambers et al., 1967), *Harpochytrium hedinii* (Travland & Whistler, 1971), *Allomyces macrogynus* and *A. neo-moniliformis* (Fuller & Olson, 1971). A nuclear cap is not present in zoospores of *Phlyctochytrium arcticum* (Chong & Barr, 1973), *Rhizophydium sphaerotheca* (Fuller, 1966) and *Olpidium brassicae* (Temmink & Campbell, 1969).

Side bodies are commonly seen closely associated with zoospore nuclei in several chytrids (see Chong & Barr, 1973). A complete side body can contain: closely appressed membranes, electron-opaque membrane-bound bodies, lipid bodies, and a mitochondrion (Fuller & Olson, 1971). Associated with the side bodies there can be the Stuben bodies which are electron-opaque membrane-bound and nonlipid containing entities. In *Phlyctochytrium arcticum* (Chong & Barr, 1973) Stuben bodies are differently shaped at different developmental stages of the zoospore. It has been suggested that one function of the Stuben bodies may be related to primary differentiation of the zoospore (Chong & Barr, 1973).

Mitochondria

In *Blastocladiella emersonii* a single, large cup-shaped mitochondrion is situated asymmetrically around part of the nuclear apparatus and around the kinetosome (Cantino et al., 1963). During germination the mitochondrion undergoes drastic changes in shape and the presence of multiple mitochondrial profiles in later stages of germination have been suggested to result from mitochondrial fragmentation

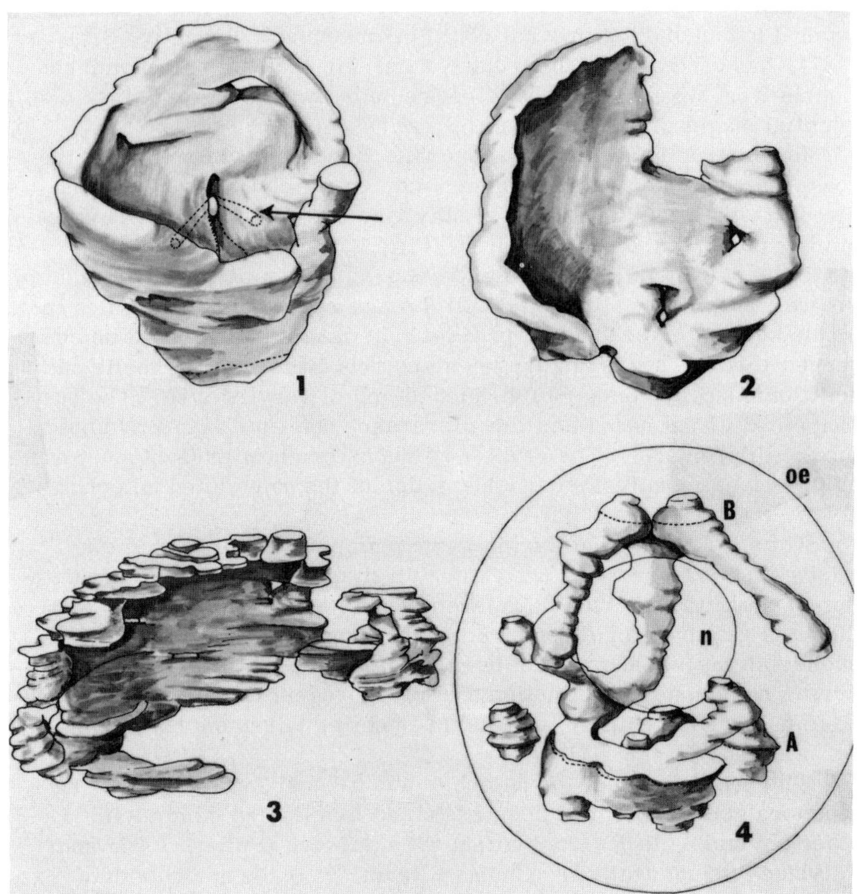

Fig. 12. Drawings of mitochondrial models reconstructed from serial sectioning of *Blastocladiella emersonii*.
1. Zoospore with channel and canals for flagellar assembly indicated by arrow. 2. Round cell. 3. Intermediate between round cell I and round cell II. 4. Round cell II. The approximate position of the nucleus (n) and the outer envelope (oe) are indicated. From Bromberg, 1974.

or division (Soll et al., 1969). However, depending on the plane of sectioning a highly branched profile could be present while in others several mitochondrial profiles could be observed. Recently, Bromberg (1974) has carefully reconstructed three-dimensional mitochondrial models from several sections of entire cells from representative sequential stages of germination of *B. emersonii*.

Reconstruction of the zoospore mitochondrion confirmed that the zoospore possessed a single mitochondrion which contained a canal through which the flagellar assembly may pass and also canals containing the rootlet of the flagellum (Fig. 12). During round cell I stage the mitochondrion retains the cup-shaped appearance, although the wall is thinner. By round cell II stage reconstruction studies indicate several mitochondrial profiles. In particular, it could be observed that there was one large extensively branched organelle and additional smaller ones---unbranched and of varying size. The cup-shape had disappeared and the appearance was now as a convoluted tube with branches. The smaller mitochondria now present are probably derived from further fragmentation of the convoluted mitochondrion.

Thus it is clear that during germination of the *Blastocladiella emersonii* zoospore there are extensive shape changes and fragmentation of the mitochondrion. Apparently neither DNA nor protein synthesis is required for these structural changes. The role of the mitochondrial genome has not yet been studied. Do the DNA units fuse during mitochondrial formation during sporogenesis and subsequently do they partition equally into the fragmenting mitochondria during germination?

The relationship of the mitochondria to other organelles during zoospore germination has been examined to a limited extent with other chytrids. In *Rhyzophydium sphaerotheca* (Fuller, 1966) and *Phlyctochytrium arcticum* (Chong & Barr, 1973) the mitochondria occur clustered around the posterior end, surrounding the kinetosome and rhizoplast in a manner not unlike that found in *Blastocladiella* and *Allomyces*. In *Olpidium brassicae* (Temmink & Campbell, 1969) the mitochondria are clustered around the nucleus. During encystment there is a loss of this defined pattern of mitochondrial distribution in all zoospores examined.

Gamma Particles

These cytoplasmic organelles were first described in

Blastocladiella emersonii by Cantino & Horenstein (1956) and have been extensively discussed by Cantino & Mack (1969), Matsumae et al., (1970) and Truesdell & Cantino (1971). The gamma particle consists of a unit membrane which encloses an ellipsoid, bowl-shaped matrix about 0.5 nm in length. These organelles are now believed to contain DNA (Myers & Cantino, 1971). Gammalike particles have been found in *B. britannica* (Cantino & Truesdell, 1971) and in the meiospore of *Allomyces macrogynus* (Olson, 1973a).

It now seems clear that the gamma particles are associated with wall synthesis in the encysting zoospore. Furthermore, the apparent close association of gamma-particles with cytoplasmic vesicles has lead to the suggestion that they may be the functional equivalents of the dictyosomes (Truesdell & Cantino, 1971). Such conclusions were based on the following observation: gamma-particles start to break down just before the new cell wall is detected and continue to break down as the cell wall is being deposited; electron dense material resembling the gamma matrix accumulates at the cell surface while simultaneously the electron dense substances of the gamma matrix of the particles is disappearing (Truesdell & Cantino, 1970).

Gamma-particles have been shown to contain chitin synthetase (Myers & Cantino, 1971). The specific activity of this enzyme can be increased 15-to 20-fold over the whole cell homogenates by centrifugation of partially purified gamma-particles through discontinuous sucrose density gradients. Light appears to affect the incorporation of chitin synthetase into gamma particles but has no effect on the total amount of this enzyme in the spore (Cantino & Myers, 1972).

SPECULATIONS ON THE ESTABLISHMENT OF POLARITY

The germination of most fungal spores is characterized by an initial phase of cell enlargement which is followed by germ tube formation and extension. In spores which show a considerable degree of swelling during germination and probably for most other spores uniform spherical wall synthesis is suggested to occur. In contrast, during germ tube formation wall formation becomes restricted to a small area of the round spore wall. This interpretation of wall synthesis during the two growth phases has been substantiated by autoradiographic evidence (Bartnicki-Garcia & Lippman, 1969; Bartinicki-Garcia, 1973).

If the vesicular hypothesis of wall growth can be considered applicable to the wall synthesis which occurs during spore germination, then it is clearly of fundamental importance to attempt to understand those factors which control the movement of vesicles and more precisely what controls their movement to the site of germ tube formation.

Of the various environmental factors which are known to affect the establishment and maintenance of the hyphal condition in fungi those which affect respiration appear to be of particular significance.

There is now considerable evidence that mitochondrial functions are required for polar or hyphal growth to occur (Storck & Morrill, 1971; Terenzi & Storck, 1969; Zorzopulos et al., 1973). In particular, Zorzopulos et al. (1973) have shown that inhibition of mitochondrial protein synthesis by chloramphenicol induces yeastlike growth under aerobic conditions. In contrast, the filamentous cells obtained under anaerobic conditions in the presence of EDTA possessed potentially active mitochondria. They have thus suggested that morphogenesis may be controlled through some type of mitochondrial/cytoplasmic interaction.

The duration of the swelling phase prior to the establishment of the polarization of wall synthesis required for germ tube emergence can vary in a number of fungi. Fletcher & Morton (1970) with *Penicillium griseofulvum* and Ekundayo & Carlile (1964) with *Rhizopus arrhizus* have shown that completion of swelling is not an essential prerequisite for germ tube formation. On the other hand, conditions have been recorded which favor the prolongation of the spherical phase and in some cases completely prevent normal germ tube production. It is perhaps significant that here again factors which affect respiration are implicated. In the absence of oxygen spores of *Mucor rouxii* germinate into spherical, yeastlike cells which reproduce by budding. Also with *Mucor rouxii* spores certain concentrations of phenethyl alcohol have no effect on the swelling phase of growth but will inhibit germ tube formation (Terenzi & Storck, 1969). In biochemical studies on the effect of phenethyl alcohol Terenzi & Storck (1969) concluded that the effect was mediated by an inhibition of respiration and enhancement of fermentation.

Bussel & Sommer (1973) have described morphological changes in *Rhizopus stolonifer* spores which appear to support the involvement of respiratory metabolism in the generation of the germ tube from the fungal spore. Anoxic conditions were shown to prevent germ tube formation without interfering with spherical cell formation. Anoxia did not inhibit dispersed (nonpolarized) wall synthesis as evidenced by the excessive thickening of the complete inner wall of the spore but did prevent the localization of wall synthesis necessary for germ tube formation. With *Aspergillus niger* spores elevated incubation temperatures prevent the establishment of polarity and this leads to the formation of giant cells which possess excessively thick walls. Preliminary evidence on the biochemical nature of this effect suggests that the oxidative phosphorylation capacity of the mitochondria of these cells may be impaired at the elevated incubation temperature. Recently Bates & Wilson (1974) have shown that in the presence of 3.33 M ethylene glycol or glycerol conidia of *Neurospora crassa* grow as single cells without germ tube formation. While it was suggested that an osmotic effect was probably operating it is perhaps significant that preliminary evidence indicated that the mitochondria of these cells possess a reduced cytochrome c content.

At this stage it is pertinent to return to the question of what causes the migration of vesicles to the sites of incipient germ tube formation. Bartnicki-Garcia (1973) has considered that since there is no evidence for microtubular guidance of vesicles the vesicles may move along electrochemical gradients. It is possible that such gradients could be set up within the cell due to the uptake and release of materials from the cell which occurs during active metabolism. One simple form of gradient which could be envisaged is that gradient set up by the movement of respiratory gases through the cytoplasm. Any permeability differences at different areas of the spore wall or alternatively any nonuniform spatial arrangement of mitochondria within the cell could conceivably result in the establishment of a polarization gradient to which the wall synthesizing apparatus of the cell could respond. That such an oxygen gradient could have pronounced effects on the movement of cell organelles is suggested by the observation in *Rhizopus stolonifer* that mitochondria migrate to the periphery of the spores during the establishment of anoxic conditions (Bussel & Sommer, 1973). It is tempting to speculate that dormant spores may be partially anoxic and that during spherical growth there may be permeability changes or changes in organelle location which in turn could create the necessary oxygen gradient. Low levels of oxygen during germination or the presence of other environmental conditions or substances which inhibit mitochondrial function could prevent the establishment of such a gradient with the result that only dispersed and not localized wall synthesis could occur.

This is, perhaps, a gross oversimplification of the concepts but it does suggest that more attention should be given to elucidating the mechanism of spherical cell wall formation. Furthermore, it will only be by a coupling of biochemical investigations with more refined ultrastructural studies that a more comprehensive understanding of fungal spore germination will be obtained.

REFERENCES

Adelman, T. G. & Lovett, J. S. (1972). Synthesis of ribosomal protein without *de novo* ribosome production during differentiation in *Blastocladiella emersonii*. Biochem. Biophys. Res. Comm. 49, 1174--1182.

Anderson, J. G. & Smith, J. E. (1971). The production of conidiophores and conidia by newly germinated conidia of *Aspergillus niger* (microcycle conidiation). J. Gen. Microbiol. 69, 185--197.

Anderson, J. G. & Smith, J. E. (1972). The effects of elevated temperature on spore swelling and germination in *Aspergillus niger*. Can. J. Microbiol. 18, 289--297.

Baker, G. E. (1944). Heterokaryosis in *Penicillium notatum*. Bull. Torrey Bot. Club 71, 367--373.

Baker, G. E. (1945). Conidium formation in species of aspergilli. Mycologia 37, 582--600.

Barnes, M. & Parker, M. S. (1966). The increase in size of mould spores during germination. Trans. Br. Mycol. Soc. 49, 487--494.

Barnes, M. & Parker, M. S. (1967). Use of the Coulter Counter to measure osmotic effects on the swelling of mould spores during germination. J. Gen. Microbiol. 49, 287--292.

Bartnicki-Garcia, S. (1969). Cell wall differentiation in the Phycomycetes. Phytopathology 59, 1065--1071.

Bartnicki-Garcia, S. (1973). Fundamental aspects of hyphal morphogenesis. In Microbial Differentiation (ed. J. M. Ashworth & J. E. Smith), pp. 245--267. London: Cambridge University Press.

Bartnicki-Garcia, S. & Lippman, E. (1969). Fungal morphogenesis: Cell wall construction in *Mucor rouxii*. Science 165, 302--304.

Bartnicki-Garcia, S., Nelson, N., & Cota-Robles, E. (1968a). Electron microscopy of spore germination and cell wall formation in *Mucor rouxii*. Arch. Mikrobiol. 63, 242--255.

Bartnicki-Garcia, S., Nelson, N., & Cota-Robles, E. (1968b). A novel apical corpuscle in hyphae of *Mucor rouxii*. J. Bacteriol. 95, 2399--2402.

Bates, W. K. & Wilson, J. F. (1974). Ethylene glycol-induced alteration of conidial germination in *Neurospora crassa*. J. Bacteriol. 117, 560--567.

Border, D. J. & Trinci, A. P. J. (1970). Fine structure of the germination of *Aspergillus nidulans* conidia. Trans. Br. Mycol. Soc. 54, 143--146.

Bracker, C. E. (1966). Ultrastructural aspects of sporangiospore formation in *Gilbertella persicaria*. In The Fungus Spore (ed. M. F. Madelin), pp. 39--60. London: Butterworths.

Bracker, C. E. (1967). Ultrastructure of fungi. Annu. Rev. Phytopath. 5, 343--374.

Bracker, C. E. (1971). Cytoplasmic vesicles in germinating spores of *Gilbertella persicaria*. Protoplasma 72, 381--397.

Brenner, D. M. & Carroll, G. C. (1968). Fine-structural correlates of growth in hyphae of *Ascodesmis sphaerospora*. J. Bacteriol. 95, 658--671.

Bromberg, R. (1974). Mitochondrial fragmentation during germination in *Blastocladiella emersonii*. Devel. Biol. 36, 187--194.

Buckley, P. M., Sommer, N. F., & Matsumoto, T. T. (1968). Ultrastructural details in germinating sporangiospores of *Rhizopus stolonifer* and *Rhizopus arrhizus*. J. Bacteriol. 95, 2365--2373.

Burguillo, P. F., Nicolas, G., & Martin, J. F. (1972). Abnormal swelling of fungal spores. Trans. Br. Mycol. Soc. 59, 512--516.

Bussel, J. & Sommer, N. F. (1973). Lomasome development in *Rhizopus stolonifer* sporangiospores during anaerobiosis. Can. J. Microbiol. 19, 905--907.

Bussel, J., Sommer, N. F., & Kosuge, T. (1969). Effect of anaerobiosis upon germination and survival of *Rhizopus stolonifer* sporangiospores. Phytopathology 59, 946--952.

Bussel, J., Buckley, P. M., Sommer, N. F., & Kosuge, T. (1969). Ultrastructural changes in *Rhizopus stolonifer* sporangiospores in response to anaerobiosis. J. Bacteriol. 98, 774--783.

Campbell, C. K. (1971). Fine structure and physiology of conidial germination in *Aspergillus fumigatus*. Trans. Br. Mycol. Soc. 57, 393--402.

Campbell, R. (1969). An electron microscope study of spore structure and development in *Alternaria brassicola*. J. Gen. Microbiol. 54, 381--392.

Campbell, R. (1970). An electron microscope study of exogenously dormant spores, spore germination, hyphae and conidiophores of *Alternaria brassicola*. New Phytol. 69, 287--293.

Cantino, E. C. & Horenstein, E. A. (1956). Gamma particles and the cytoplasmic control of differentiation in *Blastocladiella*. Mycologia 48, 443--446.

Cantino, E. C. & Mack, J. P. (1969). Form and function in the zoospore of *Blastocladiella emersonii*. 1. The γ-particle and satellite ribosome package. Nova Hed. 18, 115--147.

Cantino, E. C. & Myers, R. B. (1972). Concurrent effect of visible light on gamma particles, chitin synthetase and encystment capacity in zoospores of *Blastocladiella emersonii*. Arch Mikrobiol. 83, 203--215.

Cantino, E. C. & Truesdell, L. C. (1971). Light induced encystment of *Blastocladiella emersonii* zoospores. J. Gen. Microbiol. 69, 199-204.

Cantino, E. C., Lovett, J. S., Leak, L. V. & Lythgoe, J. (1963). The single mitochondrion, fine structure and germination of the spore of *Blastocladiella emersonii*. J. Gen. Microbiol. 31, 393--404.

Cassone, A., Simonetti, N., & Strippoli, V. (1973). Ultrastructural changes in the wall during germ tube formation from blastospores of *Candida albicans*. J. Gen. Microbiol. 77, 417--426.

Chambers, J. C., Markus, K., & Willoughby, L. G. (1967). The fine structure of the mature zoosporangium of *Nowakowskiella profusa*. J. Gen. Microbiol. 46, 135--141.

Chong, J. & Barr, D. J. S. (1973). Zoospore development and fine structures in *Phlyctochytrium arcticum* (Chytridiales). Can. J. Bot. 51, 1411--1420.

Cortat, M., Matile, P., & Wiemkin, A. (1972). Isolation of glucanase-containing vesicles from budding yeast. Arch. Mikrobiol. 82, 189-205.

Dallner, G., Siekevitz, P., & Palade, G. E. (1966). Biogenesis of endoplasmic reticulum membranes. J. Cell Biol. 30, 73--96.

Dunkle, L. D., Wergin, W. P., & Allen, P. J. (1970). Nucleoli in differentiated germ tubes of wheat rust uredospores. Can. J. Bot. 48, 1693--1695.

Ekundayo, J. A. (1966). Further studies on germination of sporangiospores of *Rhizopus arrhizus*. J. Gen. Microbiol. 42, 283--291.

Ekundayo, J. A. & Carlile, M. J. (1964). The germination of sporangiospores of *Rhizopus arrhizus*; spore swelling and germ tube emergence. J. Gen. Microbiol. 35, 261--269.

Fletcher, J. (1969). Morphology and nuclear behaviour of germination conidia of *Penicillium griseofulvum*. Trans. Br. Mycol. Soc. 53, 425--432.

Fletcher, J. & Morton, A. G. (1970). Physiology of germination of *Penicillium griseofulvum* conidia. Trans. Br. Mycol. Soc. 54, 65--81.

Florance, E. R., Denison, W. C., & Allen, T. C. (1972). Ultrastructure of dormant and germinating conidia of *Aspergillus nidulans*. Mycologia 69, 115--123.

Fuller, M. S. (1966). Structure of the uniflagellate zoospores of aquatic Phycomycetes. In The Fungus Spore (ed. M. F. Madelin), pp. 67--84. London: Butterworths.

Fuller, M. S. & Olson, L. W. (1971). The zoospore of *Allomyces*. J. Gen. Microbiol. 66, 171--183.

Garcia-Acha, I., Aguirre, M. J. R., Uruburu, F., & Villanueva, J. R. (1966). The fine structure of the *Fusarium culmorum* conidia. Trans. Br. Mycol. Soc. 49, 695--702.

Gay, J. L. & Greenwood A. D. (1966). Structural aspects of zoospore production in *Saprolegnia ferax* with particular reference to the cell and vesicular membranes. In The FungusSpore (ed. M. F. Madelin), pp. 95--110. London: Butterworths.

Gregory, P. H. (1966). The fungus spore: what it is and what it does. In The Fungus Spore (ed. M. F. Madelin), pp. 1--13. London: Butterworths.

Griffiths, D. A. (1973). Fine structure of chlamydospore germination in *Fusarium oxysporum*. Trans. Br. Mycol. Soc. 61, 7--12.

Grove, S. N. (1972). Apical vesicles in germinating conidia of *Aspergillus parasiticus*. Mycologia 64, 638--641.

Grove, S. N. & Bracker, C. E. (1970). Protoplasmic organization of hyphal tips among fungi: vesicles and Spitzenkörper. J. Bacteriol. 104: 989--1009.

Grove, S. N., Bracker, C. E., & Morré, D. J. (1968). Cytomembrane differentiation in the endoplasmic reticulum--golgi apparatus--vesicle complex. Science 161, 171--173.

Grove, S. N., Bracker, C. E., & Morré, D. J. (1970). An ultrastructural basis for hyphal tip growth in *Pythium ultimum.* Am. J. Bot. 59, 245--266.

Gull, K. & Trinci, A. P. J. (1971). Fine structure of spore germination in *Botrytis cinerea.* J. Gen. Microbiol. 68, 207--220.

Hawker, L. E. (1966). Germination: morphological and anatomical changes. In The Fungus Spore (ed. M. F. Madelin), pp. 151--162. London: Butterworths.

Hawker, L. E. & Abbot, P. M. McV. (1963). An electron microscope study of maturation and germination of sporangiospores of two species of *Rhizopus*. J. Gen. Microbiol. 32, 295--298.

Hawker, L. E. & Hendy, R. J. (1963). An electron microscope study of germination of conidia of *Botrytis cinerea*. J. Gen. Microbiol. 33, 43--46.

Hawker, L. E., Thomas, B., & Beckett, A. (1970). An electron microscope study of structure and germination of conidia of *Cunninghamella elegans* Lender. J. Gen. Microbiol. 60, 181--189.

Heale, J. B., Gafoor, A., & Rajasingham, K. C. (1968). Nuclear division in conidia and hyphae of *Verticillium albo-atrum*. Can. J. Genet. Cytol. 10, 321--340.

Heath, I. B. & Greenwood, A. D. (1970). The structure and formation of lomasomes. J. Gen. Microbiol. 62, 129--137.

Heintz, C. E. & Niederpruem, D. J. (1971). Ultrastructure of quiescent and germinated basidiospores and oidia of *Coprinus lagopus*. Mycologia 63, 745--766.

Hemmes, D. E. & Hohl, H. R. (1969). Ultrastructural changes in directly germinating sporangia of *Phytophthora parasitica*. Am. J. Bot. 56, 300--313.

Holley, R. A. & Kidby, D. K. (1973). Role of vacuoles and vesicles in extracellular enzyme secretion from yeast. Can. J. Microbiol. 19, 113--117.

Holton, R. W. (1961). Studies on pyruvate metabolism and cytochrome system in *Neurospora tetrasperma*. Plant Physiol. 35, 757--761.

Hyde, J. M. & Walkinshaw, C. H. (1966). Ultrastructure of basidiospores and mycelium of *Lenzites saepiaria*. J. Bacteriol. 92, 1218--1227.

Jansons, V. K. & Nickerson, W. J. (1970). Induction, morphogenesis and germination of the chlamydospores of *Candida albicans*. J. Bacteriol. 104, 910--921.

Jones, A. L. & Fawcett, D. W. (1966). Hypertrophy of the agranular endoplasmic reticulum in hamster liver induced by phenobarbital. J. Histochem. Cytochem. 14, 215--232.

Lessie, P. E. & Lovett, J. S. (1968). Ultrastructural changes during sporangium and zoospore differentiation in *Blastocladiella emersonii*. Am. J. Bot. 55, 220--236.

Lovett, J. S. (1963). Chemical and physical characterization of "nuclear caps" isolated from *Blastocladiella* zoospores. J. Bacteriol. 85, 1235--1246.

Lovett, J. S. (1968). Reactivation of ribonucleic acid and protein synthesis during germination of *Blastocladiella emersonii* zoospores and the role of the nuclear cap. J. Bacteriol. 96, 962--969.

Lowry, R. J. & Sussman, A. S. (1968). Ultrastructural changes during germination of ascospores of *Neurospora tetrasperma*. J. Gen. Microbiol. 51, 403--409.

McClure, W. K., Park, D., & Robinson, P. M. (1968). Apical organization in the somatic hyphae of fungi. J. Gen. Microbiol. 50, 177--182.

McCully, E. K. & Bracker, C. E. (1972). Apical vesicles in growing bud cells of heterobasidiomycetous yeasts. J. Bacteriol. 109, 922--926.

McKeen, W. E., Mitchell, N., Jarvie, W., & Smith, R. (1966). Electron microscopy studies of conidial walls of *Sphaerotheca macularis*, *Penicillium levitum* and *Aspergillus niger*. Can. J. Microbiol. 12, 427--428.

Mandels, G. R. & Darby, R. T. (1953). A rapid cell volume assay for fungitoxity using fungus spores. J. Bacteriol. 65, 16--26.

Manners, J. G. (1966). Assessment of germination. In The Fungus Spore (ed. M. F. Madelin), pp. 165--173. London: Butterworths.

Marchant, R. (1966a). Fine structure and spore germination in *Fusarium culmorum*. Ann.Bot. (Lond.) 30, 441--445.

Marchant, R. (1966b). Wall structure and spore germination in *Fusarium culmorum*. Ann. Bot. (Lond.) 30, 821--829.

Marchant, R. & White, M. F. (1966). Spore swelling and germination in *Fusarium culmorum*. J. Gen. Microbiol. 42, 237--244.

Martin, J. F., Uruburu, F., & Villanueva, J. R. (1973). Ultrastructural changes in the conidia of *Penicillium notatum* during germination. Can. J. Microbiol. 19, 797--801.

Matile, P., Cortat, M., Wiemken, A., & Frey-Wyssling, A. (1971). Isolation of glucanase containing particles from budding *Saccharomyces cerevisiae*. Proc. Nat. Acad. Sci. U.S.A. 68, 636--640.

Matsumae, A., Myers, R. B., & Cantino, E. C. (1970). Comparative numbers of γ-particles in the flagellate cells of various species and mutants of *Blastocladiella*. J. Gen. Appl. Microbiol. 16, 443--453.

Mims, C. W. (1971). An ultrastructural study of spore germination in the Myxomycete *Arcyria cinerea*. Mycologia 63, 586--601.

Mitchell, N. L. & McKeen, W. E. (1970). Light and electron microscope studies on the conidium and germ tube of *Sphaerotheca macularis*. Can. J. Microbiol. 16, 273--280.

Moore, R. T. & McAlear, J. H. (1961). Fine structure of mycota. 5.

Lomasomes previously uncharacterized hyphal structures. Mycologia 53, 194--200.

Myers, R. B. & Cantino, E. C. (1971). DNA profile of the spores of *Blastocladiella emersonii:* evidence for gamma particle DNA. Arch. Mikrobiol. 78, 252--267.

Niederpruem, D. J. & Dennen, D. W. (1966). Kinetics, nutrition and inhibitor properties of basiodiospore germination in *Schizophyllum commune*. Arch. Mikrobiol. 54, 91--105.

Niederpruem, D. J. & Wessels, J. G. H. (1969). Cytodifferentiation and morphogenesis in *Schizophyllum commune*. Bacteriol. Rev. 33, 505--535.

Nolan, R. A. & Bal, A. K. (1974). Cellulase localization in hyphae of *Achlya ambisexualis*. J. Bacteriol. 117, 840--843.

Olson, L. W. (1973a). The meiospore of *Allomyces*. Protoplasma 78, 113--127.

Olson, L. W. (1973b). Synchronized development of gametophytic germlings of the aquatic Phycomycete *Allomyces macrogynus*. Protoplasma 78, 129-144.

Olson, L. W. & Fuller, M. S. (1971). Leucine--lysine synchronization of *Allomyces* germlings. Arch. Mikrobiol. 78, 76--91.

Page, W. J. & Stock, J. J. (1971). Regulation and self-inhibition of *Microsporum gypseum* macroconidia germination. J. Bacteriol. 108, 276--281.

Park, D. & Robinson, P. M. (1970). Germination studies with *Geotrichum candidum*. Trans. Br. Mycol. Soc. 54, 83--92.

Reichle, R. & Fuller, M. S. (1967). The fine structure of *Blastocladiella emersonii* zoospores. Am. J. Bot. 54, 81--92.

Reiss, J. (1971). Lysome-like particles in *Geotrichum candidum*: A cytological study. Z. Allg. Mikrobiol. 11, 319--323.

Richmond, D. V. & Pring, P. J. (1971). Fine structure of germinating *Botrytis fabae* Sardina conidia. Ann. Bot. (Lond.) 35, 493--500.

Robb, J. (1972). Ultrastructure of *Ustilago hordei*. 1. Pregermination development of hydrating teliospores. Can. J. Bot. 50, 1253--1261.

Schmoyer, I. R. & Lovett, J. S. (1969). Regulation of protein synthesis in zoospores of *Blastocladiella*. J. Bacteriol. 100, 854--864.

Smith, J. E. & Galbraith, J. C. (1971). Biochemical and physiological aspects of differentiation in the fungi. Adv. Microbial Physiol. 5, 45--134.

Smith, J. E. & Berry, D. R. (1974). An introduction to the

biochemistry of fungal development. New York and London: Academic Press.

Soll, D. R., Bromberg, R., & Sonneborn, D. R. (1969). Zoospore germination in the water mold *Blastocladiella emersonii*. I. Measurement of germination and sequence of subcellular morphological changes. Dev. Biol. 20, 183--217.

Soll, D. R. & Sonneborn, D. R. (1971). Zoospore germination in *Blastocladiella emersonii*. 111. Structural changes in relation to protein and RNA synthesis. J. Cell Sci. 9, 679--699.

Steele, S. D. (1973). Double membraned vesicles in *Geotrichum candidum* Can. J. Microbiol. 19, 534--536.

Steele, S. D. & Fraser, T. W. (1973). Ultrastructural changes during germination of *Geotrichum candidum* arthrospores. Can. J. Microbiol. 19, 1031--1034.

Storck, R. & Morrill, R. C. (1971). Respiratory-deficient, yeast-like mutant of *Mucor*. Biochem. Genet. 5, 467--479.

Sussman, A. S. (1966). Dormancy and spore germination. In The Fungi, Vol. 2 (ed. G. C. Ainsworth and A. S. Sussman), pp. 733--764. New York and London: Academic Press.

Sussman, A. S. & Halvorson, H. O. (1966). Spores, Their Dormancy and Germination. New York and London: Harper and Row.

Sussman, A. S., Lowry, R. J., Durkee, T. L., & Maheshwari, R. (1969). Ultrastructural studies of cold-dormant and germinating uredospores of *Puccinia graminis* var. *tritici*. Can. J. Bot. 47, 2073--2078.

Temmink, J. H. M. & Campbell, R. N. (1969). The ultrastructure of *Olpidium brassicae*. 11. Zoospores. Can. J. Bot. 47, 227--231.

Terenzi, H. F. & Storck, R. (1969). Stimulation of fermentation and yeastlike morphogenesis in *Mucor rouxii* by phenethyl alcohol. J. Bacteriol. 97, 1248--1261.

Thornton R. M. (1968). Fine structure of *Phycomyces*. 1. Autophagic vesicles. J. Ultrastruct. Res. 21, 269--280.

Travland, L. B. & Whistler, H. C. (1971). Ultrastructure of *Harpochytrium hedinii*. Mycologia 63, 767--789.

Truesdell, L. C. & Cantino, E. C. (1970). Decay of gamma particles in germinating zoospores of *Blastocladiella emersonii*. Arch. Mikrobiol. 70, 378--392.

Truesdell, L. C. & Cantino, E. C. (1971). The induction and early events of germination in the zoospore of *Blastocladiella emersonii*. Cur. Top. Dev. Biol. 6, 1--44.

Tsukahara, T., Yamada, M., & Itagaki, T. (1966). Micromorphology on conidiospores of *Aspergillus niger* by electron microscopy. Jap. J. Microbiol. 10, 93--107.

Voelz, H. & Niederpruem, D. J. (1964). Fine structure of basidiospores of *Schizophyllum commune*. J. Bacteriol. 88, 1497--1502.

Van der Woude, W. J., Morré, D. J. & Bracker, C. E. (1971). Isolation and characterization of secretory vesicles in germinating pollen of *Lilium longiflorum*. J. Cell Sci. 8, 331-351.

Weisberg, S. H. & Turian, G. (1971). Ultrastructure of *Aspergillus nidulans* conidia and conidial lomasomes. Protoplasma 72, 55--67.

Wergin, W. P., Dunkle, L. D., Van Etten, J. L., St. Julian, G., & Bulla, L. A. (1973). Microscopic observations of germination and septum formation in pycnidiospores of *Botryodiplodia theobromae*. Dev. Biol. 32, 1--14.

Williams, P. G. & Ledingham, G. A. (1964). Fine structure of wheat stem rust uredospores. Can. J. Bot. 42, 1503--1508.

Wilsenach, R. & Kessel, M. (1965). The role of lomasomes in wall formation in *Penicillium vermiculatum*. J. Gen. Microbiol. 40, 401--404.

Yanagita, T. (1957). Biochemical aspects of the germination of conidiospores of *Aspergillus niger*. Arch. Mikrobiol. 26, 329--344.

Yarwood, C. E. (1936). The tolerance of *Erysiphe polygoni* and certain other powdery mildews to low humidity. Phytopathology 26, 845--859.

Zachariah, K. & Fitz-James, P. C. (1967). The structure of phialides in *Penicillium claviforme*. Can. J. Microbiol. 13, 249--256.

Zorzopulos, J., Jobbagy, A. J., & Terenzi, H. F. (1973). Effects of ethylenediaminetetraacetate and chloramphenicol on mitochondrial activity and morphogenesis in *Mucor rouxii*. J. Bacteriol. 115, 1198--1204.

DISCUSSION

Organelle Changes During Germination

Chairman: V. N. Armentrout
University of Wisconsin
Madison, Wisconsin

In reply to a question about the effects of dinitrophenol on the *Aspergillus niger* microcycle system, Dr. Smith commented that no data was yet available as to effects on growth or morphology of *A. niger* under these conditions. However, preliminary work of his group indicates no stimulation of respiration with addition of dinitrophenol. He cautioned that it would be necessary to prove that such inhibitors penetrate the cell. At present, they are developing methods to study mitochondria at a very early stage; production of protoplasts in large numbers for ease of disruption would be a useful, although difficult, method to obtain mitochondria from the spores.

Dr. Smith was asked whether substrate-level phosphorylation was sufficient for spherical growth while oxidative phosphorylation (from coupled mitochondria) was required for germ tube growth. He agreed that this was believed to be the case, but pointed out that further experiments needed to be done with gentle disruption of spherical cells. At present, the mitochondria appear to be uncoupled. Dr. Cantino reported that mitochondria isolated from *Blastocladiella* zoospores appear to be partially uncoupled. It was asked whether anyone had succeeded in isolating tightly coupled mitochondria from fungi. Dr. Smith replied that he and a co-worker had published results on the successful isolation of mitochondria from vegetative cells of *A. niger* grown at normal temperatures. The use of controlled pH and bovine serum albumin in the extraction medium was stressed in overcoming the problem of natural uncouplers present in the cell which become decompartmentalized during disruption. Dr. Storck commented that his group had also been able to isolate coupled mitochondria from *Mucor* using methods developed by Dr. Smith and his associates.

The recent work showing a single mitochondrion in the yeast cell was cited and it was asked whether a similar situation could exist in the *A. niger* spore. Also, since the cell wall of yeast shows no polar growth apart from the budding process, could there be a correlation of such mitochondrial morphology and nonpolar growth as seen in *A. niger* at high temperatures? Dr. Hashimoto indicated that serial sections of *Histoplasma* cells showed many small mitochondria and stated that the single mitochondrion concept need not apply to all fungal cells. Another participant doubted that mitochondria could play a primary role in the phenomenon of dimorphism, since both polar and nonpolar growth have been observed under anaerobic conditions where mitochondria are presumably nonfunctional. It was suggested that the high temperatures used to obtain nonpolar growth in *A. niger* may alter mitochondrial membranes which could in turn have an indirect effect related to spherical growth under those conditions. A further comment was that some spores do not show "swelling" or nonpolar growth, whereas others do, and that generalizations are dangerous.

Finally, Dr. Turian asked whether mitochondria are excluded from the vesicle of the conidiophore of *A. niger*, or whether apical vesicles disappear and mitochondria crowd into the rounded tip of the conidiophore as he has observed in *Neurospora*. Dr. Smith replied that he has not yet examined the full microcycle by electron microscopy.

CHAPTER 9

FINE STRUCTURE OF THE SPORE WALL AND GERM TUBE CHANGE DURING GERMINATION

S. Akai, M. Fukutomi, H. Kunoh, and M. Shiraishi
Kyoto University

INTRODUCTION

A great majority of fungi belong to the Eumycota and the reproduction is accomplished by forming spores of various kinds. Sexual reproduction in some aquatic Mastigomycotina is a fusion between motile gametes but in Oomycetes and Zygomycotina by conjugation between gametangia, sexually differentiated hyphal branches, resulting in a formation of thick-walled oospores or zygospores. In Ascomycotina and Basidiomycotina, however, sexual fusion of some fungi is initiated by spermatization, and in many others sexual fusion occurs between two vegetative hyphae which retain the sexual potential after they have lost the ability to produce sexual organs. Since the development of electron microscopy cytological details of fungal cells have been gradually clarified.

SPORE AND GERM TUBE WALL FINE STRUCTURE

The cell wall is a structure that gives a rigidity to cells and protects intracellular organelles. Although the function of the cell wall is not well understood, it is also important to clarify the structural detail of cell walls to interprete the mechanism of physiological function.

This paper presents a review on the fine structure of cell walls, especially of fungal spores, and their changes during germination.

FINE STRUCTURE OF CELL WALLS OF DORMANT SPORES

Sporangial Walls and Sporangiospore Walls of Oomycetes and Zygomycotina

In fungi of Mastigomycotina and Zygomycotina asexual reproduction is accomplished by a saclike sporangium borne on a tip of a specialized hyphal branch (sporangiophore). This structure is bounded by a wall derived from the sporangiophore wall and contains protoplasm with a large number of haploid nuclei. The protoplast cleaves into numerous, usually uninucleate sporangiospores. There is a crucial difference in the walls of spores between the two fungal groups, i.e., motile sporangiospores in Mastigomycotina, which are generally called zoospores, may have naked protoplasts without cell wall (Plate 1, Part 1). This is especially common in aquatic forms, and they, afterwards, encyst with a cell wall (Plate 1, Part 2). On the other hand, Zygomycotina sporangiospores are nonmotile. A wall is secreted during formation (Bartnicki-Garcia, 1968). However, some groups in Oomycetes, living parasitically and freely exposed to air, have lost their ability to form zoospores, and release the entire sporangium as a conidium. Zygomycetous sporangia in some Zygomycotina fungi are small and have a limited number of sporangiospores. These are termed sporangioles, and thus, they also behave as conidia, borne externally on a parent hypha.

Zoosporangium

The resistant sporangium of *Allomyces neo-moniliformis* is produced on the tip of a growing hypha, by deposition of a thick-pigmented wall within the original hyphal wall. The walls of these sporangia are very thick multilayered and consist of three parts: an electron-dense outer wall, cementing substances and an electron-transparent inner wall. The outer wall is pitted and consists of five layers, at least one of which is of fibrillar nature. The pits do not penetrate the outer wall. Two layers between the outer and inner walls were termed "cementing substances" by Skucas (1967), because during germination of sporangia these layers dissolved, separating the outer wall from the single-layered inner wall. The walls of fully developed sporangia are composed of six structural components: glucose, glucosamine, chitin, melanin, protein, and lipids. Comparison of the structure and composition of the wall of resistant sporangia with the wall of hyphae of *Allomyces* shows that, while the structure is very different, the composition is quite similar as only melanin and lipids are apparently absent from the hyphal walls (Skucas, 1967).

Multilayered wall of sporangia appears to be uncommon in fungi. In usual conditions sporangial walls of many phycomycetous fungi are one-, two-, or three-layered. The outer layer is electron-dense and the inner one is transparent (Chapman & Vujičić, 1965; Hohl & Hamamoto, 1967; King et al., 1968; Skucas, 1967).

In young nongerminating sporangia of *Phytophthora erythroseptica* the wall consists of two layers (Chapman & Vujičić, 1965). The outer layer is homogeneous, relatively transparent after fixation with OsO_4 or $KMnO_4$, and appears to give a cellulose reaction. The inner layer is variable in thickness and less electron-transparent than the outer layer, and has a vesiculated nature. This vesiculated nature is particularly prominent in the apical region of sporangia. Old nongerminating sporangial walls have a single layer. The vesicular layer found in young sporangia disappears and the outer electron-transparent layer decreases in thickness, and is directly apposed to the plasma membrane (Chapman & Vujičić, 1965).

Zoosporangial walls of *Phytophthora macrospora* consist of two layers. The outer layer is thin and electron-dense and the inner layer is thick and electron-transparent. From the surface view of zoosporangia, randomly arranged cellulosic microfibrils of the outer layer are observed clearly (Plate 1, Parts 3, 4) (Fukutomi, 1969).

The papillae of sporangia which are involved in the liberation of zoospores in indirect germination appear with the light microscope, as a mucilaginous thickening of the apical wall. The sporangial wall of *P. infestans* is single-layered, with an apical pore blocked by a papilla. The papillae of sporangia fixed with $KMnO_4$ have a laminated appearance and appear to be composed of protein and pectin compounds (King et al., 1968). In old sporangia of *P. erythroseptica* the papillae appear as a homogenous layer, and are finely granular after osmic acid fixation (Chapman & Vujičić, 1965). After chilling of *P. infestans*, changes occur at the apex of the sporangia where a granular layer appears between the inner surface of the papilla plug and the plasma membrane. King et al. (1968) suggested that this layer was probably formed by a lomasome and might be a site of accumulation of enzymes which degraded the papilla. Thus, the sporangial pore opens before zoospores release. A similar observation was reported for *P. erythroseptica* (Chapman & Vujičić, 1965). The terminal discharge papilla of sporangia of *Blastocladiella emersonii* is formed by a process of localized cell wall breakdown and deposition of the papilla material by secretory granules (Lessie & Lovett, 1968).

The fully expanded dormant sporangia of *Phytophthora parasitica* have two-layered walls. The outer wall appears finely granular and the inner wall is made of a homogeneous matrix with a faint longitudinal striation. The outer wall is continuous over the tip of the sporangium and confluent with the apical plug (papilla) material. The inner wall extends to the plug. The latter is composed of a fibrous material which is approximately at a right angle to the long axis of sporangia. The material of the basal plug appears to be continuous with both the outer and inner wall, and the homogeneous matrix of which it is composed is perforated with a complex, anastomosing tubular network filled with electron-dense material (Hohl & Hamamoto, 1967). In the basal plug of sporangia of *P. macrospora* tubular structures are also found (Fukutomi, 1969). Aragaki (1968) reported that sporangia of Peronosporales fungi, such as *Phytophthora*, *Bremia*, and *Albugo* were set off from sporangioshpores by callose plugs and that basal septa of terminal sporangia and septa of intercalary sporangia were mainly callose depositions whose primary function was probably to seal or plug.

In zoosporangia of *Olpidium brassicae*, the wall consists of three layers. The inner layer is electron-dense and is as thick as the two layers combined. The middle layer is striated and more electron-transparent and the thinnest outer layer is electron-dense. As the thalli mature, they form zoospores. One or more exit tube starts to be formed as a bulge in the sporangium wall. As the bulge enlarges and is pushed outward, the sporangium wall ruptures, first the outer two layers and then the inner layer (Temmink & Campbell, 1968).

The vegetative multinucleate plasmodium of *Plasmodiophora brassicae* is contained within the cytoplasm of cells of cruciferous plant roots. The plasmodium is bounded by a deep stained osmiophilic

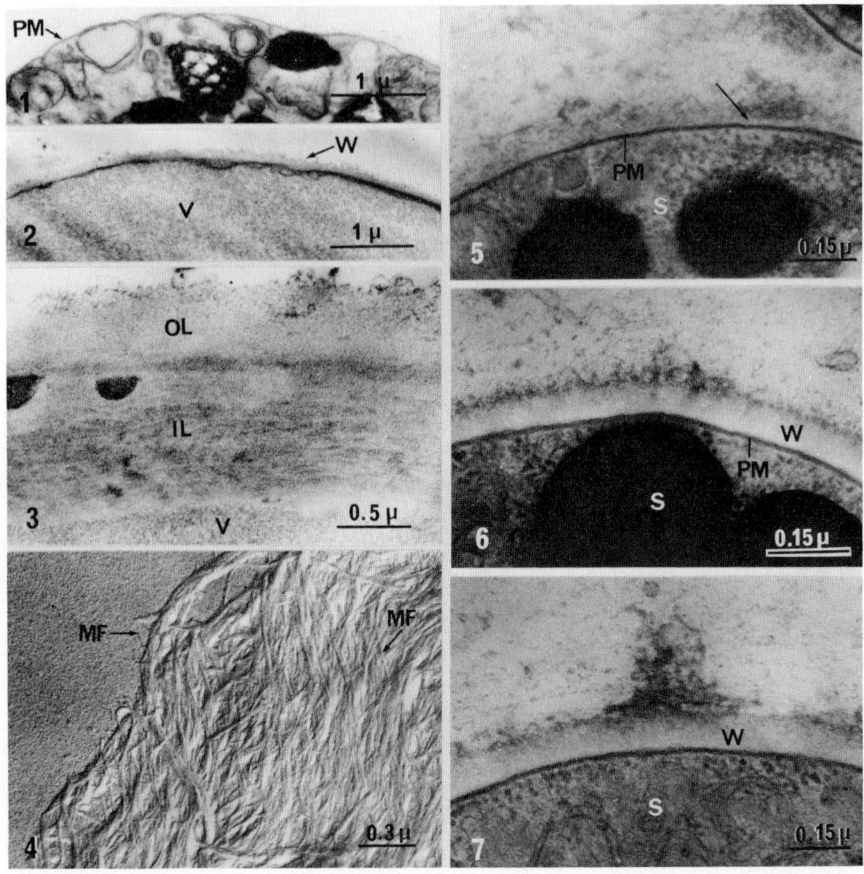

Plate 1. Electron micrographs of zoosporangial wall of *Phytophthora macrospora* and *Plasmodiophora brassicae*. Fig. 1--4. Electron micrographs of zoospore, cystospore, and zoosporangium of *P. macrospora*.

1. Plasma membrane (PM) of zoospore without cell wall. 2. Cystospore. Deposition of mucilaginous wall material (W) around the zoospore. A vacuole (V) is also shown. 3. Zoosporangium wall, showing the inner layer (IL); outer layer (OL); and a vacuole (V). 4. Microfibrils (MF) of a zoosporangial wall surface shadowed with platinum.

Fig. 5--7. Development of sprangial wall of *P. brassicae*.

5. Deposition of spine initial. Cell wall material begins to deposit between the spine initial and sporangial membrane (arrow). The

envelope, composed of two closely appressed membranes, each consisting of a pair of electron-dense layers. As sporogenesis is initiated, the outer membrane of the plasmodial envelope disintegrates, leaving the plasmodium bounded by a single membrane, the plasmodial membrane (Williams & McNabola, 1967). Vacuoles appear to be formed from the invagination of the plasmodial membrane or from the golgi body and rough-surface endoplasmic reticulum (ER). These vacuoles become aligned in planes of cleavage around each nucleus, forming the boundaries of each future sporangium. Spines are formed on the sporangial membrane as aggregates of residual vacuolar material. The sporangial wall is then deposited between the spines and sporangial membrane (Plate 1, Parts 5-7) (Akai & Shiraishi, 1974; Williams & McNabola, 1967).

Sporangiospore

In Mucoraceous fungi sporangia usually develop on the terminal of sporangiophores. The sporangia contain many encysted aplanospores. The most common type of sporangia (*Mucor*-type) possess a central dome-shaped columella, which extends into the sporangium. In *Gilbertella persicaria* (Bracker, 1966, 1968) the cleavage membrane formed from ER is transformed into the plasma membrane of spore initials. Granules are formed on the outer surface of spore plasma membranes after cleavage. The granules fuse to form a continuous spore envelope, and subsequently the spore wall is laid down inside the envelope, thus the envelope becomes the outermost layer of the spore wall. A wall is formed around the columella. This wall appears fibrillar and is structurally distinct from the spore wall. It is laid down inside the sporangiophore wall, and extends down the sporangiophore and tapers to an end.

Hawker & Abbott (1963) showed that walls of immature sporangiospores of *Rhizopus sexualis* and *R. stolonifer* were relatively thin and consisted of single layer, but mature spores possessed a thick wall of reticulate structure. However, there are many cases where sporangiospores possess two- or more-layered walls.

Sporangiospores of *G. persicaria* have bipolar appendages with

---

plasma membrane (PM) of the sporangium is also shown. 6 & 7. Development of spine and thickening of cell wall (W). The plasma membrane (PM) of the sporangium is also shown.

ribbon-like appearance. In longitudinal sections, the appendages consist of two parallel electron-dense layers and in cross sections dense layers are curled like a thin sheet partially rolled into a tube. These appendages develop before the spore walls are formed, and attach to the outermost layer of spores. A similar appendage is found on the spores of *Blakeslea trispora* and *Choanephora cucurbitarum* (Bracker, 1968).

The Kickxellaceae is a family of saprophytic Mucorales in which the asexual organ is complex and does not form a normal *Mucor*-type sporangium with columella. The asexual spore is typically elongated and is borne singly within a smooth-walled merosporangium (cylindrical sporangiole) at the apex of a pseudophialide. Pseudophialides are massed on septate or aseptate sporocladia developed on aerial hyphae. The walls of sporangiospores of *Coemansia reversa* lack spinous protuberances, evident in other species of Kickxellaceae, e.g., *Linderina pennispora* and *L. macrospora* (Young, 1968, 1973). Spore walls of the latter two species are composed of two layers, an outer spiny amorphous and fibrillar complex and an inner largely fibrillar complex. The outer layer is disrupted by treating with hot dilute alkali, whereas the inner multilayered complex of helically oriented fibrils is largely resistant to alkali-digestion (Young, 1968, 1970).

The sporangiospore walls of *Mucor rouxii* are qualitatively different from vegetative walls or sporangiophore walls of the same organism (Bartnicki-Garcia, 1968, 1969). The principal components of sporangiospore walls are a glucose polymer (S). Glucosamine is present in a much lower proportion, mostly in a triple complex with protein and melanin (Bartnicki-Garcia & Reyes, 1964, 1968b; Bartnicki-Garcia, 1969). By contrast, the polysaccharides of sporangiophore walls are primary glucosamine polymers (chitin, chitosan) (Bartnicki-Garcia & Nickerson, 1962; Bartnicki-Garcia & Reyes, 1968a, b; Barnicki-Garcia, 1969).

Cell Wall of Asexual Spores in Ascomycotina, Deuteromycotina, and Basidiomycotina

The conidium is a secondary spore formed without nuclear fusion followed by meiosis, and has been generally used for any asexual spores except sporangiospores (Talbot, 1971). Vuillemin (1910) proposed the terms of thallospores and conidia vera. Thallospores are spores which are formed on/in hypha, terminally or intercalarily.

They are not distinct structures formed *de novo*. Talbot (1971) has placed arthrospores with thallospores, but blastospores, aleuriospores, and phialospores were placed with true conidia, the latter are formed from new elements of the thallus.

Blastospore

Blastospores or bud-spores arise by budding of somatic cells of hypha or conidiophores and are not born on stipules. The genera of Hyphomycetes, such as *Candida, Cladosporium* and *Phymatotrichum* produce conidia of this type (Tubaki, 1963).

According to Djaczenko & Cassone (1971), in *Candida albicans* fixed with acrolein--TAPO*--OsO_4, the cell walls of mother cells (mature blastospore) have laminated structures with five distinguishable layers and young bud cells have wall with four layers. The outermost layer of mature cells is composed of anastomosing filaments of moderate electron opacity and granules with a certain regularity are embedded in the network of filaments. The outermost layer of mother cells does not continue into the bud. The periphery of the bud is covered only by floccular material. An amorphous matrix of high electron opacity is a major component of the second layer, which is the most electron opaque layer of the cell wall in *C. albicans*. The third layer is composed of electron opaque filaments which form bundles oriented parallel to the cell surface. The fourth layer is composed predominantly of narrow, electron-transparent spaces oriented perpendicular to the cell surface. Amorphous material of moderate electron opacity is interspersed between the spaces. The innermost layer is closely appressed to the plasma membrane and has a tortuous course. This layer is composed of homogeneous material.

A majority of reports on the ultrastructure of *C. albicans* have mentioned the presence of two layers in the cell wall. According to Djaczenko & Cassone (1971), the external layer may correspond to the second layer, because of its higher electron opacity. The internal electron-transparent layer described by many authors probably corresponds to the third and fourth layers. Cell walls prefixed with glutaraldehyde (GU) alone or mixtures of GU with acrolein and/or TAPO were homogeneous. Acrolein or TAPO alone gave poor visualization of both the cell wall and the cytoplasm of *C. albicans* cells. The

---

*TAPO, Tris-(1-aziridinyl) phosphine oxide.

cell wall of *Rhodotorula glutinis* shows a multilayered structure, overlaid with mucilage (Marchant & Smith, 1967) and *Saccharomyces cerevisiae* has a wall consisting of two layers, an outer layer and a globular inner layer (Ghosh, 1971). The wall of bud cells in *Rhodotorula rubra* have three layers when fixed with $KMnO_4$. The outer and inner layers are thin, and between the two layers there is a middle layer which consists of electron-transparent amorphous material (Marquardt, 1962).

In bud cells of *Tremella mesenterica* a capsular layer is visible outside of cell wall. The thick capsular layer of mature cells may result in part from degeneration of old wall layers. The cell walls of mature cells consist of two distinct layers. The outer layer has an outer electron-transparent zone and an inner electron-opaque zone with a fibrillar structure. These two zones are not sharply delimited, but merge into one another. Between this bizonal layer and the cytoplasmic membrane lies a thin granular electron-transparent layer (inner layer) (Bandoni & Bisalputra, 1971). The cell wall of *Paracoccidioides brasiliensis (Entomophthorales Imperfecti)* displays an outer layer composed of low-density fibrils and a thinner, denser and more homogeneous middle layer. Between the middle layer and the plasma membrane, there is a broader inner layer devoid of a distinct fibrillar structure. In *Blastomyces dermatitidis* the layers of the cell wall are not clearly observed, but the outer layer shows fibrils without a definite arrangement (Carbonell, 1967).

At an early stage in bud development of *C. albicans*, an electron-transparent encircling region (ETER) similar to those described by Marchant & Smith (1968) for *S. cerevisiae* is seen at the area where mother and bud cell wall materials come together, and the ETER appears to be continuous with the inner layer of the mother cell wall, giving the impression that the bud cell wall may have originated from beneath the mother cell wall (Shannon & Rothman, 1971).

Formation of the bud cell wall from beneath the mother cell wall has also been reported for *Rhodotorula glutinis* and *Endomycopsis platypodis* (Marchant & Smith, 1967; Kreger-Van Rij & Veenhuis, 1969). The transverse septum was formed from bud cell wall material and subsequently thickened on both sides, similar to the septum thickening described by Sentaudreu & Northcote (1969) for *S. cerevisiae*. The septum has a dolipore as in Basidiomycotina, but with no pore cap (Kreger-Van Rij & Veenhuis, 1969). In *C. albicans* the electron-transparent primary septum originates from the vicinity of the ETER (electron-light collar of Kreger-Van Rij & Veenhuis, 1969), and then electron-dense secondary septa are formed on both sides of the primary septum. The ETER will become part of the bud scar margin on the mother cell, whereas the mother cell secondary septum becomes the bud scar plug on the mother cell. The bud secondary septum will become the birth scar region on the daughter cell wall (Shannon & Rothman, 1971).

As is the case for many blastospores, in *Sporobolomyces salmonicolor, Rhodotorula glutinis*, and *Tremella mesenterica* the outer layer of the mother cell wall is ruptured at the point of budding and the inner layer bulges out as a new bud wall. The newly formed buds do not have a distinct outer wall layer seen on the mother cell (Bandoni & Bisalputra, 1971; McCully & Bracker, 1972). The fluorescent antibody technique would be helpful in determining the time of bud development and in determining the origin of various wall layers (Chung et al., 1965). Chung et al. (1965) and Jansons & Nickerson (1970) concluded that in *S. cerevisiae* and *C. albicans* bud formation did not involve incorporation of segments of cell wall from the mother cell. The budding process or blastospore formation may be a process closely resembling spore germination as observed in *Rhizopus* and other Mucoraceous fungi (Hawker & Abbott, 1963; Hawker, 1966).

Bud and birth scars have been verified in many yeast species. Budding in *S. cerevisiae* has been described as multilaterial with each successive bud developing from a new site (Streiblova & Beran, 1963). In contrast to this, successive buds of apiculate yeasts (tribe Nadsonieae) develop from a single site at each pole of the cell. Multilateral budding results in many separate bud scars around the cell surface, and in apiculate yeasts the successive bud formation results in a multiple scar. In *T. mesentrica*, both types of budding occur (Bandoni & Bisalputra, 1971), and Streiblova & Beran (1963) observed concentric ring on the apiculus of *Saccharomycodes, Nadsonia*, and other genera of the Nadsonieae.

Conidiophores of *Rosellinia buxi, R. aquila*, and other members of Xylariaceae bear relatively large and conspicuous scars. The conidia fall away leaving pores in the conidiophore wall which are plugged at the base. This plugging presumably occurs before spore detachment. The conidial scars produced by *Hypoxylon fuscum* are less conspicuous but have essentially the same form (Greenhalgh, 1967).

Besides pinkish orange moniloid conidia (macroconidia) *Neurospora* has a secondary type spore called microconidia. The first visible sign of microconidiation is a cell wall protuberance in the microconidiophore. The outer layer of the hyphal wall ruptures at this site and the incipient microconidium is forced out. Simultaneously, a collar (may correspond to ETER in *C. albicans*, stated above) is formed at the base of the developing microconidium by heavy deposition of wall material which appears to be fibrous and continuous with the wall of the young spore. After the wall of the microconidium is fully formed, the connection between the collar and spore wall is eroded faciliating release of the spore (Lowry et al., 1967).

Plate 2, Parts 1 and 4 show the surface fine structure of spore chains of *Cladosporium cucumerinum*. Sporulated materials for observation were prepared as follows: the mycelium was grown on urethane foam cubes (ca. 7 mm^3) and was incubated in a moist petri dish for sporulation, after squeezing out water. Sporulated materials were frozen instantaneously in liquid nitrogen, coated with gold, and observed with a JSM-50 A cryoscanning electron microscope. Spores budded from the mother cells forming an acropetalous long chain, sometimes branching. Small buds on the apical portion of the spores are shown in Plate 2, Part 1 (Fukutomi, 1974).

Chemical analyses of cell walls of yeasts (Ballesta & Villanueva, 1971) showed that in *Torulopsis aeris* monosaccharide constituents, glucose, and mannose were found. These sugars and galactose were also found in *Oospora suaveolens* and *Geotrichum lactis*. Only glucosamine was a common amino sugar in all three species. According to Djaczenko & Cassone (1971) the first and second layers of *C. albicans* cell walls may contain mainly mannan-protein and perhaps chitin, whereas the third and fourth layers are predominantly composed of glucan-protein.

Radulaspore

When the blastospores are abstricted, the denticles may remain upon the conidiophore and give it a rasplike appearance. For this reason blastospores are sometimes termed radulaspores (Talbot, 1971). The genera in Hyphomycetes, such as *Brachysporium, Cercospora, Fusicladium, Ramularia, Ovularia, Aureobasidium,* and *Botrytis* form radulaspore-type blastospores (Tubaki, 1963).

Aureobasidium pullulans produces radulaspore-type blastospores (Durrell, 1968). Conidia of this fungus are yeastlike cells that bud from hyphae. Their walls dissolve in 70% H_2SO_4. For production of conidia the outer wall layer of the hypha (mother cell) appears dissolved by enzyme action at the point of emergence (Durrell, 1968), and the inner wall bulges out, extruding a spore. A scar comparable to scars on yeast cells remains.

Botrytis cinerea has an electron-dense outer wall and a thick, electron-transparent, inner wall (Buckley et al., 1966; Shiraishi, 1971). The fine structure of a conidial surface is shown in Plate 2, Part 6 (Fukutomi, 1974). The surface is not smooth, but has verrucose scattering and small rounded warts on the whole surface (Akai et al., 1969). *Peziza ostracoderma,* and *Chromelosporium* produce blasto-type conidia on swollen elongate ampullae. Ampulla formation is initiated by swelling of the conidiophore apex. The denticle and associated conidia are produced by an outgrowth of the ampulla wall, i. e., the ampulla wall remains continuous and expands to form the wall of the new conidium. Mature conidial walls seem to be two-layered; the outer layer is electron-dense with a rough surface and the inner one is less electron-dense (Carmichael, 1971; Hughes & Bisalputra, 1970).

Porospore

The porospore is sometimes included in blastospores. The spores develop as a bud extruded through a distinct pore in wall of the conidiophore. Tubaki (1963) stated that genera of Hyphomycetes, such as *Alternaria, Curvularia, Helminthosporium,* and *Stemphylium* form porospore-type conidia. Ellis (1971) pointed out that in *Alternaria, Curvularia* and *Drechslera* conidia develop enteroblastically through channels (pores) in the conidiogenous cell walls.

The origin of pores in conidiophore walls of the Imperfect Fungi which produce porospores is not clear (Greenhalgh, 1967; Subramanian, 1965). Pores may represent thin areas in the conidiophore walls which are then blown out in blastospore formation. Cole (1973) thought that these pores probably were formed by autolysis of the outer layers. The distinction between porospores and blastospores is not perfect, and as Luttrell (1963) stated, these spores may not be fundamentally different.

Conidia of *Alternaria* are porospores. Dormant spores of *A. brassicicola (A. oleracea)* have very thick, heavily pigmented, melanized walls with plugged septal pores (Campbell, 1970a). The walls of young spores are quite thin and have a granular electron-dense substance concentrated in the outer region. Thus,

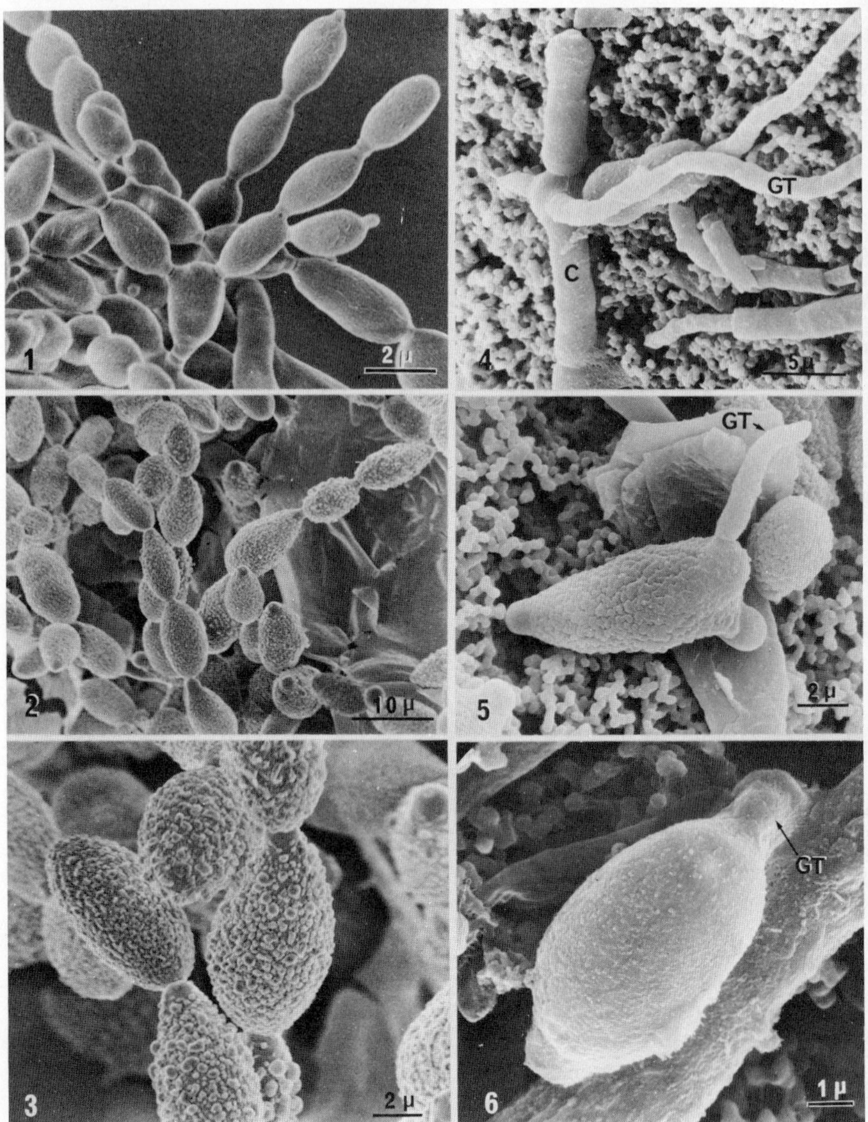

Plate 2. Scanning electron micrographs of conidia of *Alternaria solani*, *Botrytis cinerea*, and *Cladosporium cucumerinum*.
1. A cryoscanning electron micrograph of spore chains of *C. cucumerinum.* Spores budded acropetally from the mature spores

the wall develops into two ill-defined layers. The outer one has a granular texture and darker color, probably dependent upon the melanin deposition, and the inner one is electron-transparent. The outer layer is the original wall of the young spore formed when the spore is budded on, while the inner layer surrounds individual cells of spores as they are delimited by septum formation. The apical pore is originally formed as a simple perforation of the primary wall through which the secondary wall blows out as a bud. Soon after, the wall around the pore starts to develop a ring-shaped thickening called an annulus, just inside the secondary wall (Campbell, 1968).

When the septum is laid down before the secondary (inner) wall formation begins, the septal partition abuts onto the primary (outer) wall, but when the septal partition is laid down after forming the secondary wall, it will abut onto the latter. In the young septal wall the electron-transparent dividing layer between the two layers of the septal partition is clearly seen. The secondary wall is laid down on this partition. Each of the new cells is thus surrounded by its own double wall. The fully formed septum has five layers; two layers of septal partition separated by an electron-transparent layer and a layer of secondary wall deposited on each side of the partition. Melanin deposition occurs from the time of spore formation, but is restricted, excepting in the last stage of maturation, to the primary wall and septal partition of spores which become progressively more electron-dense and finally completely opaque (Campbell, 1968).

Conidial walls of *A. kikuchiana* consists of two layers, outer and inner layers similar to the walls of *A. brassicicola* (Plate 3,

---

forming a long chain, and sometimes branched. A small bud on the apical portion of spore is seen. 2 and 3. Scanning and cryoscanning electron micrographs of spore chains of *A. solani*, respectively. The verrucose surface with knobby processes is evident. 4. A cryoscanning electron micrograph of a germinating conidium (C) of *C. cucumerinum* and a germ tube wall (GT) are shown. The apical portion of conidial scar is also seen. 5. A cryoscanning electron micrograph of germinating condidia of *A. solani*. Note the germ tube wall (GT). 6. A scanning electron micrograph of germinating conidium of *B. cinerea*, showing the rough surface of a conidium with small warts, and the germ tube (GT).

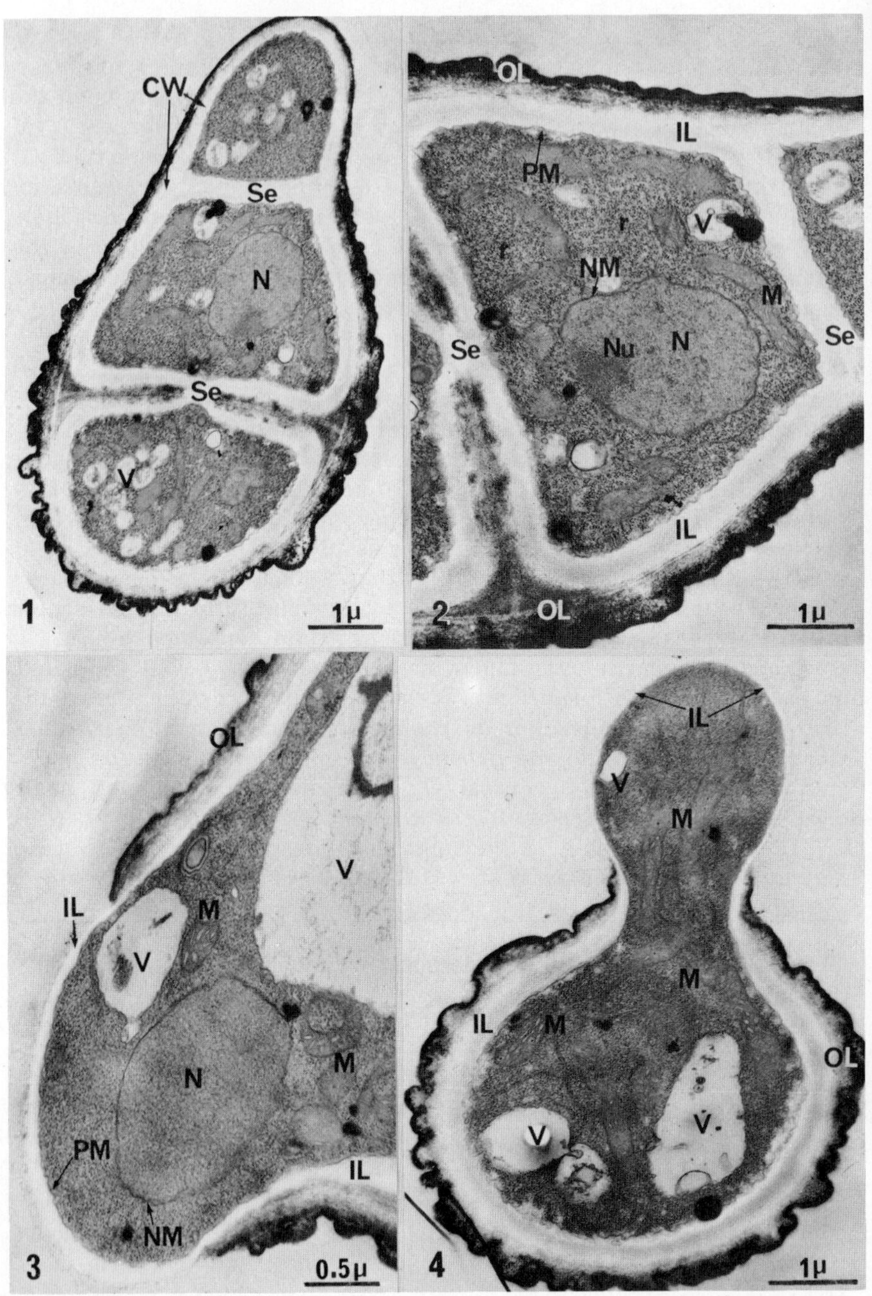
CW
Se
N
Se
V
1
1μ
OL
IL
PM
r
V
NM
M
Se
Nu
N
Se
IL
2
OL
1μ
OL
V
IL
M
V
N
M
PM
IL
NM
3
0.5μ
IL
V
M
M
IL
M
OL
V
V
4
1μ

Parts 1--4). The outer layer is densely pigmented, and has an uneven surface with knobby processes. The outermost region is extremely pigmented forming a spore coat and the granular electron-dense substances are deposited in its inside region. In the latter striated structures are seen (Plate 3, Parts 1--3). The inner layer is electron-transparent and surrounds individual cells. A slightly electron-dense zone is visible in the middle of the inner layer which separates two layers (Plate 3, Parts 2,4). The septum consists of five layers similar to septa of *A. brassicicola* (Plate 3, Parts 1,2) (Ishizaki et al., 1974). Plate 2, Parts 2, 3, 5 are surface views of spore chains of *A. solani* grown in shake culture. Verrucose surfaces of spores which are covered with knobby or warty processes, except on apical portions, are apparent (Fukutomi, 1974).

As mentioned above, in many blastospores and porospores the secondary walls of the conidiogenous (mother) cells are continuous with the primary wall of the new spore, which in turn has a secondary wall lining it. However, in *Stemphylium botryosum* the primary walls of conidiophores and conidia are continuous, until the time of conidium secession (Fanny & Carroll, 1971).

Campbell (1970b) reported that an albino strain of *A. brassicola* had a cell of hyphal and conidiophore wall structure similar to the wall structures of the wild type, except that electron-dense deposits of melanin in the wall and the annulus around the pore in the conidiophore tips were absent. Spores were produced normally, by the extrusion of the inner layer through pores in the outer layer. When spores of the albino strain were produced in submerged liquid culture which contained excess tyrosine, they developed normally. They were true porospores and had an

---

Plate 3. Electron micrographs of conidia of *Alternaria kikuchiana*. 1 and 2. A longitudinal section of a nongerminating condidium. 3 and 4. Sections of germinating conidia. The outer layer ruptured, and the wall surrounding the germ tube is continuous with the inner region of the inner layer of the conidial wall. A slightly electron-dense region is seen in the middle of the inner layer. Labels for these figures are as follows: Inner layer (IL), mitochondria (M), nucleus (N), nuclear membrane (NM), nucleolus (Nu), outer layer (OL), plasma membrane (PM), ribosomes (r), septum (Se), spore wall (CW), and vacuole (V).

electron-dense, brown substance, presumably very similar to melanin, in their primary walls and septum partitions.

Luttrell (1963) distinguished two types of conidia, euseptate and distoseptate. Campbell (1968, 1970b) reported that the wild type spores of *A. brassicicola* are apparently euseptate in that they maintain their integrity as multicellular spores, even though each cell is surrounded by its own walls. The albino spores are distoseptate.

The pigment in spore wall of soil fungi may be melanin (Durrell, 1964), and melanin may protect spores from radiation damage and from enzymic attack by microorganisms during dormancy in the soil (Kuo & Alexander, 1967). Melanin is insoluble in usual organic solvents, but may be bleached by H_2O_2, potassium chlorate, and by sodium hypochlorite (Bancroft, 1967; Gurr, 1960; McManus & Mowry, 1964; Pearse, 1961). Melanin is stained with $AgNO_3$ (Bancroft, 1967; Pearse, 1961) which darkens the spores. When these dark fungi, such as *Stemphylium ilicis*, are grown on agar containing 100 ppm hexachloracetone, approximately half of the spores were not pigmented. These light spores are killed in 4 min or less by ultraviolet light (Durrell, 1964).

The spore walls of *S. ilicis* are thick and consist of a light chitinous inner layer, often appearing laminated. Outside of this is a porous pigmented layer covered by a dense pigmented layer. The spore surface is roughened by knobby processes. Spores of *Alternaria tenuis* very closely resemble conidia of *S. ilicis*. Their porous walls, however, may have an effect on the weight of spores and their dispersion (Durrell, 1964). The spore walls of *S. sarciniforme* are lighter in color and thinner than spore walls of *S. ilicis*. The melanin is scattered as dark granules and is not encrusted in a dense surface layer. This structure affects the resistance of the spores to ultraviolet radiation, i.e., *S. ilicis* spores germinate 100% after 20 min exposure to light of 2537 A at a dosage of 10 $\mu W/cm^2$, and germinate 74% after 40 min and can withstand over 60 min exposure. *Stemphylium sarciniforme* conidia on the other hand drops to 35% germination after 20 min exposure of the same intensity, and after 40 min only 1% of the spores germinate (Durrell, 1964). Multicellular spores of *S. botryosum* have three-layered walls. The external electron-dense primary wall appears to be the outside layer of spores and has a rough surface with knobby processes. This layer is separated from a less electron-dense inner (secondary) wall by a thin electron-transparent zone.

Spores of *Humicola nigricans* have the chitinous inner layer and the pigmented outer layer with melanin granules sloughing off from the surface, and spores of *Coccosporium maculiforme* have the laminated chitinous inner wall layer and the outer layer is heavily impregnated with melanin. Spores of *Dicoccum asporum* and *Stachybotrys atra* have structures similar to those of *Humicola* and in the latter the pigment is enveloped in a black slime. Spore walls of *Torula* sp. are quite different from those of other species, and consist of an inner chitinous layer with dark bars capped over by a very thin fragile dark layer that readily shatters off. The walls of spores of *Helminthosporium sativum* are similar to spore walls of *Torula* (Durrell, 1964).

Conidial walls of *Helminthosporium oryzae* are three-layered, with a thin, electron-dense outer layer and thick, electron-transparent middle and inner layers. The inner layer encloses the conidial protoplast, and forms a spherical sac. The outer and middle layers are on the outside, and surround whole cells of spores (Horino & Akai, 1965; Matsui et al., 1962; Nozu & Matsui, 1959). Conidia of *H. sativum* also have pigmented, multilaminated walls. An electron-dense outer layer consists of two fibrillar layers, one denser than another, and an inner densely fibrillar layer (Old & Robertson, 1969, 1970).

The detail of septal partition in conidia of *H. oryzae* is very similar to that of *A. brassicicola* (Campbell, 1968) and *A. kikuchiana* (Ishizaki et al., 1974).

The structure of conidia of an albino mutant of *H. oryzae* treated with γ-ray irradiation was very similar to the structure of conidia of the original strain except for a lack of pigment in the outer layer. These spores are distoseptate, and the outer layer is broken down easily by a slight physical shock. This results in separation of individual cells. Each cell can then germinate and attack rice leaves similar to infection caused by conidia of the original strain (Masago, 1961). Pigmented wild type conidia resist lysis by soil bacteria and enzymes. This resistance may be associated with the electron-density of the surface of the cell wall which has a broad distribution of pigment (Old & Robertson, 1970).

Conidia of *Drechslera sorokiniana* develop enteroblastically through pores in the conidiogenous cell wall. These pores are probably formed by autolysis of the outer layer, and the inner layer protrudes through the pore encompassing the initial conidium. A new conidium wall begins to be formed by apposition adjacent to the plasma membrane inside the outer layer. At this stage lomasomes adjacent to this wall region are seen. At the later stage of conidium development, the outer layer remains thin and becomes relatively electron-dense, while the inner layer is poorly stained and thickens (Cole, 1973). Walls of spores of *Pityrosporum ovale* consist of a thin, electron-dense outer layer, an electron-opaque middle layer with corrugation at the inside and a dark undulating inner layer, when stained with permanganate (Kreger-Van Rij & Veenhuis, 1970).

Aleuriospore

The aleuriospores are not formed by budding, but by the apex of conidiophores becoming inflated and delimiting by a septum at an early stage. Aleuriospores have a wide plane of attachment, usually as wide as the conidiophores, whereas blastospores generally have a narrow point of attachment (Talbot, 1971). The genera of Hyphomycetes, e.g., *Acremoniella, Scopulariopsis, Clasterosporium, Humicola, Mycogone, Nigrospora*, and *Trichothecium*, produce spores of this type (Tubaki, 1963).

According to Pore & Larsh (1967), aleuriospores are formed in *Aspergillus terreus-flavipes* group species. *Aspergillus carneus*, a member of this group, produces aleuriospore-type secondary spores on the submerged mycelium (Pore et al., 1969). These spores with a wide plane of attachment to the conidiophore, found in *A. carneus*, have a four-layered wall. The outer layer is continuous with the wall of the conidiophore or hypha, and is separated from the inner layer by an electron-dense region. The inner layer resembles the outer layer in all respects, and therefore, the cell wall of aleuriospores has four layers, and are recognized by dense regions in each layer. In contrast to these spores the conidial cell wall (phialospore-type) has no electron-dense area in the middle (Pore et al., 1969). The conidium of *Acremoniella velata* possesses a relatively thick fibrous wall which is enveloped by an outer loose fibrous membrane. The fibrous nature of the inner layer is prominent near the point of attachment between conidia and conidiophores (Jones, 1968). By the scanning electron microscope observations conidia of *Acremoniella* sp. have a netlike veil on the surfaces (Jones, 1967).

In *Scopulariopsis brevicaulis* walls of differentiated conidia consist of outer and inner layers and both have the apparent fibrous nature. In the process of primary conidium formation, a new wall layer is differentiated at the conidiogenous cell apex. It appears as a thin electron-transparent zone lying between the plasma membrane and the outer wall layer. At a still later stage of development, this inner layer expands and encases the conidium initial, abutting the plasma membrane. The outer wall is initially continuous with the wall of the conidiogenous cell, but when it reaches its maximum extensibility at the base of the conidium during the development of the double septum, it fractures as a result of the internal pressure of the growing inner wall layer. The outer wall layer of subsequently formed conidia is initially continuous with the inner wall of the primary conidium, and the inner wall develops inside the outer wall (Cole & Aldrich, 1971).

In *Phoma exigua* spores initiate to differentiate by papillalike thickenings of the walls of the mother cells. At this place the plasma membrane shows a corrugation, which is apparently associated with the endoplasmic reticulum. An electron-dense layer occurs in the papilla of the mother cell wall, gradually differentiates into spore wall, and develops into a budlike protrusion. When the bud has attained the diameter of a mature spore, a light cloudy substance develops between the spore wall and the outer electron-transparent layer. This electron-transparent layer gradually disintegrates. Both substances form the slimy mass surrounding the mature spores. This process of the spore development is similar in *Phoma herbarum* and *Ascochyta bohemica* (Brewer & Boerema, 1965).

In *Ascochyta pisi* the process of spore development is somewhat different from the developmental process of *Phoma exigua*. The undifferentiated top of a mother cell develops into a thin-walled protrusion. If the mother cell previously produced spores, the top will be partly covered by the remaining parts of wall which are later visible as one or more collars (scars) at the base of the spore initial. When the thin-walled protrusion attained mother spore size, the detachment of the spores is initiated by double-septation at the conidial base and mother cell apex, so that just before detachment, an intercellular space is formed between the spore and mother cell. The portion of the wall between the detached spore and the mother cell remains attached to the mother cell, thus the collar is produced (Brewer & Boerema, 1965).

In detached thin-walled spores a septum soon develops by a thickening of the lateral wall, and the development of a new distinct wall divides the spore into two cells which are surrounded by individual walls. The mature spore produces a light cloudy substance representing the mucilaginous appearance around the spore in the pycnidium. When spores of *A. pisi* are crushed by pressure, individual cells are squeezed out of the ruptured outer wall, showing a ditoseptate nature (Brewer & Boerema, 1965). As in *Alternaria, Helmithosporium*, and *Ascochyta,* the distoseptate nature of conidia suggest a similarity of the walls of the sporangioles (merosporangium) in Mucorales.

Phialospore

Conidia which are abstricted in basipetal sequence from phialides have been termed phialospores (Talbot, 1971). The genera of Hyphomycetes, e.g., *Thielaviopsis*, *Aspergillus*, *Cephalosporium*, *Cylindrocladium*, *Fusarium*, *Graphium*, *Penicillium*, *Phialophora*, *Stachybotrys*, *Trichoderma*, and *Verticillium*, form spores of this type (Tubaki, 1963).

The cell wall of micro- and macroconidia of *Fusarium oxysporum* f. *cucumerinum* fixed with $KMnO_4$ consists of two layers, an electron-dense outer layer and a thick, electron-transparent inner one. Both the outer and inner layers of conidia fixed with OsO_4 or Gu-OsO_4, however, appear to be two-layered (Akai et al., 1969). Walls of conidia of *F. culmorum* fixed with $KMnO_4$ seem to consist at least of a triple layer of low electron density and fibrous nature (Garcia Acha et al., 1966; Marchant, 1966a, b). The inner layer surrounds individual cells forming the cell compartment of the conidium, and the outer layer forms a coat enveloping all cells of the conidium, as is seen in spores of *Alternaria* and *Helminthosporium* (Garcia Acha et al., 1966). When cultured on modified Czapek's medium containing D-arginine instead of $NaNO_3$, *Fusarium roseum* forms thick-walled, yeastlike cells. The walls of these yeastlike cells have a lamellar structure with ten or more layers between the inner and outer layers, while conidia with two-layered walls are formed when the fungus is grown on L-arginine medium (Egawa et al., 1968; Kunoh et al., 1968).

The conidiophores of *Aspergillus nidulans* have a two-layered wall (Oliver, 1972). According to Tanaka & Yanagita (1963a,b), the phialide wall of *A. niger* almost completely surrounds the first conidium except at the base, where it is broken off at a point that corresponds to the broken apex of the phialide. The walls of the first and second conidial intitials are continuous with each other and with the newly formed wall within the apex of the phialide. Subramanian (1971) showed that the outer walls of the second and subsequent conidia in a chain are continuous, thus this wall also forms the isthmus that separates adjacent conidia. The inner wall is newly formed in each conidium, and is thicker than the outer wall. In mature spores the surface lamina is a prominent, markedly corrugated envelope with the corrugations filled by electron-dense material, either continuous or formed into a network of fibrils permeated by the embedding resin. The electron-lucent wall of mature spores is thinner than in immature spores and has a more sharply defined surface (Oliver, 1972). The conidia of many species of *Aspergillus* have surfaces with characteristic rodlet patterns (Hess & Stocks, 1969).

Phialide walls of *Penicillium clavigerum, P. claviforme,* and *P. corymbiferum* are electron-transparent with thin electron-opaque surface layers. The phialide apex is closed by a hollow pluglike structure of wall material. A conidium initial is formed by distension of this plug. The conidium protoplast is cut off from the phialide protoplast by an initially thin perforate septum which is peripherally continuous with the apical plug. When fully developed, the septum becomes thicker but the perforation remains as a channel filled with electron-opaque material. In *P. clavigerum* and *P. corymbiferum* a new inner layer is present around older conidial protoplasts. This inner layer is thin around protoplasts close to the phialide and thicker around protoplasts distal from the phialide. Conidium chains are covered with a thin electron-opaque surface layer (Fletcher, 1971).

In the resting spores of *P. megasporum* the outermost covering of the spore consists of rodlets embedded in or resting on a basal layer of homogenous material. Beneath the basal layer, there is an irregularly electron-dense layer and an electron-transparent layer with a definite fibrillar appearance. Furthermore, one more transparent layer between the fibrillar layer and the plasma membrane is seen, which upon germination extends over the germtube (Remsen et al., 1967).

The surface of conidia of *Penicillium* spp., e.g., *P. chrysogenum* and *P. digitatum*, is covered by an outer layer made by ordered arrays of rodlets, which appear to vary in pattern and distribution from species to species. The rodlets consist of linear arrays of particles approximately 50Å diameter (Hess et al., 1968).

In *Trichoderma viride* conidia have two-layered walls, consisting of an electron-dense rough outer layer (epispore) and a moderately electron-dense inner layer. A species of *Pachybasium* has a thinner and more transparent wall in conidia. A distinct mucilaginous substance is characteristically present around conidia and hyphae (Hashioka et al., 1966).

In stationary culture medium, mycelium of *Verticillium albo-atrum* grows typically, but in shake culture it develops chiefly as budding yeastlike cells (Wang & Bartnicki-Garcia, 1970). The conidia have a rather smooth wall outline, but the outer surface of cell walls of the yeast form has a coarsely and irregularly granulated appearance; there is only a vague microfibrillar pattern underlying the surface granulation. The granular components were extracted by alkali and revealed a fibrillar wall fabric. The microfibrils were randomly oriented except at one of the cell poles where there was a region of circularly arranged microfibrils with a raised outer annulus and a minute central orifice or depression from which buds were formed. This central orifice may be the septal pore (Buckley et al., 1969), which resembles the bud scars of *Saccharomyces cerevisiae*. Cell walls of *V. albo-atrum* dissolved by alkali showed a heteropolysaccharide-protein complex containing mannose, galactose, glucose, glucuronic acid, glucosamine, and the common range of amino acids. The alkali-insoluble microfibrillar network was made of a β-linked glucan and chitin, and no evidence for cellulose was found (Wang & Bartnicki-Garcia, 1970). Acid treatment produced coarse microfibrills resembling those in *S. cerevisiae* walls. The hydroglucan was soluble in alkali and alkali treatment resulted in leaving an insoluble microfibrillar network composed mainly of chitin (Wang & Bartnicki-Garcia, 1970).

Chlamydospores of *Thielaviopsis basicola* are formed inside the sporophore, and the spore chain is enveloped by a distinct outer wall (Christias & Baker, 1969; Tsao & Tsao, 1970). Like the walls of hyphae and endoconidia, this outer wall of the spore chain (chain envelope) consists of two layers, an electron-dense outer layer and an electron-transparent inner layer. The chain envelope is an extension of the hyphal wall, and within each cell its own thick wall is present consisting of an electron-dense outer layer and an electron-transparent inner one, the latter directly abutts the plasma membrane. The spore wall is 2 -3 times as thick as the chain envelope, and is thickest at the corners of two adjoining cells. At these four corners, the generally electron-transparent inner layer of the spore wall extends, by wedging obliquely at about a 45-degree angle, into the thick electron-dense outer layer of the spore wall. Each wedge tapers gradually to a fine point which extends just short of the outer edge of the dark spore wall, and creates a weak junction at the rims. The wedges at the four corners of each spore are, in fact, two continuous transversely located circular bands. The spore wall may dehisce at the rim during spore germination, to form the operculum (Patrick et al., 1965; Tsao & Tsao, 1970).

Chlamydospores may also be formed from the cells of spores instead of from hyphae, e.g., from the cells of conidia in *Fusarium* (French & Nielsen, 1966; Nash et al., 1961) or in ascospores of some *Mycosphaerella* species (Talbot, 1971). Overwintering conidia of *Helminthosporium turcicum* associated with corn residue and soil may sometimes form chlamydospores within the spores (Boosalis et al., 1967). In conidia of *Alternaria porri* f. sp. *solani*, chlamydospores developed especially when the fungus contacted natural soil (Basu, 1971). The process of their development resembles the formation of chlamydospores in *Fusarium* macroconidia (Nyvall, 1970) and in conidia of *Helminthosporium turcicum* (Boosalis et al., 1967).

In cultures of fungi, addition of some substances stimulates the formation of chlamydospore. Amici & Locci (1968) described the chlamydospore formation in hypha of *Helminthosporium carbonum* in the presence of α-solanine and Ishida et al. (1968) found chlamydosporelike cells in *H. oryzae* hyphae, when cultured on a modified Czapek-D-valine medium. These cells are surrounded by a mucilaginous sheath.

Oidium

Oidia are thin-walled conidia borne in chains. Conidia of *Sphaerotheca pannosa* have a relatively thin wall, which consists of two layers. The outer layer is electron-dense, extremely thin, but not uniform and sometimes looks obscure, as if it was lacking. The inner layer is electron-transparent, relatively thick with an ill-defined stratifying structure (Akai et al., 1966). This structure is very similar to that of *S. macularis* (McKeen et al., 1966). The cell wall of conidia of *Erysiphe graminis* (Akai et al., 1968; Hossain & Manners, 1964) and *E. cichoracearum* (McKeen et al., 1967) is also two-layered. The inner layer of *E. cichoracearum* is electron-transparent, probably chitinous and possesses a few spines which project into the thicker gelatinous outer layer (McKeen et al., 1967). The outer layer of conidial walls of *E. graminis hordei* is electron-transparent, relatively thick, undulated, and not uniform in thickness. The inner layer is thin with even thickness, and has the same electron density as the outer layer (Fig. 1; Part 3) (Akai et al., 1968). The roughness of conidial cell surfaces of powdery mildew fungi has been observed by many investigators. The surfaces of *E. graminis* conidia are covered by a thin electron-dense layer which is probably mucilaginous in nature (Kunoh, 1972; Stanbridge et al., 1971) and is warty (Fig. 1; Parts 1, 2) (Kunoh & Akai, 1967) or spiny with a collar at the region of attachment (Plumb & Turner, 1972).

The existence of oidia in Hymenomycetes has long been known. The oidia of *Psathyrella coprophila* are borne on oidiophores and have filamentous appendages on the surfaces of the oidia, which taper slightly. After osmic acid fixation the filaments appear to be hollow, while after Gu-OsO_4 double fixation they appear solid (Jurand & Kemp, 1972). In the indian ink preparations examined by the light microscope the capsule, similar to those found in many bacteria, is seen as a light band which surrounds the oidia and is positive for ruthenium red. However, the relationship of the filamentous appendanges to the capsule is not clear. The capsule appears to contain a highly polmerized acid-mucopolysaccharide (Jurand & Kemp, 1972).

Aecidiospore

Aecidiospores are a dikaryotic cells. Mature aecidiospores of *Uromyces caladii* have thick strongly rough walls. The spore coat consists of a continuous wall with tubercles or cogs which project from it; between these cogs are electron-dense intercog particles that apparently fall out in the course of maturation (Moore & McAlear, 1961). Aecidiospores of *Endocronartium harknessii*, *Cronartium coleosporioides*, *C. comptoniae*, and *C. ribicola* possess smooth areas and annulated processes of various sizes with five to seven layers. The processes of *C. coleosporioides* are more prominently annulated, and have more varied shapes and sizes than those of the other three species. Processes of *E. harknessii* narrow toward the tips while the processes of *C. coleosporioides*, *C. comptoniae*, and *C. ribicola* have nearly the same width from the bottom to the tips (Hiratsuka, 1971). *C. comandrae* has smooth processes without annulations. Walkinshaw et al. (1967) who observed the aecidiospores of *C. fusiforme* showed that the aecidiospore had an extremely thick wall with striking spicules protruding from it. The wall was readily degraded by commercial chitinase, but the spicules were not affected. When the spore germinated, a thin-walled germ tube always emerged through a pore in the spore wall.

Urediospore

Urediospores are a kind of conidia borne in a uredosorus. The spore wall consists of three recognizable layers (Ehrlich & Ehrlich, 1969; Littlefield & Bracker, 1971; Thomas & Issaac, 1967; Williams & Ledingham, 1964). With both $KMnO_4$ and Gu-OsO_4 fixed specimens of *Puccinia graminis* urediospores, the external layer of spore walls appears as a thin dark line, consisting of a single strand of wall material. The middle layer is electron-lucent and may consist of pectin and other water absorbing compounds, while the more massive inner layer consists of a loose network of electron-dense microfibrils. The only visible wall in the germ pore area in spores is an external layer (Ehrlich & Ehrlich, 1969; Williams & Ledingham, 1964). In *Melampsora lini* the cell walls of urediospores are covered by a wrinkled pellicle, ca. 150-200 Å thick, including the spines. The pellicle is continuous with an accumulation of apparently the same substance in the interstices between spores. This substance forms a matrix around all the spores in a developing uredium, and is probably a hydrophobic lipid component (Littlefield & Bracker, 1971; Williams & Ledingham, 1964).

Thomas & Isaac (1967) discussed the development of echinulation in the urediospores, and suggested that the spines developed beneath the urediospore wall and moved to the surface during spore maturation. In electron micrographs, however, a young spine, which begins as a small electron-transparent body (small deposit), appears at first in the internal wall layer adjacent to the plasma membrane. This globose body then centrifugally develops as a conidial projection. In the earliest identifiable stage of spine development, endoplasmic reticulum lies near the base of the spine close to the plasma membrane (Ehrlich & Ehrlich, 1969; Littlefield & Bracker, 1971; Thomas & Isaac, 1967). During development the spore walls thicken and the spines increase in size. Centripetal growth of the wall encases the spines in the wall material. The spines progressively assume a more external position in the spore wall and finally locate at the outer surface of the wall (Littlefield & Bracker, 1971). The spines are electron-transparent, conical or pyramidal projections, with their basal portion embedded in the ringlike slightly electron-dense spore wall. The spinous surface of urediospores has been observed by scanning electron microscopy or replica methods for various kinds of rust spores (Akai et al., 1969; Arthaud, 1972; Fukutomi et al., 1969a; Manocha & Shaw, 1967; Payak et al., 1967; Stanbridge & Gay, 1969; Schwinn, 1969; Van Dyke & Hooker, 1969).

A mutant strain of *Melampsora lini* has finely verrucose urediospores. The warts on the surface of mutant spores are round electron-dense structures, shorter than the spines of the wild type. The walls are thinner in mutant spores than in wild type spores (Littlefield & Bracker, 1971). The structural components of urediospore walls have been little known, though chitin, hemicellulose, and glucomannan are thought to be basic wall constituents (Staples & Wynn, 1965).

Teliospore

Very little is known about the ultrastructure of teliospores of Ustilaginales and Uredinales. The teliospore walls of *Ustilago hordei* and *Tilletia contraversa* (Graham, 1960; Hess & Weber, 1970; Robb, 1972) consist of three major layers. The outermost layer is thinnest with moderate electron density and the external features of spores are slightly irregular. The middle layer is exceedingly electron-dense and structurally amorphous, and is very variable in thickness. The inner layer has the least density of the three and consists of fibrils oriented parallel to the circumference of spores and is embedded in a homogenous matrix. This layer has an uneven internal surface with ridges into the protoplast. That the innermost wall layer of *U. hordei* is the only fibrillar layer is in agreement with biochemical studies in *T. contraversa* which show that chitin is predominant in the inner layers but lacking in the outer layer (Graham, 1960; Robb, 1972). According to Graham (1960) the sheath of teliospores of *T. contraversa* is primarily a pectic material complex, which incorporates hemicellulose (including callose) and lipids, and the reticulum contains pectic materials, hemicelluloses, proteins, melanin pigments, and lipids. The outer endospore layer is primarily chitin but some hemicelluloses were defined histochemically. The inner endospore layer contains at least chitin, hemicellulose and pectic materials. The material cementing the reticulum to the endospore contained lipoids.

Teliospores of *Gymnosporangium haraeanum* (Plate 4, Part 1) have thick walls consisting of two layers. The outer layer is thin and electron-dense, and the inner layer consists of three zones, IL_1, IL_2, and IL_3. The thinnest innermost layer (IL_1) is electron-transparent, and abutts directly to the plasma membrane, the next layer (IL_2) is very electron-dense and the other layer (IL_3) is less electron-dense. During the early stages of germination, at the rim of two adjoining cells the innermost electron-transparent layer (IL_1) increases in thickness, while the outer layer and the inner layers (IL_2 and IL_3), especially the latter two layers decrease in thickness and then the germ tube bulges out with the loss of the IL_2- and IL_3-inner layers. The promycelium wall (germ tube) is continuous with the inner layer IL_1 of the spore wall, and at least in the early stage of germ tube development, the very thin outer layer may be recognizable. It is continuous with the spore wall (Plate 4, Part 1) (Kunoh, unpublished data).

At an early stage in development of *Puccinia chaerophylli*

teliospores, but after the division into its two spore cells, the wall is smooth and homogenous except for a thin, ill-defined, darker outer zone. Well-defined lomasomes are present between the plasma membrane and the wall. An outer dark layer is clearly differentiated and the inner zone contains electron-dense granules. Electron-transparent halos develop around the dark granules in the spore wall which is still smooth. The halos are at first more or less circular, but later extend radially, then centrifugally, toward the surface of the wall. This results in the formation of hyaline areas in the wall. At this stage when the spore cell has attained full size, the outline of the spore begins to alter and results in formation of the surface network of the mature spore. The wall shrinks in the region of the hyaline material. The reticulate ornamentation of mature teliospores develops by reabsorption of portions of the hyaline material of the original smooth wall (Henderson et al., 1972).

Cell Walls of Sexual Spores of Oomycetes, Ascomycotina, and Basidiomycotina

Cell walls of oogonia and oospores of Oomycetes

Oogamy occurs in some of the lower fungi, e.g., Oomycetes. In *Phytophthora macrospora (Sclerophthora macrospora)* the oogonial primordia are formed in the developing rice leaves of early age. The antheridial and oogonial primorida gradually develop, but it is difficult to distingush them until the hyphal knobs attain a relatively large size and the antheridial hyphae come in contact with the oogonial elements paragynously (Fukutomi & Akai, 1966; Fukutomi et al., 1969b).

Before fertilization occurs, some nuclei and a large number of scattered mitochondria are observed in the antheridium. The mitochondria are found especially around the fertilization pore. Two or three layers and masses of ER often appear along the thin wall of the pore in the antheridium. The cell wall at the fertilization pore becomes very thin but the wall surrounding the fertilization pore becomes swollen and protrudes markedly toward the inside of the antheridium and oogonium. This wall becomes several times thicker than the other areas. This was termed as the "fertilization annular swelling" (Fig. 2, Parts 1, 2) (Fukutomi et al., 1968, 1969b). The wall swelling around the fertilization pore is more remarkable in the oogonium. A fibrillar structure is observed in the annular swelling

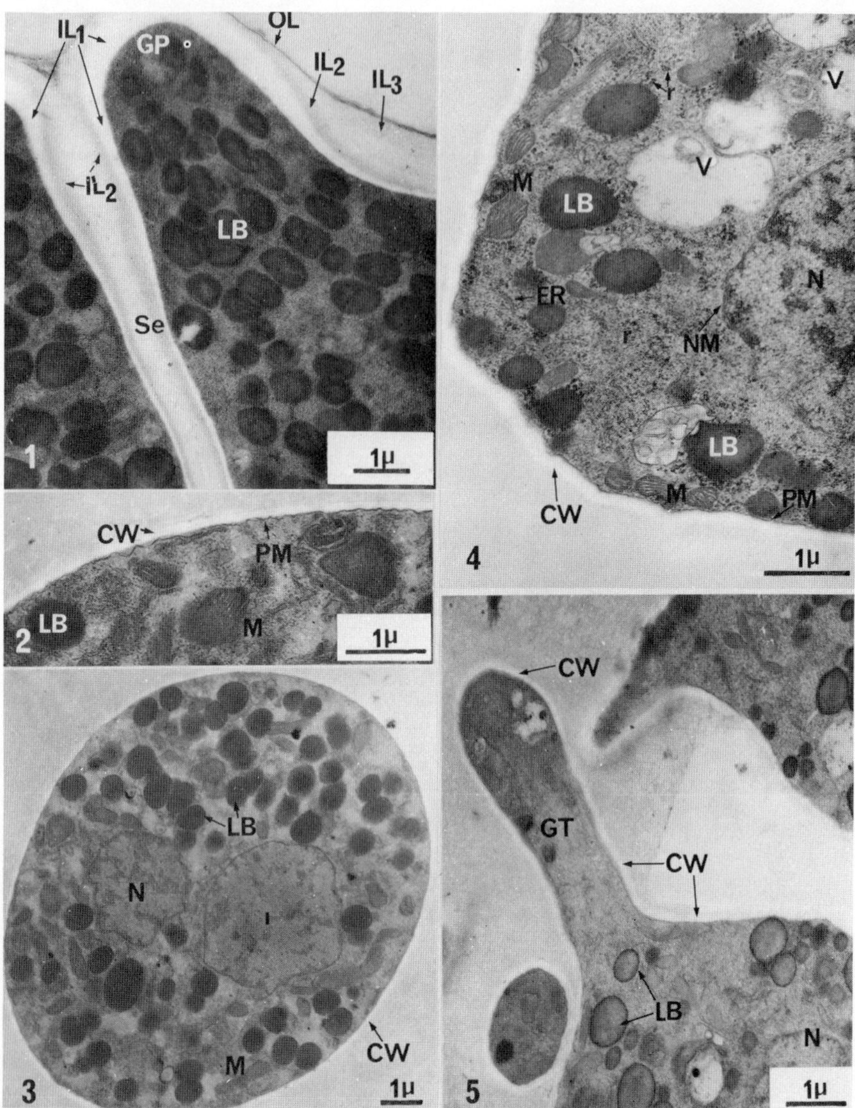

Plate 4. Electron micrographs of teliospore and sporidia of *Gymnosporangium haraeanum* and *Kuehneola japonica*.
1. An early stage of germination of a teliospore of *G. haraeanum*. 2 and 3. Sporidia of *G. haraeanum*. An electron-transparent thin, single-layered wall surrounds the sporidium. 4. A sporidium of

and appears to extend continuously to the oogonial wall (Fig. 2, Part 2). Many oogonial papillae are observed at the surface of the oogonium and the cytoplasm of the papilla is continuous with the cytoplasm of the oogonium. The thick wall of the oospore is electron-transparent, single-layered, and smooth to somewhat rough. The oogonial surface becomes wrinkled with short warts, which is thought to be due to the contraction of the oogonium after fertilization (Fig. 2, Parts 3, 4). The relation of the papillae to the short warts of the oogonial surface, and the function of the papillae are unknown (Fukutomi et al., 1971).

In postferilization stages of oospores of *Phytophthora drechsleri*, the oospores differentiate within the oogonia and become shrunken in size or become aplerotic. The thin inner walls of oospores are plastic and follow the shapes of the spherical bodies in the cytoplasm. The antheridia become empty and the oospore walls thicken (Zentmyer & Erwin, 1970).

The walls of the oogonia, antheridia, zoosporangia, and cytospores of *P. macrospora* appear to be composed of cellulose which is dissolved with 3% acetic acid. Except for the oospore wall, they show a positive reaction of cellulose with zinc-chloriodide and zinc-chlor-sulphate. The oospore wall has a positive reaction for chitin after potassium hydroxide treatment followed by the stains with the above solutions, and it was digested by chitinase. These histochemical reactions indicate that the oospore wall may be composed of chitin. In an electron microscope, the oospore wall is amorphous, while other walls of this fungus are considered to be composed of cellulose. They have a fibrillar structure. These results suggest that the cell wall synthetic metabolism of this fungus may shift from cellulose to chitin synthesis after fertilization (Fukutomi et al., 1974).

K. japonica. A thin electron-transparent single layered wall is recognizable in the circumference of the spore. 5. Germinating sporidia of *K. japonica*. A thin electron-transparent single-layered wall surrounds the germ tube. The wall is continuous with that of the sporidium. Labels for these figures are as follows: Cell wall (CW), endoplasmic reticulum (ER), germ tube (GP) (initial of promycelium), inner wall layers (IL_1, IL_2, IL_3), lipid bodies (LB), mitochondria (M), nucleus (N), nuclear membrane (NM), outer wall layer (OL), plasma membrane (PM), ribosome (r), septum (Se) and vacuole (V).

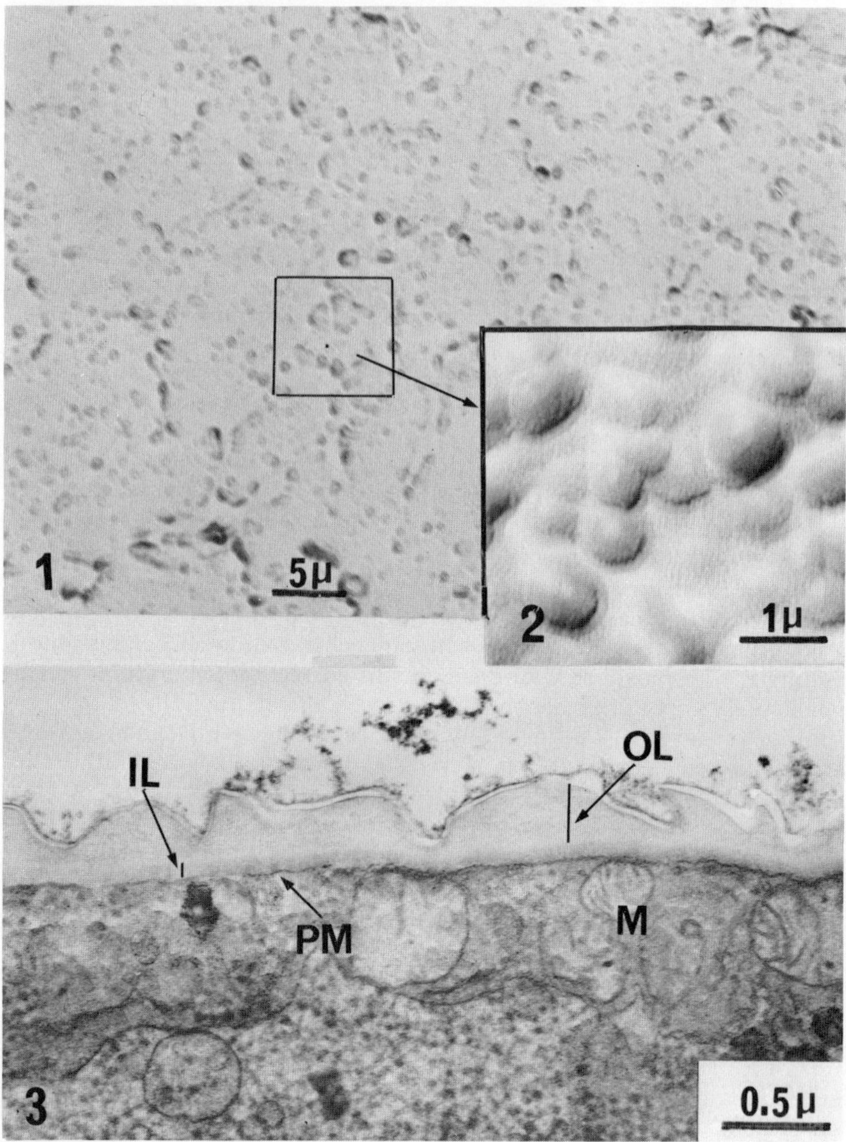

Fig. 1. Cell wall of conidia of *Erysiphe graminis hordei*.
1. A replica of a warty surface of a conidium. 2. A higher magnification of the warty structure of the conidial surface. 3. An electron micrograph of a thin section of a conidium showing the inner layer

In *P. palmivora*, young antheridia have single, thin, apparently homogenous, cellulose walls. However, in mature antheridium the walls distend and burst through oogonial enlargement and turn down to form a collar around the oogonial stalk. The walls of young oogonia, like the antheridial walls, are thin, homogeneous monolayers. Within hours, the walls of the oogonium thicken, but remains a monolayer. Soon after the oogonium has reached full size, it is cut off from its hypha by a plug, made of loosely packed vesicles which contain dense material, which are formed in the oogonial stalk (Vujičić, 1971).

The oogonial walls of *Pythium ultimum* are relatively thin. An extensive system of discontinuous ER and vesicles is observed, which may be important in the synthesis of walls of developing oogonia (Marchant et al., 1967). The oogonium is separated from the supporting hypha by forming a loose plug of material as in *P. palmivora*. This wall is poorly organized and does not resemble the more highly organized septa in other organisms. This plug appears to simply separate the oogonium from the supporting hypha. The wall thickening gives an irregular outer surface to the oospore (Marchant, 1968). Thick walled oospores are the major structures of *P. aphanidermatum* and are capable of long term survival in soil. The mature oospore wall consists of a thin outer wall (exospore) and a thick inner wall (endospore) (Stanghellini & Russell, 1973).

Ascospore

Ascospores are delimited inside the ascus by free cell formation. In *Schizosaccharomyces octosporus* (Conti & Naylor, 1960), *S. cerevisiae* (Marquardt, 1963), *Podospora arizonensis* (Alloway & Wilson, 1972), *Pyronema domesticum* (Reeves, 1967), *Ceriosporopsis halima* (Lutley & Wilson, 1972) and *Nematospora coryli* (Plurad, 1972), the spore wall is laid down between the two delimiting membranes of the newly formed spores. It is thought that double membranes are derived from ER (Bracker, 1967). These membranes become the membrane investing ascospores. The outer membrane becomes the spore membrane and the inner one becomes the plasma membrane of spores. In ascospores of *P. arizonensis* the primary wall layer

---

(IL), outer layer (OL), a mitochondrion (M), and the plasma membrane (PM).

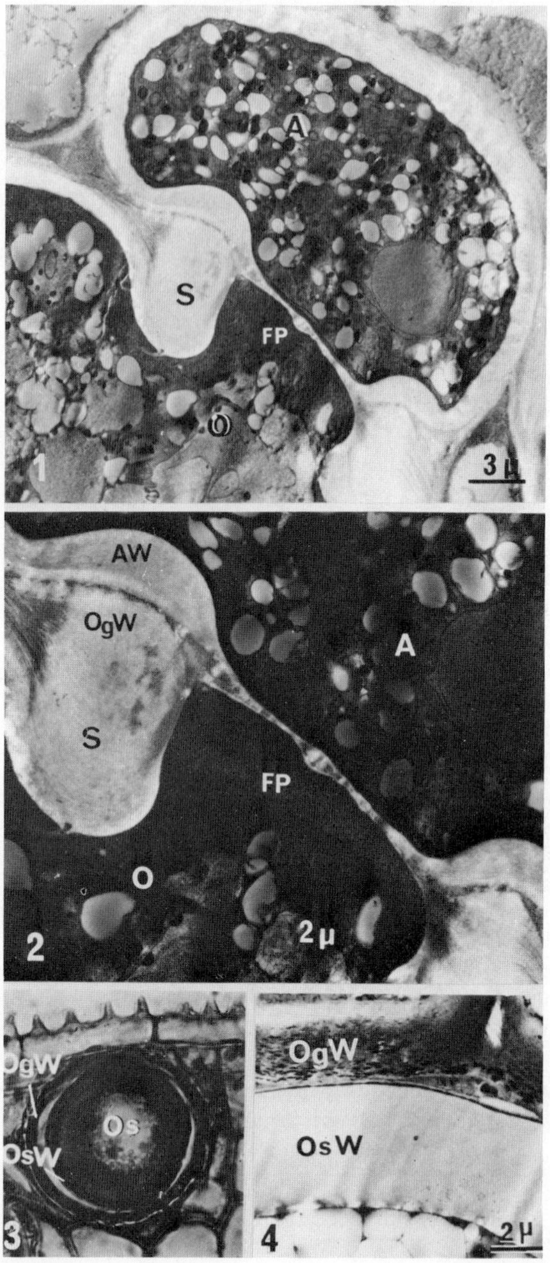
A
S
FP
O
1
3 μ
AW
OgW
A
S
FP
O
2
2 μ
OgW
Os
OsW
3
OgW
OsW
4
2 μ

develops between two delimiting membranes, and the secondary and tertiary wall layers develop within the matrix of the thickening primary layer (Alloway & Wilson, 1972). Two-celled ascospores with appendages are formed by *C. halima*. The cell walls consist of three layers (Lutley & Wilson, 1972). The first layer, the electron-dense epispore, contains acidic mucins, chitin, and protein. It has a high affinity for such stains as ruthenium red, toluidine blue and thionin. The terminal appendages grow out from apical caps of the epispore, a slightly thickened apical region of the firstly formed electron-dense layer of the wall, and is supported internally by a fibrillar network. The exospore develops external to the epispore but stains less intense. It contains acid-polysaccharides in irregular chambers which surround the spore and form a sheath at the base of each appendage. Internally, a thick electron-transparent mesospore (inner wall), consisting largely of periodic acid, Schiff (PAS)-negative polysaccharide, develops and forms individual cells (Lutley & Wilson, 1972). In the immature ascospores of *P. domesticum* elements of the ER are found close to the plasma membrane of the sporoplasm. This close association of the ER with the plasma membrane is found only during the development of the endospore (inner wall) layer. Therefore, it is suggested that this ER may play a part in ascospore wall formation (Reeves, 1967). Ascospores of *Emmonsiella capsulata* (the ascomycetous state of *Histoplasma capsulatum* and *Ajellomyces dermatitidis* (Gymnoascaceae) have walls composed of an electron-dense outer layer with warty projection and a lighter electron-transparent inner layer (Glick & Kwon-Chung, 1973).

Fig. 2. Electron and light micrographs of an antheridium, an oogonium and oospores of *Phytophthora macrospora*.
1. Fine structure of the contact area of an oogonium and an antheridium. The thin wall of the fertilization pore and thick fertilization annular swelling are evident. 2. A higher magnification of the contact area. A fibrillar structure in the fertilization annular swelling is observed. This seems to be extended to the oogonial wall. 3. A light micrograph of an oospore in the leaf tissue of a rice plant. 4. An electron micrograph of an oospore wall and an oogonial wall. Labels for these figures are as follows: Antheridium (A), antheridial wall (AW), fertilizaiton pore (FP), fertilization annular swelling (S), oogonium (O), oogonial wall (OgW), oospore (Os) and oospore wall (OsW).

Walls of dormant ascospores of *Neurospora tetrasperma* are composed of three major layers, the endosporium (endospore wall) which surrounds the protoplast, a melanized and rigid middle layer, the episporium (epispore wall), and a ribbed outer layer, the perisporium (perispore wall, ribbed coat) (Lowry & Sussman, 1968; Sussman & Halvorson, 1966). Observations with electron microscopy reveal the presence of two additional layers. One of these is relatively electron-transparent, covers the ridge of the perisporium and fills in between them. It is lamellate with a fine fibrous surface and can be chemically distinguished. The presence of this outermost layer, at least in wet spores, results in an appearance of a relatively smooth surface. The second additional wall layer is composed of a very electron-dense material lying between the episporium and the endosporium. This material appears to accumulate at the germ pore which is apparently pushed out of the way by the emerging germ tube. During germination the continuity of the germ tube wall and the endospore wall are observed, and the outermost layer of the spore wall is also continuous with the fibrous outer layer of the germ tube (Lowry & Sussman, 1968).

In ascospores of *Ascodesmis sphaerospora* the ornamented secondary wall layer (outer layer) is added outside the primary wall (inner wall). The primary wall is smooth, thick and electron-transparent and the secondary wall is electron-dense and roughened with markings which form a reticulate pattern (Moore, 1963). The outer wall (perisporium) of ascospores of *Hypoxylon rubiginosum* appear rough and spores are circumscribed with threads or fibrils at close intervals (Rogers, 1969). The ornamentation of ascospores of *Elaphomyces granulatus* results from the formation of rods within gelatinous exospores, and ascospores of *E. muricatus*, *E. anthracinus*, and *E. aculatus* have rodlike structure on the outer layer with ornamentation similar to that of spores of *E. granulatus* (Hawker, 1968).

Basidiospore

Throughout the classes of Basidiomycotina basidiospores are basically very uniform. The cell wall of basidium is rather thin and the wall of the zone where sterigmata is ever thinner (Oláh, 1973). In *Poria latemarginata*, after the discharge of basidiospores from the initial basidium, the top of the basidium disintegrates and a new young basidium develops inside the walls of the preceding basidium. The renewed growth occurs at the basal septum of the expending basidium, and results in a disintegration of the septal pore apparatus. As many as eleven basidia may develop from one point (Setliff et al., 1972).

Basidiospores are usually uninucleate but some, e.g., *Schizophyllum commune* and *Psilocybe* sp. (Madelin, 1966; Stocks & Hess, 1970; Wells, 1965), have binucleate basidiospores as a result of mitotic division. The interpretation of fine structure of basidiospore walls is quite new. Wells (1965) observed the fine structure of basidia and basidiospores of *S. commune*. According to him, the spore is initially a vesicular enlargement of the tip of sterigma. The basidiospore becomes binucleate and is soon delimited from the sterigma by an electron-transparent layer. Subsequently the electron-dense hilum is formed but the wall of the spore is continuous across the electron-dense hilum to the sterigma. A cuplike, parahilar, electron-transparent layer develops between the wall and the ectoplast of the basidiospore. It is postulated that this parahilar layer functions to strengthen the wall of the basidiospore in the areas against which force is exerted during discharge.

A thin fragile extension of the wall of the sterigma is present around the base of a basidiospore (Wells, 1965). Basidiospores are formed in a sterigma envelope which suggests a homology of basidia and asci. Basidiospore walls have one fibrous layer (Voelz & Niederpruem, 1964). Dormant basidiospores of *Psilocybe* sp. (Stocks & Hess, 1970), *P. quebecensis* (Oláh, 1973), and of *Coprinus cinereus (C. fimetarius)* (McLaughlin, 1973) possess a thick wall composed of three layers. In *Psilocybe* sp. the outer layer (ectospore) is a fibrous layer and is continuous around the entire spore. It varies in texture from a thin, hairlike layer to a covering of long loose fibers. The middle layer (epispore) is thick and electron-dense. The thickness of these two layers is constant except for the germ pore area, where the covering is very thin. The inner layer (endospore) is more electron-transparent than the middle layer (Stocks & Hess, 1970). In *P. quebecensis* the basidiospore wall consists of a thin ectospore, a thick electron-dense epispore, and an electron-transparent endospore. The ectospore appears to be continuous with the sterigma wall (Oláh, 1973). The walls of young basidiospores of *C. cinereus* consist of two layers, a dense inner one and a thin, easily lost outer one (McLaughlin, 1973), but in the spore walls of *C. atramentarius* and *C. narcoticus* five layers are distinguished (Meléndez-Howell, 1966). Griffiths (1971) described a two-layered wall in basidiospores of *Panaeolus campanulatus* a faintly fibrillar outer electron-dense layer in and an inner less electron-dense layer, composed of loosely arranged fibrils. At the germ pore, the outer layer is reduced to a thin covering and the inner cell wall is absent. In this region numerous melanin granules are observed embedded in a faintly granular matrix exterior to the plasma membrane. At the apicular region the outer wall is incomplete and projects outwards, the channel is filled with fibrillar material, apparently an extension of the inner cell wall (Griffiths, 1971).

Sporidia (basidiospores) of many rust fungi may have a thin single-layered wall. The cell walls of the sporidia of *Puccinia horiana* and *Kuehneola japonica* appear to have a single fibrous layer when fixed with $KMnO_4$. The cell walls of sporidia of both fungi are exceedingly difficult to stain, when Gu-OsO_4 fixation is employed. The cell walls, in this case, appear to consist of a homogenous electron-transparent single layer (Plate 4, Part 4). Sporidia of *Gymnosporangium haraeanum* also have an electron-transparent, single-layered wall similar to the walls of *P. horiana* and *K. japonica* (Plate 4, Parts 2, 3,). Upon germination, sporidial walls extend over the germ tube similar to the process observed in other basidiospores (Plate 4, Part 5) (Kohno et al., 1974).

STRUCTURAL CHANGES IN SPORE WALLS DURING GERMINATION

The germination of fungal spores involves the development from the dormant state of spores to some other phase in the cycle. Bartnicki-Garcia (1968) suggested three basically different mechanisms of vegetative wall formation during spore germination. Each type is characteristic of certain groups of fungi. The first type is a de novo formation of a cell wall on a naked protoplast, the second type is a de novo formation of a vegetative wall under the spore wall. The third type is the extension of the spore wall, or one of its innermost layers of wall. The latter two types are very similar to the mode of blastospore-type sporulation.

The first type of germination is observed in the encystment of zoospores of aquatic Mastigomycotina. As is evident in micrographs, (Plate 1, Part 1), the cell wall is lacking in zoospores of *P. macrospora*. However, when zoospores encyst, a mucilaginous substance deposits on the surface of the plasma membrane to form thin mucilaginous walls around the mature cystospores (Plate 1, Part 2). The mucilaginous walls of cystospores react postively to a cellulose reaction and upon germination the walls elongate and cover germ tubes (Fukutomi & Akai, 1966).

The second type of germination was first observed by Hawker & Abbott (1963) in *Rhizopus* sporangiospores. In germinating sporangiospores of *Mucor rouxii*, spore walls rupture and new walls are formed under the old spore walls which bulge out (Bartnicki-Garcia, 1969; Bartnicki-Garcia, et al., 1968). As stated above, immature sporangiospores of *Rhizopus stolonifer (R. nigricans), R. sexualis* and *R. arrhizus* are relatively thin-walled, but mature spores have a thick wall of reticulate structure. The first visible change in germination is the formation of an inner cell wall under the original spore wall. The original spore wall becomes considerably stretched and finally ruptures to allow the emergence of germ tube enveloped by a newly formed elastic inner wall (Baker & Smith, 1970; Buckley et al., 1968; Ekundayo, 1965, 1966; Hawker & Abbott, 1963; Matsumoto et al., 1969). When germinated spores of *R. stolonifer* were held for 35 days at OC, the most striking feature was the increase in thickness of the inner wall layers of spores. Evidently, wall thickening was completed during the first 35 days of cold exposure, and the cytological changes showed that metabolic activity associated with wall synthesis continued for many days after the loss of colony-forming potential (Matsumoto et al., 1969). Bussel et al. (1969) also observed that the synthesis of the inner wall of sporangiospores of *R. stolonifer* progressed under anoxia; spore metabolic activity and synthesis of the inner wall layer continued in the absence of oxygen. *Cunninghamella elegans* lacks Mucor-type, large multispored sporangia and sporangioles, and bears only conidia, single-spored sporangioles. These conidia have a two-layered wall from an early stage of development (Hawker, 1966; Hawker et al., 1970). The conidia possess a thin outer electron-dense layer corresponding to the sporangial wall and a thick, less electron-dense, inner wall corresponding to the sporangiospore wall. After germination both layers of the original spore wall are ruptured and the emerging germ tubes are enveloped in a newly formed extensible inner wall. Spores of a Mucoraceous fungus, *Gilbertella persicaria* also form a new wall during germination (Hawker, 1966).

With direct germination of sporangia of *P. parasitica* and *P. infestans*, a new wall layer deposits inside the existing sporangial wall, and this new germination wall extends to form the hyphal wall of the germ tube (Elsner et al., 1970; Hemmes & Hohl, 1969; Hohl & Hemmes, 1968). The difference in these two fungi is that the newly formed inner wall completely surrounds the sporangial protoplast in *P. infestans* (Hemmes & Hohl, 1969).

De novo formation of new wall layers within the existing wall prior to germ tube emergence has also been found in fungi other than Phycomycetous fungi, e.g., in *Fusarium culmorum* (Marchant, 1966a) and in *Aspergillus nidulans* (Border & Trinci, 1970). In germinating conidia of *A. nidulans*, a new wall layer is clearly distinguishable in conidia fixed by Kellenberger's method (Kellenberger et al., 1958), but markedly less in Gu-fixed conidia and often not at all in $KMnO_4$-fixed conidia (Border & Trinci, 1970). The walls of dormant conidia of *A. nidulans* consist of three layers. The outer layer may represent a cuticle. The inner layer is extensible, while the outer and middle layers are relatively inextensible and rupture during spore swelling, and sometimes slough off. The new layers (N_1 and N_2) are formed under the inner layer during germination. As germination proceeds, the inner layer decreases in thickness, while the newly formed N_2-layer increases in thickness. At the point of germ tube emergence the conidial wall bulges out and the middle layers become disorganized and eventually disappear. The inner layer becomes progressively thinner until it disappears from the apex of the germ tube initial. The N_2-layer alone forms the germ tube wall (Border & Trinci, 1970).

The third type of germination is found most commonly throughout the higher fungi, e.g., in conidia of *Botrytis cinerea* (Hawker & Hendy, 1963: Shiraishi, 1971), *Aspergillus oryzae* and *A. niger* (Tanaka & Yanagita, 1963a; Tanaka, 1966), *Colletotrichum lagenarium* (Akai & Ishida, 1967, 1968), *Penicillum frequentans* (Hawker, 1966) and *Alternaria kikuchiana* (Ishizaki et al., 1974) (Plate 3, Parts 3, 4), ascospores of *Saccharomyces cerevisiae* (Hashimoto et al., 1958), *Schizosaccharomyces octosporus* (Conti & Naylor, 1960) and *Neurospora tetrasperma* (Lowry & Sussman, 1968), and uredospores of *Melampsora* (Manocha & Shaw, 1967).

In the pregermination stage of oospores of *P. infestans* the endospore wall is absorbed and decreases in thickness until only the thin exospore wall remains (Romero & Erwin, 1969). In oospores of *Pythium aphanidermatum* absorption of the endospore wall was also observed (Stanghellini & Russell, 1973).

At the initial stage of germination of *A. kikuchiana*, the outer pigmented layer appears to break down and allow the passage of the germ tube (Plate 3, Parts 3, 4). The two electron-transparent regions of the inner layer coalesce at the point of germ tube emergence, and extend to become the wall of the germ tube. Sometimes, however, the innermost layer alone appears to extend over the germ tube (Plate 3, Part 4). Scanty mucilaginous substances are recognized around the germ tube (Plate 3, Part 3) (Ishizaki et al., 1974).

Conidia of *Penicillium megasporum* have walls which consist of several layers (Remsen et al., 1967). During germination, there is no essential change in the structure of the wall layers. The first signs of germination reveal an increase in thickness of the innermost layer. This area of increased thickness determines the point of germ tube emergence. This increase in thickness of the layer results in a splitting of all other wall layers. The germ tube expands through this split in the wall of the spores. The wall of the newly formed germ tube is continuous with the innermost layer of the spore. The walls of younger germ tubes are relatively electron-dense, but the walls at the tips of older germ tubes become increasingly electron-transparent and begin to deposit electron-dense material on the outer surface (Remsen et al., 1967). After germination of conidia of *Aspergillus fumigatus*, the inner wall layer gives rise to the germ tube wall by growing out through the remains of the outer wall (Campbell, 1971). In germinating ascospores of *N. tetrasperma* the walls of emerging germ tubes are continuous with the inner wall layers (endosporium) of spores and the continuity of the outermost layer (perisporium) and the fibrous layer of the germ tube wall are also seen (Lowry & Sussman, 1968).

Very few of the chlamydospores of *Thielaviopsis basicola*, even after being aged, germinated in sterile distilled water or in soil extract. The breakup of chlamydospores into individual cells or the rupturing of the outer wall of the chlamydospores appears to be necessary before germination takes place. The germ tubes always emerged at one end of the cylinderlike chlamydospore cell at the junction of the horizontal and vertical planes of the spore walls. This suggests, as stated already, a weak point around the rim at each end of the chlamydospore cell. As the germ tubes emerge, a lid or operculum is clearly visible (Patrick et al., 1965; Tsao & Tsao, 1970). Christias & Baker (1967) reported that the disrupture of chlamydospore chains could be induced by the action of the enzyme chitinase. According to them certain isolates of soil-borne Actinomycetes and bacteria that produce chitinase break chlamydospore chains of *T. basicola* into individual spores. Thus the cell becomes germinable.

The mature ascospores of *S. octosporus* are surrounded by thick walls with non-electron-dense structure. However, in the early stage of germination ascospores appear to be surrounded by two layers.

The outer layer appears to be quite electron-dense, and is apparently shed, whereas the inner layer is of lower electron density and appears to serve as the cell wall of the germ tube (Conti & Naylor, 1960). In *Psilocybe* sp. as germination of basidiospores takes place, the outer two layers of the cell wall appear to break down and the inner layer is stretched and appears to be partly dissolved. The germ tube develops into a small bubble outside of the spore. At this stage of germination a wall is readily distinguishable around the developing germ tube tip. The wall appears to be continuous with the inner layer of the old spore wall (Stocks & Hess, 1970).

CONCLUSION

In general, fungus spores may have one-, two-, or three-layered walls, but sometimes have a multilayered wall. In many cases, most of these layers, at least one of them, have a fibrous appearance. Walls of some fungi react positively to celluloselike reactions and walls of others are chitinous. The visualization of ultrastructural components in cell walls by electron microscopy may be attributable in part to the development of suitable fixatives and fixation methods.

Chemical composition of cell walls has been known to correlate with accepted taxonomic groups to some extent, e.g., some of the aquatic Phycomycetous fungi are able to be distinguished by the presence of cellulose in the cell walls, but some Ascomycetous fungi are characterized by chitin in cell walls intead of cellulose. However, as a whole, little is known about the chemical composition of spore cell walls.

Several authors have suggested that lomasomes are associated with active wall synthesis and deposition of wall materials. Lomasomes in fungi were first named by Moore & McAlear (1961). They were described as aggregates of vesicles which always appeared against the cell wall. Grove et al. (1970) presented evidence for a correlation between the biosynthetic activities and the ER—golgi apparatus—vesicle complex found in *Pythium ultimum*. Campbell (1969) reported stacks of ER membranes which resemble a simple golgi body in cells of *Alternaria brassicicola*. Cole (1973) observed that the layers of ER lay near the growing apex of a conidiogenous cell of *Drechslera sorokiniana*, and in basidiospores of *Coprinus cinerea* McLaughlin (1973) found some ER which paralleled the cell wall near the plasma membrane. The dynamics of the endoplasmic reticulum system have led several workers to speculate that ER might play a significant role in wall formation.

ACKNOWLEDGMENTS

We wish to express our sincere thanks to Professors W. M. Hess and D. J. Weber for kindly presenting the manuscript, and Mr. M. Kohno, Mie University, for his technical assistance. We are indebted to the members of the Laboratory of Plant Pathology, Kyoto University, for their valuable suggestions and to Miss Kyoko Kawamura for her kind assistance in the preparation of this manuscript.

REFERENCES

Akai, S., Fukutomi, M., & Kobayashi, N. (1969). Fine structure of conidia of *Fusarium oxysporum* f. *cucumerinum*. Ann. Phytopath. Soc. Japan 35, 351--353.

Akai, S., Fukutomi, M., & Kunoh, H. (1966). An observation on fine structure of conidia of *Sphaerotheca pannosa* (Wallr.) Lév. attacking leaves of roses. Mycopathol. Mycol. Appl. 29, 211--216.

Akai, S., Fukutomi, M., & Kunoh, H. (1968). An electron microscopic observation of conidium and hypha of *Erysiphe graminis hordei*. Mycopathol. Mycol. Appl. 35, 217--222.

Akai, S., Fukutomi, M., & Shiraishi, M. (1969). Application of a scanning electron microscope to the studies in phytopathological and mycological fields. JEOL News (Japan Electron Optics Laboratory Co. Ltd.) 7B, 22--25.

Akai, S. & Ishida, N. (1967). Electron microscopic observations of conidia germination and appressorium formation in *Colletotrichum lagenarium*. Shokubutsu Byogai Kenkyu (Forsch. Gebiet Pflanzenkrankh.), Kyoto 7, 71--72.

Akai, S. & Ishida, N. (1968). An electron microscopic observation on the germination of conidia of *Colletotrichum lagenarium*. Mycopathol. Mycol. Appl. 34, 337--345.

Akai, S. & Shiraishi, M. (1974). Electron microscopic observations on the nuclear division and resting spore formation in *Plasmodiophora brassicae* Woronin. Unpublished paper, Kyoto University.

Alloway, J. M. & Wilson, I. M. (1972). Fine structure of spore degeneration in *Podospora arizonensis*. Trans. Brit. Mycol. Soc. 58, 231--236.

Amici, A. & Locci, R. (1968). Possible phytopathological implications of the behaviour of *Helminthosporium carbonum* in presence of α-solanine. Riv. Pat. Veg. Ser. IV, 4, 51--62.

Aragaki, M. (1966). Deposition and function of callose in the Peronosporales. Phytopathology 56, 145 (Abstr.).

Arthaud, J. (1972). Les rouilles du pin maritime. Examen des spores au microscope electronique à balayage. Note preliminaire. Europ. J. For. Pathol. 2, 81--87.

Baker, J. E. & Smith, W. L., Jr. (1970). Heat-induced ultrastructural changes in germinating spores of *Rhizopus stolonifer* and *Monilinia fructicola*. Phytopathology 60, 869--874.

Ballesta, J. P. G. & Villanueva, J. R. (1971). Cell wall components of various species of yeasts. Trans. Brit. Mycol. Soc. 56, 403--410.

Bancroft, J. D. (1967). An Introduction to Histochemical Technique London: Butterworths.

Bandoni, R. J. & Bisalputra, A. A. (1971). Budding and fine structure of *Tremella mesenterica* haplonts. Can. J. Bot. 49, 27--30.

Bartnicki-Garcia, S. (1968). Cell wall chemistry, morphogenesis, and taxonomy of fungi. Annu. Rev. Microbiol. 22, 87--108.

Bartnicki-Garcia, S. (1969). Cell wall differentiation in the Phycomycetes. Phytopathology 59, 1065--1071.

Bartnicki-Garcia, S., Nelson, N., & Cota-Robles, E. (1968). Electron microscopy of spore germination and cell wall formation in *Mucor rouxii*. Arch. Microbiol. 63, 242--255.

Bartnicki-Garcia, S. & Nickerson, W. J. (1962). Isolation, composition and structure of cell walls of filamentous and yeast-like forms of *Mucor rouxii*. Biochim. Biophys. Acta 58, 102--119.

Bartnicki-Garcia, S. & Reyes, E. (1964). Chemistry of spore wall differentiation in *Mucor rouxii*. Arch. Biochem. Biophys. 108, 125--133.

Bartnicki-Garcia, S. & Reyes, E. (1968a). Chemical composition of sporangiophore walls of *Mucor rouxii*. Biochim. Biophys. Acta 165, 32--42.

Bartnicki-Garcia, S. & Reyes, E. (1968b). Polyuronides in the cell walls of *Mucor rouxii*. Biochim. Biophys. Acta 170, 54--62.

Basu, P. K. (1971). Existence of chlamydospores of *Alternaria porri* f. sp. *solani* as overwintering propagules in soil. Phytopathology 61, 1347--1350.

Boosalis, M. G., Sumner, D. R., & Rao, A. S. (1967). Overwintering of conidia of *Helminthosporium turcicum* on corn residue and in soil in Nebraska. Phytopathology 57, 990--996.

Border, D. J. & Trinci, A.P.J.(1970). Fine structure of the germination of *Aspergillus nidulans* conidia. Trans. Brit. Mycol. Soc. 54, 143--152.

Bracker, C. E. (1966). Ultrastructural aspects of sporangiospore formation in *Gilbertella persicaria*. In The Fungus Spore (ed. M. F. Madelin), pp. 39--60. London: Butterworths.

Bracker, C. E. (1967). Ultrastructure of fungi. Annu. Rev. Phytopathol. 5, 343--374.

Bracker, C. E. (1968). The ultrastructure and development of sporangia in *Gilbertella persicaria*. Mycologia 60, 1016--1067.

Brewer, J. G. & Boerema, G. H. (1965). Electron microscope observations on the development of pycnidiospores in *Phoma* and *Ascochyta* spp. Proc. Koninkl. Nederl. Akad. Wetenschappen, Ser. C, 68, 86--97.

Buckley, P. M., Sjaholm, V. E., & Sommer, N. F. (1966). Electron microscopy of *Botrytis cinerea* conidia. J. Bacteriol. 91, 2037--2044.

Buckley, P. M., Sommer, N. F., & Matsumoto, T. T. (1968). Ultrastructural details in germinating sporangiospores of *Rhizopus stolonifer* and *Rhizopus arrhizus*. J. Bacteriol. 95, 2365--2373.

Buckley, P. M., Wyllie, T. D., & DeVay, J. E. (1969). Fine structure of conidia and conidium formation in *Verticillium albo-atrum* and *V. nigrescens*. Mycologia 61, 240--250.

Bussel, J., Buckley, P. M., Sommer, N. F., & Kosuge, T. (1969). Ultrastructural changes in *Rhizopus stolonifer* sporangiospores in response to anaerobiosis. J. Bacteriol. 98, 774--783.

Campbell, R. (1968). An electron microscope study of spore structure and development in *Alternaria brassicicola*. J. Gen. Microbiol. 54, 381--392.

Campbell, R. (1969). Further electron microscope studies of the conidium of *Alternaria brassicicola*. Arch. Mikrobiol. 69, 60--68.

Campbell, R. (1970a). An electron microscope study of exogenously dormant spores, spore germination, hyphae and conidiophores of *Alternaria brassicicola*. New Phytol. 69, 287--293.

Campbell, R. (1970b). Ultrastructure of an albino strain of *Alternaria brassicicola*. Trans. Brit. Mycol. Soc. 54, 309--313.

Campbell, C. K. (1971). Fine structure and physiology of conidial germination in *Aspergillus fumigatus*. Trans. Brit. Mycol. Soc.

57, 393--402.

Carbonell, L. M. (1967). Cell wall changes during the budding process of *Paracoccidioides brasiliensis* and *Blastomyces dermatitidis*. J. Bacteriol. 94, 213--223.

Carmichael, J. W. (1971). Blastospores, aleuriospores, chlamydospores. In Taxonomy of Fungi Imperfecti (ed. B. Kendrick), pp. 50--70. Toronto and Buffalo: University of Toronto Press.

Chapman, J. A. & Vujičić, R. (1965). The fine structure of sporangia of *Phytophthora erythroseptica* Pethyb. J. Gen. Microbiol. 41, 275--282.

Christias, C. & Baker, K. F. (1967). Chitinase as a factor in the germination of chlamydospores of *Thielaviopsis basicola*. Phytopathology 57, 1363--1367.

Christias, C. & Baker, K. F. (1969). Development and macro- and micro-structure of chlamydospores of *Thielaviopsis basicola*. Phytopathology 59, 1021 (Abstr.).

Chung, K. L., Hawirko, R. Z., & Isaac, P. K. (1965). Cell replication in *Sacchromyces cerevisiae*. Can. J. Microbiol. 11, 953--957.

Cole, G. T. (1973). Ultrastructure of conidiogenesis in *Drechslera sorokiniana*. Can. J. Bot. 51, 629--638.

Cole, G. T. & Aldrich, H. C. (1971). Ultrastructure of conidiogenesis in *Scopulariopsis brevicaulis*. Can. J. Bot. 49, 745--755.

Conti, S. F. & Naylor, H. B. (1960). Electron microscopy of ultrathin sections of *Schizosaccharomyces octosporus*. III. Ascosporogenesis, ascospore structure and germination. J. Bacteriol. 79, 417--425.

Djaczenko, W. & Cassone, A. (1971). Visualization of new ultrastructural components in the cell wall of *Candida albicans* with fixatives containing TAPO. J. Cell Biol. 52, 186--190.

Durrell, L. W. (1964). The composition and structure of walls of dark fungus spores. Mycopath. Mycol. Appl. 23, 339--345.

Durrell, L. W. (1968). Studies of *Aureobasidium pullulans* (DeBary) Arnaud. Mycopath. Mycol. Appl. 35, 113--120.

Egawa, H., Tsuda, M., Ueyama, A., & Matuo, T. (1968). Formation of abnormal mycelium of *Fusarium roseum* Link. on a modified Czapek-D-amino acid medium. Experimentia 24, 403--404.

Ehrlich, M. A. & Ehrlich, H. G. (1969). Uredospore development in *Puccinia graminis*. Can. J. Bot. 47, 2061--2064.

Ekundayo, J. A. (1965). Studies on germination of fungus spores, with special reference to sporangiospores of *Rhizopus arrhizus*.

Ph. D. Thesis, Univ. of Ibadan, Nigeria, 123 pp. Cited in Hawker, 1966.

Ekundayo, J. A. (1966). Further studies on germination of sporangiospores of *Rhizopus arrhizus*. J. Gen. Microbiol. 42, 283--291.

Ellis, M. B. (1971). Porospores. In Taxonomy of Fungi Imperfecti, (ed. B. Kendrick), pp. 71--74. Toronto and Buffalo: University of Toronto Press.

Elsner, P. R., VanderMolen, G. E., Horton, J. C., & Bowen, C. C. (1970). Fine structure of *Phytophthora infestans* during sporangial differentiation and germination. Phytopathology 60, 1765--1772.

Fanny, E. & Carrol, G. C. (1971). Fine structural studies on "Poroconidium" formation in *Stemphylium botryosum*. In Taxonomy of Fungi Imperfecti, (ed. B. Kendrick), pp. 75--91. Toronto and Buffalo: University of Toronto Press.

Fletcher, J. (1971). Conidium ontogeny in *Penicillium*. J. Gen. Microbiol. 67, 207--214.

French, E. R. & Nielsen, L. W. (1966). Production of macroconidia of *Fusarium oxysporium* f. *batatas* and their conversion to chlamydospores. Phytopathology 56, 1322--1323.

Fukutomi, M. (1969). Studies on the downy of graminae plants caused by *Phytophthora macrospora* (Sacc.) S. Ito et I. Tanaka. Doctoral dissertation, Kyoto University.

Fukutomi, M. (1974). Scanning electron microscope observation of the sporulation and germination in some fungi. 8th Intern. Congress on Electron Microscopy, Canberra, Vol. II. Biological, pp. 562--563 (Australian Academy of Science).

Fukutomi, M. & Akai, S. (1966). Fine structure of the zoospores, cystospores and germ tubes of *Sclerophthora macrospora*. Trans. Mycol. Soc. Jap. 7, 199--202.

Fukutomi, M., Akai, S., & Hirata, K. (1969a). Scanning electron microscopy for the observation of the surface view of plants and fungi. Kagaku to Seibutsu (Japan) 7, 92--96.

Fukutomi, M., Akai, S., & Shiraishi, M. (1969b). Ultrastructure of antheridium and oogonium of the downy mildew fungus of rice plants. Shokubutsu Byogai Kenkyu, Kyoto, 7, 79--80.

Fukutomi, M., Akai, S., & Shiraishi, M. (1971). Fine structure of antheridia and oogonia of *Phytophthora macrospora*, the downy mildew fungus of rice plants. Mycopath. Mycol. Appl. 43, 249--258.

Fukutomi, M., Akai, S., & Shiraishi, M. (1974). Ultrastructure and composition of cell walls of sexual and asexual organs in *Phytophthora macrospora*. Unpublished paper, Kyoto University.

Fukutomi, M., Shiraishi, M., & Akai, S. (1968). Fine structure of oogonia and antheridia of the downy mildew fungus of rice plants. Ann. Phytopath. Soc. Japan 34, 177--178 (Abstr.).

Garcia Acha, I., Aguirre, M. J. R., Uruburu, F., & Villanueva, J. R. (1966). The fine structure of the *Fusarium culmorum* conidium Trans. Brit. Mycol. Soc. 49, 695--702.

Ghosh, B. K. (1971). Grooves in the plasmalemma of *Saccharomyces cerevisiae* seen in glancing sections of double aldehydefixed cells. J. Cell Biol. 48, 192--197.

Glick, A. D. & Kwon-Chung, K. J. (1973). Ultrastructural comparison of coils and ascospores of *Emmonsiella capsulata* and *Ajellomyces dermatitidis*. Mycologia 65, 216--220.

Graham, S. O. (1960). The morphology and a chemical analysis of the teliospore of the dwarf bunt fungus, *Tilletia contraversa*. Mycologia 52, 97--118.

Greenhalgh, G. N. (1967). A note on the conidial scar in the Xylariaceae. New Phytol. 66, 65--66.

Griffiths, D. A. (1971). The fine structure of basidiospores of *Panaeolus campanulatus* (L.) Fr. revealed by freeze-etching. Arch. Mikrobiol. 76, 74--82.

Grove, S. N., Bracker, C. E., & Morre, D. J. (1970). An ultrastructural basis for hyphal tip growth in *Pythium ultimum*. Am. J. Bot. 57, 245--266.

Gurr, E. (1960). Methods of analytical histology and histochemistry. Baltimore: Williams and Wilkins Co.

Hashimoto, T., Conti, S. F., & Naylor, H. B. (1958). Fine structure of microorganisms. III. Electron microscopy of resting and germinating ascospores of *Saccharomyces cerevisiae*. J. Bacteriol. 76, 406--416.

Hashioka, Y., Horino, O., & Kamei, T. (1966). Electron micrographs of *Trichoderma* and *Pachybasium*. Rept. Tottori Mycol. Inst. (Japan) 5, 18--24.

Hawker, L. E. (1966). Germination: Morphological and anatomical changes. In The Fungus Spore, (ed. E. F. Madelin), pp. 151--163. London: Butterworths.

Hawker, L. E. (1968). Wall ornamentation of ascospores of species of *Elaphomyces* as shown by the scanning electron microscope. Trans. Brit. Mycol. Soc. 51, 493--498.

Hawker, L. E. & Abbott, P. McV. (1963). An electron microscope study of maturation and germination of sporangiospores of two species of *Rhizopus*. J. Gen. Microbiol. 32, 295--298.

Hawker, L. E. & Hendy, R. J. (1963). An electron-microscope study of germination of conidia of *Botrytis cinerea*. J. Gen. Microbiol. 33, 43--46.

Hawker, L. E., Thomas, B., & Beckett, A. (1970). An electron microscope study of structure and germination of conidia *Cunninghamella elegans* Lendner. J. Gen. Microbiol. 60, 181--189.

Hemmes, D. E. & Hohl, H. R. (1969). Ultrastructural changes in direct germinating sporangia of *Phytophthora parasitica*. Am. J. Bot. 56, 300--313.

Henderson, D. M., Dudall, R., & Prentice, H. T. (1972). Morphology of the reticulate teliospore of *Puccinia chaerophylli*. Trans. Brit. Mycol. Soc. 59, 229--232.

Hess, W. M., Sassen, M. M. A., & Remsen, C. C. (1968). Surface characteristics of *Penicillium* conidia. Mycologia 60, 290--303.

Hess, W. M. & Stocks, D. L. (1969). Surface characteristics of *Aspergillus* conidia. Mycologia 61, 560--571.

Hess, W. M. & Weber, D. J. (1970). Ultrastructure of *Tilletia contraversa* teliospores as revealed by freeze-etching. Amer. J. Bot. 57, 745 (Abstr.).

Hiratsuka, Y. (1971). Spore surface morphology of pine stem rusts of Canada as observed under a scanning electron microscope. Can. J. Bot. 49, 371--372.

Hohl, H. R. & Hamamoto, S. T. (1967). Ultrastructural changes during zoospore formation in *Phytophthora parasitica*. Am. J. Bot. 54, 1131--1139.

Hohl, H. R. & Hemmes, D. E. (1968). Ultrastructure of cell wall formation during direct germination in *Phytophthora parasitica*. J. Cell Biol. 39, 168a (Abstr.).

Horino, O. & Akai, S. (1965). Comparison of the electron micrographs of conidia of *Helminthosporium oryzae* Breda de Haan and *Pyricularia oryzae* Cav. Trans. Mycol. Soc. Japan 6, 41--46.

Hossain, S. M. M. & Manners, J. G. (1964). Relationship between internal characters of the conidium and germination in *Erysiphe graminis*. Trans. Brit. Mycol. Soc. 47, 39--44.

Hughes, G. C. & Bisalputra, A. A. (1970). Ultrastructure of hypomycetes. Conidium ontogeny in *Peziza ostracoderma*. Can. J. Bot. 48, 361--366.

Ishida, N., Kunoh, H., Akai, S., Tsuda, M., Egawa, H., & Ueyama,

A. (1968). Fine structure of chlamydospore-like cell of *Helminthosporium oryzae* Breda de Haan produced on a modified Czapek-D-valine medium. Shokubutsu Byogai Kenkyu, Kyoto 7, 75--76.

Ishizaki, H., Mitsuoka, K., Kohno, M., & Kunoh, H. (1974). Effect of polyoxin on fungi (II). Electron microscopic observations of spore germ tube of *Alternaria kikuchiana* Tanaka. Ann. Phytopath. Soc. Jap., in press.

Jones, D. (1967). Examination of mycological specimens in the scanning electron microscope. Trans. Brit. Mycol. Soc. 50, 690--691.

Jones, D. (1968). An electron microscope study of the fine structure of *Acremoniella velata*. Trans. Brit. Mycol. Soc. 51, 515--518.

Jurand, M. K. & Kemp, R. F. (1972). Surface ultrastructure of oidia in the Basidiomycete *Psathyrella coprophila*. J. Gen. Microbiol. 72, 575--579.

Kellenberger, E., Ryter, A., & Séchaud, J. (1958). Electron microscope study of DNA-containing plasms. II. Vegetative and mature phage DNA as compared with normal bacterial nucleoid in different physiological states. J. Biophys. Biochem. Cytol. 4, 671--676. Cited in Ghosh, B. K. (1971).

King, J. E., Colhoun, J., & Butler, R. D. (1968). Changes in the ultrastructure of sporangia of *Phytophthora infestans* associated with indirect germination and ageing. Trans. Brit. Mycol. Soc. 51, 269--281.

Kohno, M., Nishimura, T., Kunoh, H., & Ishizaki, H. (1974). Ultrastructural observations on sporidia of *Puccinia horiana* and *Kuehneola japonica*. Unpublished paper, Mie University.

Kreger-Van Rij, N. J. W. & Veenhuis, M. (1969). A study of vegetative reproduction in *Endomycopsis platypodis* by electron microscopy. J. Gen. Microbiol. 58, 341--346.

Kreger-Van Rij, N. J. W. & Veenhuis, M. (1970). An electron microscope study of the yeast *Pityrosporum ovale*. Arch. Mikrobiol. 71, 123--131.

Kunoh, H. (1972). Morphological studies of host-parasite interaction in powdery mildew of barley, with special reference to affinity between host and parasite. Bull. Fac. Agr., Mie Univ., No. 44, 141--224.

Kunoh, H. & Akai, S. (1967). An electron microscopic observation of the surface structure of conidia of *Erysiphe graminis hordei*. Trans. Mycol. Soc. Japan 8, 77--78.

Kunoh, H., Ishida, N., Akai, S., Tsuda, M., Egawa, H., & Ueyama,

A. (1968). Fine structure of yeast-like cell of *Fusarium roseum* Link obtained from a modified Czapek-D-arginine medium. Shokubutsu Byogai Kenkyu, Kyoto 7, 77-78.

Kuo, M. J. & Alexander, M. (1967). Inhibition of lysis of fungi by melanins. J. Bacteriol. 94, 624--629.

Lessie, P. E. & Lovett, J. S. (1968). Ultrastructural changes during sporangium formation and zoospore differentiation in *Blastocladiella emersonii*. Am. J. Bot. 55, 220--236.

Littlefield, L. J. & Bracker, C. E. (1971). Ultrastructure and development of uredospore ornamentation in *Melampsora lini*. Can. J. Bot. 49, 2067--2073.

Lowry, R. J., Durkee, T. L., & Sussman, A. S. (1967). Ultrastructural studies of microconidium formation in *Neurospora crassa*. J. Bacteriol. 94, 1757--1763.

Lowry, R. J. & Sussman, A. S. (1968). Ultrastructural changes during germination of ascospores of *Neurospora tetrasperma*. J. Gen. Microbiol. 51, 403--409.

Lutley, M. & Wilson, I. M. (1972). Development and fine structure of ascospores in the marine fungus *Ceriosporopsis halima*. Trans. Brit. Mycol. Soc. 58, 393--402.

Luttrell, E.S. (1963). Taxonomic criteria in *Helminthosporium*. Mycologia 55, 643--674.

Madelin, M. F. (1966). The genesis of spores of higher fungi. In The Fungus Spore, (ed. M. F. Madelin), pp. 15--37. London: Butterworths.

Manocha, M. S. & Shaw, M. (1967). Electron microscopy of uredospores of *Melampsora lini* and of rust-infected flax. Can. J. Bot. 45, 1575--1582.

Marchant, R. (1966a). Fine structure and spore germination in *Fusarium culmorum*. Ann. Bot. 30, 441--445.

Marchant, R. (1966b). Wall structure and spore germination in *Fusarium culmorum*. Ann. Bot. 30, 821--830.

Marchant, R. (1968). An ultrastructural study of sexual reproduction in *Pythium ultimum*. New Phytol. 67, 167--171.

Marchant, R. & Smith, E. G. (1967). Wall structure and bud formation in *Rhodotorula glutinis*. Arch. Mikrobiol. 58, 248--256.

Marchant, R. & Smith, E. G. (1968). Bud formation in *Saccharomyces cerevisiae* and a comparison with the mechanism of cell division in other yeasts. J. Gen. Microbiol. 53, 163--169.

Marchant, R., Peat, A., & Banbury, G. H. (1967). The ultrastructural basis of hyphal growth. New Phytol. 66, 623--629.

Marquardt, H. (1962). Der Feinbau von Hefezellen im

Electronenmikroskop. I. Mitt. *Rhodotorula rubra.* Z. Naturforsch. 17b: 42--48.

Marquardt, H. (1963). Elektronoptischer Untersuchungen über die Ascosporenbildung bei *Saccharomyces cerevisiae* unter cytologischem und cytogenetischem Aspekt. Arch. Mikrobiol. 46, 308--320.

Masago, H. (1961). Effects of radiation on microorganisms, with special reference to the effect on *Helminthosporium oryzae.* Spec. Publ., Lab. Plant Pathol., Kyoto Univ. No. 14, 1--115.

Matsui, C., Nozu, M., Kikumoto, T., & Matsuura, M. (1962). Electron microscopy of conidial cell wall of *Cochliobolus miyabeanus.* Phytopathology 52, 717--718.

Matsumoto, T. T., Buckley, P. M., Sommer, N. F., & Shella, T. A. (1969). Chilling-induced ultrastructural changes in *Rhizopus stolonifer* sporangiospores. Phytopathology 59, 863--867.

McCully, E. K. & Bracker, C. E. (1972). Apical vesicles in growing bud cells of Heterobasidiomycetous yeasts. J. Bacteriol. 109, 922--926.

McKeen, W. E., Mitchell, N., Jarvie, W., & Smith, R. (1966). Electron microscopy studies of conidial walls of *Sphaerotheca macularis, Penicillium levitum,* and *Aspergillus niger.* Can. J. Microbiol. 12, 427--428.

McKeen, W. E., Mitchell, N., & Smith, R. (1967). The *Erysiphe cichoracearum* conidium. Can. J. Bot. 45, 1489--1496.

McLaughlin, D. J. (1973). Ultrastructure of sterigma growth and basidiospore formation in *Coprinus* and *Boletus.* Can. J. Bot. 51, 145--150.

McManus, J. F. A. & Mowry, R. W. (1964). Staining Methods, Histological and Histochemical. New York: Hoeber.

Meléndez-Howell, L. M. (1966). Ultrastructure du pore germinatif sporal dans le genre *Coprinus* Link. C. R. Acad. Sc. Paris, Ser. D, 263, 717--720.

Moore, R. T. (1963). Fine structure of mycota. I. Electron microscopy of the Discomycete *Ascodesmis.* Nova Hedwigia 5, 263--278.

Moore, R. T. & McAlear, J. H. (1961). Fine structure of mycota. 8. On the aecidial stage of *Uromyces caladii.* Phytopathol. Z. 42, 297--304.

Nash, S. M., Christou, T., & Snyder, W. C. (1961). Existence of *Fusarium solani* f. *phaseoli* as chlamydospores in soil. Phytopathology 51, 308--312.

Nozu, M. & Matsui, C. (1959). The fine structure of conidia of

Cochliobolus miyabeanus (Ito et Kurib.) Drechsler. Ann. Phytopath. Soc. Jap. 24, 135--138.

Nyvall, R. F. (1970). Chlamydospores of *Fusarium roseum* "Graminearum" as survival structures. Phytopathology 60, 1175--1177.

Oláh, G. M. (1973). The fine structure of *Psilocybe quebecensis*. Mycopath. Mycol. Appl. 49, 321--338.

Old, K. M. & Robertson, W. M. (1969). Examination of conidia of *Cochliobolus sativus* recovered from natural soil using transmission and scanning electron microscopy. Trans. Brit. Mycol. Soc. 53, 217--221.

Old, K. M. & Robertson, W. M. (1970). Effects of lytic enzymes and natural soil on the fine structure of conidia of *Cochliobolus sativus*. Trans. Brit. Mycol. Soc. 54, 343--350.

Oliver, P. T. P. (1972). Conidiophore and spore development in *Aspergillus nidulans*. J. Gen. Microbiol. 73, 45--54.

Patrick, Z. A., Toussoun, T. A., & Thorpe, H. J. (1965). Germination of chlamydospores of *Thielaviopsis basicola*. Phytopathology 55, 466--467.

Payak, M. M., Joshi, L. M., & Mehta, S. C. (1967). Electron Microscopy of urediospores of *Puccinia graminis* var. *tritici*. Phytopathol. Z. 60, 196--199.

Pearse, A. G. E. (1961). Histochemistry, Theoretical and Applied. 2nd Ed. London: J. and A. Churchill.

Plumb, R. T. & Turner, R. H. (1972). Scanning electron microscopy of *Erysiphe graminis*. Trans. Brit. Mycol. Soc. 59, 149--150.

Plurad, S. B. (1972). Fine structure of ascosporogenesis in *Nematospora coryli* Peglion, a pathogenic yeast. J. Bacteriol. 109, 927--929.

Pore, R. S. & Larsh, H. W. (1967). Aleuriospore formation in four related *Aspergillus* species. Mycologia 59, 318--325.

Pore, R. S., Pyle, C., Larsh, H. W., & Skvarla, J. J. (1969). *Aspergillus carneus* aleuriospore cell wall ultrastructure. Mycologia 61, 418--422.

Reeves, F., Jr. (1967). The fine structure of ascospore formation in *Pyronema domesticum*. Mycologia 59, 1018--1033.

Remsen, C. C., Hess, W. M., & Sassen, M. M. A. (1967). Fine structure of germinating *Penicillium megasporum* conidia. Protoplasma 64, 439--451.

Robb, J. (1972). Ultrastructure of *Ustilago boudei*. I. Pregermination development of hydrating teliospores. Can. J. Bot. 50,

1253--1261.

Rogers, J. D. (1969). *Hypoxylon rubiginosum*: cytology of the ascus and surface morphology of the ascospore. Mycopath. Mycol. Appl. 38, 215--223.

Romero, S. & Erwin, D. C. (1969). Variation in pathogenicity among single oospore culture of *Phytophthora infestans*. Phytopathology 59, 1310--1317.

Schwinn, F. J. (1969). Die Darstellung von Pilzsporen im Raster-Elektronenmikroskop. Phytopathol. Z. 64, 376--379.

Sentaudreu, R. & Northcote, D. H. (1969). The formation of buds in yeasts. J. Gen. Microbiol. 55, 393--398.

Setliff, E. C., MacDonald, W. L., & Patton, R. F. (1972). Fine structure of percurrent basidial proliferations in *Poria latemarginata*. Can. J. Bot. 50, 1697--1699.

Shannon, J. L. & Rothman, A. H. (1971). Transverse septum formation in budding cells of the yeastlike fungus *Candida albicans*. J. Bacteriol. 106, 1026--1028.

Shiraishi, M. (1971). Electron microscopical studies on the infection of leaves of host plants by *Botrytis cinerea* Pers. Doctoral Dissertation, Kyoto University.

Skucas, G. P. (1967). Structure and composition of the resistant Structure and composition of the resistant sporangial wall in the fungus *Allomyces*. Am. J. Bot. 54, 1152--1158.

Stanbridge, B. & Gay, J. L. (1969). An electron microscope examination of the surfaces of the uredospores of four races of *Puccinia striiformis*. Trans. Brit. Mycol. Soc. 53, 149--153.

Stanbridge, B., Gay, J. L., & Wood, R. K. S. (1971). Gross and structural changes in *Erysiphe graminis* and barley before and during infection. In Ecology of Leaf Surface Microorganisms (ed. T. F. Preece & C. H. Dickinson), pp. 367--379. London and New York: Academic Press.

Stanghellini, M. E. & Russell, J. D. (1973). Germination *in vitro* of *Pythium aphanidermatum* oospores. Phytopathology 63, 133--137.

Staples, R. C. & Wynn, W. K. (1965). The physiology of uredospores of the rust fungi. Bot. Rev. 31, 537--564.

Streiblova, E. & Beran, K. (1963). Types of multiplication scars in yeasts, demonstrated by fluorescence microscopy. Folia Microbiol. 8, 221--227. Cited in Bandoni, R. J. & Bisalputra, A. A. (1971).

Stocks, D. L. & Hess, W. M. (1970). Ultrastructure of dormant and germinated basidiospores of a species of *Psilocybe*. Mycologia

62, 176--191.

Subramanian, C. V. (1965). Spore types in the classification of the Hyphomycetes. Mycopath. Mycol. Appl. 26, 373--384.

Subramanian, C. V. (1971). The phialide. In Taxonomy of Fungi Imperfecti, (ed. B. Kendrick), pp. 92--119. Toronto and Buffalo: University of Toronto Press.

Sussman, A. S. & Halvorson, H. O. (1966). Spores, Their Dormancy and Germination, pp. 7--44. New York and London: Harper and Row.

Talbot, P. H. B. (1971). Principles of Fungal Taxonomy. London: MacMillan.

Tanaka, K. & Yanagita, T. (1963a). Electron microscopy on ultrathin sections of *Aspergillus niger*. I. Fine structure of hyphal cells. J. Gen. Appl. Microbiol. Tokyo 9, 101--118.

Tanaka, K. & Yanagita, T. (1963b). Electron microscopy of ultrathin sections of *Aspergillus niger*. II. Fine structure of conidia-bearing apparatus. J. Gen. Appl. Microbiol., Tokyo 9, 189--203.

Tanaka, K. (1966). Change in ultrastructure of *Aspergillus oryzae* conidia during germination. J. Gen. Appl. Microbiol., Tokyo 12, 239--246.

Temmink, J. H. M. & Campbell, R. N. (1968). The ultrastructure of *Olpidium brassicae*. I. Formation of sporangia. Can. J. Bot. 46, 951--956.

Thomas, P. L. & Isaac, P. K. (1967). The development of echinulation in uredospores of wheat stem rust. Can. J. Bot. 45, 287--289.

Tsao, P. W. & Tsao, P. H. (1970). Electron microscopic observations on the spore wall and "operculum" formation in chlamydospores of *Thielaviopsis basicola*. Phytopathology 60, 613--616.

Tubaki, K. (1963). Taxonomic study of Hyphomycetes. Annu. Rep. Inst. Ferment., Osaka 1, 25--54.

Van Dyke, C. G. & Hooker, A. L. (1969). Ultrastructure of host and parasite in interactions of *Zea mays* with *Puccinia sorghi*. Phytopathology 59, 1934--1946.

Vuillemin, P. (1910). Les conidiophores. Bull. Soc. Sci. Nancy, Ser. III, XI, 2, 129--172.

Vujičić, R. (1971). An ultrastructural study of sexual reproduction in *Phytophthora palmivora*. Trans. Brit. Mycol. Soc. 57, 525--530.

Voelz, H. & Niederpruem, D. J. (1964). Fine structure of basidiospores of *Schizophyllum commune*. J. Bacteriol. 88, 1497--1502.

Walkinshaw, C. H., Hyde, J. M., & VanZandt, J. (1967). Fine

structure of quiescent and germinating aeciospores of *Cronartium fusiforme*. J. Bacteriol. 94, 245--254.

Wang, M. C. & Bartnicki-Garcia, S. (1970). Structure and composition of walls of the yeast form of *Verticillium albo-atrum*. J. Gen. Microbiol. 64, 41--54.

Wells, K. (1965). Ultrastructural features of developing and mature basidia and basidiospores of *Schizophyllum commune*. Mycologia 57, 236--261.

Williams, P. G. & Ledingham, G. A. (1964). Fine structure of wheat stem rust uredospores. Can. J. Bot. 42, 1503--1508.

Williams, P. H. & McNabola, S. S. (1967). Fine structure of *Plasmodiophora brassicae* in sporogenesis. Can. J. Bot. 45, 1665--1669.

Young, T. W. K. (1968). Electron microscopic study of asexual spores in Kichxellaceae. New Phytol. 67, 823--836.

Young, T. W. K. (1970). Ultrastructure of the spore wall of *Linderina*. Trans. Brit. Mycol. Soc. 54, 15--25.

Young, T. W. K. (1973). Ultrastructure of the sporangiospore of *Coemansia reversa* (Mucorales). Trans. Brit. Mycol. Soc. 60, 57--63.

Zentmyer, G. A. & Erwin, D. C. (1970). Development and reproduction of Phytophthora. Phytopathology 60, 1120--1127.

DISCUSSION

Fine Structure of the Spore Wall and Germ Tube Change During Germination

Chairman: T. Hashimoto
Loyola University
Maywood, Illinois

Dr. Bartnicki-Garcia commented that *N*-acetyl glucosamine content of both oospore and oogonium walls of *Phytophthora megasperma* was only 0.43% and expressed a doubt as to the occurrence of this type of biochemical differentiation (cellulose to chitin) during oospore formation in fungi. He suggested that the authors' conclusion as drawn from the cytochemical test be confirmed by the direct biochemical analysis.

Dr. Kunoh agreed that further chemical analysis of the walls is essential for substantiating his conclusion.

In reply to a question as to the purity of chitinase used in their experiments, Dr. Kunoh stated that it was a crude preparation and that the contamination of other hydrolytic enzymes was possible.

C

Form and Function of Fungal Spores

CHAPTER **10**

RESISTANT STRUCTURES IN THE MYXOMYCETES

H. C. Aldrich and Meredith Blackwell
University of Florida

INTRODUCTION

There may be some hesitation among mycologists to embrace Myxomycetes as true fungi, and we have no definitive answer to the question of their true relationships. Olive (1970, 1975) maintains that their affinities lie among the Protozoa and presents persuasive arguments to support this view. He places the organisms usually referred to as protostelids, cellular slime molds, and plasmodial slime molds in a single class Mycetozoa, with subclasses Protostelia, Dictyostelia, Acrasia, and Myxogastria. Martin & Alexopoulos (1969) and Alexopoulos (1973) continue Martin's long-standing assertion that the acellular or plasmodial slime molds are fungi and treat them as the Class Myxomycetes within the subdivision Myxomycotina, all the other fungi being treated in the subdivision Eumycotina. They note the protostelids described by Olive and suggest that this group at present not be included in the class Myxomycetes. We are inclined to believe that at least the flagellated genera of Protostelids are related to the Myxomycetes; perhaps the others may be, too, and hopefully continued work on these organisms will clarify the relationships. Since we believe that some direct relationships do hold, and so that this chapter together with Professor Hohl's chapter will constitute a complete survey of resistant structures in the "slime molds," we will include material on Olive's subclasses Protostelia and Myxogastria. Using the

Martin & Alexopoulos system, this would then include the organisms in the class Myxomycetes and the protostelids.

Recent reviews (Alexopoulos, 1962; Gray & Alexopoulos, 1968; Olive, 1975) have included sections on myxomycete spores, microcysts, and sclerotia. In most cases, we will restrict our coverage to papers published since 1967.

Three kinds of resistant structures are formed by Myxomycetes and protostelids to survive unfavorable conditions: plasmodia form sclerotia, myxamoebae form microcysts, and plasmodia and protostelid myxamoebae can also undergo sporulation to produce haploid spores. This chapter will review the conditions inducing sclerotization and encystment and the structure, chemistry, and germination of microcysts, sclerotia, and spores. We will restrict discussion of sporulation to events occurring after spore cleavage; that process has been agreed to be beyond the score of this symposium.

METHODS OF STUDY

The few rigorous analyses of the chemical nature of spore and sclerotial walls have employed standard biochemical techniques. McCormick et al. (1970b) broke cells in a French press, purified the wall materials by repeated washings in sodium chloride and sodium lauryl sulfate solutions, and performed amino acid analyses, melanin extractions, and thin layer chromatography of acid hydrolyzed material to determine percentage composition of the walls of *Physarum polycephalum*. Ideally, such analyses should be performed on axenically grown material to eliminate contaminating bacterial polysaccharides. This may limit future progress in this area to the few species that can be cultured under bacteria-free conditions, but we hope that other studies will follow, particularly on distantly related genera.

The various electron microscope techniques have proven particularly useful in studying morphology of these structures. The scanning electron microscope (SEM) allows visualization of the surface morphology of structures such as spores (Carr, 1971). Most presently available instruments can produce resolution in the 100--200 Å range, but whether such resolution can be achieved on a day-to-day basis for the examination of biological material remains to be seen. Myxomycete spores can be shed onto double stick tape, coated with carbon and then with gold-palladium in a rotational shadowing unit, and viewed in the SEM. This usually results in spores with a collapsed appearance although Schoknecht & Small (1972) believe that dry spores in nature have this appearance anyway, and that the image thus obtained is a faithful one. They have a point, of course, because most spores of dry specimens when shaken out onto a slide and viewed with no mounting medium do appear foot ball-shaped or caved in on one side. However, the usual way of observing spores when identifying Myxomycetes is to mount them in water, whereupon they spring out to their original globular shape. Furthermore, fresh specimens from the field are frequently moist enough that the spores are globular. On balance, then, we prefer to pretreat spores so that the image obtained in the SEM is globular and directly comparable to the light microscope image. To achieve this condition, several procedures can be used. The spores are first suspended in water until they round up, usually a few minutes. They are then fixed overnight in a mixture of equal parts of 2% OsO_4 and saturated mercuric chloride solution. This mixture hardens the spores and prevents collapse during final drying (Boyde & Wood, 1969). The sample is then dehydrated through increasing concentrations of ethanol, transferred to acetone or ether, and air dried. An alternative method is critical point drying (Anderson, 1951). This works exceptionally well for many soft biological samples (e.g., Pickett-Heaps, 1973). However, we find that the hardening procedure works equally well for many kinds of material, and requires no special equipment or expertise.

Because the small size of myxomycete spores taxes the resolution of the SEM, it is worthwhile to mention another method of visualization of spore surface morphology. The single-stage surface replication technique has been elegantly applied to basidiomycete spores by Bigelow & Rowley (1968). This procedure involves coating spores with platinum and carbon, and then removing the spores with chromic acid, rendering the replica transparent to the electron beam. The replica can then be viewed in a transmission electron microscope at considerably better resolution and higher magnification than presently possible in commercially available scanning instruments.

The freeze-etch technique can also be profitably employed where equipment is available. Spores are suspended in water or 20--30% glycerol, frozen in small droplets on 3-mm discs in liquid Freon, transferred to a precooled stage in the freeze-etch unit, fractured, shadowed, thawed, and cleaned from the replicas in an acid bath (Moor & Mühlethaler, 1963). Again, replicas are viewed in a conventional transmission electron microscope. Figures 1 and 2 illustrate spores viewed with SEM and freeze-etch- TEM techniques, respectively. The SEM offers greater depth of focus and permits visualization of more of the spore surface; the freeze-etched specimen can be viewed with better resolution and planes other than the surface are visible in favorably fractured samples.

Finally, the familiar fixation and ultrathin sectioning techniques can be used in preparing spores for electron microscopy. Most of these have been adequately reviewed by Hayat (1970). Our routine methods usually include fixation in 2.5% glutaraldehyde, postfixation in 2% OsO_4, *en bloc* staining with 2% uranyl acetate in 70% alcohol, and embedding in either Mollenhauer's mixture No. 2 (Mollenhauer, 1964) or Spurr's (1969) low viscosity medium. We have recently found that stains specific for polysaccharide can be profitably applied to the thin sections (Aldrich, 1974), and will sometimes stain layers which would otherwise be electron transparent. Future possibilities which should be exploited include treatment of spores with purified enzyme preparations to selectively remove specific chemical compounds, followed by fixation, embedding, and transmission electron microscopy as just described, to determine the chemical composition of particular wall layers. Dr. Hohl and co-workers have demonstrated the utility of this technique in studies on *Dictyostelium* (Hemmes et al., 1972). Thus far, similar studies on Myxomycetes are lacking.

SPORES

Chemical Composition of Spore Wall

Application of light microscope cytochemical methods to the myxomycete spore wall has resulted in confusing and inconclusive data (Schuster, 1964; Goodwin, 1961). We have evaluated these studies elsewhere (Aldrich, 1974). Essentially, they left the controversy unresolved as to whether chitin, or cellulose, or both, occur in the spore wall of Myxomycetes. A more recent study (Nields, 1968) reports that *Physarum polycephalum* spore walls consist of an unspecified amount of chitin, 23.5% protein, 16.4% carbohydrate and 10.7% hexosamine. These data have only been published in abstract form, as far as we can determine, and their validity is therefore difficult to evaluate. The single well-documented study of spore wall composition utilizing modern biochemical methods is that of McCormick et al., (1970b) for *Physarum polycephalum*. The dark color of these spores is due to their 15.4% melanin content. Minor components of the wall

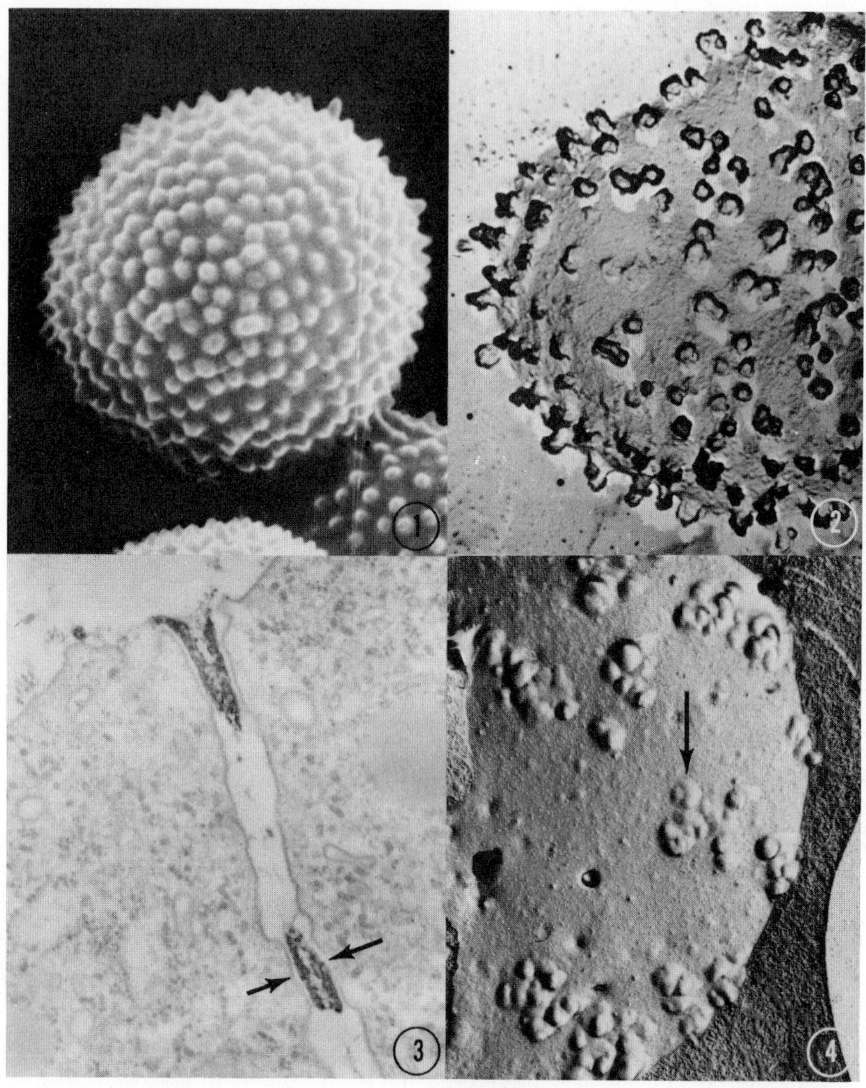

Fig. 1 *Stemonitis* sp.: Scanning electron micrograph of a spiny spore (x approx. 5300).
Fig. 2 *Fuligo septica*: Freeze-etch preparation of spore surface, showing warts. Fibrils comprising continuous wall are also visible. Note the increased resolution of this technique compared to the spore at left (x 14,800).

include amino acids (2.1%) and phosphate (1.4%), with the predominant compound and the only polysaccharide being a galactosamine polymer (81.0%). The possibility that this polymer might be an analogue of chitin was eliminated. Exhaustive enzymatic degradation trials were made, with the finding that the walls were resistant to cellulase, chitinase, lysozyme, alpha-and beta-amylases, snail digestive enzyme, pronase, trypsin, and a variety of glucanases. Further investigations of this quality on additional species are needed.

Morphological Studies

In this section, we will deal with the events that occur after the spores are cleaved out of the multinucleate sporangial proroplast. Events of spore cleavage have been dealt with separately (Aldrich, 1974), and stalk development in a member of the Physarales has also been described at the ultrastructure level (Blackwell, 1974). Most of the work to be reviewed or presented uses the electron microscope as the primary research tool. Specific topics to be dealt with include: (1) events in spore wall formation, (2) cytoplasmic phenomena preceding meiosis, (3) meiosis and related phenomena, (4) morphology of the mature spore, (5) germination.

Deposition of Spore Wall Layers

Schuster (1964) showed that the spines are the first layer of the wall to be laid down in *Didymium nigripes*. More recent studies have confirmed this in *Physarum polycephalum* (Randall & Lynch, 1974). In *Arcyria cinerea* (Mims, 1972b), *Physarum flavicomum*, and *Didymium iridis* (Aldrich, 1974) the deposition begins even as cleavage is still in progress. In the latter three species it appears that the placement of the warts is determined by areas of closely appressed

---

Fig. 3 *Didymium iridis:* Early stage of wart formation, shortly after cleavage. Membrane at top left represents periphery of sporangium. Dark material, presumably melanin, is preferentially deposited in areas where parallel membranes are closest (arrows) (x 37,400).
Fig. 4 *Physarum flavicomum:* Freeze-etched preparation showing plasma membrane of spore at same stage as in Fig. 3. Areas of membrane convolutions (arrow) show distribution of areas in which spines are being deposited. Compare with Fig. 5. (x 15,700).

membrane lining the cleavage furrow. Figure 3 illustrates an early stage in wart deposition. Freeze-etched material at this stage confirms that the distribution of membrane convolutions is similar to placement of spines on the mature spore (Figs. 4 and 5). It appears that all of the spines ultimately covering the mature *Didymium* spore are laid down in this manner, although this is admittedly difficult to determine. The situation in *Arcyria* (Mims, 1972b) differs somewhat. Larger flat-topped projections form first in the manner just described for *Didymium*. Then smaller, sharp projections form somewhat later by some undetermined mechanism. The deposition of the larger projections may be mediated by endoplasmic reticulum seen underneath the areas of deposition. The dimorphism of the ornamentation thus deposited is apparent, even in the mature spore. No such dimorphism has been noted in *Didymium iridis* and *Physarum flavicomum*, but we cannot positively say that such later deposition of some of the spines does not occur. Many species of Myxomycetes exhibit localized thickenings or areas of darker spines on the walls [e.g., *Lamproderma* (Kowalski, 1970) *Echinostelium* (Haskins, 1971) and *Arcyria nutans* (Schoknecht & Small, 1972)], and it seems likely that these areas develop in the way that Mims has described for *Arcyria cinerea*. Certainly this explains why spores in mature sporangia often show a one-to-one spine correspondence on adjacent spores (Mims, 1972b; Furtado, et al., 1971; Aldrich, 1974).

The source of the material forming the early warts in *Didymium iridis* appears to be small vesicles in the underlying cytoplasm. The vesicles may contain polysaccharide as demonstrated by the periodic acid-silver hexamine (PASH) stain procedure (Fig. 6). However, the specificity of this reaction has recently been discussed (Rambourg, 1971), and melanin is stained, in addition to polysaccharides, due to its reducing properties. It is likely that the PASH-positive material observed in Fig. 6 and by Aldrich (1974) was melanin, since the first wall layers contain this compound (Rusch, 1970). Vesicular activity continues in the peripheral cytoplasm during deposition of the continuous wall layers, but polysaccharide localization has been impossible in these vesicles. This difficulty further strengthens the possibility that the vesicle-bound, PASH-positive material seen in early wart deposition was melanin. Assuming that the dark- staining material in glutaraldehyde-osmium treated spores of *Didymium* and *Physarum* represents melanin, its deposition proceeds for one-third to one-half of the period of wall synthesis. The last-formed wall layers are almost electron transparent, although prolonged treatment with lead citrate during poststaining reveals their fibrillar nature (Fig. 27). A morphological difference between the material comprising the spines and the underlying wall layers can be seen in most species (Mims, 1971, 1972b; Haskins et al., 1971; Hung & Olive, 1972; Randall & Lynch, 1974). The warts or spines usually appear amorphous and stain darker than the underlying continuous wall, which is fibrillar (Fig. 27). This difference in morphology implies a change in wall composition during deposition, but more information is needed before definite conclusions are possible.

Cytoplasmic Phenomena Preceding Meiosis

Shortly after cleavage of the spores, a closely packed band of microtubules appears in the cytoplasm of *Didymium iridis* (Fig. 7) and persists for 10 or more h, until the meiotic spindle forms. This band has also been seen in several other species during meiotic prophase, as incidental observation while searching for synaptonemal complexes, namely *Physarum globuliferum*, *Stemonitis virginiensis*, *Fuligo septica*, *Tubifera ferruginosa*, *Arcyria incarnata*, *Arcyria cinerea*, and *Hemitrichia stipitata*. We suspect that this organelle may be ubiquitous in myxomycete spores at this stage. The cases where it has not been observed after extended observation only represent one or two species, and lability during fixation may be the reason in those cases. *Didymium iridis* is the only species studied in enough detail to ascertain precisely when the microtubule band appears. We feel it likely that it exists through the interphase preceding meiosis in most species, as it does in *Didymium*. The function of this aggregation of microtubules is unknown. The two possibilities we have suggested are (1) to expand the spores into their characteristic spherical shape after their cleavage into polyhedra during sporulation, and (2) as a precursor of the meiotic spindle. Evidence for possibility (1) is circumstantial, based on the time when the structures are observed. Evidence for possibility (2) is that when the meiotic spindle is first seen (Fig. 8) shortly after synaptonemal complexes disappear, it consists of closely appressed, almost paracrystalline microtubules which later spread to form the metaphase spindle. Since the nuclear envelope is intact as this intranuclear meiotic spindle forms, the cytoplasmic tubules would have to depolymerize, move through the nuclear pores as subunits, and re-form in the nucleoplasm.

Most myxomycete spores, after cleavage and before meiosis, contain a large central autophagic vacuole. The material in this vacuole

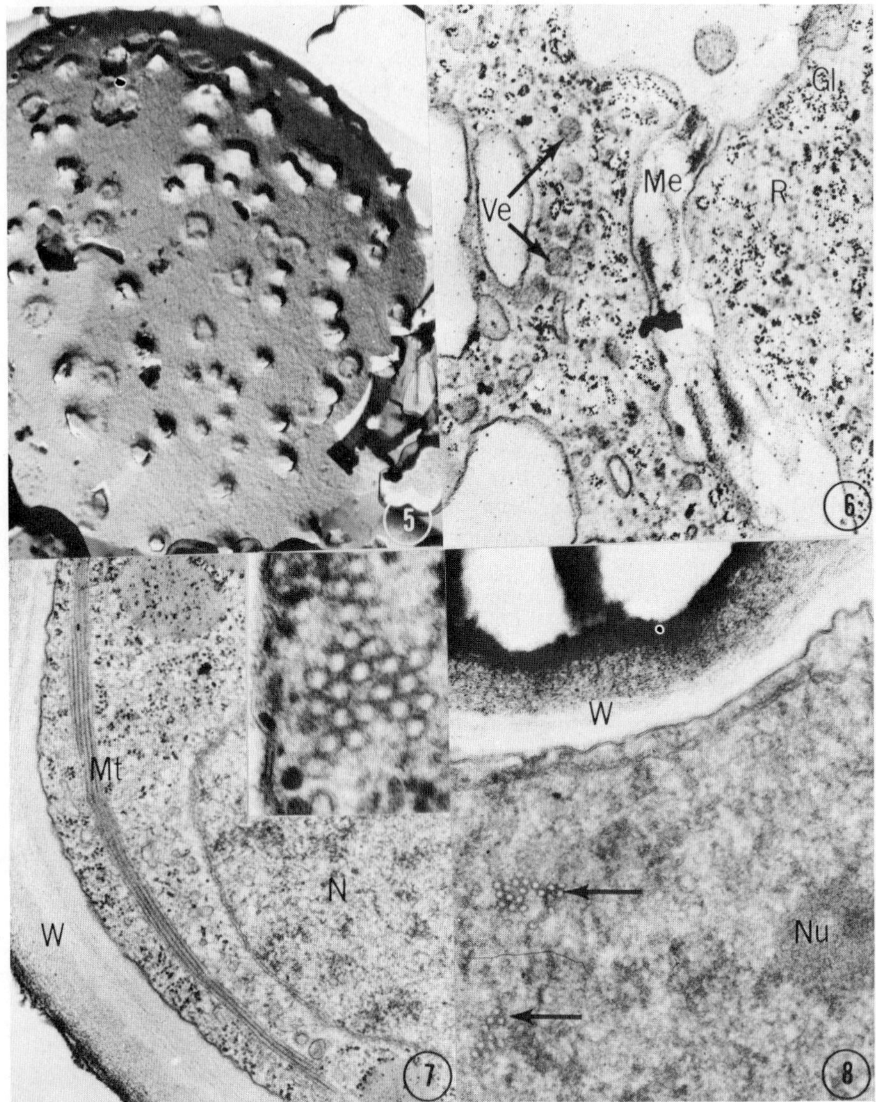

Fig. 5 *Physarum flavicomum:* 7-day old spore, freeze-etched, showing spine placement. Distribution is very similar to that of membrane blebs in Fig. 4 at time spines were being laid down. Note fibrillar nature of underlying continuous wall (x 13,100).

exhibits acid phosphatase activity (Cronshaw & Charvat, 1973), which observation attests to its digestive nature (Fig. 9). Movement of material from the cytoplasm into these vacuoles follows a sequence similar to that described by Locke & Collins (1965) during formation of protein granules of cytolysosome or autophagic vacuole origin in a butterfly. In *Didymium*, smooth endoplasmic reticulum cisternae surround a cytoplasmic mass, such as lipids, mitochondria, or ER, grow in extent by addition of vesicles of Golgi apparatus origin (Fig. 10), and finally completely isolate the mass (Fig. 11). The inner membrane recedes, the phagocytized material stains darker, and digestion and disintegration begin. At about this stage, the vacuole fuses with the large central one and the contents mix (Fig. 12). The reason for such sequences is obvious. With the spore protoplast isolated from the environment by a wall, and with considerable activity still to come, new amino acids, nucleotides, etc., are required. These are obtained by an efficient cellular level recycling plant, the large central autophagic vacuole. This vacuole appears to be found in all species critically examined at this stage.

Storage polysaccharide accumulates during the hours after spore cleavage and preceding meiosis. Glycogen has been isolated from the plasmodium of *Physarum polycephalum* (Goodman & Rusch, 1969), and takes the morphological form of the beta-particles of Revel (1964), ovoid particles about 400 x 700 Å in thin section. These particles show particulate substructure when stained with lead citrate (Fig. 13), stain heavily with the PASH polysaccharide stain (Fig. 6), and

---

Fig. 6 *Didymium iridis:* Same stage as Fig. 3, stained to demonstrate polysaccharide and melanin with PASH technique. Black globular deposits are metallic silver. Note heavy deposition over glycogen particles in cytoplasm, while ribosomes are unstained. Vesicles containing melanin (?) were probably moving to plasma membrane to contribute spine material (x 37,000).

Fig. 7 *Didymium iridis:* Band of closely packed cytoplasmic microtubules in longitudinal section. Inset shows band in cross section. Main figure (x 21,500; inset x 120,100). Mt, microtubules; W, wall; N, nucleus.

Fig. 8 *Didymium iridis:* Early stage of meiotic spindle formation. Note the similarity in appearance of the cross sectioned spindle tubules (arrows) to the microtubule band in cytoplasm in Fig. 7 (x 44,500). Nu, nucleolus.

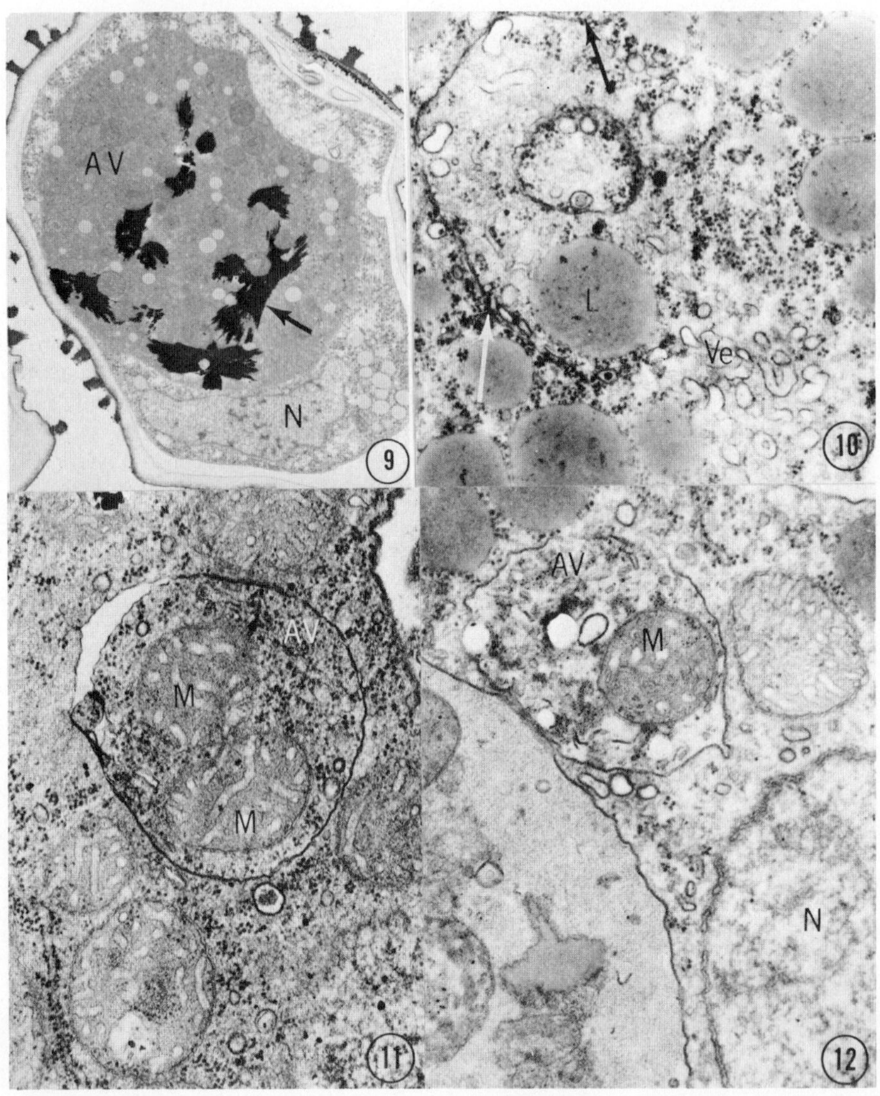

Fig. 9 *Perichaena vermicularis*: Demonstration of acid phosphatase activity in large central autophagic vacuole (AV) several hours after spore cleavage. Black deposits (arrow) localize the enzyme (x approx. 8400). (From Cronshaw & Charvat, 1973).

are removable from the section with alpha-amylase (Fig. 14). The particles can be distinguished from ribosomes by their oblong appearance. We find them commonly in all stages of the life cycle of Myxomycetes and see no reason to question their identity as glycogen. Glycogen particles often accumulate in large cytoplasmic aggregations in premeiotic spores of *Didymium* and *Arcyria* (Fig. 15), and we surmise that their confusion with ribosomes is responsible for their not being noted prominently in all studies undertaken. The glycogen particles likely represent the primary storage reserve in spores. They persist into mature spores.

Nuclear Phenomena in the Meiotic Spore

Besides the events of meiosis itself, several anomalies in the ultrastructure of the nucleus have been seen. Gustafson (1974) has noted an intranuclear crystal in *Didymium squamulosum* just prior to and concurrent with meiotic prophase (Fig. 16). It is not present at the preceding mitotic division and disappears by Metaphase I of meiosis. He believes it is not of viral origin but is uncertain of its composition and function. This crystalline structure has been seen in no other Myxomycete. During meiotic prophase in Myxomycetes, the nucleolus typically becomes eccentric. In *Didymium iridis,* as it moves against the nuclear envelope, a fibrous mass containing dark granules appears in the center of the nucleus (Fig. 17). We believe that this may be a portion of the nucleolus which remains behind as the bulk of

---

Fig. 10 *Didymium iridis*: Early stage in formation of small autophagic vacuole. Paired membranes between arrows are surrounding an area of cytoplasm. Vesicles could be contributing membrane to cisterna comprising vacuole (x 30,500).
Fig 11 *Didymium iridis*: Two mitochondria (M) sequestered into autophagic vacuole (AV). Slightly later stage than Fig. 10. No digestion apparent yet; mitochondria in autophagic vacuole show no damage. Inner membrane of sequestering cisterna is beginning to withdraw at upper left (x 32,300).
Fig. 12 *Didymium iridis*: Late autophagic vacuole, with degenerating mitochondrion. Inner membrane of sequestering cisterna no longer present. Small autophagic vacuole appears ready to contribute contents to large central one at lower left. Figures 10--12 taken from spores about 10 h past cleavage (x 33,600).

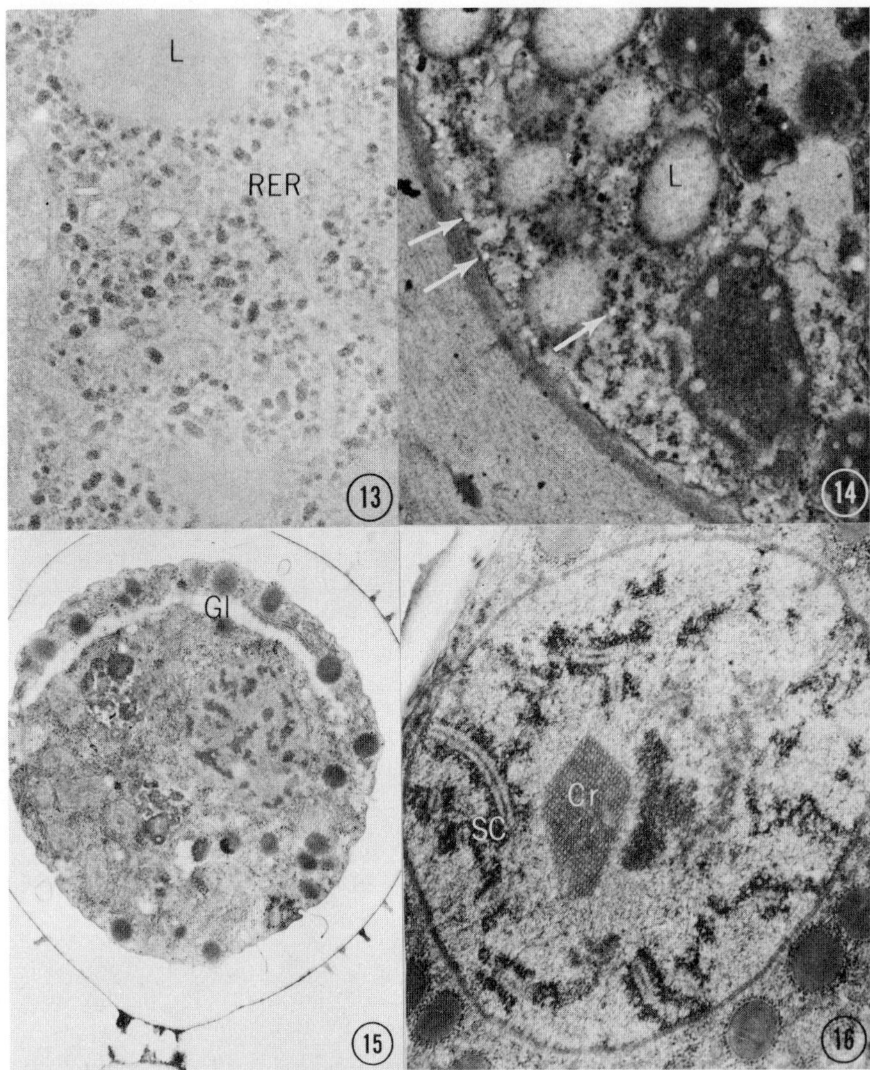

Fig. 13 *Didymium iridis:* Cytoplasm of spore treated with 0.2% ribonuclease for two hours before fixation. Ribosomes have been removed from area of rough endoplasmic reticulum and stain only lightly. Remaining dark staining granules are glycogen particles typical of all stages of Myxomycetes (x 24,000). RER, rough endoplasmic reticulum; L, lipid.

the nucleolar material moves to the periphery of the nucleus. The dark granules are removable with pronase after 48 h of digestion (Fig. 18), indicating that they are proteinaceous in nature. Attempts at ribonuclease digestion of the mass have been unsuccessful. This mass of material also disappears by the onset of Metaphase I. Nothing else is known of its nature of function. A third type of nuclear inclusion has been informally referred to in our lab as a "spaghetti-some," but we are not anxious to create the sort of monster Fuller (1966) has invented with the "rumposome" in Phycomycetes! Hence, we simply refer to the structure as a tangled mass of membranes inside the nucleus (Fig. 19). This membranous inclusion has been seen most consistently in *Physarum globuliferum*. Whether it originates at or remains attached to the nuclear envelope is unclear. Its function is unknown. Finally, lipid inclusions have occasionally been noted inside the nuclear envelope just prior to meiotic prophase (Fig. 20). Their significance is unclear.

The controversy over the position of meiosis in the myxomycete life cycle should now be familiar to most mycologists and we will resist the urge to belabor it here in detail. We would point out that most evidence from complementary techniques confirms our original thesis (Aldrich, 1967) that meiosis occurs in the spores of most species 12 to 18 h after cleavage is complete. Subsequent ultrastructure work has supported this (Aldrich & Mims, 1970; Aldrich & Carroll, 1971; Haskins et al., 1971; Randall & Lynch, 1974), but more importantly, Mohberg & Rusch (1971) used biochemical methods to confirm that postcleavage spores of *Physarum polycephalum* remain diploid until

---

Fig. 14 *Didymium iridis*: Cytoplasm of spore treated on grid after thin sectioning with 1% H_2O_2 and 0.2% alpha-amylase. Glycogen particles are completely removed from the section, leaving holes (arrows). Remaining dense granules are ribosomes (compare with Fig. 13) (x 32,200).

Fig. 15 *Arcyria cinerea*: Postmeiotic spore with crescent of tightly packed glycogen particles, probably reserve food. (From Mims, 1972b) Note also correspondence of spines on adjacent spores, shown at bottom of picture (x approx. 8000).

Fig. 16 *Didymium squamulosum*: Nuclear crystal appearing at meiosis. Typical synaptonemal complexes are present. One at left clearly ends at nuclear envelope (x 20,000). (Courtesy Dr. Ralph Gustafson) SC, synaptonemal complex; Cr, crystal.

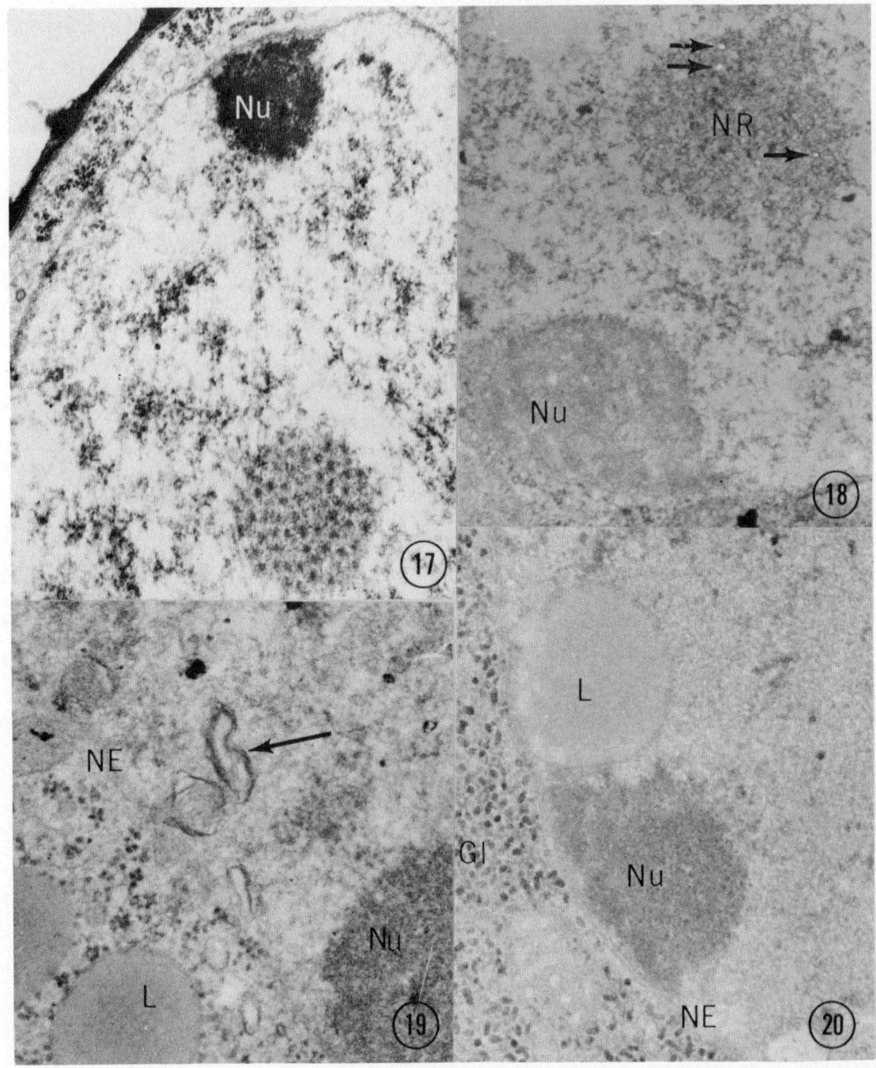

Fig. 17 *Didymium iridis*: Fibrous globular mass (lower right) containing dark granules, found just prior to meiosis. Most of the nucleolar material is at periphery of nucleus (top). Fibrous mass is thought to be nucleolar remnant (x 23,000).
Fig. 18 *Didymium iridis*: Nucleolar remnant like that in Fig. 17 after treatment on the grid with 1% H_2O_2 for 15 min, followed by digestion

after the intrasporic meiotic divisions. LeStourgeon et al., (1971) have compared the numbers of spores produced at cleavage in *Physarum flavicomum* with the number of nuclei in the predivision sporangium and found only twice the number of postcleavage spores as predivision nuclei. They reason that if the division seen before cleavage were meiotic, spore number should be about four times the number of precleavage nuclei. We reiterate that we consider satisfaction of all the following as incontrovertible cytological evidence of meiosis: (1) precleavage divisions with no synaptonemal complexes in prophase, (2) an intervening interphase with no synapsis, (3) synaptonemal complexes in spore nuclei, (4) division figures in spore nuclei, and (5) nuclear degeneration (Aldrich & Mims. 1970). Therrien & Yemma (1974) have recently raised the possibility that synaptonemal complexes taken alone do not represent compelling evidence of meiotic division, especially since Menzel & Price (1966) found complexes in anthers of haploid tomato plants. They considered them evidence of nonhomologous pairing. We agree, and affirm that the complexes in our case only indicate where meiosis would normally occur, and do not indicate that a reduction division does actually take place. That is, observation of synaptonemal complexes does not in itself rule out an apogamic situation. Therrien & Yemma (1974) have presented microspectrophometric data which show essential constancy of DNA throughout the life cycle of a "homothallic" isolate of *Didymium iridis*, Philippine-1. We have examined this isolate (Aldrich & Carroll, 1971) and found synaptonemal complexes in the spores at the usual time. To confirm Therrien's conclusion of apogamy here, we have searched our files and found a single unmistakable picture of an anphase figure in this isolate, in the same preparation in which spores were in synapsis. This suggests that meiosis was indeed completed.

---

with pronase (0.2%) for 48 h at 37°C. Nucleolus is considerably lighter, and dark granules are gone from nucleolar remnant at upper right. Arrows indicate holes left by removal of granules, evidently protein (x 9,600). NR, nucleolar remnant.

Fig. 19 *Didymium iridis*: Premeiotic spore nucleus, showing apparent proliferation of inner nuclear membrane into area of convoluted membrane (arrow) (x 41,000). NE, nuclear envelope.

Fig. 20 *Didymium iridis*: Intranuclear lipid sometimes seen before meiosis. Most of the dark particles in cytoplasm are glycogen (x 30,700).

The inconsistency between our results and those of Therrien & Yemma (1974) might be explained in two ways. Either the single anaphase figure we saw was an isolated occurrence and no divisions normally occur (our search for division figures was not an extensive one in that isolate), or the isolate has lost the ability to complete meiosis while being maintained in culture, between the time our isolate was received from Collins and the time Therrien's was received. With this single problematical exception, however, we feel that the evidence for the occurrence of meiosis in myxomycete spores is strong.

A variation on the usual meiotic scheme has recently been discovered in our laboratory (Cathcart, 1974). Some isolates of *Fuligo septica* begin and complete meiosis normally, except that synaptonemal complexes are noted as early as 6 h after spore cleavage in a few cases. Spores from these isolates germinate 2--4 h after wetting. In other isolates, however, meiosis proceeds only as far as synapsis (Fig. 21) or metaphase (Fig. 22) and then is blocked for some reason. In such cases, metaphase spindles can be seen in dry spores months or years old. When these isolates are wet for germination, about 36 h elapse before any swarm cells are seen. Cytological examination reveals that before germinating, these spores blocked at metaphase complete meiosis. Spores 2--6 h after wetting contain division figures (Fig. 23), slightly older spores contain degenerating nuclei (Fig. 24), and centrioles are seen for the first time. These observations satisfy most of our criteria for meiotic events. Cathcart (1974) also observed precleavage nuclei and postcleavage nuclei; neither type shows any svnapsis, so all the above-mentioned criteria are satisfied. Cathcart's study indicates that long germination times in *Fuligo* are always linked with a blocked meiosis when cytological examination is carried out. This observation explains the difficulty some investigators have experienced in germinating spores of certain *Fuligo* isolates, (Braun, 1971a). At our suggestion, Braun reexamined some isolates which his observations showed failed to germinate after the usual 2 h, and reported that these specimens germinated within 36 h as we had predicted. Cytological examination of this material shows condensed chromatin, but no spindle tubules or synaptonemal complexes are visible. We surmise that the nuclei are in some stage approximating prometaphase of meiosis. We have also seen metaphase figures (Fig. 25) in *Comatricha typhoides* spores which have been wet for only 1 h. This strongly suggests a meiotic interruption there, too. We have not investigated this species as carefully as Cathcart did *Fuligo septica,* however. Mohberg & Rusch (1971) have shown that the M_3cIV isolate of *Physarum polycephalum* takes 4--6 days to achieve the haploid DNA level, indicating that the process is somehow drawn out there also. Further investigation may show that the phenomenon is more common. In particular, the species of Myxomycetes which fail to germinate in culture should be examined for such abnormalities. Gilbert (1929) noted several cases in which several months of aging were necessary before high germination percentages could be obtained (*Reticularia* and *Lycogala*). An interrupted meiosis or a situation like that in the *P. polycephalum* isolate just mentioned where longer than usual is required for the meiotic divisions to be completed might explain Gilbert's results.

After the normal meiotic divisions in the spore, the uninucleate condition is established by degeneration of the other meiotic products. There are two ways in which this could occur: (1) the two sequential divisions comprising meiosis could be completed, followed by generation of three of the four resulting nuclei, or (2) one product of each division might degenerate immediately, so that tetranucleate spores never exist. Von Stosch (1935) favors the latter method. Aldrich (1967) favored the former, but had little real evidence. The difficulty is that sporangia developed in culture sometimes contain multinucleate spores. Such spores are usually larger than uninucleate ones, but are otherwise indistinguishable. *Didymium iridis* tends to form these multinucleate spores less frequently than *Physarum flavicomum*, the species used in the 1967 study. In *Didymium*, we have seen an instance of a normal sized spore containing one degenerating nucleus and another with some spindle tubules. This constitutes evidence supporting von Stosch's advocacy (1935) of possibility (2) above. Reexamination of our *Physarum flavicomum* material convinces us that no conclusion can be reliably drawn from that material because of the multinucleate spore problem. The scanty evidence available, then, favors, possibility (2).

The final event in spore maturation is centriole formation, which occurs after meiosis is complete. This appears to be a genuine case of de novo origin of centrioles. Their method of formation is poorly understood. Short, indistinct centriole-like structures can sometimes be observed among the smooth surfaced cisternae near the Golgi apparatus (Fig. 26). Whether such stages represent poorly preserved mature centrioles or forming stages is difficult to determine, however. The centriole is of the usual nine triplet structure (Aldrich, 1967, Mims, 1972a). The mature spore contains two oriented at right angles, and one often has the structures of the flagellar basal apparatus

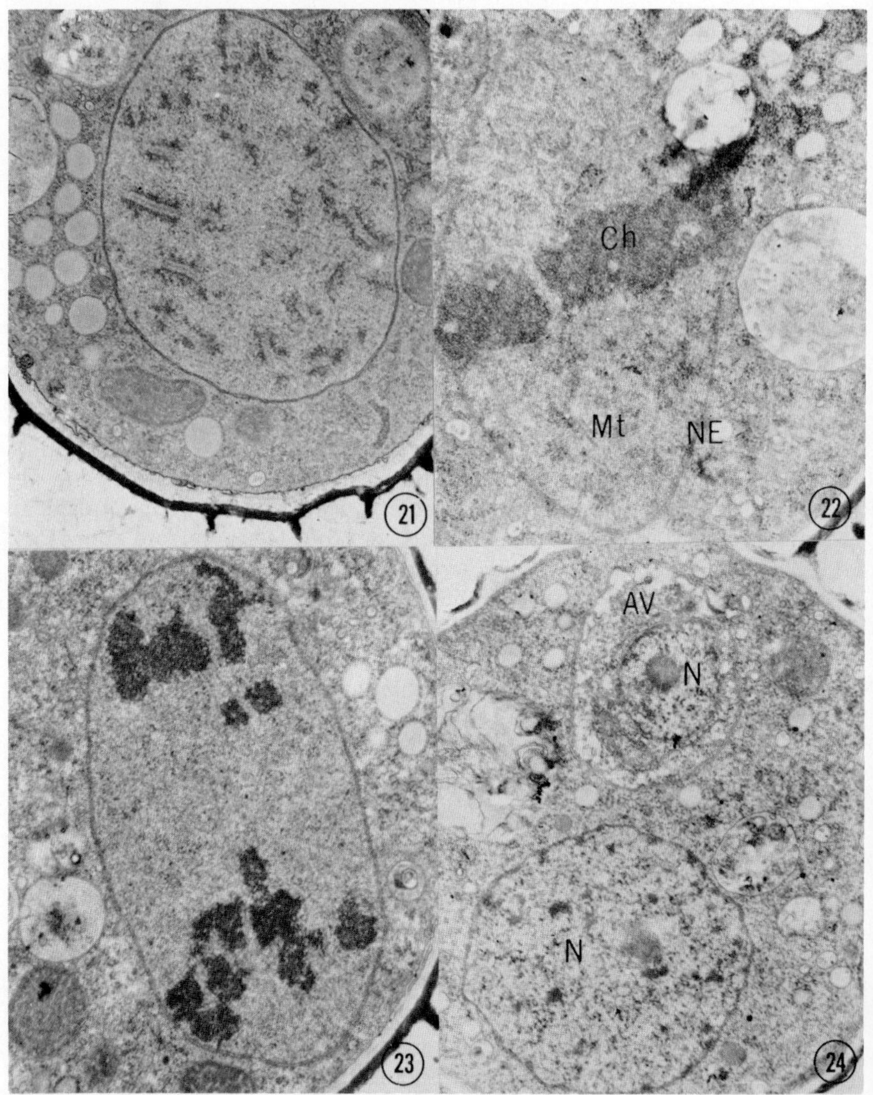

Fig. 21 *Fuligo septica*: Synapsis of meiotic prophase (zygotene) in spore 11 h past cleavage (x 11,800). Figs. 21--24 from Cathcart, 1974).

Fig. 22 *Fuligo septica*: Same isolate, same aethalium as Fig. 21, but spores 1-month-old. Metaphase figure indicates meiosis is blocked,

associated with it, presumably in preparation for germination as a swarm cell.

Morphology of the Mature Spore

Spores and capillitium of Myxomycetes have long been considered the most stable taxonomic characters (Martin & Alexopoulos, 1969). Such characteristics as spore color and ornamentation are most important. Nothing is known of the pigments responsible for spore color in the light-spored groups. Presumably the dark color in spores of the Stemonitales and Physarales is imparted by melanin, although the only biochemical verification of this possibility is in the case of *Physarum polycephalum* (McCormick et al., 1970b). There does appear to be a general correlation between the thickness of the electron dense layer in spores viewed ultrastructurally and the darkness or lightness of the wall under the light microscope. For example, the electron dense layer in *Arcyria cinerea* and *Perichaena vermicularis*, both light spored species, is very thin (Fig. 15) compared to that in dark-spored species such as *Fuligo septica* and *Didymium iridis* (Figs. 7, 27). We cannot say with certainty whether the thin electron dense layer in light colored spores represents melanin, but it seems reasonable to assume that a thin layer of this substance might result in an almost hyaline spore color under the light microscope. It also seems likely that the electron dense material is melanin, since melanin acts as a reducing agent (Rambourg, 1971) and would therefore accumulate osmium selectively.

Five types of ornamentation characterize myxomycete spores:

1. Spiny or warted (verrucose), i.e., projections may be sharp,

---

probably at Metaphase I (x 17,700). Ch, chromatin; Mt, microtubules.
Fig. 23 *Fuligo septica*: Spore from same aethalium as Figs. 21 and 22, 2 h after wetting for germination. Metaphase block has been overcome and nucleus has proceeded into anaphase (x 17,700).
Fig. 24 *Fuligo septica*: Same material as three previous figures, six hours after wetting for germination. Only a single, presumably haploid nucleus was seen. Autophagic vacuole at top contains a nucleus, probably a product of one of the meiotic divisions in the process of degenerating (x 14,800).

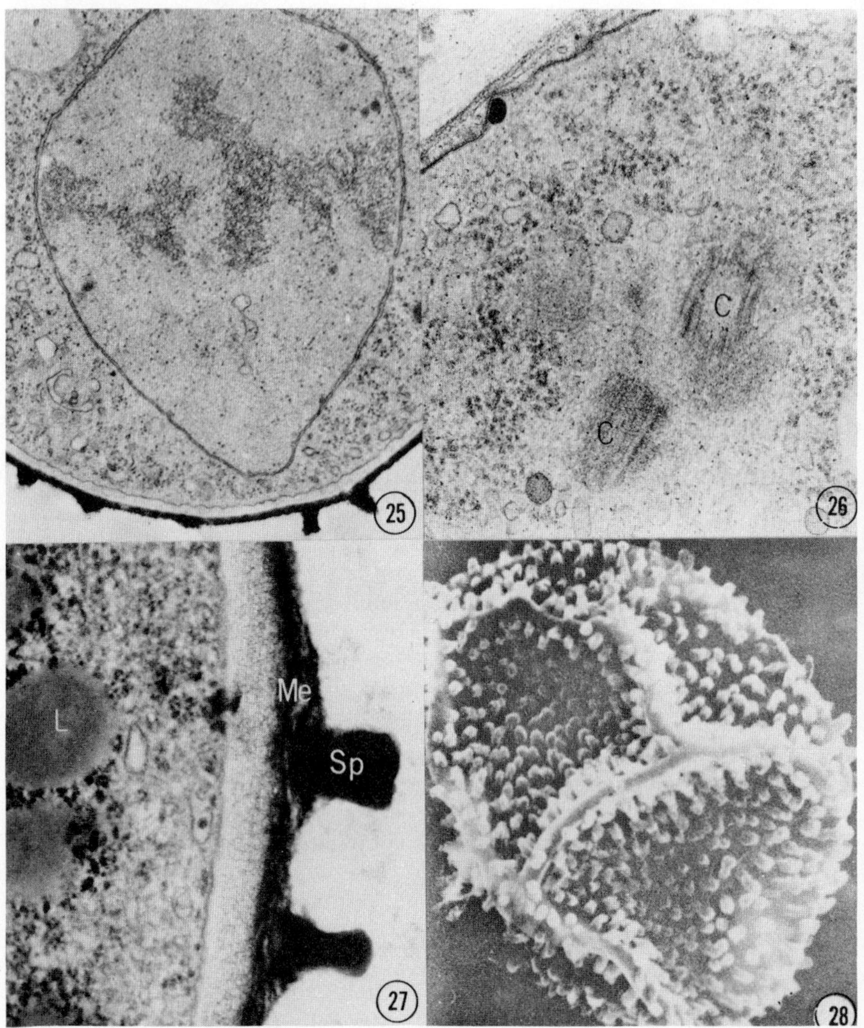

Fig. 25 *Comatricha typhoides:* Mature spore which had been wet for one hour before fixation. Metaphase figure suggests this species undergoes meiotic block as *Fuligo* does (x 23,100).
Fig. 26 *Physarum flavicomum*: Spore fixed about 2 h after meiosis was complete. The two centrioles are believed to be in formative stage. The one at left is shorter than normal, and both are somewhat

blunt, or tack-shaped.

a. Uniform. All projections identical in size and shape, for example, some *Stemonitis* species (Fig. 1).
b. Some projections or areas darker, larger, or otherwise distinct, for example, *Arcyria cinerea* (Fig. 15), *Lamproderma* (Kowalski, 1970). These specialized areas probably all form as Mims (1972b) has described.

2. Rugose, for example, *Cribraria intricata* (Schoknecht & Small, 1972); *Badhamia gracilis*, actually a combination of rugose and warted (Scheetz & Alexopoulos, 1971), Fig. 28.

3. Columnar-reticulate. Many of the Stemonitales are described as having spiny-reticulate spores. Examination of these with the scanning microscope (Fig. 29) indicates that the "spines" are actually baculae or columns supporting a raised reticulum.

4. Banded-reticulate. In this type, there are no open spaces at the base of the reticulum, hence, no "spiny" effect is created (Figs. 30, 31), for example, *Hemitrichia stipitata, Trichia favoginea* (Schoknecht & Small, 1972). Actually, this type may intergrade with the columnar-reticulate. Also, smaller ornamentation may be found in between very coarse reticulations in some cases (Schoknecht & Small, 1972). Incomplete reticulations have also been seen (Fig. 32).

5. Smooth. Several species have been described in this category, for example *Badhamia ovispora, Cribraria intricata,* and many of the Protostelids (Olive, 1967, 1975). Examination of the first with freeze-etch reveals low projections on the surface (Fig. 33).

---

indistinct (x 45,600). C, centriole.
Fig. 27 *Didymium iridis*: Mature spore in section. Spines are amorphous, dark. Underlying layer is electron dense and believed to be melanin. Inner layer is lighter but still fibrillar (x 44,800). Sp. spine (or wart).
Fig. 28 *Badhamia gracilis*: Scanning electron micrograph of mature spore, showing spines or warts and coarse rugose ornamentation (x 6,400). From Scheetz & Alexopoulos, 1971.

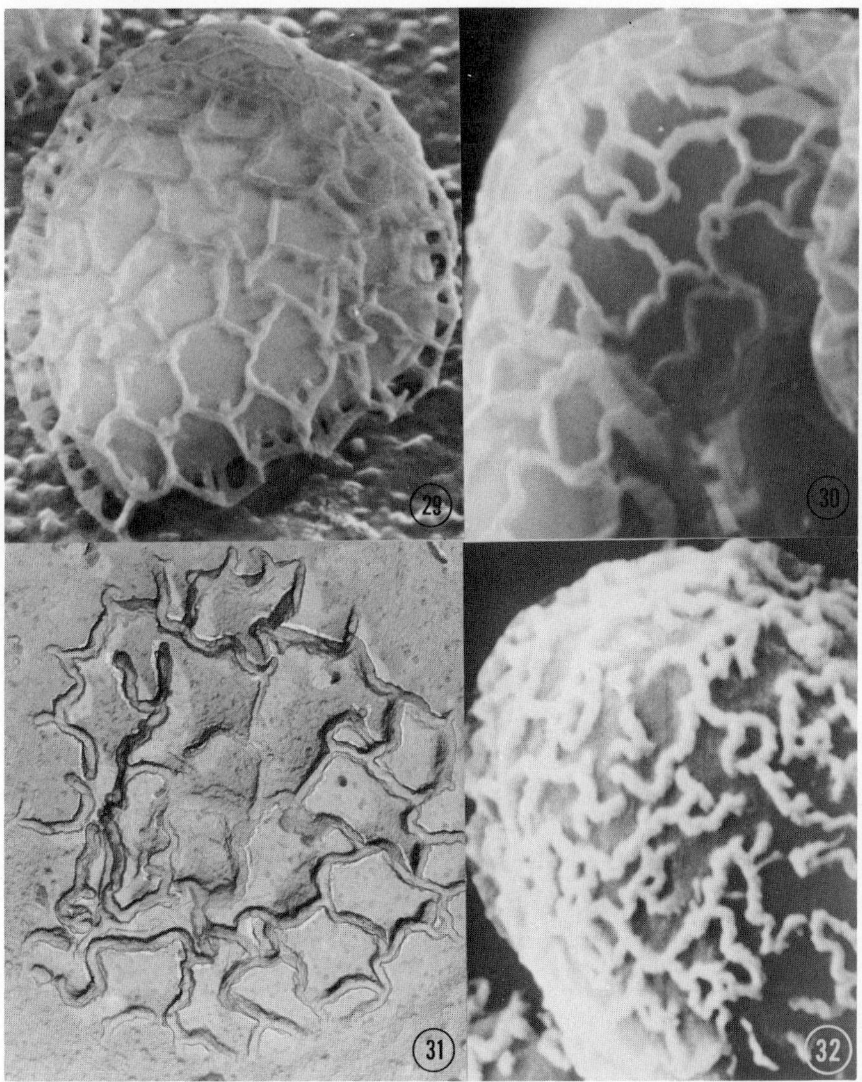

Fig. 29 *Stemonitis nigrescens*: Scanning micrograph of spore, described as spiny-reticulate based on light microscopy. Spines are here seen to be columns supporting raised reticulum (x approx. 7,400).
Fig. 30 *Hemitrichia stipitata*: Banded-reticulate spore seen with SEM.

Schoknecht & Small (1972) have found a rugose pattern in the second, and Furtado & Olive (1971) showed a thin wrinkled surface layer over the spore of *Ceratiomyxella*. We expect that when proper techniques of sufficient resolution are applied, few myxomycete spores will actually turn out to be smooth.

A problem that seems likely to arise is that as increasingly sophisticated techniques are utilized, smaller samples of spores are often examined. The investigator must decide to what extent a given spore is representative of that species or collection. Candidly, there is no good answer to this question. We would argue in favor of examination of as many samples as feasible, however, for the following reasons. Schoknecht & Small (1972) investigated this question in *Physarum cinereum*, and although they concluded that ornamentation in a given isolate tends to remain unchanged on different substrates, they find considerable variation within a collection and from one collection to another. We have observed similar wide variations in *Hemitrichia stipitata* and *H. clavata,* and have even despaired of completing an SEM comparison of these species because of the range of variation observed. Finally, Fujii & Hess (1969) found an alarming range of variation among freeze-etched conidia collected from monosporous cultures of *Penicillium digitatum* and *Aspergillus nidulans* , concluding that such variability must be assumed and taxonomic studies utilizing ultrastructural methods must be designed with extreme caution. We would extend this *caveat* to investigations of the Myxomycetes.

Spore Germination

Anyone who has handled Myxomycetes has noticed that germination percentage in the spores declines rapidly after a year or so of storage in most cases. That percentage germination does not decline to zero was clearly shown by several workers (reviewed in Gray & Alexopoulos, 1968) who succeeded in germinating 30 to 75-year-old

---

Entire base of reticulum contacts spore surface (x approx. 9,800).
Fig. 31 *Hemitrichia stipitata*: Freeze-etched sample for comparison with Fig. 30 (x 11,800).
Fig. 32 *Hemitrichia stipitata*: (SEM) Spore with discontinuous or incomplete reticulum. From different collection than Fig. 30 (x approx. 11,500).

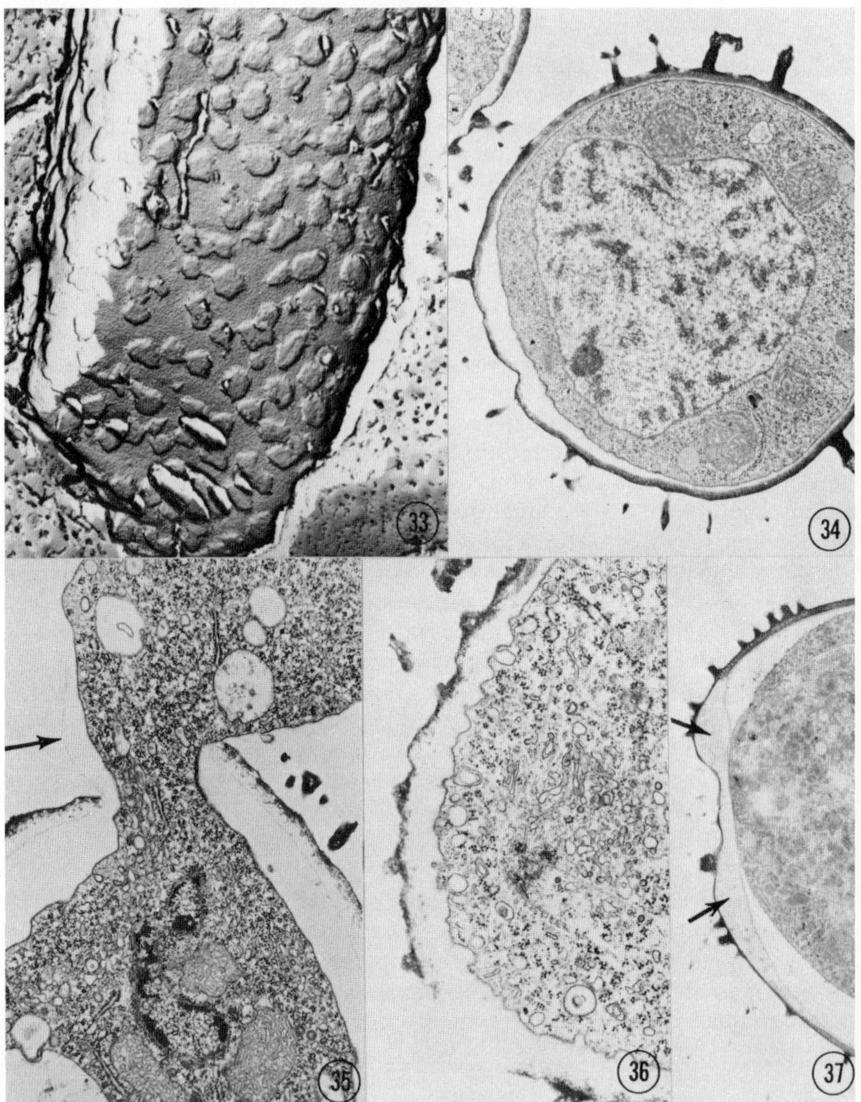

Fig. 33 *Badhamia ovispora*: Spore of one of the few ovoid-spored species. Described as smooth with light microscope, freeze-etch shows low placque-like ornamentation (x 8,900).
Fig. 34 *Stemonitis virginiensis*: Spore at meiotic prophase, illustrating thin place in wall (left) (x 8,200).

spores of several species. The factors influencing spore germination have been reviewed very completely by Martin (1940), Alexopoulos (1962), and Gray & Alexopoulos (1968), and it makes little sense to repeat their efforts here. Nothing of significance on this subject has been published since the latest of these reviews, as far as we can determine, with the exception of a short paper by Braun (1971b). He investigated the effect of aeration on germination time in *Fuligo septica* spores, and found that the time required for germination and swarm cell production was shorted considerably by this treatment.

Recent years have seen some activity in terms of ultrastructural studies of spore germination in Myxomycetes. Germination has been said to occur by two methods, namely cracking or splitting of the wall, and by formation of a small, pore-like opening (Gilbert, 1928). Change from one type to the other has been achieved by manipulating the conditions (see Gray & Alexopoulos, 1968). Unfortunately, the studies published so far employing the electron microscope have utilized only thin sectioning, and the near median sections usually studied would not distinguish very well between these two methods of germination. Thus, freeze-etching or scanning EM studies in this area would be useful.

Mims (1971) and Mims & Rogers (1973) have examined the sequence of germination events in spores of *Arcyria cinerea*, which normally germinates by the split method, and *Stemonitis virginiensis*, which germinates by the pore method. In *Arcyria*, there is a decrease in number of lipid droplets, and Mims speculates that these, as well as the glycogen particles mentioned earlier, are utilized as energy sources at this time. The inner wall layers are observed to be more elastic and to rupture last. The mature spore of *Stemonitis virginiensis* has a thin area in the outer wall layer (Fig. 34), and it

---

Fig. 35 *Stemonitis virginiensis*: Spore midway through germination. Note remnant of inner wall (arrow) pushed outward during protoplast emergence (x 8,200).
Fig. 36 *Stemonitis virginiensis*: Detail of early stage in spore germination, showing wall rupture. Note profusion of vesicles underlying and fusing with plasmalemma in area of rupture (x 20,500).
Figures 34--36 from Mims & Rogers, 1973.
Fig. 37 *Physarum melleum*: Portion of wall of mature spore, showing thin area in outer layer, underlain by greatly expanded fibrous inner layer (arrows) (x 7,600).

is this area which becomes the germination pore (Fig. 35) (Mims & Rogers, 1973). Vesicular activity is noted in the area underneath the thin region during germination (Fig. 36), but the significance and contents of the vesicles are undetermined. Whether the thin area in the wall is really circular in outline has not been determined either. Our observations indicate that *Physarum melleum* contains such a thin area in the wall (Fig. 37), and this species germinates by the split method, so it may be that occurrence of this specialized area of the wall does not necessarily correlate with the pore type of germination in all cases. Now that the chemical composition of the spore wall is known in one case (McCormick et al., 1970b), the time seems ripe for real progress to be made in elucidating the biochemical events accompanying spore germination in Myxomycetes. Hopefully answers to such questions as whether wall softening enzymes are involved, what food sources are utilized, and what the spore surface looks like during the germination process will be forthcoming soon.

MICROCYSTS

The microcyst of the Myxomycetes is not well known. This is probably due to the fact that it is a nonessential stage in the life cycle of an organism which produces a variety of other more spectacular stages. Only recently have any axenic culture techniques for myxamoebae been developed (Goodman, 1972; Henney and Asgari, 1975). For this reason it has been difficult to proceed very far in the study of any of the uninucleate, haploid stages of these organisms.

Microcysts form under cultural conditions which are favorable neither to myxamoebal-swarm cell population growth nor to plasmodial formation. This information is reviewed by Gray & Alexopoulos (1968).

There are few observations of the role of microcysts in nature. We recently collected sporangia of *Stemonitis fusca* for a classroom demonstration of spore germination. Upon wetting on an agar plate, swarm cells filled the culture dish within 10 min, while germ pores were not observed in the spores. Examination of unwet sporangia of the collection showed that numerous microcysts were present, held along with the spores by the capillitium. Thus, it is assumed that the microcysts do play an important survival role in nature.

In agar culture it is well known that myxamoebae quickly encyst in response to a variety of conditions. Myxomycete geneticists routinely store their encysted myxamoebal clones on agar slants. In fact, clones of *Protophysarum phloiogenum* microcysts which we kept at room temperature for three years remained viable. Such cultures, when lyophilized remain viable for extremely long periods (Kerr, 1965; Davis, 1965).

Recently Raper & Alexopoulos (1973) have described a species of *Didymium* in which propagation takes place, at least in culture, primarily by an elaborated encystment stage; although plasmodium and sporangium formation do occur from single amoebal culture, it is rare, and the organism regularly forms columns of amoebae. These amoebae after encystment can be transferred to repeat the column formation. This organism will be discussed further below.

Schuster (1965) has published a brief ultrastructural study of microcysts. In looking at *Physarum cinereum* and *Didymium nigripes* he found a relatively thin fibrous wall surrounding a virtually undifferentiated amoeba. He showed smooth cytoplasmic vesicles which he believed contained wall material in an early stage of cyst formation. Basal bodies, their microtubular arrays, and what appears to be glycogen layered in an artifact of fixation, were seen in the microcyst cytoplasm.

We have examined the microcysts of four additional species, *Physarum flavicomum*, *P. polycephalum* (APT-1 mutant and the Colonia isolate), *Didymium iridis*, and *Didymium* sp. (Raper & Alexopoulos, 1973). Aged clonal cultures of the first three heterothallic species and aged single amoebal cultures of the homothallic *Didymium* sp. were used. The exact cause of encystment is unknown; it may have been due to depletion of the nutrients or more likely, a change in hydrogen ion concentration from the use of *Escherichia coli* as the food organism.

As found by Schuster (1965) encysted cells contain an organellar component similar to that of non-encysted cells, including a ribosomal-coated outer nuclear membrane and basal bodies with associated microtubular arrays (Fig. 38). Several differences from myxamoebae are immediately apparent. Amoebal encystment is accompanied by an increase in the number of smooth membrane vesicles with fibrous contents of probable Golgi apparatus origin (Fig. 39). A distinct wall is present in later stages, although some vesicles were still present in the cytoplasm.

Mitochondrial profiles are circular and rarely elongated as they are in myxamoeba (Fig. 40). This may represent a different physiological response of the cells to the same fixation regime or it may be

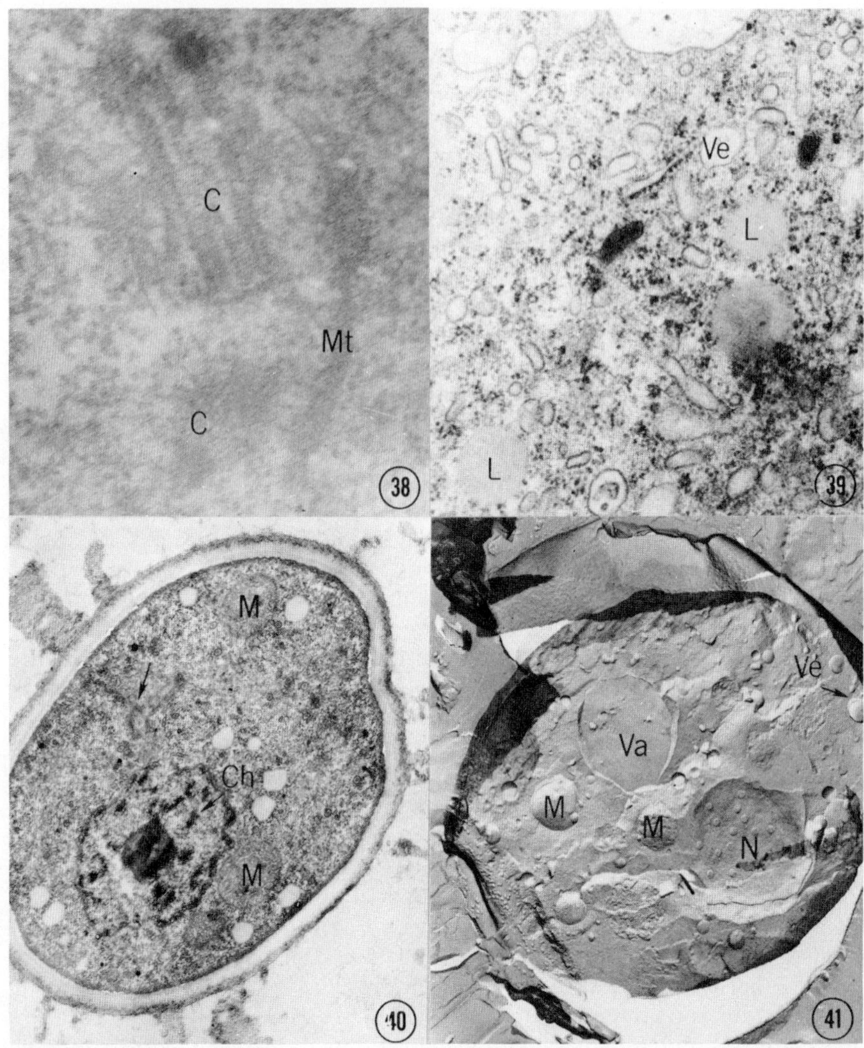

Fig. 38 *Didymium* sp. (In this and subsequent figures, this taxonomic designation refers to the unusual column-forming species described by Raper & Alexopoulos, 1973): Microcyst showing paired centrioles and some of the associated microtubular array (x 62,700).

Fig. 39 Smooth membraned vesicles of possible Golgi apparatus

a true morphological difference; glutaraldehyde-fixed, freeze-fractured microcysts also contain rounded mitochondrial profiles (Fig. 41). The microcyst plasmalemma is singular in that it is often crenulated (Figs. 40, 42) when compared to that of vegetative amoebae (Fig. 43).

The fibrous wall is of interest especially since it is variable between cells from the same cultures and between cells of different species. The intracultural differences are probably due to differences in age since the cultures in many cases were not closely synchronized. The differences between species appear to be actual interspecific differences. In *Physarum flavicomum, P. polycephalum,* and *Didymium* sp. the wall is composed of a single layer of microfibrils, randomly arranged (Figs. 42, 44). The microcyst wall of *Didymium iridis* is much thicker than those of the other three species and is made up of three layers which show differences in electron density. It is not known whether these layers represent differences in wall material composition or, merely, differences in degree of compactness of fibrils of a single component. In fact, no microcyst wall has been analyzed chemically as have been the walls of spores and sclerotial macrocysts. As Schuster (1965) found in *Didymium nigripes*, rudimentary spines are usually found in the outer wall layer of *D. iridis* (Fig. 40). The cyst wall of *D. iridis* appears far more complex than those of the other species; the layered wall and spines cause a close resemblance to the spore of this species.

Although single microcysts are formed by the *Didymium* sp. reported by Raper & Alexopoulos (1973), it is the columns of cysts, somewhat analogous to the sorocarp of acrasids, which are more striking. At the ultrastructural level the amoebal food vacuoles of this organism contain numerous myelin figures. Other species of myxomycete myxamoebae, also fed *Escherichia coli*, seldom break down food material in this manner. As the amoebae migrate to a common point to form the

---

origin in *Didymium iridis*. Some contain fibrous material (x 31,700).
Fig. 40 *Didymium iridis*: Microcyst containing paired centrioles (arrow), rounded mitochondrial profiles, and condensed chromatin. The three-layered wall bears spines on the outermost layer. Stained with alcian blue during osmium step (x 11,500).
Fig. 41 *Physarum polycephalum*, APT-1 mutant: Cross-fractured, freeze-etched microcyst prepared by spray-freezing (x 8,600).
Va, vacuole.

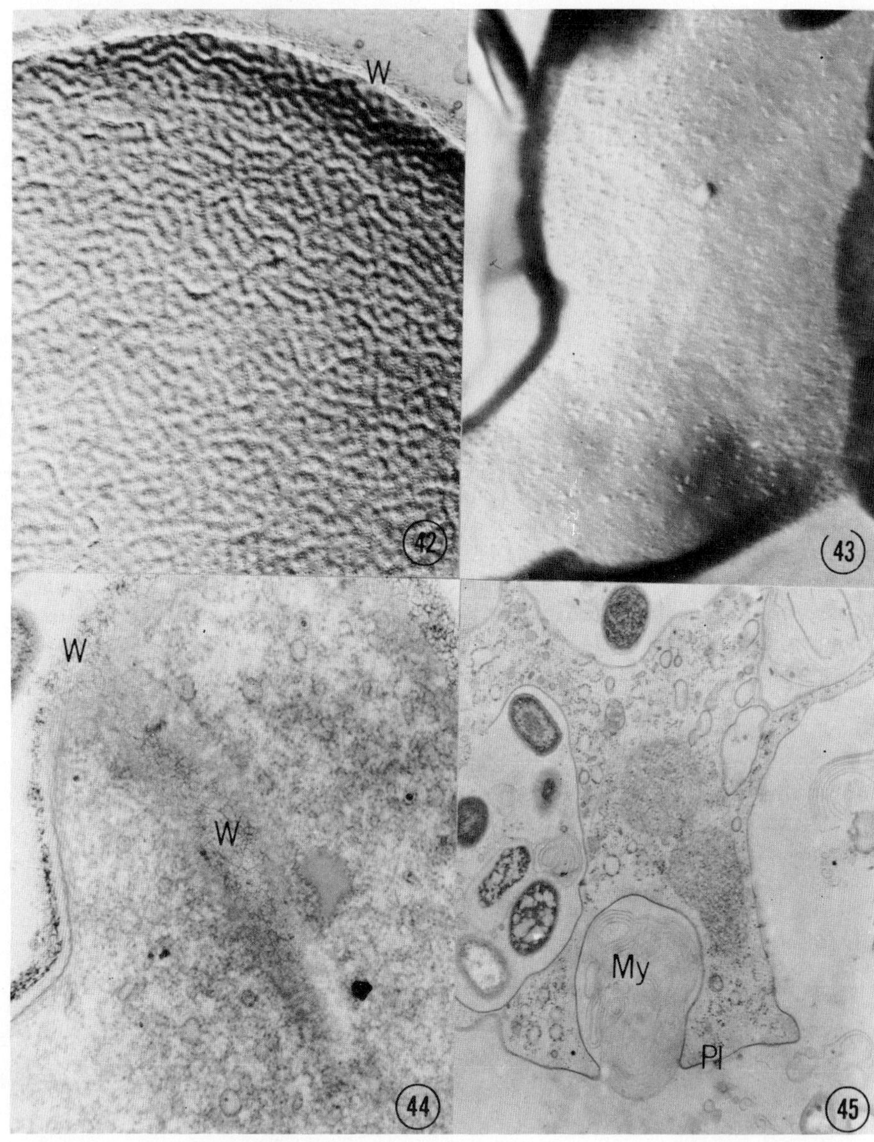

Fig. 42 *Physarum polycephalum*, Colonia isolate: Freeze-etched microcyst in surface view, showing crenulated plasmalemma and thin fibrous wall (x 25,800).

column, food vacuolar contents are extruded into the medium (Figs. 45, 46). The amoebae which soon encyst are thus embedded in and probably cemented together by the food vacuolar debris. Uningested food organisms are included in and sometimes coat the column structure (Fig. 47). No sheath such as the one found around acrasid sorocarps is present, nor is there any differentiation of the microcysts along a column. The cyst often appear irregular in shape and are seldom spherical as were the other cysts studied (Fig. 44). *Didymium* sp. has another unusual feature; it is not known to be confined to the microcysts, but so far as been found only in this stage. The nucleoli contain fine tubular structures; there are also the numerous dense granules found in many myxomycete nuclei (Figs. 48, 49). These structures are morphologically quite different from those lamellate structures found in *Acrasis rosea* (see chapter by Hohl in this volume). In another species of *Didymium*, *D. squamulosum*, a somewhat similar structure was found associated with sporic meiosis (see Fig. 16 this chapter).

Nuclear changes of microcysts are discussed below with those of the sclerotial nuclei.

Microcysts germinate by a pore to release myxamoebae or swarm cells. This is easily induced by the addition of water to old, dried cultures or by transferring the cysts to fresh medium. After wetting, cultures of *Physarum polycephalum* which had previously contained only microcysts, developed numerous swarm cells and myxamoebae. At the ultrastructural level the microcysts still present in the cultures contained many smooth membraned vesicles of possible Golgi origin and an increased amount of rough endoplasmic reticulum (Figs. 50, 51). Although rough endoplasmic reticulum is common in myxamoebae and swarm cells, it is not found once mycrocyst walls are formed until the cysts are wet and germination probably triggered.

Fig. 43 *Physarum polycephalum*, APT-1 mutant: Freeze-etched myxamoeba plasmalemma for comparision with Fig. 42. There are no crenulations evident even at high magnification (x 56,000).
Fig. 44 *Didymium* sp.: Grazing thin section of irregular microcyst wall, showing random arrangement of wall fibrils (x 24,596).
Fig. 45 Myxamoeba of *Didymium* sp.: Food vacuoles contain myelin figures which are not usually found in plasmodial slime molds. One is being extruded into the medium (x 13,400). My, myelin figure; Pl, plasmalemma.

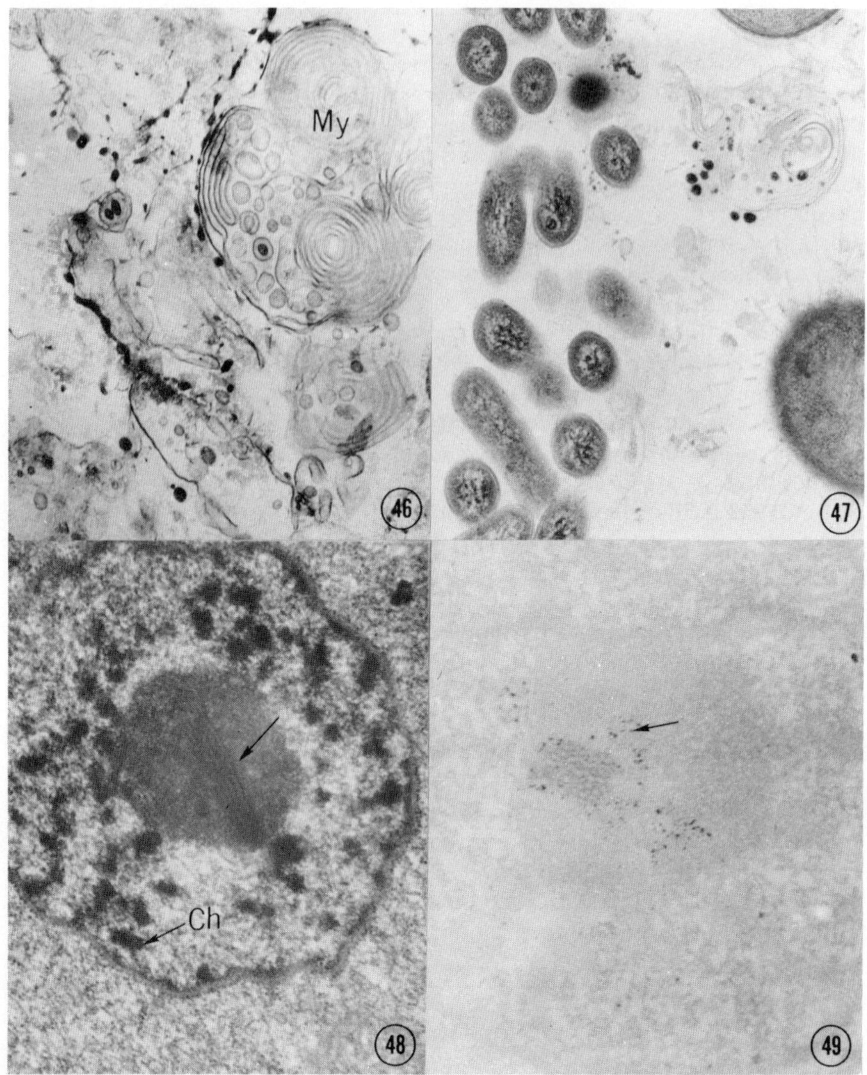

Fig. 46 *Didymium* sp.: Matrix area of columns, showing food vacuole material in which the encysting myxamoebae are embedded (x 21,800).
Fig. 47 *Didymium* sp.: Periphery of column at left with a coating of uningested bacteria (x 16,100).
Fig. 48 *Didymium* sp.: Microcyst nucleus with condensed chromatin and nucleolar inclusion (arrow) (x 21, 800).

In the flagellated protostelid, *Cavostelium bisporum*, Furtado & Olive (1970) observed that upon encystment the polarization of the swarm cell was lost, and the Golgi apparatus was altered to form a system of vesicles instead of the usual stacked cisternae. Membranes thought to be remnants of these vesicles were found between the plasmalemma and the wall. Although what they termed centrioles at this stage were retained, the associated microtubular arrays were eventually resorbed. Food vacuoles disappear from the cytoplasm before wall formation occurs. It should be pointed out that these cysts, as in many of the protostelids, vary in size and nuclear number. The protostelids which form plasmodia tend to have irregular cysts, while those which retain uninucleate protoplasts tend to have more regularly spherical cysts (Olive, 1967). It is not known which of these stages actually represent microcysts and which might actually be sclerotia.

Taylor & Kahn (1969) were able to nearly synchronize amoebal encystment in the nonflagellated protostelid *Protostelium mycophaga* by use of a technique designed for *Acanthamoeba*. They believed that cyst formation was dependent on a high concentration of inorganic monovalent cations, especially at a slightly basic pH (7.5). This did not appear to be caused by a raised osmotic pressure as was the case in *Acanthamoeba*. Based on PAS-Alcian blue and Lugol's iodine staining they found that cytoplasmic acid mucopolysaccharides became increasingly neutral, wall material was present after 3 h and appeared two layered after 27 h, at which time cysts were considered mature based on comparison with the appearance of stock cultures. No work of this sort has been done with any other species.

In *Ceratiomyxella tahitiensis*, another flagellated protostelid, two types of cysts are formed: thin-walled, uninucleate zoocysts derived from emerging protoplasts (the zoocysts give rise to flagellated cells) and thicker walled, multinucleate resting cysts which are derived from the plasmodia. The resting cyst walls are fibrous in nature and are surrounded by a fibrous mucilaginous coat; no evidence of any sort of cleavage into macrocysts is seen. Resting cysts probably represent sclerotia in the sense of this review.

No zoocyst wall was shown. An additional stage, walled with scales of probable Golgi activity, the amoebo-flagellate stage, is not

---

Fig. 49 *Didymium* sp.: Nonpoststained nucleus similar to Fig. 48, showing nucleolar inclusion in cross section. The dense granules (arrow) often seen in myxomycete nuclei are also present (x 31,000).

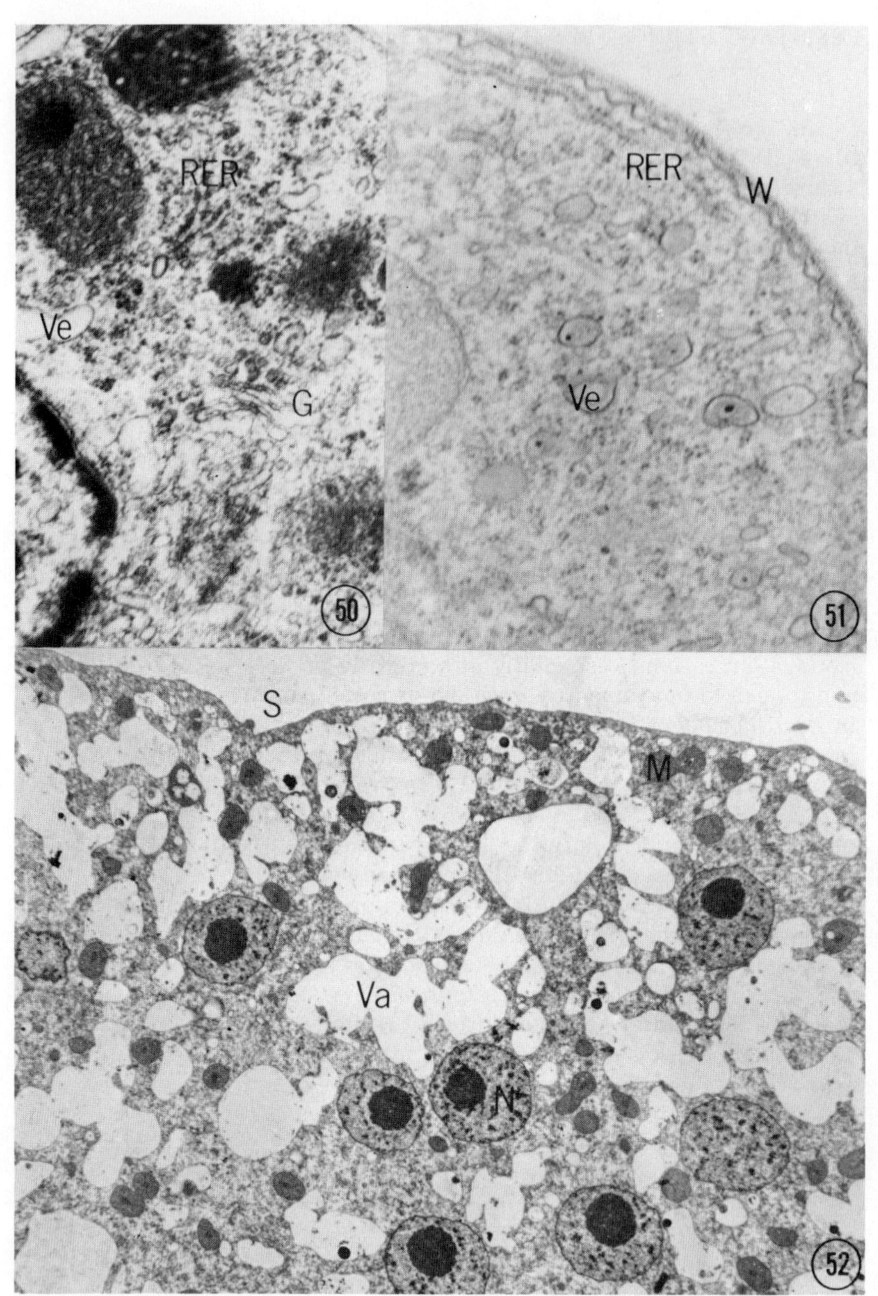
RER
Ve
G
50
RER
W
Ve
51
S
M
Va
N
52

thought to be resistant (Furtado & Olive, 1971).

SCLEROTIA

Sclerotization is a response of the plamodium to environmental factors which are favorable to neither plasmodial maintenance nor sporulation. That the ability to sclerotize affords the plasmodium a quick response to deleterious environmental conditions has long been known from observations both in nature and of cultures (see Gray & Alexopoulos, 1968, for a review of this information). Although this is a nonessential stage in the life cycle of the Myxomycetes, it is believed to play an important role in the survival of these organisms. Sclerotia of larger plasmodia are commonly found by experienced collectors; they are abundant in nature, perhaps in some instances overwintering, and certainly helping the organisms through periods when rapid drying occurs. Sporangial formation is a time consuming activity: a minium of four days' starvation, an exposure to light, the actual time of sporangial formation, a 24-h period of spore maturation are required (Daniel & Rusch, 1961; Aldrich, 1967). With sclerotization a resistant stage can form in less than 2 days (Daniel & Rusch, 1961).

Alexopoulos (1964) has discussed the rapid sporulation in moist chamber culture of some species of Myxomycetes, those which form aphanoplasmodia or protoplasmodia. It does appear that those species with aphanoplasmodia (in the broad sense) form a sclerotium which is different from that of the physaraceous species which will be discussed below. In general the macrocysts of *Stemonitis fusca* and *Arcyria cinerea* do not appear to be embedded in slime so that the cysts

---

Fig. 50 *Physarum polycephalum*: Microcyst 6 h after wetting. Note Golgi apparatus, many smooth membraned vesicles, and dispersed chromatin (x 30,800). G, Golgi apparatus.
Fig. 51 *Physarum polycephalum*: Microcyst similar to that in Fig. 50. Plasmalemma is crenulated within the single wall layer. Smooth membraned vesicles and rough ER are also present (x 28,500).
Fig. 52 *Physarum polycephalum*: Microplasmodium 8 h after induction of spherulation. Nuclear chromatin is condensed and large vacuoles appear to have fused in places (x 4,300). Micrograph by H. Kleinig. S, slime.

lie in the outline of the old plasmodium (Alexopoulos, 1964). Wollman (1966) reported this type of sclerotium in *Comatricha nodulifera,* and Nauss (1943), in *Metatrichia vesparium.* However, Wollman (1966) found that *M. vesparium* formed a sclerotium similar to that of *Physarum polycephalum.* Protoplasmodial species sometimes produce sporangia in moist chamber within 24 h. (Alexopoulos, 1964), as does *Protophysarum phoiogenum,* a species which fruits from a protoplasmodium or a small phaneroplasmodium (Blackwell, 1974). Other possible sclerotia of this type found in the protostelids have been mentioned above.

With the development of a semidefined, axenic medium for plasmodial growth of *Physarum polycephalum* (Daniel & Rusch, 1961) and the techniques for sclerotial induction (Daniel & Rusch, 1961; Chet & Rusch, 1969), much information about the biochemical events of sclerotial differentiation and some on plasmodial reconstitution has been obtained. This material has been reviewed recently by Sauer (1973), Hüttermann (1973), and Chet (1973). Lynch & Henney (1973) have discussed carbohydrate metabolism during sclerotization of another species, *P. flavicomum.*

Jump (1954) showed that sclerotia of *Physarum polycephalum* could be induced to form by desiccation, low temperatures, low PH, high osmotic pressure, sublethal concentrations of heavy metal ions, starvation, and presence of certain fungal contaminants. Plasmodial streaming stopped, the cytoplasm became gelatinized, wall material was deposited around small protoplasmic entities, the entire mass became hardened, and nuclear diameter was reduced to about half of that in the plasmodium. These individual cells of the sclerotium have been termed macrocysts, spherules, and sclerotiospores. As Sauer (1973) has pointed out, the term "spherule" should be reserved for the cells of sclerotial formed in liquid culture, and this special case of sclerotization should be called spherulation. We shall use "macrocyst" for the cells of sclerotia formed by other means.

Both spherules and macrocysts of physaraceous species are usually embedded in a large amount of slime, which upon drying, hardens and acts as a common, horny covering. Chet & Rusch (1969) found that mannitol induced spherules were formed more synchronously, but were larger and not clustered as those formed in starvation medium. It is thought that the nonclustered condition may be a result of absence of much slime.

Several ultrastructual studies have produced information on sclerotization and mature spherules and macrocysts; some of it is contradictory. Goodman & Rusch (1970) induced spherulation in a nonnutrient salts medium. Within 15 h of inoculation into the medium a 70% decrease in glycogen and an increase in vesicular structures was found. Between 18 and 24 h the Golgi apparatus was seen, these cisternae do not have quite the appearance of the well developed, extensive Golgi apparatus of the myxamoebal and swarm cell stages. They found cytoplasmic cleavage furrows centripetally formed from fusion of prealigned vesicles at about 24 h. Wall material of 34--40 Å diameter microfibrils was deposited just after cleavage. These findings agree with those of Stewart & Stewart (1961) on similar material. During development to this point only short segments of rough endoplasmic reticulum were present in the cytoplasm, but this changed to a primarily smooth system. Nuclear changes were also noted. Kleinig & Zarr (personal communication) working with spherulation in *P. polycephalum*, suggest a different method of cleavage furrow formation which occurs approximately 30 h after induction. This agrees with the results of Stiemerling (1971), who studied desiccation induced sclerotia of *Physarum confertum*. In this case it was believed that a vacuolar system formed from fused slime and food vacuoles completed the cleavage furrows to give rise to the discrete multinucleated macrocysts (Figs. 52--54).

In mature low temperature induced spherules Rhea (1966) found electron dense mitochondrial intracristal and vesicular inclusions which were not seen in plasmodia. They are morphologically like the dense bodies found in microcyst formation of *Didymium* sp. reported above, which are like the inclusion bodies of developing sporangia; these are known to be at least partially calcium (Nicholls, 1972; Daniel & Jarlfors, 1972a, 1972b; Gustafson, 1973; Blackwell, 1974).

We have thin sectioned and freeze-etched macrocysts in our own laboratory. Aged plasmodia of *P. polycephalum* (Colonia isolate) were allowed to sclerotize on agar plates. Thin sectioned and freeze fractured mature macrocysts contain both rough and smooth endoplasmic reticulum and autophagic vacuoles as found by Kleinig & Zarr (personal communication) (Fig. 55) and rough membranes underlie the plasmalemma (Figs. 56, 57). The walls are clearly of a fibrous nature (Figs. 55--59), embedded in the loosely arranged fibrous slime. At the sclerotial surface, the slime is very compacted (Figs. 58, 59). McCormick et al., (1970a) have found that this slime, a sulphated galactose polymer, during plasmodial growth and sclerotization is identical, while the spherule wall material is a galactosamine polymer (McCormick et al., 1970b).

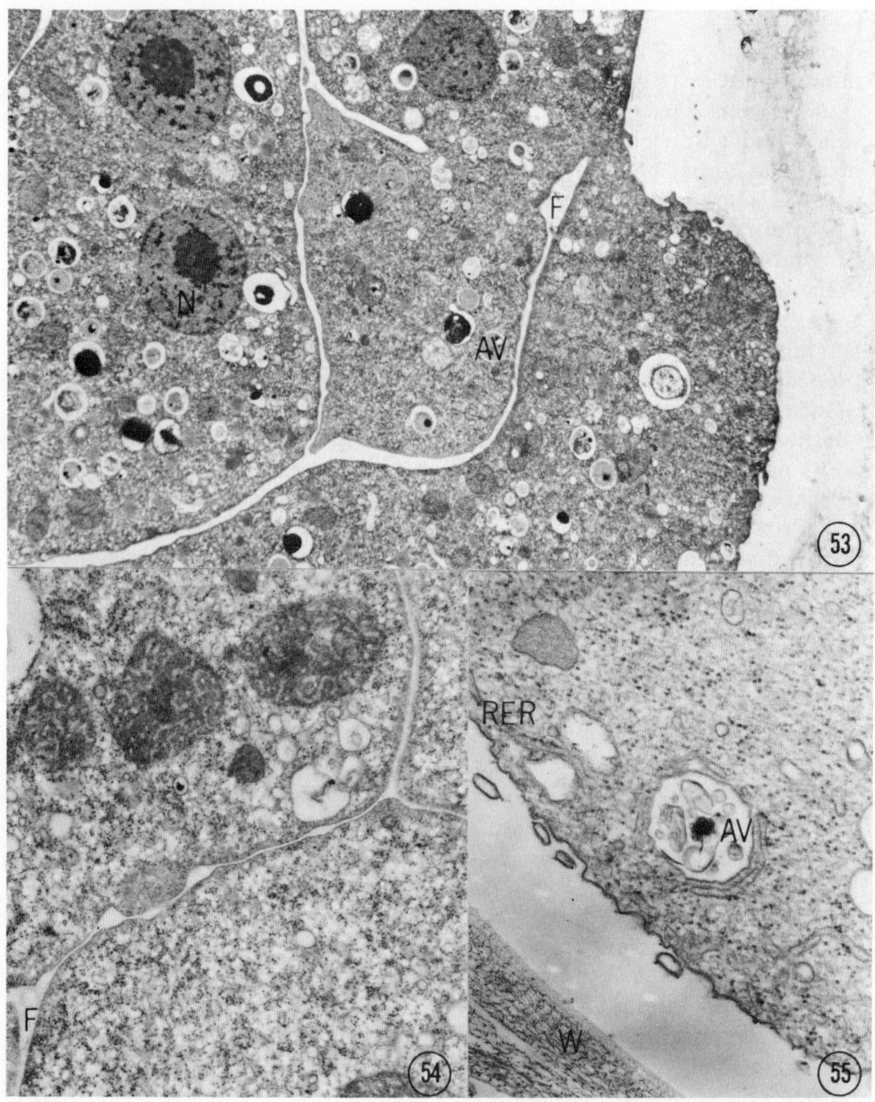

Fig. 53 *Physarum polycephalum*: Microplasmodium 30 h after induction. Cytoplasmic microfilaments may be involved in formation of these clefts, which do not seem to be formed by the fusion of any particular vesicles (x 5,900). Micrograph by H. Kleining. F, fibrils.

Nuclei of both microcysts and sclerotia have a somewhat different appearance from the nonencysted forms; the chromatin is in a far more condensed state (Figs. 40, 48, 52, 53). Figure 50 shows the nucleus of a cyst which was wet 6 h earlier. Goodman & Rusch (1970) noted less densely staining chromocenters and more easily distinguished nucleolar components in spherules of *Physarum polycephalum*. It is quite likely that these observations indicate a real change in the nuclei of these encysted forms. Mohberg & Rusch (1971) have measured amounts of nuclear DNA in different stages of the myxomycete life cycle. Although G_1 phase is not known in the cell cycle of *Physarum polycephalum*, they found half the amount of DNA in the encysted amoebae and spherules (0.3 and 0.6 pg) as in the vegetative amoebae and plasmodia (0.6 and 1--1.2). Perhaps only those cells which have recently divided enter a G_1 phase and encyst. This also correlates with the decrease in nuclear size after macrocyst wall formation (Jump, 1954). We have made an additional observation on sclerotial nuclei of *P. polycephalum*. These nuclear envelopes often have large poreless areas (Fig. 60) when compared to the more evenly distributed pores on envelopes of plasmodia (Fig. 61).

Thus, we see that Myxomycetes produce three types of resistant structures: long-lived spores, known in some cases to be the site of meiosis, and therefore the bearers of the recombinants of sexual reproduction; and microcysts and sclerotia, not so long-lived as spores, but the result of a quick response by a mass of unprotected protoplasm to harsh, often sudden, environmental stimuli. New cultural methods and techniques have greatly supplemented earlier observations and are providing a wealth of new information on these resistant structures, but knowledge of their role in nature is still largely lacking.

ACKNOWLEDGMENTS

We express our sincere appreciation to Drs. K.B. Raper and

Fig. 54 *Physarum polycephalum*: Microplasmodium after cleft formation is complete (x 16,000). (Micrograph by H. Kleinig).
Fig. 55 *Physarum polycephalum*, Colonia isolate: Macrocyst (sclerotium formed on old agar culture) showing autophagic vacuole, rough ER, and fibrous wall (x 34,300).

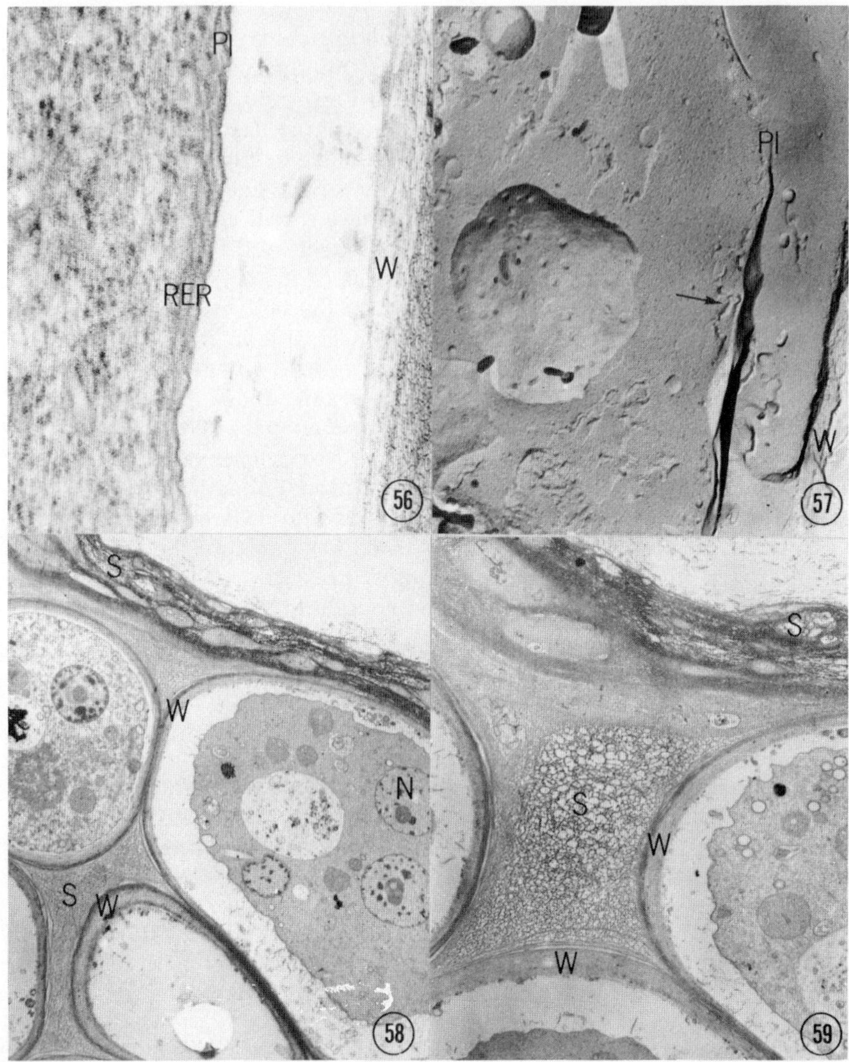

Fig. 56 *Physarum polycephalum*, Colonia isolate: Plasmalemma of mature macrocyst. Wall is fibrous. Rough ER underlies plasmalemma (x 53,800).

Fig. 57 *Physarum polycephalum*, Colonia: Cross-fractured freeze-etched cell of macrocyst. Endoplasmic reticulum (arrow) lies just

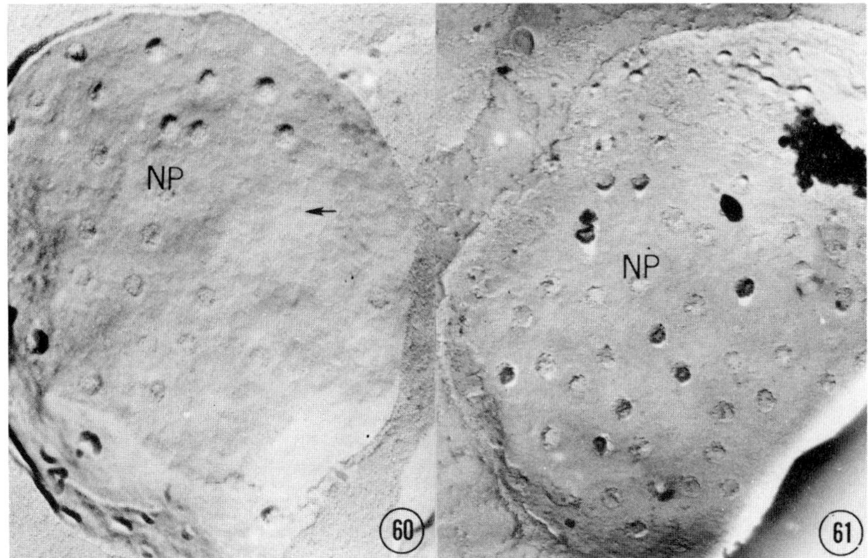

Fig. 60 *Physarum polycephalum*, Colonia: Freeze-etched macrocyst nucleus. Nuclear envelope often shows regions without pores (arrow) (x 30,000). NP, nuclear pore.
Fig. 61 *Physarum polycephalum*, Colonia: Freeze-etched plasmodial nucleus for comparison with Fig. 60. Nuclear pores are evenly distributed over surface of envelope (x 30,000).

Jennifer Dee for providing cultures, to Drs. Iris Charvat, Ralph Gustafson, Raymond Scheetz, H. Kleinig, and Charles Mims for generously furnishing unpublished data and micrographs, and to Dr. Mims for helpful discussions and critically reading the manuscript. All uncredited micrographs originated in the Biological Ultrastructure Laboratory at the University of Florida, Gainesville.

---

under plasmalemma. Fibrous wall is also visible (x 12,500).
Fig. 58 *Physarum polycephalum*, Colonia: Edge of whole sclerotium. Individual walled macrocysts are embedded in slime which is continuous over and compacted at surface. Chromatin is condensed (x 3,400).
Fig. 59 *Physarum polycephalum*, Colonia: Same as Fig. 58, showing fibrous nature of slime and macrocyst walls (x 5,800).

REFERENCES

Aldrich, H. C. (1967). Ultrastructure of meiosis in three species of *Physarum*. Mycologia 59, 127--148.

Aldrich, H. C. (1974). Spore cleavage and the development of wall ornamentation in two Myxomycetes. Proc. Iowa Acad. Sci. 81, 28--36.

Aldrich, H. C. & Carroll, G. C. (1971). Synaptonemal complexes and meiosis in *Didymium iridis:* A reinvestigation. Mycologia 63, 308--316.

Aldrich, H. C. & Mims, C. W. (1970). Synaptonemal complexes and meiosis in Myxomycetes. Amer. J. Bot. 57, 935--941.

Alexopoulos, C. J. (1962). The Myxomycetes II. Bot. Rev. 29, 1--78.

Alexopoulos, C. J. (1964). The rapid sporulation of some Myxomycetes in moist chamber. Southwest. Nat. 9, 155--159.

Alexopoulos, C. J. (1973). Myxomycetes. In The Fungi: An Advanced Treatise, Vol. IV B, pp. 39--60 (eds. G. C. Ainsworth, F. K. Sparrow, & A. S. Sussman). New York: Academic Press.

Anderson, T. F. (1951). Techniques for the preservation of three dimensional structure in preparing specimens for the electron microscope. Trans. N.Y. Acad. Sci., Ser. II, 13, 130.

Bigelow, H. E. & Rowley, J. R. (1968). Surface replicas of the spores of fleshy fungi. Mycologia 60, 869--887.

Blackwell, M. (1974). A study of sporophore development in the myxomycete *Protophysarum phloiogenum*. Arch. Microbiol., 99, 331--344.

Boyde, A. & Wood, C. (1969). Preparation of animal tissues for surface SEM. J. Microsc. 90, 221--249.

Braun, K. L. (1971a). Spore germination time in *Fuligo septica*. Ohio J. Sci. 71, 304--309.

Braun, K. L. (1971b). Effects of aeration on the germination of *Fuligo septica* spores. Mycologia 63, 669--671.

Carr, K. E. (1971). Applications of scanning electron microscopy in biology. Internat. Rev. Cytol. 30, 183--256.

Cathcart, M. E. (1974). An ultrastructural study of sporulation and spore germination in *Fuligo septica*. M. S. Thesis, University of Florida, 30 pp.

Chet, I. (1973). Changes in ribonucleic acid during differentiation. Ber. Deutsch. Bot. Ges. 86, 77--92.

Chet, I & Rusch, H. P. (1969). Induction of spherule formation in *Physarum polycephalum*. J. Bacteriol. 100, 673--678.

Cronshaw, J. & Charvat, I. (1973). Localization of beta-glycerophosphatase activity in the myxomycete *Perichaena vermicularis*. Can. J. Bot. 51, 97--101.

Daniel, J. W. & Jarlfors, U. (1972a). Plasmodial ultrastructure of the myxomycete *Physarum polycephalum*. Tissue Cell 4, 15--36.

Daniel, J. W. & Jarlfors, U. (1972b). Light induced changes in the ultrastructure of a plasmodial myxomycete. Tissue Cell 4, 405--426.

Daniel, J. W. & Rusch, H. P. (1961). The pure culture of *Physarum polycephalum* on a partially defined soluble medium. J. Gen. Microbiol. 25, 47--59.

Davis, E. E. (1965). Preservation of myxomycetes. Mycologia 57, 986--988.

Fujii, R. K., & Hess, W. M. (1969). Surface characteristics of conidia from monosporous cultures of *Penicillium digitatum* and *Aspergillus nidulans* var. *echinulatus*. Can. J. Microbiol. 15, 1472--1473.

Fuller, M. S. (1966). Structure of the uniflagellate zoospores of aquatic Phycomycetes. In The Fungus Spore, (ed. M. F. Madelin), 338 pp. London: Butterworths.

Furtado, J. S. & Olive, L. S. (1970). Ultrastructural studies of protostelids: The cyst stage of *Cavostelium bisporum*. Protoplasma 70, 379--387.

Furtado, J. S. & Olive, L. S. (1971). Ultrastructure of the protostelid *Ceratiomyxella tahitiensis* including scale formation. Nova Hedwigia 21, 537--576.

Furtado, J. S., Olive, L. S., & Jones, S. B. (1971). Ultrastructural studies of protostelids: The fruiting stage of *Cavostelium bisporum*. Mycologia 63, 132--143.

Gilbert, F. A. (1928). A study of the method of spore germination in Myxomycetes. Am. J. Bot. 15, 345--352.

Gilbert, F. A. (1929). Factors influencing the germination of myxomycetous spores. Amer. J. Bot. 16, 280--286.

Goodman, E. M. (1972). Axenic culture of myxamoebae of the myxomycete *Physarum polycephalum*. J. Bacteriol. 111, 242--247.

Goodman, E. M. & Rusch, H. P. (1969). Glycogen in *Physarum polycephalum*. Experientia 25, 580.

Goodman, E. M. & Rusch, H. P. (1970). Ultrastructural changes during spherule formation in *Physarum polycephalum*. J. Ultrastr. Res. 30, 172--183.

Goodwin, D. C. (1961). Morphogenesis of the sporangium of

Comatricha. Am. J. Bot. 48, 148--154.

Gray, W. D. & Alexopoulos, C. J. (1968). Biology of the Myxomycetes. New York: Ronald Press.

Gustafson, R. A. (1973). The life cycle of *Didymium squamulosum*. A physiological and ultrastructural study. Ph.D. Thesis, Univ. of Texas, Austin.

Gustafson, R. A. (1974). Personal communication.

Haskins, E. F. (1971). Sporophore formation in the Myxomycete *Echinostelium minutum* de Bary. Arch. Protistenk. 113, 123--129.

Haskins, E. F., Hinchee, A. A., & Cloney, R. C. (1971). The occurrence of synaptonemal complexes in the slime mold *Echinostelium minutum* de Bary. J. Cell Biol. 51, 898--903.

Hayat, M. A. (1970). Principles and Techniques of Electron Microscopy Vol. I. New York: Van Nostrand Reinhold.

Hemmes, D. E., Kojima-Buddenhagen, E. S., Hohl. H. R. (1972). Structural and enzymatic analysis of the spore wall layers in *Dictyostelium discoideum*. J. Ultrastr. Res. 41, 406--417.

Henney, H. R. & Asgari, M. (1975). Growth of the haploid phase of the myxomycete *Physarum flavicomum* in defined minimal medium. Arch. Microbiol. 102, 175--178.

Hung, C. Y. & Olive, L. S. (1972). Ultrastructure of the spore wall in *Echinostelium*. Mycologia 64, 1160--1163.

Huttermann, A. (1973). Biochemical events during spherule formation of *Physarum polycephalum*. Ber. Deutsch. Bot. Ges. 86, 55--76.

Jump, J. A. (1954). Studies on sclerotization in *Physarum polycephalum*. Am. J. Bot. 41, 561--567.

Kerr, N. S. (1965). A simple method of lyophilization for the long-term storage of slime molds and small soil amoebae. BioScience 15, 469.

Kowalski, D. T. (1970). The species of *Lamproderma*. Mycologia 62, 621--673.

LeStourgeon, W. M., Bohstedt, C. F., & Thimell, P. (1971). Supportive evidence for postcleavage meiosis in *Physarum flavicomum*. Mycologia 63, 1002--1012.

Locke, M. V. & Collins, J. V. (1965). The structure and formation of protein granules in the fat body of an insect. J. Cell Biol. 26, 257--368.

Lynch, T. J. & Henney, H. R. (1973). Carbohydrate metabolism during differentiation (sclerotization) of the myxomycete

Physarum flavicomum. Arch. Mikrobiol. 90, 189--198.
Martin, G. W. (1940). The Myxomycetes. Bot. Rev. 6, 356--388.
Martin, G. W. & Alexopoulos, C. J. (1969). The Myxomycetes Iowa City: University of Iowa Press.
McCormick, J. J., Blomquist, J. C., & Rusch, H. P. (1970a). Isolation and characterization of an extracellular polysaccharide from *Physarum polycephalum*. J. Bacteriol. 104, 1110--1118.
McCormick, J. J., Blomquist, J. C., & Rusch, H. P. (1970b). Isolation and characterization of a galactosamine wall from spores and sphaerules of *Physarum polycephalum*. J. Bacteriol. 104, 1119--1125.
Menzel, M. Y. & Price, J. M. (1966). Fine structure of synapsed chromosomes in F_1 *Lycopersicon esculentum-Solanum lycopersicoides* and its parents. Am. J. Bot. 53, 1079--1086.
Mims, C. W. (1971). An ultrastructural study of spore germination in the Myxomycete *Arcyria cinerea*. Mycologia 63, 586--601.
Mims, C. W. (1972a). Centrioles and Golgi apparatus in postmeiotic spores of the Myxomycete *Stemonitis virginiensis*. Mycologia 64, 452--456.
Mims, C. W. (1972b). Spore wall formation in the Myxomycete *Arcyria cinerea*. Trans. Brit. Mycol. Soc. 59, 477--581.
Mims, C. W. & Rogers, M. A. (1973). An ultrastructural study of spore germination in the Myxomycete *Stemonitis virginiensis*. Protoplasma 78, 243--254.
Mohberg, J. & Rusch, H. P. (1971). Isolation and DNA content of nuclei of *Physarum polycephalum*. Exp. Cell Res. 66, 305--316.
Mollenhauer, H. H. (1964). Plastic embedding mixtures for use in electron microscopy. Stain Tech. 39, 111.
Moor, H. & Muhlethaler, K. (1963). Fine structure in frozen-etched yeast cells. J. Cell Biol. 17, 609--628.
Nauss, R. (1943). Observations on the culture of *Hemitrichia vesparium* with special reference to its black plasmodial color. Bull. Torrey Bot. Club 70 152--163.
Nicholls, A. V. (1972). The effects of starvation and light on intramitochondrial granules in *Physarum polycephalum*. J. Cell Sci. 10, 1--14.
Nields, H. W. (1968). Hexosamine formation in the Myxomycete *Physarum polycephalum*. J. Cell Biol. 39, 100a (abstract).
Olive, L. S. (1967). The Protostelida---A new order of Mycetozoa. Mycologia 59, 1--29.
Olive, L. S. (1970). The Mycetozoa: A revised classification. Bot.

Rev. 36, 59--89.

Olive, L. S. (1975). The Mycetozoans. New York: Academic Press. 293 pp.

Pickett-Heaps, J. D. (1973). Stereo-scanning electron microscopy of desmids. J. Microsc. 99, 109--116.

Rambourg, A. (1971). Morphological and histochemical aspects of glycoproteins at the surface of animal cells. Int. Rev. Cytol. 31, 57.

Randall, L. P. & Lynch, D. L. (1974). Spore ultrastructure of the myxomycete *Physarum polycephalum*. Am. J. Bot. 61, 513--524.

Raper, K. B. & Alexopoulos, C. J. (1973). A myxomycete with a singular myxamoebal encystment stage. Mycologia 65, 1284--1295.

Revel, J. P. (1964). Electron microscopy of glycogen. J. Histochem. Cytochem. 12, 104--114.

Rhea, R. P. (1966). Electron microscopic observations on the slime mold *Physarum polycephalum* with specific reference to fibrillar structures. J. Ultrastruct. Res. 15, 349--379.

Rusch, H. P. (1970). Some biochemical events in the life cycle of *Physarum polycephalum*. In Advances in Cell Biology, Vol. 1, pp. 297--327 (eds. D. M. Prescott, L. Goldstein & E. McConkey). New York: Appleton-Century Crofts.

Sauer, H. W. (1973). Differentiation in *Physarum*. Soc. Gen. Microb. Symp. 23, 375--405.

Scheetz, R. W. & Alexopoulos, C. J. (1971). The spores of *Badhamia gracilis* (Myxomycetes). Trans. Am. Micros. Soc. 90, 473--475.

Schoknecht, J. D. & Small, E. B. (1972). Scanning electron microscopy of the acellular slime molds (Mycetozoa=Myxomycetes) and the taxonomic significance of surface morphology of spores and accessory structures. Trans. Am. Micros. Soc. 91, 380--410.

Schuster, F. (1964). Electron microscope observations on spore formation in the true slime mold *Didymium nigripes*. J. Protozool. 11, 207--216.

Schuster, F. (1965). Ultrastructure and morphogenesis of solitary stages of true slime molds. Protistologica 1, 49--62.

Spurr, A. R. (1969). A low-viscosity epoxy resin embedding medium for electron microscopy. J. Ultrastr. Res. 26, 31.

Stewart, P. A. & Stewart, B. T. (1961). Membrane formation during sclerotization of *Physarum polycephalum* plasmodia. Exp. Cell Res. 23, 471--478.

Stiemerling, R. (1971). Die Sklerotisation von *Physarum confertum*. Cytobiologie. 3, 127--136.

Taylor, P. & Kahn, A. J. (1969). Initiation of encystment in the mycetozoan *Protostelium mycophaga*. J. Protozool. 16, 152--154.

Therrien, C. D. & Yemma, J. J. (1974). Comparative measurements of nuclear DNA in a heterothallic and a self-fertile isolate of the Myxomycete *Didymium iridis*. Am. J. Bot. 61, 400--404.

Von Stosch, H. A. (1935). Untersuchungen Über die Entwicklungsgeschichte der Myxomyceten: Sexualität und Apogamie bei Didymiaceen. Planta 23, 623--656.

Wollman, C. (1966). Cultural studies of selected species of Myxomycetes. Ph.D. Thesis, Univ. of Texas, Austin.

DISCUSSION

Resistant Structures in the Myxomycetes

Chairman: David A. Cotter
Southeastern Massachusetts University
North Darthmouth, Massachusetts

Dr. Alfred Sussman suggested that colchicine might be used to distinguish between alternative functions for microtubules. He felt that the early addition of colchicine might prevent microtubule formation in the cytoplasm of the plasmodial slime mold. This treatment should arrest meiosis. The removal of the colchicine should then allow the microtubules to migrate into the nucleus as suggested by Dr. Aldrich and meiosis should then continue. Dr. Sussman also felt that colchicine might be employed to determine if microtubules function to produce cell shape in the plasmodial slime molds.

Dr. Aldrich answered that the production of polygonal spores after colchicine treatment would produce evidence for his hypothesis. However, he pointed out that colchicine either does not penetrate the cells or at least does not affect myxomycete nuclei. He stated that cold treatments or high hydrostatic pressure might block microtubule formation and thus produce the effects suggested by Dr. Sussman. Dr. Aldrich also indicated that pronase did not digest the microtubular bands in the cytoplasm; this was not surprising since many kinds of microtubules are not digestible with pronase.

CHAPTER 11 MYXOMYCETES

H. R. Hohl
University of Zürich

INTRODUCTION

Cellular slime molds do not form true plasmodia during their development. Instead, the myxamoebae aggregate to form multicellular pseudoplasmodia which are ultimately transformed either into stalked sori bearing the spores or into macrocysts. Alternatively each individual myxamoeba may directly become a microcyst in the absence of any cell aggregation (Fig. 1; for further details of the life cycle see Bonner, 1967). Thus the cellular slime molds possess a unicellular cycle involving the formation and germination of microcysts and a multicellular cycle involving the transformation of cell masses into spore-bearing sorocarps or macrocysts, both of which give rise again to myxamoebae upon germination.

For heuristic reasons it is useful to divide the complex developmental sequences of these cycles into those concerned with differentiation of myxamoebae to microcysts, spores, or stalk cells, and those dealing with cell-to-cell interactions leading to a specific spatial arrangement of the various cell types in the mature sorocarp or macrocyst. There exists an impressive and rapidly expanding literature on these aspects of temporal and spatial differentiation which has been reviewed in a series of papers (e.g., Gerisch, 1968; Francis, 1969; Sussman & Sussman, 1969; Ashworth, 1971; Bonner, 1971; Newell, 1971; Garrod & Ashworth, 1973; Gregg & Badman, 1973; Wright, 1973; Killick & Wright, 1974). These investigations have lead to sophisticated models of development in the cellular slime molds, with emphasis placed on the formation of the sorocarp.

In this chapter we will concentrate on the formation and

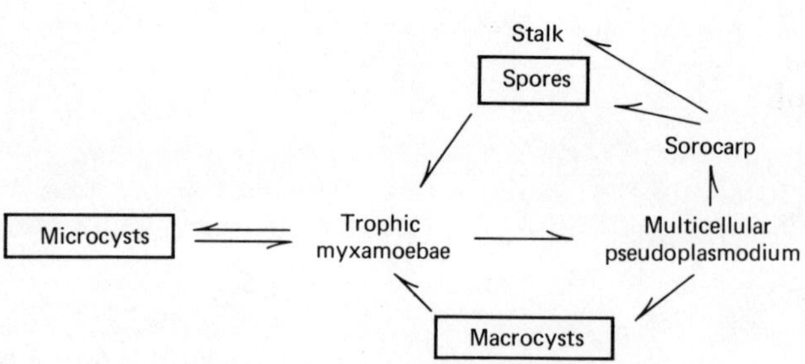

Fig. 1 Generalized life cycle of a cellular slime mold.

germination of spores, microcysts, and macrocysts, that is, on the temporal aspects of differentiation. Not all the species produce all three of these resting stages but some do. Emphasis will be placed on the correlation of ultrastructural and biochemical features, and also on a comparative approach to these structures, each of which is very distinct and occupies its characteristic position in the life cycle.

The study of these organisms is facilitated by the fact that the entire developmental cycle takes place after the vegetative stage is completed. This means that development can be studied separately from events such as feeding, growth, and cell division. Furthermore, these organisms share with the zoosporic fungi the advantage of forming naked protoplasts as a natural stage of their life cycle; that is, spore or cyst wall formation may be particularly well studied.

MYXAMOEBAE

Before dealing with the formation of the resting stages it is appropriate to include a few remarks about the myxamoebae from which all these structures are derived. The ultrastructure of the myxamoebae has been described for various species: *Acrasis rosea*

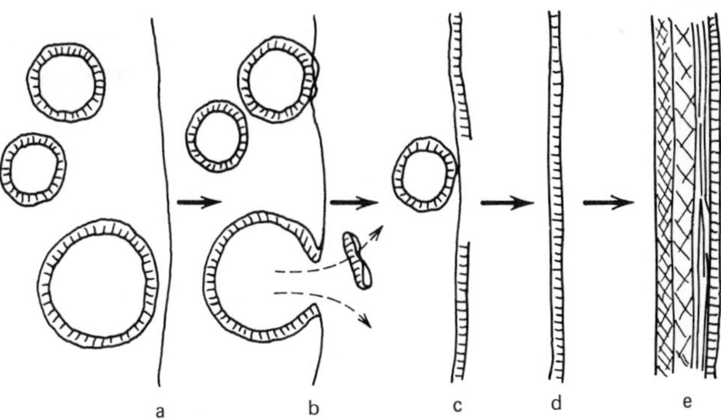

Fig. 2 Spore wall formation in *Dictyostelium*. (a) Prespore vacuoles at cell periphery, (b) prespore vacuoles fusing with plasma membrane, (c) incomplete, (d) completed first wall layer, (e) formation of additional wall layers.

(Hohl & Hamamoto, 1968, 1969a), *Acytostelium leptosomum* (Hohl et al., 1968), *Dictyostelium discoideum* (Gezelius & Ranby, 1957; Gezelius 1959, 1961, 1971; Mercer & Shaffer, 1960; Hohl, 1965; Hohl & Hamamoto, 1969b; Maeda & Takeuchi, 1969; Gregg & Badman, 1970; George et al., 1972), *D. mucoroides* and *D. purpureum* (Hohl, unpublished), *Polysphondylium pallidum* (Hohl, 1965; Hohl et al., 1970) and *P. violaceum* (Joss, unpublished manuscript).

The myxamoebae display a typical eucaryotic ultrastructure essentially similar to that of soil amoebae. Among the species investigated, *A. rosea* is distinguished from all others by its disk-shaped instead of tubular mitochondrial cristae and by the extensive lamellar inclusions in its nucleolus (Hohl & Hamamoto, 1968). Furthermore, it does not develop the prespore cell specific vacuoles (PV) characteristically formed during the development of the other species. According to Bonner (1967), *A. rosea* has the *Sappinia*-type myxamoeba while all the others are of the *Dictyostelium*-type, a difference obviously reflected in its ultrastructure.

Surprisingly no report on the ultrastructural aspects of mitosis has been published so far. In a recent article (Roos, 1975) this deficiency is made up for one species, *P. violaceum*. In this species the nuclear envelope appears to remain intact throughout mitosis, the spindle apparatus is built from a bundle of parallel microtubules and terminates at the nuclear envelope with the spindle poles marked by small, dense spindle pole bodies (Plate 1, Fig. 6). Kinetochore microtubules have also been observed. This line of investigation needs to be further pursued and might provide important clues for possible ancestors of this group among the Protozoa or Fungi.

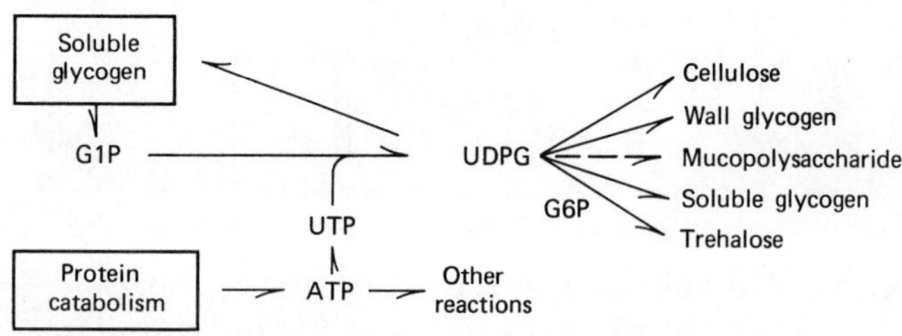

Fig. 3 Pathways of spore polysaccharide formation in *Dictyostelium* (adapted from Wright, 1973).

SPORES

Spore Formation

By a mechanism not entirely understood the presumptive spore cells become located in the posterior, the presumptive stalk cells in the anterior section of the pseudoplasmodium. During culmination the prestalk cells contribute to the formation of the cellulose sheath surrounding the stalk and finally turn into highly vacuolated, dead stalk cells after having surrounded themselves with a cellulose wall. Each prespore cell will develop into a single, uninucleate spore surrounded by a multilayered spore case.

It is possible to disturb the normal morphogenetic pattern without interfering with normal cell differentiation, for example, by treating the system with EDTA (Gerisch, 1961) or by producing aberrant mutants (Sonneborn et al., 1963). One may even induce unaggregated myxamoebae to turn into stalk cells by exposing them to cAMP (Bonner, 1970). But the absence of reports of spore differentiation outside the confines of a pseudoplasmodium indicates that it depends strictly on a multicellular condition.

Prespore cells become structurally distinguishable from prestalk cells by the presence of prespore vacuoles (PV) as shown by Hohl & Hamamoto (1969b) and Maeda & Takeuchi (1969). These vacuoles are formed in increasing numbers in prespore cells after aggregation is completed (Gregg & Badman, 1970; Müller & Hohl, 1973) and are intimately involved in spore wall formation (Hohl & Hamamoto, 1969b). They are most likely the site of the polysaccharide spore antigens demonstrated by Takeuchi (1963) with the use of immunofluorescence techniques. By this technique the fate of the prespore cells under a variety of experimental conditions has been studied (Sakai, 1973): fragmentation of the pseudoplasmodium leads to the interconversion of prestalk and prespore cells, that is, each fragment differentiates into a normal pseudoplasmodium, a phenomenon already studied by Raper (1940). The conversion is inhibited at low temperature, in the presence of cycloheximide or Actinomycin D. LiCl inhibits the transition from prestalk cells to prespore cells although the reverse event is not affected. Isolated prespore cells eventually lose their PV, slowly when not dividing, rapidly when induced to divide (Gregg, 1971).

The PV are approximately 0.7 μm in diameter. In species such as *P. pallidum* and *P. violaceum* several vacuoles may be interconnected

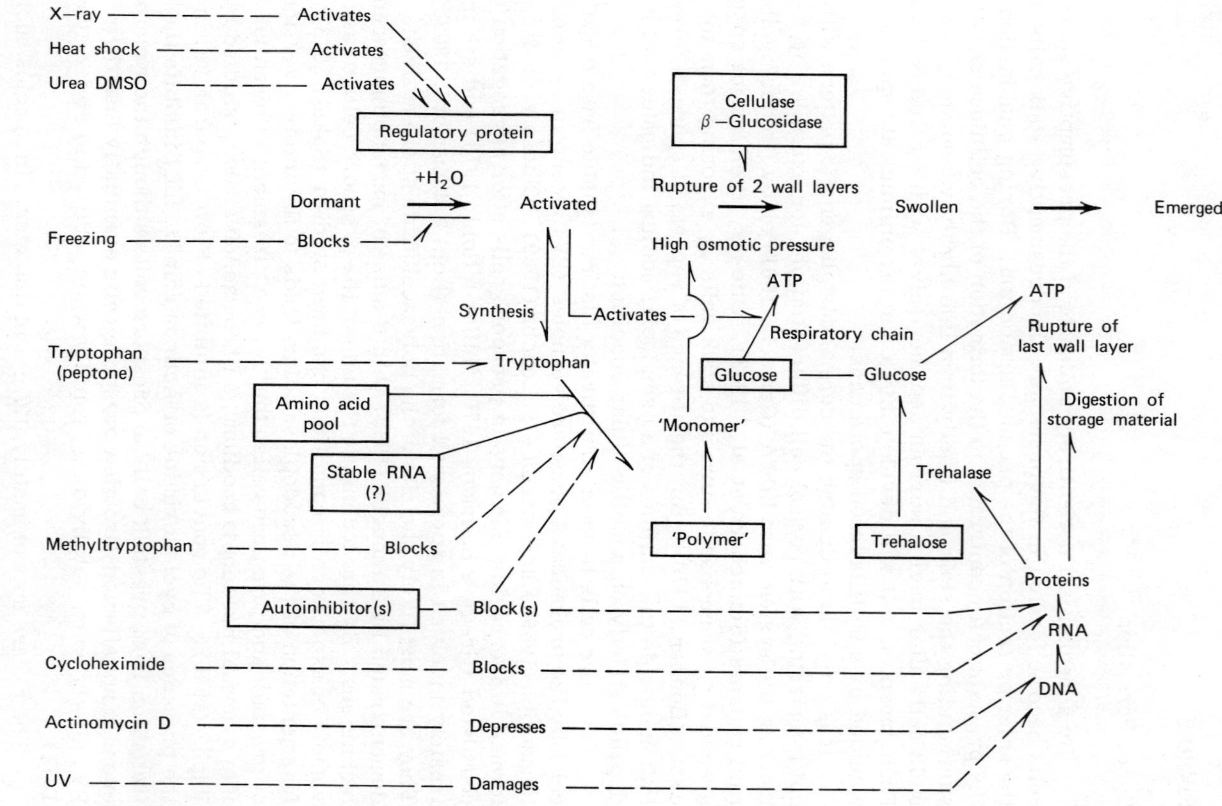
X–ray
Activates
Heat shock
Activates
Urea DMSO
Activates
Regulatory protein
Cellulase
β–Glucosidase
$+H_2O$
Dormant
Activated
Rupture of 2 wall layers
Swollen
Emerged
Freezing
Blocks
High osmotic pressure
ATP
Synthesis
Activates
Respiratory chain
ATP
Rupture of
last wall layer
Tryptophan
(peptone)
Tryptophan
Glucose
Glucose
Amino acid
pool
'Monomer'
Digestion of
storage material
Stable RNA
(?)
Trehalase
Methyltryptophan
Blocks
'Polymer'
Trehalose
Proteins
Autoinhibitor(s)
Block(s)
RNA
Cycloheximide
Blocks
DNA
Actinomycin D
Depresses
UV
Damages

through open bridges. The membrane-bound vesicles are lined on their inside with a thin, electron dense layer of material (Fig. 7) which is morphologically identical to the first, outermost spore wall layer. When the spore wall is formed late in culmination, the PV move to the periphery of the cells, and their membranes fuse with the plasma membrane (Fig. 2 & 7). Their contents are released to the environment by evagination, while their membranes become part of, and apparently replace, the plasma membrane (Hohl & Hamamoto, 1969b). The inner dense lining of the vacuoles now faces the outside and covers the incipient spore as a first layer (Fig. 2).

The bulk of material released by the PV consists of slime which differs from the slime produced during migration by its lack of fibrillar components (Hohl & Jehli, 1973). There is good evidence that in addition to the slime several enzymes are released by this same bulk mechanism. Among these is UDP-galactose polysaccharide transferase, an enzyme present only in prespore cells (Newell et al., 1969) which participates in the formation of the spore wall mucopolysaccharide (White & Sussman, 1963; Sussman & Lovgren, 1965) represented by the dense, inner lining of the PV. UDP-galactose-4-epimerase (Telser & Sussman, 1971) and acid phosphatase (Gezelius, 1972) might well be other enzymes secreted in this fashion.

In summary, the PV performs the following roles in spore differentiation: (1) the prefabrication of a wall layer in the cell interior and its subsequent release to the cell surface, (2) the localization of at least part of the wall-producing machinery in a defined cellular compartment, (3) the bulk elimination of slime material and superfluous enzymes, and (4) the apparent replacement of the plasma membrane by the membranes of the PV (Hohl & Hamamoto, 1969b).

Once the outermost wall layer has formed, two more layers, making up the bulk of the spore case, are rapidly produced (Fig. 8 & 9). The heavy middle layer, made essentially of cellulose, is added by an unknown mechanism, perhaps by polymerization of cellulose microfibrils at the plasma membrane (Hohl et al., 1968; Hohl & Hamamoto, 1969b). Even less is known about the formation of the third, innermost layer which is sensitive to cellulase and pronase

---

Fig. 4 Diagram summarizing aspects of spore germination in *Dictyostelium* as discussed in the text. Framed compounds: present in dormant spores; dashed lines: presumed action of external factors.

(Hemmes et al., 1972).

The biochemical aspects of differentiation in *D. discoideum* are under intensive scrutiny. Since carbohydrates constitute the main components of the differentiated state, special attention has been paid to the metabolism of these compounds. The major carbohydrates found in mature spores are formed according to the scheme outlined in Fig. 3.

Soluble glycogen appears to be the main source of all the polysaccharides and also of trehalose. Thus, the total glucose residue content of the glycogen present before differentiation just about equals that of the carbohydrates of the mature spores (Rosness & Wright, 1974), while at the same time gluconeogenesis remains at a very low level (Baumann & Wright, 1968; Cleland & Coe, 1968). The extensive degradation of protein during differentiation (Gregg & Bronsweig, 1956a,b) thus does not seem to provide carbon skeletons for polysaccharides but rather is involved in energy production for development. The breakdown is characterized at the cellular level by the presence of numerous autophagic vacuoles (Gezelius, 1972; George et al., 1972). Endogenous energy production is essential in view of the total absence of exogenous energy provisions during development.

While the main biochemical steps leading to the various end products are generally agreed upon, vigorous debate continues over the mode, levels, and relative importance of the controls governing

---

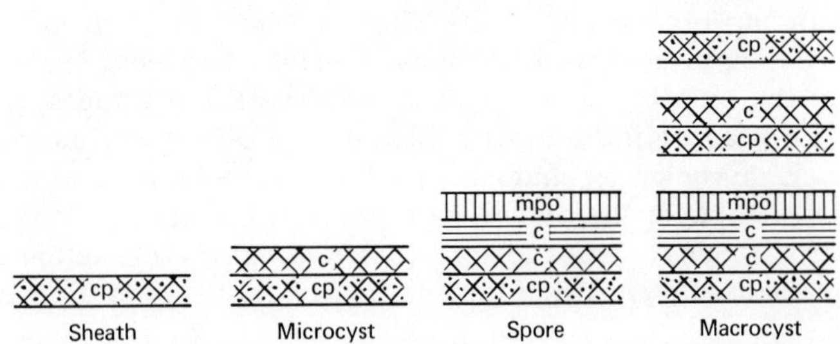

Fig. 5 Diagram summarizing the architecture of the slime sheath and of the walls of microcysts, spores and macrocysts. Tentative homologies between wall layers are marked: cp, cellulose and protein; c, cellulose; mpo, mucopolysaccharide.

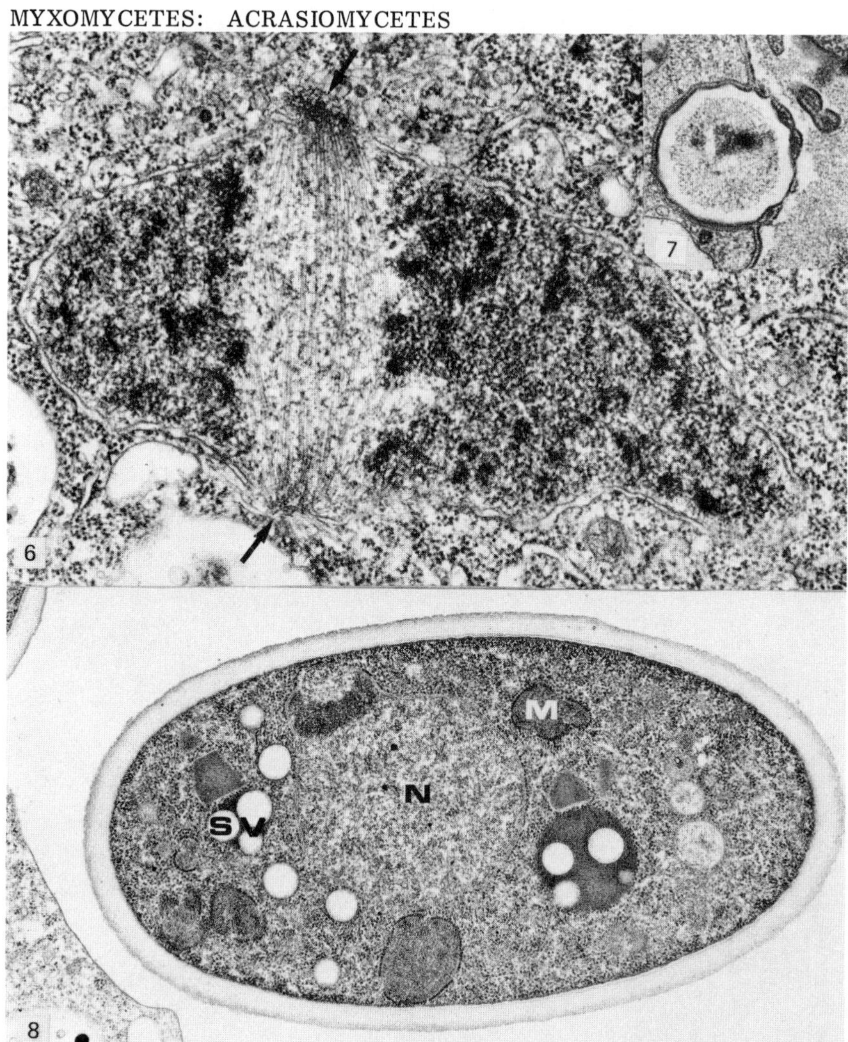

Figs. 6--8. Fig. 6. Metaphase nucleus of *Polysphondylium violaceum* with intranuclear spindle and spindle pole bodies (arrows). Photograph provided by Dr. U.-P. Roos (x 32,000). Fig. 7. Prespore vacuole fusing with plasma membrane during early stage of spore wall formation (x 24,000). Fig. 8. Mature spore of *Dictyostelium discoideum*. Nucleus (N), mitochondria (M), and spore vesicles (SV), (x 17,600). Figs. 7 and 8 from Hohl & Hamamoto (1969b).

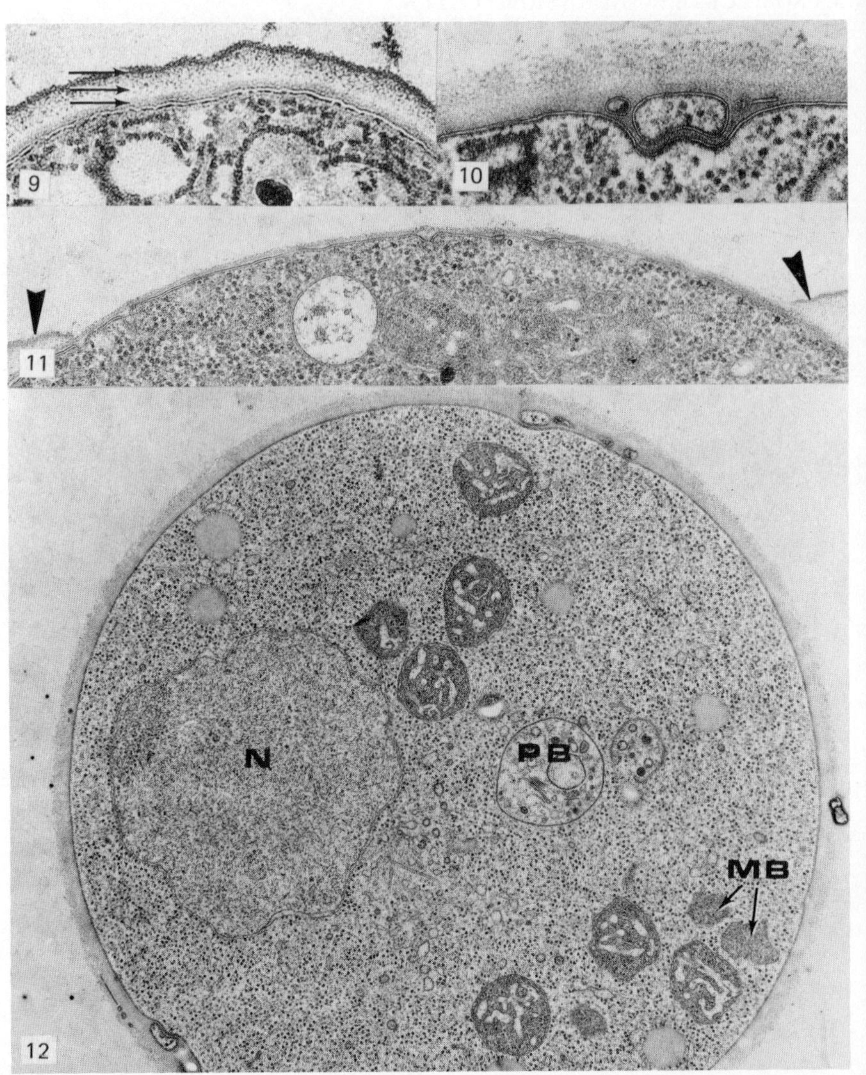

Figs. 9--12. Fig. 9. Mature spore wall of *Dictyostelium discoideum* with three distinct layers. From Hohl & Hamamoto (1969b), (x 53,600). Fig. 10. Microcyst wall with a dense inner and a fibrillar outer layer; cytoplasmic material entrapped in dense layer, (x 45,600). Fig. 11. Swollen spore with ruptured outer layers (arrows) and a cytoplasm retained by innermost wall layer only, (x 36,000). Fig. 12. Mature microcyst. Nucleus (N), polyvesicular bodies (PB), microbodylike structures (MB), (x 14,400).

development. Controls must be exerted to provide for temporal as well as spatial organization, that is, the substrates and enzymes must be present at the right time and place to permit the orderly construction of the mature fruiting body. A discussion of the intensive work going on in this field lies beyond the scope of this paper. Suffice it to say that there exist two major camps. Sussman's group (e.g., Sussman & Sussman, 1969; Loomis, 1970a; Newell & Sussman, 1970; Newell et al., 1972) adheres to the model of a "developmental program" according to which the necessary array of developmentally important enzymes are produced through a series of time-ordered transcriptional and translational events. Their main evidence derives from the modified patterns of enzyme activities in morphogenetically deranged mutants and from drug inhibition studies.

Wright (Wright, 1968, 1973; Killick & Wright, 1974) suggests that rates of synthesis or levels of enzyme activities usually are not the critical variables which set the pace for the entire process, since *in vivo* these rates and levels normally exceed those needed to account for the metabolic fluxes observed. According to her, substrate availability, end product inhibition and rates of enzyme turnover more often represent the critical variables in development.

For more information on this controversy, which has important bearings on the basic concepts of control in developmental systems, the reviews of Francis (1969), Newell (1971), Bonner (1971) and Ashworth (1971) should be consulted. A particularly valuable insight may be gained by comparing the opposite views in respect to the one enzyme sufficiently studied to allow such a comparison, uridine diphosphoglucose pyrophosphorylase (Franke & Sussman, 1973; Gustafson et al., 1973).

Spore Germination

The mature spore of *D. discoideum* (Fig. 8 & 9) is surrounded by a three-layered spore wall (Hohl & Hamamoto, 1969b; Cotter et al., 1969). The outermost layer most likely consists of an acid mucopolysaccharide (White & Sussman, 1963) while the innermost layer contains cellulose and protein (Hemmes et al., 1972). The heavy middle layer is composed largely if not entirely of cellulose with the microfibrils on its outer face oriented parallel and at right angles to the main axis of the spore while on its inner face they are randomly oriented.

The cellulose of the spore wall is made of alpha- and beta-cellulose

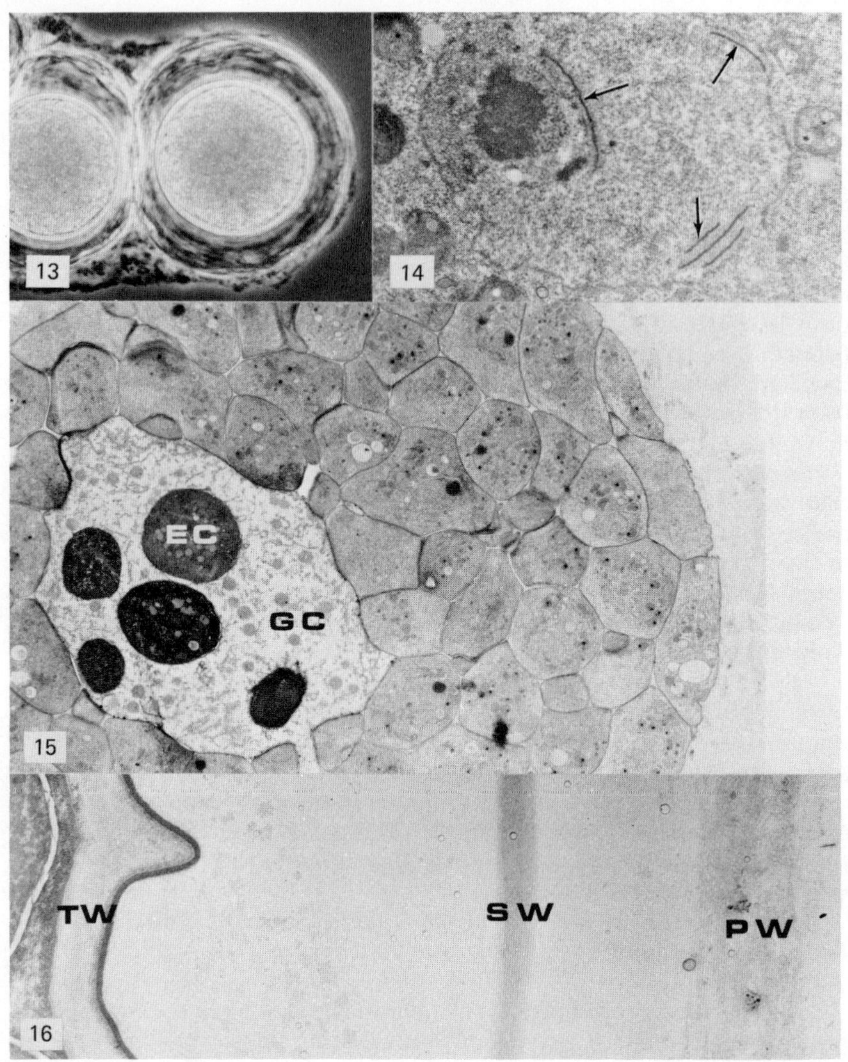

Figs. 13--16. Macrocysts. Fig. 13. Mature macrocyst of *Polysphondylium violaceum*, from Nickerson & Raper (1973a), (x 400). Fig. 14. Nucleus from developing macrocyst of *P. violaceum*, with axial elements (arrows) indicative of a meiotic prophase (x 12,800). Fig. 15. Developing macrocyst with central giant cell (GC) containing several endocytes (EC) surrounded by the aggregate of myxamoebae (x 3,360). Fig. 16. Mature macrocyst with primary (PW),

(Rosness & Wright, 1974) and must also be tightly associated with a heavily branched glucan (glycogenlike material) according to Ward & Wright (1965) and Rosness & Wright (1974). Both the intimate association of cellulose with other glucans and with proteins, as indicated by these investigations, are interesting aspects of cell wall biochemistry which need further clarification.

In addition to the cell wall carbohydrates (Fig. 3) the spore contains soluble glycogen (Rosness & Wright, 1974), trehalose (Clegg & Filosa, 1961; Ceccarini & Filosa, 1965; Ceccarini, 1967; Rosness & Wright, 1974), and some glucose (Ceccarini & Filosa, 1965).

Spores of *D. discoideum* and other but not all members of the Dictyosteliaceae (Cotter & Raper, 1968c) are constitutively dormant. They do not germinate when placed in distilled water or buffer solutions. The dormancy of the spores appears to be maintained by several distinct means. There is good evidence for the existence of a regulatory protein which has to be "turned on" during activation of the spore (Cotter, 1973a). Besides this the spore contains one or several specific selfinhibitors (Russell & Bonner, 1960; Snyder & Ceccarini, 1966; Ceccarini & Cohen, 1967; Cotter & Raper, 1968b; Bacon et al., 1973; Obata et al., 1973) one of which appears to act by inhibiting protein synthesis (Bacon & Sussman, 1973). Furthermore, other substances such as sugars present in the sorus slime may prevent germination by virtue of maintaining the spores in an environment of high osmotic pressure (Cotter, 1973b). Thus at least three mechanisms are concerned with keeping the cell in its resting state; a state which by its resistance to desiccation, freezing and thawing, and temperatures above 50°C assures a higher degree of survival for the organism.

Germination in the species best studied, *D. discoideum*, follows a sequential pattern common to many other fungi and also comparable to bacterial endospores. Its main stages are activation, swelling and emergence. According to Cotter & Raper (1968b) activation refers to the stage between the time spores are exposed to the activation event and the first sign of spore swelling. Swelling encompasses the events from the end of activation until the outer two wall layers have split longitudinally. This stage is accompanied by a loss in refractility, the formation of one or more contractile vacuoles (water expulsion

---

secondary (SW) and tertiary wall (TW) (x 22,400). Figs. 14--16 from Erdos et al., 1972).

vesicles) and the loss of heat and cold resistance. During emergence the myxamoeba escapes from the spore case leaving it behind. The term is equivalent to the "out-growth of bacterial endospores" or "germtube formation" in fungi.

The ultrastructural changes accompanying germination have been described by Cotter et al. (1969). No changes can be noted during activation, but during swelling the mitochondria lose their dense, crenate appearance and their peripheral lining of ribosomes, while the tubular cristae become prominent again. The stacks of flat cisternae have inflated to large, roundish vacuoles containing floccular material and what appears to be remnants of cytoplasmic material. The dilation of these flat cisternae might possibly be the mechanism whereby the osmotic pressure necessary for spore swelling is created. Since the osmotic pressure is a function of the concentration of the dissolved, osmotically active compounds, the creation of a high pressure within small compartments such as these vacuoles would be much more economical than raising the concentration of the solutes in the entire cell.

During the same period the numerous electron transparent vesicles surrounded by elements of the rough ER (the "spore vesicles" of Gregg & Badman, 1970) are often still associated with dark, lipid-like material and perhaps represent some storage material, such as trehalose. The elements of the ER enveloping the electron transparent vesicles disappear during swelling. Together with numerous proteinaceous crystals contained within tightly apposed cisternae of the rough ER, the dark material and the electron transparent vesicles disappear during emergence, and are absent in the emerged myxamoebae.

During swelling the two outer wall layers rupture and the cells are retained by the innermost wall layer (Fig. 11) only. As shown by Cotter & Raper (1968a, 1970) protein synthesis is required for emergence of the myxamoebae from this innermost layer, indicating that formation of one or several enzymes needed for dissolving the restraining wall layer may be one of the important controls for myxamoeba emergence.

Spore germination occurs in the absence of any external energy source, and heat-activated spores will germinate provided they are suspended in water which has been well aerated (Cotter & Raper, 1968a). According to these authors a 10 mM potassium phosphate buffer at pH 6.5 is optimal.

In the following section an attempt is made to correlate the available data on spore germination and to summarize our present understanding of the situation in a diagram. This diagram (Fig. 4) is necessarily rather crude but should help to point out some of the salient features of spore germination and to focus on some of those links in the system, about which still very little is known.

Spore activation may be achieved by heat shock or the addition of peptone (Cotter & Raper, 1966, 1968a), by gamma irradiation (Hashimoto & Yanagisawa, 1970; Khoury et al., 1970; Hashimoto, 1971) or by adding protein denaturing agents such as dimethyl sulfoxide (DMSO) or urea (O'Connell, 1975). According to the multistage theory of spore activiation by Cotter (1973a) and Cotter et al. (1975) heat shock and other activation treatments induce a partial helix-coil transition in a "dormant" regulatory protein, located in the mitochondria (Cotter & George, 1975), which is responsible for restricted oxidative phosphorylation. The induction leads to a metastable state of the protein which then decays to a relaxed state functional in spore swelling by allowing ATP production to commence. The rate of oxygen consumption rises sharply after activation (Bacon & Sussman, 1973).

Some of the amino acids contained in peptone, particularly L-tryptophan, L-phenylalanine and L-methionine, at 10 mM each, largely replace the peptone mixture as activator (Cotter & Raper, 1966). Tryptophan is the most interesting one as it has not been found---at least in its free form---in any developmental stage (Krivanek & Krivanek, 1959), and appears only in activated spores (Bacon & Sussman, 1973). It is synthesized by activated spores from exogenously added glucose (Bacon & Sussman, 1973) but free glucose is also known to be present inside dormant spores (Ceccarini, 1967). Alternatively it might be produced through protein degradation, evidence for which has been presented (Cotter et al., 1969; Cotter & Raper, 1970).

Only the level of tryptophan, of all the amino acids tested, decreases during germination (Bacon & Sussman, 1973). These authors suggest that the amino acid, which accumulates in spores inhibited by the self inhibitor N, N-dimethylguanosine (Bacon et al., 1973), is incorporated into an enzyme necessary for spore swelling. Its direct involvement in spore swelling is also indicated by the observation that of all the amino acid antagonists studied only methyltryptophan inhibits swelling to some degree (Cotter & Raper, 1970). What then might be the identity of such a hypothetical "spore swelling enzyme"? Trehalase activity is very low in dormant spores and increases sharply during germination (Ceccarini, 1967; Cotter & Raper, 1970). According to Ceccarini & Filosa (1965) trehalose is the main energy source during germination. But Cotter & Raper (1970) have pointed out that the increase in trehalase specific activity occurs only at the end of spore swelling and during emergence of the myxamoebae. Thus, this enzyme is not likely to be a spore swelling enzyme nor to provide the means for energy production during the early stages of germination. It is more probably that the initial source of energy is obtained from the small amount of glucose present in dormant spores (Ceccarini, 1967; Cotter & Raper, 1970), while trehalose-derived glucose is used during emergence. The situation is not entirely clear, however, as Killick & Wright (1972) using different preparative methods have found a considerable trehalase activity in dormant spores.

It is possible that such a spore swelling enzyme produces an increased osmotic pressure by degrading some polymer to monomers (e.g., proteins to amino acids or glycogen to glucose). Since soaking dormant spores does not lead to swelling this hypothetical enzyme may be formed only after activation, when tryptophan becomes available. Furthermore, swelling proceeds in the presence of RNA-synthesis inhibitors indicating that a stable mRNA might possibly participate in this process (Bacon & Sussman, 1973). In fact, small amounts of polyribosomes have been observed in dormant spores (Feit et al., 1971). Once formed, the enzyme would then produce the high concentration of monomers necessary to swell the spores. As indicated earlier, this process might conceivably take place in vacuoles, thus minimizing the quantities of osmotically active compounds necessary to achieve swelling. A concomitant inflation of vacuoles has been observed not only in spores (Cotter et al., 1969) but also in swelling microcysts (Hohl et al., 1970). In this connection it is noteworthy that spores can be prevented from entering the swelling stage by any osmotically active compound added in sufficiently high concentration (Cotter, 1973b).

Increase in osmotic pressure alone might not be sufficient to rupture the outer spore wall layers. This process is probably aided by two enzymes present in the dormant spore, a cellulase and a beta-glucosidase (Rosness, 1968).

According to the scheme presented in Fig 4, once the spore has attained the required levels of ATP and osmotic pressure it proceeds to completion of the swelling stage. This stage is reached in the presence of inhibitors of RNA or protein synthesis (Cotter & Raper, 1966; Bacon & Sussman, 1973). Even cells whose DNA has been damaged by UV irradiation are capable of progressing to the swollen stage (Cotter, 1973a).

The emergence of the myxamoebae must involve removal of the retaining innermost wall layer. As emergence is prevented by inhibiting RNA (Bacon & Sussman, 1973) or protein synthesis (Cotter & Raper, 1966) it is probable that an enzyme system has to be formed which is capable of dissolving this layer. From the observation that this layer contains cellulose and protein (Hemmes et al., 1972) and that cellulase is present in dormant spores (Rosness, 1968) one is tempted to conclude that a proteolytic enzyme is synthesized which controls the final step of emergence. Treating swollen spores, inhibited from emerging by the presence of cycloheximide (Cotter et al., 1969), with cellulase and pronase will liberate a good portion of the myxamoebae (Hemmes et al., 1972) lending further evidence to this notion.

Apart from the postulated proteolytic enzyme and the trehalase whose specific activity increases sharply during emergence it must be assumed that a series of other enzymes are being formed or activated, some presumably induced if nutrients are available. These events will eventually lead to DNA replication and resumption of the vegetative condition, thus ending the events of spore germination. Very little is known about the switch to the "vegetative" genome.

The summary diagram (Fig. 4) emphasizes several points. The events from activation to the swollen stage are under control of specific factors and conditions which, with the exception of the activation factor(s), are "prepackaged" (Newell, 1971) in the mature spore. Towards the end of the swollen stage the cell enters a more conventional type of metabolism and thus becomes susceptible to inhibitors of RNA and protein synthesis. The enzyme trehalase apparently does not play a decisive factor in spore germination of *Dictyostelium* until the late swelling stage.

Figure 4 also points to some major unsolved questions. What are the immediate consequences of activation? Does it affect only one, or, as suggested in the diagram, two or more metabolic pathways simultaneously? What is the actual link between tryptophan and spore swelling, and is a stable RNA really involved in this process (Sussman & Douthit, 1973)? Finally there remains the problem of the transition from the emerged myxamoeba to a truly vegetative cell.

MICROCYSTS

Microcyst formation is a unicellular event involving the differentiation of individual myxamoebae from the trophic to a resting stage which differs from both spores and macrocysts (Fig. 1). It occurs in several members of the Acrasiomycetes (Blaskovics & Raper, 1957), but has been studied to an appreciable degree only in *P. pallidum*. The discussion will, therefore, be restricted to this species.

Several factors favor encystment: increased osmolarity of the medium, presence of specific ions, suppression of aggregation, and lack of some essential nutrients. Volatile compounds, given off by the myxamoebae, such as NH_3 may also be involved (Lonski, 1973). The most synchronous and rapid encystment is obtained by first exposing myxamoebae to starvation and then transferring them to a medium with increased osmolarity (Toama & Raper, 1967a; Hohl et al., 1970).

Optimal concentrations are provided by 0.08 *M* KCl or iso-osmolar concentrations of sucrose or glucose (Toama & Raper, 1967a). Of the various salts tested KCl proved to be the best but the rate of encystment may be increased by adding divalent cations such as Mg^{12+} or Ca^{2+}. This together with the observation that NaCl does not induce encystment indicates that ions such as K^+, Mg^{2+} or Ca^{2+} may be used in the process itself. Ca^{2+} also stimulates stalk formation (Maeda, 1970).

In liquid culture, shaking the myxamoebae prevents their aggregation. This effect is achieved on solid media by using higher concentrations of KCl (Toama & Raper, 1967a). When using solid substrates it is also necessary to keep the cultures in the dark, as light is known to stimulate sorocarp formation (e.g., Kahn, 1964) and thereby depress encystment.

From the observation that microcysts eventually form when cells are grown and continuously kept in the growth medium (O'Day & Francis, 1973) one might conclude that the change from low to high osmolarity per se is not a prerequisite for induction of encystment. Rather, this process is induced as a consequence of depletion of one or several factors essential for vegetative growth. Although the mechanism of induction is unknown, a clue is provided by the work of Githens & Karnovsky (1973b) who found that both starvation and increased osmolarity depress phagocytosis and thus possibly trigger microcyst formation.

Once myxamoebae are transferred to the encystment medium a lag of approximately 7 h occurs before encystment is morphologically recognizable as a rounding up of the cells followed by wall formation (Hohl et al., 1970). The process is completed within 5--6 h after the end of the lag period.

At the electron microscope level (Hohl et al., 1970) there is first an increase of finely fibrillar material in the cytoplasmic matrix. When the cells have rounded up, cytoplasmic microprojections may protrude from the generally smooth surface. The fibrillar material becomes more concentrated at the cell periphery where small vesicles measuring about 50--100 nm have also been observed, some of which fuse with the plasmamembrane and apparently release their contents to the outside. The presence of large vacuoles filled with floccular or fibrous material, microbodylike structures, spiny vesicles, and autophagic vacuoles suggest a high intracellular activity.

The cell wall is first recognizable, after about 9 h as a loose, fluffy cover of fibrillar material surrounding the entire cell. The wall becomes progressively firmer and eventually consists of two layers (Fig. 10), an inner electron denser and an outer, somewhat looser and fibrillar one (Hohl et al., 1970; Bühlmann, 1971). Not infrequently cytoplasmic inclusions and what appear to be remnants of autophagic vacuoles become entrapped in the microcyst wall.

The cytoplasm of the mature cyst (Fig. 12) generally resembles that of trophic myxamoebae. Some notable exceptions include the absence of food vacuoles which are replaced by a cluster of mainly autophagic vacuoles and polyvesicular bodies, and the occurrence of darkly staining material which is found in roundish masses or lining the cell periphery. In particular, the cytoplasm, and especially the mitochondria, does not exhibit the dense, dehydrated appearance typical of spores, which might explain, for example, the higher susceptibility of microcysts to heat (Cotter & Raper, 1968c).

The mature cyst wall is composed of cellulose and associated glycogenlike material, protein, and lipid (Toama & Raper, 1967b). While the presence of cellulose has already been reported by Blas- & Raper (1957), the high amounts of lipids (15--20% of wall material) and protein (about 30% of wall material) are rather surprising but have been substantiated by Bühlmann (1971). The latter author has obtained considerable evidence that the outer wall layer is made of cellulose microfibrils while the inner layer contains a complex of protein and glucose polymers (amorphous cellulose or glycogenlike material). It may be assumed that the lipids are also situated in this inner layer.

Biochemical investigations on microcyst formation are scarce.

While DNA content per cell stays constant throughout encystment, dry weight drops by one third, protein content by about 25% and RNA content by about 60% (Githens & Karnovsky, 1973a). The loss in dry weight is correlated with a noticeable increase in activity of several lysosomal enzymes: alphamannosidase (O'Day, 1973a), acetylglucosaminidase (O'Day, 1973b), and acid phosphatase (O'Day & Riley, 1973). The activity of these enzymes eventually drops, most likely as a result of the observed extracellular release of these enzymes. The initial increase in enzyme activity is not simply a result of starvation as it does not occur in starved myxamoebae (O'Day, 1973c), but is a part of microcyst development just as these same enzymes partake in fruiting body formation (Loomis, 1969, 1970b; Gezelius, 1972). The increase but not the release of these enzymes is dependent on concomitant protein synthesis as judged from experiments with cycloheximide.

The question remains, however, how critical these enzymes really are for encystment. It has been shown (O'Day, 1973a,b; O'Day & Riley, 1973) that a cystless mutant displays an enzyme activity pattern generally different from the wild type. This might indicate, but does not prove that these enzymes are part of the developmental program controlling morphogenesis. Particularly in the case of alkaline phosphatase it could be demonstrated (O'Day & Francis 1973) that depressing the enzyme activity with beryllium does not significantly alter the encystment process. These authors point out that the best way to solve this problem would be by using mutants entirely lacking the enzyme in question.

The results obtained so far would agree with the notion that the activity pattern of the lysosomal enzymes reflects the production and release of the autophagic vacuoles seen with electron microscopy (Hohl et al., 1970). Since encystment proceeds in the absence of an external energy source, energy must be provided by consumption of cytoplasmic material. Although the enzymes tested would thus be linked to encystment via the energy requirement, one could easily imagine that this link is not stringent as some of these enzymes could perhaps be replaceable. It might be more fruitful to use enzymes with a priori functions more closely related to the construction of the cyst, such as wall synthesizing enzymes.

In contrast to spores, microcysts can germinate without being activated (Cotter et al., 1968c). Germination occurs when the cysts are suspended in dilute buffer solutions, but not in 0.1 *M* sucrose buffer, indicating that dormancy may be induced by external factors.

Excystment is accompanied by a swelling of the cluster of vacuoles observed in the mature cyst and the disappearance of the peripheral cytoplasmic lining of electron dense material (Hohl et al., 1970). For a limited period the cisternae of the rough endoplasmic reticulum appear dilated and filled with amorphous material. Next, a loosening of the cell wall texture becomes apparent and the myxamoebae finally emerge by what might best be described as localized dissolution of the very loose cyst wall. The wall does not completely dissolve, however, and is eventually left behind as a thin empty shell (Blaskovics & Raper, 1957).

The biochemical events of excystment have only recently been investigated. It has been known for several years that protein synthesis is required as shown by reversible cycloheximide inhibition (Cotter & Raper, 1968c). The inhibition is effective only within the first couple of hours of germination (O'Day, 1974) and is thus coincident with the dilation of the endoplasmic reticulum. While application of the drug during this early period inhibits emergence, the initial swelling phase is not affected. According to the same author (O'Day, 1974; Gwynne & O'Day, personal communication), there is good evidence that ^{3}H-uridine incorporation, which increases drastically after $1\frac{1}{2}$--2 h, is almost completely stopped by actinomycin D. The sequence of morphological events during germination and the normal accumulation of alkaline phosphatase activity, both sensitive to cycloheximide, are unaffected by this drug. In addition, the incorporation of ^{3}H-leucine increases long before significant amounts of ^{3}H-uridine are incorporated. These data suggest that new RNA synthesis is not required for germination and indicate that alkaline phosphatase may be made by stable RNA.

During germination the activity of several lysosomal enzymes in the extracellular medium increases sharply (O'Day, 1974). This increase correlates well with loosening and partial dissolution of the wall, suggesting that some of these enzymes might be involved in this process.

MACROCYSTS

Macrocyst formation represents an alternative pathway of development (Fig. 1) presently known to occur in certain strains of species belonging to two genera, *Dictyostelium* and *Polysphondylium*. The structure produced is a large, heavy walled cyst (Fig. 13) of multicellular origin, first mentioned by Raper (1951) and described in detail by Blaskovics & Raper (1957). The size of macrocysts varies from 15--100 μm among the species investigated. Having suffered from benign neglect for some time, the macrocysts have gained prominence over the past few years as evidence accumulates that their formation and germination represents the hitherto unknown sexual cycle of these organisms.

Macrocyst formation in competent strains is greatly enhanced by keeping the cultures in the dark (Hirschy & Raper, 1964; Weinkauff & Filosa, 1965; Filosa & Dengler, 1972; Nickerson & Raper, 1973a), exposing them to elevated temperatures (Blaskovics & Raper, 1957; Nickerson & Raper, 1973a), omitting phosphates from the medium (Nickerson & Raper, 1973a), or incubating the cultures at very high humidity (Weinkauff & Filosa, 1965; Filosa & Chan, 1972; Nickerson & Raper, 1973a). The entire process of macrocyst formation may be achieved with a good degree of synchrony in shaken submerged cultures (Filosa & Dengler, 1972). In addition to the above factors, or possibly connected with them, a volatile compound or compounds produced by the aggregating cells might also enhance macrocyst formation (Weinkauff & Filosa, 1965). Reversing the above conditions will favor sorocarp production, which means that changing environmental parameters will alter the ratio of sorocarps and macrocysts ultimately produced (Nickerson & Raper, 1973a).

Up to late aggregation no significant differences between sorocarp formation and macrocyst production have been observed (Blaskovics & Raper, 1957; Nickerson & Raper, 1973a), except that aggregations destined to form macrocysts are generally smaller. As a first step towards macrocyst formation the polarization of the cell mass, a prerequisite of sorocarp induction (Gerisch, 1960), is suppressed. The myxamoebae remain in heaps which become surrounded by a loose fibrillar sheath very similar to the slime sheath surrounding the migrating pseudoplasmodium (Blaskovics & Raper, 1957: , Erdos et al., 1972; Hohl & Jehli, 1973). One or several macrocysts may develop from the cell population contained within the fibrillar sheath (termed primary wall by Erdos et al., 1972). Although many detailed features of macrocysts, such as the three walls surrounding the mature cyst, have been observed and described by Blaskovics & Raper (1957), the cytoplasmic events occurring during development have only been elucidated accurately by using electron microscopic techniques. The following account is based primarily on the studies of Erdos et al. (1972) on *P. violaceum* and those of Filosa & Dengler (1972) on *D. mucoroides*.

Within the cell mass enclosed by the primary wall one or several giant cells can be observed, each serving as the focal point of a future macrocyst. This giant cell begins to enlarge further by ingesting the peripheral myxamoebae (Fig. 15). Each phagocytized cell or endocyte (Blaskovics & Raper, 1957) is enclosed in a vacuole equivalent to the food vacuole of trophic myxamoebae. The giant cell (or cytophagic cell of Filosa & Dengler, 1972) is at first binucleate but at the onset of its phagocytic activity becomes uninucleate with a greatly enlarged nucleus (Erdos et al., 1972) whose diameter is approximately double that of aggregative myxamoebae.

There is ample evidence that the endocytes are broken down and eventually digested, remaining for a long time as characteristic granules in the cytoplasm of the mature cyst. The breakdown of endocytes coincides with the appearance of the brownish color characteristic of the maturing macrocyst.

Presumptive evidence for the occurrence of meiosis at about this time, when most of the endocytes have fragmented, has been obtained by Erdos et al. (1972). Within the nucleus of the cytophagic cell, single or paired axial elements of the synaptonemal complex have been observed (Fig. 14). Subsequently the giant cell becomes multinucleate and the nuclei return to their normal size. The entire sequence from the binucleate to the uninucleate and finally the multinucleate stage is strong evidence for a sexual process.

During the same stage of development the number of lipoidal inclusions increases greatly. As the cyst matures fixation becomes more difficult and substructures are hard to identify. When all the surrounding myxamoebae have been consumed by the giant cell its cytoplasmic membrane has become the plasma membrane of the maturing macrocyst. At this stage the secondary wall forms. Ultrastructurally it is equivalent to the microcyst wall (Hohl et al., 1970), though considerably thicker. It is rigid and does not collapse as the macrocyst protoplast shrinks away from it during further differentiation.

In the last phase a third, pliable wall develops on the surface of the giant cell with a trilaminar appearance strongly resembling that of the spore wall of the *Dictyosteliaceae* (Hohl & Hamamoto, 1969b; Hemmes et al., 1972). This third wall resembles the spore wall not only structurally, but also functionally as will be shown in the discussion on macrocyst germination.

The mature macrocyst has a cytoplasm filled with numerous granules representing remnants of the endocytes, lipoidal inclusions, and above all, numerous nuclei. The entire protoplast is surrounded by three different walls (Fig. 16). While the first wall is produced by the multicellular stage just as is the case in the structurally similar slime sheath surrounding the migrating pseudoplasmodium, the second and third wall are formed by the single protoplast of the giant cytophagic cell, a situation comparable to that of the structurally homologous (Fig. 5) microcyst and spore walls respectively which are also formed around individual cells.

Germination of macrocysts has for a long time been a puzzle which, despite the limited successes obtained by Blascovics & Raper (1957) has only recently been satisfactorily elucidated. With the establishment of proper germination conditions for a strain of *D. mucoroides* and exploration of many other strains and species by Nickerson & Raper (1973b) the road was paved for the ultrastructural study reported by Erdos et al. (1973a).

The minimum period of dormancy of macrocysts varies among strains and species, and has been reported to range from almost zero in *D. minutum* to about 50 days in *P. violaceum*. Normally it requires at least a few weeks. Factors favoring macrocyst germination include ageing of cysts, lowering incubation temperatures to below 20°C and exposing cultures to light. The siutation is rather complex, however, as the various parameters are all interrelated, young cysts requiring more light and lower temperatures than older ones, and the rate of germination increases with age (Nickerson & Raper, 1973b).

The germination of macrocysts is first indicated by a swelling of the contracted internal cell mass within the secondary wall. This is the only stage in germination which is reversible (Nickerson & Raper, 1973b). Swelling is followed by a slow tumbling movement within the cytoplasm. As the movement becomes increasingly turbulent the dark coloration gradually fades, leaving an essentially colorless mass of granular cytoplasm. This is followed by the reappearance of individual cells as well as contractile vacuoles. The final stage begins with the partial dissolution and rupture of the heavy secondary wall and ends with the emergence of the myxamoebae.

In most cases myxamoebae were released soon after the secondary wall was broken but the tertiary wall still could prevent liberation of the cells. In the absence of bacteria these myxamoebae are capable of immediate aggregation followed by formation of either new macrocysts or fully differentiated sorocarps, a situation comparable to microcycle sporogenesis in bacteria (Vinter & Slepecky, 1965) or secondary sporangia formation in *Phytophthora* (Hemmes & Hohl, 1973).

At the ultrastructural level the first sign of germination is expressed as a splitting of the tertiary wall. The innermost layer remains intact and retains the protoplast. The entire protoplast then cleaves into large uninucleate segments larger than myxamoebae. These proamoebae (Erdos et al., 1973a) proceed to digest the larger part of their endocyte granules which probably coincides with the disappearance of the brownish color at the light microscope level. As a next step, the proamoebae divide several more times, presumably accompanied by concurrent nuclear divisions, to form the myxamoebae. The cleavage process is progressive rather than simultaneous and similar to the one described for *Didymium* by Schuster (1964). However, the mechanism of cleavage, if by furrow or coalescence of vesicles, remains unsolved.

If macrocysts are placed on media containing 150 µg/ml of cycloheximide, germination will proceed to this point but myxamoebae will not emerge (Erdos et al., 1973a). This situation is homologous to spore germination where the application of this drug also prevents dissolution of the innermost wall layer and with it release of myxamoebae (Cotter et al., 1969). By analogy the secondary wall becomes loose and fluffy during germination just as in the case of microcyst germination (Hohl et al., 1970).

The strong ultrastructural evidence of sexuality, together with the fact that only a few strains form macrocysts (i.e., are self compatible) has prompted investigations into the possible occurrence of mating types. Clark et al. (1973) and Francis (personal communication) report mating in four out of five species tested, namely *D. purpureum*, *D. discoideum*, *P. violaceum*, and *P. pallidum*. *Dictostelium mucoroides* is self compatible and has no mating types as judged from the strains so far checked, whereas *D. giganteum* a near relative, does have mating types. While *D. purpureum* and *D. discoideum* possess two mating types, the situation in *P. violaceum* is complicated by the presence of two syngens with two mating types each.

In a similar study of Erdos et al. (1973b) the situation has been explored in more detail for *D. discoideum*. Of 32 strains tested only one proved to be self compatible. Twenty-four strains will form macrocysts when paired with an alternate compatible strain, while five strains formed no macrocysts under any circumstances investigated. The two last strains are self incompatible but will form macrocysts when paired with either one of the two mating types. No macrocysts form when the two strains are paired with each other.

While the authors state that it is premature to talk of heterothallism until the genetics of the mating system is known they, as well as Clark et al. (1973), do suggest the existence of a single-locus, two-allele system. The discovery in *D. discoideum* of the two ambisexual strains may, however, also indicate that a multiallelic system similar to that found in some Basidiomycetes, exists in some cases.

More definitive information on the genetics of mating types must await a solution to the curious situation that in the only species which can be germinated satisfactorily, that is, *D. mucoroides*, no mating types are known, while in the species with mating types the degree of germination is still inadequate. Yet there is hardly any doubt that the development of the cellular slime molds will soon become amenable to genetic analysis.

SYNOPSIS

Common to all three resting stages described in this chapter is the ability to develop and germinate in the absence of nutrients. In fact, the absence of essential nutrients provides one of the main triggers for their formation, and the presence of high levels of nutrients may, at least in the case of microcysts, prevent germination. Apart from this, microcysts, spores, and macrocysts differ in most respects, as each appears to fulfill a particular role in the life cycle of this extraordinary group of microorganisms.

Microcysts represent a transient resting stage not requiring activation for subsequent germination. Formed by individual myxamoebae, they can differentiate under conditions unfavorable for spore or macrocyst production, for example, low cell density or the inhibition of cell aggregation. Similarly, the environmental factors favoring macrocyst or spore formation differ greatly. Compared with microcysts, spores are more resistant to adverse environmental conditions and their germination is, with notable exceptions (Cotter & Raper, 1968c), controlled by a constitutive dormancy. While spore germination may be induced immediately following culmination, macrocysts normally require a resting period of several weeks and can survive for several months (Nickerson & Raper, 1973b). Considering their sexual nature, their resistance, and longivity, macrocysts undoubtedly represent the most important single resting stage of the entire life cycle.

In considering the possible ecological significance of these various structures, one is tempted to argue that a strain capable of forming all three resting stages would have optimal means for survival and should be isolated more frequently from soil samples than strains of more limited potential. Yet despite the apparent advantages no obvious correlation between resting stage formation and abundance in soil has been noted.

Furthermore, hardly any information is available on the formation of the various structures under natural conditions. One simply has to assume that microcysts and spores are actually formed in the soil. The occurrence of macrocysts in nature is indicated by the close proximity of complementary mating types, some of which have been isolated from the same square meter of forest ground or even from the same soil sample (Clark et al., 1973). Obviously the relative ecological significance of spores and cysts can presently only be guessed at, and is in urgent need of experimental clarification.

From a developmental point of view the comparison of the various structures reveals two striking points. One concerns the formation of the macrocyst. Its three walls correspond structurally and, judging from the few data available, also functionally to the sum of the spore wall, the microcyst wall and the slime sheath, as indicated in Fig. 5. This implies that during macrocyst differentiation groups of genes become activated which are normally engaged in sheath, microcyst, or spore formation. The novel combination of segments from alternate developmental pathways helps to create an entirely new structure, the macrocyst.

The second point in question carries with it some phylogenetic implications. Despite the incomplete data a comparison of the different walls reveals a striking resemblance among various wall layers. A tentative result of such a comparison is shown in Fig. 5. Obviously much more information is needed before definite parallels can be drawn. Should the diagram prove essentially correct this would mean that in the cellular slime molds there exist several resting stages with increasingly complex wall architectures. In particular, the more complex walls would retain the wall layers of the simpler stage while adding new ones. The situation suggests that during the evolution of the Acrasiomycetes the different wall types, and with it the different resting stages, developed in a time-ordered sequence. This could mean that the most primitive cellular slime molds were enclosed by only a slime sheath, and that the ability to form microcysts, then spores, and finally macrocysts, represent major stepping stones of cellular slime mold phylogeny.

ACKNOWLEDGEMENTS

I would like to thank Drs. Erdos, Nickerson, Raper, and Roos for providing photographs as indicated in the text, and Drs. Cotter, Killick, O'Day, Rosness, and Wright for letting me use manuscripts in press and/or unpublished results. Dr. G. Michalenko kindly helped in editing the manuscript.

REFERENCES

Ashworth, J. M. (1971). Cell development in the cellular slime mould *Dictyostelium discoideum*. Symp. Soc. Exp. Biol. 25, 27--49.

Bacon, C. W. & Sussman, A. S. (1973). Effects of the self-inhibitor of *Dictyostelium discoideum* on spore metabolism. J. Gen. Microbiol. 76, 331--344.

Bacon, C. W. , Sussman, A. S., & Paul, A. G. (1973). Identification of a self-inhibitor from spores of *Dictyostelium discoideum*. J. Bacteriol. 113, 1061--1063.

Baumann, P. & Wright, B. E. (1968). The phosphofructokinase of *Dictyostelium discoideum*. Biochemistry, 7, 3653--3661.

Blaskovics, J. C. & Raper, K. B. (1957). Encystment stages of *Dictyostelium*. Biol. Bull. 113, 58--88.

Bonner, J. T. (1967). The Cellular Slime Molds, 2nd ed., p. 205, Princeton, New Jersey: Princeton U. P.

Bonner, J. T. (1970). Induction of stalk cell differentiation by cyclic AMP in the cellular slime mold *Dictyostelium discoideum*. Proc. Nat. Acad. Sci. U.S.A. 65, 110--113.

Bonner, J. T. (1971). Aggregation and differentiation in the cellular slime molds. Annu. Rev. Microbiol. 25, 75--92.

Bühlmann, M. (1971). Morphologische und chemische Charakterisierung isolierter Microcystenwaende von *Polysphondylium pallidum*. Diplomarbeit, Universitaet Zuerich.

Ceccarini, C. (1967). The biochemical relationship between trehalase and trehalose during growth and differentiation in the cellular slime mold, *Dictyostelium discoideum*. Biochim. Biophys. Acta 148, 114--124.

Ceccarini, C. & Filosa, M. (1965). Carbohydrate content during development of the slime mold, *Dictyostelium discoideum*. J. Cell Comp. Physiol. 66, 135--140.

Ceccarini, C. & Cohen, A. (1967). Germination inhibitor from the cellular slime mould *Dictyostelium discoideum*. Nat. (Lond.) 214, 1345--1346.

Clark, M. A., Francis, D., & Eisenberg, R. (1973). Mating types in cellular slime molds. Biochem. Biophys. Res. Commun. 52, 672--678.

Clegg, J. & Filosa, M. (1961). Trehalose in the cellular slime mold *Dictyostelium mucoroides*. Nat. (Lond.) 192, 1077--1078.

Cleland, S. V. & Coe, E. L. (1968). Activities of glycolytic enzymes during the early stages of differentiation in the cellular slime mold *Dictyostelium discoideum*. Biochim. Biophys. Acta 156, 44--50.

Cotter, D. A. (1973a). Spore germination in *Dictyostelium discoideum*. I. The thermodynamics of reversible activation. J. Theor. Biol. 41, 41--51.

Cotter, D. A. (1973b). Autoinhibition of spore germination in the slime mold, *Dictyostelium discoideum*. Abstr. Annu. Meet. Am. Soc. Microbiol., G 83.

Cotter, D. A. & George, R. P. (1975). Germination and mitochondrial damage in spores of *Dictyostelium discoideum* following supraoptimal heating. Arch. Microbiol., 103, 163--168.

Cotter, D. A., Morin, J. W., & O'Connell, R. W. (1975). Spore germination in *Dictyostelium discoideum*. II. Effects of dimethyl sulfoxide on post-activation lag as evidence for the multistate model of activation. In press.

Cotter, D. A., Miura-Santo, L. Y., & Hohl, H. R. (1969). Ultrastructural changes during germination of *Dictyostelium discoideum* spores. J. Bacteriol. 100, 1020--1026.

Cotter, D. A. & Raper, K. B. (1966). Spore germination in *Dictyostelium discoideum*. Proc. Nat. Acad. Sci. U.S.A. 56, 880--887.

Cotter, D. A. & Raper, K. B. (1968a). Factors affecting the rate of heat-induced spore germination in *Dictyostelium discoideum*. J. Bacteriol. 96, 86--92.

Cotter, D. A. & Raper, K. B. (1968b). Properties of germinating spores of *Dictyostelium discoideum*. J. Bacteriol. 96, 1680--1689.

Cotter, D. A. & Raper, K. B. (1968c). Spore germination in strains of *Dictyostelium discoideum* and other members of the *Dictyosteliaceae*. J. Bacteriol. 96, 1690--1695.

Cotter, D. A. & Raper, K. B. (1970). Spore germination in *Dictyostelium discoideum*: Trehalase and the requirement for protein synthesis. Dev. Biol. 22, 112--128.

Erdos, G. W., Nickerson, A. W., & Raper, K. B. (1972). Fine structure of macrocysts in *Polysphondylium violaceum*. Cytobiology, 6, 351--366.

Erdos, G. W., Nickerson, A. W., & Raper, K. B. (1973a). The fine structure of macrocyst germination in *Dictyostelium mucoroides*. Dev. Biol. 32, 321--330.

Erdos, G. W., Raper, K. B., & Vogen, L. K. (1973b). Mating types and macrocyst formation in *Dictyostelium discoideum*. Proc. Nat. Acad. Sci. U.S.A. 70, 1828--1830.

Feit, I. N., Chu, L. K., & Iverson, R. M. (1971). Appearance of polyribosomes during germination of spores in cellular slime mold *Dictyostelium purpureum*. Exp. Cell Res. 65, 439--444.

Filosa, M. F. & Chan, M. (1972). The isolation from soil of macrocyst-forming strains of the cellular slime mould *Dictyostelium mucoroides*. J. Gen. Microbiol. 71, 413--431.

Filosa, M. F. & Dengler, R. E. (1972). Ultrastructure of macrocyst formation in the cellular slime mold, *Dictyostelium mucoroides*: Extensive phagocytosis of amoebae by a specialized cell. Dev. Biol. 29, 1--16.

Francis, D. (1969). Time sequences for differentiation in cellular slime molds. Quart. Rev. Biol. 44, 277--290.

Franke, J. & Sussman, M. (1973). Accumulation of uridine diphosphoglucose pyrophosphorylase in *Dictyostelium discoideum via* preferential synthesis. J. Mol. Biol. 81, 173--185.

Garrod, D. & Ashworth, J. M. (1973). Development of the cellular slime mould *Dictyostelium discoideum*. Symp. Soc. Gen. Microbiol 23, 407--435.

George, R. P., Hohl, H. R., & Raper, K. B. (1972). Ultrastructural development of stalk-producing cells in *Dictyostelium discoideum*, a cellular slime mould. J. Gen. Microbiol. 70, 477--489.

Gerisch, G. (1960). Zellfunktionen und Zellfunktionswechsel in der Entwicklung von *Dictyostelium discoideum*. I. Zellagglutination und Induktion der Fruchtkoerperpolaritaet. Roux. Arch. Entwicklungsmech. 152, 632--654.

Gerisch, G. (1961). Zellfunktionen und Zellfunktionswechsel in der Entwicklung von *Dictyostelium discoideum*. III. Gentrennte Beeinflussung von Zelldifferenzierung und Morphogenese. Roux Arch. Entwicklungsmech. 153, 158--167.

Gerisch, G. (1968). Cell aggregation and differentiation in *Dictyostelium*. Curr. Top. Dev. Biol. 3, 157--197.

Gezelius, K. (1959). The ultrastructure of cells and cellulose

membranes in *Acrasiae*. Exp. Cell Res. 18, 425--453.
Gezelius, K. (1961). Further studies in the ultrastructure of *Acrasiae*. Exp. Cell Res. 23, 300--310.
Gezelius, K. (1971). Acid phosphatase localization in myxamoebae of *Dictyostelium discoideum*. Arch. Mikrobiol. 75, 327--337.
Gezelius, K. (1972). Acid phosphatase localization during differentiation in the cellular slime mold *Dictyostelium discoideum*. Arch. Mikrobiol. 85, 51--76.
Gezelius, K. & Ranby, B. G. (1957). Morphology and fine structure of the slime mold *Dictyostelium discoideum*. Exp. Cell Res. 12, 265--289.
Githens III, S. & Karnovsky, M. L. (1973a). Biochemical changes during growth and encystment of the cellular slime mold *Polysphondylium pallidum*. J. Cell Biol. 58, 522--535.
Githens III, S. & Karnovsky, M. L. (1973b). Phagocytosis by the cellular slime mold *Polysphondylium pallidum* during growth and development. J. Cell Biol. 58, 536--548.
Gregg, J. H. (1971). Developmental potential of isolated *Dictyostelium* myxamoebae. Dev. Biol. 26, 478--485.
Gregg, J. H. & Badman, W. S. (1970). Morphogenesis and ultrastructure in *Dictyostelium*. Dev. Biol. 22, 96--111.
Gregg, J. H. & Badman, W. S. (1973). Transitions in differentiation by the cellular slime molds. In Developmental Regulation. Aspects of Cell Differentiation (ed. S. J. Coward), pp. 85--106. New York and London: Academic Press.
Gregg, J. H. & Bronsweig, R. D. (1956a). Biochemical events accompanying stalk formation in the slime mold, *Dictyostelium discoideum*. J. Cell Comp. Physiol. 48, 293--300.
Gregg, J. H. & Bronsweig, R. D. (1956b). Dry weight loss during culmination of the slime mold, *Dictyostelium discoideum*. J. Cell Comp. Physiol. 47, 483--488.
Gustafson, G. L., Kong, W. Y., & Wright, B. E. (1973). Analysis of uridine diphosphate-glucose pyrophosphorylase synthesis during differentiation in *Dictyostelium discoideum*. J. Biol. Chem. 248, 5188--5196.
Hashimoto, Y. (1971). Effect of radiation on the germination of spores of *Dictyostelium discoideum*. Nat. (Lond.) 231, 316--317.
Hashimoto, Y. & Yanagisawa, K. (1970). Effect of radiation on the spore germination of the cellular slime mold *Dictyostelium discoideum*. Rad. Res. 44, 649--659.
Hemmes, D. E. & Hohl, H. R. (1973). Mitosis and nuclear

degeneration: simultaneous events during secondary sporangia formation in *Phytophthora palmivora*. Can. J. Bot. 51, 1673--1675

Hemmes, D. E., Kojima-Buddenhagen, E. S., & Hohl, H. R. (1972). Structural and enzymatic analysis of the spore wall layers in *Dictyostelium discoideum*. J. Ultrastr. Res. 41, 406--417.

Hirschy, R. A. & Raper, K. B. (1964). Light control of macrocyst formation in *Dictyostelium purpureum*. Bacteriol. Proc. (Abstr.), p. 27.

Hohl, H. R. (1965). Nature and development of membrane systems in food vacuoles of cellular slime molds predatory upon bacteria. J. Bacteriol. 90, 755--765.

Hohl, H. R. & Hamamoto, S. T. (1968). Lamellate structures in the nucleolus of the cellular slime mold *Acrasis rosea*. Pac. Sci. 22, 402--407.

Hohl, H. R. & Hamamoto, S. T. (1969a). Ultrastructure of *Acrasis rosea*, a cellular slime mold during development. J. Protozool. 16, 333--344.

Hohl, H. R. & Hamamoto, S. T. (1969b). Ultrastructure of spore differentiation in *Dictyostelium*: The prespore vacuole. J. Ultrastr. Res. 26, 442--453.

Hohl, H. R., Hamamoto, S. T., & Hemmes, D. E. (1968). Ultrastructural aspects of cell elongation, cellulose synthesis, and spore differentiation in *Acytostelium leptosomum*, a cellular slime mold. Am. J. Bot. 55, 783--796.

Hohl, H. R. & Jehli, J. (1973). The presence of cellulose microfibrils in the proteinaceous slime track of *Dictyostelium discoideum*. Arch. Mikrobiol. 92, 179--187.

Hohl, H. R., Miura-Santo, L. Y., & Cotter, D. A. (1970). Ultrastructural changes during formation and germination of microcysts in *Polysphondylium pallidum*, a cellular slime mould. J. Cell Sci. 7, 285--306.

Kahn, A. J. (1964). The influence of light on cell aggregation in *Polysphondylium pallidum*. Biol. Bull. 127, 85--96.

Khoury, A. T., Deering, R. A., Levin, G., & Altman, G. (1970). Gamma-ray-induced spore germination of *Dictyostelium discoideum*. J. Bacteriol. 104, 1022--1023.

Killick, K. A. & Wright, B. E. (1972). Trehalose synthesis during differentiation in *Dictyostelium discoideum*. IV. Secretion of trehalase and the *in vitro* expression of trehalose-6-phosphate synthetase activity. Biochem. Biophys. Res. Commun. 48, 1476--1481.

Killick, K. A. & Wright, B. E. (1974). Regulation of enzyme activity during differentiation in *Dictyostelium discoideum*. Annu. Rev. Microbiol. 28, 139--166.

Krivanek, J. O. & Krivanek, R. C. (1959). Chromatographic analysis of amino acids in the developing slime mold, *Dictyostelium discoideum* Raper. Biol. Bull. 116, 265--271.

Lonski, J. (1973). Ph. D. Thesis, Princeton Univ.

Loomis, W. F., Jr. (1969). Acetyl glucosaminidase, an early enzyme in the development of *Dictyostelium discoideum*. J. Bacteriol. 97, 1149--1154.

Loomis, W. F., Jr. (1970a). Temporal control of differentiation in the slime mold, *Dictyostelium discoideum*. Exp. Cell Res. 60, 285--289.

Loomis, W. F., Jr. (1970b). Developmental regulation of α - mannosidase in *Dictyostelium discoideum*. J. Bacteriol. 103, 375--381.

Maeda, Y. (1970). Influence of ionic conditions on cell differentiation and morphogenesis of cellular slime molds. Dev. Growth Differ. 12, 217--227.

Maeda, Y. & Takeuchi, I. (1969). Cell differentiation and fine structures in the development of the cellular slime molds. Dev. Growth Differ. 11, 232--245.

Mercer, E. H. & Shaffer, B. M. (1960). Electron microscopy of solitary and aggregated slime mould cells. J. Biophys. Biochem. Cytol. 7, 353--356.

Müller, U. & Hohl, H. R. (1973). Pattern formation in *Dictyostelium discoideum*: Temporal and spatial distribution of prespore vacuoles. Differentiation, 1, 267--276.

Newell, P. C. (1971). The development of the cellular slime mould *Dictyostelium discoideum*: A model system for the study of cellular differentiation. Essays Biochem. 7, 87--126.

Newell, P. C. & Sussman, M. (1970). Regulation of enzyme synthesis by slime mold cell assemblies embarked upon alternative developmental programs. J. Mol. Biol. 49, 627--637.

Newell, P. C., Ellingson, J. S., & Sussman, M. (1969). Synchrony of enzyme accumulation in a population of differentiating slime mold cells. Biochim. Biophys. Acta 177, 610--614.

Newell, P. C., Franke, J., & Sussman, M. (1972). Regulation of four functionally related enzymes during shifts in the developmental program of *Dictyostelium discoideum*. J. Mol. Biol. 63, 373--382.

Nickerson, A. W. & Raper, K. B. (1973a). Macrocysts in the life cycle of the *Dictyosteliaceae*. I. Formation of the macrocysts. Am. J. Bot. 60, 190--197.

Nickerson, A. W. & Raper, K. B. (1973b). Macrocysts in the life cycle of the *Dictyosteliaceae*. II. Germination of the macrocysts. Am. J. Bot. 60, 247--254.

Obata, Y., Abe, H., Tanaka, Y., Yanagisawa, K., & Uchiyama, M. (1973). Isolation of a spore germination inhibitor from a cellular slime mold *Dictyostelium discoideum*. Agr. Biol. Chem. 37, 1989--1990.

O'Connell, R. W. (1975). M. S. thesis, Southeastern Massachusetts Univ.

O'Day, D. H. (1973a). α-Mannosidase and microcyst differentiation in the cellular slime mold *Polysphondylium pallidum*. J. Bacteriol. 113, 192--197.

O'Day, D. H. (1973b). Intracellular and extracellular acetylglucosaminidase activity during microcyst formation in *Polysphondylium pallidum*. Exp. Cell Res. 79, 186--190.

O'Day, D. H. (1973c). Intracellular localization and extracellular release of certain lysosomal enzyme activities from amoebae of the cellular slime mould *Polysphondylium pallidum*. Cytobios 7, 223--232.

O'Day, D. H. (1974). Intracellular and extracellular enzyme patterns during microcyst germination in the cellular slime mold *Polysphondylium pallidum*. Dev. Biol. 36, 400--410.

O'Day, D. H. & Francis, D. W. (1973). Patterns of alkaline phosphatase activity during alternative developmental pathways in the cellular slime mold, *Polysphondylium pallidum*. Can. J. Zool. 51, 301--310.

O'Day, D. H. & Riley, L. J. (1973). Acid phosphatase activity and microcyst differentiation in the cellular slime mold *Polysphondylium pallidum*. Exp. Cell Res. 80, 245--249.

Raper, K. B. (1940). Pseudoplasmodium formation and organization in *Dictyostelium discoideum*. J. Elisha Mitchell Sci. Soc. 56, 241--282.

Raper, K. B. (1951). Isolation, cultivation and conservation of simple slime molds. Quart. Rev. Biol. 26, 169--190.

Roos, U. P. (1975). Mitosis in the cellular slime mold *Polysphondylium violaceum*. J. Cell Biol. 64, 480--491.

Rosness, P. A. (1968). Cellulolytic enzymes during morphogenesis in *Dictyostelium discoideum*. J. Bacteriol. 96, 639--645.

Rosness, P. A. & Wright, B. E. (1974). *In vivo* changes of cellulose, trehalose and glycogen during differentiation of *Dictyostelium discoideum*. Arch. Biochem. Biophys. 164, 60--72.

Russell, G. K. & Bonner, J. T. (1960). A note on spore germination in the cellular slime mold *Dictyostelium mucoroides*. Bull. Torrey Bot. Club 87, 187--191.

Sakai, Y. (1973). Cell type conversion in isolated prestalk and prespore fragments of the cellular slime mold *Dictyostelium discoideum*. Dev. Growth Differ. 15, 11--19.

Schuster, F. (1964). Electron microscope observations on spore formation in the true slime mold *Didymium nigripes*. J. Protozool. 11, 207--216.

Snyder, H. M. & Ceccarini, C. (1966). Interspecific spore inhibition in the cellular slime moulds. Nat. (Lond.) 209, 1152.

Sonneborn, D. R., White, G. J., & Sussman, M. (1963). A mutation affecting both rate and pattern of morphogenesis in *Dictyostelium discoideum*. Dev. Biol. 7, 79--93.

Sussman, A. S. & Douthit, H. A. (1973). Dormany in microbial spores. Annu. Rev. Plant Physiol. 24, 311--352.

Sussman, M. & Lovgren, N. (1965). Preferential release of the enzyme UDP-galactose polysaccharide transferase during cellular differentiation in the slime mold, *Dictyostelium discoideum*. Exp. Cell Res. 38, 97--105.

Sussman, M. & Sussman, R. (1969). Patterns of RNA synthesis and of enzyme accumulation and disappearance during cellular slime mould cytodifferentiation. Symp. Soc. Gen. Microbiol. 19, 403--435.

Takeuchi, I. (1963). Immunochemical and immunohistochemical studies on the development of the cellular slime mold *Dictyostelium mucoroides*. Dev. Biol. 8, 1--26.

Telser, A. & Sussman, M. (1971). Uridine diphosphate galactose-4-epimerase, a developmentally regulated enzyme in the cellular slime mold *Dictyostelium discoideum*. J. Biol. Chem. 246, 2252--2257.

Toama, M. A. & Raper, K. B. (1967a). Microcysts of the cellular slime mold *Polysphondylium pallidum*. I. Factors influencing microcyst formation. J. Bacteriol. 94, 1143--1149.

Toama, M. A. & Raper, K. B. (1967b). Microcysts of the cellular slime mold *Polysphondylium pallidum*. II. Chemistry of the microcyst wall. J. Bacteriol. 94, 1150--1153.

Vinter, V. & Slepecky, R. A. (1965). Direct transition of outgrowing bacterial spores to new sporangia without intermediate cell division. J. Bacteriol. 90, 803--807.

Ward, C. & Wright, B. E. (1965). Cell-wall synthesis in *Dictyostelium discoideum*. I. *In vitro* synthesis from uridine diphosphoglucose Biochemistry, 4, 2021--2027.

Weinkauff, A. M. & Filosa, M. F. (1965). Factors involved in the formation of macrocysts by the cellular slime mold, *Dictyostelium mucoroides*. Canad. J. Microbiol. 11, 385--387.

White, H. J. & Sussman, M. (1963). Polysaccharides involved in slime-mold development. II. Water-soluble acid mucopolysaccharide(s). Biochim. Biophys. Acta 74, 179--187.

Wright, B. E. (1968). An analysis of metabolism underlying differentiation in *Dictyostelium discoideum*. J. Cell Physiol. 72, Suppl. 1, 145--160.

Wright, B. E. (1973). Critical Variables in Differentiation, p. 109. Englewood Cliffs, N. J.: Prentice-Hall.

MYXOMYCETES: ACRASIOMYCETES

DISCUSSION

Myxomycetes: Acrasiomycetes

Chairman: David A. Cotter
Southeastern Massachusetts University
North Dartmouth, Massachusetts

Dr. Lichtwardt questioned the validity of the term spores for resting stages of the cellular slime molds as the number of propagules actually decreases during differentiation. To him the term cyst would be more appropriate.

Dr. Hohl in reply argued in favor of the term spore as these structures possess, in contrast to the microcysts, most other salient features of fungal spores such as requirement for activation (dormancy), presence of autoinhibitors, swelling stage, and thick, complex walls.

Dr. Weber pointed out the stability of freshly emerged myxamoebae and wondered whether anyone had compared the physical strength of plasma membranes from myxamoebae and fungal protoplasts.

Dr. Hohl was not aware of any such studies, but noted that there is considerable membrane turnover during aggregation when most of the food vacuoles are expelled and again during culmination when the prespore vacuoles move to the cell periphery and release their content to the environment. This could mean that plasma membranes from different developmental stages possess greatly different properties.

Dr. Lovett inquired about possible mechanisms that could account for resorption of the surplus membrane material accumulating during vacuole release.

Dr. Hohl replied that work on this problem is in progress involving plasma membrane markers but that it is not known yet whether resorption occurs at the molecular level, that is, by resorption of individual lipoproteins, or involves processes such as pinocytosis.

Dr. Aldrich asked for more information about the fuzzy coat lining the inner face of the plasma membrane and the membranous trappings observed in the cyst wall.

Dr. Hohl specified that in cysts of *Polysphondylium pallidum* and *Acytostelium leptosomum* the darkly staining peripheral material may occur as fuzzy coat or assume more complex shapes such as stacks of layers or rows of hollow spheres. Similar material in form of amorphous masses may be present in the cytoplasm. The nature of the material is unknown. The membranous material trapped in the cyst wall originates from food vacuoles that are expelled during wall formation or from microprojections of the cell periphery being caught in the wall during encystment. It is of interest to note, however, that even in the absence of these inclusions the wall still has a high protein and lipid content.

Dr. Eveleigh inquired how it was demonstrated that the prespore vacuoles emptied their enzymatic complement to the environment during spore wall formation.

Dr. Hohl pointed out that at present there is a large amount of circumstantial evidence supporting this view but that only recently the prespore vacuoles have been isolated by Takeuchi's group thus opening the way for a direct analysis of their content.

Dr. Hashimoto was interested in the exact nature of the intimate association of proteins and polysaccharides and Dr. Bartnicki-Garcia requested a review of the evidence for true glycogen--cellulose complexes in the spore walls.

Referring mostly to the work carried out in Dr. Wright's laboratory Dr. Hohl recalled the need for dissolving first the cellulose before the glycogen can be demonstrated but added that neither cellulose-glycogen nor polysaccaride--protein bondings had been analyzed in detail. There is evidence for covalent bonds between cellulose and protein in walls of higher plants.

CHAPTER 12

FORM AND FUNCTION IN CHYTRIDIOMYCETE SPORES

E. C. Cantino and G. L. Mills
Michigan State University

*We have interpreted this title rather literally in that we have restricted our presentation to those associations between structure and function which occur during the lifetime of a zoospore---that is, between the time it is released as a motile cell (but not before, hence we do not consider sporogenesis) and the time it ceases to exist, that is, encysts (but not beyond, hence we do not consider spore germination, a process to which other contributors to this volume have addressed themselves).
†This appendange has been added to the title originally assigned to us for two reasons: first, while new accounts of fine structure among Chytridiomycete spores are now accumulating rapidly in the literature, what grasp we presently have of relationships between form and function in them still stems for the most part from studies of *Blastocladiella emersonii*; and second, this occasion being the 25th

FORM AND FUNCTION IN CHYTRIDIOMYCETE SPORES

---

anniversary of the unveiling of this water fungus in 1949 (Cantino, 1951), we are sacrificing 12 minutes of our allotted time to permit *B. emersonii* to make her cinematic debut in Utah via a film (Cantino, 1973) which illustrates---better than any words from us---the colorful nature of her charm.

FORM AND FUNCTION IN CHYTRIDIOMYCETE SPORES

INTRODUCTION

If we simply had to compile a description of what has been learned in the past ten years about the fine structure of Chytridiomycete zoospores, along with what has been discovered about their activities---or even if we were to go one step further, to point up similarities and differences among them and perhaps deduce, thereby, new insights on phylogenetic relationships among the water molds---the task would at least be productive for the newly blended facts would surely lead to new perceptions. But this not being the purpose of our symposium, at least as we interpret it, we have restricted ourselves in the following pages to an examination of <u>relationships</u> between form and function in Chytridiomycete spores, and to discussions of conclusions and ideas about them based primarily upon substantial experimental evidence. Within these guidelines, the information that qualifies for inclusion in this review is rather limited in scope.

Consider, as an introductory illustration of this outlook, what is known today about the delimiting boundary around a zoospore. At one time or another, at least a few mycologists must have watched Chytridiomycete spores shrink in hypertonic media and swell in hypotonic solutions. It is probably not unreasonable, therefore, to suppose that such zoospores ". . . are constantly battling with the aquatic environment to maintain osmotic balance because they are not provided with a cell wall," as we (Truesdell & Cantino, 1971) have done for the zoospores of *Blastocladiella emersonii*. But does it follow with comparable justification that the plasma membrane must be the primary implement used by a zoospore to carry on this battle? Probably not. What is really known about the structural and functional aspects of the region that delimits the Chytridiomycete spore from its surroundings---a region which might be expected to serve quite favorably as a vehicle for discussions of relationships between form and function? Discouragingly little!

To begin with, remembering that it was not so many years ago that man rediscovered how many chromosomes he really possessed in his somatic cells, we should not feel too timid to ask: Do Chytridiomycete zoospores <u>have</u> a plasma membrane? From our own observations, we do know that the *B. emersonii* zoospore is bounded by something about 80---100 Å thick made up of two dark lines separated by an intervening light line. The literature provides, at best, no more than a handful of articles on posteriorly uniflagellate spores wherein a careful search through the photographs reveals an occasional segment of a section through a similarly trilaminar arrangement. Perhaps it is safe to assume that Chytridiomycete zoospores are surrounded by a "unit membrane," but it is really an act of faith in the predictive value of comparative biology. And what of the topic before us, form and function; what about the operation of this plasma membrane? For example, does it possess that impressive and important property of being differentially permeable? We know of no critical direct evidence involving Chytridiomycete zoospores that bears conclusively on this question.

In summing up his reflections on a recent conference involving membrane structure, Hechter (1972) pointed out, "The elucidation of the molecular principles underlying the structural and functional properties of biological membranes has emerged as one of the cardinal problems of contemporary biology, in consequence of the recognition that many of the major activities of living cells involve reactions associated with membranes." So, perhaps chytridiomycetologists need not be unduly alarmed at their lack of knowledge of relationships between form and function in the plasma membrane, and about its connection with the form and function of a zoospore, for even the celebrated cell biologists cannot say a great deal more, at least with any kind of certainty, about whether or not there are any invariant principles underlying membrane functions and structures. But for the purposes of our review, this discussion serves to illustrate the first in a sizeable list of structures that have been described in one Chytridiomycete zoospore or another, but about which little if anything can be said relating form to function.

THE FLAGELLUM

It might be supposed that the flagellum which seems to propel the Chytridiomycete zoospore around would also constitute a good focal point for any consideration of form and function. But what (besides intuition and, again, comparative biology) is there to suggest that the flagellum is in fact the agent directly responsible for the movements of the zoospore?

It is common experience, of course, that when a zoospore is seen to interrupt its rapid swimming momentarily for some reason, often there is also a corresponding transient reduction or even stoppage of its flagellar beat, but not always! Conversely, when zoospores propel themselves about more slowly in amoeboid fashion, their flagella often exhibit little movement, but sometimes they will lash about vigorously. And yet, when a zoospore settles down before encysting, its flagellum may vibrate rapidly back and forth for some seconds without providing any propulsive force at all. If commonplace observations such as these argue for the direct role of the flagellum in the swimming process, it is hardly an air tight case.

Perhaps a somewhat more direct line of evidence stems from our observations (unpublished) that populations of zoospores of *B. emersonii*, deflagellated mechanically (Cantino et al., 1968), but just as viable as their nondeflagellated counterparts (and presumably capable of exhibiting normal encystment kinetics; Soll & Sonneborn, 1969b), are incapable of exhibiting the typical motions of a swimming zoospore. Probably other Chytridiomycete zoospores, if deflagellated without introducing other kinds of damage, would similarly lose their capacities for locomotion.

Yet, there must be better ways than frontal onslaughts on the integrity of zoospores to find out what they are doing with their tails. Fortunately, there are clear signs that such indignities will give way to less destructive analytical approaches as newer tools and fresh skills are brought to bear from other disciplines. Miles & Holwill (1969), for example, have not only concluded that the turning movements of a *B. emersonii* zoospore may be caused by asymmetric activity of its flagellum, and that coordination of the planar waves it propagates may be affected by stimuli other than those derived from its mechanical properties, they have even specified its wave parameters: wavelength, 18.4 ± 1.0 μm; amplitude, 2.4 ± 0.4 μm; frequency, 84 ± 10 Hz; and velocity, 97 ± 23 μm;/sec. How quickly this type of probing will generate basic understanding of the processes responsible for flagellar movements---and zoospore movements---in the Chytridiomycetes remains to be seen.

Interestingly, according to Koch's (1968a) studies of some eight chytrids, the length of the flagellum is not a function of zoospore body size, which tends to rule out from further consideration some possible relationships between flagellar dimensions and function that might otherwise have been envisioned.

The evidence about internal structure, starting with Koch's (1956) examinations of *Chytridium* and *Rhizophidium* and ending with the miscellaneous profiles of various Chytridiomycetes presently scattered throughout the literature, points to the likelihood that all Chytridiomycete flagella (exclusive of their whip lashes) possess the well-known 9 + 2 arrangement. Unfortunately, there is no significant experimental evidence derived from studies of Chytridiomycete spores to suggest if (and, if so, how) this internal plan relates to the way the flagella beat out their planar wave patterns (Koch, 1958; Miles & Holwill, 1969).

Is There a Relationship Between Flagellar Activity and the Presence of Lipid Globules?

If the driving force behind flagellar movements is chemical, and we can't imagine what else it could logically be, what is its nature? In recalling an early observation by Berdan that there might have been some attachment between the flagellum and a lipoid body in the spores of *Cladochytrium*, Koch (1958) said his own observations suggested that the lipoid body might play a direct role in the action of the flagellum. Regrettably, this was not accompanied by any indication of what the evidence was, and the basis for Koch's opinion has not been spelled out since then as far as we know; hence, this intriguing curiosity must be put aside for lack of evidence.

Along a different, but still lipoidal vein, Bernstein (1968) turned up a correlation of potential interest during a study of the zoospores of *Rhizophlyctis*: namely, that the smaller the zoospore, the greater was the number of lipoid globules per zoospore. Unfortunately, no direct connection between lipid and flagellar activity was indicated from this investigation.

There is also the zoospore of *Blastocladiella emersonii*, wherein the proximity of lipid globules to flagellar apparatus suggests a possible role for lipid in flagellar activity. All the lipid globules in this cell are tightly arranged in a "side body" complex (Cantino & Truesdell, 1970) to which the base of the flagellum appears to be anchored (Fig. 1; see subsequent section on side bodies). As zoospores swim about (for 5--10 h, for example) these lipid globules become progressively less prominent and harder to find in random thin sections through such cells (Suberkropp & Cantino, 1973).* As a matter of fact, the total lipid gets used during swimming at a rate of 0.14--0.19

*In an early attempt to guess the possible nature of the internal energy pool used by a zoospore, Cantino & Lovett (1964) suggested that since the refractile lipid granules in a spore that had been swimming for "some hours" apparently did not decrease in number, they were probably ". . . not used in quantity. . ." Although the observations were essentially correct, the conjecture was wrong.

pg x h^{-1} x $spore^{-1}$ (Suberkropp & Cantino, 1973; Mills et al., 1974). Hence, with a starting level of total lipid between ca. 3.0 pg/spore (Mills & Cantino, 1974; Mills et al., 1974) and 4.8 pg/spore (Suberkropp & Cantino, 1973), it is not surprising to find when hourly percentage losses are calculated (Table 1) that while the lipid globules (Fig. 1) become less prominent and harder to find in random thin sections, they nonetheless remain visible when an entire zoospore is examined microscopically *in vivo* (Cantino & Lovett, 1964) after various periods of swimming. More importantly, the triglycerides, which are believed (Mills et al., 1974) to be the major components of the lipid globules in the side body complex, decrease most pronouncedly during 5 h of swimming (Table 2). The foregoing observations are not inconsistent with the possibility that lipid is used for flagellar activity. However, zoospores also consume protein and polysaccharide during the same 5 h period at rates of 0.6 pg x h^{-1} x $spore^{-1}$ (Suberkropp & Cantino, 1973) or a little less (Lodi & Sonneborn, 1974) and 0.1--0.2 pg x h^{-1} x $spore^{-1}$ (Suberkropp & Cantino, 1973) respectively. The question still remains, therefore: Is it lipid metabolism that drives flagellar activity?

Another line of evidence bearing on this question involves the aerobic activities of swimming zoospores.* The O_2 consumed by them (QO_2 at 22°C = 9; QO_2 also = µl x h^{-1} x 10^{-7} cells; Suberkropp & Cantino, 1972, 1973 is probably needed for various things (osmotic regulation, maintenance of the highly ordered state of their organelle complexes, etc.) in addition to flagellar motion. However, deflagellated but fully viable and otherwise apparently unharmed (Cantino et al., 1968) spores consume ca. 15--20% less 0_2 (Suberkropp & Cantino, 1972). This suggests that about 15--20% of the 0_2 consumed by a swimming spore is employed to support flagellar activity under aerobic conditions.

The crucial issue then becomes: Is this ca. 15--20% of the total O_2 used for lipid metabolism? Direct evidence bearing on this question

*We make a distinction here between aerobically and anaerobically swimming spores. Several observations (Cantino et al., 1968) suggest that anaerobic metabolism of a glycogenlike polysaccharide can provide energy used for flagellar activity. While the evidence was not so compelling that lipid could be ruled out preferentially over polysaccharide, semi-anaerobiosis did not cause spores to stop swimming. Hence, if lipid drove flagellar movements under these conditions, it would probably have been <u>via</u> a non-0_2-consuming process.

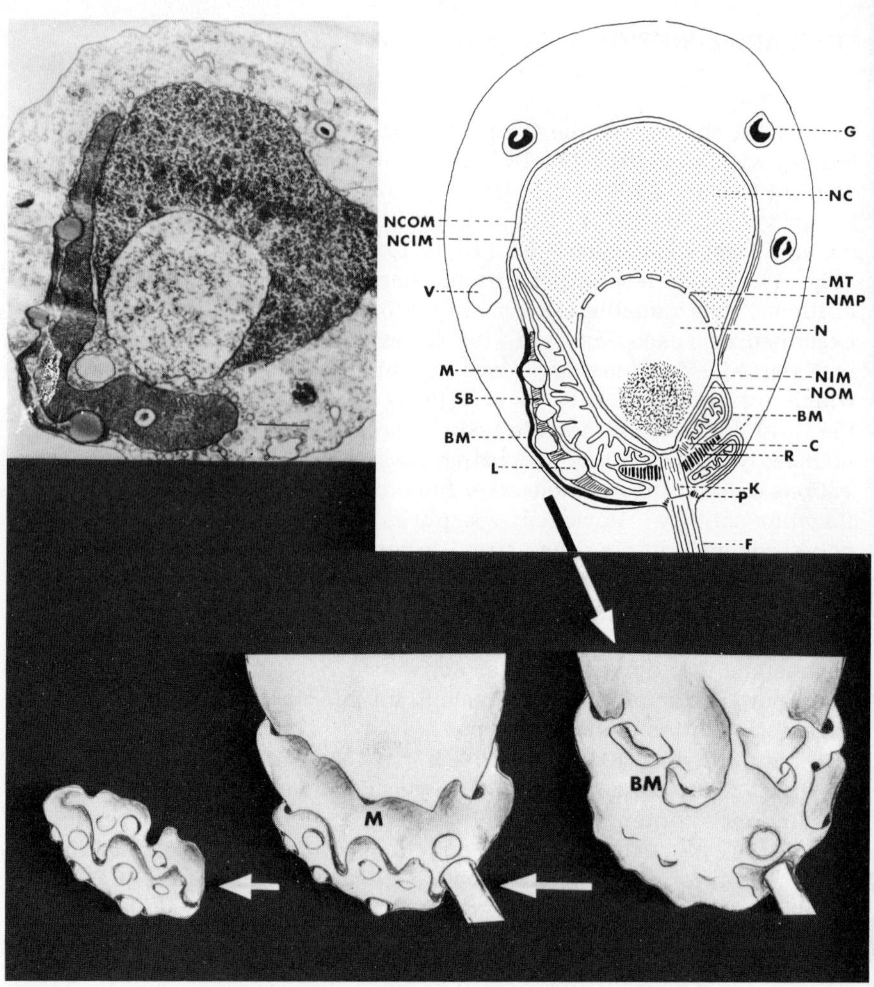

Fig. 1 Structure of the *B. emersonii* zoospore. Top right: composite diagram of cell components; NCOM, nuclear cap outer membrane; NCIM, nuclear cap inner membrane; V, vacuole; M, mitochondrion; SB, SB-matrix (= symphyomicrobody); BM, backing membrane; L, Lipid globule; F, flagellum; P, prop; K, kinetosome; R, rootlet; C, centriole; NOM, nuclear outer membrane; NIM, nuclear inner membrane; N, nucleus; NMP, nuclear membrane pore; MT, microtubule (occur in triplets); NC, nuclear cap; G, gamma particle. Top left: Longitudinal section through a zoospore (derived from dark grown OC plant on Difco PYG agar) fixed in OsO_4 and Ur acetate (fixation II, Truesdell & Cantino, 1970); The marker represents 1 μm. Bottom:

Table 1. Average Hourly Percentage Loss of Total Lipid in *B. emersonii* Zoospores During a 5-h Swimming Period

Av. loss of lipid / Initial lipid, zero time [a] (%)	Av. loss of lipid / Wt (pg) of spore, zero time [b] (%)	Reference
3.9 (4.9)[c]	0.40	d
4.7	0.30	e

[a] (pg x h^{-1} x $spore^{-1}$ /pg x $spore^{-1}$) x 100.
[b] (pg x h^{-1} x $spore^{-1}$/ 47) x 100.
[c] value in parentheses based on a dichromate assay procedure; other values based on dry weight assays.
[d] Suberkropp & Cantino, 1973.
[e] Mills et al., 1974.

Table 2. The Changes in Total Neutral Lipid and Triglycerides in *B. emersonii* Zoospores During a 5-h Swimming Period

	Amount (mg/10^{10} spores)	
	0 h	5 h
Neutral lipid	5.95 (6.37)[a]	3.69 (2.69)
Triglycerides	5.16 (5.53)	2.62 (1.91)

[a] First value based on dry weight; value in parentheses based on assays of ^{14}C-labeled lipids.

is not available. However, assuming that lipid is completely oxidized aerobically to CO_2 and H_2O, applying the data of Suberkropp and Cantino (1972, 1973) for lipid utilization and O_2 uptake, and accepting the conclusion (Mills et al., 1974) that the lipid globules in the zoospore are made up essentially of triglycerides, calculations show (Table 3) that the 15--20% of the total O_2 consumed is approximately the amount required to completely oxidize the quantity of neutral lipid actually consumed by a zoospore. What is the likelihood that such a one-to-one relationship would have occurred by chance alone? We do not know, but we suspect it is pretty low.

A related matter of importance here concerns the question: If lipid is in fact the underlying source of chemical energy behind

---

three dimensional organization (Cantino & Truesdell, 1970) of the side body pictured as if it were being progressively dismantled (from right to left) so as to yield (far left) an isolated SB--lipid complex (i.e., a symphyomicrobody--lipid complex; see section on microbodies); legends as for top right.

Table 3. Quantity (Theoretical) of 0_2 used for Complete Oxidation of the Amount of Neutral Lipid Actually Consumed by Swimming Zoospores of *B. emersonii*

No.	Calculation	Reference
(1)	Complete oxidation of tripalmitin (theoretical), i.e., $O_2\downarrow$ x μg tripalmitin^{-1} = 2.015 μl (2.88 μg)	a
(2)	Total lipid↓ x h^{-1} x spore^{-1} = 0.19 pg	b
(3)	Weight of a spore = 47 pg	b
(4)	QO_2 (μl $O_2\downarrow$ x h^{-1} x mg^{-1}) = 9	b
(5)	$O_2\downarrow$ x h^{-1} x spore^{-1} = 0.423 pℓ	b
(6)	$O_2\downarrow$ x pg total lipid^{-1} = 2.226 pℓ	(5)/(2)
(7)	Neutral lipid (triglycerides)↓ x h^{-1} x spores^{-1} = 0.038 x 10^{-6} μg	20% of (2)
(8)	$O_2\downarrow$ x h^{-1} x spore^{-1} due to flagellar activity = 0.0635 x 10^{-6} to 0.0846 x 10^{-6} pℓ	15--20% of (5)
(9)	$O_2\downarrow$ due to flagellar activity/neutral lipid↓ = 1.67--2.23 μl O_2	(8)/(7)
(10)	Therefore, $O_2\downarrow$ from (1) is approximately = $O_2\downarrow$ from (9)	

[a]Hawk et al., 1951.
[b]Suberkropp & Cantino, 1972, 1973.

flagellar activity in *B. emersonii*, does the transducing mechanism depend upon the close physical proximity of the lipid globules to the flagellar apparatus? In the model (Cantino & Truesdell, 1970) illustrating this relationship (Fig. 1), the kinetosome of the flagellum is anchored by a single rootlet to a cup shaped mitochondrion; the lipid globules in the side-body complex lie along the outer surface of the long arm of the mitochondrion, most of them in intimate contact with it. The pertinent facts that bear on this discussion are as follows. When a zoospore is chilled to ca. 4°C, it stops consuming O_2 (Cantino et al., 1969); at such reduced temperatures, the backing membrane (BM; Fig. 1) that holds the SB--lipid complex and the mitochondrion in place fragments, and a few of the lipid globules are released and dispersed into the cytoplasm (Shaw & Cantino, 1969; Cantino et al., 1969). Yet when such spores are brought back to 22°C, they begin to consume O_2 again at the same rate as spores that have not been chilled (Cantino, 1969; Cantino et al., 1969). This suggests that, if lipid is in fact the chemical energy source that supports flagellar activity, either (a) it need not be in its final position next to the mitochondrion, or (b) the relative amount of it dispersed is insignficant.

Is There a Relationship Between Flagellar Activity and the Presence of Mitochondria?

Since the sheathless end of the flagellum in *B. emersonii* is rooted (Fig. 1) in a single mitochondrion (Cantino et al., 1963; Cantino & Truesdell, 1970), an arrangement recently verified by Bromberg (1974), might this linkage constitute part of the transducing device for converting chemical energy into the energy of flagellar motion? If this idea has any merit, it must be reconciled with the fact that when a zoospore of *B. emersonii* is chilled, its single mitochondrion becomes spherical and displaced to one side (Shaw & Cantino, 1969), although it does remain secured to the kinetosome via the banded rootlet (see Fig. 7 in Shaw & Cantino, 1969). Chilled spores quickly encyst when they are brought to 22°C; if the mitochondrion normally plays a direct role in flagellar activity, the cessation of swimming might be due either to the translocation of some of the lipid globules (described above), or to a cold-induced change in the mitochondrion, or to both of them.

At first glance, the notion is appealing; how much more elegantly could one design a coupling among the cell's only appendage that performs an obvious mechanical task, its presumed power unit, and its lipoidal fuel supply? *Blastocladiella britannica's* anatomical attributes are basically similar (Cantino & Truesdell, 1971a), hence it could possess a similar transmission system. And, although no other Chytridiomycete is presently known to be endowed with an identical apparatus that so completely surrounds the kinetosome at the base of the flagellum, some species come very close to it. In the zoospores of *Coelomomyces*, a single mitochondrion also encircles the kinetosome, albeit only partially; however, a rootlet is missing (Martin, 1971; Madelin & Beckett, 1972). In the meiospores (Olson, 1973a) and zoospores (Fuller & Olson, 1971) of *Allomyces*, there are many mitochondria, but a particularly large one is generally found near the base of the flagellum; in meiospores, it not only partially surrounds the kinetosome, but a striated rootlet is also present. In the zoospores of *Olpidium*, too, there are several mitochondria (Temmink & Campbell, 1969a), but one of them surrounds the end of the kinetid (rhizoplast + kinetosome + centriole at the end of the flagellum) like a collar. If the striated rhizoplast in *Olpidium* is something like the striated rootlet in *Blastocladiella*, could it be that these two water molds have evolved a particularly efficient energy transducing system for flagellar activity?

There are, however, a large number of Chytridiomycetes with no such apparent physical connection between flagellum and mitochondrion, including *Phlyctochytrium, Nowakowskiella, Monoblepharella, Oedogoniomyces,* and so on (Umphlett & Olson, 1967; Chambers et al., 1967; Fuller & Reichle, 1968; Reichle, 1972; Chong & Barr, 1973). If the relationship has functional significance, it does not apply to all Chytridiomycetes.

Implicit in the foregoing arguments, of course, is the assumption that mitochondria in Chytridiomycete zoospores, especially those in the spores of *B. emersonii*, consume oxygen and are functionally similar to mitochondria in other organisms, that is, that they carry the conventional enzymatic machinery that mediates a controlled aerobic oxidation of metabolic substrates. But is such an assumption justifed? It has become abundantly clear that there are metabolically different classes of microbodies in animals and plants (Tolbert, 1971; Richardson, 1974), and that diversity can even be detected within the same organism at different developmental stages. So who can say what the mitochondria seen in sections of pickled zoospores of *Phlyctochytrium* or *Monoblephrella* or *Allomyces* can actually do? And in particular, is there any direct evidence that the mitochondria in the zoospores of *B. emersonii* can, in fact, consume oxygen? Horgen & Griffin (1969) and LéJohn et al. (1969) have prepared "mitochondrial pellets and mitochondrial fractions," respectively, from them, and demonstrated that they exhibit enzymatic activities. However, the microscopic character of the ingredients in their pellets and fractions was not defined, morphological evidence was not provided that mitochondria were actually present, and whether or not the preparations had the capacity to consume oxygen was not mentioned.

We take this opportunity, therefore, to report that in our laboratory fairly clean suspensions of microscopically definable mitochondria, capable of consuming O_2 and of exhibiting some respiratory control with additions of adenosine diphosphate (ADP) have been isolated from *B. emersonii* zoospores (Suberkropp & Cantino, unpublished data). Achieving suitably high breakage of zoospores without concomitant mitochondrial damage was a major obstacle. Since the single mitochondrion is so relatively large (Fig. 1), some techniques which lyse cells adequately also tend to injure the mitochondrion, whereas others which yield intact mitochondria often do not rupture enough cells. For example, if osmotic shock techniques (Myers & Cantino, 1971) involving certain concentrations of sucrose and mannitol are used, whenever cell breakage is high enough to permit isolation of mitochondria the respiratory control with α-ketoglutarate is close to one (i.e., no effect of added ADP). Homogenates prepared according to LéJohn et al. (1969) did not, in our hands at least, yield any respiratory control (ratios never over 1); however, a modification of their method yielded better results.* The resultant mitochondrial fraction was resuspended and used to determine its O_2 uptake capacity and respiratory control ratios at 22°C. The representative tracings (Fig. 2) for results obtained with NAD-linked substrates such as α-ketoglutarate and glutamate show low but readily detectable levels of respiratory control (ratios of 2.5--3.5); secondary additions of ADP always caused a secondary stimulation of the O_2 uptake level. On the other hand, with the flavin-linked substrate, succinate, there was no respiratory control. This, together with the fact that these mitochondrial preparations also utilized NADH, may mean (Lehninger, 1951; Chappell & Hansford, 1969) that many mitochondria were damaged. In spite of their fairly satisfactory appearance (Fig. 3), it is likely that, due to their large size, they were damaged in the initial homogenization procedure; a gentler method of cell disruption will have to be devised before this work continues.

In the meantime, however, it is quite clear that the single mitochondrion in the zoospore of *B. emersonii*, an organelle with quite conventional morphological characteristics, also possesses the capacity to consume O_2. In all probability, therefore, it is responsible for the O_2 uptake exhibited by zoospores *in vivo* (Suberkropp & Cantino, 1972, 1973) and for their susceptibility to tricarboxylic acid cycle inhibitors such as arsenite and malonate (McCurdy & Cantino, 1960).

*A very concentrated spore suspension (ca. 5 x 10^8 spores/ml) in 0.73 M mannitol, 10 mM KCl, 10 mM K phosphate, ph 6.8, 2 mM EGTA, and 1% bovine serum albumin was passed through a 27 guage hypodermic needle 12 times to obtain thorough breakage of the spores. Phase microscopic observations suggested little damage, although almost all mitochondria had become spherical. Centrifugation of homogenates according to LéJohn et al. (1969) and Horgen and Griffin (1969) was longer than necessary to pellet intact mitochondria. Our final centrifugation sequence at 0--4°C was 3500g for 4 min to pellet whole spores, large debris, and mitochondria; 400g for 8 min with one wash of the resuspended pellet to separate mitochondria from whole cells and intact nuclear apparatii; and 3500g for 4 min to pellet mitochondria and eliminate debris.

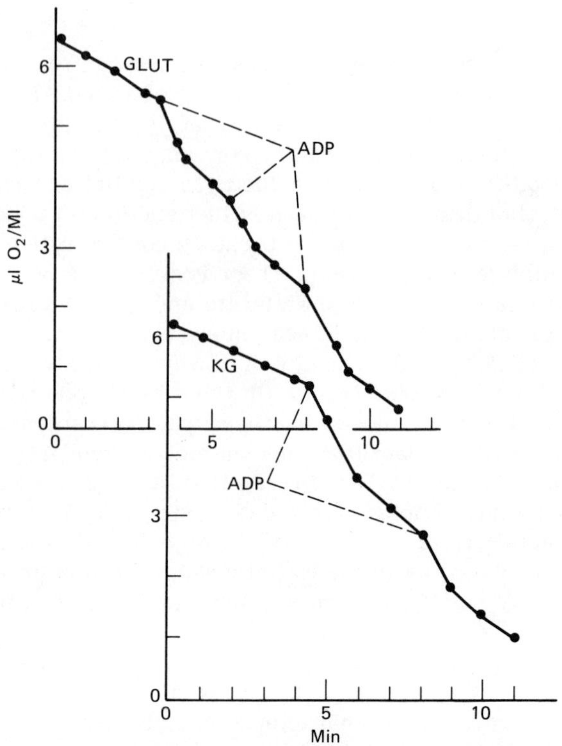

Fig. 2 Oxygen uptake by mitochondrial fractions from *B. emersonii* zoospores. Suspensions were equilibrated to constant endogenous rate (set as zero time), and additions (5 mM glutamate, 5 mM α-ketoglutarate, and 0.1 mM ADP) were then made as shown. About 5 x 10^9 zoospores were processed as described in text, one fourth of the final preparation being used for each experiment illustrated (from Suberkropp & Cantino, unpublished data).

THE NUCLEAR CAP: RIBOSOMAL ORGANELLE PAR EXCELLENCE

The nuclear cap has been recognized (Sparrow, 1960) for many years as a regular and prominent feature of the Blastocladiales. In the zoospores of *Blastocladiella emersonii*, the nuclear cap is entirely surrounded by a double membrane and covers the anterior nuclear hemisphere (Cantino et al., 1963). Such nuclear caps have been

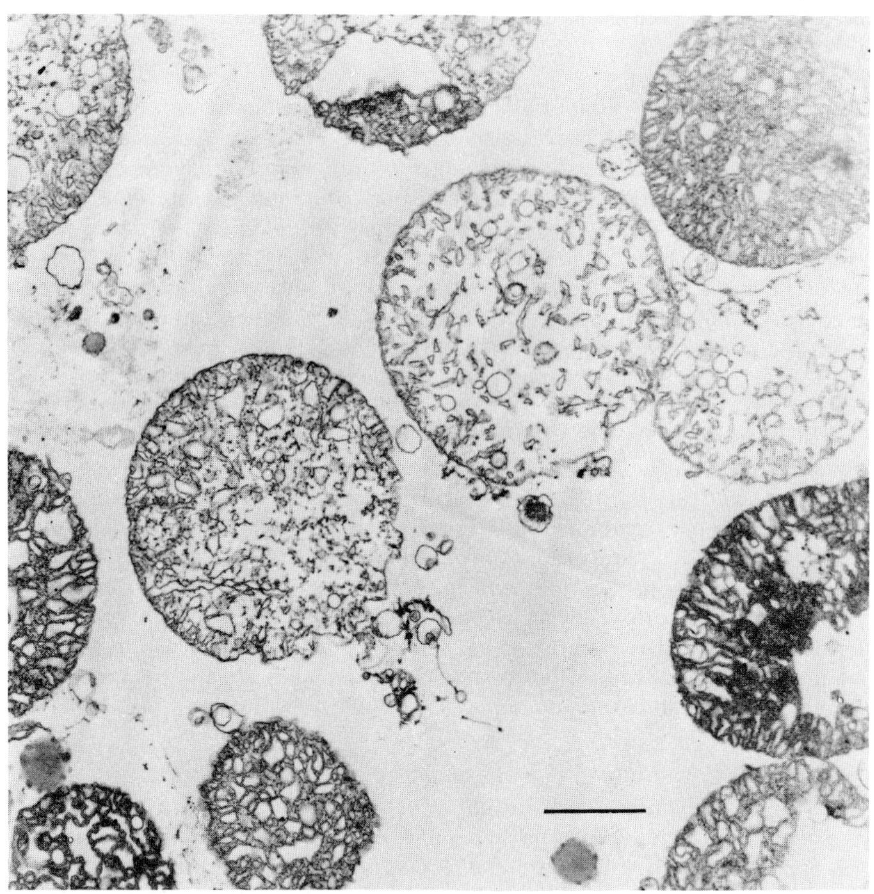

Fig. 3 The appearance of mitochondria isolated (see text) from *B. emersonii* zoospores and further purified via a linear (30--60%) sucrose gradient; the band collected at 1.23 g/cm^3 was fixed in 0.5% glutaraldehyde and 0.5% OsO_4, and postfixed in 1% OsO_4, each for 1 h at 4°C, dehydrated, and embedded in Spurr's medium. The marker represents 1 μm.

---

isolated and analyzed (Lovett, 1963); their contents, consisting of granules with the classic appearance of ribosomes, have also been analyzed chemically and identified as 83S ribosomes. Hence, in the typical swimming zoospore of *B. emersonii*, the nuclear cap is a packet that confines almost all* of the cell's ribosomes.

What is the function of the nuclear cap? A swimming zoospore of *B. emersonii* accumulates only insignificant amounts of RNA and DNA and, at best, exceedingly small quantities of protein (Lovett, 1968; Soll & Sonneborn, 1971a), its metabolic activities being mainly if not exclusively oxidative and degradative (Suberkropp & Cantino, 1972, 1973; Lodi & Sonneborn, 1974; Mills et al., 1974). But shortly after a zoospore retracts its flagellum and encysts (see section on encystment), the exposed segment of the double membrane around the nuclear cap begins to fragment.† The dispersal of its ribosomes commences therewith, and the process continues until they are scattered throughout the cytoplasm (Lovett, 1968; Soll & Sonneborn, 1969a; Truesdell & Cantino, 1971). The evidence suggests that when protein synthesis begins during spore germination, it takes place on these ribosomes---ribosomes which were formed during growth of the preceeding parental generation, and then kept clustered within the nuclear cap during the intervening motile stage in this organism's life cycle. Lovett's (1968) results also indicate that such early protein synthesis depends upon the presence of a cryptic mRNA in the spore being held in readiness, so to speak, to help mediate the process. Hence it seems as if the zoospore can logically be looked upon as a

*Adelman & Lovett (1972) err in concluding that the nuclear cap contains ". . . the entire complement of zoospore ribosomes." Firstly, the single mitochondrion in the zoospore of *B. emersonii* contains ribosomelike particles (Truesdell & Cantino, 1971) identical in appearance to nuclear cap ribosomes. And secondly, nuclear caps do not invariably ". . . remain intact in the discharged zoospore. . ." (Adelman & Lovett, 1972); additional membrane bound aggregates of ribosomelike particles (in effect small satellite ribosome packages situated either in the zoospore cytoplasm or in mitochondrial "chambers"), although rarely seen in swimming zoospores, can be commonly fabricated by amoeboid spores (Shaw & Cantino, 1969; Cantino, 1969; Cantino & Mack, 1969) and recovered in cell-free preparations (Cantino, unpublished). Their significance, if any, is unknown.

†Although there is circumstantial evidence suggesting an alteration of the membrane between the nuclear cap and nucleus (Cantino & Truesdell, 1972) after a spore is triggered to encyst, and for possible involvement of the mitochondrion (Truesdell & Cantino, 1971) in fragmentation of the nuclear cap double membrane after encystment, there are no clues about the mechanisms that might be involved.

cell preprogrammed with templates to carry on the synthesis of materials that are needed after spore encystment (see section on the gamma particle for related information).

This led to the conclusion (Lovett, 1963, 1968) that two important functions of the nuclear cap might be to protect ribosomes during the motile zoospore stage thereby conserving them for use when *B. emersonii* began growing again, and to exercise a regulatory function by spacially separating from the enclosed ribosomes one or more components necessary for their function. Subsequently, Schmoyer & Lovett (1969) showed that charged tRNAs and aminoacyl-tRNA synthetases were detectable both inside and outside the nuclear cap, and that there was an unidentified material, also present inside and outside of the cap, which blocked ribosome function by inhibiting the binding of amino acyl-tRNA and peptide bond formation. However, removal of the ribosomes from the nuclear cap did not release the inhibitor, as would be expected if compartmentalization per se had restricted ribosome function. This led to a reevaluation of previous ideas, and to the conclusion that mere enclosure of ribosomes in a nuclear cap did not in itself prevent protein synthesis, and that the lack of ribosome activity may have been due to the presence of the unidentified inhibitor---a concept that Soll & Sonneborn (1971a) also favor from their own studies of amino acid uptake by zoospores of *B. emersonii*.

The latest evidence (Adelman & Lovett, 1974) suggests that the elusive inhibitor is a substance of low molecular weight, possibly a nucleotide. It is hypothesized that the inhibitor regulates polypeptide synthesis by preventing translation of the enigmatic mRNA in swimming zoospores, or by blocking the degradation of inactive 80s ribosomes until it is released during their encystment or germination. To prove all this, efforts are being made to isolate and identify the inhibitor; however, as far as we know it has not yet been accomplished.

Does the role of the nuclear cap, as it is presently envisioned for *B. emersonii*, have any general applicability to the zoospores of other Chytridiomycetes? As stated previously, early light-microscopic studies led to the belief that nuclear caps were typical ingredients of blastocladeaceous zoospores. Recent electron microscopic studies have tended to confirm this point of view. In *B. britannica* (Cantino & Truesdell, 1971a) and other members of the Blastocladiales such as *Allomyces* (Blondel & Turian, 1960; Moore, 1968; Fuller & Olson, 1971; Olson, 1973a; etc.), *Blastocladia* (Fuller & Calhoun, 1968), and *Coelomomyces* (Martin, 1971; Madelin & Becket, 1972; Whisler et al., 1972), nuclear caps apparently enclosing essentially all of a planont's ribosomes are present. Furthermore, during encystment of *Allomyces* meiospores (Olson, 1973b), fragmentation of the nuclear cap envelope and dispersal of its ribosomes apparently takes place before a significant amount of protein is synthesized, more or less as occurs in *B. emersonii*. Perhaps it will not be surprising, then, if the basic picture, as presently portrayed for *B. emersonii* from the studies done in Lovett's and Sonneborn's laboratories, turns out to be substantially the same for all the members of the Blastocladiales.

However, the unmistakable fact remains: The enclosure of a zoospore's ribosomes within some sort of bundle is not an indispensible requirement for all posteriorly uniflagellate fungi. It has been known (Koch, 1958) for many years that the motile cells of many Chytridiomycetes lack any obvious nuclear cap. Recent electron microscopic studies demonstrate that *Olpidium* (Temmink & Campbell, 1969a) and *Phlyctochytrium* (Chong & Barr, 1973) bear naked aggregates of ribosomelike* particles which are widely distributed in the cytoplasm, while in others such particles may be enclosed only partially within loose networks of double membranes (Fuller, 1966; Fuller & Reichle, 1968). However, there are also some not belonging to the Blastocladiales in which ribosomelike particles are almost completely confined within nuclear caplike structures, examples being *Harpochytrium* (Travland & Whisler, 1971), *Oedogoniomyces* (Reichle, 1972), and *Nowakowskiella* (Chambers et al., 1967).

It can be concluded, therefore, as LéJohn and Lovett (1966) emphasized almost ten years ago for *Rhizophlyctis* in particular, that in many Chytridiomycete spores, whatever the control system operating to prevent protein synthesis may be, a nuclear cap cannot be part of the regulatory device as it is in *B. emersonii* and *R. rosea*. This assumes, of course, that protein synthesis is not going on in these various Chytridiomycete spores, as is the case in *Rhizophlyctis*. The assumption could be wrong.

*It is a telling commentary on our capacity for self-persuasion and the hectic climate of research that nowadays investigators of zoospore fine structure routinely, and with apparent confidence, identify ribosomes by their appearance in thin sections; the fact remains that ribosomes have been isolated, and then characterized by chemical and centrifugal methods, only from *B. emersonii* (Lovett, 1963) and *R. rosea* (LéJohn & Lovett, 1966).

SIDE BODIES, STÜBEN BODIES, MICROBODIES, AND SYMPHYOMICROBODIES BY SYMPHYOGENESIS

The first electron microscopic study of thin sections through the zoospores of *B. emersonii* (Cantino et al., 1963; Lovett, 1963) revealed the true nature of the so-called side body, a theretofore je ne sais quoi that had been seen (Stüben, 1939; Couch & Whiffen, 1942) by way of light microscopy in swarmers from other species of the Blastocladiaceae. The largest structural component of the side body turned out to be a single giant mitochondrion. Adjacent to it were some much smaller lipid granules interspersed among even thinner profiles of unknown material, the combination seemingly held in place by a double membrane and resembling a sort of shallow sac or blister upon the much more massive mitochondrion. Later on, the unidentified profiles among the lipid granules were labeled sb (Lessie & Lovett, 1968) to identify them on electron photomicrographs. Although sb granules were at first treated as if they were individual entities,* subsequent work with serial sections showed (Cantino & Truesdell, 1970) that the sb profiles in swimming zoospores literally represented segments of a much larger continuous structure; the latter was then labeled the SB matrix. In short (Fig. 1), the lipid granules (which are not unit membrane bounded) are dispersed along the outer surface of the long "arm" of the mitochondrion, where they are usually in intimate contact with it. Molded against them is the SB matrix and its limiting unit membrane, a somewhat spoon-bowl shaped organelle with an irregularly wavy rim confined to a region around the lipid bodies. The SB matrix consists of a granular to amorphous substance of moderate electron density. A sheet of double membrane, the backing membrane (BM), covers the SB--lipid--mitochondrial complex and is attached at several places to (i.e., is continuous with) the outer unit membrane of the nuclear apparatus. The two unit membranes of the BM are placed exceptionally close to each other wherever they lie adjacent to the SB matrix, and the central region between them stains intensely with OsO_4, UO_2^{2+}, or Pb^{2+}.

Within the past few years, structural aggregates resembling either the whole side body ensemble in *B. emersonii* (i.e., a BM-bound

*Chong & Barr (1973) are mistaken in repeatedly attributing to Lessie & Lovett (1968) the conclusion that sb granules were interconnected to form a matrix; Lessie and Lovett did suggest, however, the possibility of <u>mitochondrial</u> fusions.

SB--lipid--mitochondrial complex), or just its SB--lipid grouping, have been seen in the zoospores of *B. britannica* (Cantino & Truesdell, 1971a) and other Chytridiomycetes. Especially noteworthy among them is the side-body complex described by Martin (1971) in the zoospores of *Coelomomyces punctatus*, wherein nonmembrane bound lipid globules are positioned between, and strongly indent, a single mitochondrion on one side and a unit-membrane bound, undulate rimmed, SB matrix on the other; the diameter of the lipid bodies even exceeds the thickness of the SB matrix, as is true in *B. emersonii*. A somewhat comparable arrangement of solitary mitochondrion and lipid bodies partially embedded in profiles of electron dense material resembling the SB matrix of *B. emersonii* apparently also occurs in the motile planonts of two other species of *Coelomomyces*, *C. psorophorae* (Whisler et al., 1972) and *C. indicus* (Madelin & Beckett, 1972).

Similarly, something resembling a side body complex has been observed in the zoospores of *Allomyces* (Fuller & Olson, 1971), wherein the apparent equivalents of the membrane-bound SB matrix in *B. emersonii* were named Stüben bodies.* And recently, Olson (1973a)

*Fuller & Olson (1971) give several reasons for introducing a new term, Stüben bodies, to denote structures in *Allomyces* that resemble the side body granules in *B. emersonii*. One argument is that priority or description identifies these structures as being an integral part of the "seitenkörper" discovered by Stüben in 1939. However, the drawings (e.g., Fig. 64) of Couch & Whiffen (1942), who reexamined Stüben's side body by light microscopy in several species of *Blastocladiella*, and their observations that "adhering to the side body were . . . fat bodies," attest convincingly to our conclusion (Cantino et al., 1963) that what Stüben saw was almost certainly the giant mitochondrion, not the much smaller SB-matrix. If priority of description is at issue, then surely it is the mitochondrion, not the SB matrix, that should be renamed a Stüben body! Fuller and Olson also rationalize that because similar granules are found in all phases of the *Allomyces* life cycle (and in *Phlyctochytrium*), a more general name is called for. The reasoning escapes us, however, for not only is Stüben no more general than sb, but there is no connection between Stüben's visualization of a tear-drop shaped blob in the zoospore of *Blastocladiella* and the granules seen recently in the thallus of *Allomyces*. Perhaps a more useful generalization, however, could have been based upon the fact that the granular profiles in both *Allomyces* and *Blastocladiella* do resemble microbodies.

concluded from studies of serial sections that there is a side body complex in the meiospores of *Allomyces* with a three dimensional organization that resembles the one in *B. emersonii*. However, one particular difference is worth noting. In *Allomyces*, the membranes around the Stüben bodies ". . . can be traced to be continuous with the outer membrane of the nuclear envelope . . ." (Olson, 1973a). Furthermore, Olson's sketch (Fig. 21) depicts the membrane around the Stüben bodies as being continuous with the membrane around the side body complex. But in *B. emersonii*, the unit membrane around the SB matrix is not traceable to either the membrane around the nucleus or the nuclear cap, nor is it continuous at any point with the double membrane (The BM, Fig. 1) around the side body complex. If this dissimilarity should reflect variations in the origin of the SB matrix and Stüben bodies, and conseqeuntly differences in function, such a variation would be important.

Finally, although *Harpochytrium* zoospores do not possess a fully blastocladiacean side body complex in that it lacks the mitochondrion, they do contain (Travland & Whisler, 1971) a tight association of lipid like globules, a single "rumposome" (Fuller, 1966), and profiles of electron dense granular material resembling the SB matrix in *B. emersonii* and the Stüben bodies in *Allomyces*.

What happens to SB matrices and Stüben bodies when zoospores encyst has not been established in any great detail. In the meiospores of *Allomyces*, where the side body complex seems to be almost identical (Olson, 1973a) to that (Cantino & Truesdell, 1970) in the swarmers of *B. emersonii*, one of the first events in encystment is the disassembly of the side body complex (Olson, 1973b). In the process, the filled granular structures in the side body (i.e., the apparent equivalents of the SB matrix in *B. emersonii*) do not disintegrate but are dispersed to the cytoplasm in the cyst. In the zoospores of *B. emersonii*, studies (Cantino & Truesdell, unpublished data) of serial sections through spores fixed at 2, 4, and 15 min after induction of synchronous encystment, show that the SB matrix remains membrane bound and intact---although it becomes somewhat branched---in cysts 5--10 min old.

Understandably, the important question has arisen repeatedly: What is the function of these various bodies enclosed by single membranes, composed of a moderately electron opaque substance, and closely associated with lipid bodies and a single mitochondrion---or alternatively, a rumposome?

The Membrane Bound SB Matrix in the Side Body of a B. emersonii Zoospore is a "Symphyomicrobody"

Over the past few years, a large literature has accumulated about "microbodies." These organelles have been found in higher animals and plants, protozoans, algae, and fungi (de Duve, 1969; Hruban & Rechcigl, 1969; Tolbert, 1971). Morphologically, microbodies have some or all of the following characteristics (Mollenhauer et al., 1966; Frederick et al., 1968; Hruban & Rechcigl, 1969): a round or elongate shape; a single ca 60Å--80Å limiting membrane; a close association with endoplasmic reticulum and/or lipid bodies; a finely granular matrix of varying electron density; and the presence of an electron opaque inclusion, a crystalloid structure, or both, embedded in the ground substance. The matrix is homogenous after $KMnO_4$ fixation, but may have slightly fibrillar texture after OsO_4 fixation. The size ranges from 0.3 to 1.5 μm at its maximum diameter. Physiologically, microbodies have been divided into peroxisomes and glyoxysomes (de Duve, 1969). Peroxisomes are respiratory bodies, play a role in glycolate metabolism and photorespiration, and contain enzymes that produce and degrade H_2O_2 (de Duve, 1969; Tolbert, 1971). Glyoxysomes are frequently found near lipid bodies (Vigil, 1970; Mendgen, 1973; Silverberg & Sawa, 1973) and function in the conversion of fats to carbohydrates (Beevers, 1969; Tolbert, 1971).

Morphologically, the SB matrix in the *B. emersonii* zoospore resembles a microbody in some respects: It is delimited by a single unit membrane, has a somewhat electron opaque matrix, and is always found in close association with lipid bodies, a common characteristic of glyoxysomes. But in comparison with the average microbody, the SB matrix is enormous in size although this could conceivably be misleading since the size of the SB matrix is based on a three dimensional model derived from studies of serial sections (Cantino & Truesdell, 1970) whereas the sizes reported for most microbodies have been based on two dimensional measurements of random thin sections. Another difference concerns the fact that many microbodies are closely associated with endoplasmic reticulum, while the zoospores of *B. emersonii* do not possess (Cantino et al., 1963) an endoplasmic reticulum. However, an association between sb granules and endoplasmic reticulum may occur when zoospores are being formed, because endoplasmic reticulum is present at that time (Lessie & Lovett, 1968). Since microbodies are thought to be formed from or by endoplasmic reticulum (Matsushima, 1972; Vigil, 1973), this may become one important factor in establishing whether or not the SB matrix can be considered to be a microbody.

Physiological evidence for the identity of microbodies has been gotten at the cellular level using ultracytochemical techniques. Although many enzymes have been used as markers for microbodies, only catalase seems to occur in almost all of them (de Duve, 1969); it can be demonstrated ultracytochemically (Novikoff & Goldfischer, 1968; Frederick & Newcomb, 1969), and it has been used to establish the presence of microbodies in a number of different organisms (Beard & Novikoff, 1969: Frederick & Newcomb, 1969; Vigil, 1970; Matsushima, 1972; Stewart et al., 1972). We have applied the procedure of Beard & Novikoff (1969) to fresh suspensions of *B. emersonii* zoospores. Incubation in the DAB/H_2O_2 medium produced a very dense deposit of osmium black over the SB matrix (Fig. 4) but nowhere else. Hence, it appears as if the SB matrix contains catalase, and that therefore it may be a sort of microbody.

Lessie & Lovett (1968) have shown that the cytoplasm in a growing plant of *B. emersonii* contains numerous, small, irregularly shaped, "sb granules" with moderately electron dense content bounded by a single membrane; that these granules, along with lipid granules, surround each nucleus during sporogenesis; and that both sb and lipid granules eventually align themselves alongside the mitochondrion in the finished zoospore while it is still inside the sporangium.* Since it has been further shown (Cantino & Truesdell, 1970) that, in free swimming zoospores, such sb granules are interconnected into a single, large, unit membrane bound SB matrix (see Fig. 1), it now seems to us an inescapable conclusion that the SB matrix is formed by symphyogenesis, in other words, that it is produced by the union of formerly separate elements (Fig. 5). And, since we have herewith provided evidence that the SB matrix probably possesses catalase activity, and that it is the only organelle in the zoospore that does so, the SB matrix can properly be called a symphyomicrobody. The building blocks from which it is constructed, that is, the sb granules of Lessie & Lovett (1968), are therefore almost certainly microbodies.

The glyoxysome plays a central role in gluconeogenesis by virtue of its capacity to burn up fat via β-oxidation and to mediate the

*Madelin & Beckett (1972) report that a somewhat similar sequence of events, involving lipid droplets and flattened sacs of granular material slightly denser than the cytoplasm, occurs during the genesis of a side body complex in the planonts of *Coelomomyces*.

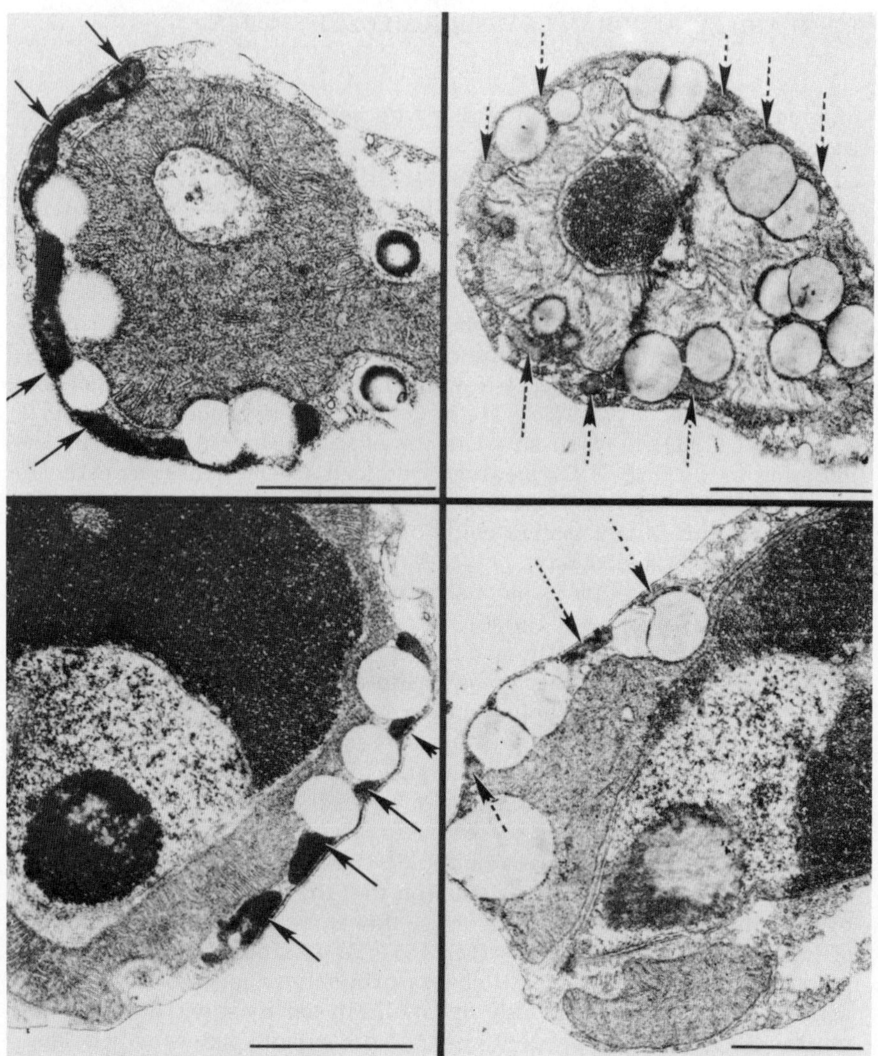

Fig. 4 The response of the SB-matrix in zoospores of *B. emersonii* to the Beard and Novikoff (1969) DAB test for catalase activity. Zoospores were fixed in cacodylate buffered 2% glutaraldehyde for 2 h at 4°C, rinsed 12 h in a 0.1 M cacodylate buffer, pH 6.9, containing 0.22 M sucrose, incubated in the DAB oxidation medium at pH 9 and 37°C for 60 min (and others in appropriate control solutions with KCN or without substrate), rinsed, postfixed with 2% OsO_4 for 1 h at 4°C, rinsed again, dehydrated and embedded in Spurr's medium. Note

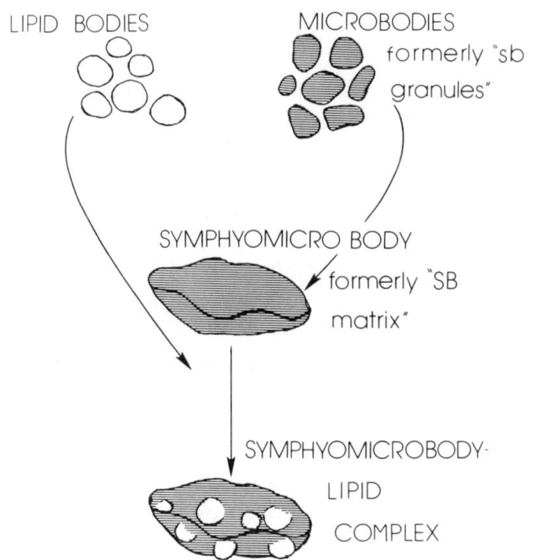

Fig. 5 Formation of the symphyomicrobody (by symphyogenesis) and the symphyomicrobody--lipid complex in the zoospore of *B. emersonii*.

---

glyoxylate cycle via the key enzymes isocitrate lyase and malate synthetase, thus bringing about net synthesis of succinate from acetyl CoA; the succinate then finds its way to mitochondria where it is further metabolized (Beevers, 1969; Tolbert, 1971; Richardson, 1974). A key point in this operation is the fact that continuous activity of the glyoxysome requires the cooperation of outside enzymes (Müller et al., 1968; Beevers, 1969); therefore a metabolic interplay---a movement of C_4 and C_6 molecules back and forth---must occur between glyoxysomes and mitochondria.

Early preparations of *B. emersonii* spores exhibited isocitrate lyase activity (McCurdy & Cantino, 1960); our work in progress has verified the presence of this enzyme in spore populations, using newer assay procedures now commonly used by others (e.g., Huang & Beevers, 1971, 1973; Bieglmayer et al., 1973). However, before the symphyomicrobody itself can be assayed for this enzyme, purified preparations free of contaminating material will obviously have to be

---

the dense deposit (solid arrows) in the SB-matrices, and the lack of it (dotted arrows) in the controls without substrate.

obtained. To this end, we are attempting to establish procedures for isolating symphyomicrobodies in sufficient quantity for analyses. Preliminary work in progress suggests* that isocitrate lyase activity may be associated with it. However, enough information is not yet available about the other enzyme activities and metabolic properties of the symphyomicrobody in the zoospore of *B. emersonii* to establish firmly whether or not this organelle should be considered a kind of glyoxysome or peroxisome, or both, or neither one.

If the symphyomicrobody behaves like the conventional glyoxysome, it presumably functions during swimming rather than encystment, for in the latter process there is no decrease in total spore lipid (Mills et al., 1974). Yet, its method of origin (by symphyogenesis) and its decrease in size (while the spore is swimming and consuming lipid; Suberkropp & Cantino, 1973; Mills et al., 1974) are not characteristics that we have encountered in the literature on microbodies; from these two points of view, the *B. emersonii* symphyomicrobody apparently encompasses something more than what is circumscribed by the present day stereotype of the glyoxysome. However, if it is a glyoxysome, then its placement (Fig. 1) within the sacciform side body complex (in intimate contact with both the spore's single mitochondrion on the one hand, and its total complement of lipid globules on the other) illustrates what appears to be a highly efficient arrangement for carrying out the mission of a glyoxysome.†

RETRACTION OF THE FLAGELLUM: AXONEMAL TRANSLOCATION

When a Chytridiomycete zoospore stops swimming and transforms itself into a cyst, its tail vanishes. Is there some understandable mechanism behind this disappearing act? Koch (1968b) has described in detail some painstaking observations of some 17 species of Chytridiales and Blastocladiales, from which he distilled the conclusion that there are four recognizably distinct patterns of flagellar retraction to

*Homogenates of *B. emersonii* have an average specific activity of 0.014, μmole x min^{-1} x mg^{-1} protein for isocitrate lyase. The enzyme, enriched 10-fold, is associated with a particulate fraction which has a buoyant density of 1.28 g/cm^3 on a linear sucrose gradient.

†A tightly knit nucleus--microbody--chloroplast association in another cell in which there is also only one DAB positive microbody has been described (Stewart et al., 1972) in the alga *Klebsormidium*.

be found among them. He labeled them "lash-around," "body-twist," "vesicular," and "straight-in." Since that time, there have been additional reports of other Chytridiomycetes also pulling in their flagella by methods resembling the first three processes. Since no reasonable mechanism, based either on experimental evidence or just the plain force of logic, has been offered to explain these first three retraction phenomena, the subject probably should not qualify for further discussion here.

However, since Koch (1968b) has included extensive descriptions about flagellar retraction in our strain of *B. emersonii*, and since we, too, have investigated in detail the way it retracts its flagellum---and have proposed a mechanism to account for it as well---we feel compelled to provide some comments on this matter. Although our own observations have been described (Cantino & Hyatt, 1953; Cantino et al., 1963), expanded (Cantino et al., 1968), and summarized (Truesdell & Cantino, 1971), this is an appropriate opportunity to reflect upon our ideas and those of others with a broader perspective.

For *B. emersonii*, the end of a zoospore's existence coincides with the beginning of its encystment, and this involves a very rapid and highly coordinated sequence of events (Truesdell & Cantino, 1971; Soll & Sonneborn, 1969a). Among them are two phenomena, verified repeatedly (Cantino et al., 1968) and captured on 16-mm film (Cantino, 1973) since they were first observed (Cantino et al., 1963), which in any viable spore of *B. emersonii* are intrinsically linked to one another: namely, flagellar retraction and rotation of the nucleus and its associated nuclear cap (the nuclear apparatus). When the time for encystment draws near, the spore stops swimming and becomes more spherical (Fig. 6). Simultaneously, the flagellum straightens out, vibrates rapidly, and then stops. The nuclear apparatus shifts slightly toward the spore's anterior, then begins to rotate in the direction of the short arm of the mitochondrion. As rotation proceeds, the flagellum sweeps into an arc; meanwhile, the axoneme is being translocated into the main spore body. The nuclear apparatus continues turning until the entire axoneme has been retracted through a fixed point of entry on the spore's surface. Thus, the body of the spore does not rotate simultaneously; it simply becomes progressively more spherical until it is cystlike in shape, after which the initial cyst wall is detectable in electron micrographs (Lovett, 1968; Soll & Sonneborn, 1969a; Truesdell & Cantino, 1971). Koch (1968b) also reported having seen flagellar retraction in *B. emersonii*, essentially as we have described it above, a process he termed

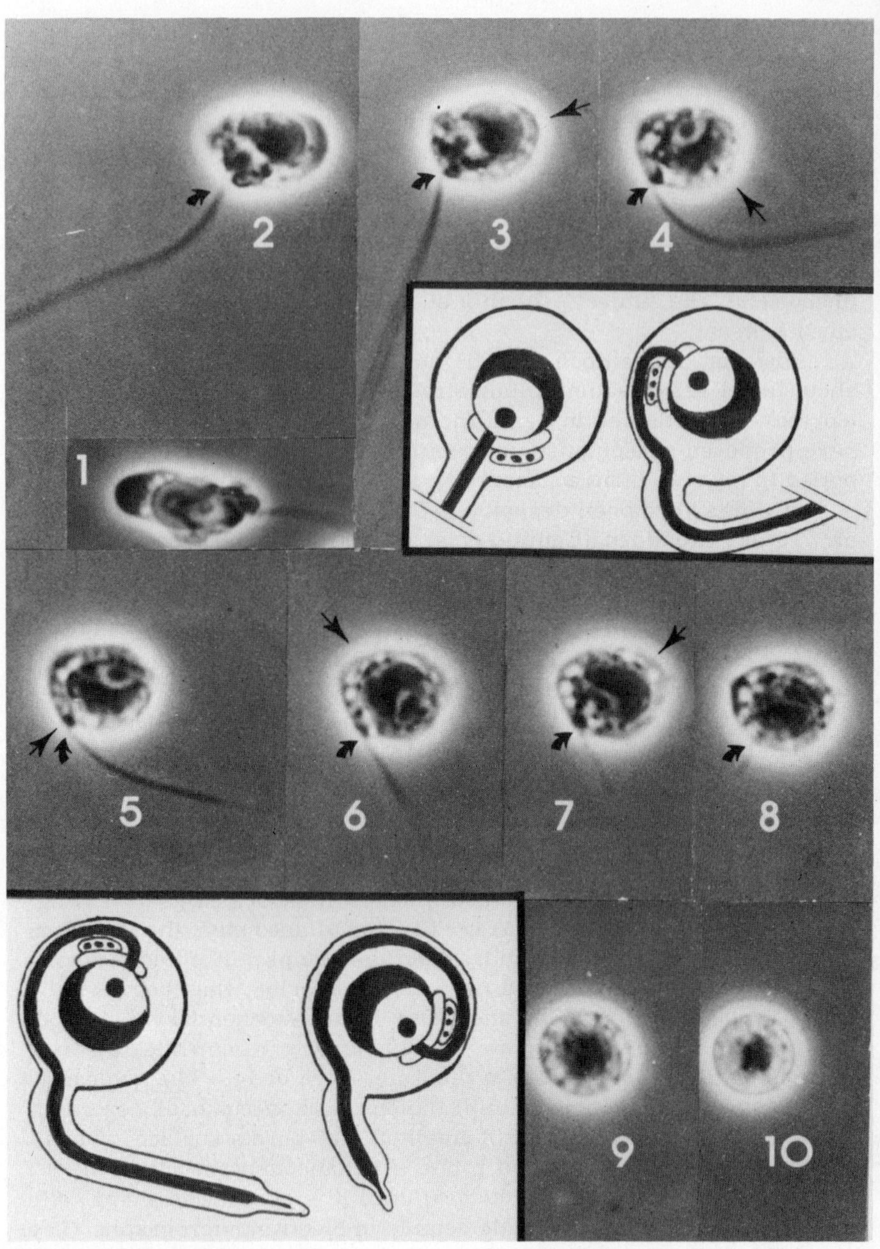

Fig. 6 Rotation of the nuclear apparatus during flagellar retraction by the zoospore of *B. emersonii*. Photographs x 2,350. Pictures

straight-in retraction.

What is the mechanism for such a flagellar retraction with its concurrent rotation of the nuclear apparatus (hereafter nuclear rotation)? We have concluded (Truesdell & Cantino, 1971) that it is a purely mechanical process. The flagellum is composed of an axoneme and its membranous sheath. During retraction and nuclear rotation, the axoneme coils up along the inside periphery of the encysting spore. After the axoneme has been withdrawn, the membranous sheath is no longer associated with it. The sheath is continuous with the plasma membrane before retraction; apparently it is retained as part of the plasma membrane after retraction. Since translocation of the axoneme occurs only during nuclear rotation, and vice versa, the two structures probably remain connected throughout the process. In such a linked device, the force responsible for the simultaneous translocation and rotation could theoretically be applied at either "end" of the system. An internal force causing nuclear rotation would, in effect, "wind in" the axoneme; conversely, a force applied to the axoneme would, in turn, be transmitted to the kinetosome (which abuts upon the nucleus) and thereby push the nuclear apparatus in its circular

---

were taken about 15 seconds apart; all spores are in their original positions on the slide. The straight arrow points to the broad top side of the nuclear cap which lies anterior and adjacent to the nucleus (light colored) with nucleolus (dark spherical body) therein. As the amoeboid spore (1) begins to round up (2), its outstretched flagellum vibrates back and forth without propelling the spore away (3); immediately thereafter, the nuclear apparatus begins to rotate clockwise (4) while the flagellum sweeps around in an arc and simultaneously begins to move into the spore (5); as rotation continues, the flagellum gets progressively shorter (6--8) until it is withdrawn *in toto* (9,10). Note that the point (curved arrow) at which the flagellum makes contact with the periphery of the spore remains fixed at the same position throughout the sequence (for additional details, see Cantino et al., 1968). Explanatory diagrams. Flagellar retraction mechanism: the short elongated body containing three dots represents the symphyomicrobody with its attached lipid globules; the longer body next to it is the mitochondrion which lies adjacent to the crescent-shaped nuclear cap and its adjoining nucleus. The circumference of a rounded spore about to retract its flagellum is approximately equal to or slightly greater than the length of the flagellum.

path. Our observations (Cantino et al., 1968; Truesdell & Cantino, 1971) support the latter interpretation.

During encystment, the spore becomes spherical, that is, its surface area is decreasing with respect to its volume; furthermore, its volume, as seen microscopically and as measured quantitatively (Truesdell & Cantino, 1971), is also decreasing. Thus, forces are operating which oppose any extension of the cell surface; the latter includes the membrane extending outward around the axoneme. These forces, therefore, cause the flagellar membrane to assume a new position confluent with the contour of the cell, which is becoming increasingly spherical.* Axonemal translocation and nuclear rotation will follow.†

If the force causing nuclear rotation is transmitted via the axoneme, as visualized in our model, nuclear rotation should never be detected in spores which have been deflagellated (but not otherwise harmed). Experiments (Cantino et al., 1968) have shown this to be the case. Throughout encystment of such mechanically deflagellated spores, there is very little movement of the nuclear apparatus.

Finally, if the spore's membrane must contract to bring about flagellar retraction, is there any evidence that it acquires this capacity before the spore rounds up? Certain observations such as the development of flocculation competence, effects of certain oxidants and reductants, release of fluorescent material, and so on (Cantino et al., 1968), could be interpreted as being consistent with the idea that the plasma membrane does change chemically or physically

*We have obtained a few micrographs of zoospores actually caught in the process of retracting their flagella; they show that the nuclear cap, nucleus, and axoneme remain connected and arranged along an arc that follows the contour of the spore surface, just the arrangement anticipated from our model of flagellar retraction. Furthermore, the plasma membrane of a retracting spore is irregularly pleated in such micrographs, which may be a manifestation of surface forces that come into play during retraction, possibly by way of an enfolding process (as contrasted with an actual reduction by uniform contraction).

†There is evidence that binding sites must be broken before rotation can occur; when such sites are broken, the nuclear apparatus--axonemal assemblage would presumably float without restraint and thereby be free to rotate. The way in which the spore is triggered to encyst, however, does not relate directly to the mechanism of flagellar retraction; see section on encystment.

before encystment. But, the only direct evidence that the plasma membrane may be altered for its special task is the *in vivo* observation (Cantino et al., 1968; Truesdell & Cantino, 1971) that flagellar retraction is preceeded by formation of vesicles which migrate through the cytoplasm to the cell surface where they seem to fuse with it.

In summary, a body of circumstantial evidence supports the view that in *B. emersonii* contraction of the plasma membrane drives flagellar retraction, and that this is an integral part of the overall process of encystment in any zoospore which can continue to develop into a germling, that is, any viable zoospore. We have not uncovered any experimental evidence in our own quarters that would argue against the mechanism proposed above for flagellar retraction in *B. emersonii*, nor are we aware of any arguments made by others against this mechanism.

There is another issue, however, that cannot be set aside quite so easily, and it concerns the several alternate ways in which *B. emersonii* also seems to be able to dispose of its flagellum. Koch (1968b) has provided an abundance of photographic documentation for them in *B. emersonii*, including vesicular, body-twist, and lash-around. The latter is the only method that Fuller (1966) and Reichle & Fuller (1967) were able to observe or recognize in *B. emersonii*. From this perspective, Koch (1968b) concluded that since he had seen *B. emersonii* retract its flagellum by all of these methods, the apparently contradictory reports by Cantino et al. (1963, 1968), Fuller (1966), and Reichle & Fuller (1967) regarding flagellar retraction in this one organism were not really in conflict. This outlook constitutes the issue we wish to discuss now.

From one point of view, there is no disagreement, for we, too, have observed *B. emersonii* zoospores retract their flagella by all four methods---under certain conditions! The important question, in our judgement, really is: when a zoospore exhibits lash-around, or body-twist, or vesicular retraction, does it also annihilate itself in the process of such indulgences?

Consider lash-around retraction in *B. emersonii*, as illustrated in Koch's (1968b, Figs. 28A--C) photographs. By our standards, these spores are exceedingly bloated in appearance and decidedly not of normal dimensions (compare Truesdell & Cantino, 1971, Figs. 3A--D). Furthermore, they do not seem to be decreasing in size in the manner of zoospores undergoing nuclear rotation during flagellar retraction* (compare with our Fig. 6). Being nearly transparent in their highly swollen states, their insides are revealed with pleasurable clarity, but they are decidedly not happy spores. We, too, have put *B. emersonii* zoospores in various controlled experimental situations where they behave like those described by Koch; one of them involves incubating spores in reduced temperatures (Truesdell & Cantino, 1971). Under such conditions, spores will gradually swell and become spherical. Extensive water uptake during swelling apparently lowers cytoplasmic viscosity, for particles within them begin to exhibit rapid motion. The single mitochondrion enlarges and becomes globose. Eventually, flagellar activity gets to be increasingly erratic and eventually stops as the flagellum straightens out and extends directly away from the spore; then, as the flagellum wraps itself around the spore, it is absorbed.

What happens, exactly, as the flagellum disappears during such lash-around retraction in *B. emersonii*? Fuller (1966), apparently basing his conclusions on studies of seemingly swollen spores (see his Fig. 16), felt that his observations supported a mechanism suggested by Koch in that the flagellum wrapped itself around the body of the zoospore with a subsequent "fusion" of the flagellar sheath and the cell membrane; Reichle and Fuller (1967) reached a similar conclusion. But, according to Koch (1968b), "Actual fusion of the flagellum with the outer membrane of the planont body has not been seen by anyone . . . ."

Our own position on this matter, based solely on observations of *B. emersonii*, is this: As the flagellum wraps around the spore, the flagellar sheath pulls away from one side of the axoneme; simultaneously, it becomes part of the plasma membrane, as illustrated in Fig. 7. In this fashion, the axoneme is absorbed without any fusion of membranes! The axoneme automatically finds itself coiled up in the cytoplasm of the spore as the flagellar sheath progressively, but rapidly, peels away from the side of the axoneme where its sheath is confluent with the plasma membrane around the spore body proper. Because of the great difference in flagellar and spore body surface areas, the relative amount of flagellar membrane that is translocated

*In the sketches (Cantino et al., 1963, Fig. 4) accompanying our first description of nuclear rotation and translocation of the flagellum, overall cell sizes were not shown in correct proportions to one another; from this particular point of view they are misleading and should be disregarded.

to the rest of the cell surface is quite small. Unfortunately, there is presently no experimental evidence which might help explain what causes the peeling off of the flagellar membrane and thereby drives the consequent illusory translocation of the axoneme in lash-around retraction in *B. emersonii*.

We come, now, to the consequences of lash-around retraction as we have observed it in *B. emersonii*. Complete absorption of the axoneme may take from a few seconds to a few minutes; it seems to be related to the rate at which a spore is swelling. The axoneme is often visible within the spore, pushing against the plasma membrane and sometimes distorting its spherical contour. Usually, the spore continues to swell until it bursts; the cytoplasm is discharged violently

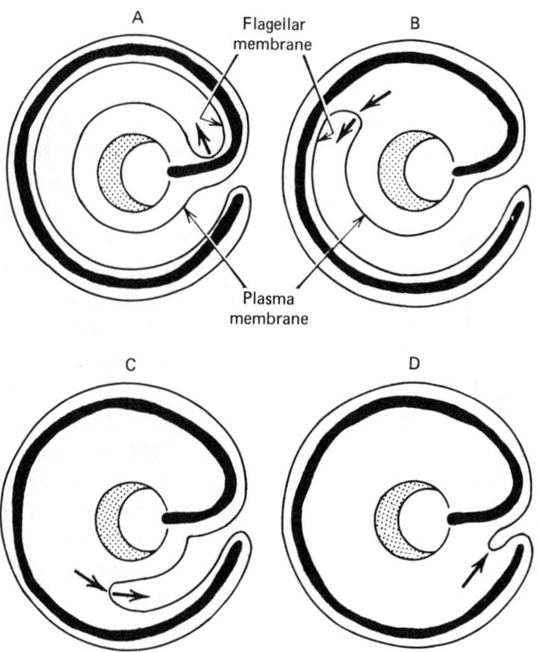

Fig. 7 The mechanism of "lash-around" flagellar retraction in *B. emersonii*. Proportions of various cell parts are highly exaggerated to facilitate this illustration of the way in which the flagellum can be "absorbed" without a concomitant fusion of membranes.

into the suspending medium, and the flagellar axoneme uncoils from its original position within the spore to become readily visible.

Koch (1968b) does not provide specific evidence that the swollen *B. emersonii* spores depicted in his figures which exhibited lash-around retraction, or that others like them, ever germinated or had the capacity to do so. Similarly, no evidence is given by Reichle and Fuller (1967) that their "slightly flattened" (but also, apparently, somewhat swollen) *B. emersonii* zoospores, as pictured in their Fig. 1, were viable. It is our contention that while lash-around flagellar retraction can indeed occur in *B. emersonii* zoospores, the process is generally lethal and the cell does not survive.

Koch (1968b) also protrayed body-twist retraction in *B. emersonii* which we, too, have observed in chilled swollen spores. But again, the pictures provided by Koch (Figs. 20--28) to illustrate this method of flagellar absorption invariably show spores that are swollen. In sharp contrast, his pictorial evidence (Fig. 1C, D) for straight-in flagellar retraction, which involves rotation of the nuclear apparatus, shows spores that are not swollen.

In conclusion, we would like to reemphasize a point made earlier (Truesdell & Cantino, 1971). Zoospores of *B. emersonii* can swell for various reasons, not only because of exposure to low temperatures. It is not clear what the conditions may have been that caused swelling in the spores pictured by Koch (1968b) and Reichle and Fuller (1967). But, flagellar absorption associated with such swollen spores differs in an important way from flagellar retraction in nonswollen spores. In the former, the nuclear apparatus does not rotate, the spores increase in volume, flagellar absorption and zoospore encystment are not intrinsically associated, and the cells usually die. In the latter, the nuclear apparatus does rotate, the spores decrease in volume, rotation of the nuclear apparatus and zoospore encystment are intrinsically associated, and the cysts that ensue are viable and germinate.

Finally, what can be said about the mechanisms of flagellar retractions in Chytridiomycetes other than *B. emersonii*? We emphasize that our reservations, as expressed above, regarding the significance of the various retraction methods applies only to *B. emersonii*, which we have studied extensively; we do not imply that lash-around retractions are necessarily fatal in all other Chytridiomycetes. Koch (1968b) has provided detailed descriptions of how flagella are retracted in numerous chytrids, and brief accounts have appeared since then for *Rozella allomycis* (Held, 1973), *Phlyctochytrium arcticum* (Chong & Barr, 1973), *Olpidium brassicae* (Temmink & Campbell,

1969b), and for the meiospores of *Allomyces macrogynus* (Olson, 1973b). However, no facts or theories to account for the mechanisms involved have been forthcoming, and there is little if any compelling evidence to suggest that zoospores generally survive all of these gesticulations. Clearly, some experimental data and some positive conclusions are needed.

However, Held (1973) has turned up the provocative observation that in all species for which suitable cross sections through encysted spores are available, the fibrils in the retracted axonemes are similarly oriented relative to the cell surface. Held does not suggest a mechanism for retractions based on this observation, nor do we, but we do wonder if this might simply mean that all of the sections which happened to provide the evidence for Held's idea might also have come from spores which had encysted by one particular process only (could it have been the straight-in method?). Even though chytridiomycete zoospores can use more than one method of retraction, it could be---as, we maintain, is true for *B. emersonii*---that only one of them will generally yield viable cysts. As far as we can tell, no one has done the necessary experimental work to rule this possibility in or out.

ZOOSPORE ENCYSTMENT

The encystment of a zoospore coincides with the end of its life history; it constitutes a distinctive kind of relationship between form and function. To appreciate this relationship, at least three things are needed: an adequate knowledge of the structural details in non-encysting zoospores (and they must be zoospores that have not unknowingly or inadvertently been triggered to encyst); suitable methods for inducing synchronous encystment in zoospore populations and for specifying the degree of synchrony in precise terms such that the significance of any changes in structure can be meaningfully guaged; and familiarity with any alterations in structure that occur during encystment. These requirements have been met in recent studies of *Blastocladiella emersonii* in several laboratories and they deserve coverage in this chapter. However zoospore encystment in this fungus, including a treatment of zoospore behavior and fine structural changes during encystment, environmental influences upon zoospores, the kinetics of encystment, and other matters, has been reviewed in some depth very recently (Truesdell & Cantino, 1971). Furthermore, one phase of the encystment process, flagellar retraction, has already been examined in the preceding section. And another particularly important feature, the role of gamma particles in encystment, will be singled out for separate discussion in the next section. Hence, we refer the interested reader to the afore mentioned review article for full details and appropriate literature citations. We do, however, take this opportunity to select for special emphasis a few aspects of encystment in *B. emersonii* and to call attention to recent observations of encystment in other fungi.

The Event of Encystment

As we (Truesdell & Cantino, 1971) have pointed out, the zoospore of *B. emersonii* ". . . contains an exceptionally tight and highly ordered arrangement of membrane-bound organelles. During encystment, this subcellular assemblage is swiftly disorganized by a cascade of changes: the flagellum is retracted; the nuclear apparatus rotates; the cell becomes spherical, loses volume, and forms a cyst 'wall'. Substantial vesiculation accompanies the process. This succession of events is temporally coordinated and spatially integrated. Axonemal translocation and rotation of the nuclear apparatus results from structural--mechanical transformations requiring no special energy sources or mechanisms other than those needed for the other processes associated with encystment. In the more camouflaged operation of cell wall synthesis, decay of gamma particles generates vesicles that fuse with the plasma membrane and thereby alter it; presumably they bring enzymes and/or structural components to the cell surface to lay down the foundation for synthesis of the cyst wall. Moreover, vesicles fusing with the plasmalemma contribute, on the one hand, to the volume decrease associated with encystment; on the other hand, by depositing new cyst wall material, they may be generating the surface forces needed for the change in cell shape and translocation of the axoneme."

A particularly important point, one that cannot be stressed enough because it bears so directly upon any understanding we may eventually achieve about the mechanism of encystment, has been rightly emphasized in one way (Soll & Sonneborn, 1971a, 1971b) or another (Cantino et al., 1968; Truesdell & Cantino, 1971); it is that the events of encystment, as outlined above, occur in the absence of any significant amount of protein synthesis (Lovett, 1968; Soll & Sonneborn, 1971a, 1971b) and, therefore, that they are not controlled by any underlying mechanism requiring substantial protein synthesis. This means that if proteins are needed for any of the structural changes and rearrangements in encystment, such proteins are presumably already there, preformed, in a zoospore before it encysts. It forces one to call into question whether or not differential protein synthesis can always provide, as Soll & Sonneborn (1971b) have aptly put it, an exclusive and sufficient mechanism for all kinds of cellular differentiation. The zoospore of *B. emersonii* is serving as a useful vehicle for examining this question.

Interestingly, a somewhat similar picture may be developing about the encystment process in *Allomyces*, a closely related blastocladiaceous mold with some historical seniority over *Blastocladiella*. The available evidence suggests that the zoospores of *A. arbuscula* (Burke et al., 1972) and *A. neo-moniliformis* (Olson & Fuller, 1971) do not synthesize protein until some time after encystment. Similarly, in the meiospores of the latter species, protein synthesis apparently does not accompany disassembly of various structures associated with encystment (Olson, 1973b). While two different species of aquatic fungi newly arrived on the scene is hardly a flock, it does encourage us to suppose that a general pattern may be evolving which could lead to a fuller understanding of the interesting "fast differentiations" associated with encystment in Chytridiomycete spores.

The Trigger for Encystment

From the foregoing discussion, two important questions arise. The first is: How is encystment triggered? There are several ways of inducing synchronous encystment in populations of *B. emersonii* zoospores; those most thoroughly studied involve controlling population density (Truesdell & Cantino, 1971) or placing spores in the cold (Cantino et al., 1969; Truesdell & Cantino, 1971), white light (Cantino & Truesdell, 1971b; Cantino & Myers, 1972), inorganic salts (Cantino et al., 1968; Soll & Sonneborn, 1969a, 1969b, 1972), and sulfonic acid dyes (Truesdell & Cantino, 1971).

In vivo the onset of encystment is preceded by at least two visible events: the acquisition of a capacity for spore floculation (which may be due to changes in the plasma membrane), and formation of vesicles which migrate to the cell surface and disappear (a phenomenon related to the activities of Gamma particles; see next section).

Electron microscopic investigations (Lovett, 1968; Soll & Sonneborn, 1969a; Truesdell & Cantino, 1971) have clarified some of the major structural events that may be associated with the triggering of encystment in *B. emersonii*. Within minutes after induction of encystment, myelinlike profiles appear in the vicinity of the porous nuclear membrane; they are thought (Cantino & Truesdell, 1972) to reflect membrane alterations that occur when spores are triggered to encyst. However, the most consistent and well-defined change (Truesdell & Cantino, 1971) detected after induction but before flagellar retraction occurs in the undifferentiated part of the backing membrane (BM, Fig. 1). The plasma membrane is closely associated with the electron scattering portion of the BM; it is continuous, in turn, with the outer unit membrane of the nuclear cap. The undifferentiated BM fragments rapidly; the differentiated portion does not break down as quickly although it, too, is also consumed slowly via vesicle formation. The BM has never been seen in sections through cells that have undergone normal nuclear rotation and flagellar retraction. Hence the BM seems to function as a 'binding site' that must first be broken before rotation of the nuclear apparatus can occur. Furthermore, there are visible props (Fig. 1) apparently connecting the kinetosome with the plasma membrane at the base of the flagellum (as there are in *B. britannica*; Cantino & Truesdell, 1971a) which presumably must be disengaged. There are also triplets of microtubules that extend upward from the kinetosome along the outside of the nuclear apparatus (Fig. 1) which deserve consideration (Shaw & Cantino, 1969; Myers & Cantino, 1974) as anchoring devices that must be defeated in some way, for they have never been seen (Cantino & Truesdell, unpublished data) in spores that have retracted their flagella; there is little doubt, apparently (Bardele, 1973) that a primary function of microtubules in other cell types is to serve as a cytoskeleton.*

The release of such mechanical connections holding various parts of the cell in position appears to be essential for encystment in that it is required for nuclear rotation. If all binding sites are ruptured,

*However, in this connection we are confronted by a distracting contradiction. On the one hand, it is thought (Weisenberg, 1972) that calcium may regulate the assembly of microtubules by virtue of its powerful capacity to prevent repolymerization of microtubule sub unit proteins. On the other hand, calcium is a potent encystment inhibitor (Suberkropp & Cantino, 1972) for prechilled *B. emersonii* spores which would otherwise ordinarily encyst; that is, calcium ensures regeneration of the motile state, which presumably involves regeneration of microtubule triplets (assuming that chilling, an effective encystment trigger, disrupts the spore's microtubule triplets).

the nuclear-axonemal assemblage is presumably free to rotate without restraint. It is not known, however, how the various triggering agents bring about the membrane alterations involved in the breaking of such binding sites.

Opportunely, other Chytridiomycetes are now entering onto this scene. In *Allomyces neo-moniliformis*, zoospore encystment can be triggered (Olson & Fuller, 1971) with a solution of L-leucine and L-lysine which bring about paralysis of flagella. Flagellar withdrawal may take place then, but more typically it occurs when the zoospores are then transferred to rich media. Of the various events occurring during synchronous encystment of meiospores (Olson, 1973b), one of the first is disassembly of the side body complex; the backing membrane disintegrates, this being followed by depolymerization of cytoplasmic microtubule triplets.* The parallelism with the way things happen in *B. emersonii* is impressive.

In *Rozella*, encystment can be triggered (Held, 1972) by providing zoospores with a host (e.g., *A. arbuscula*), reduced temperatures, elevated pressures, Cu ions, or colchicine. Then, when the ". . . swimming zoospore loses its motility and becomes ovate before encystment, the nucleus slips from its position directly anterior to the mitochondrion to a dorsal one, and consequently the lipid sac is pushed to a median position; this arrangement is still maintained in the naked cyst. . . . The fate of the backing membrane of the zoospore is not clear, although it may be retained in the cyst. . ." (Held, 1973). Since most of the treatments used to induce encystment in *Rozella* reportedly also degrade cytoplasmic microtubules in Heliozoa and some other organisms, Held (1972) suggests that its flagellated state may depend upon the intactness of its skeletal structure, or, more specifically, that encystment may be induced as a consequence of a modification of the lining of microfilaments around the proximal part of the flagellum, the microtubules radiating from the kinetosome and centriole, and the circle of props connecting the base of the flagellum to

*Although we have not considered in this article events that occur after the initial encystment of a zoospore, it seems to us of special interest (in view of the possible function and fate of the microtubule aggregates in *Blastocladiella, Allomyces*, and *Rozella*) to note that Olson (1972, 1973c), who has also studied the fate of microtubule triplets, concluded that a pool of protein derived from them is conserved and used later on by the spindle apparatus in the first mitotic division.

the plasmalemma.

The second cardinal question that arises from observing the events of encystment is: By what means are such events prevented in a nonencysting spore? There are probably interlocking ways by which a zoospore keeps some things shut down. A simple, on-or-off inhibitor-mediated process could underlie one or more of them, for Schmoyer & Lovett (1969) and Adelman & Lovett (1974) provide evidence for an internal inhibitor that suppresses protein synthesis in *B. emersonii*, while Cantino et al. (1968), Truesdell & Cantino (1971), and Soll & Sonneborn (1972) provide various kinds of arguments that unknown substances released by *B. emersonii* spores are capable of regulating encystment. Yet, the possible existence of such chemical agents provokes new questions, and one of them is especially meaningful; as we asked recently (Truesdell & Cantino, 1971), "Should the unidentified material which decreases percent encystment in a zoospore population be viewed as an encystment inhibitor or as a <u>zoospore stabilizer</u>?" Independently, Soll & Sonneborn (1972) seemed to be thinking along a similar vein when they cited unpublished evidence for the existence of a factor, recovered from suspensions of *B. emersonii* zoospores, which prevented germination and which they referred to as a zoospore maintenance factor. It is worth reemphasizing that the significance of this provocative question goes well beyond the problem in semantics that it poses. Believing it to be of extraordinary significance, yet not wanting to rephrase (simply for the sake of being nonrepetitive) points we have already put as best we could, we quote:

> Suppose it could be shown unequivocally that the encystment 'inhibitor' actually stabilizes zoospores in the sense of buffering them against adverse environments. For example, if the 'inhibitor' prolonged the time that swimming spores could withstand some new external stress, or the proportion of them that lysed after the environmental change took place, would not 'stabilizer' be the more appropriate term to use? . . . In any case, returning to the question of causality, the intricate character of a zoospore's internal architecture may also be operating to prevent encystment; unfortunately, it is exceedingly hard to demonstrate . . . This inquiry into the enigmatic nature of encystment also calls for brief consideration of some aspects of the kinetics of encystment. Spores induced with low temperatures and sulfonic acid azo dyes display similar kinetics, while spores induced with KCl

show a much greater mean time of encystment. The evidence suggests that the difference in behavior must result from an increase in the number and/or the average rates of encystment 'processes' which ensue after triggering. When the reason for this difference is uncovered, an important aspect of encystment will have been resolved. But in the meantime, a comparison of similarities among induction methods may also be fruitful. First, as far as we can tell, they all yield essentially the same sequence of structural changes associated with encystment, from which it can be argued that all induction methods must affect pathways that converge at some common step or process. Second, they all involve the placement of a spore under great stress. During cold incubation, the cell eventually becomes precariously balanced on the verge of lysis as it takes up water and approaches its "elastic limit". In sulfonic acid azo dyes, the effects are not so dramatic, but experience shows that spores in contact with these dyes are more labile to the effects of other environmental factors, such as cold incubation and fixation. In KCl, spores are also being pushed toward their limits; in 50 mm KCl, there is a little lysis, and at higher concentrations lysis is substantial. . . . One way of harmonizing these observations into a unified concept of the trigger mechanism is to conceptualize the induction step as a perturbation of the delicate balance of cellular controls in a spore. A momentary imbalance could evoke emergence of the new set of interrelated processes which comprise encystment. If this accurately represents what takes place, then it would be most enlightening to find answers to the question: what specific cellular controls will, when disturbed, lead to breakdown of other control mechanisms?

Diverse induction methods may disrupt different control mechanisms. Disengagement of some of them will be sufficient to induce encystment. Disengagement of others will not, and may, instead, cause autolysis or cell death. For example, cold incubation may be the type of trigger that interferes in a nondiscriminating fashion with control processes in the zoospore. If it modulates those critical things that suppress encystment without tampering with

> those that cause cell death, the spore will be induced to encyst. We can also suppose that when the different induction methods act on dissimilar controls, the results may pivot the breakdown of others. Depending on what is first attacked, the sequence of subsequent disjunctions will probably vary, as will their rates. Thus, the identity of the first control mechanism to be attacked will establish the rate of subsequent events; this will be reflected in the value of $\bar{T}$ (mean time of encystment) for the particular induction method used. (Truesdell & Cantino, 1971).

THE GAMMA PARTICLE: TYPE SPECIMEN FOR A NEW CLASS OF CYTOPLASMIC ORGANELLE, THE ENCYSTOSOME

The gamma particle is a ca. 0.5 μm cytoplasmic organelle discovered (Cantino & Horenstein, 1956) in the zoospores of *B. emersonii*. The main body of the gamma particle is exceedingly electron opaque with most fixatives; it resembles a hemi-ellipsoidal bowl with its acute ends removed (Fig. 8, top left) and is enclosed by a surrounding unit membrane. Studies of zoospore populations monitored during synchronous encystment have shown that the gamma particle decays in a characteristic fashion and that it apparently plays a cause-and-effect role in the formation of the cyst wall. The principle events (Fig. 8) are the release of vesicles from the gamma particle surrounding membrane which fuse with the plasma membrane, and the simultaneous decay of the gamma particle matrix via genesis of a tier of 80 nm vesicles on its inner surface which fuse with, and thus replenish, the gamma particle surrounding membrane. The case that has been developed for the function of the gamma particle in *B. emersonii* involves its frequency (number of gamma particles/cell) in zoospores derived from different phenotypes; its three dimensional shape; its de novo appearance during sporogenesis and its abrupt decay during zoospore encystment; its staining properties; its isolation in sufficient quantity and cleanliness to permit chemical analyses; the detection of DNA and RNA in it; the enzyme components it does and does not have, especially the absence of catalase and phosphatase and the presence of chitin synthetase (but not the preceding enzymes in the biosynthetic path to chitin, i.e., L-glutamine: D-fructose 6-phosphate aminotransferase, and acetyl-CoA: 2-amino-2-deoxy-D-glucose 6-phosphate N-acetyl transferase), and the relationship of the chitin synthetase to encystment; its apparent distinctiveness relative to lysosomes and microbodies; and the effects of visible light upon it. While it is impossible for us to view this summary with complete detachment, it is probably not an exaggeration to conclude that no other cytoplasmic organelle in the aquatic fungi has been studied to such an extent.

The facts thus far discovered about the gamma particle in *B. emersonii* have also led us to discuss the question of its autonomy and continuity as a cytoplasmic DNA organelle; the significance of its DNA and its possible role in encystment; and whether or not it might represent a third kind of genetic system. And with respect to the function of the gamma particle in a zoospore, in particular, the facts have led us to the idea that just as the nuclear cap may function to "protect" intact ribosomes for use in early development (see section on nuclear cap), so the gamma particle may serve as a protective package for sequestering other key ingredients in the organism, and as a device for transferring them from one generation to the next by way of the organism's only reproductive vehicle, the zoospore. These key ingredients could have special roles to play as soon as growth begins and, therefore, be located exclusively in the gamma particle for protection. Such ingredients include the chitin synthetase, a 4S RNA, and a DNA satellite, the lightest one present in the zoospore.

Since current knowledge of the gamma particle in *B. emersonii* has been reviewed in depth so recently (Truesdell & Cantino, 1971; Cantino & Myers, 1974; Myers & Cantino, 1974), we by-pass, albeit reluctantly, any further discussion of this subject now, and suggest that the aforementioned sources be consulted for futher information about the gamma particle and its role in intracellular interactions in *B. emersonii*.

There is one topic, however, of special relevance to this symposium that does merit further discussion here; namely, might the gamma particle in *B. emersonii* be a highly specialized form of some subcellular structures ubiquitously distributed among the zoospores of the water fungi? Encystment occurs universally among all viable zoospores of the aquatic Phycomycetes; hence if gamma particles play an integral role in encystment, might they not also belong to a family of widespread organelles involved in the encystment of all Chytridiomycete zoospores? When a gamma particle is defined in the most general of terms, that is, as a highly electron opaque particle contained in a vesicle, then similar structures have indeed been pictured, albeit not always indicated, in the motile cells of many species of Chytridiomycetes, not to mention other aquatic fungi, examples being

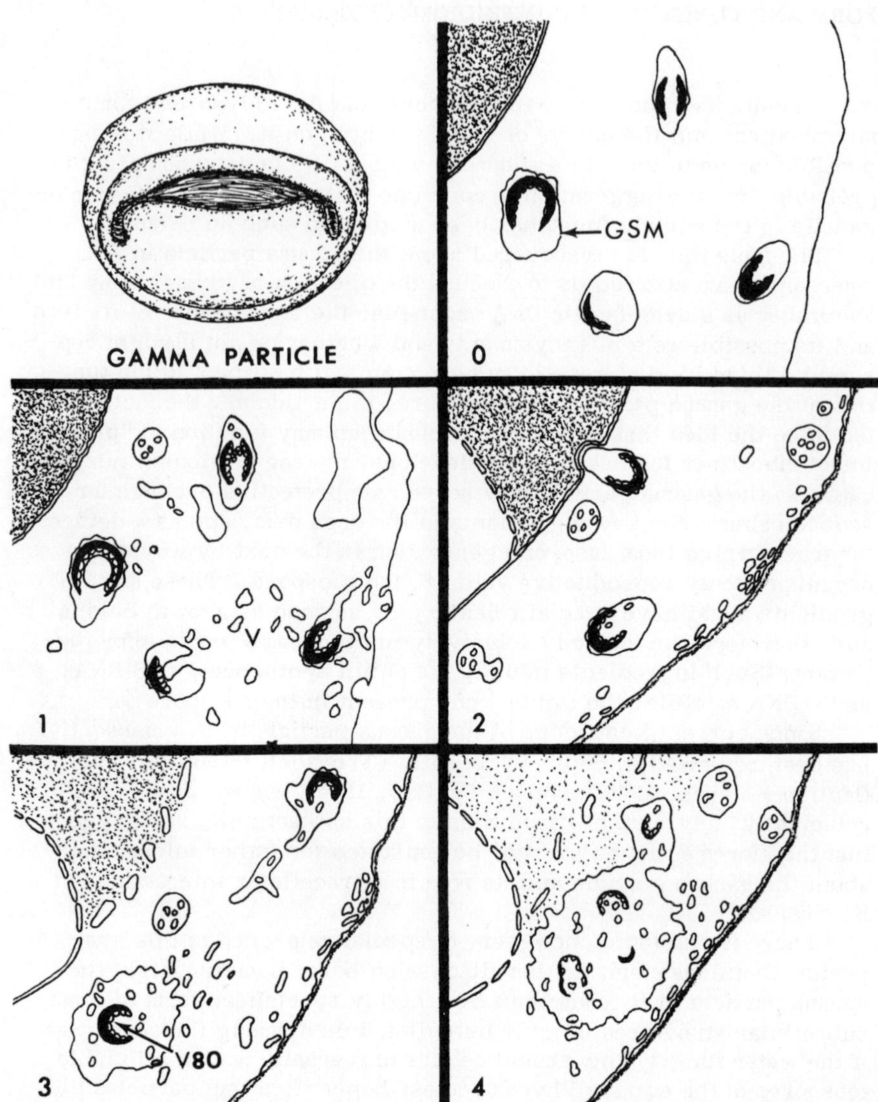

Fig. 8 The Gamma particle (top left), and successive stages in its decay during zoospore encystment (from Truesdell & Cantino, 1970). The nearly spherical (0) Gamma particle surrounding membrane (GSM) becomes amorphic (1) and begins to pinch off vesicles (V) which migrate to the plasma membrane. Simultaneously the Gamma particle matrix generates a tier of 80 nm vesicles (V80) lined up on its inner surface. The V80 fuse with the GSM, thus adding membrane

Monoblepharella (Fuller & Reichle, 1968) *Rhizidiomyces* (Fuller & Reichle, 1965), *Rhizophlyctis* (Chambers & Willoughby, 1964), and *Nowakowskiella* (Chambers et al., 1967).

In particular, let's consider the members of the Blastocladiales. Gamma particle like bodies have been seen, stained, and counted (Matsumae et al., 1970) in the motile cells of *B. britannica, B. simplex,* and *B. cystogena*. Ultrastructurally, the particles in the spores of *B. britannica* (Cantino & Truesdell, 1971a) differ from the gamma particles of *B. emersonii*; they never yield typical horseshoe-like profiles when sectioned, and they are recognizably distinct in structure. But, they do resemble them in size, the number per cell, their strongly osmiophyllic nature, and the possession of a network of membranous material that extends into the cytoplasm and buds off vesicles. Considering the fact that, in spite of the close relationship between *B. emersonii* and *B. britannica*, their respective particles do differ morphologically, variation in the structure of such organelles among other species in this order would not seem to be either unexpected or unreasonable. Among them, *Allomyces* appears to be in certain ways the most closely related to *B. emersonii*, and some of its motile cells contain particles that most nearly resemble gamma particles. Longitudinal sections through a zoospore of *A. macrogynus* (Fig. 1 in Fuller & Calhoun, 1968), and several sections through zoospores of *Allomyces* x *javanicus* (Moore, 1968; for example, Figs. 16 and 20) contain profiles of organelles that strikingly resemble the gamma particles in *B. emersonii*. Although Fuller & Olson (1971; p. 181) concluded that gamma bodies such as those seen in the spores of *B. emersonii* had not been observed in the zoospores of *A. macrogynus* and *A. neo-moniliformis*, Robertson (1972, p. 264) reported seeing gamma particles similar to those described in *B. emersonii* scattered throughout the zoospores of a "new" *Allomyces*. Then most recently, Olson made the interesting observation (1973a) that the meiospores and mitospores of *A. macrogynus* resemble one

---

components to the GSM faster than it loses them. As these processes unfold, deposition of the cyst wall begins (2). The expanding Gamma particles with their decaying matrices coalesce with one another (3); as the matrices are gradually depleted via V80 formation, the coalesced Gamma particles take on the appearance (4) of multivesicular bodies which, finally, bear no resemblance to the original Gamma particles (1).

another except that meiospores contain bodies similar to gamma particles of *B. emersonii*. Although the matrix of the *Allomyces* particle is apparently spherical, hence the reason, we presume, for the absence of typical horseshoe profiles in the pictures shown, the matrix is said to change during "mobilization" (i.e., it starts to produce small vesicles); thus it begins to resemble, in both behavior and morphology, the gamma particles in *B. emersonii*. In a second paper (Olson, 1973b) it is shown that among the early events in meiospore encystment, electron dense material in the gamma particles is mobilized; the amount decreases rapidly, and what remains produces vesicles which bleb off internally and contain electron dense material. Since these vesicles are found in the cytoplasm, and near the plasmalemma where they apparently fuse with it, Olson is tempted to speculate that electron dense material is used for wall synthesis, and that the vesicles contribute the transport mechanism.

Coelomomyces punctatus is also a member of the Blastocladiales. In the anterior region of its motile cells, there are "inclusion bodies" (Martin, 1971) bound by single membranes; they vary in size, as do their contents, and they are very electron dense. Apparently they do not contain the three-holed, boat-shaped matrix so characteristic of the gamma particles in *B. emersonii*; rather, they more nearly resemble the gammalike particles in *B. britannica* in the sense that their cores are not of definite and regular shape, but, rather, are very similar to the bodies in *B. emersonii* (See Fig. 7) that result from the multiple fusions of decaying gamma particles (Truesdell & Cantino, 1970).

Outside of the Blastocladiales, *Olpidium brassicae* (Temmink & Campbell, 1969a) should be mentioned because its zoospores contain multivesicular bodies (MVB) which resemble the MVB generated by a *B. emersonii* zoospore (see Fig. 8) as its gamma matrix buds off small vesicles, and which are also thought to migrate to sites of cyst wall deposition (Temmink & Campbell, 1969b). Another Chytridiomycete we wish to single out for special mention because of the electron dense gammalike particles contained in its motile cells is *Rozella* (Held, 1973). In it, Held has seen not only gammalike granules but also MVB extending from vacuoles containing gammalike granules. The gammalike granules are apparently common in *Rozella* zoospores, but the accompanying vesicles and MVB are not. However, vesicles are released from the electron dense granules after flagellar retraction, at the time of wall formation. The MBV appear to migrate to the periphery, the site of wall deposition. Hence, the gammalike particles and the MVB appear to be the same organelles at different stages of development, just as are the corresponding organelles in *B. emersonii.**

Finally, it is of general interest that the zoospores of some non-Chytridiomycete water fungi also contain gammalike particles, for example, the "dense-body vesicles" in the zoospores of *Saprolegnia ferax* and *Dictyuchus sterile* (Gay et al., 1971). It is thought that dense-body vesicles may serve as centers of membrane formation during zoospore germination and that, in this sense at least, they may resemble the gamma particles of *B. emersonii*. Coincidentally, it is interesting, especially in view of our own earlier stand (Truesdell & Cantino, 1971) and the position we are taking here (see below), that Gay et al. (1971) believe dense-body vesicles are sufficiently characteristic to deserve a status comparable to that of lysosomes and microbodies of higher plants and animals, but nonetheless distinct from and not necessarily homologous with any of them. Hopefully the dense body vesicles in the Saprolegniales, and those in *Allomyces* and *Rozella*, will be isolated in sufficient quantities for comparative analyses of their properties.

It is quite apparent that there is interspecific, not to mention intraspecific, variation in the appearance of sections through the electron-dense, unit membrane-bound organelles thus far detected in the zoospores of aquatic fungi, and which we have here tried to compare with the gamma particles in *B. emersonii*. When the functions of these organelles, especially those thought to be involved in encystment, have been established using spore suspensions sufficiently synchronized such that zoospore activities can be meaningfully quantified, when their origins have been ascertained, and when these same organelles have been isolated and their chemical and enzyme compositions partially established, it should become possible to decide in no uncertain terms whether or not they are all homologous structures with analogous properties.

In the meantime, we believe that it is no longer premature, having documented numerous details of its structure and behavior, its chemistry and enzymology, to give the gamma particles of *B. emersonii* a

*Held also suggested a possible function for the large vacuoles into which the gamma particles in both *B. emersonii* and *Rozella* eventually decay. His discussion went beyond the limits we set for the subject matter to be covered in this article, but the reader interested in the activities of subcellular organelles during germination of fungal spores will find it profitable to examine Held's inviting ideas on this matter.

familial name. We propose to have it serve as the cornerstone for the erection of a family of single membrane bound cytoplasmic organelles* found in the zoospores of aquatic fungi which we now name encystosomes. It is not expected that all encystosomes will be structurally identical, but we do expect that they will share certain properties, hence we define the encystosome as follows: a cytoplasmic, single membrane bound organelle, containing nucleic acid and chitin synthetase, which is required for Chytridiomycete zoospore encystment, one of its direct and essential functions being the production of vesicles which fuse with the plasma membrane during the initial deposition of the cyst wall; in the process, the matrix of the encystosome may be gradually "used up," hence its original identity can eventually be lost.

ACKNOWLEDGEMENTS

Unpublished work on *B. emersonii* was supported by general research grants AI 10568 (N.I.H.) and GB 42425 (N.S.F.) to E.C.C.

REFERENCES

Adelman, T. G. & Lovett, J. S. (1972). Synthesis of ribosomal protein without *de-novo* ribosome production during differentiation in *Blastocladiella emersonii*. Biochem. Biophys. Res. Commun. 49, 1174--1182.

Adelman, T. G. & Lovett, J. S. (1974). Evidence for a ribosome-associated translation inhibitor during differentiation of *Blastocladiella emersonii*. Biochim. Biophys. Acta 335, 236--245.

Bardele, C. F. (1973). Struktur, biochemie und funktion der mikrotubuli. Cytobiologie 7, 442--488.

Beard, M. E. & Novikoff, A. B. (1969). Distribution of peroxisomes (microbodies) in the nephron of the rat. J. Cell Biol. 42, 501--518.

*In a recent symposium on control of organelle development, Lewis (1970) commented, "What is an organelle? No speaker has been rash enough to define one. For the purpose of this symposium it may be useful to define an organelle as a membrane bound body with specialized enzymic functions and genetic information (DNA)." Even by this rather rigid standard our prototype qualifies to be called an organelle.

Beevers, H. (1969). Glyoxysomes of castor bean endosperm and their relation to gluconeogenesis. Ann. N.Y. Acad. Sci. 168, 313--324.

Bernstein, L. B. (1968). A biosystematic study of *Rhizophlyctis rosea* with emphasis on zoospore viability. J. Elisha Mitchell Sci. Soc. 84, 84--93.

Bieglmayer, C., Graf, J., & Ruis, H. (1973). Membranes of glyoxysomes from castor-bean endosperm. Eur. J. Biochem. 37, 553--562.

Blondel, B.& Turian, G. (1960). Relation between basophilia and fine structure of cytoplasm in the fungus *Allomyces macrogynus* Em. J. Biophys. Biochem. Cytol. 7, 127--134.

Bromberg, R. (1974). Mitochondrial fragmentation during germination in *Blastocladiella emersonii*. Dev. Biol. 36, 187--194.

Burke, D. J., Seale, T. W., & McCarthy, B. J. (1972). Protein and ribonucleic acid synthesis during the diploid life cycle of *Allomyces arbuscula*. J. Bacteriol. 110, 1065--1072.

Cantino, E. C. (1951). Metabolism and morphogenesis in a new *Blastocladiella*. Antonie v. Leeuwenhoek, 17, 325--362.

Cantino, E. C. (1969). The gamma particle, satellite ribosome package, and spheroidal mitochondrion in the zoospore of *Blastocladiella emersonii*. Phytopathology, 59, 1071--1076.

Cantino, E. C. (1973). Developmental Pathways of the Water Mold, *Blastocladiella emersonii*. 16 mm film, 11 min., color. BFA Educational Media, 2211 Michigan Ave., Santa Monica, California, 90404.

Cantino, E. C. & Horenstein, E. A. (1956). Gamma and the cytoplasmic control of differentiation in *Blastocladiella*. Mycologia 48, 443--446.

Cantino, E. C. & Hyatt, M. T. (1953). Phenotypic "sex" determination in the life history of a new species of *Blastocladiella, B. emersonii*. Antonie v. Leeuwenhoek 19, 25--70.

Cantino, E. C. & Lovett, J. S. (1964) Non-filamentous aquatic fungi: model systems for biochemical studies of morphological differentiation. In Advances in Morphogenesis, Vol. 3 (ed. M. Abercrombie & J. Brachet), pp. 33--93. New York and London: Academic Press.

Cantino, E. C., Lovett, J. S., Leak, L. V., & Lythgoe, J. (1963). The single mitochondrion, fine structure, and germination of the spore of *Blastocladiella emersonii*. J. Gen. Microbiol. 31, 393--404.

Cantino, E. C. & Mack, J. P. (1969). Form and function in the zoospore of *Blastocladiella emersonii* I. The gamma particle and satelite ribosome package. Nova Hedwigia 18, 115--148.

Cantino, E. C. & Myers, R. B. (1972). Concurrent effect of visible light on gamma particles, chitin synthetase, and encystment capacity in zoospores of *Blastocladiella emersonii*. Arch. Mikrobiol. 83, 203--215.

Cantino, E. C. & Myers, R. B. (1974). The gamma-particle and intracellular interactions in *Blastocladiella emersonii*. Brookhaven Symp. Biol. 25, 51--74.

Cantino, E. C., Suberkropp, K. F., & Truesdell, L. C. (1969). Form and function in the zoospore of *Blastocladiella emersonii* II. Spheriodal mitochondria and respiration. Nova Hedwigia 18, 149--158.

Cantino, E. C. & Truesdell, L. C. (1970). Organization and fine structure of the side body and its lipid sac in the zoospore of *Blastocladiella emersonii*. Mycologia 62, 548--567.

Cantino, E. C. & Truesdell, L. C. (1971a). Cytoplasmic gamma-like particles and other ultrastructural aspects of zoospores of *Blastocladiella britannica*. Trams. Br. Mycol. Soc. 56, 169--179.

Cantino, E. C. & Truesdell, L. C. (1971b). Light-induced encystment of *Blastocladiella emersonii* zoospores. J. Gen. Microbiol. 69, 199--204.

Cantino, E. C. & Truesdell, L. C. (1972). Myelin-like "artifacts" in the zoospores of *Blastocladiella emersonii*. Trans. Br. Mycol. Soc. 59, 129--132.

Cantino, E. C., Truesdell, L. C., & Shaw, D. S. (1968). Life history of the motile spore of *Blastocladiella emersonii*: A study in cell differentiation. J. Elisha Mitchell Sci. Soc. 84, 125--146.

Chambers, T. C. & Willoughby, L. G. (1964). The fine structure of *Rhizophlyctis rosea*, a soil Phycomycete. J. R. Microsc. Soc. 83, 355--364.

Chambers, T. C., Markus, K., & Willoughby, L. G. (1967). The fine structure of the mature zoosporangium of *Nowakowskiella profusa*. J. Gen. Microbiol. 46, 135--141.

Chappell, J. B. & Hansford, R. G. (1969). Preparation of mitochondria from animal tissues and yeasts. In Subcellular Components: Preparation and Fractionation (ed. G. D. Birnie and S. M. Fox), pp. 77--91. London: Butterworths.

Chong, J. & Barr, D. J. S. (1973). Zoospore development and fine structures in *Phlyctochytrium arcticum* (Chytridiales). Can. J. Bot. 51, 1411--1420.

Couch, J. N. & Whiffen, A. J. (1942). Observations on the genus *Blastocladiella*. Am. J. Bot. 29, 582--591.

de Duve, C. (1969). Evolution of the peroxisome. Ann. N. Y. Acad. Sci. 168, 369--381.

Frederick, S. E. & Newcomb, E. H. (1969). Cytochemical localization of catalase in leaf microbodies (peroxisomes). J. Cell Biol. 43, 343--353.

Frederick, S. E., Newcomb, E. H., Vigil, E. L., & Wergin, W. P. (1968). Fine-structural characterization of plant microbodies. Planta (Berl.) 81, 229--252.

Fuller, M. S (1966). Structure of the uniflagellate zoospores of aquatic Phycomycetes. In The Fungus Spore (ed. M. F. Madelin), pp. 67--84. London: Butterworths.

Fuller, M. S. & Calhoun, S. A. (1968). Microtubule--kinetosome relationships in the motile cells of the Blastocladiales. Z. Zellforsch. 87, 526--533.

Fuller, M. S. & Olson, L. W. (1971). The zoospore of *Allomyces*. J. Gen. Microbiol. 66, 171--183.

Fuller, M. S. & Reichle, R. (1965). The zoospore and early development of *Rhizidiomyces apophysatus*. Mycologia. 57, 946--961.

Fuller, M. S. & Reichle, R. E. (1968). The fine structure of *Monoblepharella* sp. zoospores. Can. J. Bot. 46, 279--283.

Gay, J. L., Greenwood, A. D,. & Heath, I. B. (1971). The formation and behavior of vacuoles (vesicles) during oosphere development and zoospore germination in *Saprolegnia*. J. Gen. Microbiol. 65, 233--241.

Hawk, P. B., Oser, B. L., & Summerson, W. H. (1951). Practical Physiological Chemistry, pp. 1323. New York: Blakiston.

Hechter, O. (1972). Reflections on general membrane structure: The conference in review. Ann. N. Y. Acad. Sci. 195, 506--519.

Held, A. A. (1972). Fungal zoospores are induced to encyst by treatments known to degrade cytoplasmic microtubules. Arch. Mikrobiol. 85, 209--224.

Held, A. A. (1973). Encystment and germination of the parasitic chytrid *Rozella allomycis* on host hyphae. Can. J. Bot. 51, 1825--1836.

Horgen, P. A. & Griffin, D. H. (1969). Cytochrome oxidase activity in *Blastocladiella emersonii*. Plant Physiol. 44, 1590--1593.

Hruban, Z. & Rechcigl, M. Jr. (1969). Microbodies and Related Particles. 296 pp. New York and London: Academic Press.

Huang, A. H. C. & Beevers, H. (1971). Isolation of microbodies from plant tissues. Plant Physiol. 48, 637--641.

Huang, A. H. C. & Beevers, H. (1973). Localization of enzymes within microbodies. J. Cell Biol. 58, 379--389.

Koch, W. J. (1956). Studies of the motile cells of Chytrids I. Electron microscope observations of the flagellum, blepharoplast and rhizoplast. Am. J. Bot. 43, 811--819.

Koch, W. J. (1958). Studies of the motile cells of chytrids II. Internal structure of the body observed with light microscopy. Am. J. Bot. 45, 59--72.

Koch, W. J. (1968a). Studies of the motile cells of chytrids. IV. Planonts in the experimental taxonomy of aquatic Phycomycetes. J. Elisha Mitchell Sci. Soc. 84, 69--83.

Koch, W. J. (1968b). Studies of the motile cells of chytrids. V. Flagellar retraction in posteriorly uniflagellate fungi. Am. J. Bot. 55, 841--859.

Lehninger, A. L. (1951). Phosphorylation coupled to oxidation of dihydrodiphosphopyridine nucleotide. J. Biol. Chem. 190, 345--359.

LéJohn, H. B., Jackson, S. G., Klassen, G. R., & Sawula, R. V. (1969). Regulation of mitochondrial glutamic dehydrogenase by divalent metals, nucleotides, and α-ketoglutarate. J. Biol. Chem. 244, 5346--5356.

LéJohn, H. B. & Lovett, J. S. (1966). Characterization of ribonucleic acids from *Rhizophlyctis rosea* zoospores. Planta (Berl.) 71, 283--290.

Lessie, P. E. & Lovett, J. S. (1968). Ultrastructural changes during sporangium formation and zoospore differentiation in *Blastocladiella emersonii*. Am. J. Bot. 55, 220--236.

Lewis, D. (1970). Conclusions: Organelles as membrane complexes. Sympos. Soc. Exp. Biol. 24, 497--501.

Lodi. W. R. & Sonneborn, D. R. (1974). Protein degradation and protease activity during the life cycle of *Blastocladiella emersonii*. J. Bacteriol. 117, 1035--1042.

Lovett, J. S. (1963). Chemical and physical characterization of "nuclear caps" isolated from *Blastocladiella* zoospores. J. Bacteriol. 85, 1235--1246.

Lovett, J. S. (1968). Reactivation or ribonucleic acid and protein synthesis during germination of *Blastocladiella* zoospores and the role of the ribosomal nuclear cap. J. Bacteriol. 96, 962--969.

Madelin, M. F. & Beckett, A. (1972). The production of planonts by thin-walled sporangia of the fungus *Coelomomyces indicus*, a

parasite of mosquitoes. J. Gen. Microbiol. 72, 185--200.
Martin, W. W. (1971). The ultrastructure of *Coelomomyces punctatus* zoospores. J. Elisha Mitchell Sci. Soc. 87, 209--221.
Matsumae, A., Myers, R. B., & Cantino, E. C. (1970). Comparative numbers of gamma particles in the flagellate cells of various species and mutants of *Blastocladiella*. J. Gen. Appl. Microbiol. 16, 443--453.
Matsushima, H. (1972). The microbody with a crystalloid core in tobacco cultured cell clone XD-6S. III. Developmental studies on the microbody. J. Electr. Microsc. 21, 293--299.
McCurdy, H. D., Jr. & Cantino, E. C. (1960). Isocitritase, glycine-alanine transaminase, and development in *Blastocladiella emersonii*. Plant Physiol. 35, 463--476.
Mendgen, K. (1973). Microbodies (glyoxysomes) in infection structures of *Uromyces phaseoli*. Protoplasma 78, 477--482.
Miles, C. A. & Holwill, M. E. J. (1969). Asymmetric flagellar movement in relation to the orientation of the spore of *Blastocladiella emersonii*. J. Exp. Biol. 50, 683--687.
Mills, G. L. & Cantino, E. C. (1974). Lipid composition of the zoospores of *Blastocladiella emersonii*. J. Bacteriol. 118, 192--201.
Mills, G. L., Myers, R. B., & Cantino, E. C. (1974). Lipid changes during development of *Blastocladiella emersonii*. Phytochemistry 13, 2653--2657.
Mollenhauer, H. H., Morré, D. J., & Kelley, A. G. (1966). The widespread occurrence of plant cytosomes resembling animal microbodies. Protoplasma 62, 44--52.
Moore, R. T. (1968). Fine structure of mycota 13. Zoospore and nuclear cap formation in *Allomyces*. J. Elisha Mitchell Sci. Soc. 84, 147--165.
Müller, M., Hogg, J. F., & de Duve, C. (1968). Distribution of tricarboxylic acid cycle enzymes and glyoxylate cycle enzymes between mitochondria and peroxisomes in *Tetrahymena pyriformis*. J. Biol. Chem. 243, 5385--5395.
Myers, R. B. & Cantino, E. C. (1971). DNA profile of the spore of *Blastocladiella emersonii*: evidence for gamma-particle DNA. Arch. Mikrobiol. 78, 252--267.
Myers, R. B. & Cantino, E. C. (1974). The gamma particle. A study of cell-organelle interactions in the development of the water mold *Blastocladiella emersonii*. pp. 117. Basel: S. Karger.
Novikoff, A. B. & Goldfischer, S. (1968). Visualization of microbodies for light and electron microscopy. J. Histochem. Cytochem.

16, 507.

Olson, L. W. (1972). Colchicine and the mitotic spindle of the aquatic Phycomycete *Allomyces*. Arch. Mikrobiol. 84, 327--338.

Olson, L. W. (1973a). The meiospore of *Allomyces*. Protoplasma 78, 113--127.

Olson, L. W. (1973b). Synchronized development of gametophytic germlings of the aquatic Phycomycete *Allomyces macrogynus*. Protoplasma 78, 129--144.

Olson, L. W. (1973c). A low molecular weight colchicine binding protein from the aquatic Phycomycete *Allomyces neo-moniliformis*. Arch. Mikrobiol. 91, 281--286.

Olson, L. W. & Fuller, M. S. (1971). Leucine--lysine synchronization of *Allomyces* germlings. Arch. Mikrobiol. 78, 76--91.

Reichle, R. E. (1972). Fine structure of *Oedogoniomyces* zoospores, with comparative observations on *Monoblepharella* zoospores. Can. J. Bot. 50, 819--824.

Reichle, R. E. & Fuller, M. S. (1967). The fine strucutre of *Blastocladiella emersonii* zoospores. Am. J. Bot. 54, 81--92.

Richardson, M. (1974). Microbodies (glyoxysomes and peroxisomes) in plants. Sci. Progr. Oxf. 61, 41--61.

Robertson, J. A. (1972). Phototaxis in a new *Allomyces*. Arch. Mikrobiol. 85, 259--266.

Schmoyer, I. R. & Lovett, J. S. (1969). Regulation of protein synthesis in zoospores of *Blastocladiella*. J. Bacteriol. 100, 854--864.

Shaw, D. S. & Cantino E. C. (1969). An albino mutant of *Blastocladiella emersonii*: comparative studies of zoospore behavior and fine structure. J. Gen. Microbiol. 59, 369--382.

Silverberg, B. A. & Sawa, T. (1973). An ultrastructural and cytochemical study of microbodies in the genus *Nitella* (Characeae). Can. J. Bot. 51, 2025--2032.

Soll, D. R. & Sonneborn, D. R. (1969a). Zoospore germination in the water mold, *Blastocladiella emersonii*. I. Measurement of germination and sequence of subcellular morphological changes. Dev. Biol. 20, 183--217.

Soll, D. R. & Sonneborn, D. R. (1969b). Zoospore germination in the water mold, *Blastocladiella emersonii*. II. Influence of cellular and environmental variables on germination. Dev. Biol. 20, 218--235.

Soll, D. R. & Sonneborn, D. R. (1971a). Zoospore germination in *Blastocladiella emersonii*, III. Structural changes in relation to protein and RNA synthesis. J. Cell Sci. 9, 679--699.

Soll, D. R. & Sonneborn, D. R. (1971b). Zoospore germination in *Blastocladiella emersonii*: Cell differentiation without protein synthesis? Proc. Nat. Acad. Sci. U.S.A. 68, 459--463.

Soll, D. R. & Sonneborn, D. R. (1972). Zoospore germination in *Blastocladiella emersonii*. IV. Ion control over cell differentiation. J. Cell Sci. 10, 315--333.

Sparrow, F. K., Jr. (1960). Aquatic Phycomycetes. 1187 pp. Ann Arbor: Univ. of Michigan Press.

Stewart, K. D., Floyd, G. L., Mattox, K. R., & Davis, M. E. (1972). Cytochemical demonstration of a single peroxisome in a filamentous green alga. J. Cell Biol. 54, 431--434.

Stüben, H. (1939). Über Entwicklungsgeschichte und Ernährungsphysiologie eines neüen niederen Phycomyceten mit Generationswechsel. Planta, Archiv. Wiss. Bot. 30, 353--383.

Suberkropp, K. F. & Cantino, E. C. (1972). Environmental control of motility and encystment in *Blastocladiella emersonii* zoospores at high population densities. Trans. Br. Mycol. Soc. 59, 463--475.

Suberkropp, K. F. & Cantino, E. C. (1973). Utilization of endogenous reserves by swimming zoospores of *Blastocladiella emersonii*. Arch. Mikrobiol. 89, 205--221.

Temmink, J. H. M. & Campbell, R. N. (1969a). The ultrastructure of *Olpidium brassicae*. II. Zoospores. Can. J. Bot. 47, 227--231.

Temmink, J. H. M. & Campbell, R. N. (1969b). The ultrastructure of *Olpidium brassicae*. III. Infection of host roots. Can. J. Bot. 47, 421--424.

Tolbert, N. E. (1971). Microbodies---peroxisomes and glyoxysomes. Annu. Rev. Plant Physiol. 22, 45--74.

Travland, L. B. & Whisler, H. C. (1971). Ultrastructure of *Harpochytrium hedinii*. Mycologia 63, 767--789.

Truesdell, L. C. & Cantino, E. C. (1970). Decay of gamma particles in germinating zoospores of *Blastocladiella emersonii*. Arch. Mikrobiol. 70, 378--392.

Truesdell, L. C. & Cantino, E. C. (1971). The induction and early events of germination in the zoospore of *Blastocladiella emersonii*. In Current Topics In Developmental Biology, Vol. 6 (ed. A. A. Moscona & A. Monroy), pp. 1--44. New York and London: Academic Press.

Umphlett, C. J. & Olson, L. W. (1967). Cytological and morphological studies of a new species of *Phlyctochytrium*. Mycologia 59,

1085--1096.

Vigil, E. L. (1970). Cytochemical and developmental changes in microbodies (glyoxysomes) and related organelles of castor bean endosperm. J. Cell Biol. 46, 435--454.

Vigil, E. L. (1973). Plant microbodies. J. Histochem. Cytochem. 21, 958--962.

Weisenberg, R. C. (1972). Microtubule formation in vitro in solutions containing low calcium concentrations. Science 177, 1104--1105.

Whisler, H. C., Shemanchuk, J. A. & Travland, L. B. (1972). Germination of the resistant sporangia of *Coelomomyces psorophorae*. J. Invert. Path. 19, 139--147.

DISCUSSION

Form and Function in Chyridiomycete Spores With Emphasis on *Blastocladiella emersonii*

Chairman: D. E. Hemmes
University of Hawaii
Honolulu, Hawaii

In response to a question about the presence of various enzymes in the symphyomicrobodies, Professor Cantino replied that isocitrate lyase has long been known to be present in zoospores of *Blastocladiella emersonii* and is probably localized in the symphyomicrobodies. Attempts are now underway to isolate these organelles and to track them by their isocitrate lyase activity.

Questions dealing with gamma particles (encystosomes) centered on their nucleic acid content and their role in encystment. Professor Cantino stated that studies of gamma particle DNA and RNA thus far have dealt only with isolated gamma particles. Details as to when DNA or RNA are incorporated into the gamma particle or as to what happens to the gamma particle DNA after encystment cannot be answered yet. Gamma particle DNA is thought to be 20--25% double stranded.

The gamma matrix appears to be a highly compacted mat of either membrane or material that produces membrane. The membrane cannot be resolved until the matrix starts to unravel and then, as the gamma core decays, vesicles are produced. All of this activity occurs in the absence of protein synthesis.

It was previously shown that populations of orange zoospores had an <u>average</u> of only eight gamma particles (i.e., a very low gamma particle count), and that these spores were very poorly viable, an observation supporting the idea that gamma particles are required for encystment.

On-going studies include protein assays on the gamma particle membranes and the plasma membrane immediately before, during, and after encystment with synchronous cultures. Professor Cantino emphasized that a major goal is to substantiate all microscopical observations chemically and enzymatically.

CHAPTER 13

FORM AND FUNCTION IN ZYGOMYCETE SPORES

S. N. Grove

Goshen College

FORM AND FUNCTION IN ZYGOMYCETE SPORES

INTRODUCTION

This chapter on zygomycete spores is primarily a summary of information on structure and biochemistry which has appeared since the First Symposium with emphasis on the more recent work. Several excellent reviews are available on a variety of subjects which relate to this discussion of zygomycete spore form and function (Bartnicki-Garcia, 1968a, 1973; Bergman et al., 1969; Bracker, 1967; Gooday, 1973; Hess, 1973). The reader will be referred to these reviews for discussions of pertinent topics.

Spore form is interpreted here as including structure and composition while function is discussed relative to metabolic activities and the role of spores in the life cycle of an organism. That role as discussed in this chapter is one of re-establishing the mycelium by germination and subsequent hyphal growth. Consequently, a section on morphogenesis presents information on the mechanisms involved in fungal growth and related phenomena.

An examination of the literature on the Zygomycetes shows that mucoraceous species have received the most attention while entomophthoraceous organisms, although potentially very interesting, have seldom been studied. As a result information on the following aspects of the Mucorales is included in this chapter; spore surfaces, wall structure of quiescent and germinated spores, protoplasmic components of quiescent and germinated spores, wall composition, lipid composition, and metabolic phenomena associated with germination. In contrast, a very limited treatment of the Entomophthorales will be found.

The original micrographs used in this chapter were obtained by the methods of Grove & Bracker (1970), Grove (1972) and Oujezdsky et al., (1973).

QUIESCENT AND GERMINATED ZYGOMYCETE SPORES

Structure

Asexual spores

The development of scanning microscopic techniques has stimulated interest in spore surface characteristics so that several laboratories have now collected this type of information. The surface features of spore walls of several mucoraceous fungi have been examined by a variety of electron microscope techniques (see Chapter by Hess & Weber for a discussion of techniques). Appendages and surface striations are revealed in scanning electron micrographs of sporangiospores of *Gilbertella persicaria* (Fig. 1) and *Choanephora trispora* (Fig. 2). The spines and other surface structures of stylospores and sporangioles of *Mortierella* species were shown by Chien et al., (1974). When observed by the same technique spores of *Cunninghamella elegans* exhibit surface spines with blunt tips and disklike bases (Hawker et al., 1970). In cross-section these spines are hollow and are more or less circular. The delicate nature of the attachment is evident from the ease with which the spines are detached during processing for transmission electron microscopy. Patterns of surface spines have also been shown for kickxellaceous spores from several genera (Young, 1968); but the smooth sporangiospores of *Coemansia reversa* apparently lack the protuberances evident on the other species of the Kickxellaceae (Young, 1973). For three dry-spored species of *Piptocephalis* Young (1968, 1969) found surface patterns of warts and ridges but the surfaces of a slime-spored species were relatively smooth. Young (1968) also showed by the use of carbon replicas that spore walls of many mucroaceous spores are spiny. A discussion of the possible significance of many of the above observations was presented by Hawker et al., (1970).

Additional information about surface features was obtained from sections and freeze-etch replicas of *Rhizopus arrhizus* sporangiospores (Hess & Weber, 1973). They show deep furrows and prominent ridges in the ungerminated condition. However, these features disappear during germination. These characteristics are shared by other mucoraceous species (Figs. 1--3, 12). Apparently most of these surface elaborations contribute to the dispersive potential of spores, for such ornamentations so far have not been found on germ tubes or hyphae which are relatively smooth.

The walls of ungerminated mucoraceous spores have proven to be composed of at least two different morphological entities, an outer electron-opaque portion and an inner relatively electron-translucent region (Figs. 3, 4, 8, 10, 11). This is also shown in electron micrographs of a variety of representative fungi (Bartnicki-Garcia et al., 1968b; Bracker, 1971; Buckley et al., 1968; Hawker et al., 1970; Hess & Weber, 1973; Young, 1971). In *Rhizopus* sp the electron-opaque outer region is uneven in width (Buckley et al., 1968; Hess & Weber, 1973) corresponding to the surface furrows and ridges

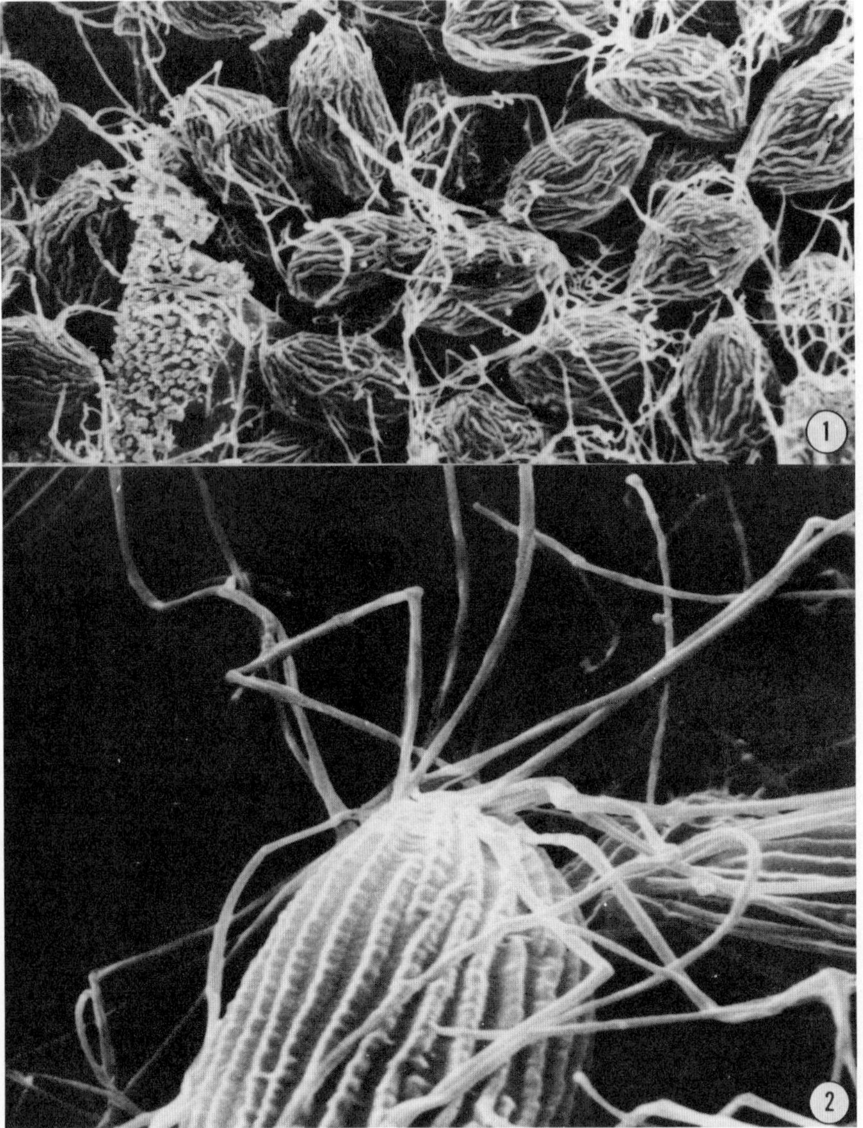

Key to labeling: IW - inner wall, L - lipid body, M - mitochondrion, MT - Microtubule, N - nucleus, SW - spore wall, V - vesicle, VA - vacuole.

Fig. 1 A scanning electron micrograph of sporangiospores of

illustrated by Hess & Weber (1973). In *Mucor, Gilbertella*, and *Cunninghamella* the uneven nature of this surface component is much less prominent. (Bartnicki-Garcia et al., 1968b; Bracker, 1971; Hawker et al., 1970).

During germination stage I (Bartnicki-Garcia et al., 1968b) the uneven nature of the spore wall is reduced. This is probably due to the spherical growth of the spore and the associated stretching of the spore wall (Figs. 3, 4, 12, 14). When compared to those of ungerminated spores electron micrographs of germinated spores prepared by freeze-etching show that the furrows and ridges almost disappear in *R. arrhizus* (Hess & Weber, 1973). However, thin sections of germinated *R. arrhizus* (Buckley et al., 1968; Hess & Weber, 1973) and *R. stolonifer* (Buckley et al., 1968) (Figs. 8, 10) reveal that the outer electron-opaque region of the spore wall retains some of its uneven distribution.

For all mucoraceous fungi examined so far a new inner wall that is different from either of the original spore wall components is deposited within the original spore wall during germination (Figs. 5, 7, 8, 10, 15--18). Thin sections of germinated spores from a variety of organisms show the general nature of this wall (Bartnicki-Garcia et al., 1968b; Bracker, 1971; Buckley et al., 1968; Hawker et al., 1970; Hess & Weber, 1973; Young, 1971), while a study by Bracker and Halderson (1971) illustrates and discusses negatively contrasted fibrils, 20--71 Å diameter, which are characteristic of this inner wall in *G. persicaria*. Similar investigations on this inner wall have not been reported for other mucoraceous fungi. However, autoradiographic evidence shows that this inner wall is synthesized in an evenly disperse pattern over the entire spore periphery in *M. rouxii* (Bartnicki-Garcia & Lippman, 1969). Presumably these features are of generl occurrence among this group of fungi, but verification of this awaits further investigation.

Extensive information is now available to support the concept that the wall of the emerging germ tube is continuous with the new inner wall formed in germination stage I. Autoradiographic data

---

Gilbertilla persicaria (5--13 x 4.5--11 μm). (Courtesy of E. E. Butler).

Fig. 2 Part of a sporangiospore of *Choanephora trispora* (14 μm) showing striations which are not visable with the light microscope. (Courtesy of E. E. Butler).

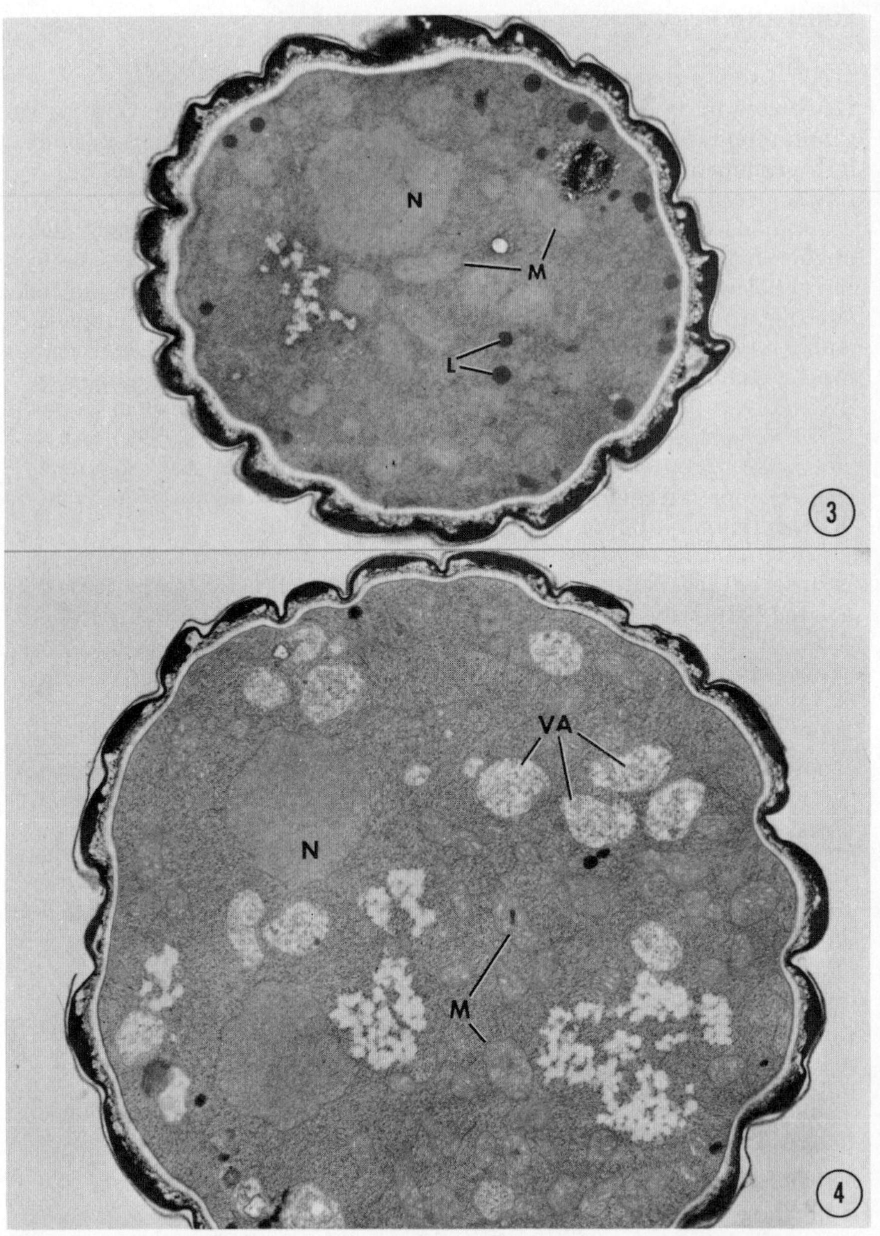

Fig. 3 Ungerminated spore of *Choanephora cucurbitans* showing a three layered spore wall. The thin outer wall layer is greatly folded,

shows that prior to germ tube protrusion the pattern of wall synthesis gradually becomes polarized on a small portion of the spherical cell of *M. rouxii* (see Bartnicki-Garcia, 1973, for discussion). This region of the stage I cell becomes the apex of the emerging germ tube. Thus, the conversion to germination stage II requires the establishment of polarized wall growth where previously only a uniform distribution of wall growth existed. Electron micrographs of *M. rouxii* (Bartinicki-Garcia et al., 1968b) and numerous other mucoraceous fungi (Bracker, 1971; Bracker & Halderson, 1971; Buckley et al., 1968; Hawker et al., 1970; Hess & Weber, 1973; Young, 1971; and Figs. 5, 7, 8, 10, 15--18) show the direct continuity between germ tubes and the inner wall of germinated spores.

Since the wall is deposited by the living protoplasm it is necessary to look within the cell for structures associated with wall formation. When prefixation stresses are reduced electron microscopic evidence from *G. persicaria* indicates that cytoplasmic components are also polarized during the initiation and growth of the germ tubes (Bracker, 1971). Cytoplasmic vesicles similar to the secretory vesicles found at the apices of somatic hyphae (Girbardt, 1969; Grove & Bracker, 1970; Hess & Weber, 1973) are present from the earliest detected stages of germ tube initiation throughout subsequent growth of the hyphae. Germlings of other mucoraceous fungi show similar cytoplasmic polarization (Figs. 5--10, 15--18) if the necessary precautions

---

but is continuous over the spore surface. This layer apparently corresponds to the sporangial wall of organisms with multispored sporangia. The middle layer is of variable thickness with prominent regions composed of electron-opaque material. The inner portions of this layer are more electron-translucent. The inner wall is of more uniform thickness and is the most electron-translucent part of the wall. The intracellular components include nucleus (N), mitochondria (M), and lipid bodies (L). Note that membranes and ribosomes are not easily distinguished (x 13,000).

Fig. 4 A spore of *C. cucurbitans* after initiation of germination shown at the same magnification as Fig. 3. Spherical growth has occurred. The outer wall layer no longer covers the entire surface. The membranes and ribosomes are easily seen at this stage. Numerous vacuoles (VA) containing granular material are present and may have formed at the expense of the lipid bodies which are no longer observed. Mitochondrion (M), nucleus (N) (13,000).

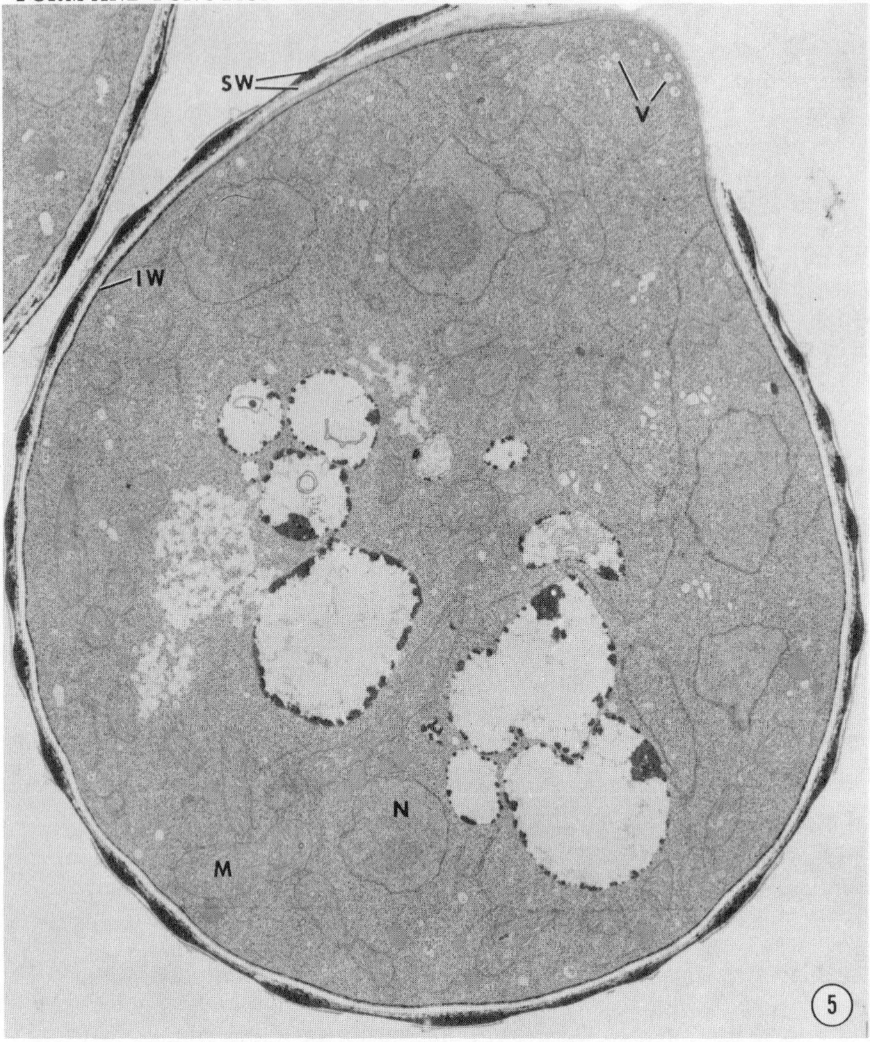

Fig. 5 A later stage of germination that shows the site of germ tube initiation in *C. cucurbitans*. The spore wall (SW) has expanded and no longer shows the deep infolding of earlier stages. A new inner layer (IL) has formed and is continuous with the wall of the germ tube that protrudes through a discontinuity in the spore wall. The vacuoles at this stage are much larger. The apex of the germ tube contains a cluster of cytoplasmic vesicles (V). Small groups of additional vesicles are also found in other areas of the cytoplasm. Mitochondrion

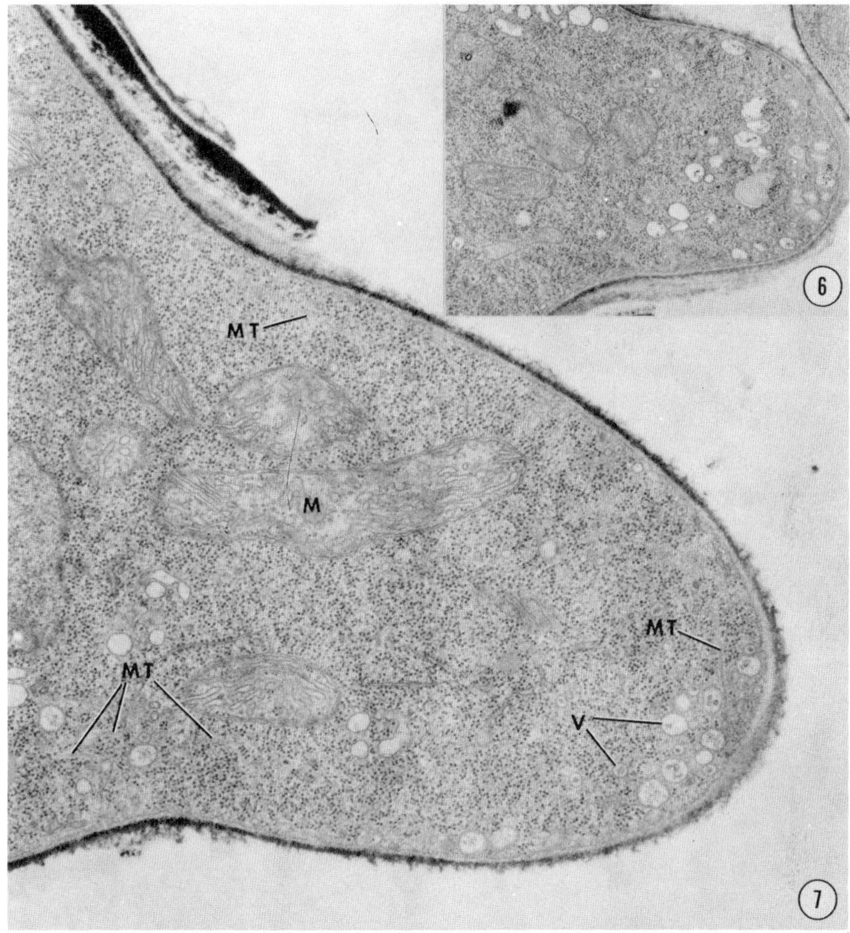

Fig. 6 A slightly later stage of germination in *C. cucurbitans* showing numerous vesicles in the germ tube apex (13,800).
Fig. 7 Part of a *C. cucurbitans* germling showing both tip and subapical vesicles. Microtubules (MT) are also present in the germ tube region. Some of these are adjacent to groups of vesicles (V). Mitochondion (M) (x 21,600).

---

are taken to reduce prefixation stresses. The significance of the vesicles found in these germlings and their role in wall formation and

---

(M), nucleus (N) (x 10,000).

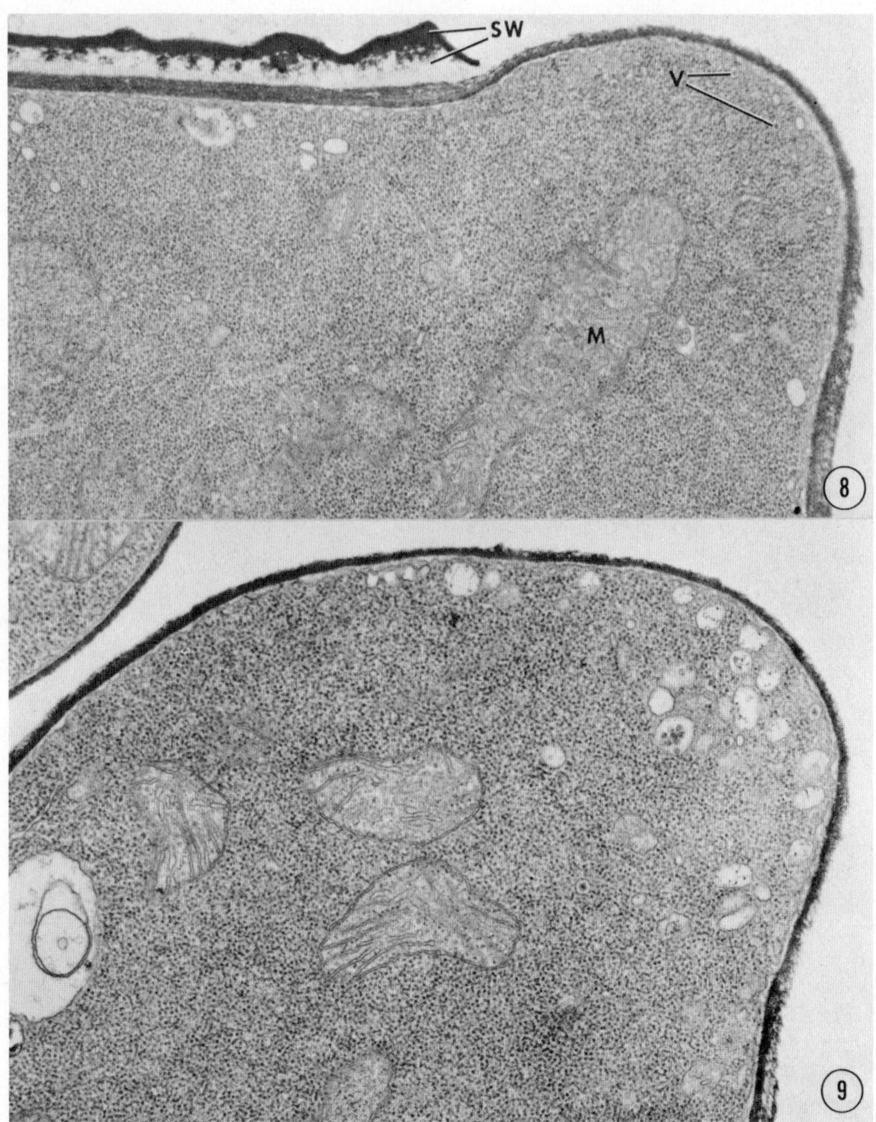

Fig. 8 Part of a germinating spore of *Rhizopus stolonifer* showing an early stage in germ tube emergence. Numerous vesicles (V) are located near the apex. Other vesicles are also evident in the subapical cytoplasm. Spore wall (SW), mitochondrion (M) (x 25,400).
Fig. 9 Part of an emerging germ tube of *R. stolonifer* which contains a large cluster of apical vesicles (x 29,500).

hyphal morphogenesis is probably to secrete wall precursors and enzymes in the region of wall synthesis as has been recently discussed by Bartnicki-Garcia (1973).

Quiescent mucoraceous spores possess many of the intracellular components typical of growing fungal cells, but often have qualitative and quantitative variations (Bartnicki-Garcia et al., 1968b; Buckley et al., 1968; Hawker et al., 1970; Hess & Weber, 1973). During the germination of these spores many changes occur in the existing components and new ones arise. In *M. rouxii* (Bartnicki-Garcia, 1968b), *Rhizopus* sp. (Buckley et al., 1968; Hess & Weber, 1973), *Cunninghamella elegans* (Hawker et al., 1970), *Choanephora cucurbitans* (Figs. 3--5) and *Phycomyces blakesleeanus* (Figs. 12, 15) a system of vacuoles develops while numerous lipid bodies present in the quiescent spore gradually disappear. Changes also occur in the shape and abundance of mitochondria (Hawker et al., 1970; Hess & Weber, 1973). In *R. arrhizus* mitochondria are spherical and contain few cristae in ungerminated spore cytoplasm, but they become larger and polymorphic and contain abundant cristae during the germination process (Hess & Weber, 1973). Figures 11 and 12 show ungerminated spores of *P. blakesleeanus* which appear to have no lack of cristae although the mitochondrial profiles are quite different from those in germlings (Figs. 15--18). In these fungi the number of mitochondrial cristae may not reflect the mitochondrial activity since anaerobic hyphae of *M. genevensis* possess well defined cristae (Clark-Walker, 1972). The endomembrane system (endoplasmic reticulum--golgi apparatus--cytoplasmic vesicle complex) is sparse in the quiescent spore, but it is frequently observed in germinated spores. Micrographs of *M. rouxii* (Bartnicki-Garcia et al., 1968b) and *Rhizopus* sp. (Buckley et al., 1968; Hess & Weber, 1973) show increased amounts of endoplasmic reticulum in germinated spores. Bracker (1971) showed that in *G. persicaria* the endomembrane system is a prominent component of germinated spores. This is also true for three additional mucoraceous species (Figs. 5--10, 15--18). The vesicles derived from the golgi apparatus have a role in wall formation as mentioned above and discussed by Bracker (1971) and Bartnicki-Garcia (1973).

Where ribosomes have been preserved in mucoraceous fungi (Bracker, 1971; Buckley et al., 1968; Hess & Weber, 1973; Young, 1971; and Figs. 5--10, 15--18) they are unusually abundant compared to other fungi (Grove & Bracker, 1970). In fact, they are so closely spaced in the cytoplasm that one cannot be sure if they are associated

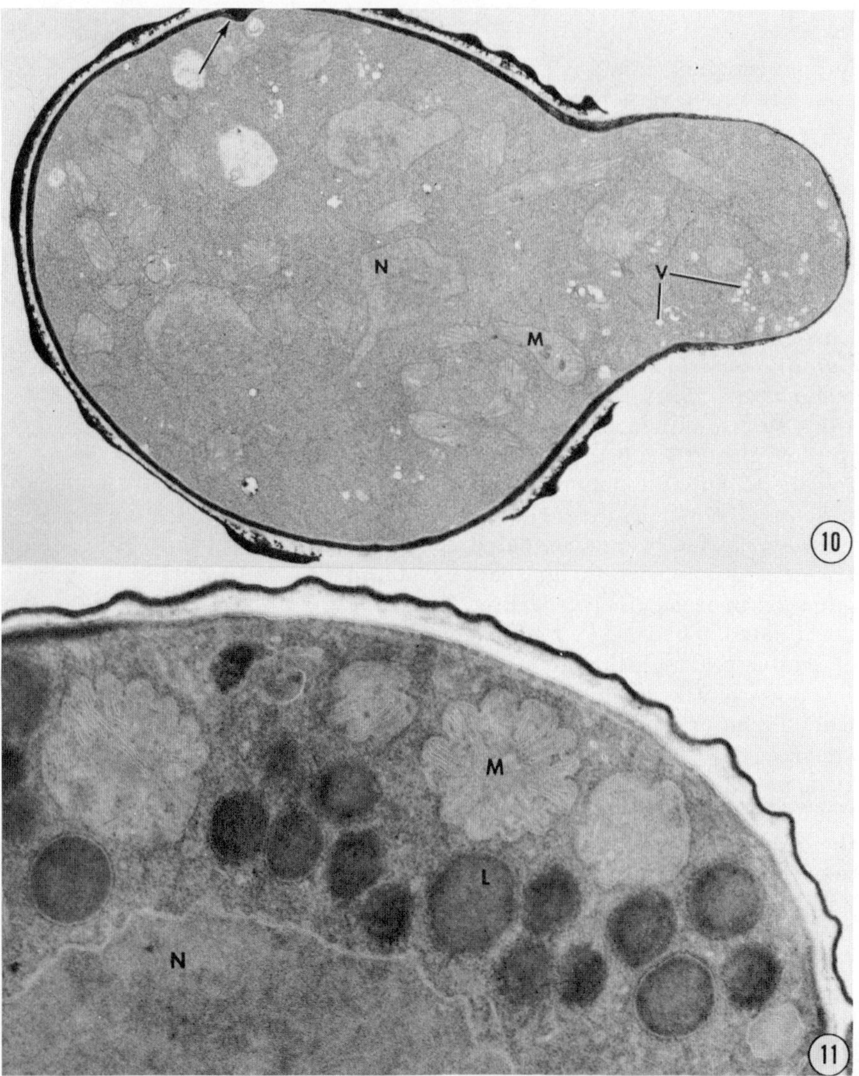

Fig. 10 A young germling of *R. stolonifer* showing several breaks in the spore wall. Through one break a germ tube has grown. The new inner wall at another break contains an acculumation of material (arrow) resembling the apical corpuscles found in *Mucor rouxii*. Note that cytoplasmic vesicles (V) are found near the wall in the germ tube but none are near the wall at the other sites of ruptured spore

as polyribosomes or not.

When studied by freeze-etching techniques membrane surfaces in sporangiospores of *Phycomyces* (Tewari et al., 1971), *M. plumbeus* (Jones, 1972), and *R. arrhizus* (Hess & Weber, 1973) show a variety of features not observed with other techniques. For example, the distribution of intramembrane particles varied among the mitochondrial membranes, with the lowest population density observed in the outer membrane, the highest on crista membranes, and the inner membrance having an intermediate density (Tewari et al., 1971). The particle size varied with the fracturing temperature. Particles of 50--60 Å were observed at -100°C or -115°C and particles of 80--100 Å were observed at -150°C. No damage in the membrane structure was noticed with or without glycerol pretreatment to reduce the size of the ice crystals formed during the freezing process.

Potential problems exist, however, since Hess & Weber (1972, 1973) offer evidence for the production of artifacts between the cell wall and the plasma membrane of *R. arrhizus* following glycerol pretreatment. They concluded that suitable controls must be employed when such antifreeze agents are used. In the same study similar particles were found in plasma membranes of quiescent spores, germinated spores, and germ tubes of *R. arrhizus*. This suggests that no change in plasma membrane structure accompanies the various developmental stages. Jones (1972) observed unusual ridges on the inner surface of the nuclear membrane which may correspond to grooves on the outer surface in *M. plumbeus*, but no significance can yet be attached to this phenomenon. Thus far the freeze-etching techniques have not been widely used for mucoraceous spores and represent a potential source of additional information about these cells.

The Entomophthorales are potentially an interesting group but recent information about spore form and function in these organisms is scarce. Many species have been associated with various diseases of vertebrates and invertebrates. Especially promising as biological

---

wall. Mitochondrion (M), nucleus (N) (x 9,100).

Fig. 11 Part of an ungerminated spore of *Phycomyces blakesleeanus* showing the two layered wall and some cytoplasmic components. Due to the electron-opaque nature of the spore cytoplasm the membranes are often scalloped and they contain numerous cristae. The lipid bodies are coated with granules (see Figs. 13 and 14) presumed to be ferritin. Nucleus (N) (x 28,800).

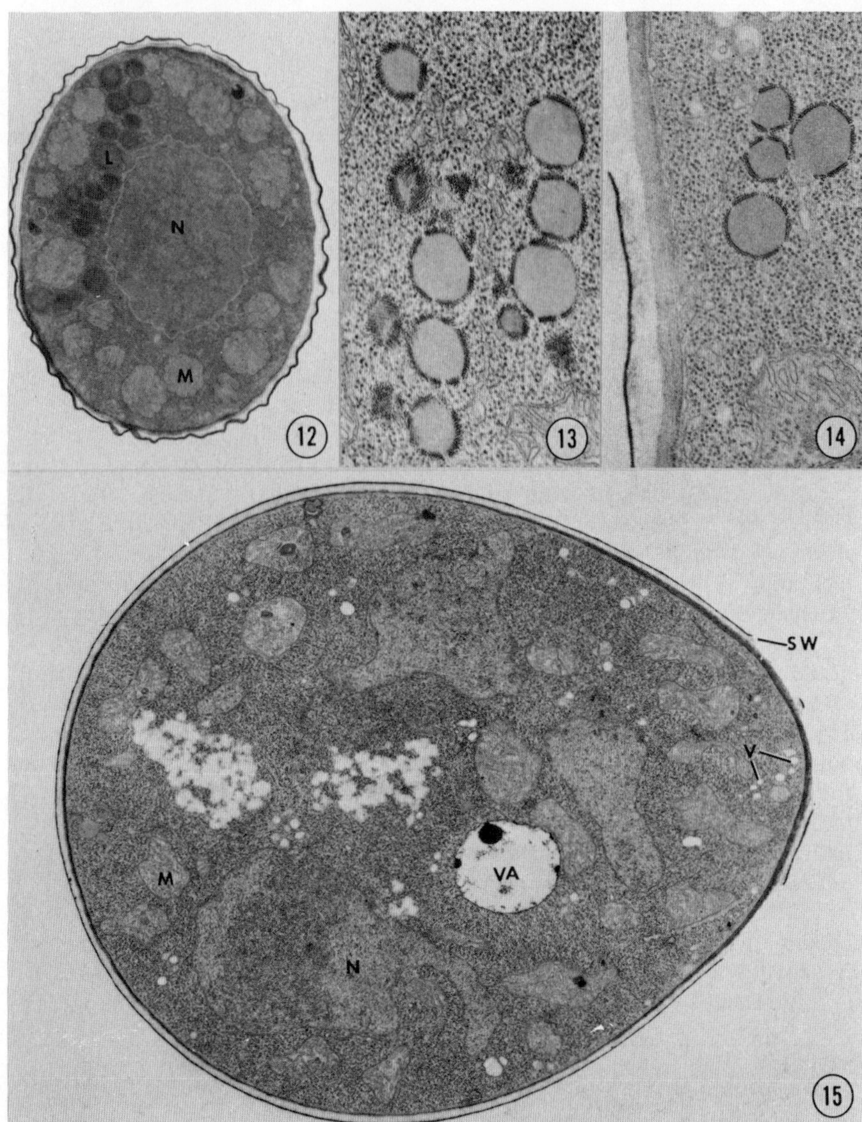

Fig. 12 A section of an ungerminated *P. blakesleeanus* spore showing the distribution of cell components. One nucleus (N), numerous mitochondrial (M), and lipid bodies (L) are evident. (See Fig. 15 for comparison with a germling at the same magnification) (x 9,600). Figs. 13 and 14 Lipid bodies in the cytoplasm of very young

controls are species of *Entomophthora* which infect insects (Tyrrell & MacLeod, 1972). When conidia isolated from infected hosts are placed in a culture medium they germinate, and when the germ tubes grow to about 100--150 the protoplasm is released from the tip as a viable protoplast. The protoplasts proliferate rapidly and pass through a complex set of developmental stages. The significance of these observations to the life cycle of *Entomophthora* sp. is not yet understood and provides a fertile field for further investigations.

Zygospores

The wall and surface structure of zygospores of several mucoraceous species have been studied. Scanning electron micrographs of zygospores from *G. persicaria* (Fig. 21) show the surface characteristics of both the spore itself and the suspensors. In a developmental study Hawker & Beckett (1971) illustrate the surface features of mature zygospores of *R. sexualis* as well as details of wall structure. The outer layer of wall in this organism can be removed by peroxide treatment leaving the inner translucent layer which is highly birefringent (Fig. 19, 20). Histochemistry of the zygospores of both *Mucor mucedo* and *R. sexualis* indicates that the outer wall layer contains melanin, chitin chitosan, while the inner wall is composed chiefly of chitin (Gooday, personal communication).

The cytoplasm of mature zygospores is highly specialized containing, in addition to the usual components, various storage inclusions. Mature zygospores of *R. sexualis* (Hawker & Beckett, 1971) contain massive amounts of stored material in the form of lipid bodies and glycogen filled regions. Collectively these features along with wall features are consistent with the apparently resistent nature of zygospores.

---

germlings of *P. blakesleeanus*. The angular arrangement of ferritin around the lipid bodies suggests that this layer has a crystalline structure (x 32,000 and x 28,000, respectively).

Fig. 15 A young germling of *P. blakesleeanus*. A new inner wall has formed beneath the spore wall which has been ruptured by the emerging germ tube. A small cluster of vesicles (V) is near the apical wall and others are in scattered small groups throughout the cytoplasm. Mitochondrial (M) and nuclear (N) profiles are of variable shapes. Vacuole (VA) (x 9,600).

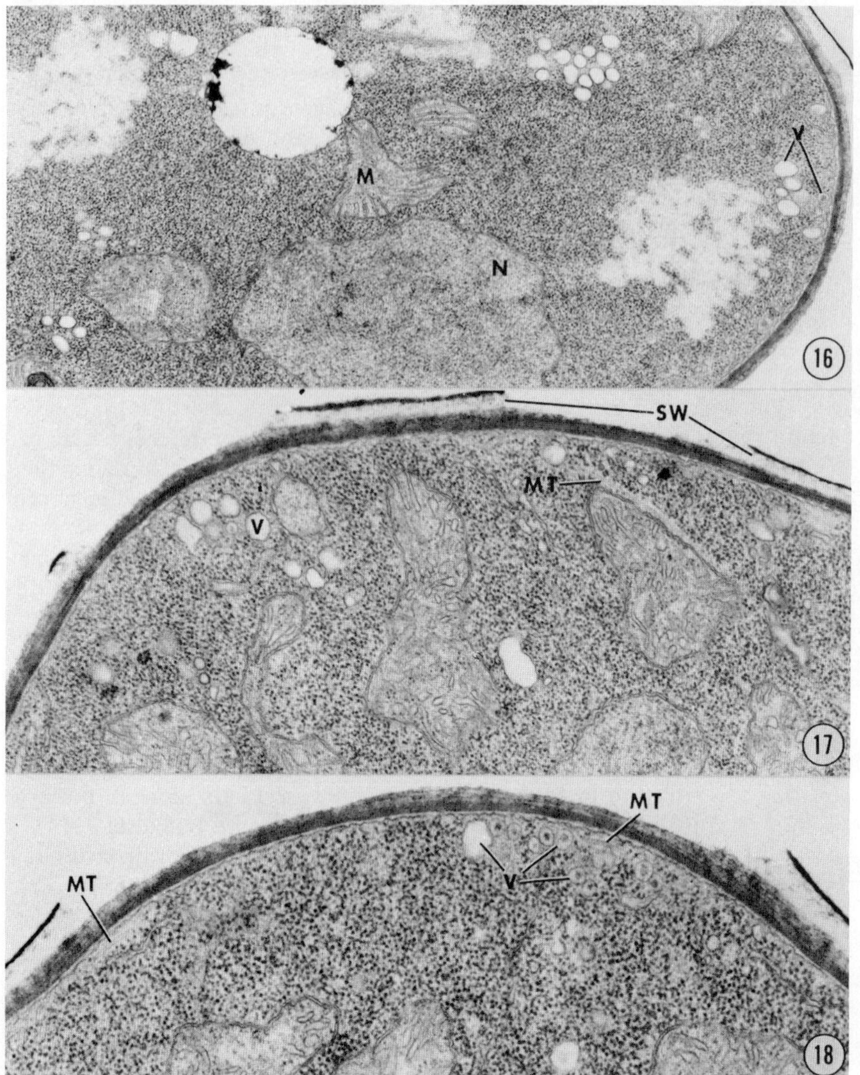

Figs. 16--18 Germ tubes of young germlings of *P. blakesleeanus* showing that each apex contains a small cluster of vesicles (V). Occasional microtubules (MT) are also seen near the germ tube apices. Mitochondrion (M), spore wall (SW) (x 18,400, x 28,800, x 33,600).

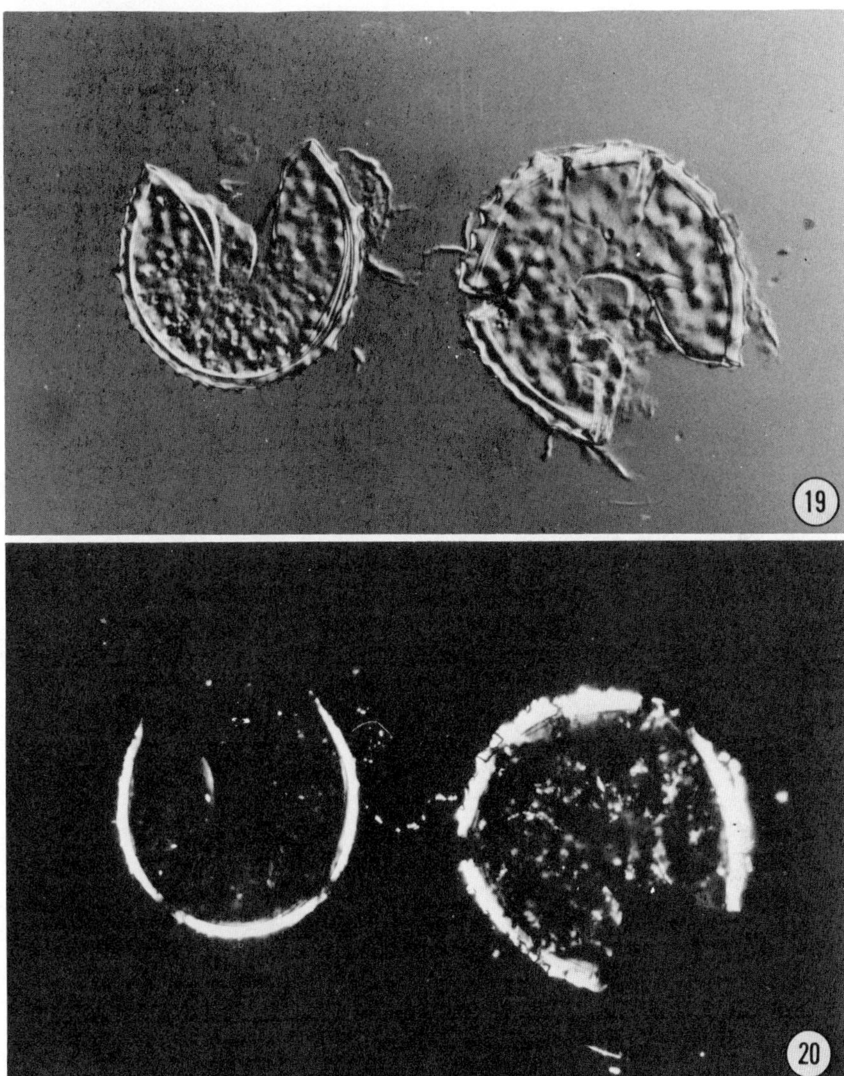

Figs. 19 and 20 Photomicrographs showing inner zygospore walls of *Rhizopus sexualis* after the outer wall has been disrupted with peroxide. Interference contrast optics show details of the wall fragments (Fig. 19). Polarizing optics show the highly birefringent nature of this wall (Fig. 20). (Courtesy of G. W. Gooday).

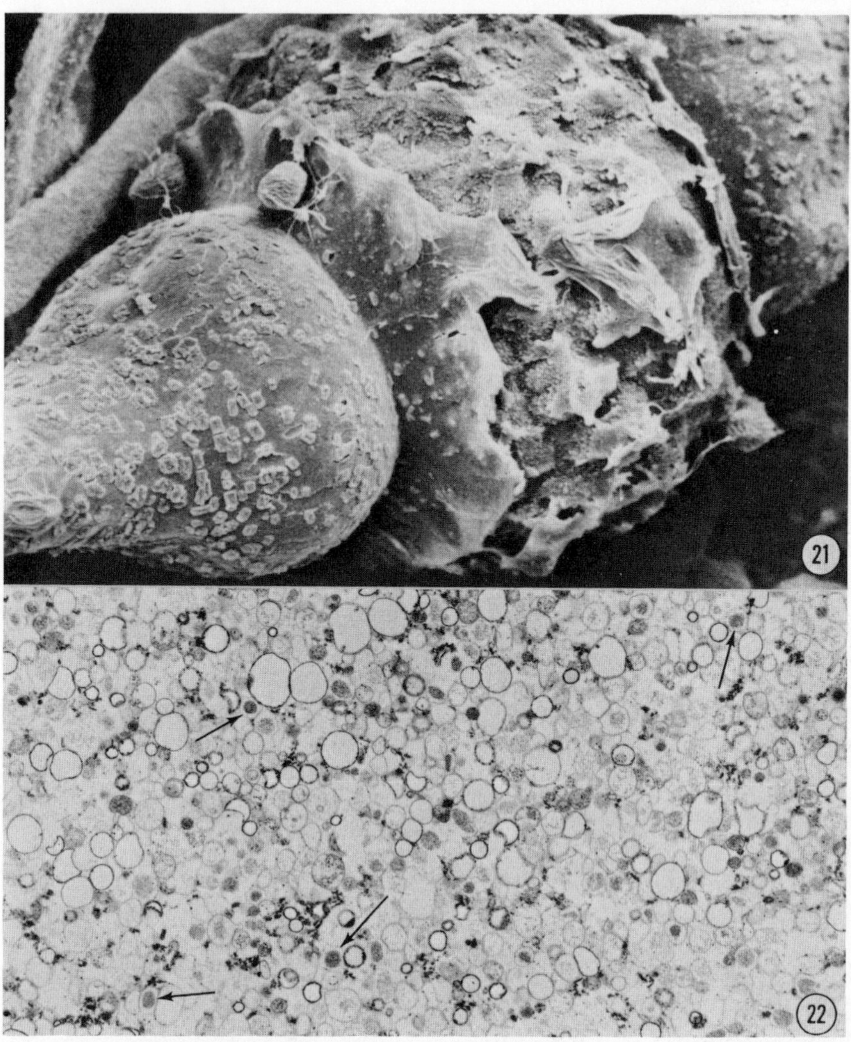

Fig. 21 Zygospore of *Gilbertella persicaria* (70--80 μm diam) as seen by scanning electron microscopy. (Courtesy of E. E. Butler).
Fig. 22 A thin section of a membrane fraction from a homogenate of *Gilbertella persicaria* germlings. This fraction is enriched for apical vesicles and was obtained by a combination of differential filtration, differential centrifugation, and density gradient centrifugation. The apical vesicles are recognized by their granular or fibrillar inclusions (arrows). Structures which appear similar to the vesicle inclusions

Knowledge about the conditions required for induction of germination in zygospores is scarce so that little information is available concerning this process. The unparalleled study by Mosse (1970 a--c) of probable zygospores of an *Endogone* species may be the only recent information on the germination of these specialized cells. The mature resting spore is enclosed by six distinct layers (Mosse, 1970c). The outermost layer (I) is continuous with the subtending hypha and is sloughed off at maturity. Next is a colored layer (II) which is naturally electron-opaque, does not blacken in osmium, and may consist of up to six distinct bands. The bands have a curved fibrillar substructure, alternating with narrow bands of tangentially oriented fibrils. A structurally similar layer (III) may also contain several bands, but it is electron-translucent and lacks color in the light microscope. The inner three layers (IV--VI) are separated from the outer three layers (I--III) by a three-layered membrane about 140 Å thick. The region of layer III adjacent to this membrane shows a regular periodic structure oriented perpendicular to the membrane. Layer IV is not homogeneous, but consists of alternating electron-opaque and electron-translucent segments. The innermost wall layer (VI) is electron-translucent and appears structureless.

In the mature resting spore the cytoplasm contains numerous storage components and small amounts of other organelles (Mosse, 1970b). Much of the cytoplasmic region is filled by oil bodies and a few electron-opaque structures interpreted as polysaccharide storage bodies. Nuclei, mitochondria, and other cell components are limited to small interstices between oil bodies. Another interesting component of the spore cytoplasm in this fungus, which typically forms vesicular-arbuscular mycorrhizae with a variety of plant hosts, is a bacteria-like organism. These organisms lie free in the cytoplasm, but they cause no apparent disturbance of their host. During the resting state they form extensive colonies in the host cytoplasm, but no information is available concerning their functional relationship to the fungus.

Prior to germination the resting spore undergoes some major changes (Mosse, 1970a,c). Specialized regions develop that contain dividing nuclei. The spore wall splits between layers IV and V, allowing cytoplasm from the spore interior to accumulate in the

---

but are not surrounded by a membrane may represent broken vesicles (x 15,000). (From unpublished work of Grove, Marlowe, and Szaniszlo).

separation. Radial walls are formed which create peripheral compartments containing cytoplasm. Germ tubes subsequently arise from these peripheral compartments and penetrate the outer wall layers. Consequently these peripheral compartments may be analogous to spores and the whole structure may be a sporangium formed when the zygospore (resting spore) germinates.

Composition

Several contributions to understanding wall composition in the Zygomycetes have appeared since the excellent review by Bartnicki-Garcia (1968a). Gooday (1973) investigated the chemistry of zygospores from *Mucor mucedo* and *Rhizopus sexualis*. He found that sporopollenin is formed as a component only of the outer layer of the wall in *M. mucedo*. This material is deposited as a result of an oxidative polymerization of carotenoids. Formation of sporopollenin may be a role for the increased carotenogenesis observed during sexual reproduction in this organism. In *R. sexualis*, which does not accumulate carotenoids, no sporopollenin was found. Additional information from histochemistry of *Mucor* and *Rhizopus* zygospores indicates that the outer layer of the wall contains melanin, chitin, and chitosan, while the inner layer is chiefly chitin (Gooday, personal communication). An excellent discussion of cell wall composition in the Mucroales is given by Gooday (1973).

When the cell walls of carotenogenic and noncarotenogenic forms of *Blakeslea trispora* were compared Chenouda (1972) found quantitative differences in the chemical composition. The yellow cell walls contained larger amounts of neutral and amino sugars than walls of white cells, while the white cell walls contained more readily extractable lipid. This is interpreted as a demonstration that environmental conditions may influence the synthesis of cell wall since the white hyphae are typical of submerged cultures while the yellow hyphae are formed when the fungus is exposed to air.

An examination of the carbohydrates in the hyphal walls of selected members of the Entomophthorales (Hoddinott & Olsen, 1972) shows that the genera differ qualitatively, while species of *Entomophthora* differ only quantitatively. Chitin and glucan seem to be major components of the wall while chitosan was not detected. On this basis the Entomophthorales should be removed from the chitin-chitosan group and assigned to a chitin-glucan group distinct from the Mucorales in the classification scheme of Bartnicki-Garcia (1968a).

A microelectrophoretic examination of the fatty acid and hydrocarbon constituents of the surface and wall lipids of selected fungal spores (Fisher et al., 1972) yielded evidence for surface lipids on sporangiospores of *R. stolonifer*. The maximum distribution of surface alkanes falls in the C_{21}--C_{25} range while wall alkanes ranged from C_{16}--C_{32}. No coincidence was observed in the distribution patterns or the major peaks of *n*-alkanes in walls compared with surface fractions. By the same methods no surface lipids were found on the walls of *M. rouxii* sporangiospores.

The lipid content and the fatty acid composition of sporangiospores and mycelium of mucoraceous fungi is also of interest because they represent structural and energy storage materials (see Bergman et al., 1969 for information on *Phycomyces*). Sumner & Morgan (1969) examined several *Mucor* sp. and one of *Rhizopus* sp. and found that although in each species the spores contained less lipid it was of a more highly saturated nature than the mycelium. Weete et al., (1970) found that in *R. arrhizus* polar lipids composed 44% of the lipid fraction from the mycelium while triglycerides represented 20%, sterols 17%, and free fatty acids 12%. The saturated fatty acid composition in the spores was 17% palmitic (C_{16}), 20% stearic (C_{18}), and 8% arachidic (C_{20}), whereas the unsaturated fatty acids were 42% oleic (C_{18}: 1) and 8% linoleic (C_{18}: 2). The amounts of these substances differed in the mycelia as follows: 18% palmitic, 11% stearic, 16% arachidic, 29% oleic, and 16% linoleic.

The changes in lipid composition during the germination of sporangiospores and subsequent growth of the mycelium are also of interest because they may contribute to the understanding of the role of lipids in the life cycle of these fungi. During the first 2 h of germination in *R. arrhizus* the total lipids decreased, and then increased during later periods of an 8-h study (Gunasekaran et al., 1972a). Comparisons of the levels of total, polar, and neutral lipids suggest that lipids are used for carbon and energy functions during the early stages of germination. Subsequently the neutral lipids are utilized and polar lipids are synthesized. After 72 h of growth the highest concentration of lipids is reached, thereafter both total dry weight and total lipid declines (Gunasekaran et al., 1972b). Weete et al., (1973) also showed that during a six day growth cycle the total lipids varied from 2.2 to 15.3% with the maximum at the 72 h. Sporangiospores contained 2.6% lipids while sterols represented 1.8--9.1% of the total lipids during the growth period. From these studies it is apparent that information about lipid compositon is not meaningful unless it is accompanied by precise information on the age and stage of growth of the fungus.

Ferritin is associated with lipid bodies in *Phycomyces blakesleeanus* (Bergman et al., 1969). It has been isolated and characterized by David and Easterbrook (1971). They found that ferritin is selectively incorporated into sporangiospores and suggested that it functions as a storage from of iron required in the early stages of spore germination. They demonstrate that the ferritin is located on the surface of lipid bodies where it forms paracrystalline monolayers. However, micrographs shown here suggest a more complex arrangement of ferritin, possibly paracrystalline bilayers which may be rigid enough to empart an angular profile to the lipid bodies (Figs. 11, 13, 14). The level of ferritin iron is regulated by the amount of iron in the growth medium. Evidence for ferritin has also been found in other spores (Buckley et al., 1968). where it presumably would have a similar function.

Metabolism

Although mature sporangiospores of *Phycomyces blakesleeanus* exhibit dormancy, they can be activated by several treatments and germinate by a series of developmental stages defined as swelling, vacuolization, and germ tube emergence (Figs. 12, 15--18; reviewed by Bergman et al., 1969). The biochemical reactions associated with activation have been the subjects of many recent investigations which have attempted to determine the relative contributions of increased glycolytic activity and trehalose breakdown (Delvaux, 1973; Furch, 1971, 1972; Rudolph, 1973; Rudolph & Furch, 1970; Van Assche et al., 1972). Among other things heat treatment activates trehalose activity (Van Assche et al., 1972) causing a decrease in trehalase with accumulation of glucose and activation of certain glycolytic enzymes (Rudolph & Furch, 1970; Delvaux, 1973). Although the direct connection between the trehalase activity and the breaking of dormancy has not been established, it is clear that during activation the glycolytic pathway is stimulated first and followed later by activation of the pentose phosphate pathway (Delvaux, 1973). Increased amounts of cytochrome and cytochchrome oxidase accompany an increase in respiration rate during germination (Keyhani et al., 1972). Van Assche et al., (1973) provide evidence that RNA and protein synthesis increase immediately after activation and suggest that RNA is present in the ungerminated spores. RNA synthesis decreases during the first nuclear divison and DNA synthesis begins only after germ tubes are present (See Chapter by Sussman for further discussion of dormancy and activation).

The initiation of and rapid increase in protein synthesis during fungal spore germination has also been investigated in *Rhizopus stolonifer*. Transfer RNA isolated from germinated and ungerminated spores was shown to possess aminoacyl-acceptor activity for all 20 commonly found amino acids when assayed with aminoacyl-tRNA synthetase from germinated spores (Van Etten et al., 1969). However, Merlo et al., (1972) found quantitative changes in one of the two isoaccepting species of lysyl-tRNA and in valyl-tRNA. In addition a quantitative change occurred in isoleucyl-tRNA. In a further contribution to understanding spore germination Gong & Van Etten (1972) found that germinated spores of *R. stolonifer* contain three soluble DNA-directed RNA polymerases while ungerminated spores contain only two. RNA polymerase I was present in both, but it differed slightly in germinated spore fractions. RNA polymerase II was present only in germinated spores and first appeared near the end of the spherical growth stage (stage I, swelling). Thus, both qualitative and quantitative changes in RNA polymerases accompany germination of *R. stolonifer* sporangiospores.

The functions of zygomycete spores are influenced by various environmental parameters which have been used by several investigators in the study of fungi. The effects of nutrition (Balasubramanian & Srivastava, 1972) and temperature (Streets & Ingle, 1972) on germination and growth of mucoraceous sporangiospores have been investigated. The resistance of resting spores of *Entomophthora* sp. to low temperature (Krejzova, 1971) and chemicals (Krejzova, 1973) has been shown but knowledge of the metabolic implications of this resistance is lacking. In a study of the effect of chemical sensitization on the repair of potentially lethal irradiation injury Sommer et al, (1971) suggested a relationship to spore metabolism to explain reduced potential for repairs. However, direct information is lacking and this provides a fertile field for study.

The conditions under which resting spores of *Entomophthora pyriformis* germinate are not known and as a first step in understanding the dormancy of these cells Tyrrell (unpublished) has examined various enzymes concerned with glucose metabolism. Of fourteen enzymes of the Emden-Meyerhof-Parnas and pentose phosphate pathways detected in mycelia only nine were detected in the spores. The activities of these nine enzymes were generally much lower than their counterparts from the mycelia. These deficiencies suggest that glucose metabolism by these pathways does not occur at significant levels in the dormant spores.

Progress could potentially be made in understanding spore function by the use of mutants. But recessive mutants of *P. blakesleeanus* have been difficult to obtain due to the typically multinucleate nature of the sporangiospores (Cerda-Olmedo & Reau, 1970). The discovery that on an acidified minimal medium the usually small (0.3%) fraction of uninucleate spores can be increased to 4.5% will aid in obtaining recessive mutants (Reau, 1972). Fractions containing over 80% uninucleate spores are obtained by low speed gradient centrifugation. The spores are fully viable and produce homokaryotic mycelia.

A limited amount of information is available about carbohydrate metabolism in zygomycetes. For example, the utilization of maltose by *Mucor rouxii* requires oxygen (Flores-Carreon et al., 1969) and is mediated by an inducible wall-bound β-glucosidase (Fores-Carreon et al., 1970; Reyes & Ruiz-Herrera, 1972). The glycoprotein with this enzymatic activity is capable of hydrolyzing maltose and starch to yield glucose (Flores-Carreon & Ruiz-Herrera, 1972). Apparently maltose does not penetrate into the fungus to be used as a carbon source, rather it is first hydrolyzed to glucose by the wall-bound enzyme (Reyes & Rhiz-Herrera, 1972).

Morphogenesis

Mucor rouxii has been the subject of extensive investigation into the mechanisms of hyphal morphogenesis (Bartnicki-Garcia, 1968b, Bartnicki-Garcia & Lippman, 1969, 1972a,b; McMurrough & Bartnicki-Garcia, 1971; McMurrough et al., 1971). Bartnicki-Garcia (1973) has recently provided an excellent review of hyphal morphogenesis, and consequently the discussion here will be brief.

The control of dimorphism in *M. rouxii* as influenced by environmental conditions may involve a control over the initiation of apical growth (Bartnicki-Garcia, 1968b). By adjusting the levels of carbon dioxide, oxygen, and hexoses one may induce growth of the yeast or hyphal form. Autoradiographic evidence shows that wall formation is distributed in different patterns for the two types of morphogenesis (Bartnicki-Garcia & Lippman, 1969). The hyphal wall is deposited primarily at the apex while in spherical cells wall formation is dispersed uniformly over the surface. Chitin synthetase activity and chitin deposition are also localized preferentially at the tips of hyphae

(McMurrough et al., 1971) where the enzyme is closely associated with the wall. Additional chitin synthetase activity is associated with a microsome fraction, but the identity and origin of the actual chitin-synthesizing particles are unknown (McMurrough & Bartnicki-Garcia, 1971).

The most likely candidates for these enzyme laden particles are the numerous secretory vesicles which accumulate in the apices of growing germ tubes and hyphae of mucoraceous fungi (Dargent, 1972; Girbardt, 1969; Grove & Bracker, 1970; Hess & Weber, 1973; Syrop, 1973) (See Figs. 5--10, 15--18). These structures have not yet been reported in *M. rouxii* but this may be due to prefixation stresses which prevent the preservation of the delicate apical organizaiton as demonstrated by Bracker (1971) for *G. persicaria*. In germ tubes of *M. rouxii* which lack apical vesicles, an apical corpuscle has been observed (Bartnicki-Garcia, 1968b; Bartnicki-Garcia et al., 1968a), but its nature and origin are as yet unclear. Similar structures are not usual components of growing hyphae or germ tubes of related fungi which contain apical vesicles (Figs. 5--10, 15--18) (Bracker, 1971).

Turgor pressure may provide the driving force for the expansion of the apical wall during hyphal growth (Bartnicki-Garcia, 1973). This is a reasonable hypothesis since hyphal tips have an osmotic pressure significantly higher than that of the surrounding medium (Adebayo et al., 1971). However, it is not likely that turgor alone can cause the expansion and a concurrent lysis of the wall may be needed (Bartnicki-Garcia, 1973).

Circumstantial evidence for lytic enzyme activity in the tips of growing hyphae comes from careful observations of the bursting of growing tips when they are subjected to various environmental changes (Bartnicki-Garcia & Lippman, 1972b). Bartnicki-Garcia (1973) suggests that this lytic activity may be carried by the system of vesicles which is associated with wall growth. A model has been proposed to integrate all the essential components of wall growth (Bartnicki-Garcia, 1973) and it may be best illustrated in the highly polarized growth of fungal hyphae.

This model for wall formation relies heavily on the vesicles as carriers of the various enzymatic and structural components of wall formation. The exact function of such vesicles, however, must wait until they can be isolated and biochemically characterized (Bartnicki-Garcia, 1973). Preliminary biochemical analysis of isolated fractions rich in vesicles (Fig. 22) suggest that the vesicles are associated with polysaccharides which are qualitatively similar to those found in the hyphal wall in *G. persicaria* (Grove et al., 1972). No informaiton is yet available on the enzymatic activities associated with these vesicle fractions, but if the model is correct one would expect both lytic and synthetic enzymes as prominent components.

CONCLUSIONS

The examples cited in this chapter show that the bulk of work on zygomycete spores and their germination has been concerned with mucoraceous species. But even in this group we are far from understanding the controls over activation and germination of the delicate balance between synthesis and lysis in hyphal growth. Additional effort should also be focused on the entomophthoraceous fungi since they potentially can serve as biological controls for some economically important pests. The use of pesticides is becoming unacceptable because of economic and environmental limitations. Consequently, biological controls are becoming increasingly important. Thus, future investigations of zygomycetes holds promise of not only increasing our understanding of basic biological principles but of contributing to the impact of mycology on the "relevant" problems of our society.

REFERENCES

Adebayo, A. A., Harris, R. G., & Gardner, W. R. (1971). Turgor pressure of fungal mycelia. Trans. Br. Mycol. Soc. 57, 145--151.

Balasubramanian, K. A. & Srivastava, D. N. (1972). Relationships of nutrition to germination and pathogenicity of spores of *Rhizopus stolonifer*, Ind. J. Exp. Biol. 13, 445--447.

Bartnicki-Garcia, S. (1968a). Cell wall chemistry, morphogenesis, and taxonomy of fungi. Annu. Rev. Microbiol. 22, 87--108.

Bartnicki-Garcia, S. (1968b). Control of dimorphism in *Mucor* by hexoses: Inhibition of hyphal morphogenesis. J. Bacteriol. 95, 1586--1594.

Bartnicki-Garcia, S. (1973). Fundamental aspects of hyphal morphogenesis. In Microbial Differentiation, Vol, XXIII, Symp. Soc. Gen. Microbiol. (ed. J. M. Ashworth & J. E. Smith), pp. 245--267. Cambridge, England: Cambridge U. P.

Bartnicki-Garcia, S. & Lippman, E. (1969). Fungal morphogenesis. Cell wall construction in *Mucor rouxii*. Science 165, 302--304.

Bartnicki-Garcia, S. & Lippman, E. (1972a). Inhibition of *Mucor rouxii* by polyoxin D: Effects on chitin synthesis and morphological development. J. Gen. Microbiol. 71, 301--309.
Bartnicki-Garcia, S. & Lippman, E. (1972b). The bursting tendency of hyphal tips of fungi: Presumptive evidence for a delicate balance between wall synthesis and wall lysis in apical growth. J. Gen. Microbiol. 73, 487--500.
Bartnicki-Garcia, S., Nelson, N., & Cota-Robles, E. (1968a). A novel apical corpuscle in hyphae of *Mucor rouxii*. J. Bacteriol. 92, 2399--2402.
Bartnicki-Garcia, S., Nelson, N., & Cota-Rables, E. (1968b). Electron microscopy of spore germination and cell wall formation in *Mucor rouxii*. Arch. Mikrobiol. 63, 242--255.
Bergman, K., Burke, P. V., Cerda-Olmedo, E., David, C. N., Delbruck, M., Foster, K. W., Goodel, E. W., Heisenberg, H., Meissner, G., Zalokar, M., Dennison, D. S., & Sheapshire, W., Jr. (1969). Phycomyces. Bacteriol. Rev. 33, 99--157.
Bracker, C. E. (1976). Ultrastructure of fungi. Annu. Rev. Phytophathol. 5, 343--374.
Bracker, C. E. (1971). Cytoplasmic vesicles in germinating spores of *Gilbertella persicaria*. Protoplasma 72, 381--397.
Bracker, C. E. & Halderson, N. K. (1971). Wall fibrils in germinating sporangiospores of *Gilbertella persicaria* (Mucorales). Arch. Mikrobiol. 77, 366--376.
Buckley, P.M., Sommer, N. F., & Malsumoto, T. T. (1968). Ultrastructural details in germinating sporangiospores of *Rhizopus stolonifer* and *Rhizopus arrhizus*. J. Bacteriol. 95, 2365--2373.
Cerda-Olmedo, E. & Reau, P. (1970). Genetic classification of the lethal effects of various agents on heterokaryotic spores of *Phycomyces*. Mutation Res. 9, 369--384.
Chenouda, M. S. (1972). Chemical composition of the cell walls of carotenogenesis and non-carotenogenesis forms of (-) strain of *Blakeslea trispora*. J. Gen. Appl. Microbiol. 18, 143--153.
Chien, C.-Y., Kuhlman, E. G., & Gams, W. (1974). Zygospores in two *Mortierella* species with "Stylospores." Mycologia 66, 114--121.
Clark-Walker, G. D. (1972). Development of respiration and mitochondria in *Mucor genevensis* after anaerobic growth: Absence of glucose repression. J. Bacteriol. 109, 399--408.
Dargent, R. (1972). Sur l'ultrastructure des hyphes en croissance du *Mucor mucedo* L. et la differenciation de leurs ramifications. C. R. Acad. Sci. Paris. 275, 1485--1488.

David, C. N. & Easterbrook, K. (1971). Ferritin in the fungus *Phycomyces*. J. Cell Biol. 48, 15--28.

Delvaux, E. (1973). Some aspects of germination induction in *Phycomyces blakesleenanus* by an ammonium--acetate pretreatment. Arch. Mikrobiol. 88, 273--284.

Fisher, D. J., Holloway, P. J., & Richmond, D. V. (1972). Fatty acid and hydrocarbon constituents of the surface and wall lipids of some fungal spores. J. Gen. Microbiol. 72, 71--78.

Flores-Carreon, A., Reyes, E., & Ruiz-Herrera, J. (1969). Influence of oxygen on maltose metabolism by *Mucor rouxii*. J. Gen. Microbiol. 59, 13--19.

Flores-Carreon, A., Reyes, E., & Ruiz-Herrera, J. (1970). Inducible cell wall-bound α-glucosidase in *Mucor rouxii*. Biochem. Biophys. Acta 222, 354--360.

Flores-Carreon, A. & Ruiz-Herrera, J. (1972). Purification and characterization of α-glucosidase from *Mucor rouxii*. Biochim. Biophys. Acta 258, 296--505.

Furch, B. (1971). Uber die Beeinflussung des Ruhesstofwechsels der Sporangiospores von *Phycomyces blakesleeanus*. Arch. Mikrobiol. 80, 293--299.

Furch, B. (1972). Zur Warmeaktivierung der Sporen von *Phycomyces blakesleeanus*. Das Auftreten von Garungen unter aeroben Bedingungen. Protoplasma 75, 371--379.

Girbardt, M. (1969). Die Ultrastrukture der Apikalregion von Pilzhyphen. Protoplasma 67, 413--441.

Gong, C.-S. & Van Etten, J. L. (1972). Changes in soluble ribonucleic acid polymerases associated with the germination of *Rhizopus stolonifer* spores. Biochim. Biophys. Acta 272, 44--52.

Gooday, G. W. (1973). Differentiation in the Mucorales. In Microbial Differentiation, Vol. XXIII, Symp. Soc. Gen. Microbiol. (ed. J. M. Ashworth & J. E. Smith), pp. 269--294. Cambridge, England: Cambridge U. P.

Grove, S. N. (1972). Apical vesicles in germinating conidia of *Aspergillus parasiticus*. Mycologia 64, 638--641.

Grove, S. M. & Bracker, C. E. (1970). Protoplasmic organization of hyphal tips among fungi: Vesicles and Spitzenkörper. J. Bacteriol. 104, 989--1009.

Grove, S. M., Marlowe, J. D., & Szaniszlo, P. J. (1972). Isolation and partial chemical characterization of apical secretory vesicles from growing hyphae of the fungus, *Gilbertella persicaria*. J. Cell Biol. 55, 99a.

Gunasekaran, M., Weber, D. J., & Hess, W. M. (1972a). Changes in lipid composition during spore germination of *Rhizopus arrhizus*. Trans. Br. Mycol. Soc. 59, 241--248.

Gunasekaran, M., Weber, D. J., & Hess, W. M. (1972b). Total, polar and neutral lipids of *Rhizopus arrhizus* Fischer. Lipids 7, 430--434.

Hawker, L. E. & Beckett. A. (1971). Fine structure and development of the zygospore of *Rhizopus sexualis* (Smith) Callen. Phil. Trans. R. Soc. Lond. 263, 21--100.

Hawker, L. E., Thomas, B., & Beckett, A. (1970). An electron microscope study of structure and germination of conidia of *Cunninghamella elegans* Lendener. J. Gen. Microbiol. 60, 181--189.

Hess, W. M. (1973). Ultrastructure of fungal spore germination. Shokubutsu Byogai Kenkyn (Forsch. Gebiet Pflanzenkrankh) Kyoto 8, 71--84.

Hess, W. M. & Weber, D. J. (1972). Freezing artifacts in Frozen-etched *Rhizopus arrhizus* sporangiospores. Mycol. 64, 1164--1166.

Hess, W. M. & Weber, D. J. (1973). Ultrastructure of dormant and germinated sporangiospores of *Rhizopus arrhizus*. Protoplasma 77, 15--33.

Hoddinott, J. & Olsen, O. A. (1972). A study of the carbohydrates in the cell walls of some species of the Entomophthorales. Can. J. Bot. 50, 1675--1679.

Jones, D. (1972). An unusual feature of the nuclear membrane in freeze-etched sporangiospores of *Mucor plumbeus*. Trans Br. Mycol. Soc. 59, 175--178.

Keyhani, J., Keyhani, E., & Goodgal, S. H. (1972). Studies on the cytochrome content of *Phycomyces* spores during germination. Eur. J. Biochem. 27, 527--534.

Krejzova, R. (1971). Resistance and germinability of resting spores of some species of the genus *Entomophthora*. Ces. Mykol. 25, 231--238.

Krejzova, R. (1973). The resistance of cultures and dried resting spores of three species of the genus *Entomophthora* to ajatin and the viability of their resting spores after long term storage on the refrigerator. Ces. Mykol. 27, 107--111.

McMurrough, I. & Bartnicki-Garcia, S. (1971). Properties of a particulate chitin synthetase from *Mucor rouxii*. J. Biol. Chem. 246, 4009--4016.

McMurrough, I., Flores-Carreon, A., & Bartnicki-Garcia, S. (1971). Pathway of chitin synthesis and cellular localization of chitin synthetase in *Mucor rouxii*. J. Biol. Chem. 246, 3999--4007.

Merlo, D. J., Roker, H., & Van Etten, J. L. (1972). Protein synthesis during fungal-spore germination. II. Analysis of transfer ribonucleic acid from germinated and ungerminated spores of *Rhizopus stolonifer*. Can. J. Microbiol. 18, 949--956.

Mosse, B. (1970a). Honey-coloured, sessile *Endogone* spores. I. Life history. Arch. Mikrobiol. 70, 167--175.

Mosse, B. (1970b). Honey-coloured, sessile *Endogone* spores. II. Changes in fine structure during spore development. Arch. Mikrobiol. 74, 129--145.

Mosse, B. (1970c). Honey-coloured, sessile *Endogone* spores. III. Wall structure. Arch. Mikrobiol. 74, 146--159.

Oujezdsky, K. B., Grove, S. N., & Szaniszlo, P. J. (1973). Morphological and structural changes during the yeast-to-mold conversion of *Phialophora dermatitidis*. J. Bacteriol. 113, 468--447.

Reau, P. (1972). Uninucleate spores of *Phycomyces*. Planta (Berl.) 108, 153--160.

Reyes, E. & Ruiz-Herrera, J. (1972). Mechanism of maltose utilization by *Mucor rouxii*. Biochim. Biophys. Acta 273, 328--335.

Rudolph, H. (1973). Über die Wirkung von D-Glycerat-3-phosphat und D-Glycerat-2-phosphat-Zusätzen auf Sporenhomogenate von *Phycomyces blakesleeanus*. Z. Pflanzenphysiol. 67, 212--222.

Rudolph, H. & Furch. B. (1970). Untersuchungen zur Aktivierung von Sporenhomogenaten durch Warmebehandlung. Arch. Mikrobiol. 72, 175--181.

Sommer, N. F., Dupuy, P., & Rabalu, A. (1971). Effect of chemical sensitization on repair of potentially lethal irridiation injury in *Rhizopus stolonifer* sporangiospores. Radiation Bot. 11, 363--366.

Streets, B. W. & Ingle, M. B. (1972). The effect of temperature on spore germination and growth of *Mucor miehei* in submerged culture. Can. J. Microbiol. 18, 975--979.

Sumner, J. L. & Morgan, E. D. (1969). The fatty acid composition of sporangiospores and vegetative mycelium of temperature adapted fungi in the order Mucorales. J. Gen. Microbiol. 59, 215--221.

Syrop, M. (1973). The ultrastructure of the growing regions of aerial hyphae of *Rhizopus sexualis* (Smith) Callen. Protoplasma 76, 309--314.

Tewari, J. P., Malhotra, S. K., & Tu, J. C. (1971). A study of the structure of mitochondrial membranes by freeze-etch and freeze-fracture techniques. Cytobios 4, 97--119.

Tyrrell, D. & MacLeod, D. M. (1972). Spontaneous formation of protoplasts by a species of *Entomophthora*. J. Invert. Pathol. 19, 354--360.

Van Assche, J. A & Carlier, A. R. (1973). The pattern and nucleic acid synthesis in germinating spores of *Phycomyces blakesleeanus*. Arch. Mikrobiol. 93, 129--136.

Van Assche, J. A., Carlier, A. R., & Dekeersmaeker, H. H. (1972). Trehalase activity in dormant and activated spores of *Phycomyces blakesleeanus*. Planta 103, 327--333.

Van Etten, J. L., Koski, R. K., & El-olemy, M. M. (1969). Protein synthesis during fungal spore germination. IV. Transfer ribonucleic acid from germinated and ungerminated spores. J. Bacteriol. 100, 1182--1186.

Weete, J. D., Lawler, G. C., & Laseter, J. L. (1973). Total lipid and sterol composition of *Rhizopus arrhizus*: Identification and metabolism. Arch. Biochem. Biophys. 155, 411--419.

Weete, J. D., Weber, D. J., & Laseter, J. L. (1970). Lipids of *Rhizopus arrhizus* Fischer. J. Bacteriol. 103, 536--540.

Young, T. W. K. (1968). Electron microscopic study of asexual spores in Kickxellaceae. New Phytol. 67, 823--836.

Young, T. W. K. (1969). Asexual spore morphology in species of *Piptocephalis*. New Phytol. 68, 359--362.

Young, T. W. K. (1971). Ultrastructure of the wall of the germinating sporangiospore of *Linderina pennispora* (Mucorales). Ann. Bot. 35, 183--191.

Young, T. W. K. (1973). Ultrastructure of the sporangiospore of *Coemansia reversa* (Mucorales). Trans. Br. Mycol. Soc. 60, 57--63.

DISCUSSION

Form and Function in Zygomycete Spores

Chairman: G. C. Lawler
Brigham Young University
Provo, Utah

The major portion of the discussion of this paper centered on the vesicles commonly observed in germinating spores and growing hyphae of the Zygomycetes. In response to questions on the origin and apparent dicotomy of vesicular types, Dr. Grove stated that, based on his observations, he felt that there were two distinct types of vesicles and that, although the origin of the small vesicles is completely unknown, the large ones appear to come from the golgi apparatus. When asked to comment on the role of vesicles in spherical growth, Dr. Grove stated that the earliest stage in which vesicles are recognizably associated with the plasma membrane is when the germ tube is beginning to emerge, but that this may only reflect a change in the distribution of vesicles during germination. It was suggested that autoradiography might be of value in determining relationships between vesicles and the biochemical changes taking place during cell wall synthesis. Dr. Bartnicki-Garcia presented data on the rate of glucosamine deposition in the cell wall of germinating spores of *Mucor rouxii* which were interpreted as indicating that a spatial redistribution of chitin synthetase was taking place during polar growth. It was then suggested that vesicles would be an excellent means to effect such a redistribution. Dr. Grove commented that if the basic unit of chitin synthetase activity was the vesicle, one would expect, as has been reported,a random distribution of vesicles during spherical growth and a polar distribution during germ tube formation. It was then proposed that in Dr. Grove's movie of vesicular movement during hyphal tip growth, the vesicles appeared to be moving away from the tip as fast as they were moving towards it and that the film would, therefore, not support an argument for vesicular involvement in cell wall synthesis. Dr. Grove responded that, based on numerous observations of these movies, it appears that the vesicles first move up the perifery of the hyphae towards the apex and that once at the apex they begin a swirling motion during which a small percentage of the vesicles is caught up in a backflow type of cyclosis and moved away from the growing tip. Dr. Hawker commented that observations on two types of vesicles in the developing zygospore of *Rhizopus sexualis* led her to the conclusion that it was extremely likely that these vesicles carried wall material and enzymes to and from sites of synthesis and degradation. A question concerning the behavior of vesicles at branching was then raised. Dr. Grove stated that once vesicles reached the growing tip they do not come back, but that those en route to the tip could be sidetracked to the site of the new growing tip. He also observed that as the new tip grows out it eventually establishes a vesicle producing subapical cytoplasm of its own. It was then asked if the number of vesicles present in a hypha reflected the growth rate of that hypha. Dr. Grove responded that the number of vesicles is probably a good indication of the speed of growth of any particular hyphae or germ tube.

A question was posed as to whether there are any indications that spore wall degradation products are utilized in the synthesis of new cell wall during germination. Dr. Eveleigh commented that in the thick walled resting spores of *Entomophthora* sp. the spore wall typically occupies about 50% of the total cell volume before germination and after elongation of the germ tube there is only a very thin ghost of a spore wall remaining. Dr. Grove was then asked to express his feelings as to whether lysis or a tearing of the outer spore wall takes place as the germ tube emerges. He stated that he felt that spherical growth of the inner wall causes a physical separation or rupture of the outer wall. The next question was concerned with the number of nuclei in a sporangiospore. Dr. Eveleigh commented that in the genus *Entomophthora* the number of nuclei per sporangiospore, or conidium, varies from one to a dozen or more, depending upon the species. It was then asked if the presence of a nucleus determines where a branch point is going to occur. Dr. Grove responded that this is probably not true for the Zygomycetes, but because there are so many nuclei in each hypha it would be difficult to prove either way. Dr. Grove was asked if he had ever observed microtubules oriented parallel to the long axis of the germ tube near the growing tip. He answered that, although some of the electron micrographs he used in his presentation showed microtubules perpendicular to the long axis, he had often observed them parallel to the long axis of the growing tip. Questions were then posed concerning the identification of certain electron-opaque bodies as glycogen particles and the labeling of the outer spore wall of *Choanephora* sp. as sporangial wall. Dr. Grove responded that depending upon preparation procedures certain electron-opaque particles may be identified as glycogen because similar appearing particles have traditionally been called glycogen, and that the identification of the outermost spore wall layer of *Choanephora* sp. as sporangial wall was based on previous descriptions of the sporangiospores of *Choanephora* sp. and *Cunninghamella*, sp.

CHAPTER 14

SOME ASPECTS OF THE FORM AND FUNCTION OF OOMYCETE SPORES

S. Bartnicki-Garcia and D. E. Hemmes
University of California

FINE STRUCTURE

The number of studies on the ultrastructure of sporangia and zoospores of Oomycetes has increased greatly since the last Fungus Spore Symposium (Colhoun, 1966; Gay & Greenwood, 1966). We know now much more about the cytology of sporangial formation, sporangial aging, sporangial germination by zoosporogenesis or by germ tube formation, zoospore morphology, zoospore encystment, and cyst germination. Studies on gametangial interaction have become available recently. As to the more resistant spores, chlamydospores and oospores, little is known. There are no good electron micrographs of dormant oospores of Oomycetes mainly because of the difficulty in embedding and fixing these cells for electron microscopy.

By far, the asexual spores of two genera of Peronosporales, *Phytophthora* and *Pythium*, and two genera of Saprolegniales,

Saprolegnia and *Aphanomyces*, have received the most attention and will be compared here. In addition, results of recent unpublished findings on the fine structure of chlamydospores and gametangia of *Phytophthora* will be summarized.

Sporangia

Development

Sporangial development in the Peronosporales involves a sizable expansion of the sporangial initial by cytoplasmic inflow from the supporting mycelium. This is dramatically clear in time-lapse cinematography* of *Phytophthora palmivora*. The expansion phase is less apparent in *Saprolegnia* where the sporangium is only slightly larger in diameter than the parental hypha (Gay & Greenwood, 1966). By contrast, *Aphanomyces euteiches* does not develop distinct zoosporangia, delimited by septa; instead, when induced to sporulate, the entire thallus becomes a continuous sporangium and the protoplasm is cleaved into numerous nonmotile primary spores (Hoch & Mitchell, 1972a).

Distinct landmarks in sporangial development in *Phytophthora*, *Pythium*, and *Saprolegnia* are the differentiation of the basal plug (Fig. 1), the apical papilla (Fig. 2), and, in *Phytophthora palmivora*, secondary wall thickening (Christen & Hohl, 1972).

Cleavage vesicles and flagella develop prior to induction of zoosporogenesis in sporangia of *Phythophthora* (King & Butler, 1968; Elsner et al., 1970; Williams & Webster, 1970; Hemmes & Hohl, 1973). Mature sporangia (Fig. 3) are thus immediately ready for indirect germination once the requirements of temperature and moisture are met. In contrast, flagella are not observed in sporangia of *Pythium middletonii* (Heintz, 1971; Heintz & Bracker, unpublished), *Saprolegnia ferax* (Gay & Greenwood, 1966; Gay et al., 1971), or *Aphanomyces euteiches* (Hoch & Mitchell, 1972b) until after induction of zoosporogenesis.

---

*Time lapse 16-mm movie on *Phytophthora palmivora* made by L. Clouet, B. Boccas, J. Chevaugeon, and A. Ravise from the Laboratorie de Cryptogamie de l'Universite de Paris-Sud, France 1968.

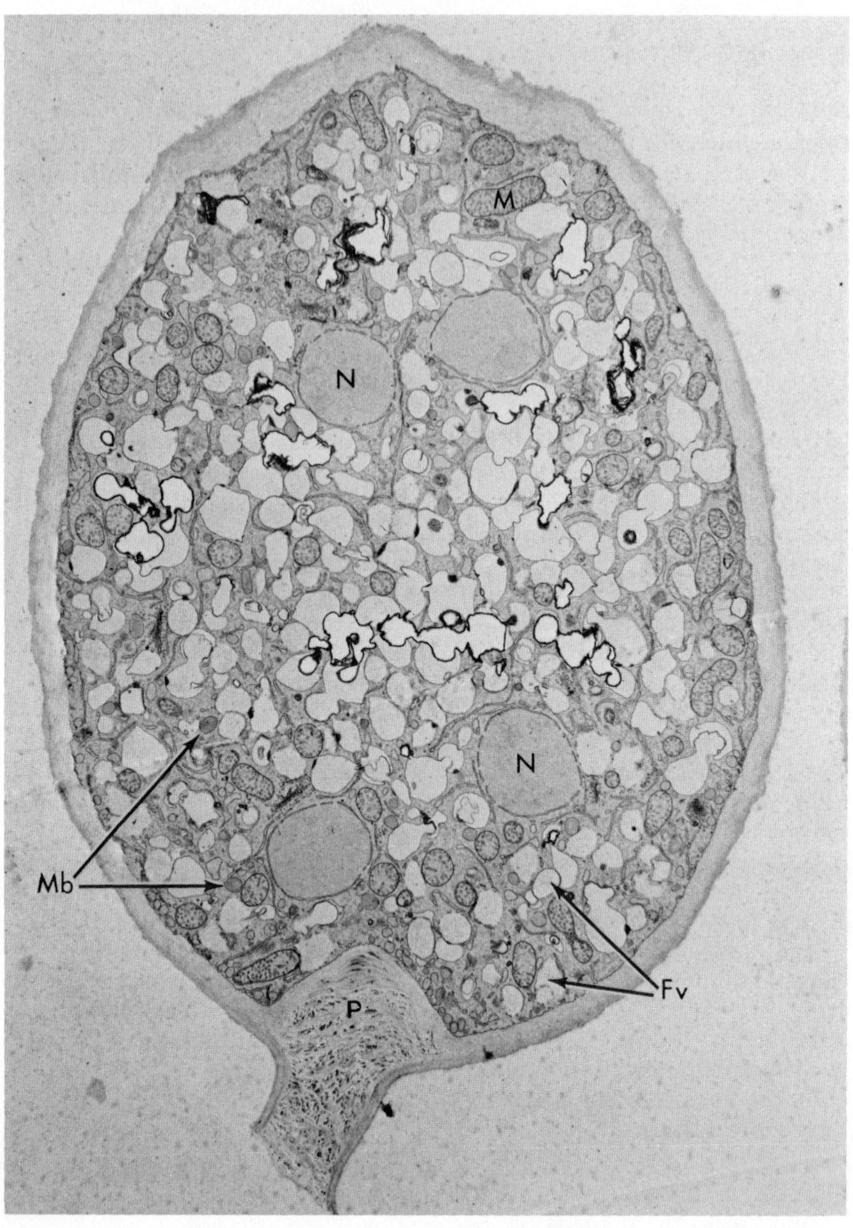

Fig. 1 Fully expanded sporangium of *Phytophthora palmivora* (formerly *P. parasitica*) before the development of flagella. Nucleus (N); microbodylike structures (Mb); fibrillar vacuoles (Fv) equal to

Zoosporogenesis (indirect germination)

Intriguingly, each of the four genera considered here has been reported to differ as to the origin of the cleavage membranes. It remains to be seen if these discrepancies are all genuine or are, for the most part, divergent interpretations of the same basic cleavage mechanism.

Hohl & Hamamoto (1967) and Elsner et al. (1970) found evidence that cleavage vesicles of vacuoles originated from dictyosomes in *Phytophthora*. Fusion of these vesicles with themselves and with the central vacuole cleaved the cytoplasm into uninucleate zoospores (Hohl & Hamamoto, 1967; Elsner et al., 1970; Williams & Webster, 1970).

In *Pythium*, zoosporogenesis is preceded by the formation of a bubblelike vesicle at the tip of a relatively long tube which develops from the distal end of the sporangium. The sporangial protoplast then flows through the tube into the vesicle where cleavage takes place. When zoosporogenesis is induced in *Pythium middletonii*, the Golgi apparatus begins to proliferate and the number of dictyosomes increases (Heintz, 1971; Heintz & Bracker, unpublished). The cytokinetic membrane, which ultimately becomes the plasma membrane of the zoospores, is formed by the coalescence of plasma membranelike vesicles and cisternae derived from paranuclear dictyosomes. Continued addition of dictyosome-derived membrane to the cytokinetic apparatus results in the establishment of furrows which partition the protoplasm into individual zoospores. Cleavage in *Pythium* spp. differs from that described in *Phytophthora* in that cytokinetic membranes are produced after inducement of zoosporogenesis and do not take the form of cleavage vesicles aligned along presumptive cleavage planes. Flagella are first observed after the protoplast is discharged from the sporangium into the vesicle.

In *Saprolegnia ferax*, zoosporogenesis is characterized by the fusion of a large number of small vesicles with dense contents to form larger vacuole-like vesicles (cleft-vesicles) which, by their positions, delimit the uninucleate spore initials at the periphery of the

---

peripheral vesicles in the zoospore. The plane of sectioning included the basal plug (P) but not the apical plug (see Fig 2) (x 5,390). Glutaraldehyde-$KMnO_4$ fixation. (From D. E. Hemmes, Ph.D. Dissertation, 1970.)

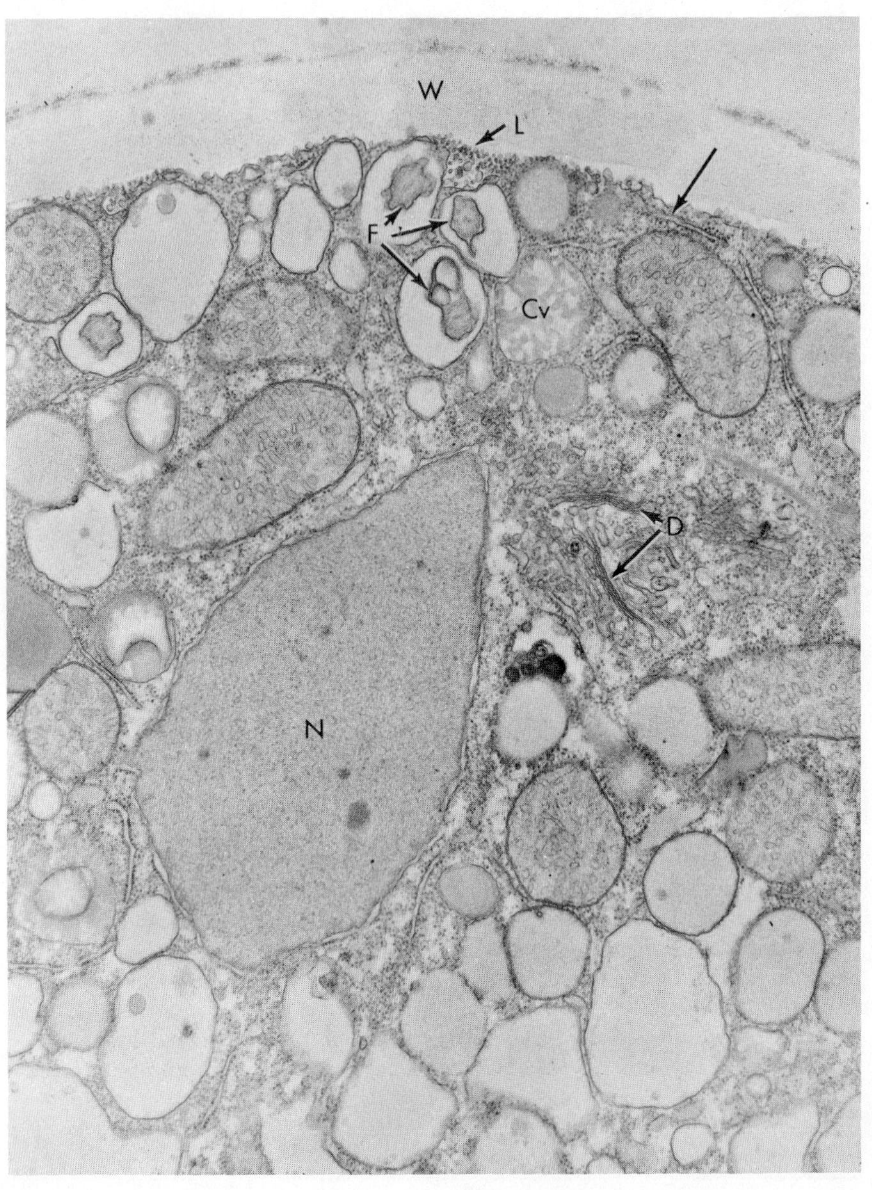

Fig. 2 Apical plug or papilla of a sporangium of *Phytophthora palmivora* (formerly *P. parasitica*). Sporangium wall (W). (x 11,650.) (From Hemmes & Hohl, 1969.)

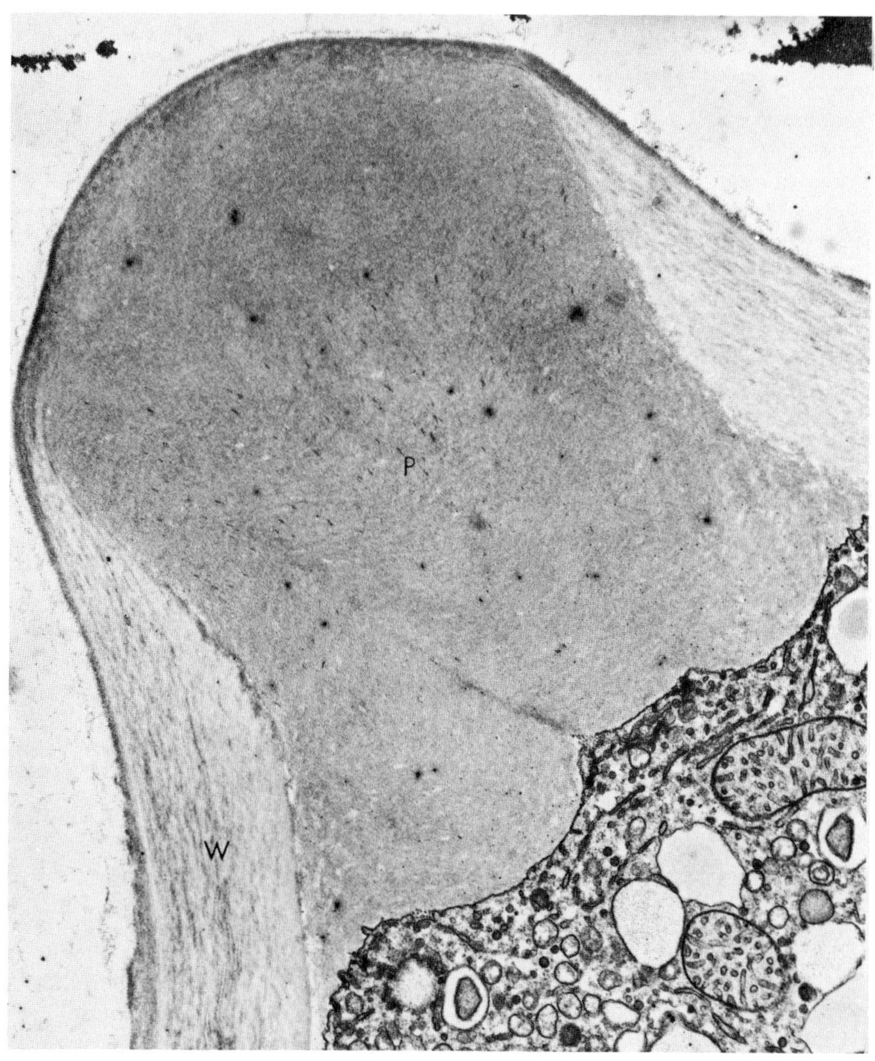

Fig. 3 Detail of a fully expanded sporangium of *Phytophthora palmivora* after the development of flagella. Lomasome (L); sporangial wall (W); flagella (F); nucleus (N); crystalline finger-print vacuole (Cv); dictyosomes (D). Glutaraldehyde-OSo_4 fixation. (x 14,630.) (From D. E. Hemmes, Ph.D. Dissertation, 1970.)

sporangium (Gay & Greenwood, 1966; Gay et al., 1971). Ultimately, the cleft-vesicles fuse to form an extensive and continuous tonoplast deeply evaginated between the lobes of the zoospore initials. During this process the plasmalemma remains closely associated with the sporangium wall and shows no evidence of invagination or other changes. The newly developed tonoplast extends to contact and fuse with the plasmalemma, thereby separating the spores. The plasmalemma of each spore is thus derived partly from the tonoplast and partly from the sporangial plasmalemma. There is no evidence to connect the precursors of the cleft-vesicles with preformed membrane systems of the endoplasmic reticulum or the Golgi bodies. Although vesicles similar to those of the dense-body type have been illustrated in the other Oomycetes (see Gay et al., 1971), they have not been implicated in the cleavage process.

The cleavage of spores within the sporangium of *Aphanomyces euteiches* appears to require no new membrane synthesis to delineate the spores (Hoch & Mitchell, 1972b). Primary spore formation does not involve the coalescence of cleavage vesicles, but utilizes instead the existing plasmalemma and tonoplast of the hyphae that are present prior to induction of spore formation. Flagella are not observed during the development of the primary spores in the hyphae or in the newly encysted primary spores. After flagella develop in the encysted primary spore, a localized bulge appears in the wall, the wall ruptures, and the incipient zoospore is extruded.

Direct Germination and Aging

Under certain conditions, such as incubation at elevated temperatures or in fruit extracts, zoosporogenesis does not occur in sporangia of *Phytophthora* spp. (Vujičić & Colhoun, 1966; Hemmes & Hohl, 1969). Instead, germ tubes grow directly through the sporangial wall or through the sporangial papilla (direct germination) (Fig. 4). Since cleavage vesicles and flagella are already present in a mature sporangium of *Phytophthora palmivora* (Hemmes & Hohl, 1973), one of the first steps in direct germination is the reversal of elements needed for zoosporogenesis, for example, the breakdown of the no longer needed flagella. The flagellar vacuole and axonemal membranes are disrupted leaving the flagellar axonemes naked in the cytoplasm where they eventually degenerate (Hemmes & Hohl, 1973). At approximately the same time, a germination wall is laid down in apposition to the primary sporangial wall; later, the germ tube wall arises from this germination wall. Direct germination can be regarded as a form of intrasporangial encystment particularly when there is partial or even total cytoplasmic cleavage but zoospores are not released. In such cases, the zoospores or multinucleate masses of cytoplasm encyst and germinate within the sporangium.

The same sequence of events, flagellar degeneration and germ tube formation, occurs spontaneously in old sporangia of *Phytophthora* (King, et al., 1968; Hemmes & Hohl, unpublished), though the sequence requires several days rather than several hours as in direct germination. Only one or two thin germ tubes (sporangiophores) are produced near the papilla in these old sporangia. The sporangial cytoplasm flows into the germ tubes which subsequently expand and differentiate into secondary sporangia (Hemmes & Hohl, 1973).

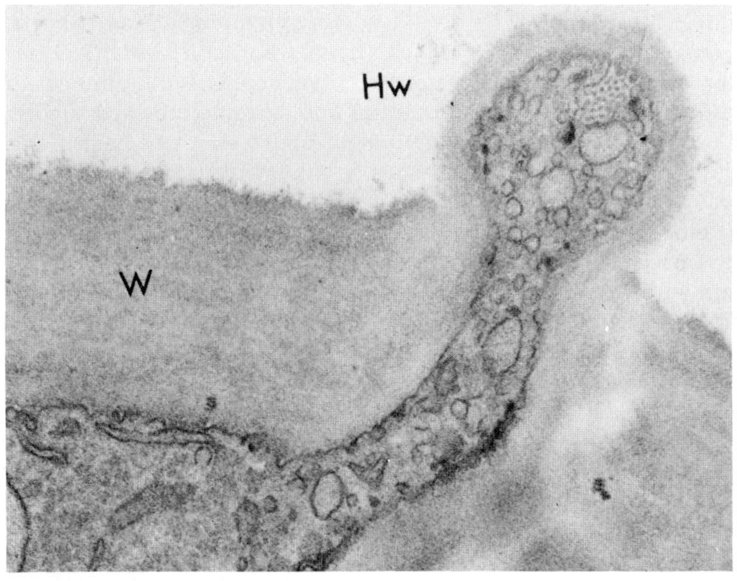

Fig. 4 Direct germination of a sporangium of *Phytophthora palmivora* (formerly *P. parasitica*). Germ tube (Hw) has perforated the sporangial wall (W). (x 35,000.) (From Hemmes & Hohl, 1969.)

Zoospores

General Morphology

Zoospores are reniform to pyriform in shape with the anterior end tapered. Scanning electron micrographs (Desjardins et al., 1973a) confirmed the interference phase contrast study of Ho et al., (1968a) showing that the zoospore of *Phytophthora* (and other Oomycetes) possesses a lengthwise groove (Fig. 5); flagella are inserted in the deep part of the groove near the middle. Both flagella have lateral appendages. The tinsel flagellum has characteristic and prominent mastigonemes; the whiplash flagellum is not hairless, as was generally believed, but bears fine lateral hairs, as first found in *Saprolegnia ferax* (Manton et al., 1951) and more recently in *Phytophthora spp*. (Vujičić et al., 1968; Reichle, 1969a; Desjardins et al., 1969). In the latter, fine hairs are also found at the tip of the whiplash flagellum (Desjardins et al., 1973b).

Longitudinal or transverse sections of zoospores of *Phytophthora parasitica* (Reichle, 1969a); *P. megasperma* var. *sojae* (Ho et al., 1968b); *P. palmivora* (Fig. 6), *Pythium aphanidermatum* (Grove, 1971) and *Aphanomyces euteiches* (Hoch & Mitchell, 1972b) show a striking similarity in fine structure. The centrally positioned nucleus is pyriform in shape with its pointed end extended toward the kinetosomes. An anterior tinsel flagellum and posterior whiplash flagellum are attached to the two kinetosomes in a slightly raised area within the groove. An elaborate system of rootlets and microtubules anchor the kinetosomes in the cytoplasm. Single microtubules extend from the kinetosomes into the cytoplasm and along the narrow part of the nucleus while rootlets composed of a band of microtubules extend parallel to the groove from each kinetosome (Reichle, 1969a; Heath & Greenwood, 1971; Grove, 1971; Hoch & Mitchell, 1972b).

A large contractile vacuole (or water expulsion vesicle apparatus) (Grove, 1971) is found at the anterior end of the zoospore and is partially surrounded by a Golgi apparatus made up of several elongated dictyosomes. Numerous coated vesicle profiles are continuous with the contractile vacuole membranes. Mitochondria, rough endoplasmic reticulum, microtubules within cisternae of endoplasmic reticulum, microbodies, and membrane-free ribosomes are dispersed within the cytoplasm. There is also an abundance of vesicles with striated inclusions (Fig. 6,25) that have received various designations: vesicles with crystalline contents (Ho et al., 1968b; Reichle, 1969a),

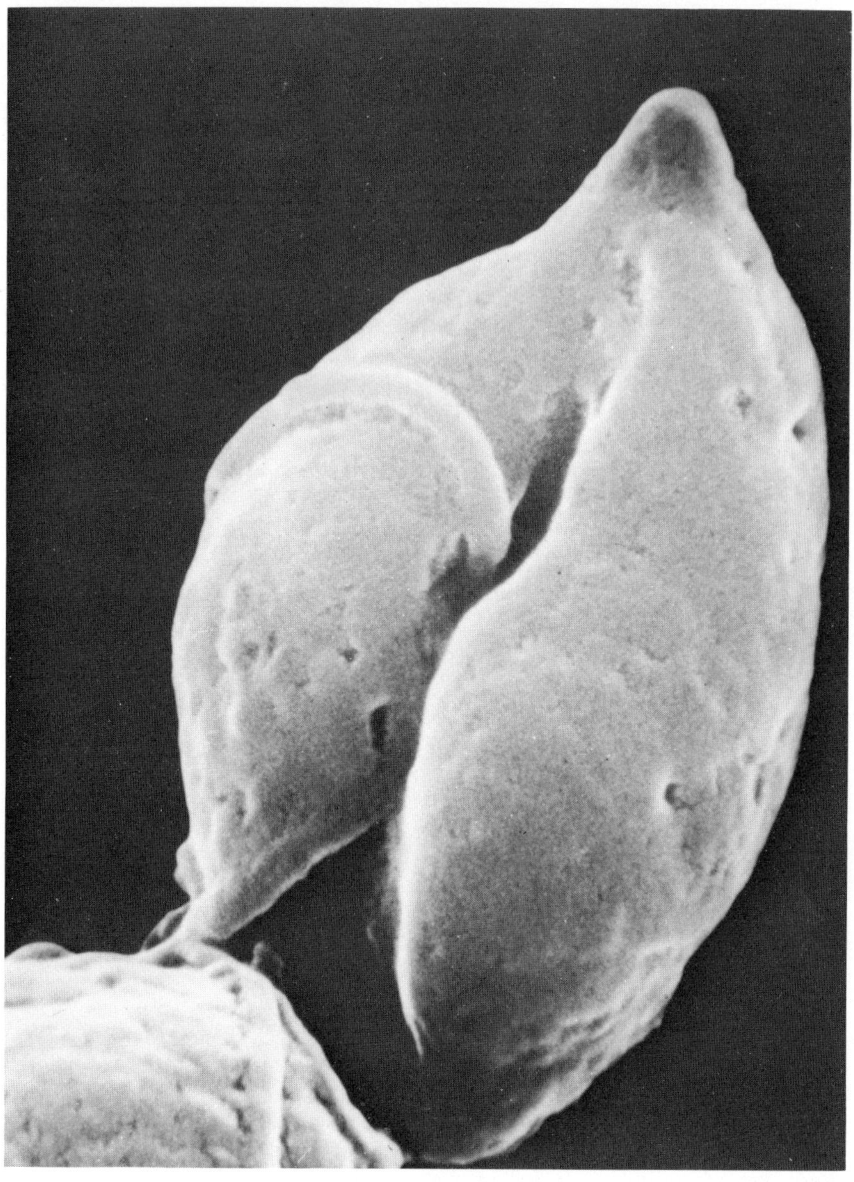

Fig. 5 Scanning electron micrograph of a zoospore of *Phytophthora palmivora* (x 16,425). (From Desjardins et al, 1973a.)

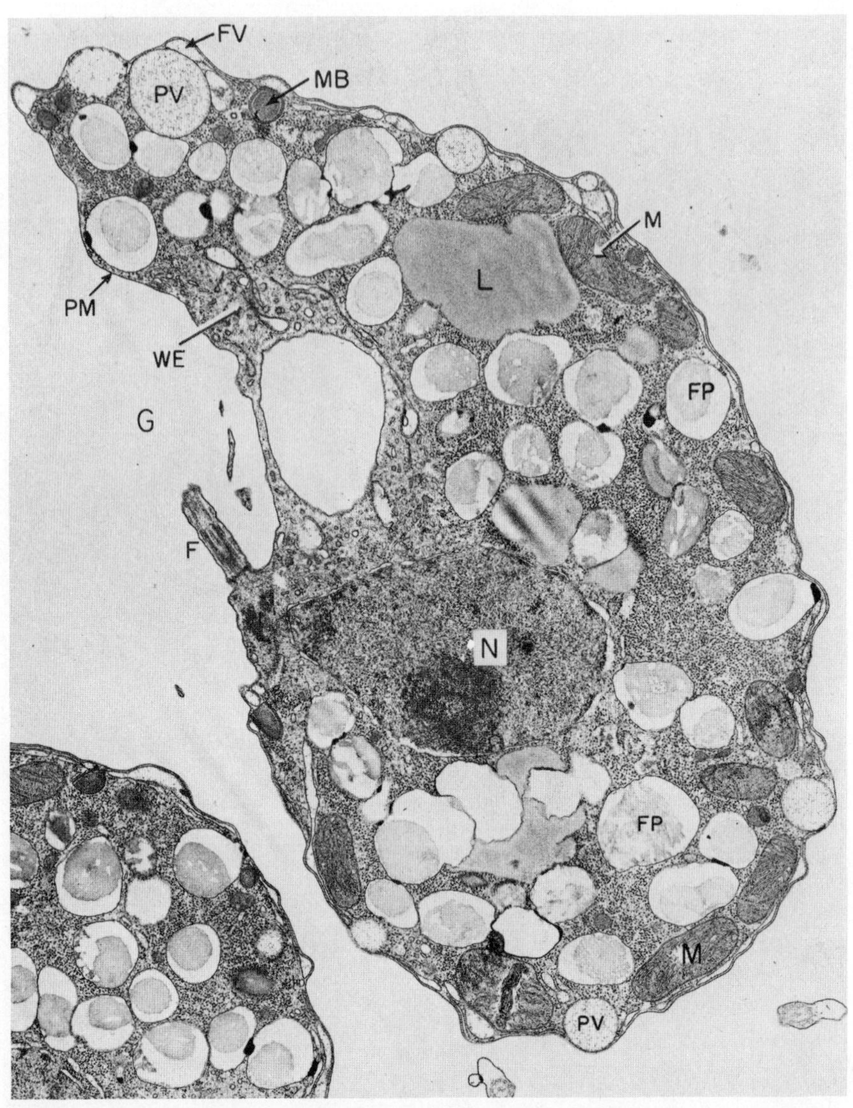

Fig. 6 Median longitudinal section of a zoospore of *Phytophthora palmivora*. Peripheral vesicles (PV); flattened vesicles (FV); microbodylike structure (MB); finger-print vacuoles (FP); nucleus (N); mitochondrion (M); water expulsion vesicle apparatus (WE); lipid-like body (L); flagellum (F); plasmalemma (PM); groove region (G). (Hemmes & Bartnicki-Garcia, previously unpublished.) x 22,200.

liposomes (Williams & Webster, 1970), finger-print vacuoles (Grove, 1971), and vesicles with dense bodies (Gay & Greenwood, 1966).

Ribosome-free cisternae (flattened vesicles) line the periphery of the zoospores of *Phytophthora*, *Pythium*, and *Aphanomyces* except in the ventral (groove) region and are interspersed with large, 0.5 µm in diameter, spherical vesicles called peripheral vesicles by Grove (1971), fibrillar vacuoles by Hemmes and Hohl (1971), and vesicles containing granular-fibrillar material by Reichle (1969a) (Fig. 6). Additional organelles are observed at the cell periphery in other Oomycete zoospores, for instance, vesicles containing an electron-opaque cortex and center in *Aphanomyces euteiches* (Hoch & Mitchell, 1972b) and "bars" in *Saprolegnia* (Heath & Greenwood, 1970). Cytoplasmic ribosomes and other organelles are excluded from the cytoplasm surrounding the Golgi apparatus, the water expulsion vesicle, and the peripheral cisternae.

Cysts

Cytological aspects of zoospore encystment are discussed below. When the cyst of *Phytophthora palmivora* reaches maturity, the dictyosomes which are initially arranged in a Golgi complex near the nucleus are dispersed toward the cell periphery and vesicles can be seen budding off the dictyosome cisternae (Hemmes & Hohl, 1973). These vesicles are similar to the apical vesicles found at the tips of growing hyphae (Grove et al., 1970; Grove & Bracker, 1970). In all oomycetous cysts studied, the germ tube arises from a point on the cyst surface under which a large cluster of such vesicles can be found (Heath et al., 1971; Hemmes & Hohl, 1971; Grove, 1971; Hoch & Mitchell, 1972b) (Fig. 7).

Chlamydospores

Our knowledge of the fine structure of Oomycete chlamydospores is limited to a study of the relatively thin-walled chlamydospores of *Phytophthora cinnamomi* formed in grapelike clusters (Hemmes & Wong, unpublished) (Figs. 8--10). Despite some superficial resemblance between sporangia and chlamydospores, viewed by light microscopy, major differences are immediately apparent between chlamydospores and sporangia of *Phytophthora* examined under the electron microscope. First, as would be expected, structures necessary for zoosporogenesis are absent (papillae, flagella, and cleavage vesicles).

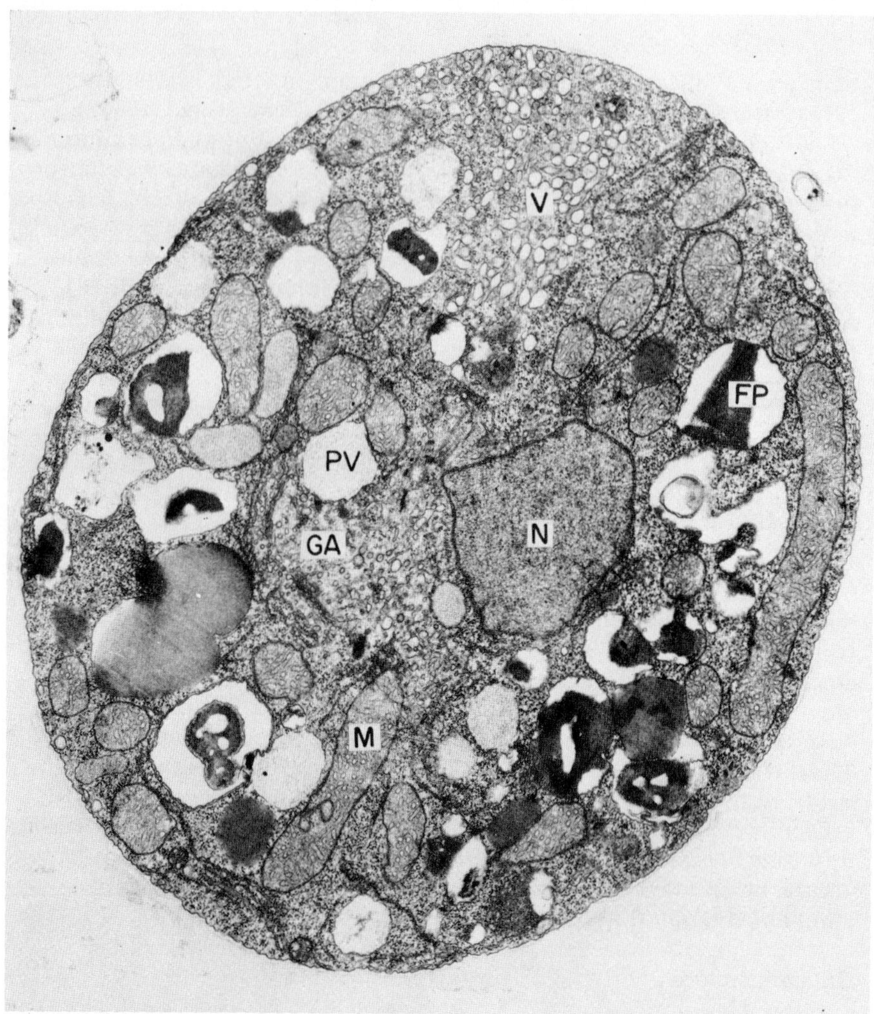

Fig. 7 Cyst of *Pythium aphanidermatum* prior to germ tube formation. A cluster of vesicles (V) marks the site where the germ tube would have emerged. The few remaining peripheral vesicles (PV) from the zoospore state are now near the center. Golgi region (GA); Fingerprint vacuole (FP); Mitochondrion (M); Nucleus (N). The cyst wall does not show on this print. (x 15,200). (Courtesy of Grove, 1971.)

Secondly, the contents of the "finger-print" vacuoles, which are major constituents of sporangia of *Phytophthora*, are condensed into highly electron dense, spherical globules (Figs. 8 and 9) and do not show the characteristic finger-print pattern seen in sporangia. In further contrast to the sporangium (compare Fig. 1), these vacuoles occupy a much larger proportion of the chlamydospore volume, squeezing nuclei, mitochondria, and other organelles into small pockets between the vacuoles. Spherical vesicles closely resembling the fibrillar vacuoles described by Hemmes and Hohl (1971), or peripheral vesicles (Grove, 1971), for sporangia and zoospores, are abundant in the chlamydospores of *P. cinnamomi* (Figs. 9, 10).

The chlamydospore walls of *P. cinnamomi* are homogeneous and similar to sporangial walls. No layering, as seen in oogonial walls of *Phytophthora capsici* (compare Fig. 8 and Fig. 16) was observed. Chlamydospore germination is similar to direct sporangial germination in that a germination wall is formed under the primary chlamydospore wall which later extends into the germ tube wall.

Gametangia and Oospores

Gametangia and oospores of Oomycetes are challenging subjects for ultrastructural studies because of the technical difficulties and many facets of their development that have remained unexplored or controversial (Dick and Win-Tin, 1973); for instance, the development of amphigynous gametangia (in *Phytophthora*), the time and site of meiosis (gametic vs zygotic), the presence and configuration of fertilization tubes, the mode of cleavage of the oogonial cytoplasm into one or more oospheres, the existence and origin of the periplasm, and the number and genesis of the various wall layers.

The original interpretation of amphigyny in *Phytophthora*, (Dastur, 1913; Pethybridge, 1913; Murphy, 1918) which had been challenged by Stephenson & Erwin (1972), has now been substantiated by electron microscopy of *P. palmivora* (Vujičić, 1971) and *P. capsici* gametangia (Hemmes & Bartnicki-Garcia, 1975). In amphigynous species of *Phytophthora*, there is the unusual penetration of the male (antheridial) initial by the female (oogonial) hypha (Fig. 11). This initial gametangial interplay, however, does not bring about cytoplasmic fusion. Plasmogamy occurs later when a fertilization tube, generated by the antheridium, penetrates the oogonium (Fig. 12). Electron microscopy of *P. capsici* gametangia (Hemmes & Bartnicki-Garcia, 1975) also confirmed the formation and subsequent degeneration of a periplasmic layer around the oosphere (Figs. 17, 18), and provided details of the structure and development of the many different types of cell walls (antheridium, oogonium, oogonial-stalk,

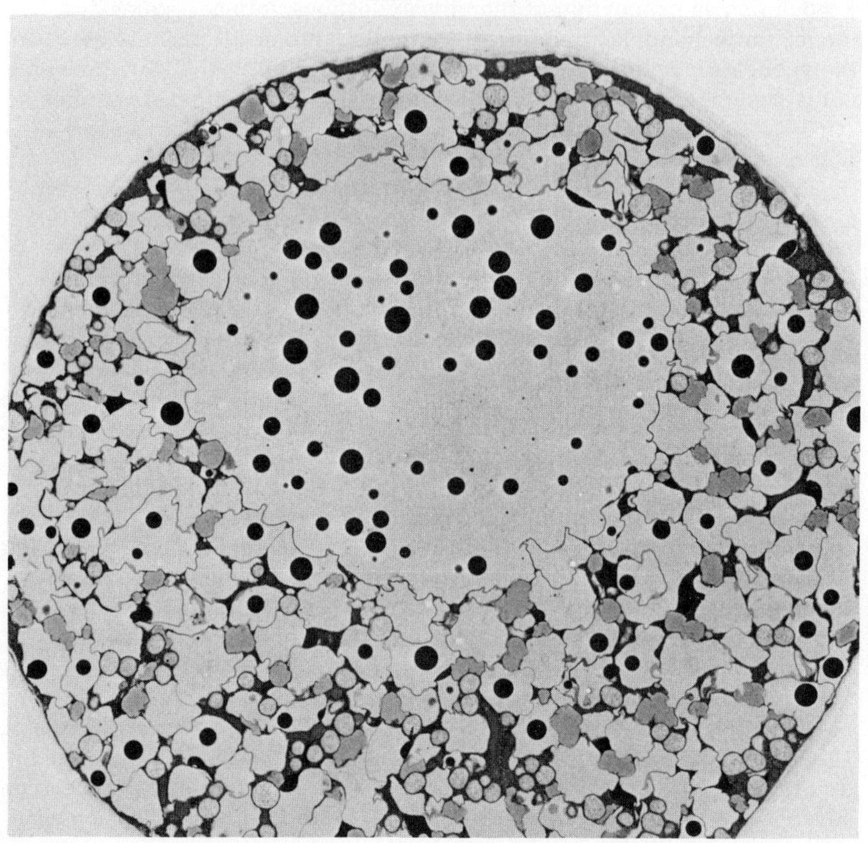

Figs. 8 and 9 Resting chlamydospore of *Phytophthora cinnamomi*. Note dense spherical inclusions in "fingerprint" vacuoles (FP) and numerous "peripheral" vesicles (PV). (These names correspond to equivalent organelles found in other cellular stages where the appearance and position of these structures, respectively, matches their designations.) Nucleus (N); Chlamydospore wall (W). (Fig. 8, x 4,000; Fig. 9, x 11,480.) (Hemmes & Wong, previously unpublished.)

fertilization-tube, outer oospore (exospore), and inner oospore (endospore) (Figs. 13--18). Dictyosome-derived vesicles play a prominent role in many aspects of wall formation during sexual morphogenesis in *Phytophthora capsici*, such as growth of the oogonial stalk, growth of the fertilization tube, oogonial expansion, and thickening of the oogonial wall (Hemmes & Bartnicki-Garcia, 1975) (Figs. 11--15). Vesicles were also found in abundance under the developing oogonial wall of *Phytophthora palmivora* (Vujičić, 1971) and

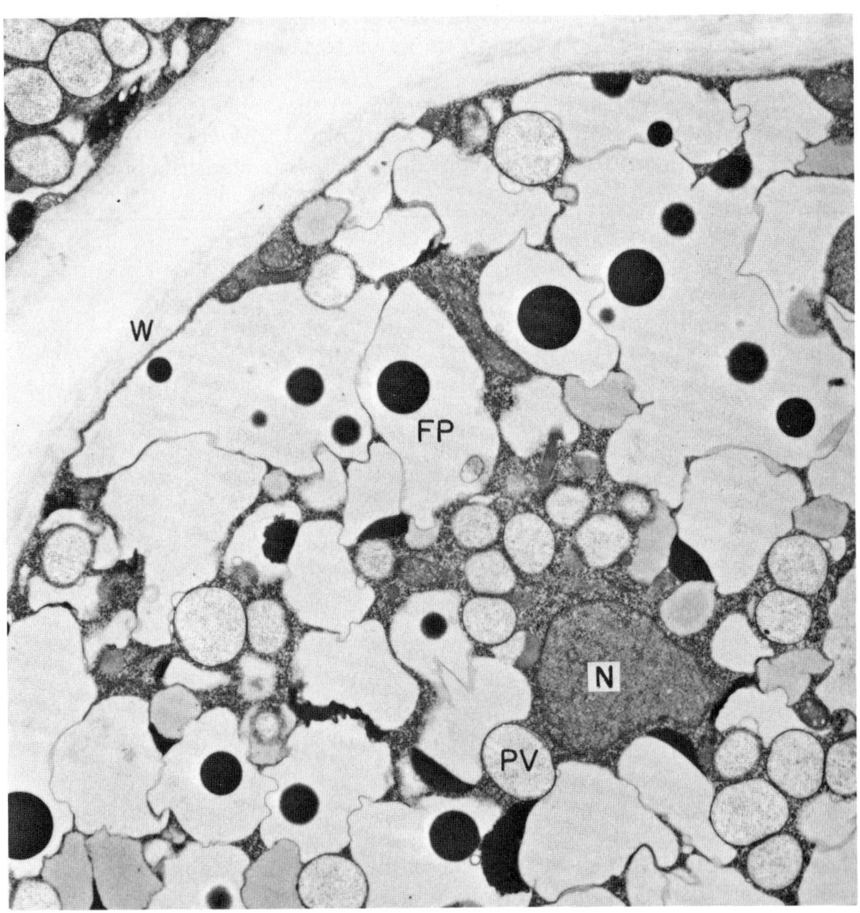

Saprolegnia furcata (Heath et al., 1971). During oosporogenesis of *P. capsici*, drastic changes in the overall appearance of the cytoplasm take place. The gametangial cytoplasm, originally rich in rough endoplasmic reticulum, dictyosomes, ribosomes, mitochondria, and dictyosome derived vesicles changes into a cytoplasm where enlarged finger print vacuoles and lipidlike bodies assume the major portion of the oospore interior (Figs. 13--18).

In *Pythium ultimum*, Marchant (1968) also found extensive changes in cytoplasmic organization of the oogonium resulting in a mature oospore filled with large "storage" organelles. However, in contrast to our findings on *Phytophthora capsici* (Figs. 12, 17, and 18), he concluded that neither a periplasm nor a fertilization tube were formed in *Pythium ultimum*.

Ultrastructural studies of oogenesis in the Saprolegniaceae showed that meiosis occurs in the gametangia (Howard & Moore, 1970; Ellzey, 1974). The oogonium of *Saprolegnia ferax* is cleaved into oospheres

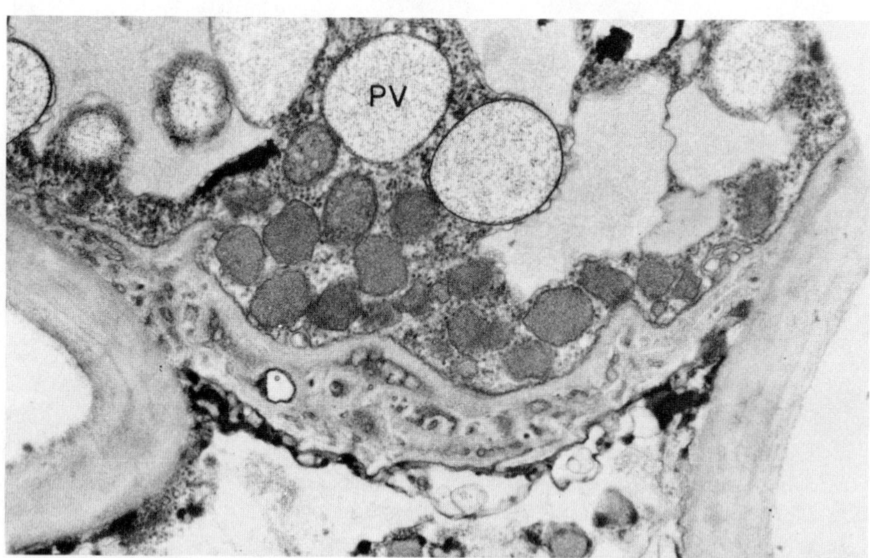

Fig. 10 Septum (plug) at the base of a chlamydospore of *Phytophthora cinnamomi* with associated electron-dense vesicles and "peripheral" vesicles (PV) (x 20,500). (Hemmes & Wong, previously unpublished.)

by a process analogous to that which occurs in zoosporogenesis (Gay et al., 1971). The mature saprolegniaceous oospore contains a large vacuole-like body (named ooplast by Howard & Moore, 1970); this corresponds to what had been erroneously considered to be the oosphere cytoplasm. The cytoplasm proper may surround the ooplast partially or entirely, or may lie outside of it. Accordingly, a revised classification of oospore types in the Saprolegniaceae was proposed (Howard, 1971).

FUNCTIONAL ASPECTS

Role of β-Glucans in the Life Cycle of Oomycetes, Mainly *Phythophthora*

The role of proteins and nucleic acids in living cells is well recognized and has been the subject of extensive studies. Polysaccharides, however, have not commanded a similar degree of attention. This relative neglect is particularly critical in organisms such as fungi where polysaccharides play a decisive role in growth and differentiation by being the main structural components of the cell walls.

The Oomycetes are a good illustrative case of the importance of polysaccharide biochemistry in the life cycle of fungi: here a single group of polysaccharides, the β-glucans, plays a preponderant role in the activities of these fungi.

The significance of β-glucans is both quantitative and qualitative. First, β-glucans constitute a major portion of the cell mass, one half of the total dry weight in some stages. Second, β-glucans, being the principal cell wall constituents, are the central ingredient for cell wall construction and morphogenesis. The sheer abundance of β-glucan in the oomycetous cells indicates that these organisms must spend a major portion of their metabolic activities making and breaking β-glucosidic linkages.

Cell Wall Glucans

Comparative studies on the composition of cell walls isolated from various life cycle stages of *Phytophthora* (Table 1) show that glucose invariably constitutes roughly 80--90% of the wall dry wt. Two different kinds of β-glucans have been recognized: cellulose, the characteristic β-1,4-linked polymer and a highly insoluble, branched glucan joined mainly by β-1,3 links with β-1,6 links at the

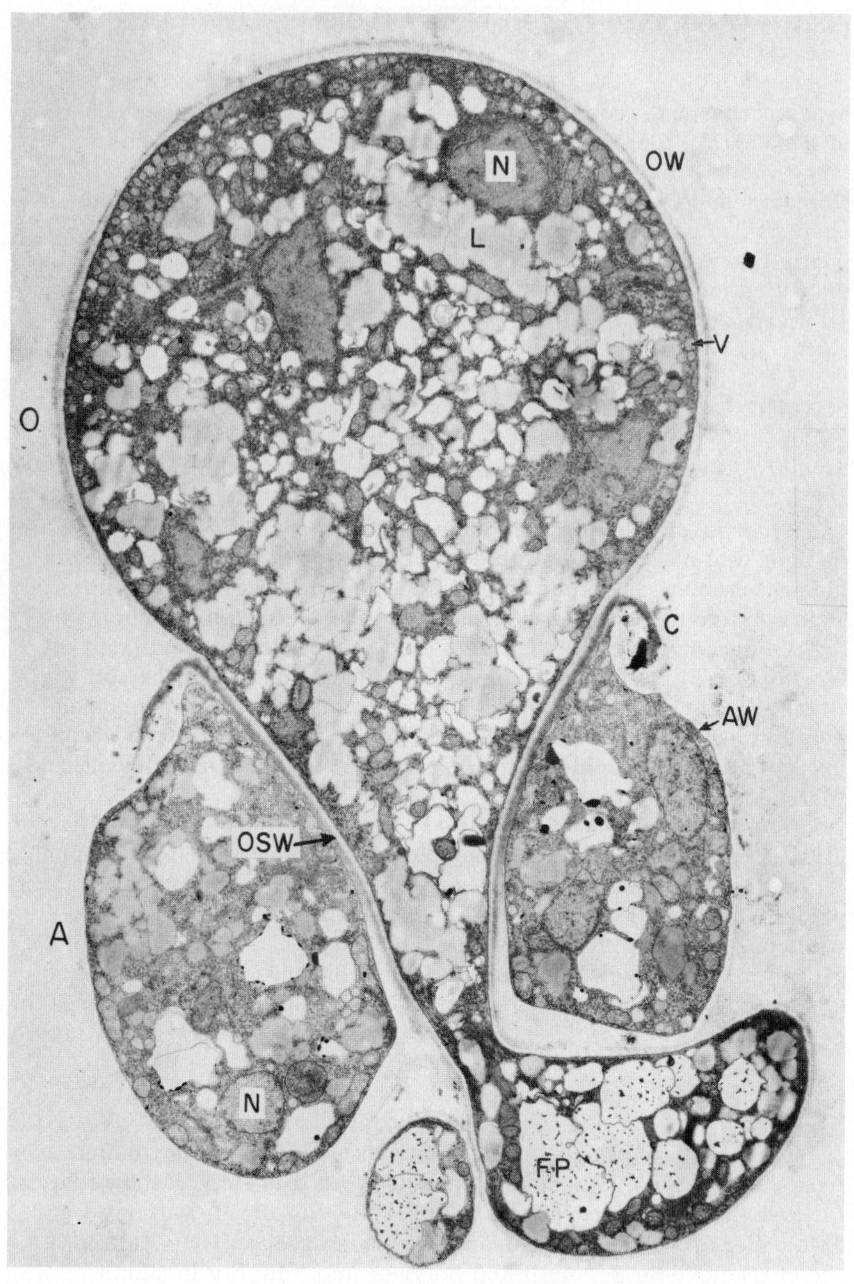

N
OW
L
V
O
C
AW
OSW
A
N
FP

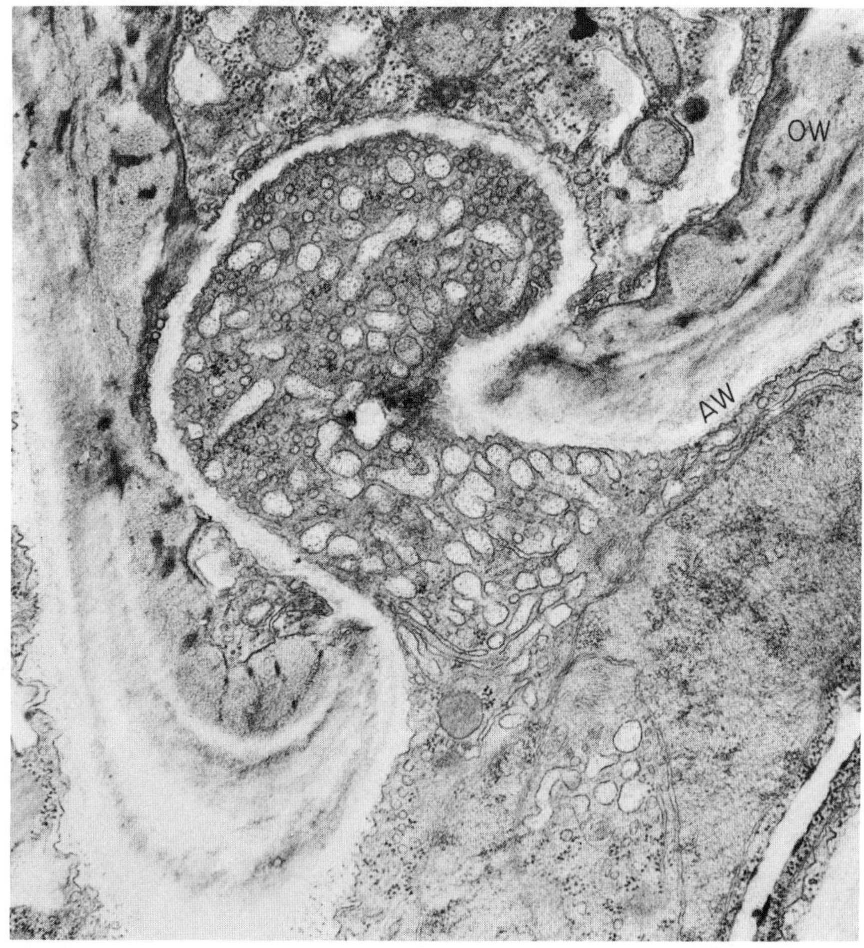

Fig. 12 An antheridial fertilization tube penetrating the base of an oogonium of *Phytophthora capsici*. The tube is filled with dictyosome-derived vesicles of two different sizes. Oogonial wall (OW); antheridial wall (AW). (x 24,000.) (Hemmes & Bartnicki-Garcia, 1975).

Fig. 11 Longitudinal section through a young expanding oogonium of *Phytophthora capsici* and its antheridium (A). The oogonium (O) initial grew through the antheridium rupturing the oogonial stalk wall (OSW) and the antheridial wall (AW) to form a curled wall collar (C). Note the profusion of vesicles (V) under the oogonial wall (OW). Fingerprint vacuoles (FP); Lipidlike droplets (L); Nuclei (N) (x 6,100). (Hemmes & Bartnicki-Garcia, 1975).

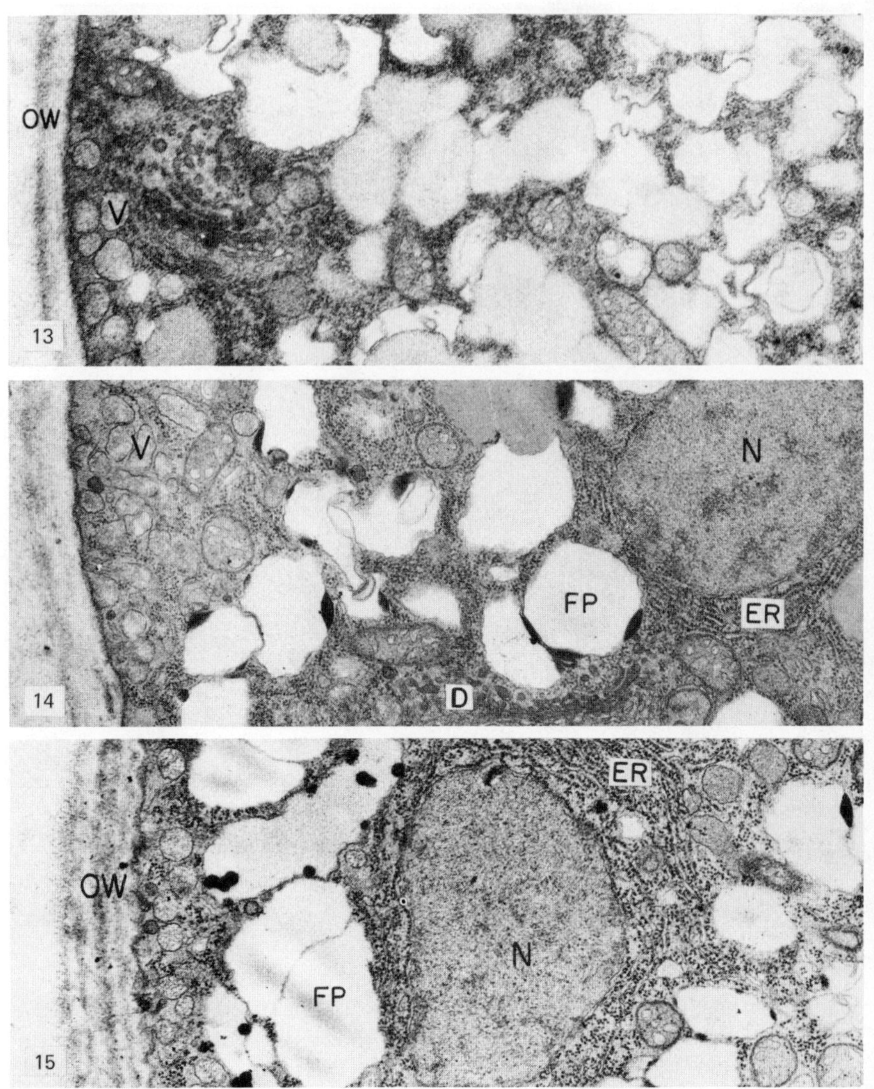

Figs. 13--18 Oosporogenesis in *Phytophthora capsici*. Details of the periphery of a differentiating oogonium showing the sequence of changes in cytoplasmic organization and cell wall development. Figure 13, young expanding oogonium (same as in Fig. 11). Oogonial wall (OW); wall-destined vesicles (V). Figure 14, oogonium near the end of expansion phase. Dictyosome (D); nucleus (N); vesicles (V); fingerprint vacuole (FP); rough endoplasmic reticulum (ER). Figure

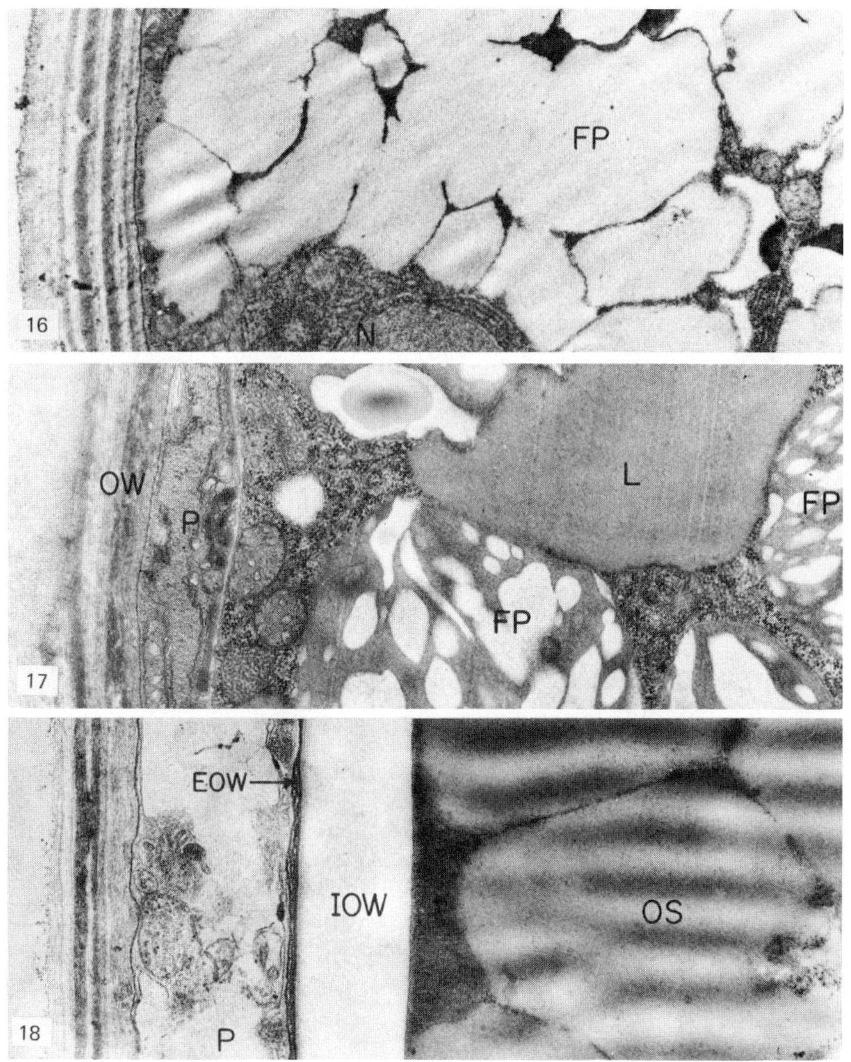

15, wall thickening in a fully expanded oogonium. (OW) oogonial wall; (FP) fingerprint vacuole; (N) nucleus; (ER) endoplasmic reticulum. Figure 16, oogonium with fully thickened multilayered wall. (FP) fingerprint vacuole. Figure 17, oosphere delimitation with formation of periplasm (P) between oogonial wall (OW) and oosphere. (L) lipid; (FP) fingerprint vacuole. Figure 18, periplasm (P) degeneration and oospore formation. Note the thin and multilaminated outer (continued on the next page)

branching residues; the latter also appears to contain some β-1,4-links (Zevenhuizen & Bartnicki-Garcia, 1969). This insoluble non-cellulosic β-glucan is the major component of the wall. These two types of β-glucans occur in the walls of all genuine members of the Oomycetes examined. They include representatives of the Peronosporales, Saprolegniales and Leptomitales (Bartnicki-Garcia, 1966; Aronson et al., 1967; Novaes-Ledieu & Jimenez-Martinez, 1968; Sietsma et al., 1969; Pao & Aronson, 1970). The only apparent exception to this uniformity of wall composition in the Oomycetes (Bartnicki-Garcia, 1968, 1970) would be in the family Thraustochytriaceae of the Saprolegniales. This unique assembly of organisms was placed in the Oomycetes mainly because of their biflagellate zoospore (Sparrow, 1943). Recently, an examination of *Schizochytrium aggregatum* and *Thraustochytrium* sp. by Darley et al., (1973) revealed that their walls have a structural organization and chemical composition entirely unlike that of other Oomycetes and the rest of the fungi (there were no β-glucans; instead, the walls were rich in D- and L-galactose polymers). These and other features (Darley et al., 1973; Goldstein, 1973), strongly dispute the validity of placing the Thraustochytriaceae among the Oomycetes and make their reclassification elsewhere mandatory.

Even though the proportion of β-glucans in the cell wall does not change markedly during the life cycle of *Phytophthora* (Table 1), sharp changes in the architecture of the wall may be detected in the electron microscope; they are indicative of changes in both the fine structure and proportions of the β-glucans. A case in point is shown in Figs. 19--22. The isolated cyst wall of *Phytophthora palmivora* is a thin fabric of tightly knit glucan microfibrils; both the external and internal surfaces have a microfibrillar texture.* Upon germination,

*In addition to the microfibrillar wall, the cyst envelope contains an outer, amorphous, patchy coat that is lost during wall isolation (sonication or alkali-extraction).

oospore wall or "exospore" (EOW), the thick electron-transparent inner oospore wall (IOW) and the large storage bodies filling the oospore (OS) interior to the near exclusion of other cytoplasmic organelles. Note changes in size and appearance of fingerprint vacuoles (FP) from young oogonium to oosphere (Figs. 14--17). (Hemmes & Bartnicki-Garcia, 1975.) (x 18,500).

Table 1. Chemical Composition of the Walls of *Phytophthora*

Component	Hypha		Sporangium	Cyst	Oospore/Oogonium
	1[a]	2[b]	3[c]	3[c]	4[d]
Hexose[f]	88.0	86.0	93.5	92.5	77.7
Protein	3.5	4.2	1.7	2.4	10.66
Lipid: Free	0.3	0.2	e	e	5.68
Bound	2.1	0.9	e	e	5.64
Amino sugar	0.3	0.3	0.2	0.1	0.43
Phosphorus	0.13	e	0.05	0.098	0.55

[a]*Phytophthora cinnamomi*, from Bartnicki-Garcia (1966).
[b]*Phytophthora parasitica*, Ibid.
[c]*Phytophthora palmivora*, from Tokunaga and Bartnicki-Garcia (1971b).
[d]*Phytophthora megasperma* var. *sojae*. (Lippman et al., 1974).
[e]Not measured.
[f]Acid hydrolysis showed that glucose was the main hexose (>95%); mannose was also present.

the outer texture of the wall changes drastically; from its base to its growing apex, the germ tube displays a non-fibrillar outer texture (Tokunaga & Bartnicki-Garcia, 1971b) (Figs. 19 and 20). This non-fibrillar component is mainly an alkali-insoluble β-1,3-glucan that can be selectively removed by digestion with an exo-β-1,3-glucanase (Figs. 21 and 22). This exposes the inner microfibrillar skeleton of the germ tube wall which is a continuation of the cyst wall fabric* (Tokunaga & Bartnicki-Garcia, 1971b). Obviously, to understand the process of zoospore germination, we must pay attention not only to the initial formation of a microfibrillar cyst wall but also to the development of a more complex double-layered wall.† This dual or multiple nature seems to be a universal character of hyphal walls (Hunsley & Burnett, 1970; see Bartnicki-Garcia, (1973a).

Storage of β-Glucans in the Cytoplasm

The cytoplasm of Oomycetes contains no glycogen (Zevenhuizen

*Hegnauer & Hohl (1973) believe that microfibrillar layer of the germ tube is not a simple extension of the cyst wall but a new separate entity.

†Triple-layered, if one includes an outermost amorphous "fluffy" coat covering the germ tube (Hegnauer & Hohl 1973).

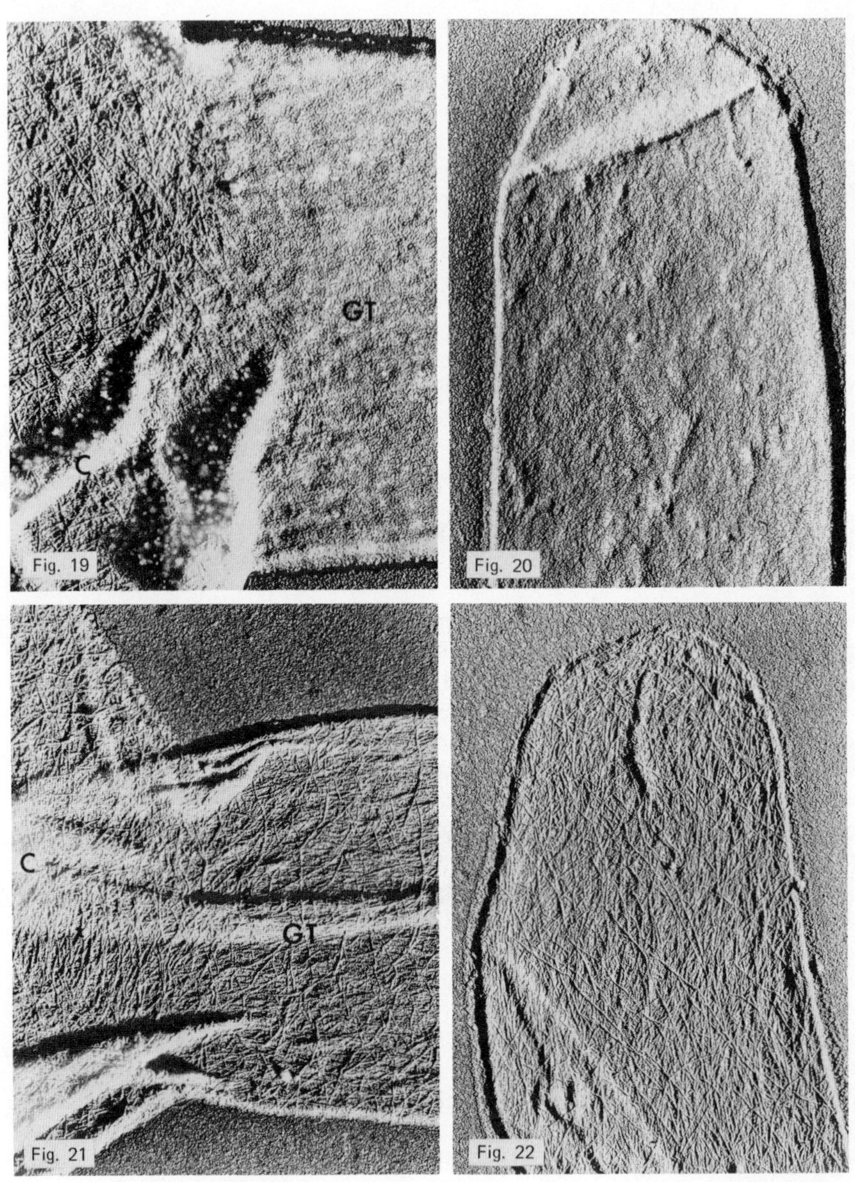

Figs. 19--22 Change in wall architecture during cyst germination of *Phytophthora palmivora*. Figure 19, germ tube wall (GT) arising from the cyst wall (C). Figure 20, germ tube wall apex. Figures 21 and

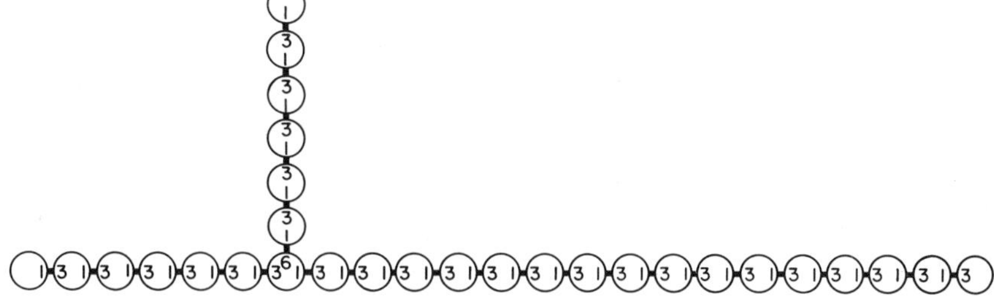

Fig. 23 Mycolaminaran structure. A plausible model for the mycolaminaran isolated from the zoospore of *Phytophthora palmivora*. The total number of glucose residues and the types of β-linkages shown were experimentally determined (Wang & Bartnicki-Garcia, 1974) but the position of the branching residue and, hence, the length of the side chain is not known.

& Bartnicki-Garcia, 1970). Instead these fungi employ as storage polysaccharides, β-1,3-glucans similar to algal laminaran (Phaeophyta and Chrysophyta) for which the designation mycolaminaran has been proposed (Wang & Bartnicki-Garcia, 1974). These highly soluble glucans were first isolated from the mycelium of *Phytophthora cinnamomi* (Zevenhuizen & Bartnicki-Garcia, 1970) and later found in the other species of *Phytophthora* (Wood et al., 1971; Wang & Bartnicki-Garcia, unpublished results) and in other genera of Oomycetes (Faro, 1972a,b). A detailed characterization of the molecular structure of mycolaminarans from *Phytophthora palmivora* revealed that these were small polymers ($\overline{DP}$ 29--36), made exclusively of glucose units, with the main chain linked by β-1,3 bonds and one or two branches at C-6 (Fig. 23); the number of branches and the molecular size depends on the developmental stage of the fungus (Wang & Bartnicki-Garcia, 1974). Other marked changes in the quality and quantity of mycolaminarans accompany sporulation and spore germination. Whereas in the vegetative mycelium of *Phytophthora* all mycolaminaran molecules are neutral, in the sporangia and in the subsequent stages of asexual development (zoospores and cysts),

---

22, same as Figs. 19 and 20 after digestion with an exo-β-1,3-glucanase which removes the non-fibrillar outer layer of the germ tube wall and exposes the inner microfibrillar texture (x 19,800). (Tokunaga & Bartnicki-Garcia, 1971b.)

most of the molecules become phosphorylated (Wang & Bartnicki-Garcia 1973) (Fig. 24). Up to five different species of mycolaminaran phosphate can be separated by ion-exchange column chromatography. Unknown is the reason why the bulk of the mycolaminaran is phosphorylated during sporulation.

Flux of β-Glucans During the Life Cycle

Figure 25 illustrates the changes in β-glucan composition that occur throughout the life cycle of *Phytophthora*. This diagram was

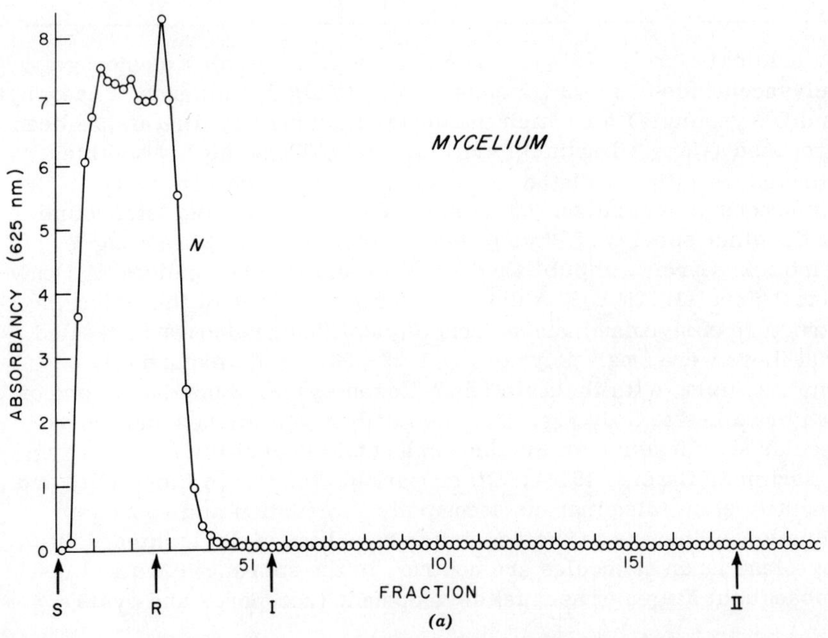

Fig. 24 Separation of mycolaminarans of *Phytophthora palmivora* by DEAE-Sephadex columns. (a) The mycelium contains only neutral (N) mycolaminaran that is not adsorbed onto the column. (b) The zoospores contain mainly a family of mycolaminaran phosphates (PI-PV). (Wang & Bartnicki-Garcia, 1973.)

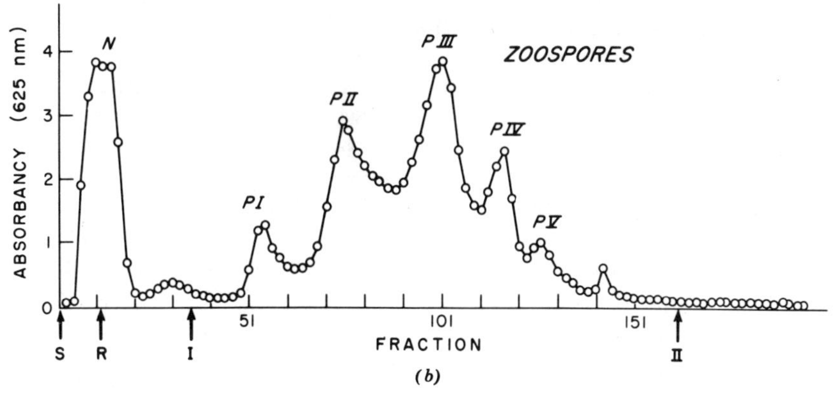

(b)

assembled from data of three different species (Wang & Bartnicki-Garcia, unpublished observations) and would require some minor readjustment when full data for a single organism is collected.

During vegetative mycelium development, a large amount of β-glucan, slightly in excess of 50% cell dry weight, is synthesized (33% cytoplasmic mycolaminaran and 18% wall glucans). When the fungus sporulates, it mobilizes massive quantities of cytoplasm into the expanding sporangia. Very likely, the vast reserves of mycelial mycolaminaran are translocated into the sporangia where they are available to support sporulation. They may be the main or even the sole source of glucose to manufacture the sporangial wall and fulfill other metabolic needs particularly if sporulation occurs during mycelial starvation, which is an effective way of inducing sporulation in some *Phytophthora* species (Menyonga & Tsao, 1966; see review by Zentmyer & Erwin, 1970). In the sporangium, the relative proportion of cytoplasmic to wall glucans of the mycelium is reversed. The sporangial wall, a much thicker structure than the hyphal wall, now accounts for more than 25% of the cell wall while the cytoplasmic glucan amounts to only 18%. Moreover, two thirds of the cytoplasmic glucan is phosphorylated. It is not known if mycelial mycolaminaran is phosphorylated during sporulation or if the mycolaminaran phosphates are formed de novo.

In the next developmental stage, zoospores are formed and liberated and leave behind the sizable sporangial wall. In the zoospore, the relative content of cytoplasmic mycolaminaran (neutral and phosphorylated) is higher but this increase is almost all due to the removal of the sporangial wall from the cell weight balance. About 18% is mycolaminaran phosphate and 6% neutral mycolaminaran. Changes in the relative proportions of the various mycolaminaran phosphates can be detected in the transition from sporangium to zoospore (Wang & Bartnicki-Garcia, 1973).

Upon encystment (in a nutrient-free environment), about one half of the total mycolaminaran is consumed (Tokunaga & Bartnicki-Garcia, 1971a; Wang & Bartnicki-Garcia, unpublished). Some is probably used for the synthesis of cyst wall glucan which accounts for about 3--4% of the cyst dry weight; the rest is probably catabolized.

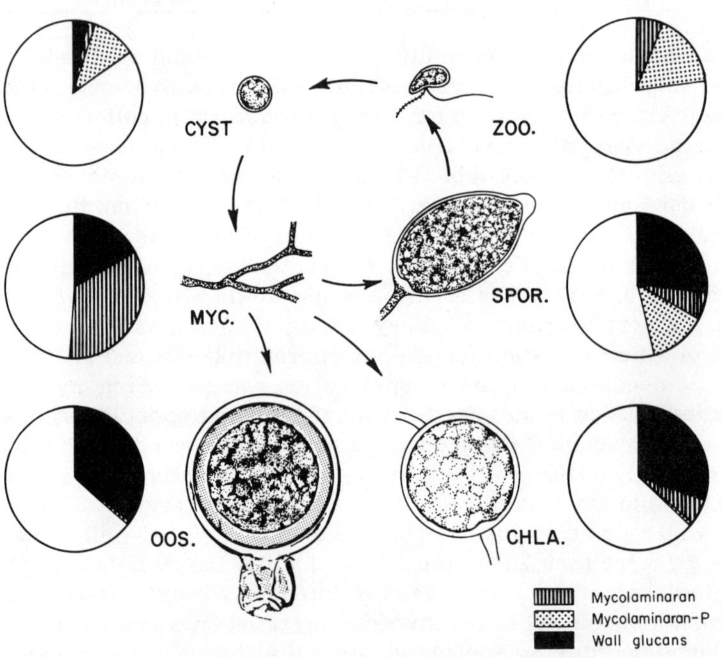

Fig. 25 Changes in β-glucan content during the life cycle of *Phytophthora*. Circles represent total cell dry weight. Zoospores, cysts, mycelium and sporangia of *P. palmivora;* oospores of *P. megasperma* var. *sojae*; chlamydospores of *P. parasitica*. (Wang & Bartnicki-Garcia, previously unpublished.)

The lowest total glucan content (15%) recorded was in the cyst stage; however, there is likely to be a further decrease when the cyst germinates, as it can (Barash et al., 1965; Hemmes & Hohl, 1972), in the absence of nutrients. For mycelium development, however, a supply of exogenous nutrients is required and eventually the fungus will regain its high level of cellular β-glucans.

Sexual sporulation also depends largely on a reservoir of mycelial mycolaminaran; this conclusion is reached from the observation that gametangial development, particularly the expanding oogonium, receives a massive influx of mycelial cytoplasm. The mycolaminaran thus mobilized is probably all ultilized for the complex sequence of sexual differentiation; in the end, the mature oospore contains only a trace of mycolaminaran in its cytoplasm. Presumably, the oospore/oogonium wall glucan, which amounts to 35% of the cell dry weight (Lippman et al., 1974), is largely or exclusively derived from mycelial mycolaminaran.

Even though the dormant oospore does not contain mycolaminaran as a storage polysaccharide, it does rely on β-glucans as reserve material for germination. This conclusion is based on the well known observation (Blackwell, 1949; Zentmyer & Erwin, 1970) that the inner wall of a germinating oospore is markedly eroded suggesting an extensive utilization of oospore wall glucan.

The asexual resting spore, or chlamydospore, resembles the sporangium in that it stores a significant amount of mycolaminaran (Fig. 5), but it differs in amount and in having only a trace of mycolaminaran phosphate. As in the sporangia and the sexual structures, there is also a large investment in the formation of wall glucans.

Information on wall composition and storage polysaccharides of other genera of Oomycetes is rather meager. Nevertheless, the available data indicate that the kinetic scheme of β-glucan distribution described for the life cycle of *Phytophthora* may be representative of Oomycetes in general. In *Achlya*, Faro (1972a,b) found that the growing mycelium stores a laminaranlike polysaccharide similar if not identical to the mycolaminaran of *Phytophthora*. This cytoplasmic glucan is extensively utilized during the formation of sex organs. The utilization begins shortly after the appearance of hormones A and B in the medium and before gametangial branches are produced. After the branches are formed, there is an additional period of net cytoplasmic glucan accumulation followed by a consumption period that begins prior to oospore formation.

β-Glucan degradation has been invoked as the target process

for the operation of the steroidal sex hormones of *Achlya*. Hormone A (antheridiol) is believed to induce the formation of a wall-softening cellulase presumably needed for branch formation (Thomas & Mullins, 1967; Mullins & Ellis, 1974). Faro (1972b) suggested that hormones A and B may reverse the repressive effect of exogenous glucose on β-1,3-glucanase. Presumably, this enzyme is needed for the utilization of mycolaminaran during sexual differentiation.

In conclusion, the biochemical machinery of an Oomycete is largely committed to the production of different types of β-glucans used as pillars of cell structure and as reservoirs of C and energy for differentiation (sporulation and spore germination). Understanding the biochemistry of morphogenesis of these fungi demands an account of the enzymes involved in the synthesis and degradation of β-glucans and their regulatory mechanisms. Of particular importance are the subcellular processes responsible for the spatial synthesis of wall β-glucans. Electron microscopy shows that Golgi-derived vesicles are associated with the many processes of cell wall formation: the apical growth of tubular cells, and the uniform expansion of spheroidal cells (Grove et al., 1970; Heath et al., 1971; Hemmes & Bartnicki-Garcia, 1975). These vesicles are likely candidates for the spatial displacement of catalysts and/or ingredients for β-glucan metabolism.

Zoospore Encystment in *Phytophthora*

This is one of the simplest and experimentally most attractive steps in the life cycle of *Phytophthora* or any flagellated fungus. The key event in encystment is the acquisition of a cyst wall. The process of encystment, however, encompasses a number of other well known changes (Waterhouse, 1962; Hickman, 1970): loss of motility, settling down, loss of flagella, rounding off. These and other recent findings on the fine structure and biochemistry of encysting zoospores, permit us to recognize the following steps in the encystment process:

Cessation of Motility

This is perhaps the first sign of impending encystment. The swimming zoospore becomes sluggish and travels shorter straight distances before turning abruptly. Eventually, it displays irregular, jerky motions or rotates slowly in small circles. The factor(s) that triggers spontaneous encystment is not well understood. In an undisturbed suspension, individual zoospores start encystment over a long and variable period. The average swimming time in a population is subject to great fluctuation depending on environmental factors. Usually, most zoospores of *Phytophthora* spp. encyst within 1 h, but they may swim for hours or even days (Ho & Hickman, 1967; Bimpong & Clerk, 1970). Many stimuli, physical and chemical, can trigger rapid encystment. Vigorous agitation has been used in our laboratory as a convenient and clean procedure to synchronize zoospore encystment (Tokunaga & Bartnicki-Garcia, 1971a). In 1--2 min of vortexing, essentially all zoospores in a suspension form a recognizable cyst wall.

Loss of Flagella

As the zoospores come to rest, their flagella are either shed or withdrawn into the zoospore body. Although shedding is more common, both mechanisms of flagellar disappearance can be found in the same population of *Phytophthora parasitica* and the details are discussed by Reichle (1969b).

Discharge of Peripheral and Flattened Vesicles and Formation of a Cyst Coat

The numerous peripheral vesicles and other structures (Figs. 6, 26) lying in close proximity to the plasmalemma fuse with it (Fig. 27) and in so doing, the contents of the vesicles are extruded. Kinetic studies showed that the contents of the vesicles are discharged within 30 sec after encystment is triggered (Sing, 1974). Major changes in the surface of the cell ensue from this discharge. The material discharged by the peripheral structures forms an amorphous coat on the cyst surface (Grove, 1971; Hemmes & Hohl, 1971); this is not the cyst wall proper, though it was described as a wall in these publications. Conceivably, it may be a matrix that provides either physical support and/or enzymes for the subsequent elaboration of a microfibrillar cyst wall. The latter is formed inside and/or under the amorphous coat. The amorphous component of the cyst envelope is easily lost during wall isolation but can be seen in carbon replicas of native cysts as amorphous patches (Desjardins et al., 1973a; Hegnauer & Hohl, 1973). It can also be seen in thin sections of mature cysts of *P. palmivora*, outside the microfibrillar wall, as a "fluffy coat" (Hegnauer & Hohl, 1973).

Acquisition of Adhesiveness

One of the most important properties of a zoospore is its ability to adhere firmly to the solid surface upon which it settles down to encyst. Adhesion insures that the zoospore, having reached a favorable substratum by chemotaxis (Zentmyer, 1970; Hickman, 1970) would remain firmly adhered to it. Obviously this behavior has profound ecological value for both saprophytes and parasites. A study of adhesion of zoospores of *Phytophthora palmivora* revealed that the adhesive phase is extremely short. Swimming zoospores are not adhesive but become so in the initial stage of encystment, before a cyst wall is made (Sing & Bartnicki-Garcia, 1972). Once the cyst matures, the cells lose the ability to attach themselves to solid substrata. Most likely, the material discharged by the peripheral vesicles is the

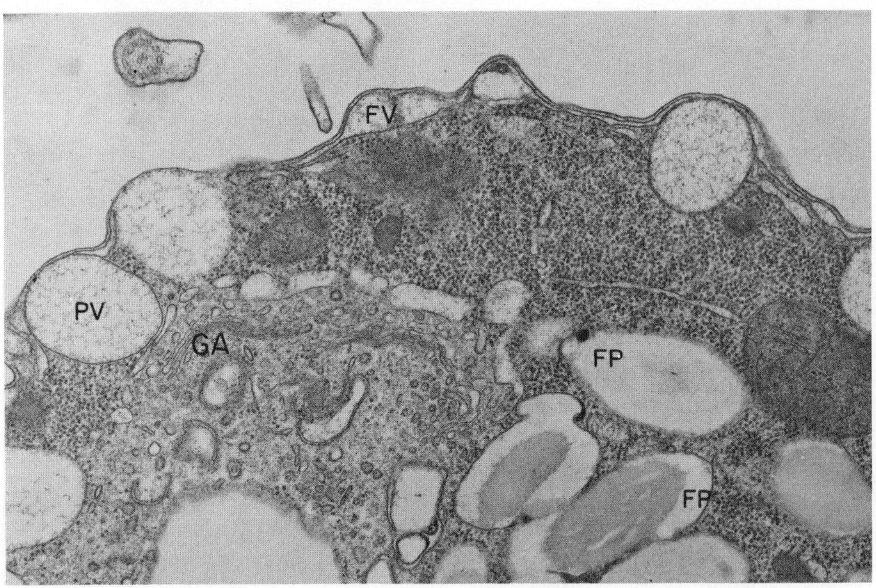

Fig. 26 Detail of the cell periphery of a zoospore of *Phytophthora palmivora*. Peripheral vesicles (PV); flattened vesicles (FV): fingerprint vacuoles (FP); and (GA) golgi apparatus. (Hemmes & Bartnicki-Garcia, previously unpublished). x 33,000.

cement that glues the zoospore to a surface. This can be deduced from Fig. 28, a fortuitous section of a zoospore that has just landed on a plastic surface. After encystment, the cell remains firmly adhered to the solid surface, be that plastic, glass, leaf, etc., and resists the shearing force of a strong water current. In mature cysts, the adhesive material appears as an electron dense substance located between the wall and the solid surface (Fig. 29) (Sing & Bartnicki-Garcia, 1975). The adhesive material appears to be a glycoprotein; it binds concanavalin A and is digested by trypsin; the digestion products contain mannose, glucose and galactose (Sing, 1974). We have been tempted to name the peripheral vesicle an "adhesosome;" yet, such specific designation must await assessment of another

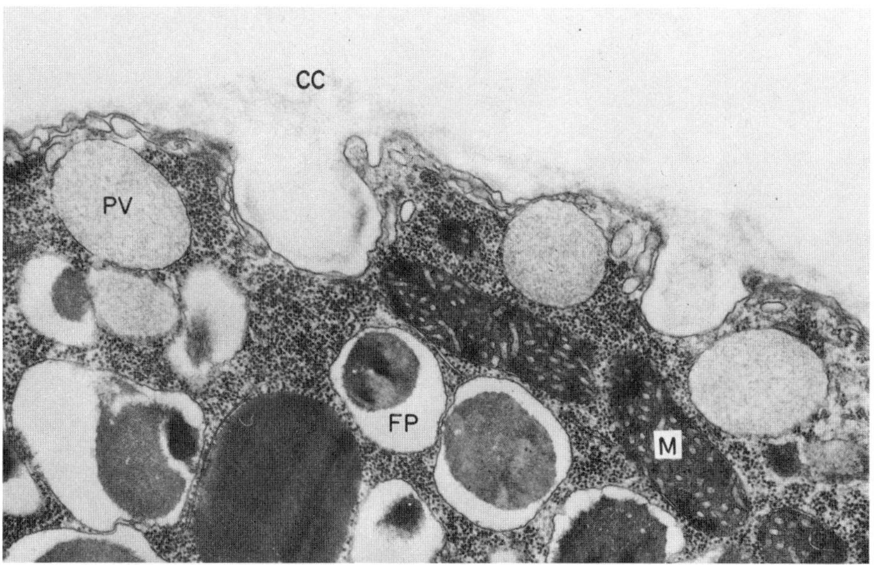

Fig. 27 Detail of the cell periphery of a zoospore of *Pythium aphanidermatum* at the onset of encystment. Note the discharge of peripheral vesicles (PV) with formation of a cyst coat (CC). Fingerprint vacuoles (FP); mitochondria (M) (x 24,420). (Courtesy of Grove, 1971.)

plausible and perhaps more basic function for these organelles, namely, cyst wall formation (Grove, 1971; Hemmes & Hohl, 1971). Conceivably, the primary function of the peripheral vesicle might be microfibril formation and/or binding the nascent microfibrils to the cell surface and to one another (i.e., serving as a matrix for wall

---

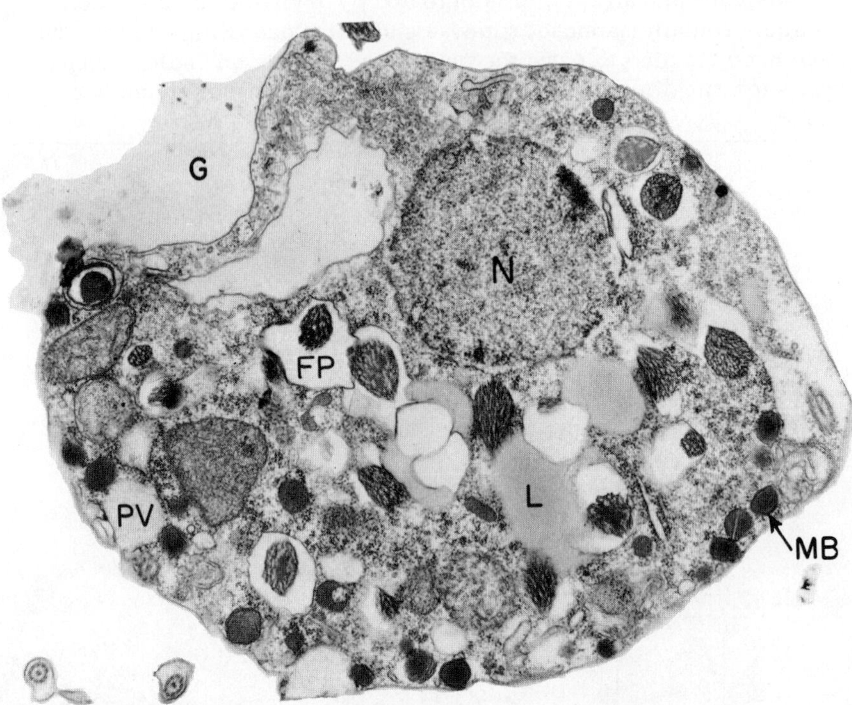

Fig. 28 Initial phase of adhesion of an encysting zoospore of *Phytophthora palmivora* to a plastic film surface. Note that the cell is anchored through the freshly discharged contents of a peripheral vesicle. Peripheral vesicles (PV); microbody-like structure (MB); finger-print vacuoles (FP); nucleus (N); lipidlike body (L); groove region (G). (x 13,440.) (Sing & Bartnicki-Garcia, 1975.)

neogenesis). This presumed interfibrillar cement would also serve, secondarily, to bind the cell to an external surface if the zoospore happened to be in contact with it at the time of encystment.

Cyst Wall Fibril Formation

There is no evidence for preformed cyst wall microfibrils in the zoospore cytoplasm (see Bartnicki-Garcia, 1973b). Chemical studies showed only the presence of a trace (0.2%) of wall glucans in the unencysted zoospore, of *Phytophthora palmivora* in contrast to the 3--4% in the cyst. Furthermore, it would seem highly unlikely that preassembled microfibrils, packed into vesicles, could upon secretion, organize themselves into a continuous, seamless, tightly knit fabric of long microfibrils. It is also unlikely, despite the rapidity of encystment, that the wall glucans would be formed simply by further polymerization of soluble mycolaminaran molecules. The wall contains more than half of its glucose residues joined by β-1,4-linkages and there is no evidence for these links in the zoospore cytoplasm; in other words, the cyst wall cellulose plus the noncellulosic glucan, which also contains some 1,4-linkages, must be synthesized de novo during encystment. Most likely, the mycolaminaran consumed during encystment (both phosphorylated and neutral) is, to a large extent, depolymerized into glucose units and then repolymerized into new β-glucans (exolaminaranase activity can be detected in the zoospore (Holten & Bartnicki-Garcia, 1972). There is, however, the possibility that some of the mycolaminaran may serve as primer for the formation of the noncellulosic glucan.

Microfibril elaboration probably starts immediately after the discharge of peripheral vesicles but it takes approximately 1 min to produce an alkali- or detergent-resistant cyst wall recognizable under the light microscope. In the electron microscope, such incipient wall appears as a loose fibrillar meshwork plastered with amorphous material (Fig. 30). Wall glucan formation continues with linear kinetics for about 10 min and then stops to give the microfibrillar envelope shown in Fig. 31.

Although the peripheral and flattened vesicles lying under the zoospore plasmalemma seem to be responsible for the formation of the amorphous (coat) and microfibrillar (wall) components of the cyst envelope, the participation of other cell structures in these events cannot be excluded; particularly, those which suffer major changes or disappear during encystment, namely, vesicles of the water

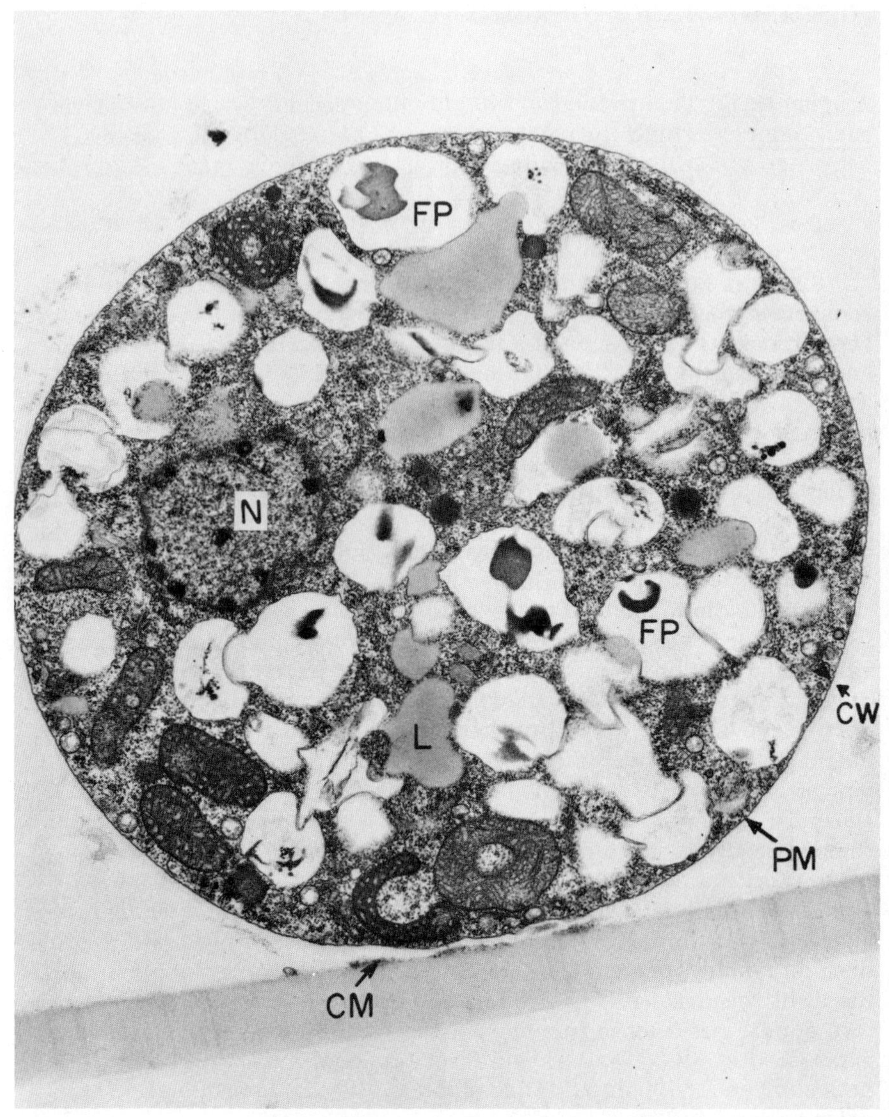

Fig. 29 Mature cyst of *Phytophthora palmivora* adhered to a plastic film surface. The electron-dense cementing material (CM) appears between the plastic film surface and the electron transparent cyst wall (CW). Finger-print vacuoles (FP); nucleus (N); lipidlike body

expulsion apparatus and the vesicles with an electron opaque cortex and core described by Hoch & Mitchell (1972b) for *Aphanomyces euteiches*. The latter may be extruded and become part of the cyst envelope as occurs with the "bars" in *Saprolegnia* (Heath & Greenwood, 1970). In the zoospores of *Phytophthora palmivora*, but rarely in the cysts, one can also find vesicles with an electron-dense core and cortex separated by a thin transparent space (Figs. 6,28); these have been described as microbodylike organelles (Hemmes & Hohl, 1971) and bear certain resemblance to the "parastrosomes" of *Phytophthora capsici* sporangia (Williams & Webster, 1970); their role in encystment, if any, remains to be determined.

Change of Shape

During encystment, the zoospore gradually loses its pyriform shape (Fig. 5) and becomes spherical. The rounding off process begins while the flagella are still active (Reichle, 1969b) but the deep groove on the zoospore persists as a depression even after a well defined cyst wall is made and does not disappear until the cyst is mature (Desjardins et al., 1973a). The shape of the zoospore is probably maintained by a microtubular skeleton which becomes progressively dismantled during encystment.

The cyst wall plays a morphogenetically passive role as it follows largely or entirely the contour of an existing cell. This is in contrast with the germ tube, or hypha, where a morphogenetically active role can be adduced for the cell wall (Bartnicki-Garcia, 1973a).

All of the preceding steps in the encystment of *Phytophthora palmivora* zoospores can take place in the presence of cycloheximide (Hemmes & Hohl, 1971; Sing & Bartnicki-Garcia, unpublished results). This indicates that the zoospore is preprogrammed with the complete battery of enzymes required for encystment. The next developmental step (cyst germination) is inhibited by cycloheximide indicating the need for new protein synthesis. The latter is necessary, among other things, to initiate the flow of dictyosome-derived vesicles to the cell surface (Hemmes & Hohl, 1971; Grove, 1971).

Functionally, the encystment of the biflagellate oomycetous zoospore and the uniflagellate chytridiomycetous zoospore (see chapter

---

(L); plasmalemma (PM). (x 14,880.) (Sing & Bartnicki-Garcia, 1975).

Fig. 30 Incipient cyst wall of *Phytophthora palmivora* isolated by extraction of whole cells with 0.5% sodium dodecyl sulfate. Note amorphous material associated with the developing microfibrillar wall (x 36,000). (Sing & Bartnicki-Garcia, previously unpublished.)

12) are similar. Both zoospores have a preformed complement of encystment enzymes, they synthesize a wall de novo from internal reserves, and they become adhesive prior to or during encystment. Yet, morphologically, the overall internal appearance and fine structure of organelles presumably involved in encystment are considerably different. Conspicuously, there are no γ-particles ("encystomes") in the oomycetous zoospore; conversely, there are no "peripheral" vesicles in uniflagellate zoospores (see chapter 12). A disparity in the presumed subcellular basis of encystment of Chytridiomycetes and Oomycetes need not be surprising. These two classes of fungi differ greatly in various basic properties, including cell wall chemistry, and probably belong to two entirely separate phylogenetic lines (see Bartnicki-Garcia, 1970). By parallel evolution, they may have developed superficially similar solutions to the problem of wall neogenesis that are different, however, in biochemical and cytological details.

ACKNOWLEDGMENTS

Experimental work co-authored by S.B.-G. which was published after 1972 or was hitherto unpublished, was supported in part by an NSF research grant (GB-32165). The experimental work on chlamydospores (Hemmes & Wong) included herewith was supported by NIH Minority Schools Biomedical support grant No. 5 S06 RR-08073.

REFERENCES

Aronson, J. M., Cooper, B. A., & Fuller, M. S. (1967). Glucans of oomycete cell walls. Science 155, 332--335.

Barash, I., Klisiewicz, J. M., & Kosuge, T. (1965). Utilization of carbon compounds by zoospores of *Phytophthora drechsleri* and their effect on motility and germination. Phytopathology 55, 1257--1261.

Bartnicki-Garcia, S. (1966). Chemistry of hyphal walls of *Phytophthora*. J. Gen. Microbiol. 42, 57--69.

Fig. 31 Fragment of a mature cyst wall of *Phytophthora palmivora* isolated by sonication. (x 32,400) (Hemmes & Bartnicki-Garcia previously unpublished.)

Bartnicki-Garcia, S. (1968). Cell wall chemistry, morphogenesis, and taxonomy of fungi. Annu. Rev. Microbiol. 22, 87--108.

Bartnicki-Garcia, S. (1970). Cell wall composition and other biochemical markers in fungal phylogeny. In Phytochemical Phylogeny (ed. J. B. Harborne), pp. 81--103. London: Academic Press.

Bartnicki-Garcia, S. (1973a). Fundamental aspects of hyphal morphogenesis. In Microbial Differentiation Vol. 23, Symp. Soc. Gen. Microbiol. (ed. J. M. Ashworth & J. E. Smith), pp. 245--267. Cambridge, England: Cambridge U. P.

Bartnicki-Garcia, S. (1973b). Cell wall genesis in a natural protoplast: The zoospore of *Phytophthora palmivora*. In Yeast, Mould and Plant Protoplasts (ed. J. R. Villanueva, I. Garcia-Acha, S. Gascon, & F. Uruburu), pp. 77--91. London: Academic Press.

Blackwell, E. (1949). Terminology in *Phytophthora*. Mycological Papers No. 30. Kew, Surrey: The Commonwealth Mycological Institute.

Bimpong, C. E. & Clerk, G. C. (1970). Motility and chemotaxis in zoospores of *Phytophthora palmivora* (Butl.) Butl. Ann. Bot. 34, 617--624.

Christen, J. & Hohl, H. R. (1972). Growth and ultrastructural differentiation of sporangia in *Phytophthora palmivora*. Can. J. Microbiol. 18, 1959--1964.

Colhoun, J. (1966). The biflagellate zoospore of aquatic Phycomycetes with particular reference to *Phytophthora* spp. The Fungus Spore (ed. M. F. Madelin). pp. 85--94 Butterworths.

Darley, W. M., Porter, D., & Fuller, M. S. (1973). Cell wall composition and synthesis via Golgi-directed scale information in the marine eucaryote, *Schizochytrium aggregatum* with a note on *Thraustochytrium* sp. Arch. Mikrobiol. 90, 89--106.

Dastur, J. F. (1913). On *Phytophthora parasitica* sp. nov. A disease of the castor oil plant. Mem. Dep. Agric. India, 5, 177--231.

Desjardins, P. R., Wang, M. C., & Bartnicki-Garcia, S. (1973a). Electron microscopy of zoospores and cysts of *Phytophthora palmivora*: Morphology and surface texture. Arch. Mikrobiol. 88, 61--70.

Desjardins, P. R., Zentmyer, G. A., Chen, D., DeWolfe, T. A., Klotz, L. J., & Reynolds, D. A. (1973b). Flagellar hairs on zoospores of *Phytophthora* Species: Tip hairs on the whiplash flagellum. Experientia 29, 240--241.

Desjardins, P. R., Zentmyer, G. A., & Reynolds, D. (1969). Electron microscopic observations of the flagellar hairs of *Phytophthora palmivora* zoospores. Can. J. Bot. 47, 1077--1079.

Dick, M. W. & Win-Tin. (1973). The development of cytological theory in the Oomycetes. Biol. Rev. 48, 133--158.

Ellzey, J. T. (1974). Ultrastructural observations of meiosis within antheridia of *Achlya ambisexualis*. Mycologia 66, 32--47.

Elsner, P. R., VanderMolen, G. E., Horton, J. C., & Bowen, C. C. (1970). Fine structure of *Phytophthora infestans* during sporangial differentiation and germination. Phytopathology 60, 1765--1772.

Faro, S. (1972a). A soluble beta-1,3-glucan found in selected genera of Oomycetes. J. Gen. Microbiol. 72, 393--394.

Faro, S. (1972b). The role of a cytoplasmic glucan during morphogenesis of sex organs in *Achlya*. Am. J. Bot. 59, 919--923.

Gay, J. L. & Greenwood, A. D. (1966). Structural aspects of zoospore production in *Saprolegnia ferax* with particular reference to the cell and vacuolar membranes. In The Fungus Spore (ed. M. F. Madelin), pp. 95--110. London: Butterworths.

Gay, J. L., Greenwood, A. D., & Heath, I. B. (1971). The formation and behavior of vacuoles (vesicles) during oosphere development and zoospore germination in *Saprolegnia*. J. Gen. Microbiol. 65, 233--241.

Goldstein, S. (1973). Zoosporic marine fungi (Thraustochytriaceae and Dermocystidiaceae). Annu. Rev. Microbiol. 27, 13--26.

Grove, S. N. (1971). Protoplasmic correlates of hyphal tip initiation and development in fungi. Ph. D. thesis, Purdue University.

Grove, S. N. & Bracker, C. E. (1970). Protoplasmic organization of hyphal tips among fungi: vesicles and Sptizenkorper. J. Bacteriol. 104, 989--1009.

Grove, S. N., Bracker, C. E., & Morré, D. J. (1970). An ultrastructural basis for hyphal tip growth in *Pythium ultimum*. Am. J. Bot. 57, 245--266.

Heath, I. B., Gay, J. L., & Greenwood, A. D. (1971). Cell wall formation in the Saprolegniales: Cytoplasmic vesicles underlying developing walls. J. Gen. Microbiol. 65, 225--232.

Heath, I. B. & Greenwood, A. D. (1970). Wall formation in the Saprolegniales. II. Formation of cysts by the zoospores of *Saprolegnia* and *Dictyuchus*. Arch. Mikrobiol. 75, 67--79.

Heath, I. B. & Greenwood, A. D. (1971). Ultrastructural observations of the kinetosomes, and Golgi bodies during the asexual life cycle of *Saprolegnia*. Z. Zellforsch. 112, 371--389.

Hegnauer, H. & Hohl, H. R. (1973). A Structural comparison of cyst and germ tube wall in *Phytophthora palmivora*. Protoplasma 77, 151--163.

Heintz, C. E. (1971). An ultrastructural analysis of zoosporogenesis in *Pythium middletonii*. Abstra. First Intern. Mycol. Congr. p. 64.

Hemmes, D. E. & Bartnicki-Garcia, S. (1975). Electron microscopy of gametangial interaction and oospore development in *Phytophthora capsici* Arch. Microbiol. 103, 91--112.

Hemmes, D. E. & Hohl, H. R. (1969). Ultrastructural changes in directly germinating sporangia of *Phytophthora parasitica*. Am. J. Bot. 56, 300--313.

Hemmes, D. E. & Hohl, H. R. (1971). Ultrastructural aspects of encystation and cyst-germination in *Phytophthora parasitica*. J. Cell Sci. 9, 175--191.

Hemmes, D. E. & Hohl, H. R. (1972). Flagellum degeneration in the fungus *Phytophthora palmivora* (formerly *Phytophthora parasitica*). J. Gen. Microbiol. 73, 345--351.

Hemmes, D. E. & Hohl, H. R. (1973). Mitosis and nuclear degeneration: simultaneous events during secondary sporangia formation in *Phytophthora palmivora*. Can. J. Bot. 51, 1673--1675.

Hickman, C. J. (1970). Biology of *Phytophthora* zoospores. Phytopathology 60, 1128--1135.

Ho, H. H. & Hickman, C. J. (1967). Asexual reproduction and behavior of zoospores of *Phytophthora megasperma* var. *sojae*. Can. J. Bot. 45, 1963--1981.

Ho, H. H., Hickman, C. J., & Telford, R. W. (1968a). The morphology of zoospores of *Phytophthora megasperma* var. *sojae* and other Phycomycetes. Can. J. Bot 46, 88--89.

Ho, H. H., Zachariah, K., & Hickman, C. J. (1968b). The ultrastructure of zoospores of *Phytophthora megasperma* var. *sojae*. Can. J. Bot. 46, 37--41.

Hoch, H. C. & Mitchell, J. E. (1972a). The ultrastructure of *Aphanomyces euteiches* during asexual spore formation. Phytopathology, 62, 149--160.

Hoch, H. C. & Mitchell, J. E. (1972b). The ultrastructure of zoospores of *Aphanomyces euteiches* and of their encystment and subsequent germination. Protoplasma 75, 113--138.

Hohl, H. R. & Hamamoto, S. T. (1967). Ultrastructural changes during zoospore formation in *Phytophthora parasitica*. Am. J. Bot. 54, 1131--1139.

Holten, V. Z. & Bartnicki-Garcia, S. (1972). Intracellular β-glucanase activity of *Phytophthora palmivora*. Biochim. Biophys. Acta 276, 221--227.

Howard, K. L. (1971). Oospore types in the Saprolegniaceae. Mycologia 63, 679--686.

Howard, K. L. & Moore, R. T. (1970). Ultrastructure of oogenesis in *Saprolegnia terrestris*. Bot. Gaz. 131, 311--336.

Hunsley, D. & Burnett, J. H. (1970). The ultrastructural architecture of the walls of some hyphal fungi. J. Gen. Microbiol. 62, 203--218.

King, J. E. & Butler, R. D. (1968). Structure and development of flagella of *Phytophthora infestans*. Trans. Brit. Mycol. Soc . 51, 689--697.

King, J. E., Colhoun, J., & Butler, R. D. (1968). Changes in the ultrastructure of sporangia of *Phytophthora infestans* associated with indirect germination and ageing. Trans. Brit. Mycol. Soc. 51, 269--281.

Lippman, E., Erwin, D. C., & Bartnicki-Garcia, S. (1974). Isolation and chemical composition of oospore-oogonium walls of *Phytophthora megasperma* var. *sojae*. J. Gen. Microbiol. 80, 131--141.

Manton, I., Clarke, B., & Greenwood, A. D. (1951). Observations with the electron microscope on a species of *Saprolegnia*. J. Exp. Bot. 2, 321--331.

Marchant, R. (1968). An ultrastructural study of sexual reproduction in *Pythium ultimum*. New Phytol. 67, 167--171.

Menyonga, J. M. & Tsao, P. M. (1966). Production of zoospore suspensions of *Phytophthora parasitica*. Phytopathology 56, 359--360.

Mullins, J. T. & Ellis, E. A. (1974). Sexual morphogenesis in *Achlya*: ultrastructural basis for the hormonal induction of antheridial hyphae. Proc. Nat. Acad. Sci. U.S.A. 71, 1347--1350.

Murphy, P. A. (1918). The morphology and cytology of the sexual organs of *Phytophthora erythroseptica* Peth. Ann. Bot. (London) 32, 115--153.

Novaes--Ledieu, M. & Jimenez-Martinez, A. (1968). The structure of cell walls of Phycomycetes. J. Gen. Microbiol. 54, 407--415.

Pao, V. M. & Aronson, J. M. (1970). Cell wall structure of *Sapromyces elongatus*. Mycologia 62, 531--541.

Pethybridge, G. H. (1913). On the rotting of potato tubers by a new species of *Phytophthora* having a method of sexual reproduction hitherto undescribed. Sci. Proc. R. Dublin Soc. 13, 463--468.

Reichle, R. E. (1969a). Fine structure of *Phytophthora parasitica* zoospores. Mycologia 61, 30--51.

Reichle, R. E. (1969b). Retraction of flagella by *Phytophthora parasitica* var. *nicotiana* zoospores. Arch. Mikrobiol. 66, 340--347.

Sietsma, J. H., Eveleigh, D. E., & Haskins, R. H. (1969). Cell wall composition and protoplast formation of some oomycete species. Biochim. Biophys. Acta. 184, 306--317.

Sing, V. O. (1974). Cytological and biochemical basis of cellular adhesion in zoospores of *Phytophthora palmivora*. Ph.D. Thesis, Univ. of California, Riverside.

Sing, V. O. & Bartnicki-Garcia, S. (1972). Adhesion of zoospores of *Phytophthora palmivora* to solid surfaces. Phytopathology 62, 790.

Sing, V. O. & Bartnicki-Garcia, S. (1975). Adhesion of zoospores of *Phytophthora palmivora*: electron microscopy of cell attachment and cyst wall fibril formation. J. Cell. Sci. 18, 123--132.

Sparrow, F. K. (1943). The Aquatic Phycomycetes, Exclusive of the Saprolegniaceae and Pythium. Ann Arbor: Univ. Michigan Press.

Stephenson, L. W. & Erwin, D. C. (1972). Encirclement of the oogonial stalk by the amphigynous antheridium in *Phytophthora capsici*. Can. J. Bot. 50, 2439--2441.

Thomas, D. Des S. & Mullins, J. T. (1967). Role of enzymatic wall-softening in plant morphogenesis: Hormone induction of Achlya. Science 156, 84--85.

Tokunaga, J. & Bartnicki-Garcia, S. (1971a). Cyst wall formation and endogenous carbohydrate utilization during synchronous encystment of *Phytophthora palmivora* zoospores. Arch. Mikrobiol. 79, 283--292.

Tokunaga, J. & Bartnicki-Garcia, S. (1971b). Structure and differentiation of the cell wall of *Phytophthora palmivora*: Cysts, hyphae and sporangia. Arch. Mikrobiol. 79, 293--310.

Vujičić, R. (1971). An ultrastructural study of sexual reproduction in *Phytophthora palmivora*. Trans. Brit. Mycol. Soc. 57, 525--530.

Vujičić, R. & Colhoun, J. (1966). Asexual reproduction in *Phytophthora erythroseptica*. Trans. Br. Mycol. Soc. 49, 245--254.

Vujičić, R., Colhoun, J., & Chapman, J. A. (1968). Some observations on the zoospores of *Phytophthora erythroseptica*. Trans. Br. Mycol. Soc. 51, 125--127.

Wang, M. C. & Bartnicki-Garcia, S. (1973). Novel phosphoglucans from the cytoplasm of *Phytophthora palmivora* and their selective occurrence in certain life cycle stages. J. Biol. Chem. 248, 4112--4118.

Wang, M. C. & Bartnicki-Garcia, S. (1974). Mycolaminarans: storage β-1,3-D-glucans from the cytoplasm of the fungus *Phytophthora palmivora*. Carbohyd. Res. 37, 331--338.

Waterhouse, G. M. (1962). The zoospore. Trans. Br. Mycol. Soc. 45, 1--20.

Williams, W. T. & Webster, R. K. (1970). Electron microscopy of the sporangium of *Phytophthora capsici*. Can. J. Bot. 48, 221--227.

Wood, F. A., Singh, R. P., & Hodgson, W. A. (1971). Characterization of a virus-inhibiting polysaccharide from *Phytophthora infestans*. Phytopathology 61, 1006--1009.

Zentmyer, G. A. (1970). Tactic responses of zoospores of *Phytophthora*. In Root Diseases and Soil-borne Pathogens (ed. Toussoun, T. A., Bega, R. V., & Nelson, P. E.), pp. 109--111. Berkeley: Univ. of California Press.

Zentmyer, G. A. & Erwin, D. C. (1970). Development and reproduction of *Phytophthora*. Phytopathology 60, 1120--1127.

Zevenhuizen, L. P. T. M. & Bartnicki-Garcia, S. (1969). Chemical structure of the insoluble hyphal wall glucan of *Phytophthora cinnamomi*. Biochemistry 8, 1496--1502.

Zevenhuizen, L. P. T. M. & Bartnicki-Garcia, S. (1970). Structure and role of a soluble cytoplasmic glucan from *Phytophthora cinnamomi*. J. Gen. Microbiol. 61, 183--188.

DISCUSSION

Form and Function in Oomycetes

Chairman: G. L. Mills
Michigan State University
East Lansing, Michigan

The question was asked if Oomycete zoospores would encyst in the presence of a protein synthesis inhibitor as do the zoospores of *Blastocladiella*, and also whether or not they would encyst in the presence of chloramphenicol. Dr. Bartnicki-Garcia replied that they provided a parallel situation since both types of zoospores could encyst without protein synthesis, that is, in the presence of cycloheximide, but he was unsure about the effects of chloramphenicol.

He was also asked if encystment was susceptible to arrest by treatment with cyanide or azide. He did not know if encystment could be arrested by the use of metabolic poisons but answered that some respiration must occur since part of the glucan is degraded into smaller noncarbohydrate fragments.

In reply to the question about whether or not the mycolaminaran is incorporated into the cell wall formed by the oogonium or if it becomes part of the wall after some chemical transformation, Dr. Bartnicki-Garcia stated that the material was probably first hydrolyzed to glucose then rebuilt into the wall glucans. He also indicated that there was no evidence of incorporation of preformed long chains into the glucans although some mycolaminaran may serve as a primer molecule.

The comment was made that the material of the fingerprint vacuoles is highly reactive to a number of different treatments and could possibly be a phosphorylated glucan. Dr. Bartnicki-Garcia responded that the material of the fingerprint vacuoles must have some crystalline properties and speculated that the fingerprint vacuoles might be the structures, since they are so ubiquitous and abundant in the cell, which store the mycolaminaran that accounts for up to 50% of the dry weight of the cytoplasm.

In answer to the question whether the fungus itself could dissolve the noncellulosic $\beta 1 \rightarrow 3, 1 \rightarrow 6$ polysaccharide of the cell wall, Dr. Bartnicki-Garcia replied no. He thought that once the walls were formed the fungus no longer had access to the glucan components (except for the inner oospore wall).

The question was asked of Dr. Hemmes if microfilaments or microtubules were ever found in the cytoplasm of the cells with the nuclear inclusions. Dr. Hemmes stated that there are very few microtubules in the cytoplasm. He also added that the nuclear inclusions were found only after fertilization had taken place, then all of the nuclei have them.

The comment was made that something unusual occurs at meiotic prophase in all of the Phycomycetes so far investigated, since it is unusually difficult to demonstrate synaptonemal complexes ultrastructurally. Dr. Hemmes replied that there is some controversy as to where meiosis takes place. He stated that meiotic figures can be seen in the oogonium but he has seen nothing other than degenerating nuclei in the antheridia.

Dr. Bartnicki-Garcia was asked if *Phytophthora* uses a cellulase to get into plants. He replied that it probably did but he did not know the mechanism that allowed the cellulase to pass through the cellulosic cell wall of the fungus without destroying it.

CHAPTER 15 FORM AND FUNCTION IN BASIDIONYCETE SPORES

W. M. Hess and D. J. Weber
Brigham Young University

INTRODUCTION

The first half of this chapter is concerned primarily with the biochemistry of dormant and germinating basidiomycete spores, particularly carbohydrate, protein, lipid, and nucleic acid metabolism, respiration, inhibitors and stimulators of germination and physical factors. The second half of this chapter is concerned with ultrastructure of basidiomycete spores. In the second half of the chapter ultrastructural details of developing, dormant and germinated *Tilletia caries* and *T. controversa* teliospores are also discussed in order to demonstrate the kinds of information which may be obtained with electron microscopy of one kind of spore.

BIOCHEMISTRY OF BASIDIOMYCETE SPORES

The basidiomycetes represent the largest and most highly evolved class of fungi with approximately 15,590 species. Basidiomycetes are also a major food source as is indicated by the yearly consumption of 225 million pounds of mushrooms in the United States. On the other hand, many basidiomycetes, such as rusts and smuts, are major plant pathogens. Complex life cycles with many different types of spores are common in basidiomycetes. The spore types include basidiospores, pycniospores, aeciospores, uredospores, teliospores, conidia, and chlamydospores. These spores have not all been investigated equally and probably 80% of the biochemical research has been done with uredospores or teliospores. Only a very limited number of the 15,590 species have been investigated extensively by biochemical methods. Only four species of Uredinales, two species of Ustilaginales, and one specie of Hymenomycetes have been investigated extensively.

Since the first fungal spore symposium in 1965, considerable research has been conducted on molecular biology of fungi (Table 1). Biochemical research has been predominately with Uredinales (rusts). Hymenomycetes and Ustilaginales (smuts) have also received significant attention, but very little research has been conducted with the Gastromycetes (Table 1).

Table 1. Distribution of Physiological Research Papers[a] on Basidiomycete Spores

	Ured.	Ustil.	Hymen.	Gast.
Carbohydrate metabolism	32	3	13	---
Protein metabolism	29	---	2	---
Lipid metabolism	24	6	9	---
Nucleic acid metabolism	27	1	8	---
Respiration	11	---	5	2
Inhibition & stimulation	41	9	14	1
Physical factors	35	5	4	2
	199	24	55	5

[a]From Abstracts of Mycology 1965 to 1973.

General reviews of spore germination which include physiological and morphological aspects of basidiomycetes have been written by Shaw (1964), Sussman (1965), Staples & Wynn (1965), Allen (1965), Madelin (1966), Burnett (1968), and Merrill (1970).

Teliomycetes

Uredinales (rusts)

The approximately 400 species of rusts have been a favorite subject for biochemical investigations of spores, possibly because of their economic significance in cereal diseases and because the uredospores can be collected in sufficient quantities for biochemical analyses. The life cycles of many rusts contain as many as five morphologically distinct types of spores: pycniospores, aeciospores, uredospores, teliospores, and basidiospores. Almost all of the biochemical investigations have been conducted with uredospores or teliospores.

With the development of axenic cultures of rusts (Williams, 1971; Williams et al., 1966; Raghunath, 1970; Kuhl et al., 1971; Hollis, 1973) it is now possible to obtain spores from rust cultures. However, the major source of uredospores for biochemical research is still from field or greenhouse sources.

Storage of rust spores

The proper storage of quantities of spores for biochemical research requires special attention. When uredospores of *Puccinia graminus tritici* were mixed and stored with different types of soil, spore germination ceased in 62 days (Orr & Tippetts, 1971). Ozone and even atmospheric ions appear to also have a detrimental effect on storage and germination of rust spores (Hibben & Stotzky, 1969; Sharp, 1967). However, the overwintering of uredospores of *Pucciniastrum agrimoniae* did not affect their viability (Ouellette, 1969). In fact, low temperatures appear to be the best storage conditions for *P. penniseti* (5--8°C) (Misra & Prasada, 1971) and for *P. graminis tritici* (4°C) (Wynn et al., 1966).

Several species of rust spores have been stored successfully after a freeze-dry treatment (Roncadori & Matthew, 1966), but the storage of rust uredospores in liquid nitrogen appears to be the most promising method as spores are still viable after ten years of storage (Loegering et al., 1966; Kilpatrick et al., 1971; Prescott & Kernkamp, 1972; Heagle & Moore, 1970). On the other hand the viablility of rust pycniospores was lost after about one year in liquid nitrogen. However, this viability is still higher than for other storage methods (Bugbee & Kernkamp, 1966).

As the uredospores aged, the rate of germination decreased and ability to assimilate exogenous substrates decreased (Wynn et al., 1966). However, this parameter was not affected by storage temperature. The levels of glutamic acid, malonic acid, and pyrrolidone carboxylic acid decreased during storage, and the capacity for assimilation of glucose and acetate decreased. Of the substrates assimilated, a higher percentage of the components were incorporated into amino acids and organic acids than into polysaccharides, lipids, and proteins.

Powell (1971) collected aeciospores from *Cronartium comandrae* as the pustules matured and found that germination was high for the initial 2--4 weeks after collection, but was lower several weeks later, which indicates that the age of spores is an important factor in germination. The highest germination percentage of uredospores of *P. gramis* occurred when they were harvested directly onto a nutrient medium (Melching et al., 1969). It is obvious that harvest and storage conditions need be taken into account in physiological studies of spores.

Effect of light on the germination of spores

Day length and light intensity had little effect on the germination of fresh spores of *Puccinia helianthi*, but higher light intensity had an adverse effect on the germination of spores that had been stored for two months at -16°C (Sood & Sackston, 1971 and 1972).

Hydration of the uredospores of *Puccinia recondita* during the germination process increased the light sensitivity. Uredospores of *P. recondita* that were produced in darkness were more sensitive to light (7600 lux) than spores that were produced in the light (Zadoks & Groenewegen, 1967).

Uredospores of *Puccinia cynodontis* germinated well over a range of temperatures from 1.5 to 35°C, but inhibition of germination occurred when the light intensity exceeded 200 foot candles (Vargas et al., 1967).

Germination of uredospores of *Puccinia graminis tritici* was reduced by near ultraviolet and far-red light (Maddison & Manners, 1973; Calpouzos & Chang, 1971). Maddison & Manners (1973) proposed that the nucleic acids and proteins of spores were affected by high light intensity.

The physiological bases for the effect of light on spores remains to be answered.

Effect of Temperature on the Germination of Spores

The time required for the release of basidiospores of *Cronartium fusiforme* was adversely related to temperature and the rate of release was directly related to the temperature of incubation (Snow, 1968). The germination rate of uredospores of *Puccinia helianthi* stored at -16°C varied in three different races. However, there was no effect upon germination when the spores were obtained fresh (Sood & Sackston, 1971, 1972). Tollenaar & Houston (1966a and 1966b) found that temperature during uredospore production of *Puccinia striiformis* governed the maximum germination temperature. The maximum temperature and the minimum temperature of germination were obtained when spores were formed at 24.5 and 0.5^{o}C.

Powell & Morf (1966) determined the pH and temperature maximum for a number of rusts. Uredospores of *P. recondita* remained viable but ungerminated on dry wheat foliage for 45 days at 5--9°C, indicating that the deposit of the spore on the leaf and infection are not necessarily simultaneous. Eversmeyer & Burleigh (1968) and Mederick & Sackston (1972) observed that the minimal temperature for germination was between 2--5°C, the optimum was 10--25°C, and the maximal was 30--35°C for *Puccinia sorghi*. The percent germination of rust spores on water agar varied with the temperature. The optimum temperature was between 15--30°C (Kushalappa & Hedge,

1971). Low temperature depressed uredospore germination of *P. graminis*. However, spores that had been in cold dormancy were able to germinate when activated at 40°C (Topolovskii, 1966).

When air-dried uredospores were exposed to temperatures below freezing, the germinability was reduced even after bringing the spores back to room temperature for a period of time. Germination was fully restored by heat shock. However, this cold dormancy could not be restored once the spores contacted water. The cold shock apparently transformed the spores into a super sensitive state in relation to water. The primary cause of damage to cold dormant spores when exposed to water appears to be irreversible permeability damage followed by metabolic injury (Maheshwari & Sussman, 1971).

While the assumption is that the major effect of temperature is on the enzymatic processes, the detailed evidence is yet to be obtained for most fungal spores.

Self-Inhibitors

Self-inhibitors in rust spores have the obvious advantage that they prevent germination of the spores in the sorus. Inhibitors also prevent rapid germination of all of the spores at the same time, which in turn increases the chances of survival of the rust over fluctuating environmental conditions. Inhibitors also favor a wider distribution of the spores which tends to increase the survival rate for rusts.

Several aromatic compounds have been reported to inhibit germination of uredospores. The most inhibitory compounds were coniferyl alcohol and ferulic acid (Chigrin et al., 1969 & 1970; Bahadur & Sinha, 1966). Phloroglucinol, a type of phenolic compound, has been reported as a self-inhibitor in *Puccinia sorghi* (Roy, 1968). French and Gallimore (1972a & b) treated uredospores of *Puccinia graminis* with high concentrations of nonanal, which normally stimulates germination of spores in water, and observed reduced spore germination. Kapooria (1970 & 1971) observed the inhibition of germination of uredospores of *Puccinia penniseti* with different fungal exudates. Gibberellic acid and canavanine were inhibitory to germination of spores of *Puccinia coronata* (Natio et al., 1972).

Allen et al. (1971) purified a self-inhibitor from *P. graminis tritici*. The inhibitor was identified as methyl-4-hydroxy-3-methoxycinnamate (methyl ferulate) (Macko et al., 1971). Another self-inhibitor, methyl-3,4-dimethoxycinnamate was isolated from uredospores of *Uromyces phaseoli, Puccinia helianthi, P. antirrhini, P. arachidis*, and *P. sorghi* (Macko et al., 1970; 1971; 1972; Roy, 1968; Foudin &

Macko, 1974). Allen (1972) reported that the *cis* not the *trans* isomer of methyl ferulate and methyl-3,4-dimethoxycinnamate inhibited the germination of the uredospores of wheat rust. One hundred molecules of the nonyl alcohol are required to overcome the inhibition of the methyl-*cis*-ferulate and restore germination to 50% (Macko et al., 1972). Methyl-*cis*-ferulate appears to block the digestion of the germ tube pore, which prevents germination (Hess et al., 1974). These results reflect the excellent progress being made on the physiology of self-inhibitors.

Effects of Surface Films on the Germination of Spores

Woodbury & Stahmann (1970) concluded that an oxidative reaction occurred in water films on uredospores of *Puccinia graminis tritici* and that the reaction was accelerated by light. The oxidized products that were formed in the film of water inhibited uredospore germination. When spores were germinated in large volumes of water, the percent of germination varied with the surface area. The inhibitor formed was apparently retained at the spore surface. Therefore, hydration of spores appears to result in increased germination, and an increased film area. Storage of spores at low humidity resulted in decreased germination, decreased film area, and a delay in the prevention of film reactions. Nair (1971) observed that under thin films of water and water temperature of 20--24°C, teliospores of *Ravenelia hobsoni* germinated at the end of four hours and germination was completed in 8 h. If the thickness of the film of water was changed, then abnormalities in spore germination occurred. The swelling of crown rust uredospores was associated with viability (Jones & Haburn, 1969). These results suggest that the effect of light in a storage condition may be mediated by surface film material.

Stimulators of Rust Uredospore Germination

Babayan & Khachatryan (1972) germinated uredospores of the wheat rust in several extracts of plants. The spores germinated only in extracts from apricot leaves. They concluded that the main factor for germination of uredospores was low humidity. A significant self-stimulation for germination of uredospores of *Puccinia striiformis* was obtained from leaves with viable spores collected in the field (up to 5,400 spores per cm) (Tollenaar & Houston, 1966b). The teliospores of *Puccinia carthami* germinated poorly or not at all on water agar medium, but germination on water agar was stimulated by supplements from leaves derived from safflower (Klisiewicz, 1973).

Swendsrud & Calpouzos (1970) reported that extracts from uredospores of *Puccinia recondita* stimulated the germination of pycnidiospores of *Darluca filum*. Wiese & Daly (1967) used nonyl alcohol to stimulate uredospores in order to investigate metabolic processes accompanying germination. They concluded that the germination process depends upon factors other than self-inhibitors or stimulators. Klisiewicz (1973) exposed the teliospores of *P. carthami* to volatile substances from safflower and observed an increase in spore germination over the nonexposed spores. The highest germination (90%) was obtained when the spores were exposed to the volatile substances from the safflower seeds. Among several non-host plants, only one volatile substance was effective in stimulating germination.

French & Gallimore (1972a & b) observed that compounds ranging from saturated and unsaturated linear hydrocarbons to esters with nitrogen and sulfur derivatives stimulated spore germination of *Puccinia graminis*. The stimulators ranged from water insoluble to water soluble compounds and included volatile and nonvolatile compounds. Nonanol was stimulatory at 0.04 parts per million (French & Gallimore, 1971). Nonanol and nonanal were able to stimulate germination of the uredospores even in the rust pustules (French & Gallimore, 1972a). Cunningham (1966) investigated the germination of teliospores of *Frommea obtusa* and found that they germinated readily on water agar as soon as they were mature, but that they lost this capacity in about two months after the spores had remained dry at room temperature. The longer the teliospores were dried, the longer it took for them to germinate.

Farkas & Ledingham (1959) suggested that self-inhibited uredospores have primarily respiration of fatty acid oxidation and the enzymes of carbohydrates are more or less inactive. Substances that overcame self-inhibition stimulated oxygen consumption, intensified the utilization in the endogenous fatty acids, and stimulated carbohydrate metabolism.

The present need is not to find more stimulators of spore germination, but rather to determine the nature and mode of action of the stimulators.

Carbohydrate Metabolism in Rust Spores

The association of carbohydrate breakdown with energy

production indicates that carbohydrate metabolism is a significant area for investigation to help elucidate spore germination. While many investigations have been conducted which have been concerned with the utilization of sugars by uredospores, the basic enzymological evidence about carbohydrate pathways is still meager.

The oxygen consumption of uredospores of *Puccinia corroni* and *Puccinia graminis tritici* increased as the time of incubation increased, while the addition of the exogenous substrates had little or no effect (Naito & Tani, 1967; White & Ledingham, 1961). The respiration of rust spores did not show any significant change when glucose was added to the incubation media (Caltrider et al., 1963).

The endogenous respiration of germinating uredospores of *P. graminis tritici* may be determined by two methods. When determined manometrically, the values were about 0.5 to 0.7 times greater than when they were measured polarographically (Williams & Allen, 1967; Bush, 1968). However, Maheshwari & Sussman (1970) found that manometric measurements of respiration rates of dormant uredospores of *P. graminis tritici* were higher than polarimetric measurements, but as germination began, the situation was reversed. By use of the oxygen electrode they divided germination into four phases. Phase I was the development of respiratory activities when the uredospores came in contact with the liquid, Phase II was a decline in respiratory rate as germination was initiated, Phase III was marked by an increase in respiratory rate which paralleled the rate of germination and germ tube elongation, Phase IV was a gradual decline in the respiratory rate which commenced after the germination was complete. The respiration rate of uredospores of *Puccinia coronata* in the distilled water declined during incubation (Naito & Tani, 1967). The respiration quotient in the early stages was about 0.5 and in the late stage was 1.0. This indicates a change from lipid metabolism to carbohydrate metabolism. Chigrin et al. (1970) found that respiration coefficients were indicative of lipid utilization at the beginning of germination periods, but later the R_Q was indicative of carbohydrate metabolism. They suggested that a copper containing terminal oxidase (perhaps polyphenol oxidase) was functioning in a significant role in the oxygen utilization in the uredospores. Lipid metabolism is common in rusts such as *P. graminis* and *Uromyces phaseoli*. The lipid content decreased during incubation (Shu et al., 1954; Caltrider et al., 1963).

Daly et al. (1967) have shown that the *P. graminis tritici* uredospores utilized some sugars and some sugar alcohols. The respiration quotient was a rough guide of the type of endogenous substrate being utilized. Ungerminated uredospores of *P. graminis tritici* had an R_Q of 0.6 (Farkas & Ledingham, 1959). The R_Q was 0.4 for *Puccinia rubigovera* (Staples, 1957). These R_Q values indicate that lipids are being oxidized. Respiration of the spores was not blocked by glycolysis inhibitors, whereas the Kreb cycle respiration was blocked by inhibitors (Shu et al., 1954).

In aeciospores of *Cronartium fusiforme*, the loss of germination appeared to be a function of water uptake and was not associated with the reduction of the ability of the spores to consume oxygen (Walkinshaw, 1968a and 1968b).

Evidence for Glycolysis in Uredospores of Rusts

Many of the enzymes associated with glycolysis, including triose isomerase and aldolase, were detected in uredospores of *Puccinia graminis tritici* by Shu & Ledingham (1956). Caltrider et al. (1963) detected the presence of most of the enzymes of the Embden Meyerhof pathway (EMP) in the spores of *P. graminis tritici*. All of the EMP enzymes could not be detected. For example, alcohol and lactic acid dehydrogenases were absent from spores of *P. graminis tritici* and *Uromyces phaseoli*. These results suggested that the corresponding alcohols are probably not produced by these fungi. In most spores, the anaerobic conditions of the EMP are probably not utilized for the production of alcohol or lactic acid. Pyruvic decarboxylase and dehydrogenase have not been demonstrated in spores of any of these rusts. The difficulty of the assay rather than the absence of the enzyme may be the reason for these results. Most fungal spores have all of the enzymes necessary to form acetyl CoA from glucose.

Evidence for HMP in Uredospores of Rusts

The early evidence for the hexose monophosphate pathway (HMP) was based on the presence of a NADH-linked dehydrogenase in *Puccinia graminis tritici* uredospores (Shu & Ledingham, 1956). Extracts of ungerminated spores also converted glucose-6-phosphate to pento-phosphates and other components of the HMP. Caltrider et al. (1963) showed the presence of all the HMP enzymes in spores of *P. graminis tritici* and the entire enzyme complement was also present in uredospores of *Uromyces phaseoli*. Enzymes of the Embden Meyerhof Pathway (EMP) and HMP declined after an initial rise at the beginning of germination of *P. graminis tritici* uredospores (Burger & Reisener, 1970).

Evidence for the Citric Acid Cycle in Uredospores of Rusts

The citric acid cycle (CAC) is the main pathway for oxidizing the acetate carbons which are derived from the Embden Meyerhof Pathway (EMP) such as acetyl CoA. Caltrider et al. (1963) demonstrated the presence of CAC enzymes: aconitase, isocitric dehydrogenases, succinic dehydrogenases, fumarase, and malic dehydrogenases, in uredospores of *Puccinia graminis tritici* and *Uromyces phaseoli*. However, alpha ketoglutaric dehydrogenases could not be found and this may be due to the difficulty of the assay. The component organic acids of this system were shown to be present in uredospores of *P. graminis tritici* and in *Puccinia rubigovera* (Caltrider et al., 1963). *Puccinia graminis tritici* uredospores fully oxidized the citric acid intermediates (White & Ledingham, 1961). Staples (1957) demonstrated the presence of keto acids, except for isocitrate, in uredospores of *P. graminis tritici*. The operation of the citric acid cycle in *P. graminis tritici* uredospores has also been supported by labeling studies with acetate, proponate, and valerate. The labeling patterns are consistent with the operation of the citric acid cycle (Burger et al., 1972; Reisener & Jager, 1967). The electron-transport system appears to be present in all the spores examined and it is usually sensitive to cyanide and azide (Klusak, 1970). Many of the CAC enzymes have been found particularly in spores of *P. graminis tritici*, *U. phaseoli* and *P. graminis rubigovera* (Staples, 1957; Caltrider et al., 1963). The endogenous oxygen consumption of *P. graminis* and *U. phaseoli* uredospores was very low and did not increase by adding exogenous glucose or succinate.

Evidence for the Glyoxylate Pathway in Uredospores of Rusts

Frear & Johnson (1961) obtained evidence for the glyoxylate pathway in *Melampsora lini* spores. Caltrider et al. (1963) obtained enzymatic evidence for a functioning glyoxylate pathway in uredospores of *P. graminis tritici* and *U. phaseoli*. Reisener & Jager (1967) analyzed different metabolic pathways in uredospores of *P. graminis tritici* and concluded that the CAC was probably more important than the glyoxylate cycle in the germination process.

Evidence for CO_2 fixation during germination

Stallknecht & Mirocha (1970) found the amount of CO_2 fixed by uredospores of *U. phaseoli* was greatest during the initial stages of germination and decreased rapidly during the first two hours of germination. There was no difference in CO_2 fixation in the light or dark. The CO_2 was incorporated into low molecular weight metabolites, RNA, and other macromolecules such as proteins and polysaccharides. Rick & Mirocha (1968) found that the malic enzyme from uredospores of *U. phaseoli* was specific for NADP and needed magnesium ions for activity. However, Phosphoenolpyruvate carboxylase could not be demonstrated in the crude spore. Macko & Novacky (1966) observed changes in the isoenzymes of malic dehydrogenase in uredospores of *Puccinia recondita*.

Fungal spores are not lacking in pathways to metabolize carbohydrates, however, the mechanism of metabolic regulation is an area which needs further research.

Changes in Carbohydrate Metabolism During Uredospore Germination

The uredospores of *Puccinia graminis tritici* can utilize a number of externally supplied carbohydrates, amino acids, and volatile fatty acids. However, the level of metabolic actitivy is very low. Mannose, mannitol, glucose, sucrose, and fructose were utilized better than other carbohydrates (Shu & Ledingham, 1956). The uredospores of flax rust had a glucose metabolic pool which was about 8% of the permanent cell carbon (Turel & Ledingham, 1959). Reisener (1962) observed the formation of trehalose and polyols in uredospores of *P. graminis tritici*. Wynn (1968) followed the increase in glucan synthesis during germination in rust spores; Wynn & Gajdusek (1968) found that glucomannan proteins were rapidly synthesized during the first two hours of germination of uredospores of *Uromyces phaseoli*, whereas degradation was associated with germ tube elongation. It was postulated that germinating uredospores have a developmental control mechanism for transferring a portion of the endogenous reserves to a reservoir of polysaccharides.

Van Sumere et al. (1957) obtained evidence for a hemicellulase, cellulase, and a pectinase in uredospores of *P. graminis*. When esterases in the uredospores of *Hemileia vastatrix* were removed, the germination rate decreased (Raphaela-Musumeci et al., 1972). No changes occurred in total dry weight of *Puccinia coronata* uredospores during germination (Naito & Tani, 1967). The dormant uredospores contained crude fat (34%), mannitol (16%), and polysaccharides (10%)

A gradual decrease in the fatty acids occurred during the first 24 h, whereas the nonsaponifiable lipid increased slightly 16 h after incubation. The majority of mannitol and free sugar disappeared 5 h after the beginning of the germination process, whereas the oligosaccharide content increased about twofold and polysaccharides were at the highest concentration 16 h after incubation. The water soluble polysaccharides (not starch or glycogen) disappeared from the germ tubes when elongation ceased. Reisener (1962) found that the spore walls of *Puccinia graminis tritici* contained N-acetylglucosamine and polymeric mannose whereas germ tube walls contained N-acetylglucosamine, polymeric glucose, and polymeric mannose.

About half of the substrate consumed by uredospores of three rusts was converted to carbon dioxide during germination (Staples et al., 1962). Uredospores of *P. graminis tritici* incorporated ^{14}C glucose into carbohydrates (mannitol, arabitol, and glycerol) and free amino acids (Burger et al., 1972). Alanine, tyrosine, and glutamic acid also showed high ^{14}C incorporation. The labeling pattern suggests that glucose is metabolized by the pentose pathway as well as by the Embden Meyerhof pathway and the citric acid cycle. However, only 12% of the added label could be accounted for by soluble products. Most of the label was incorporated into the germ tube cell walls. The major constituents of the germ tube walls were glucose, mannose, glactose, and glucosamine.

Trocha & Daly (1974) and Trocha et al. (1974) found that amino acids constituted 6% of the carbon in uredospores of *P. graminis* and 16% of the carbon in germ tube walls. Trocha & Daly (1974) found a glucan and a gluco mannan in *Uromyces phaseoli* uredospores. In the germ tube walls they found that the polysaccharides were more methylated and glucosamine was present. Joppien et al. (1972) found that the walls of uredospores of *P. graminis tritici* were composed mainly of protein, lipids, and neutral sugars such as mannose, glucose, and glactose. The carbon content of the germ tube walls (polymetric glucose and mannose) was slightly higher than the carbon content of the spore walls (mannose). The germ tube walls contained twice as much nitrogen (hexosamine) and lipid as the spore walls. After studying the utilization of internal labeled substrates during germination in *P. graminis tritici*, Daly et al. (1967) concluded that internal pools of carbohydrates and lipids were important in the germination process.

Lipid Metabolism

Chemical analysis of uredospores of *Puccinia graminis tritici* before and after germination indicated that the lipid compounds were the major endogenous substrate. Some nitrogenous material, presumably protein, was also utilized (Shu et al., 1954). Fatty acids (C_3--C_{12}) inhibited oxygen consumption of aeciospores of *Cronartium fusiforme* whereas C_{16}--C_{18} fatty acids stimulated oxygen consumption (Walkinshaw, 1968a, b). Walkinshaw also concluded that alpha decarboxylation of fatty acids was functioning in the lipid breakdown process during germination rather than beta oxidation.

Tulloch & Ledingham (1960a & b) found, by gas chromatography, that palmitic acid was a major saturated fatty acid in rusts. Tulloch et al. (1959) and Tulloch & Ledingham (1960) isolated an unusual fatty acid, *cis*-9,10-epoxy octadecanoate from rusts which constituted 28% of the fatty acid fraction. Baker & Strobel (1965) determined that alkanes ranged from C_{19} to C_{31} and were present in spores of *Puccinia stratifomis*. Hougen et al. (1958) found that egost-7-enol was the major sterol and that beta and gamma carotene were also present in uredospores of *P. graminis*. The major sterols in *P. graminis tritici* were cholesterol, stigmast-7-enol and stigmasterol, or ergost-7-enol (Novak et al., 1972). The level of cholesterol increased during germination whereas the level of stigmast-7-enol decreased. The major sterols of *Uromyces phaseoli* spores were 7--24 (28) stigmastadien-38-ol and 7-stigmasten-38-ol (Lin et al., 1972). Both of the sterols were synthesized during the germination process. During germination of flax rust uredospores the total lipids did not decrease appreciably until after the fourth hour of germination. During the first and second hour of germination, about half of the *cis*-9,10-epoxy octadecanoate was converted to the free form (Jackson & Frear, 1967; 1968). The nonsaponifiable lipids and phospholipids increased gradually during the 8 h germination period. The principle sterols of flax rust uredospores were stigmast-7-enol and stigmasta-7,24 (28) dienol with stigmast-5,7-dienol as a minor component (Jackson & Frear, 1968).

The utilization of oxygen in *P. graminis tritici* uredospores was stimulated by the short chained fatty acids. Acetic acid was the most stimulatory and valerate the least (Reisener et al., 1961). Peruanskii & Run'kovskaia (1966) found at least seven triglycerides, five lecithins, five cephalins, and six sphingomyelins in the lipids of *P. graminis tritici*. Lipase from *P. graminis tritici* was a very active enzyme with a temperature optimum of 15°C, which corresponds to the optimum germination temperature (Knoche & Horner, 1970). The phospholipids of *U. phaseoli* were phosphatidylcholine and phosphatidylethanolamine, and diphosphatidylglycerol, cardiolipin, and phosphatidylinositol were present as minor components. In germinating spores, the diphosphatidylglycol was three times higher in the germ tube wall preparation than in walls of resting spores.

Laseter et al. (1973) investigated the volatile terpenoids of aeciospores of *C. fusiforme*. Laseter & Valle (1971) found alkanes from C_{18} to C_{35} and free fatty acids ranging from C_{14} to C_{24}. The total free fatty acids from the surface was different than fatty acids obtained from a total analysis of the fungal spore. The major terpenoids were alpha pinene, beta pinene, delta-3-carene, myrcene, linonene, beta phellandrene, and delia-torpinene (Laseter et al., 1973). They also found a number of mono-terpene alcohols of which the terpine-4-ol was a predominant material. The sesquiterpenes were beta farnesene and beta citronellol.

While the identification and changes in lipids during germination of spores are valuable information, there is a pressing need for experimental evidence on enzymatic processes relating to lipid breakdown and synthesis. The physiological functions of several metabolic lipid components also remain to be elucidated.

Nucleic Acid Metabolism

The spores of basidiomycetes have not been favored subjects for nucleic acid investigations. Hollomon (1970) reviewed nucleic acid synthesis during fungal spore germination. Some base ratios of nucleic acids of fungi including some basidiomycetes were summarized by Storck & Alexopoulos (1970). Fungal spores have several forms of DNA and RNA. Mitochondrial DNA in uredospores of *Uromyces phaseoli* had a high GC content and a buoyant density in CsCl that was less than or nearly equal to that of nucleic acid DNA (Staples, 1974). In most other fungi, the base composition of mitochondrial DNA is not as variable as the base composition of nucleic acid DNA. Staples & Yaniv (1973a) found that labeled adenosine was incorporated only into mitochondrial DNA. Ramakrishnan & Staples (1968a, 1968b, 1971) investigated the changes in ribonucleic acid during uredospore differentiation. The uredospores of *U. phaseoli* contained the same complement of nucleic acids throughout germination and differentiation that was found in the resting spores.

The incorporation of uridine into template RNA occurred only during the differentiation of the germ tube. They suggested that the formation of the rust infection structures may depend upon synthesis of messenger RNA (Ramakrishnan & Staples, 1970b). The template

RNA had sedimentation values of 4S and 19S. In a cell-free system prepared from *E. coli*, the isolated fungal RNA stimulated incorporation of amino acids into protein. The template was not destroyed by DNAse and the RNA was rich in AMP (Ramakrishnan & Staples, 1970a). Uredospores of *U. phaseoli* virtually lost their capacity to incorporate exogenous uridine into RNA during germination when kept longer than four days in cold storage (Ramakrishnan & Staples, 1971).

Stallknecht & Mirocha (1971) observed that $^{14}CO_2$ was incorporated into messenger RNA, ribosomal RNA, and soluble RNA. Chakravorty & Shaw (1972) concluded that ribosomal RNA (28S, 18S, and 5S) was synthesized in germinating uredospores of *Melampsora lini* during the initial 6 h of hydration. During this period, the newly transcribed ribosomal RNA molecules were associated with ribosomal 60S and 40S components and with the monomeric ribosomes (60S). The newly formed ribosomes actively engaged in protein synthesis and could be recovered from the polyribosomal region in a sucrose density gradient with centrifugation. Extracts of uredospores of *M. lini* taken during germination all possessed DNA, ribosomal RNA, and soluble RNA (Quick, 1973). The DNA levels were not markedly changed during the first 12 h of germination (Quick, 1973).

Pavgi & Singh (1970) found that the diploid nucleus migrated in the promycelium during the germination of teliospores of *Entyloma eleocharidis*. During the germination of aeciospores of *Cronartium comandrae* and *C. comptoniae* no nuclear divisions occurred and the two nuclei migrated toward the tip of the germ tube. Powell & Hiratsuka (1969) concluded that the aeciospores of *C. comandrae* produced two germ tubes which is characteristic of host alternating races, whereas the pine to pine races only produced one germ tube. Kapooria (1971) observed that teliospores of *Uromyces fabae* produce five-celled promycelia. Four cells of the promycelia are functional, but the lower cell is non functional. The chromosome counts in the promycelia and the basidiospores indicated that there were four chromosomes grouped in two pairs of different sizes. Raghunath (1967) observed that the haploid number of chromosomes of *Tolyposporium christensenii* was two and that typical meiosis involved the formation of bivalent pairs with "chiasmata" comparable to those found in higher plant nuclei. Hiratsuka & Maruyama (1968) observed the presence of one instead of two nuclei in the germ tube during germination and suggested that an unusual nuclear cycle existed for *Peridermium ephedrae*. The nuclear behavior of teliospores of *Puccinia praninirana* and of *Uromyces orientalis* has also been investigated (Kotwal & Kulkarni,

1971; Pawar & Kulkarni, 1971).

Ribosomes in Germinating Uredospores

Staples et al. (1966) found that ribosomes of uredospores of *Uromyces phaseoli* sedimented in a single boundary and were capable of incorporating *in vitro* polyuridylic acid directed phenylalanine incorporation. They investigated the messenger RNA template in relation to ribosomes of uredospores of *U. phaseoli* (Staples, 1968). Staples et al. (1970) concluded that ribosomes from uredospores of *U. Phaseoli* that had germinated for 4 h were partly bound by cellular membranes. Approximately 57% of the ribosomes were aggregated into polysomes. Staples et al. (1971) investigated germinating uredospores of *U. phaseoli* in relation to RNA turnover rate and ribosomes (Staples & Yaniv, 1973b). Yaniv & Staples (1968) found that the activity of ribosomes was activated by hydration but began to decline after 2 h of activity. The activity of the ribosomes from germinated uredospores was inhibited by a deoxycholate wash, but the ribosomes from a resting or hydrated spore were not affected by this detergent treatment. Evidence also exists that polyribosomes are present in the uredospores of *U. phaseoli* (Staples et al., 1966). A RNA fraction which has the properties of messenger RNA was isolated from uredospores of *U. phaseoli* (Ramakrishnan & Staples, 1969). Only one RNA polymerase was detected in germinating uredospores of *U. phaseoli* (Manocha, 1973). Normally, RNA from fungi are relatively unstable. The RNA polymerase that was isolated by Manocha (1973) was unaffected by Cycloheximide and Amanitin, but was inhibited 58 and 80% by Actinomycin D and Rifampin, respectively. The sensitivity of the enzyme to Rifampin indicates that it might be of mitochondrial origin although Manocha (1973) suggested it is of nuclear origin. Germination of uredospores of *U. phaseoli* may present an exception to the theory that tRNA is synthesized early during the germination process since tRNA synthesis was not observed until after germ tube formation (Ramakrishnan & Staples, 1967, 1970b; Staples, 1968; Stallknecht & Mirocha, 1971). Ramakrishnan & Staples (1970a) reported that the germinating uredospores of *U. phaseoli* incorporated very little precursor into the mRNA fraction whereas the germinating spores were induced to form appressoria when mRNA synthesis occurred. These findings are consistent with the observation that Actinomycin D had no effect on the germination of *Puccinia graminis mustriosi* uredospores, but did inhibit appressoria formation (Dunkel et al., 1969).

In *U. phaseoli* uredospores, RNA synthesis begins 90 min after the initiation of germination whereas transfer RNA and messenger RNA synthesis were detected only after germ tubes appeared (Ramakrishnan & Staples, 1967, 1968a, 1969). In most fungi the evidence suggests that the DNA remains unchanged during spore germination, and that DNA synthesis is not required for germination. However, with RNA systhesis during spore germination there is considerable variation among the different species. It is still not resolved whether fungus spores contain preexisting mRNA. However, there is evidence that dormant spores of many fungi contain mRNA. The manner in which this mRNA is conserved in the spores is not known. Although RNA synthesis generally accompanies spore germination, the question of whether the dormant spores contain sufficient mRNA to form the germ tube cannot be answered for most fungi. Present data suggests that some spores require RNA synthesis for germination and others do not. There are many questions concerning RNA synthesis that still remain to be answered.

Amino Acid Metabolism

During germination of uredospores of *Puccinia graminis tritici*, arginine was taken up better than alanine and glycine, but the efficiency of the amino acid uptake was not as high as the uptake of glucose or valerate (Reisener, 1969). Reisener (1969) concluded that the capacities for taking up amino acids was not a limiting factor in protein synthesis, and that a large amount of the arginine, alanine, and glycine was also metabolized to carbon dioxide. These results indicate the presence of normal amino acid pathways. Smith (1965) observed similar results with uredospores of *Puccinia helminthi*. Snow & Jones (1967) concluded that the amino acids were being secreted from germinating uredospores of *Puccinia coronata* and were transported into the cell.

Jackson et al. (1970) found that many of the aromatic compounds were not taken up or metabolized by *Puccinia recondita tritici*. However, some enzymes such as shikimate dehydrogenase and tyrosine ammonia lyase were present.

The interaction of various amino acid pools in relation to germination needs to be investigated further.

Protein Synthesis

The protein synthesizing systems in spores of rusts have been investigated by several individuals. Staples & Bevigian (1967) prepared an amino acid incorporation system from uredospores of *Uromyces phaseoli*. The incorporation system required an ATP generating system, GTP, and magnesium, and the incorporation was completely inhibited by RNAse. The aminoacyl transfer RNA binding enzymes (Transferase I and II) have been purified from *U. Phaseoli* (Yaniv & Staples, 1971). Transferase I (MW 145,000) catalyzed polyuridylic acid directed and GTP dependent binding of fungal tRNA to ribosomes and was required with transferase II for the synthesis of peptides from aminoacyl tRNA. Reisinger (1967) found a decrease in all free and bound amino acids when uredospores of *Puccinia graminis tritici* germinated and only glucosamine increased. They also concluded that there was no net protein synthesis and that any release of nitrogen was used for glucosamine synthesis. Through use of labeled substrates, they concluded that the uredospores had the capabilities of synthesizing amino acids. Although they did speculate that protein synthesis may be limited by reduced synthesis of some amino acids. Yaniv & Staples (1969) determined the characteristics of the transfer of amino acids from the aminoacyl transfer of RNA to polypeptides of uredospores of *U. phaseoli*. They found that for maximum activity, the reaction required 4 m mole of magnesium, 40 mg/ml of polyuridylic acid, 2 mole of mercaptoethanol and ribosomes. The requirement for GTP was not absolute.

By use of enzymatic hydrolysis, Trocha & Daly (1970) found that, in contrast to previous reports, the amount of amino acids released by proteases was at least 20--25% during the first 5 h of germination, but decreased after 16--24 h. They concluded that the spores of obligate parasites have a functional protein synthesizing system which is involved in germination. Using electrophoretic methods, Macko et al. (1967) and Novacky & Macko (1966) followed the protein and enzymatic changes during germination of uredospores from *P. graminis*. Dunkel (1969) found that RNA and protein synthesis inhibitors blocked formation of infection structures of *P. graminis tritici*. Shipton & Fleischmann (1969) used disc electrophoresis to distinguish races of *Puccinia coronata avenae*. Eyal et al. (1967) also used protein bands to distinguish several races of *Puccinia recondita*. Acid phosphatase, esterase, and polyphenol oxidase were located both in the walls and in the cytoplasma of hydrated uredospores of *P. graminis tritici*, but alcohol dehydrogenase was only in the wall and malic dehydrogenase was mainly in the cytoplasm. Cheung & Barber (1971) concluded that the spore wall proteins were readily diffusable and are probably connected with the spore germination process.

The information on protein synthesis in fungal spores is a good example of the excellent progress that is being made with the molecular aspects of spore germination.

Ustilaginales

Spores of the approximately 700 species of smuts have also been biochemically investigated, although not as extensively as rust spores.

Dormancy and self-inhibition

Smut spores have extremely thick walls, particularly the chlamadyospores (teliospores) which are almost impermeable to organic solvents. Dormancy in smut spores undoubtedly could be due, to some extent, to the thickness of the wall and exclusion of hydration.

Singh & Trione (1969) reported the presence of a self-inhibitor in *Tilletia caries* and *Tilletia controversa* (Macko et al., 1964). Further research has shown that the self-inhibitor may be extracted by methanol (Trione, 1973). Nielsen (1962) identified two trimethylammonium compounds in *Tilletia* species which he suggested were precursors for the trimethylamine, an odor primarily associated with *Tilletia* teliospores.

In *T. caries* trimethylamine was reported to be a natural inhibitor of germination (Ettel & Halbsguth, 1963). Trione (1973) concluded that the bunt teliospores were regulated by different mechanisms of self-inhibitors which were different from those commonly found in rust uredospores. Water-soluble inhibitors of germination were detected in teliospores of *T. controversa* and these are considered to be the cause of dormancy.

Factors involved with stimulating germination

Schauz (1970) investigated the effects of storage temperatures on the germination of uredospores of *Tilletia caries*. Duran & Safeeulla (1968) pointed out that chilling is required to trigger germination in smut spores found in northern areas, whereas chilling was not required for smuts from the southern areas.

Duran & Safeeulla (1968) distinguished three basic germination types: (1) those where high germination occurred on a variety of media over a wide temperature range without any need of pretreatment, (2) those that required substantial chilling to germinate and still have very low percentages of germination, (3) those that were unaffected by chilling, but gave about 1% germination. Light does not appear to enhance germination of the wheat smut teliospores (Trione, 1973).

Carbohydrate metabolism of smut spores

Caltrider & Gottlieb (1966) studied the effects of several sugars and their relative abilities to stimulate germination in teliospores of several smuts. The germination of teliospores of *Ustilago maydis* was high with sucrose, raffinose, and melezitose, but was low with glucose and fructose. Invertase was present in the teliospores. There did not appear to be a relationship between the oxygen consumption with the different carbon sources and their relative abilities to stimulate germination. Enzymatic analysis and amino acid analoque experiments suggested that the Embden Meyerhof pathway (EMP) was the major pathway early in germination. Although there was some evidence for the hexose monophosphate pathway (HMP) and citric acid cycle (CAC) enzymes being present.

Caltrider et al. (1963) investigated the synthesis of enzymes associated with EMP and HMP in *U. maydis* during germination and found several inactive enzymes in the dormant spores. *Ustilago maydis* spores may be an example of a nonfunctional system in which all of the EMP enzymes are present, but carbon dioxide was not produced under anaerobic conditions (Caltrider et al., 1963). Of course, the system may be operative within the cells with the accumulation of pyruvic and lactic acids. *Tilletia caries* spores have all the HMP enzymes and the carbon dioxide from 1-^{14}C-glucose is incorporated more rapidly than the glucose labeled in the 3 and 4 carbons (Newburgh & Cheldelin, 1958). Ungerminated teliospores of *U. maydis* did not have all of the HMP enzymes. The HMP was more active in *T. caries* teliospores than the EMP (Newburgh & Cheldelin, 1958).

The Entner Doudoroff pathway is another system for metabolizing glucose and there appears to be some evidence that it is present in the teliospores of *T. caries* (Newburgh & Cheldelin, 1958).

Lipid content and metabolism of smut spores

The hydrocarbon patterns appeared to be similar in three species of *Tilletia*. In contrast benzene and methanol fractions gave distinct gas chromatography patterns of the three species (Laseter et al., 1968). Weete et al. (1969) and Laseter et al. (1968) analyzed the lipids of spores of *T. caries*, *T. controversa*, and *T. foetida*. They found that spores of *T. foetida* were the highest in saturated fatty acids. Those of *Tilletia controversa* were second, followed by spores of *T. caries*. With unsaturated fatty acids, the sequence was reversed. Of the total lipids present in spores of *T. controversa*, the distribution was 35% phospholipids, 52% free fatty acids, and 26% bound fatty acids (Trione & Ching, 1971). The major fatty acids were $C_{18:2}$, $C_{18:1}$, and $C_{18:0}$. Factors that reduced the permeability of the wall also stimulated the uredospores of *T. caries* to germinate about 5 h sooner. Tween 20 and dimethylsulfoxide increased the permeability, but also had an adverse affect upon germination.

The hydrocarbon fraction of *Ustilago maydis* contained predominantly C_{25}, C_{27}, and C_{29} alkanes, and natural methylesters ranged from C_{16} to C_{20}. The free fatty acids contained acids which ranged from about C_{12} to C_{20} (Laseter et al., 1968). Oro et al. (1966) reported alkanes which ranged from C_{14} to C_{37} in fungal spores. Gunasekaran et al. (1971) found that the content of polar lipids (84% of total lipids) was higher in *U. maydis* spores as compared to those of *Ustilago bullata* (53%). All other lipid classes were higher in spores of *Ustilago bullata*. Hess et al. (1974) identified a range of unusual compounds in spores of *T. caries*, and Gorin et al. (1960) found that erythrulose was an effective precursor in the synthesis of 4-0-β-D-mannopyranosyl-D-erythritol in smuts.

Nucleic acid synthesis

Teliospores of *Ustilago maydis* contained an unusual RNA, but after 12 h of germination, normal RNA appeared. Gottlieb et al. (1968) suggested that this spore RNA was complexed with protein. Lin et al. (1971) found that teliospores of *U. maydis* contained cytoplasmic and mitochondrial RNA. Spores incubated for two hours or longer formed functional transfer RNA.

At the lower temperatures, about 32°C, a DNA mutant of *U. maydis* grows as filamentous cells rather than as budding cells. Analysis of this material indicated that there was a polygene that was the structural gene for DNA precursors and that the single recessive mutation in that gene resulted in a temperature-sensitive phenotype (Jeggo et al., 1973). Raghunath (1969) concluded that the mature spores of *Georgefischeria riveae* were uninucleate and diploid. Jensen & Kiesling (1971) studied nuclear conditions of teliospores of *Ustilago hordei*, and Becerusuc (1973) used doses of X-radiation to produce mutants in smuts.

While considerable research progress has been made on smut spores, a more correlated approach on a single species would probably be much more productive.

Hymenomycetes

The Hymenomycetes can be divided into several groups such as Phragmobasidiomycetidae, Holobasidiomycetidae, and the Phyllophorales. The most thoroughly investigated Hymenomycete is *Schizophyllum commune*. Because of their use as food, the Agaricales have also been investigated.

Germination conditions

The germination of spores of *Fomes igniarius* var. *populinus*, a causal agent of a disease on quaking aspen, was stimulated by glucose (Manion & French, 1968, 1969). Manion and French found that basidiospores placed in sap-wood wounds of aspens germinated readily if the wounds were fresh, but in older wounds, the germination rate was lower. In a 6-day-old wound no germination occurred. When Manion and French (1968) further analyzed the components present in the sap-wood, they found that the major carbohydrate in the sap-wood was glucose, and that germination could be readily obtained when glucose was added. If the uptake of glucose was inhibited, germination was reduced.

The germination of basidiospores from *Polyporus tomentosus* was improved after three months storage at $-18^{o}C$ (Whitney, 1966). However, the viability of the basidiospores was reduced greatly after three months of storage at $0^{o}C$. The best germination of *P. tomentosus* was obtained with extracts of inner barks of black spruce (Whitney & Bohyachuk, 1971). Malt extract agar was the best medium for germination of basidiospores of *Pholiota aurivella*. The spores could be stored for 3--6 months at 3--$9^{o}C$ (LaVallee & Lortie, 1971). The penetration of the germ tube occurred through the end pore of *Polyporus dryophilus vulpinus* (Morton & French, 1970).

Brown & Merrill (1973) found that up to 78% germination of basidiospores of *Fomes applanatus* could be obtained if the artificial media contained glucose and certain volatiles from colonies from *Ceratocystis fagacearum*. Propyl ethyl, isovaleric acid, isoamyl alcohol, IAA, Kinetin, sodium acetate, ethanol, or vitamins did not stimulate germination. Ross (1969) concluded that the basidiospores of *Fomes annosus* would probably not be able to survive more than 1 h at 45°C and 90% relative humidity. Scheld & Perry (1970a, 1970b) found that germination of basidospores of *Lenzites sepiaria* was stimulated by hydroxy-methyl fural.

Van Alfen (1969) isolated a hot water fraction from horse dung that stimulated germination in basidiospores of *Psilocybe*. The isolated compound had properties similar to sodium glycocholate. Olah (1973) observed that an accumulation of polysaccharides occurred during germination in *Psilocybe quebecensis*.

Carbohydrate metabolism

The RQ for *Schizophyllum commune* was near 1, which suggested that the oxidation of carbohydrates was the functional pathway (Niederpruem, 1964). The ungerminated spores of *S. commune* appeared to have equally active Embden Meyerhof pathways (EMP) and hexose monophosphate pathways (HMP). At 4 h after incubation, the HMP was predominant but after 8--12 h of incubation the EMP appeared to play a greater role. The oxygen comsumption of *S. commune* was relatively high with an 0Q of 10--20 (Niederpruem, 1964). The oxygen consumption was stimulated by the addition of sucrose, some hexoses, and some xyloses, but not by glucose (Niederpruem & Hafix, 1964). The respiration of *S. commune* basidiospores which utilized glucose was inhibited by azide, cyanide, Antimycin A, and Atabrine, indicating that the normal electron transport system was present in these spores (Niederpruem, 1964; Ratts et al., 1964). Components of the citric acid cycle were also shown to be present in spores of *S. commune* (Wessels, 1959). Niederpruem (1964) demonstrated the presence of isocitrate dehydrogenase and fumerase in the basidiospores of this fungus.

Basidiospores of *Schizophyllum commune* have been investigated by feeding radioactive glucose, mannitol, and arabitol (Aitken & Niederpruem, 1972). Glucose was taken up rapidly and was not affected by antibiotics such as cycloheximide which blocks protein synthesis. With mannitol or arabitol, the uptake process was interferred with by the presence of cycloheximide, which indicates that protein synthesis was necessary for uptake to occur. The respiratory rates for either arabitol or mannitol increase markedly upon germination in glucose--asparagine minimal cultural media. Aitken & Niederpruem (1973) also investigated the fate of labeled glucose as compared to labeled mannitol and arabitol. The amount of labeled CO_2 released was greater when 1-^{14}C glucose was added to germinating spores as compared to 6-^{14}C-glucose. Unlabeled ethanol inhibited the $^{14}CO_2$ evolution when 3,4-^{14}C-glucose was used as a substrate with germinating spores. This suggests that glucose is metabolized by the common pathways present in organisms and indicates that the hexose monophosphate pathway is functioning as well as the Embden Meyerhof pathway (EMP). With ^{14}C-arabitol, this compound was converted primarily to trehalose. Mannitol, however, was not converted to trehalose. Niederpruem & Dennen (1966) grouped the sugars that were utilized by basidiospores of *S. commune* into four categories: (1) those utilized between 15 and 20 h, (2) those used between 30 and 60 h, (3) those used between 30 h and 7 days, and (4) those in which there was no activity even after 7 days. The results indicated that the basidiospores used classical sugars, hexoses and pentoses, during the first part of the germination. During the 30 to 60 h period, many of the sugar alcohols were used. During the 30 h to 7 day period, the disaccharides and some pentoses were utilized. Compounds that had no activity included fucose, rhamnose, inositol, and similar sugar compounds.

Scheld & Perry (1970) concluded that the basidiospores of *Lenzites sepiaria* rapidly utilized glucose by the EMP, and acetate or succinate by the citric acid cycle. Labeled substrates were also incorporated into RNA and protein at low rates as soon as they were added to the media. During the germination process, glucose was used primarily for the cellular material, and very little was converted to CO_2. Aitken & Niederpruem (1970) found, during basidiospore germination in *S. commune*, that a light phase occurred during which the cellular reserves, trehelose, mannitol, and arabitol were depleted (Niederpruem & Jersild, 1967). The branching pattern of basidiospores of *S. commune* indicated a linear relationship between the length of the main axis and the number of cells in the main axis (Volz & Niederpruem, 1968).

Hymenomycete lipid metabolism

Losel (1967) found that basidiospores of *Agaricus bisporus* were stimulated by vapors which diffused from dilute solutions of very short-chain fatty acids, especially isovaleric acid. She concluded that the stimulatory effect was not a pH effect, but was due to the direct entry of these compounds into the metabolic pathways. Stocks & Hess (1970) found a dung extract similar to sodium glycocholic acid that stimulated germination of a species of *Psilocybe*, and O'Sullivan and Losel (1971) compared the effect of a number of organic compounds on the germination of basidiospores of *A. bisporus* and concluded that they were stimulated by isocaporic acid, but not by straight chain C_5--C_{11} fatty acids, or by the amino acids, leucine, and isoleucine. Cholesterol at a concentration of one part per million was inhibitory. The lipid reserve of the spore comprised mono-, di-, and triglycerides, free fatty acids, and sterols. The phospholipid fraction was small and phosphatidyl-choline was the most predominant phospholipid. Wasowska (1971) investigated the effect of different substrates and the pH of germination of spores of *A. bisporus*. Cheng (1969) also studied the stimulatory effect of spore germination of *A. bisporus* and Rast and Bachofen (1967) observed the fixation of carbon dioxide in *A. bisporus*. Singer (1972) investigated the absorption of cotton blue into spores of basidiomycetes. Those spores that strongly absorbed the cotton blue he called cyanophilous. Singer (1972) found that while dye absorption was useful in some taxonomic roles, the separation was certainly not definitive.

Mills & Eilers (1973) observed constitutional dormancy in basidiospores of *Coprinus radiatus*. Germination was induced by heat and certain chemicals. Spores were heated at 45° for 4 h, resulting in higher germination. When spores were heated in the presence of chemicals, the germination rate went up to 88%.

Niederpruem & Hafiz (1964) reported that oxygen consumption was stimulated by an addition of sucrose, some xyloses, but not by glucose. The R_Q for *Schizophyllum commune* was near 1, which suggests the oxidation of carbohydrates (Niederpruem, 1964). Respiration of spores of *Corpinus lagopus* was stimulated only slightly by glucose or fructose (Heintz & Niederpruem, 1970).

Mannitol and arabitol were identified in basidiospores of *S. commune*. Arabitol was a more favorable carbon source since it disappears first during germination (Niederpruem & Hunt, 1967). Basidiospores of *S. commune* contained mannitol dehydrogenase. This enzyme was not active in ungerminated spores, but the enzyme was present during germination as was indicated by the increased oxygen consumption when mannitol was supplied (Niederpruem & Hafiz, 1964).

Gastromycetes

Almost no biochemical research on spores of gastromycetes has been done. Bushnell (1974a, 1974b) analyzed the lipids of *Lycoperdon perlatum* and found long chain aldehydes in the spores. He also found branched chain hydrocarbons in the spores of *Calvatia gigantea*.

Spore research in the basidiomycetes has made considerable progress in identifying natural components and observing the changes that occur during germination. It is obvious that many species need to be studied in detail biochemically to elucidate questions which have been raised by studies conducted to date. There is a particular need for more intense studies of gastromycete spores. Some questions that need to be answered in the future are: (1) What are the metabolic controls that regulate development during germination? (2) Can metabolic systems such as protein systems and nucleic acid metabolism be used to evaluate evolution of fungal spore systems? (3) What controls the synthesis of the metabolic controls such as the self-inhibitors?

ULTRASTRUCTURE OF BASIDIOMYCETE SPORES

Very few of the basidiomycete (basidiomycotina) fungi have been studied intensively with biochemical procedures. In contrast, approximately 200 species have been investigated by ultrastructural procedures during the last 7--8 years. Since the first international fungal spore symposium in 1966, publications which have dealt with ultrastructure of dormant and germinated basidiomycete spores have been reported for many families. Approximately 33% of the electron microscopy studies have been with Uredinales, 24% with Ustilaginales and 43% with the Hymenomycetes and Gasteromycetes.

Since authors who have written other chapters for this book have discussed ultrastructural details about spore walls and cytoplasm this discussion of basidiomycete spore ultrastructure is very general. However, specific details of some smut fungi and developing, dormant, and germinated *Tilletia caries* and *T. controversa* teliospores are discussed to clarify the kinds of morphological information which may be obtained with electron microscopy.

Electron Microscopy Procedures Employed

When the first international fungal spore symposium was held, scanning electron microscopy was not available for fungal spore investigations. During the last few years, scanning electron microscopy has become a very important research tool for investigations of fungal spores. Almost 40% of the publications which have been concerned with ultrastructure of dormant and germinated basidiomycete fungal spores since 1966 have involved the use of scanning electron microscopy and freeze-etching has been used in approximately 14% of the studies (Table 2). Also, complementary fracture freeze-etching is available in many laboratories, although it has not been used yet for investigations of basidiomycete spores.

It is much easier to prepare fungal spore samples for examination with scanning electron microscopy than it is to prepare suitable freeze-etch replicas. Studies which involve freeze-etching and scanning electron microscopy for the same study have not been conducted with Basidiomycete spores (Table 2), although Cole & Ramirez-Mitchell (1974) investigated freeze-etch replicas of conidia of an imperfect fungus with scanning electron microscopy.

Very few investigations have been conducted in attempts to correlate structure with function. Most of the emphasis has been upon structure. There is a need for more investigations which involve histochemistry at the ultrastructural level and correlation of structure with function.

Use of Electron Microscopy for Taxonomy

Replicas have been used with transmission electron microscopy for investigations of fungal spores (Bigelow & Rowley, 1968; Pegler, 1969), but most investigators now use scanning electron microscopy or transmission electron microscopy with thin sections or freeze-etch replicas for taxonomic investigations. Hymenomycetes and gasteromycetes have been investigated by all of these procedures (Besson-Antoine & Kuhner, 1972; Grand & Moore, 1971; Hashioka, 1971; Pegler & Young, 1972a, 1972b, 1972c; and Bronchart & Demoulin, 1973). Ultrastructure atlases are available for some species (Pegler & Young, 1972a, 1972b).

During the mid 1960's, Hess et al. (1966) used freeze-etching to investigate conidia of Penicillia. These investigators pointed out that conidia of Penicillia have surface rodlet patterns. Similar rodlets were reported for conidia of *Aspergillus* (Hess & Stocks, 1969) and other fungi (Hess et al., 1972). Lycoperdon basidiospores also have rodlets

Table 2. Electron Microscopy Procedures Employed Since 1966 for Investigations of Dormant and Germinated Basidiomycete Spores in Studies With Major Emphasis Upon Ultrastructure.

Procedures employed	Percent of studies reported
Thin sectioning	40.0
Thin sectioning and freeze-etching	7.0
Freeze-etching	7.0
Scanning electron microscopy	30.0
Scanning electron microscopy and thin sectioning	8.0
Scanning electron microscopy, thin sectioning and freeze-etching	1.0
Replica procedures	7.0
Scanning electron microscopy and freeze-etching	0.0
Complementary fracture freeze-etching	0.0

(Bronchart & Demoulin, 1971; Hess et al. 1972). Investigations with two species of *Aspergillus* indicated that rodlet patterns are not consistent enough to use as taxonomic characters even when monosporous cultures are used (Fujii & Hess, 1969). To date, attempts have not been made to use rodlet patterns of *Lycoperdon* basidiospores for taxonomic purposes.

Surface characteristics of spores of many rust fungi have been reported to have contributed significantly to the taxonomic relationships of rust fungi. Thin sectioning has been used to observe reticulate ornamentation (Henderson et al., 1972a), and spines and refractive granules (Henderson et al., 1972b). Littlefield (1971) and Littlefield & Bracker (1971) used various electron microscopy procedures to elucidate spine development of *Melampsora* uredospores, and scanning electron microscopy to demonstrate that wild type *Melampsora* uredospores were echinulate with sharply pointed, usually straight, spines. Whereas mutant uredospores of *Melampsora* were finely verrucose, ornamented with rounded, wartlike projections and were usually spineless.

Thomas & Isaac (1967) used electron microscopy to determine the sequence of events in the formation of spines on the uredospore walls of *Puccinia graminis tritici* uredospores, and Corlett (1970) used scanning electron microscopy to elucidate surface structures of echinulate uredospores of *Puccinia coronata*. Brandenburger & Schwinn (1971) also used scanning electron microscopy to demonstrate the different types of surface structures of teliospores and uredospores of rust fungi. They demonstrated warts, caves, ridges and spines. Many other investigators have used scanning electron microscopy for investigations of rust fungi (Holm et al., 1970; Hiratsuka, 1971; Grand & Moore, 1972; and Jones, 1971).

Many features of spores of the Ustilaginales have also been examined by various electron microscopy procedures (Hashioka et al., 1966; Dumitras, 1969, 1970; Durrieu & Rajeriarison, 1968; Jones, 1968; Kukkonen, 1969; Savulescu et al., 1968; Zogg & Schwinn, 1971; and Allen et al., 1971), although a relatively small number of spores of the Ustilaginales have been examined with electron microscopy.

Electron Microscopy of Spore Cytoplasm

Many interesting features of cytoplasm of spores of hymenomycetes and gasteromycetes have been discovered by the use of electron microscopy during the past few years. The use of freeze-etch replicas has made it possible to observe invaginations in plasma membranes. Sassen et al. (1967) were the first to demonstrate that fungal spores have invaginations. They have now been reported for several fungi, but it is normally necessary to prepare freeze-etch replicas to visualize them. It is also interesting that some fungi do not have them. A notable example is *Rhizopus arrhizus* (Hess & Weber, 1973). The invaginations were first reported for basidiomycete fungi by Stocks & Hess (1970). Subsequently other investigators who used freeze-etching in their studies have also reported them (Griffiths, 1971; Bronchart & Demoulin, 1970). Only a few kinds of fungal spores have been investigated with freeze-etching. For this reason it is not known whether this feature will be a common feature in all groups of fungi. They have also been observed in *Tilletia* teliospores (Allen et al., 1971) and *Lycoperdon* basidiospores (Bronchart & Demoulin, 1970), and in groups of fungi other than the basidiomycetes (W. M. Hess, unpublished results).

The use of freeze-etching has also made it possible to visualize other interesting features of fungal spores. Bronchart & Demoulin (1970) demonstrated that lomasomes were present in mature basidiospores of *Hypholoma fasciculare*, which indicates that lomasomes are not a fixation artifact, and Bauer & Tanaka (1968) demonstrated a regular pattern on both sides of the outer mitochondrial membrane of mature *Basidiobolus* spores. They demonstrated bands consisting of 5--10 parallel ridges, each of which were 20 nm wide. Features such as these are visible only with freeze-etch replicas.

Figs. 1--4 Freeze-etch replicas of dormant *Ustilago* teliospores. Fig. 1. A portion of a surface of a *U. bromniveri* teliospore, x 11, 480. Fig. 2. A portion of a *U. bromniveri* teliospore showing the plasma

Chang & Tanaka (1971) demonstrated membranous structures in germ tubes of *Volvariella* basidiospores. These structures consisted of three types of configurations: polyvesicular, myelinlike, and multitubular. The formation of these structures appeared to be mediated by the endoplasmic reticulum. Numerous microtubules were observed by McLaughlin (1973) during formation of basidia of *Coprinus* and *Boletus*. These observations and similar observations indicate that significant differences in spore structure and function exist from family to family, from genus to genus, from species to species, and even in spore types of the same fungus. Heintz & Niederpruem (1970, 1971) pointed out that significant morphological and physiological differences exist between oidia and basidiospores of *Coprinus lagopus*.

Spore Ultrastructure

Freeze-etching has provided a means to investigate aspects of spore morphology which were not previously possible to investigate. Dormant smut teliospores are very difficult to investigate by standard thin sectioning techniques, although after smut teliospores germinate, thin sectioning techniques may be used for ultrastructural investigations. Freeze-etching provides a means to view minute details of spore surface structures, surface views of membranes, and detailed information about organelle morphology (Figs. 1--8). Morphological differences between teliospores of different species of fungi may be investigated in detail. When teliospores of three species of *Ustilago* were examined, it was obvious that spore surface characteristics differ and that invagination patterns in plasma membrane surfaces also differ significantly (Figs. 1--5). The invaginations in the plasma membranes which are visible in freeze-etch replicas (Figs. 2, 4, 5) may be species or subspecies specific, however, investigations have not been conducted to date to determine whether this suggestion has

---

membrane where the spore wall has been fractured away, x 12,300. Figure 3. A portion of a surface of a *U. bullata* teliospore, x 12,300. Figure 4. A portion of a *U. bullata* teliospore showing the plasma membrane where the spore wall has been fractured away, x 11,800. Note the characteristic invaginations in the plasma membranes in Figs. 2 and 4.

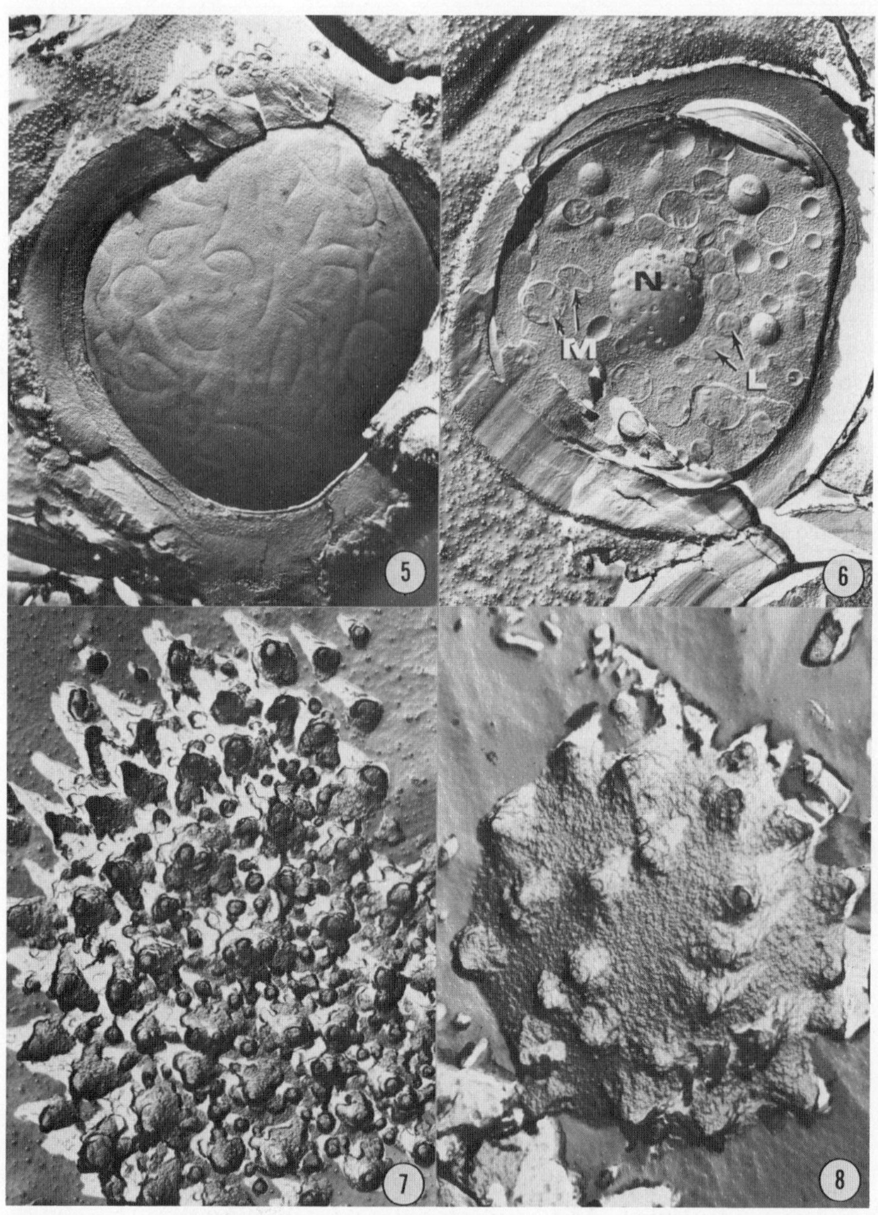

Figs. 5--8 Freeze-etch replicas of dormant grass smut teliospores. Figure 5. A portion of a *Ustilago hordei* teliospore showing the cross fractured spore wall and the plasma membrane where the wall has

validity. Certainly, teliospore surface details differ significantly from species to species.

When teliospores are cross fractured, it is possible to study spore wall layers, the relative thickness of spore walls of different species, and surface and cross fractured views of spore organelles (Fig. 6). Nuclei, mitochondria, and lipid bodies are readily evident in cross fractured teliospores (Fig. 6). Also, with freeze-etch replicas, organelle morphology is much more reliable because the only preparative techniques necessary for dormant spores is to freeze them very rapidly prior to fracturing and preparation of replicas.

It is obvious that if spore surface details differ significantly from species to species within the same genus, that significant differences will also be evident from genus to genus (Figs. 7--8). Although samples are relatively easy to prepare for scanning electron microscopy, particularly when compared with procedures necessary for preparation of freeze-etch replicas, scanning electron microscopy does not provide the detail which is available with freeze-etch replicas. However, commercial scanning electron microscopes are now available which have resolution in the range of 5 nm. For spore surface investigations, scanning electron microscopy in this range of resolution will almost approach the resolution obtainable with freeze-etch replicas. Since transmission electron microscopes have considerably better resolution, it is possible that freeze-etch replicas will still provide more information when examined with transmission electron microscopy. Also, it is difficult to get reliable information about cytoplasm with scanning electron microscopy since it is necessary to freeze dry and fracture specimens before examination with scanning electron microscopy. The processes necessary for freeze drying and fracturing introduce artifacts, particularly when compared to the relatively artifact free replicas obtained with freeze-etching. It is obvious that more reliable morphological and histochemical information may be obtained when scanning electron microscopy, thin sectioning, and freeze-etching

---

been fractured away, x 9,840. Note the characteristic invaginations in the plasma membrane. Figure 6. A cross fractured *U. hordei* teliospore showing the cross fractured spore wall, a nucleus (N), mitochondria (M), lipids (L) and other organelles, x 9,020. Figure 7. A portion of a surface of a *Sphaeclotheca reiliana* teliospore, x 13,120. Figure 8. A portion of a surface of a *Ustilago maydis* teliospore, x 9,840.

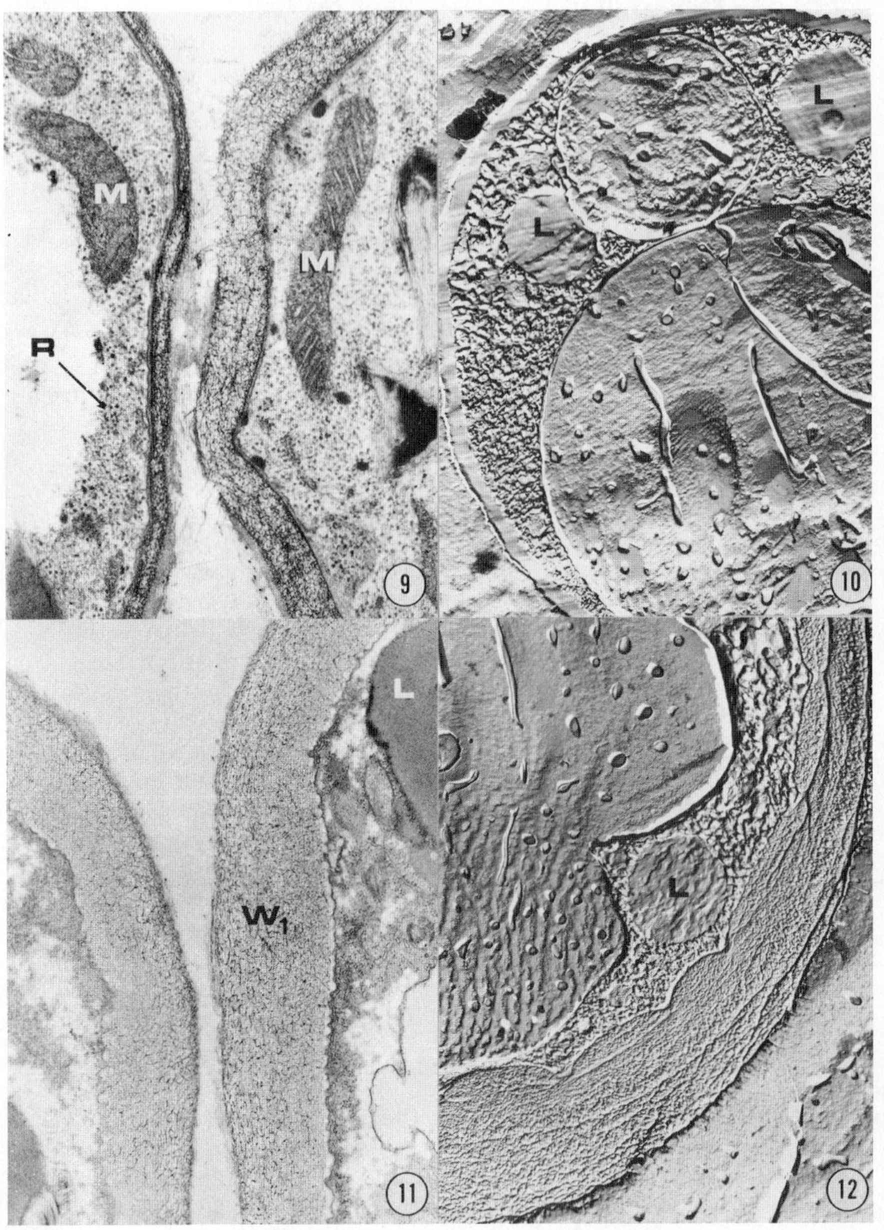

Figs. 9--12 Developing *Tilletia* teliospores. Figure 9. Developing *T. caries* teliospores showing ribosomes (R), mitochondria (M) and

are all utilized to investigate the same tissue system. Of course, it is not possible to utilize all three of these procedures on all dormant and germinated fungal spores, primarily because of the thick spore wall layers.

Ultrastructure of *Tilletia* Teliospores

When developing dormant and germinated *Tilletia caries* and *Tilletia controversa* teliospores are investigated with thin sectioning, freeze-etching, and scanning electron microscopy, it is possible to characterize changes which take place in spore walls and cytoplasmic structures during stages in the life cycles of these fungi. Each electron microscopy procedure provides a uniquely different type of information. Although the biochemistry, time of host infection, and physiology of the teliospores of these two species of fungi differ significantly, they appear, based upon studies conducted to date, to be morphologically identical.

Ultrastructural Changes During Development of *Tilletia* Teliospores

Tilletia caries and *T. controversa* teliospores germinate in the soil and infect young wheat seedlings. The hyphal cells grow through the wheat plants to the developing wheat kernels. The host tissue inside wheat kernels is replaced by fungal tissue, which becomes converted almost entirely to teliospores. As the teliospores develop, very thin spore walls are formed initially and mitochondria and ribosomes are readily evident (Fig. 9). Lipid bodies are also readily evident (Fig. 10) and developing spores appear to have a relatively high concentration of water, as is evidenced by the fracture pattern obtained with freeze-etch replicas (Fig. 10). As the spores continue to

---

early stages of spore wall development, x 19,680. Figure 10. Freeze-etch replica of a cross-fractured *T. controversa* teliospore during an early stage of development showing lipids (L) and other organelles, x 6,560. Figure 11. Developing *T. controversa* teliospores showing the developing outer spore wall layer (W_1) and lipid (L), x 14,760. Figure 12. Freeze-etch replica of a cross-fractured *T. controversa* teliospore showing lipid (L) and invaginations of the inner surface of the developing spore wall where the W_2 spore wall layer is deposited, x 6,560.

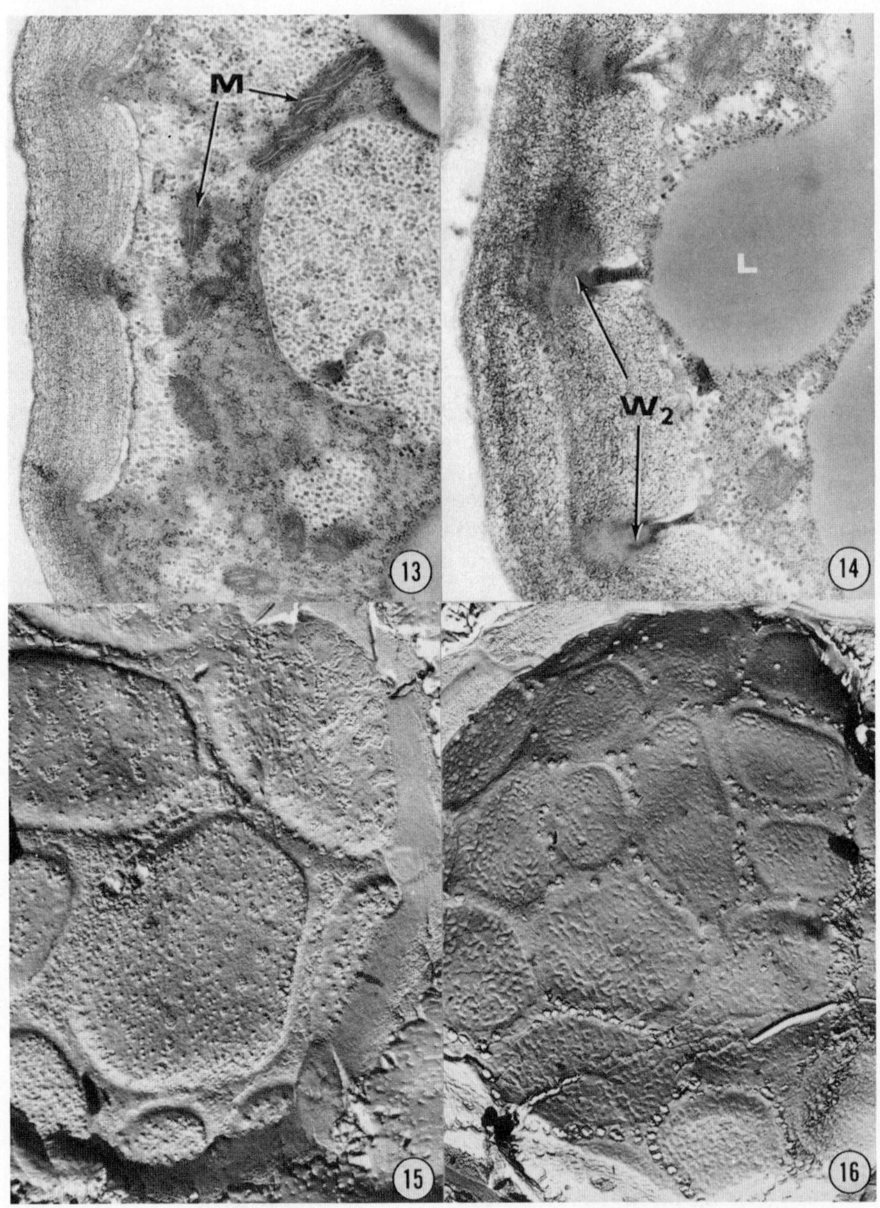

Figs. 13--16 Developing *Tilletia controversa* teliospores. Figure 13. Developing teliospore showing mitochondria (M) and invaginations in the plasma membrane where electron dense spore wall material appears

develop, the teliospore walls become much thicker, although distinct layering is not evident during the early stages of development (Fig. 11). As spore wall development continues, invaginated areas become distinct both with freeze-etching and thin sectioning (Figs. 12--14). As *Tilletia* teliospore walls continue to develop, electron dense W_2 wall material is obviously deposited at the regions of the invaginations in the plasma membrane (Figs. 13--14).

If spores are fractured so that surface views of the interface between the membrane and the cell wall are exposed, the reticulated pattern, which is indented into the cell wall, is visible (Figs. 15--16). In the reticulated regions of developing spore wall membranes, irregular areas are evident, which indicate that materials are being deposited through the membrane much more rapidly at the sites of the reticulations than in the other areas of the wall (Fig. 16). As the spores become more mature, and as the W_2 wall layer becomes continuous around the spore, the reticulated areas are still seen (Fig. 17). However, after the W_2 wall deposition is complete, or almost complete, the reticulated pattern of the plasma membrane is no longer evident (Fig. 18). After deposition of the outside (W_1) layer, and the relatively electron dense W_2 wall layer, the partition layer is deposited after which the W_3 wall layer is deposited (Fig. 19). Figure 19 shows a spore which is almost mature and contains irregular areas in the plasma membrane where wall material is possibly being deposited into the W_3 wall layer. After all of the wall layers have been deposited, the plasma membrane contains characteristic invaginations, as can be seen in Fig. 20. These invaginations are visible after the cytoplasm has been fractured away from the plasma membrane, and spore wall layers. If the spore wall layers are fractured away from the plasma membrane, the invaginations may also be seen (Fig. 21). However, when the spore wall layers are fractured away from the plasma membrane, the invaginations are invaginated (Fig. 21), but when the cytoplasm is fractured away from the wall, the invaginations

---

to be deposited, x 14,760. Figure 14. Developing teliospore showing lipid (L) and deposited electon dense (W_2) wall material, x 14,760. Figure 15. Freeze-etch replica of a portion of a teliospore showing the plasma membrane during the initial deposition of W_2 wall material, x 9,840. Figure 16. Freeze-etch replica of a portion of a teliospore showing the plasma membrane during active deposition of W_2 wall material, x 5,330.

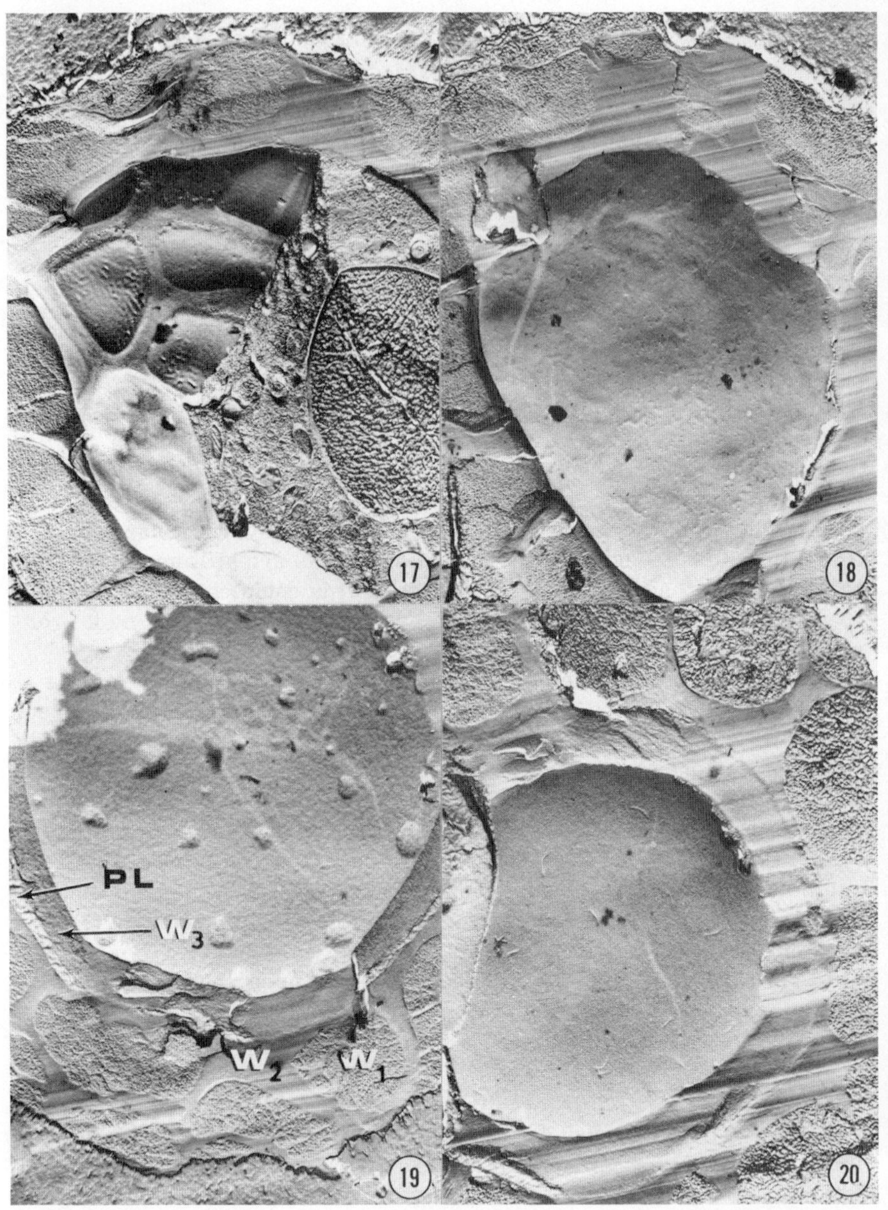

Figs. 17--20 Freeze-etch replicas of developing *Tilletia controversa* teliospores. Figure 17. A teliospore showing a portion of the plasma membrane after deposition of most of the W_2 wall layer, x 5,740.

appear as ridges (Fig. 20).

When *Tilletia caries* and *T. controversa* teliospores are exposed by surface fractures, a characteristic reticulation is evident (Fig. 22). This reticulated pattern reflects the structure if the W_2 wall layer is viewed with thin sectioning (Fig. 23) or with freeze-etch replicas (Fig. 24).

Ultrastructure of Dormant *Tilletia caries* and *T. controversa* Teliospores

Since the reticulated pattern is distinctly evident in freeze-etch replicas of dormant *Tilletia* teliospores (Fig. 22) and since thin sections of spore walls (Fig. 23) and cross fractured spore walls (Fig. 24) usually do not show significant indentations of the outer spore wall layer, it is possible that some of the more fibrous W_1 spore wall material is fractured away during the freeze-etching (fracturing) process. It must also be recognized that there is a considerable amount of variation in not only the reticulated pattern, but also in the amount of depression of the W_1 layer between the reticulations of the W_2 wall layer. In most instances, however, when cross fractured spore walls are viewed, this indentation is minimal. However, most surface fractures reveal a distinct indentation as is shown in Fig. 22.

When *Tilletia* teliospores are viewed with scanning electron microscopy (Fig. 26) the indentation is much more distinct. However, it is very likely that the scanning electron microscopy image is not accurate, primarily because emissions from the surfaces of spores may not be from the W_1 layer, particularly since this layer appears to be very fibrous with both thin sectioning and freeze etching techniques (Figs. 23--24). It is possible that with scanning electron microscopy the W_1 wall layer may not be visualized because of the chemical composition and lack of density. However, there is no other experimental evidence to elucidate this question.

Figure 18. A portion of a teliospore showing the plasma membrane during deposition of the partition layer, x 5,740. Figure 19. A portion of a teliospore showing the plasma membrane after depositon of most of the W_3 wall layer. The other wall layers (W_1, W_2, and PL) are also shown, x 5,740. Figure 20. A portion of a teliospore showing the plasma membrane of a mature spore, x 5,740. Note that the characteristic invaginations are present.

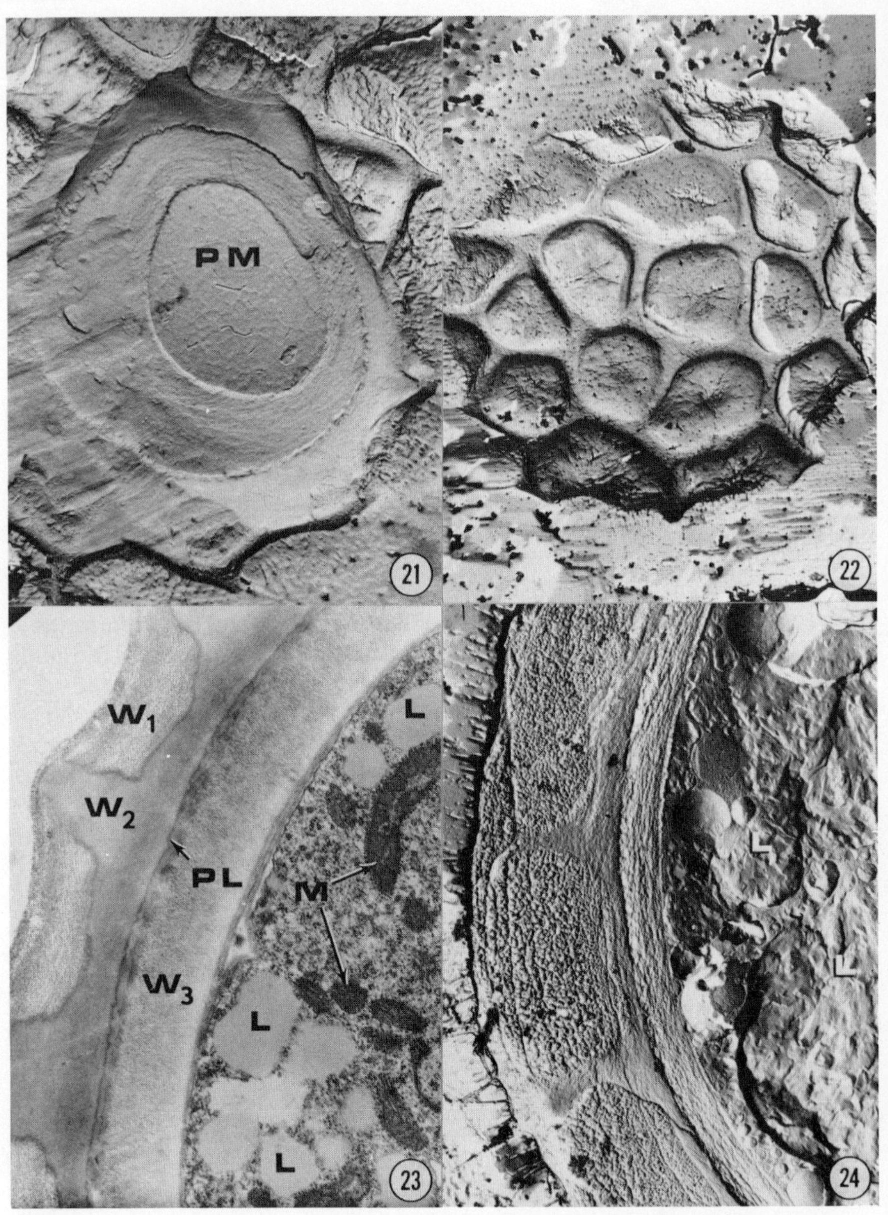

Figs. 21--24 *Tilletia* teliospores. Figure 21. Freeze-etch replica of a dormant *T. caries* teliospore showing the plasma membrane (PM) where the wall layers were fractured away, and the characteristic

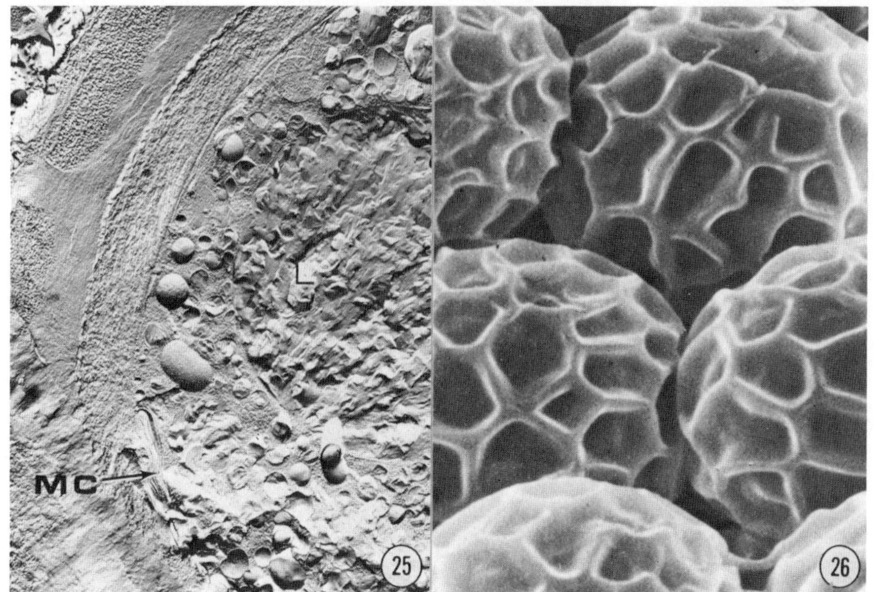

Figs. 25--26 *Tilletia* teliospores. Figure 25. Freeze-etch replica of a germination *T. caries* teliospore showing the characteristic spore wall layers, lipid (L) and a membrane complex (MC), x 7,200. Figure 26. Scanning electron micrograph of *T. controversa* teliospores, x 2,880.

After fungal spores germinate, all of the spore wall layers which are present in the dormant spores are still distinctly present in the germinated spores (Fig. 25) and structural modifications are not evident in the spore walls except where germ tube initiation is seen, and those areas are very localized. Portions of dormant teliospore surfaces are shown in Figs. 27 and 28. The ridges of the reticulations

---

invaginations in the plasma membrane, x 5,330. Figure 22. Freeze-etch replica of a dormant *T. caries* teliospore showing the characteristic reliculated pattern, x 4,100. Figure 23. A portion of a *T. caries* teliospore during early stages of germination showing the relatively thick wall layers (W_1, W_2, and W_3) the partition layer (PL) mitochondria (M) and lipids (L) in the cytoplasm, x 15,580. Figure 24. Freeze-etch replica of a dormant *T. controversa* teliospore showing the characteristic wall layers and abundant lipids (L), x 7,380.

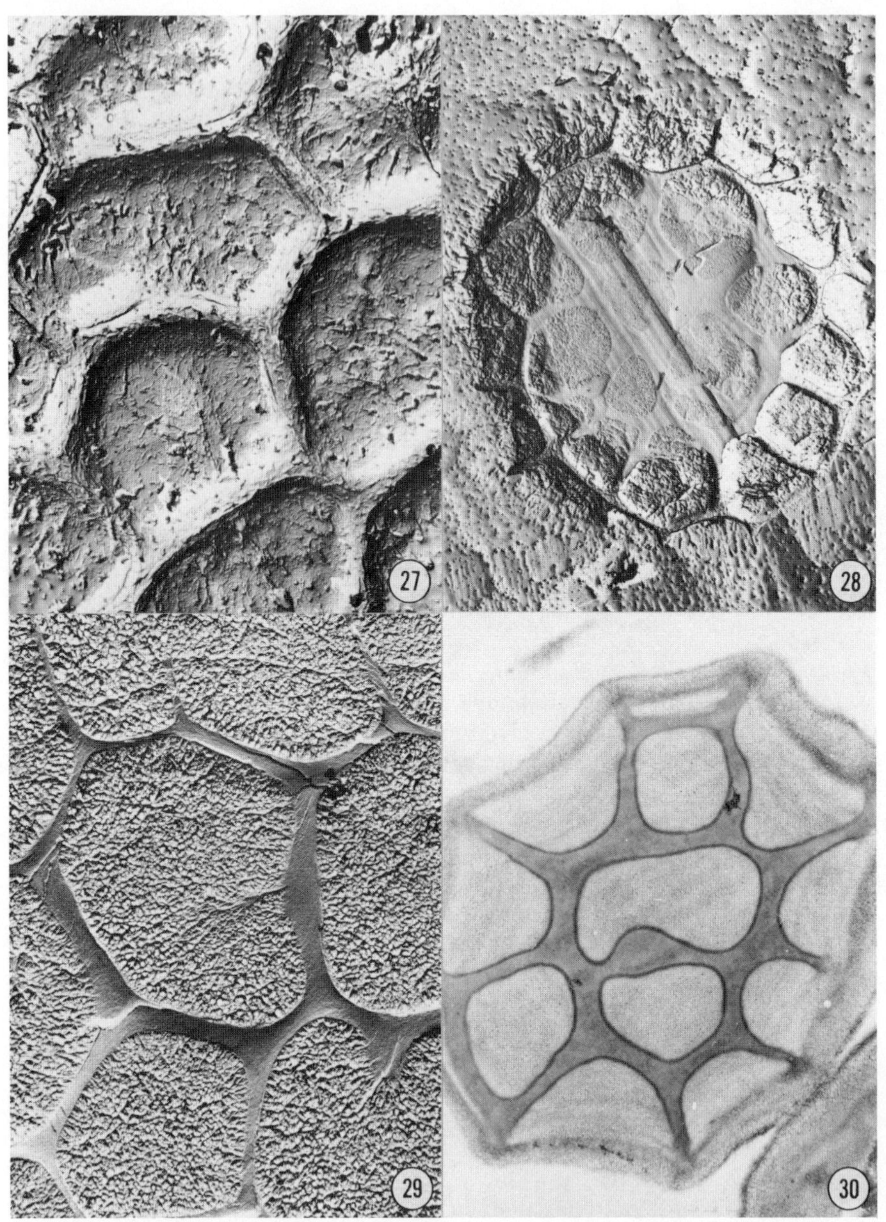

Figs. 27--30 Dormant *Tilletia controversa* teliospores. Figure 27. Freeze-etch replica of a portion of a surface of a teliospore showing the characteristic reticulated pattern, x 7,380. Figure 28. Freeze-etch

of dormant spores appear to be relatively rigid. As surface layers of teliospores are fractured away (Fig. 28) it is obvious that the relatively less electron dense W_2 wall material between the reticulated areas is more fibrous than the reticulated areas (Figs. 28--30). The electron dense W_2 wall layer is distinctly denser than the W_1 wall material when viewed with freeze-etching (Fig. 29) and with thin sectioning (Fig. 30). When teliospores are obliquely fractured, the relatively fibrous W_1 wall material, the relatively dense W_2 wall material, the distinctly undulating partition layer, and the relatively dense W_3 wall layer are distinct (Fig. 31).

Ultrastructure of Germinated *Tilletia caries* and *T. controversa* Teliospores

As was mentioned above, the spore wall layers which are present in dormant spores are unchanged in germinated spores. However, the W_1 layer sometimes is not distinct in either dormant or germinated spores (Fig. 32). However, the W_2 layer, the partition layer, and the W_3 layer have been observed in all the spores examined to date. After spores germinate, significant changes take place in spore organelles. *Tilletia* teliospores contain large amounts of lipids and unidentified organelles which are possibly associated with the glyoxalate cycle (Allen et al., 1971). These unidentified organelles change significantly in shape and content upon germination. Also, when germinated spores are viewed with thin sectioning techniques lipids appear to fuse together (Fig. 33), and when spores are germinated in distilled water unusual aggregates of particles are evident on the plasma membrane (Fig. 34) which are not present when spores are germinated on agar surfaces and they are not present in mature dormant spores. Possibly these aggregates of particles on or in the membrane are a morphological alteration which reflect an abnormal metabolic process.

As the germination process proceeds, germ tubes which are surrounded by new wall material develop rapidly and contain distinct lipids, mitochondria, endoplasmic reticulum, and microbodies. These

---

replica of a teliospore which has been cut to expose portions of the outer two spore wall layers, x 3,690. Figure 29. Freeze-etch replica of a cross fractured spore showing portions of the two outer spore wall layers, x 7,380. Figure 30. Thin section of the outer two spore wall layers showing the charateristic reticulated pattern of the more electron dense, W_2 layer, x 5,330.

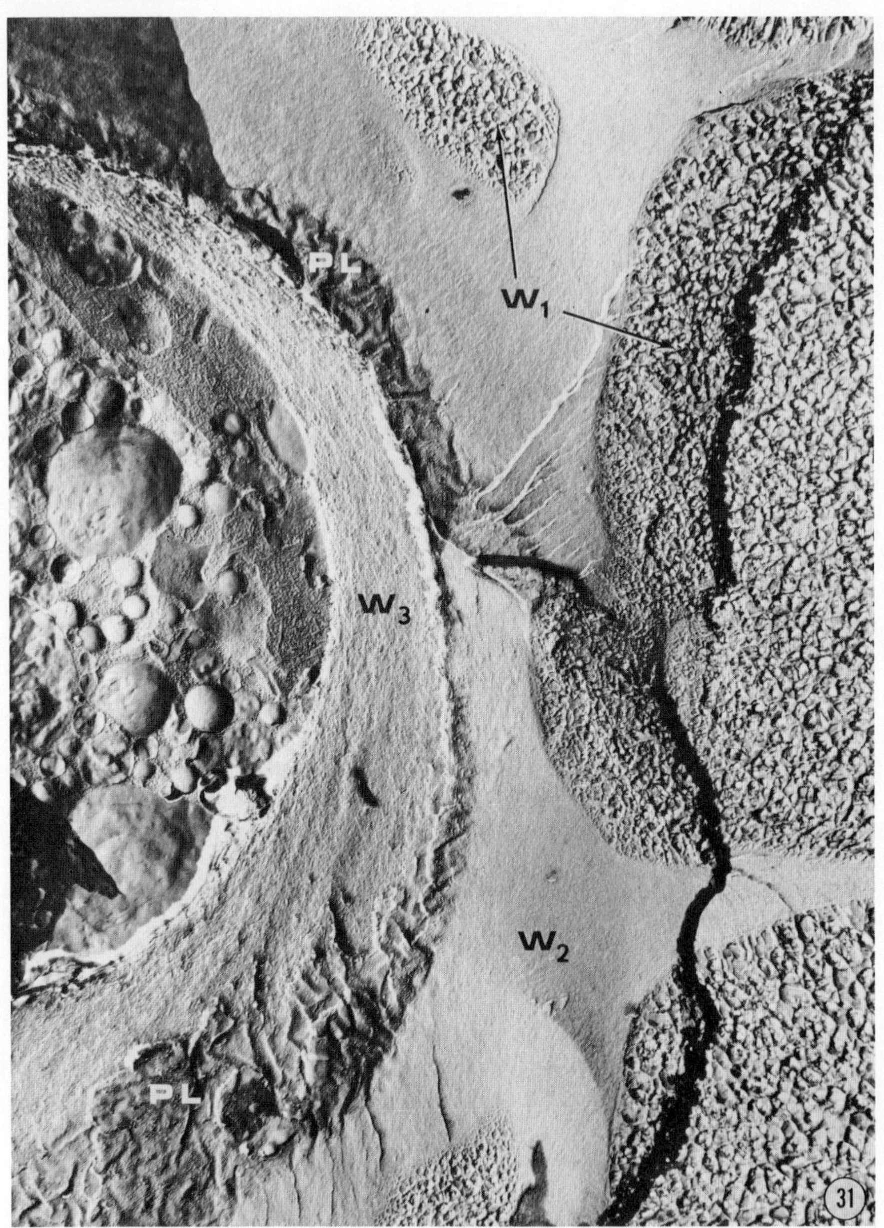

Fig. 31 A freeze-etch replica of a dormant *Tilletia controversa* teliospore showing the outer wall layers (W_1 and W_2), the partition layer (PL), the inner wall layer (W_3), and organelles in the cytoplasm, x 18,860.

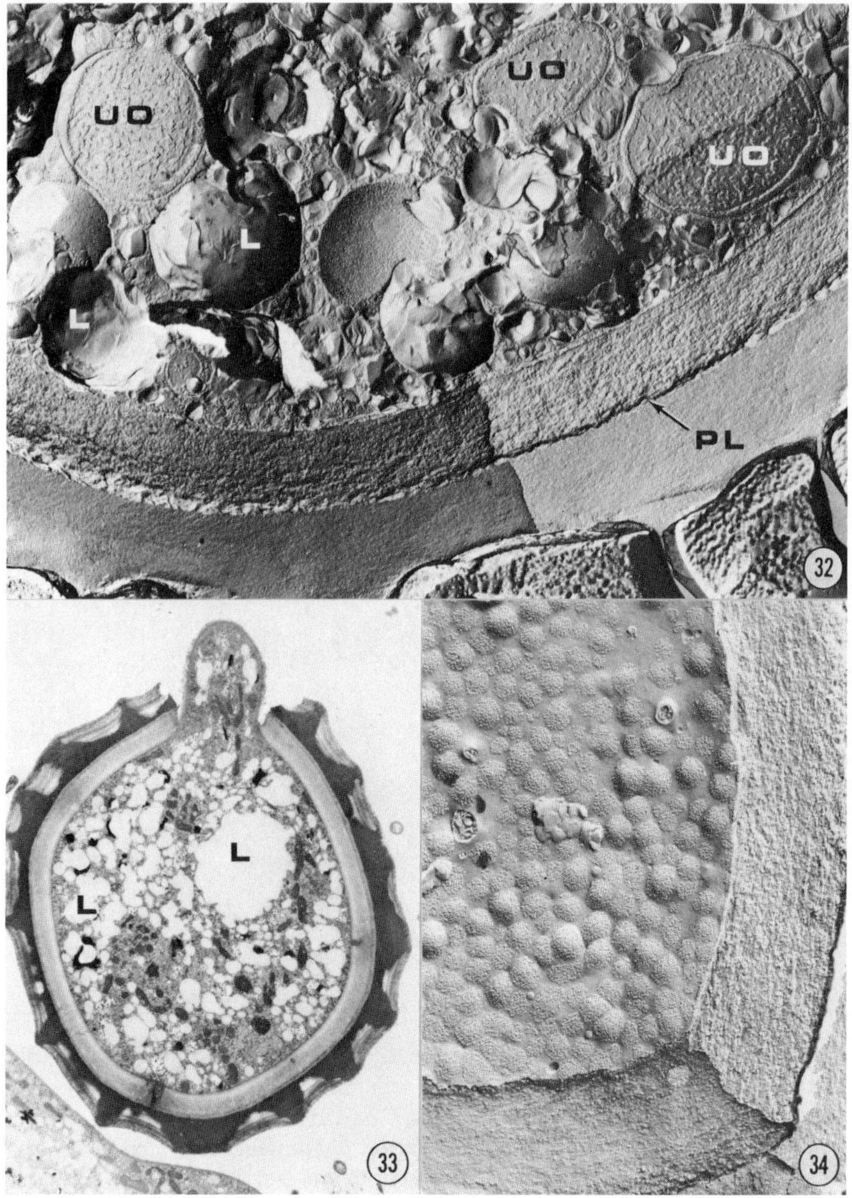

Figs. 32--34 Germinated *Tilletia caries* teliospores. Figure 32. Freeze-etch replica of a cross-fractured teliospore showing the spore wall which appears to be void of the W_1 layer. The partition layer

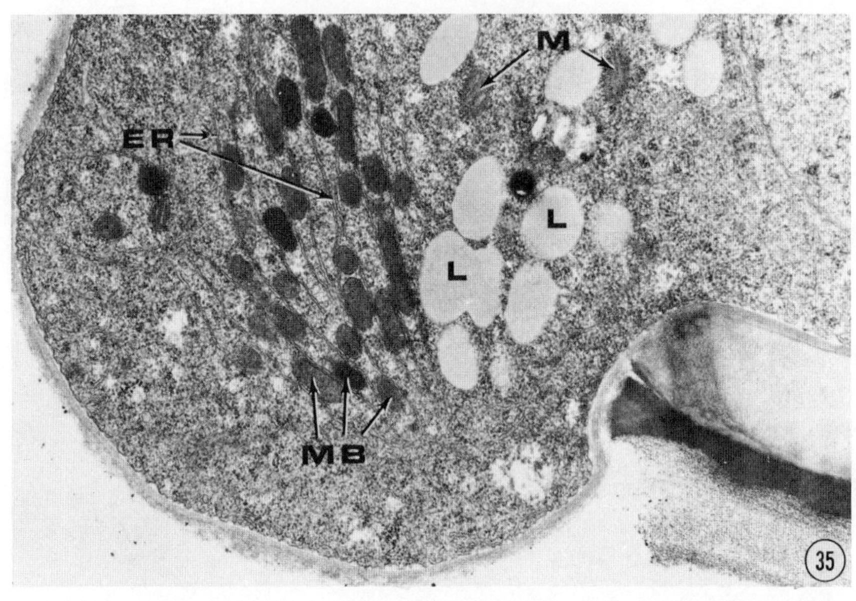

(PL) is readily evident, and unidentified organelles (UO) are present in the cytoplasm, x 11,200. Figure 33. Thin section of a teliospore showing the characteristic wall layers and lipids (L), x 3,600. Figure 34. Freeze-etch replica of a portion of the wall and membrane of a teliospore showing aggregates of membrane particles, x 16,000.

Figs. 35--36 Germinated *T. caries* teliospores. Figure 35. A portion of a germ tube showing lipids, (L), mitochondria (M), endoplasmic reticulum (ER) and organelles which appear to be microbodies (MB),

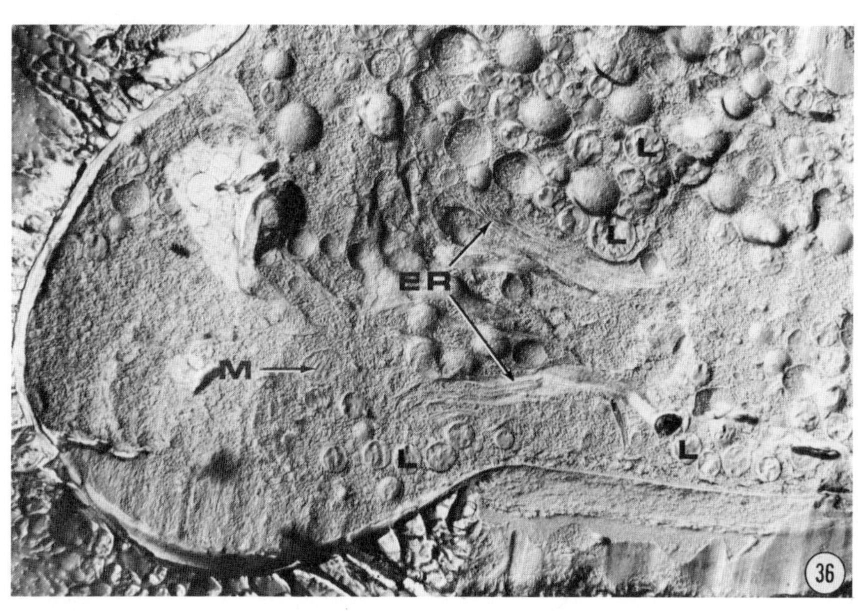

x 16,000. Figure 36. Freeze-etch replica of a portion of a germ tube showing lipids (L), a mitochondrion (M), endoplasmic reticulum (ER), and other organelles, x 8,800.

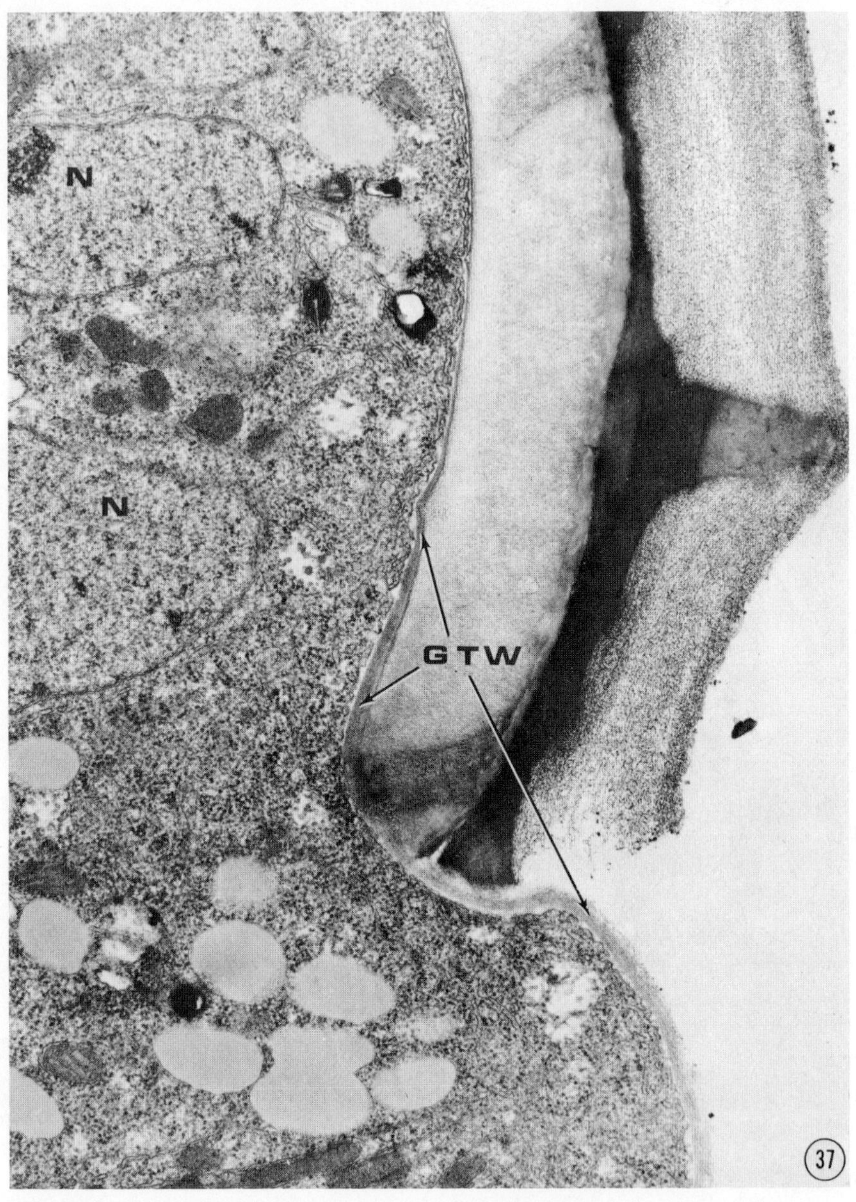

Fig. 37 A portion of a germinated *Tilletia caries* teliospore showing the newly developed germ tube wall (GTW), nuclei (N) and other organelles, x 21,200.

organelles are readily evident with thin sections and freeze-etch replicas (Figs. 35--36). Although endoplasmic reticulum is not evident in dormant spores, germinated spores have distinct endoplasmic reticulum (Figs. 35--36) and membrane complexes are often observed (Fig. 25). The developing germ tube wall is commonly evident adjacent to the old W_3 wall layer (Fig. 37).

As the germination process is initiated in teliospores, several areas in the spore wall will initiate the development of new wall material and the dissolution of the old wall layers (Fig. 38). Several of these sites may be observed in a single germinating teliospore. After

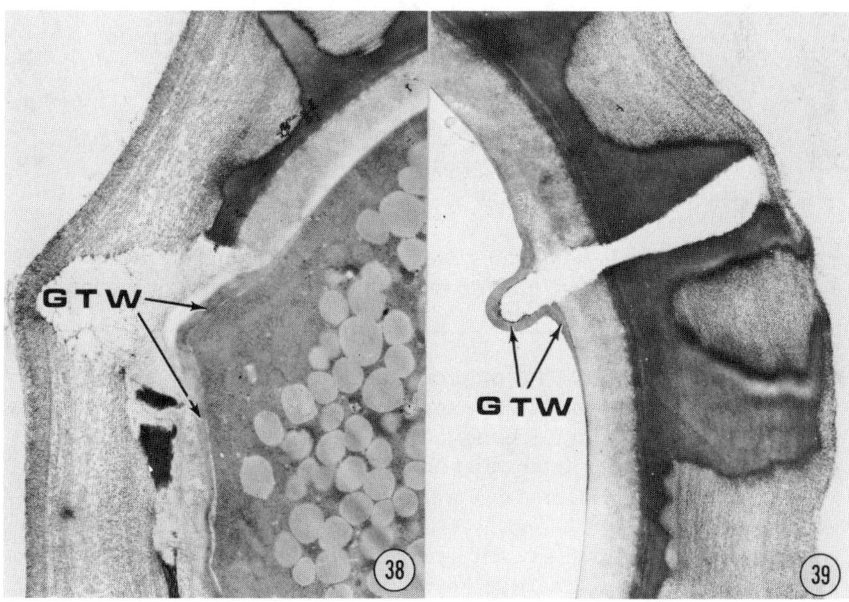

Figs. 38--39 Germinated *Tilletia controversa* teliospores. Figure 38. A portion of a teliospore showing the newly developed germ tube wall (GTW) and the dissolution of the other spore wall layers, x 10,000. Figure 39. A teliospore wall showing the newly developed germ tube wall (GTW) and the dissolution of the other spore wall layers. Germination at another site in the wall resulted in germ tube development and exit of the spore cytoplasm x 10,000.

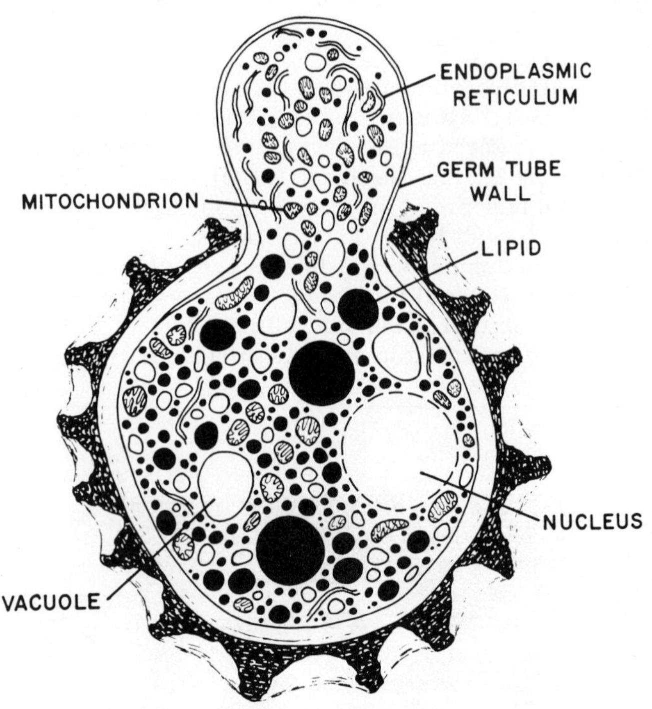

Fig. 40 Diagram showing the general distribution of organelles of *Tilletia caries* and *Tilletia controversa* teliospores during germination.

---

the germination process is complete and after the cytoplasm has left the spore, distinct areas in the spore wall are still evident where new germ tube walls began to develop and where areas in the old spore wall were almost completely dissolved (Fig. 39).

Figure 40 is a diagram of a germinated spore showing the spore wall layers, a nucleus, a vacuole, mitochondria, the new germ tube wall, distinct lipid bodies, and endoplasmic reticulum, all of which are readily distinguishable in germinated teliospores. Since histochemical investigations to positively identify microbodies have not been conducted, these organelles were omitted from this diagram.

An understanding of the developmental process of *Tilletia* teliospores, the structure of dormant teliospores, and the changes which take place as *Tilletia* teliospores germinate, may help to elucidate similar processes and phenomena which take place in other basidiomycete spores, although developmental processes and changes which take place during germination are assumed to be as diverse from species to species as spore morphology is from species to species.

ACKNOWLEDGMENTS

This research was supported in part by NIH Career Development Award No. 1-K4-GM-29, 976 to W. M. Hess, Research Corporation, and Brigham Young University Research Division.

REFERENCES

Aitken, W. B. & Niederpruem, D. J. (1970) Ultrastructural changes and biochemical events in basidiospore germination of *Schizophyllum commune*. J. Bacteriol. 104, 981--988.

Aitken, W. B. & Niederpruem, D. J. (1972). Isotopic studies of carbohydrate metabolism during basidiospore germination in *Schizophyllum commune:* I. Uptake of radioactive glucose and sugar alcohols. Arch. Mikrobiol. 82, 173--183.

Aitken, W. B. & Niederpruem, D. J. (1973). Isotopic studies of carbohydrate metabolism during basidiospore germination in *Schizophyllum commune:* II. Changes in specifically labeled glucose and sugar alcohol utilization. Arch. Mikrobiol. 88, 331--344.

Allen, J. V., Hess, W. M. & Weber, D. J. (1971). Ultrastructural investigations of dormant *Tilletia caries* teliospores. Mycol. 63, 144--156.

Allen, P. J. (1965). Metabolic aspects of spore germination in fungi. Annu. Rev. Phytopathol. 3, 313--342.

Allen, P. J. (1972). Specificity of the cis-isomers of inhibitors of uredospore germination in the rust fungi. Proc. Natl. Acad. Sci. U.S.A. 69, 3497--3500.

Allen, P. J., Strange, R. N., & Elnaghy, M. A. (1971). Properties of germination inhibitors from stem rust uredospores. Phytopathology 61, 1382--1389.

Babayan, A. A. & Khachatryan, G. A. (1972). O faktorakh vliyayushchikh na prorastanie uredospor stevlevoi rzhavchiny pshenitsy. Biol. Zh. Arm. 25, 60--68.

Bahadur, P. & Sinha, S. (1966). Self-inhibition and self-stimulation of uredospore germination in *Uromyces-ciceris-arietini*. Proc.

Ind. Sci. Congr. 53, 222.

Baker, K. & Strobel, G. A. (1965). Lipids of the cell wall of *Puccinia striiformis* uredospores. Proc. Montana Acad. Sci. 25, 83--86.

Bauer, H. & Tanaka, K. (1968). Ultrastructure of mitochondria and crystal containing bodies in mature Ballistospores of the fungus *Basidiobolus ranarum* as revealed by freeze-etching. J. Bacteriol. 96, 2132--2137.

Beceresc u, D. (1973). The influence of X-radiations on the growth of the dikaryotic mycelium and the behavior of haploid strains of certain species of *Ustilago*. Rev. Roum. Biol. Ser. Bot. 18, 119--124.

Besson-Antoine, M. & Kühner, M. R. (1972). L'ornementation sporale des Rhodophyllacees (Agaricales) et Son développement. C. R. Acad. Sci. Paris D 275, 543--548.

Bigelow, H. E. & Rowley, J. R. (1968). Surface replicas of the spores of fleshy fungi. Mycol. 60, 869--887.

Brandenburger, W. & Schwinn, F. J. (1971). Über Oberflächenfeinstrukturen von Rostsporen eine raster-elektronenmikroskopische untersuchung. Arch. Mikrobiol. 78, 158--165.

Bronchart, R. & Demoulin, V. (1970). Mise en evidence par le cryodecapage de lomasomes dans la basidiospore de *Hypholoma fasiculare* (huds. ex Fr.) Kummer. Planta 94, 229--232.

Bronchart, R. & Demoulin, V. (1971). Ultrastructure de la paroi des basidiospores de *Lycoperdon* et de *Schleroderma* (Gastéromycètes) comparée à celle de quelques autres spores de champignons. Protoplasma 72, 179--189.

Bronchart, R. & Demoulin, V. (1973). Ultrastructure de la paroi sporale des gastéromycètes-*Mycenastrum corium* et *Abstroma retuculatum* en rapport avec leur position systématique. Bull. Soc. R. Bot. Belgique 106, 267--272.

Brown, T. S., Jr. & Merrill, W. (1973). Germination of basidiospores of *Fomes applanatus*. Phytopathology 63, 547--550.

Bugbee, W. M. & Kernkamp, M. F. (1966). Storage of pyciniospores of *Puccinia graminis secalis* in liquid nitrogen. Plant Dis. Report 50, 576--578.

Burger, A., Prinzing, A., & Reisener, H. J. (1972). Untersuchungen ueber den Glucosestoffwechsel keimender Uredosporen von *Puccinia graminis* var. *tritici*. Arch. Mikrobiol. 83, 1--16.

Burger, A. & Reisener, H. J. (1970). Ueber die Veraenderung der Aktivitaeten einiger enzyme in den uredosporen von *Puccinia graminis* var. *tritici* waehrend der Keimung. Arch. Mikrobiol.

72, 260--266.

Burnett, J. H. (1968). Fundamentals of Mycology. New York: St. Martin.

Bush, L. (1968). Measurement of oxygen consumption of wheat stem rust uredospores with an oxygen electrode *(Triticum aestivum)*. Phytopathology 58, 752--754.

Bushnell, J. L. & Weber, D. J. (1974a). Lipid changes in spores of *Ustilago maydis* during germination. Unpublished data.

Bushnell, J. L. & Weber, D. J. (1974b). Lipid components in spores of *Lycoperdon perlatum* and *Calvatia gigantea*. Unpublished data.

Calpouzos, L. & Chang, H. (1971). Fungus spore germination inhibited by blue and far red radiation. Plant Physiol. 47, 729--730.

Caltrider, P. G. & Gottlieb, D. (1966). Effect of sugars on germination and metabolism of teliospores of *Ustilago maydis*. Phytopathology 56, 479--484.

Caltrider, P. G., Ramachandran, S., & Gottlieb, D. (1963). Metabolism during germination and function of glyoxylate enzymes in uredospores of rust fungi. Phytopathology 53, 86--92.

Chakravorty, A. K. & Shaw, M. (1972). Ribosmal RNA synthesis and ribosome formation during germination of flax-rust uredospores. Physiol. Plant Pathol. 2, 1--6.

Chang, S. & Tanaka, K. (1971). An electron microscope study of complex membranous structures in the basidomycete, *Volvariella volvacea*. Cytologia (Tokyo) 36, 639--651.

Cheng, S. (1969). Study on stimulation of spore germination in mushroom. (*Agaricus bisporus* (Lange) Sing.) (Chinese). J. Hort. Soc. China 15, 14--19.

Cheung, D. S. M. & Barber, H. N. (1971). Uredospore wall proteins of wheat stem rust: Localization and enzymatic activities. Arch. Mikrobiol.77, 239--246.

Chigrin, V. V., Bessmel'tseva, L. M., Polyakova, G. D., & Aleshin, E. P. (1970). Aktivnost' fermentov v pokoyashchikhsya i prorastayushchikh uredosporakh *Puccinia graminis* F. *tritici*. Sel'-skokhoz Biol. 5, 591--594.

Chigrin, V. V., Bessmel'tseva, L. M., & Rozum, L. V. (1969). Toksichnost'fenol'nykh soedinenii dlya prorastayushchikh uredospor steblevoi rzhavchiny pshenitsy *Puccinia graminis Pers*. f. sp. *tritici* Eriks. et E. Henn. Mikol Fitopatol 3, 243--248.

Cole, G. T. & Ramirez-Mitchell, R. (1974). Comparative scanning electron microscopy of *Penicillium* conidia subjected to critical point drying, freeze drying and freeze-etching. In Scanning

Electron Microscopy/1974, Part II (ed. O. Johari & I. Corvin), pp. 367--374. Chicago: IIT Research Institute.

Corlett, M. (1970). Surface structure of the urediniospores of *Puccinia coronata* f. sp. *avenae*. Can J. Bot. 48, 2159--2161.

Cunningham, J. L. (1966). Germination of teliospores of the rust fungus *Frommea obtusa*. Mycol. 58, 494--496.

Daly, J. M., Knoche, H. W., & Wiese, M. V. (1967). Carbohydrate and lipid metabolism during germination of uredospores of *Puccinia graminis tritici*. Plant Physiol. 42, 1633--1642.

Dumitras, L. (1969). Electron microscopic researches on surface markings of teliospores of some *Tilletia* species parasitic on cereals and grasses. Rev. Roum. Biol. Ser. Bot. 14, 123--126.

Dumitras, L. (1970). Electron microscopy of the surface of teliospores of some species of *Tilletia* parasite on spontaneous Gramineae. Rev. Roum. Biol. Ser. Bot. 15, 303--306.

Dunkle, L. D., Maheshwari, R., & Allen, P. J. (1969). Infection structures from rust uredospores: Effect of RNA and protein synthesis inhibitors. Science 158, 481--482.

Duran, R. & Safeeulla, K. M. (1968). Aspects of teliospore germination in some North American smut fungi. Mycol. 60, 231--243.

Durrieu, M. G. & Rajeriarison, C. (1968). L'ornementation sporale des *Ustilago* parasites des Polygonacees (observations en microscopic électronique). C. R. Acad. Sci. Paris D, 267, 1940--1942.

Ettel, G. E. & Halbsguth, W. (1963). Über die wirkung von trimethylamin calciumnitrat und licht bei der keimung ver brandsporen von *Tilletia tritici* (Bjerk) winter. Beitr. Biol. Pfl. 39, 451--488.

Eversmeyer, M. G. & Burleigh, J. R. (1968). Effect of temperature of the longevity of *Puccinia recondita* f. sp. *tritici* uredospores on dry wheat foliage. Plant Dis. Rep. 52, 186--188.

Eyal, Z., Peterson, J. L., & Cappellini, R. A. (1967). Protein content and cell-wall weight of uredospores of five races of *Puccinia recondita* Rob. ex Desm. as affected by light and temperature. Bull. Torrey Bot. Club 94, 63--67.

Farkas, G. L. & Ledingham, G. A. (1959). The relation of self-inhibition of germination to the oxidative metabolism of rust uredospores. Can J. Microbiol. 5, 141--151.

Foudin, A. S. & Macko, V. (1974). Identification of the self-inhibitor and some germination characteristics of peanut rust uredospore germination. Phytopathology, 64, 990--993.

Frear, D. S. & Johnson, M. A. (1961). Enzymes of the glyoxylate cycle in germinating uredospores of *Melampsora lini* (pers.) lev.

Biochim. Biophys. Acta. 47, 419--421.
French, R. C. & Gallimore, M. D. (1971). Effect of some nonyl derivatives and related compounds on germination of uredospores. J. Agr. Food Chem. 19, 912--915.
French, R. C. & Gallimore, M. D. (1972a). Stimulation of germination of uredospores of stem rust of wheat in the pustule by *n*-nonanal and related compounds. J. Agr. Food Chem. 20, 421--423.
French, R. C. & Gallimore, M. D. (1972b). Effect of pretreatment with stimulator and water vapors on subsequent germination and infectivity of uredospores of *Puccinia graminis* var. *tritici*. Phytopathology 62, 116--119.
Fujii, R. K. & Hess, W. M. (1969). Surface characteristics of conidia from monosporous cultures of *Penicillium digitatum* and *Aspergillus nidulans* var. *echinulatus*. Can. J. Microbiol. 15, 1472--1473.
Gorin, P. A. J., Haskins, R. H., & Spencer, J. F. T. (1960). Biochemistry of the *Ustilaginales* Bioassimilation of L-Erythrulose in 4-0-β-D-Mannopyranosyl-D-Erythritol. Can. J. Biochem. Physiol. 38, 953--956.
Gottlieb, D., Roa, M. V., & Shaw, P. D. (1968). Changes in ribonucleic acid during the germination of teliospores of *Ustilago maydis*. Phytopathology 58, 1593--1597.
Grand, L. F. & Moore, R. T. (1971). Scanning electron microscopy of basidiospores of species of Strobilomycetaceae. Can. J. Bot. 49, 1259--1261.
Grand, L. F. & Moore, R. T. (1972). Scanning electron microscopy of *Cronartium* spores. Can. J. Bot. 50, 1741--1742.
Griffiths, D. A. (1971). The fine structure of basidiospores of *Panaeolus campanulatus* (L.) Fr. revealed by freeze-etching. Arch. Mikrobiol. 76, 74--82.
Gunasekaran, M., Bushnell, J. L. & Weber, D. J. (1971). Comparative studies on lipid components of *Ustilago bullata* and *Ustilago maydis* spores. Chem. Pathol. Pharmacol. 3, 621--628.
Hashioka, Y. (1971). Scanning and transmission electronmicrographs of shiitake, *Lentinus edodes* (Berk.) Sing. Rep. Tottori Mycol. Inst. 9, 1--10.
Hashioka, Y., Ikegami, H., & Horino, O. (1966). Fine structure of the rice false smut chlamydospores in comparison with that of the cereal smut spores. Res. Bull. Fac. Agr. Gifu. Univ. 22, 40--49.
Heagle, A. S. &. Moore, M. B. (1970). Germinability and longevity of teliospores of *Puccinia coronata* f. sp. *avenae*. Phytopathology

60, 617--681.

Heintz, C. E. & Niederpruem, D. J. (1970). Ultrastructure and respiration of oidia and basidiospores of *Coprinus lagopus* (sensu Buller). Can. J. Microbiol. 16, 481--484.

Heintz, E. E. &. Niederpruem, D. J. (1971). Ultrastructure of quiescent and germinated basidiospores and oidia of *Coprinus lagopus*. Mycol. 63, 745--766.

Henderson, D. M., Eudall, R, & Prentice, H. T. (1972a). Morphology of the reticulate teliospores of *Puccinia chaerophylli*. Trans Br. Mycol. Soc. 59, 229--232.

Henderson, D. M. , Prentice, H. T., & Eudall, R. (1972b). The morphology of fungal spores: III. The aecidiospores of *Puccinia poarum*. Pollen Spores 14, 17--24.

Hess, W. M., Bushnell, J. L., & Weber, D. J. (1972). Surface structures and unidentified organelles of *Lycoperdon perlatum* Pers. Basidiospores. Can. J. Microbiol. 18, 270--271.

Hess, W. M., Sassen, M. M. A., & Remsen, C. C. (1966). Surface structure of frozen-etched *Penicillium* conidiospores. Naturwissenschaften 53, 708.

Hess, W. M. & Stocks, D. L. (1969). Surface characteristics of *Aspergillus conidia*. Mycol. 61, 560--571.

Hess, W. M. & Weber, D. J. (1973). Ultrastructure of dormant and germinated *Rhizopus arrhizus* sporangiospores. Protoplasma 77, 15--33.

Hess, S., Weber, D. J., & Hess, W. M. (1974). Unusual lipid components in *Tilletia caries*. Unpublished data.

Hess, S. L., Allen, P. J., Nelson, D., & Lester, H. (1975). Mode of action of methyl *cis*-ferulate, the self-inhibitor of stem rust uredospore germination. Physiol. Plant Pathol. 5, 107--112.

Hibben, C. R., & Stotzky, G. (1969). Effects of ozone on germination of fungus spores. Can. J. Microbiol. 15, 1187.

Hiratsuka, Y. (1971). Spore surface morphology of pine stem rusts of Canada as observed under a scanning electron microscope. Can. J. Bot. 49, 371--372.

Hiratsuka, Y. & Maruyama, P. J. (1968). Nuclear condition of the germ tubes (of aeciospores) of *Peridermium ephedrae*. Mycol. 60, 437--438.

Hollis, C. A., Schmidt, R. A. & Kimbrough, J. W. (1973). Axenic culture of *Cronartium fusiforme*. Phytopathology 62, 1417--1419.

Hollomon, D. W. (1970). Ribonucleic acid synthesis during fungal spore germination. J. Gen. Microbiol. 62, 75.

Holm, L. Dunbar, A. & Hofstein, A. (1970). Studies on the fine structure of aeciospores: II. Sv. Bot. Tidskr 64, 380--382.

Hougen, F. W., Craig, B. M. & Ledingham, G. A. (1958). The oil of wheat stem rust uredospores. I. The sterol and carotenes of the unsaponifiable matter. Can. J. Microbiol. 4, 521--529.

Jackson, A. O., Samborski, D. J., & Rahringer, R. (1970). Metabolism of alicyclic acids and phenylpropanoids by uredospores of wheat stem rust and wheat leaf rust. Can. J. Bot. 48, 1085--1091.

Jackson, L. L. & Frear, D. S. (1967). Lipids of rust fungi. I. Lipid metabolism of germinating flax rust uredospores. Can. J. Biochem. 45, 1309--1315.

Jackson, L. L. & Frear, D. S. (1968). Lipids of rust fungi. II. Stigmast-7-enol and stigmasta-7,24(28)-dienol in flax rust uredospores. Phytochemistry 7, 651--654.

Jeggo, P. A., Unrau, P., Banks, G. R., & Holliday, P. (1973). A temperature sensitive DNA polymerase mutant of *Ustilago maydis*. Nat. New Biol. 242, 14--16.

Jensen, L. L. & Kiesling, R. L. (1971). Nuclear condition in germinating teliospores of *Ustilago hordei*. Proc. N. Dak. Acad. Sci. 24, 1--6.

Jones, D. (1968). Surface features of fungal spores as revealed in a scanning electron microscope. Trans. Br. Mycol. Soc. 51, 608--610.

Jones, D. R. (1971). Surface structure of germinating *Uromyces dianthi* uredospores as observed by scanning electron microscope. Can. J. Bot. 49, 2243.

Jones, J. P. & Haburn, M. A. (1969). Swelling phenomena of germinating crown rust uredospores. Phytopathology 59, 1034.

Joppien, S., Burger, A., & Reisener, H. J. (1972). Untersuchungen ueber den chemischen Aufbau von Sporenund Keimschlauchwaenden der Uredosporen des Weizenrostes (*Puccinia graminis* var. *tritici*). Arch. Mikrobiol. 82, 337--352.

Kapooria, R. G. (1971). A cytological study of promycelia and basidiospores and chromosome number in *Uromyces fabae*. Neth. J. Plant Pathol. 77, 91--96.

Kapooria, R. G. & Sharma, M. C. (1970). Studies on the antifungal properties of the exudates of fungi: I. Fungal exudates inhibiting uredospores germination of *Puccinia penniseti* Zimm. Agra. Univ. J. Res. Sci. 17, 167--169.

Kilpatrick, R. A., Harmon, D. L. Loegering, W. Q., & Clark, W. A. (1971). Viability of uredospores of *Puccinia graminis* f. sp. *tritici* stored in liquid nitrogen. Plant Dis. Rep. 55, 871--873.

Klisiewicz, J. M. (1973). Effect of volatile substances from safflower on germination of teliospores of *Puccinia carthami*. Phytopathology 63, 795.

Klusak, H. (1970). Respiration and character of terminal oxidation of yellow rust spores (*Puccinia striiformis* West). Phytopathol. Z. 68, 55--62.

Knoche, H. W. & Horner, T. L. (1970). Some characteristics of a lipase preparation from the uredospores of *Puccinia graminis tritici*. Plant Physiol. 46, 401--405.

Kotwal, S. N. & Kulkarni, U. K. (1971). Nuclear behavior in germinating teliospores of *Puccinia prainiana*. Ind. Phytopathol. 24, 182--186.

Kuhl, J. L., Maclean, D. J., Scott, K. J., & Williams, P. G. (1971). The axenic culture of *Puccinia* species for uredospores: Experiments on nutrition and variation. Can. J. Bot. 49, 201--209.

Kukkonen, I. (1969). The spore surface in the Anthracoidea section Echinosporae (Ustilaginales). A study with light and electron microscopy. Ann. Bot. Fenn. 6, 269--283.

Kushalappa, A. C. & Hedge, R. K. (1971). Studies on maize rust *Puccinia sorghi*, in Mysore State: I. Effect of temperature on uredospore germination on water agar and detached host leaf. Indian Phytopathol. 24, 759--764.

Laseter, J. L., Hess, W. M., Weete, J. D. Stocks, D. L., & Weber, D. J. (1968). Chemotaxonomic and ultrastructural studies on three species of *Tilletia* occurring on wheat. Can. J. Micro. 14, 1149.

Laseter, J. L. & Valle, R. (1971). Organics associated with the outer surface of airborne uredospores. Environ. Sci. Technol. 5, 631--634.

Laseter, J. L., Weete, J. D., & Walkinshaw, C. H. (1973). Volatile terpenoids from aeciospores of *Cronartium fusiforme*. Phytochemistry 12, 387--390.

Laseter, J. L., Weete, J. D., & Weber, D. J. (1968). Alkanes, fatty acid methyl esters, and free fatty acids in surface area of *Ustilago maydis*. Phytochemistry 7, 1177--1181.

Lavallee, A. & Lortie, M. (1971). Quelques observations sur la germination et la viabilité des basidiospores du *Pholiota aurivella*. Phytoprotection 52, 112--118.

Lin, F. K., Davies, F. L., Tripathi, R. K., Raghu, K., & Gottlieb, D. (1971). Ribonucleic acids in spore germination of *Ustilago maydis*.

Phytopathology 61, 645--648.

Lin, H. K., Langenbach, R. J., & Knoche, H. W. (1972). Sterols of *Uromyces phaseoli* uredospores. Phytochemistry 11, 2319--2322.

Littlefield, L. J. (1971). Scanning electron microscopy of uredospores of *Melampsora lini*. J. Microsc. (Paris) 10, 225--228.

Littlefield, L. J. & Bracker, C. E. (1971). Ultrastructure and development of uredospore ornamentation in *Melampsora lini*. Can J. Bot. 49, 2067--2073.

Loegering, W. Q., Harmon, D. L., & Clark, W. A. (1966). Storage of uredospores of *Puccinia graminis tritici* in liquid nitrogen. Plant Dis. Rep. 50, 502--506.

Lösel, D. M. (1967). The stimulation of spore germination in *Agarius bisporus* by organic acids. Ann. Bot. 31, 417--425.

Macko, V. & Novacky, A. (1966). Isozymes of malate dehydrogenase in uredospores of *Puccinia recondita* Rob. ex Desm. Biologia 21, 460--462.

Macko, V., Novacky, A., & Skrobal, M. (1964). Inhibition in "slide germination test" caused by *Tilletia controversa* spores extract. Biologia 19, 869--870.

Macko, V., Novacky, A., & Stahmann, M. A. (1967). Protein and enzyme patterns from uredospores of *Puccinia graminis* var. *tritici*. Phytopath. Z. 58, 122--127.

Macko, V., Staples, R. C., Allen, P. J., & Renwick, J. A. A. (1971). Identification of the germination self-inhibitor from wheat stem rust uredospores. Science 173, 835--836.

Macko, V., Staples, R. C., Gershon, H., & Renwick, J. A. A. (1970). Self-inhibitor of bean rust uredospores: Methyl-3,4-dimethoxycinnamate. Science 170, 539--540.

Macko, V., Staples, R. C., Renwick, J. A. A., & Pirone, J. (1972). Germination self-inhibitors of rust uredospores. Physiol. Plant Pathol. 2, 347--355.

Maddison, A. C. & Manners, J. G. (1973). Lethal effects of artificial ultraviolet radiation on cereal rust uredospores. Trans. Br. Mycol. Soc. 60, 471--494.

Madelin, M. F. (1966). The Fungus Spore. London: Butterworths.

Maheshwari, R. & Sussman, A. S. (1970). Respiratory changes during germination of uredospores of *Puccinia graminis* f. sp. *tritici*. Phytopathology 60, 1357--1364.

Maheshwari, R. & Sussman, A. S. (1971). The nature of cold-induced dormancy of uredospores of *Puccinia graminis tritici*. Plant Physiol. 47, 289--295.

Manion, P. D. & French, D. W. (1968). Inoculation of living aspen trees with basidiospores of *Fomes igniarius* var. *populinus*. Phytopathology 58, 1302--1304.

Manion, P. D. & French, D. W. (1969). The role of glucose in stimulating germination of *Fomes igniarius* var. *populinus* basidiospores. Phytopathology 59, 293--296.

Manocha, M. S. (1973). DNA-dependent RNA-polymerase in germinating bean rust uredospores. Can. J. Microbiol. 19, 1175--1177.

McLaughlin, D. J. (1973). Ultrastructure of sterigma growth and basidiospore formation in *Coprinus* and *Boletus*. Can. J. Bot. 51, 145--150.

Mederick, F. N. & Sackston, W. E. (1972). Effects of temperature and duration of dew period on germination of rust uredospores on corn leaves. Can. J. Plant Sci. 52, 551--557.

Melching, J. S., Fischer, J. A., & Stanton, J. R. (1969). Influence of spore collection and deposition methods on germinability of uredospores of *Puccinia graminis tritici*. Phytopathology 59, 1558.

Merrill, W. (1970). Spore germination and host penetration by heart rotting Hymenomycetes. Annu. Rev. Phytopathol. 8, 281--300.

Mills, G. L. & Eilers, F. I. (1973). Factors influencing the germination of basidiospores of *Coprinus radiatus*. J. Gen. Microbiol. 77, 393--401.

Misra, A. & Prasada, R. (1971). Studies on uredospore germination and incubation period of bajra rust caused by *Puccinia penniseti*. Ind. J. Mycol. Plant Pathol. 1, 103--107.

Morton, H. L. & French, D. W. (1970). Attraction toward and penetration of *Polyporous dryophilus* var. *vulpinus* basidiospores by hyphae of the same species. Mycol. 62, 714--720.

Nair, K. R. G. (1971). Germination of the teliospores of *Ravenelia hobsoni*. Trans Br. Mycol. Soc. 57, 344--348.

Naito, N. Lee, M., & Tani, T. (1972). Inhibition of germination and infection structure formation of *Puccinia coronata* uredospores by plant growth regulators and antimetabolites. Kagawa Daiyaku Nogakubu Gakuzyutu Hokoku 23, 51--56.

Naito, N. & Tani, T. (1967). Respiration during germination in uredospores of *Puccinia coronata*. Ann. Phytopathol. Soc. Jap. 33, 17--22.

Newburgh, R. W. & Cheldelin, V. H. (1958). Glucose oxidation in mycelia and spores of the wheat smut fungus *Tilletia caries*. J. Bacteriol. 76, 308--311.

Niederpruem, D. J. (1964). Respiration of basidiospores of

Schizophyllum commune. J. Bacteriol. 88, 210--215.

Niederpruem, D. J. & Dennen, D. W. (1966). Kinetics, nutrition and inhibitor properties of basidiospore germination in *Schizophyllum commune.* Arch. Mikrobiol. 54, 91--105.

Niederpruem, D. J. & Hafiz, A. (1964). Polyol metabolism during basidiospore germination in *Schizophyllum commune.* Plant Physiol. Proc. Suppl. 39, vi.

Niederpruem, D. J. & Hunt, S. (1967). Polyols in *Schizophyllum commune.* Am. J. Bot. 54, 241--245.

Niederpruem, D. J. & Jersild, R. A. (1967). Nuclear behavior in multinucleate *Schizophyllum commune* germlings. Arch. Mikrobiol. 61, 223--231.

Nielsen, J. (1962). Trimethylammonium compounds in *Tilletia* spp. Can. J. Microbiol. 41, 335--339.

Novacky, A. & Macko, V. (1966). Electrophoresis of the proteins from uredospores of *Puccinia graminis* var. *tritici* Erikss. et. Henn., physiologic races 21 and 111. Naturwissenschaften 53, 281.

Nowak, R., Kim, W. K., & Rohringer, R. (1972). Sterols of healthy and rust-infected primary leaves of wheat and non-germinated and germinated uredospores of wheat stem rust. Can. J. Bot. 50, 185--190.

Oláh, G. M. (1973). The fine structure of *Psilocybe quebecensis.* Mycopathol. Mycol. Appl. 49, 321--338.

Oro, J., Laseter, J. L. & Weber, D. J. (1966). Alkanes in fungal spores. Sci. 154, 399--400.

Orr, G. F. & Tippetts, W. C. (1971). Deterioration of uredospores of wheat stem rust under natural conditions. Mycopathol. Mycol. Appl. 44, 143--148.

O'Sullivan, J. & Lösel, D. M. (1971). Spore lipids and germination in *bisporus.* Arch. Mikrobiol. 80, 27.

Ouellette, G. B. (1969). Preuve de la viabilite des urediniospores du *Pucciniastrum agrimoniae* apres repos hivernal. Can. J. Bot. 43, 497--498.

Pavgi, M. S. & Singh. R. A. (1970). Teliospore germination, cytology, and development of *Entyloma eleocharidis.* Cytologia (Tokyo) 35, 391--401.

Pawar, I. S. & Kulkarni, U. K. (1971). Nuclear behavior in the germinating teliospores of *Uromyces orientalis.* Sci. Cult. 36, 514--515.

Pegler, D. N. (1969). Ultrastructure of basidiospores in agaricales in relation to taxonomy and spore discharge. Trans. Br. Mycol.

Soc. 52, 491--513.

Pegler, D. N. & Young, T. W. K. (1972a). Basidiospore form in the British species of *Galerina* and *Kuehneromyces*. Kew Bull. 27, 483--500.

Pegler, D. N. & Young, T. W. K. (1972b). Basidiospore form in the British species of *Inocybe*. Kew Bull. 26, 499--538.

Pegler, D. N. & Young, T. W. K. (1972c). Reassessment of the Bondarzewiaceae (Alphyllophorales). Trans. Br. Mycol. Soc. 58, 49--58.

Peruanskii, Y. V. & Run'kovskaya, N. G. (1966). O fraktsiyakh lipidov uredospor steblevoi rzhavchniy pshchenitsy. Dokl. Akad. Nauk. SSSR 169, 962--964.

Powell, J. M. (1971). Daily germination of *Cronartium comandrae* aeciospores. Can. J. Bot. 49, 2123--2127.

Powell, J. M. & Hiratsuka, Y. (1969). Nuclear condition and germination characteristics of the aeciospores of *Cronartium comandrae* and *C. comptoniae*. Can. J. Bot. 47, 1961--1963.

Powell, J. M. & Morf, W. (1966). Temperature and pH requirements for aeciospore germination of *Peridermium stalactiforme* and *P. harknessii* of the Cronartium coleosporioides complex. Can. J. Bot. 44, 1597--1606.

Prescott, J. M. & Kernkamp, M. F. (1972). Genetic stability of *Puccinia graminis tritici* in cryogenic storage. Plant Dis. Rep. 55, 695--696.

Quick, W. A. (1973). Nucleic acid changes during germination and axenic development of flax rust (*Melampsora lini*). Physiol. Plant Pathol. 3, 419--430.

Raghunath, T. (1967). Behavior of the diploid nucleus in germinating teliospores of *Tolyposporium christensenii*. Bull Torrey Bot. Club 94, 191--193.

Raghunath, T. (1969). Nuclear behavior in the germinating teliospores of the smut *Georgefischeria riveae*. Caryologia 22, 223--228.

Raghunath, T. (1970). Studies in teliospore germination and artificial culture of *Tolyposporium christensenii* inciting long smut of lemon grass. J. Univ. Poona 38, 110--116.

Ramakrishnan, L. & Staples, R. C. (1967). Some observations on RNAs and their synthesis in germinating bean rust uredospores. Phytopathology 57, 826--827.

Ramakrishnan, L. & Staples, R. C. (1968a). Template activity in nucleic-acids from germinating bean rust uredospores. Phytopathology 58, 886.

Ramakrishnan, L. & Staples, R. C. (1968b). Template activity and RNA synthesis from germinating bean rust uredospores. Phytopathology 58, 1064.

Ramakrishnan, L. & Staples, R. C. (1969). Evidence for a template RNA in differentiated bean rust uredospores. Phytopathology 59, 1045.

Ramakrishnan, L. & Staples, R. C. (1970a). Evidence for a template RNA in resting uredospores of the bean rust fungus. Contrib. Boyce Thompson Inst. 24, 197--202.

Ramakrishnan, L. & Staples, R. C. (1970b). Changes in ribonucleic acids during uredospore differentiation. Phytopathology 60, 1087--1091.

Ramakrishnan, L. & Staples, R. C. (1971). Effect of storage on the rate of incorporation of uridine into ribonucleic acid by germinating uredospores. Phytopathology 61, 244--245.

Raphaela Musumeci, M., Nicholson, R. L., Moraes, W. B. C., & Kuc', J. (1972). Observacoes sobre uma atividade esterasica associada aos uredosporos de *Hemileia vastatrix* (Berk et Br.). Biologico 38, 148--150.

Rast, D. & Bachofen, R. (1967). Carboxylation reactions in *Agaricus bisporus*; I. The endogenous CO_2 acceptor. Arch. Microbiol. 57, 392--405.

Ratts, T., Hafiz, A., Niederpruem, D. J., & Egbert, P. (1964). Physiological studies on basidiospore germination in *Schizophyllum commune*. Bacteriol. Proc. p. 111.

Reisener, H. J. (1962). Formation of trehalose and polyols by wheat stem rust *Puccinia graminis tritici* uredospores. Can. J. Biochem. Physiol. 40, 1248--1251.

Reisener, H. J. (1967). Untersuchungen uber den Aminosaure-und Proteinstoffwechsel der Uredosporen des Weizenrostes (*Puccinia graminis* var. *tritici*). Arch. Mikrobiol. 55, 382--397.

Reisener, H. J. (1969). Ueber den Stoffwechsel von Alanin, Glykokoll und Arginin in den Uredosporen von *Puccinia graminis* var. *tritici* waehrend der Keimung. Arch. Mikrobiol. 69, 101--113.

Reisener, H. J. & Jager, K. (1967). Untersuchungen uber die quantitative Bedeutung einiger Stoffwechselwege in den Uredosporen von *Puccinia graminis* var. *tritici*. Planta 72, 265--283.

Reisener, H. J., McConnell, W. B. & Ledingham, G. A. (1961). Effect of short-chain fatty acids on respiration of wheat stem rust uredoapores. Can. J. Microbiol. 7, 865--868.

Rick, P. D. & Mirocha, C. J. (1968). Fixation of carbon dioxide in the dark by the malic enzyme of bean and oat stem rust uredospores. Plant Physiol. 43, 201--207.

Roncadori, R. W. & Matthews, F. R. (1966). Storage and germination of aeciospores of *Cronartium fusiforme* from *Pinus elliottii*, *Pinus taeda*. Phytopathology 56, 1328--1329.

Ross, E. W. (1969). Thermal inactivation of conidia and basidiospores of *Fomes annosus*. Phytopathology 59, 1798--1801.

Roy, M. K. (1968). Studies on the identification of a self inhibitor of uredospore germination in *Puccinia sorghi*. (Abstr.) Pap. Int. Symp. Plant Pathol. p. 44.

Sassen, M. M. A., Remsen, C. C., & Hess, W. M. (1967). Fine structure of *Penicillium megasporum* conidiospores. Protoplasma 64, 75--88.

Savulescu, A., Becerescu, D., Dumitras, L., & Ploaie, P. (1968). Electron microscopic research on teliospores of the genea *Ustilago* and *Tilletia*. Ann. Microbiol. Enzymol. 18, 45--56.

Schauz, L. (1970). Controlling the germination of smut spores and sporidium development in *Tilletia caries* (DC) Tul. through outside factors. I. The effect of storage temperature on the fungus spores on the course of germination. Wheat. Bakteriol. Parasit. Infekt. Hyg. Abstr. 2 Naturwiss. 124, 739--742.

Scheld, H. W. & Perry, J. J. (1970a). Induction of dimorphism in the basidiomycete *Lenzites sepiaria*. J. Bacteriol. 102, 271--277.

Scheld, H. W. & Perry, J. J. (1970b). Basidiospore germination in the wood-destroying fungus *Lenzites sepiaria*. J. Gen. Microbiol. 60, 9--21.

Sharp, E. (1967). Atmospheric ions and germination of uredospores of *Puccinia striiformis*. Sci. 156, 1359--1360.

Shaw, M. (1964). The physiology of rust uredospores. Phytopath. Zeit. 50, 159--180.

Shipton, W. A. & Fleischmann, G. (1969). Disc electrophoresis of proteins from uredospores of races of *Puccinia coronata* f. sp. *avenae*. Phytopathology 59, 883.

Shu, P. & Ledingham, G. A. (1956). Enzymes related to carbohydrate metabolism in uredospores of wheat stem rust. Can. J. Microbiol. 2, 489--495.

Shu, P., Tanner, K. G., & Ledingham, G. A. (1954). Studies on the respiration of resting and germinating uredospores of wheat stem rust. Can. J. Bot. 32, 16--23.

Singer, R. (1972). Cyanophilous spore walls in the Agaricales and Agaricoid basidiomycetes. Mycol. 64, 822--829.

Singh, J. & Trione, E. J. (1969). Self inhibition of the germination of bunt teliospores of *Tilletia caries* and of *Tilletia controversa*. Phytopathology 59, 15.

Smith, J. E. (1965). Glutamine synthesis in uredospores of *Puccinia helianthi* Schw. Can. J. of Microbiol. 11, 381--383.

Snow, G. A. (1968). Basidiospore production by *Cronartium fusiforme* as affected by suboptimal temperatures and preconditioning of teliospores. Phytopathology 58, 1541--1546.

Snow, J. P. & Jones, J. P. (1967). The sequence of release of amino-acids from germinating crown rust spores. Phytopathology 57, 832.

Sood, P. N. & Sackston, W. E. (1971). Studies on sunflower rust. VII. Effect of light and temperature during spore formation on the germinability of fresh and stored uredospores of *Puccinia helianthi*. Can. J. Bot. 49, 21--25.

Sood, P. N. & Sackston, W. E. (1972). Studies on sunflower rust. XI. Effect of temperature and light on germination and infection of sunflowers by *Puccinia helianthi*. Can. J. Bot. 50, 1879--1886.

Stallknecht, G. F. & Mirocha, C. J. (1970). Carbon dioxide fixation by uredospores of *Uromyces phaseoli* and its incorporation into cellular metabolites. Phytopathology 60, 1338--1342.

Stallknecht, G. F. & Mirocha, C. J. (1971). Fixation and incorporation of CO_2 into ribonucleic acid by germinating uredospores of *Uromyces phaseoli*. Phytopathology 61, 400--405.

Staples, R. C. (1957). The organic acid composition and succinoxidase activity of the uredospore of the leaf and stem rust fungi. Contrib. Boyce Thompson Inst. 19, 19--31.

Staples, R. C. (1968). Protein synthesis by uredospores of the Bean-D rust fungus. Neth. J. Plant Pathol. 74, 25--36.

Staples, R. C. (1974). Synthesis of DNA during differentiation of bean rust uredospores. Physiol. Plant Pathol. 4, 415--424.

Staples, R. C., App, A., McCarthy, W. J., & Gerosa, M. M. (1966). Some properties of ribosomes from uredospores of the bean rust fungus. Contrib. Boyce Thompson Inst. 23, 159--164.

Staples, R. C. & Bedigian, D. (1967). Preparation of an amino acid incorporation system from uredospores of the bean rust fungus. Plant Res. Contrib. 23, 345--347.

Staples, R. C., Ramakrishnan, L., Yaniv. Z. (1970). Membrane-bound ribosomes in germinated uredospores. Phytopathology 60, 58--62.

Staples, R. C., Syamananda, R., Kao, V., & Block, R. (1962). Comparative biochemistry of obligately parasitic and saprophytic fungi. II. Assimilation of C^{14}-labeled substrates by germinating spores. Contr. Boyce Thompson Inst. 21, 345--362.

Staples, R. C. & Wynn, W. K. (1965). The physiology of uredospores of the rust fungi. Bot. Rev. 31, 537--564.

Staples, R. C. & Yaniv, Z. (1973a). Synthesis of DNA in differentiating bean rust uredospores. Phytopathology 63, 207.

Staples, R. C. & Yaniv, Z. (1973b). Spore germination and ribosomal activity in the rust fungi. II. Variable properties of ribosomes in the uredinales. Physiol. Plant Pathol. 3, 137--145.

Staples, R. C., Yaniv, Z., Ramakrishnan, L., & Lipetz, J. (1971). Properties of ribosomes from germinating uredospores. Morphological and Biochemical Events in Plant Parasite Interaction (ed. S. Akai & S. Ouchi), pp. 59--90. Tokyo: Phytopathology Society of Japan.

Stocks, D. L. & Hess, W. M. (1970). Ultrastructure of dormant and germinated basidiospores of a species of *psilocybe*. Mycologia 62, 176--191.

Storck, R. & Alexopoulos, C. J. (1970). Deoxyribonucleic acid of fungi. Bacteriol. Rev. 34, 126--154.

Sussman, A. S. (1965). Physiology of dormancy and germination in the propagules of crytogamic plants. Encycl. Plant Phys. 15, Part 2, 931--1025.

Swendsrud, D. P. & Calpouzos, L. (1970). Rust uredospores increase the germination of pycindiospores of *Darluca filum*. Phytopathology 60, 1445--1447.

Thomas, P. L. & Isaac, P. K. (1967). The development of echinulation in uredospores of wheat stem rust. Can J. Bot. 45, 287--289.

Tollenaar, H. & Houston, B. R. (1966a). Effect of temperature during uredospore production and of light on *in vitro* germination of uredospores from *Puccinia striiformis*. Phytopathology 56, 787--790.

Tollenaar, H. & Houston, B. R. (1966b). *In vitro* germination of uredospores of *Puccinia graminis* and *P. striiformis* at low spore densities. Phytopathology 56, 1036--1039.

Topolovskii, V. A. (1966). Vliyanie kompleksa iskusstvenno sozdavaemykh faktorov vneshnei sredy na vskhozhest' uredospor nekotorykh rzhavchinnykh gribov. Vestnik Sel'skokhoz Nauk 1, 44--50.

Trione, E. J. (1973). The physiology of germination of *Tilletia* teliospores. Phytopathology 63, 643--648.

Trione, E. J. & Ching, T. (1971). Lipids in teliospores and mycelium of the dwarf bunt fungus, *Tilletia* controversa. Phytochemistry 10, 227--229.

Trocha, P. & Daly, J. M. (1970). Protein and ribonucleic acid synthesis during germination of uredospores. Plant Physiol. 46, 520--526.

Trocha, P. & Daly, J. M. (1974). Cell walls of germinating uredospores. II. Carbohydrate polymers. Plant Physiol. 53, 527--532.

Trocha, P., Daly, J. M., & Langenbach, R. J. (1974). Cell walls of germinating uredospores. I. Amino acid and carbohydrate constituents. Plant Physiol. 53, 519--526.

Tulloch, A. P., Craug, B. M., & Ledingham, G. A. (1959). The oil of wheat stem rust uredospores. II. The isolation of *cis*-9,10-epoxyoctadecanoic acid, and the fatty acid composition of the oil. Can. J. Microbiol. 5, 485--491.

Tulloch, A. P. & Ledingham, G. A. (1960a). The component fatty acids of oils found in spores of plant rusts and other fungi. Can. J. Microbiol. 6, 425--434.

Tulloch, A. P. & Ledingham, G. A. (1960b). The component fatty acids of oils found in spores of plant rusts and other fungi, Part II. Can. J. Microbiol. 8, 379--387.

Turel, F. L. M. & Ledingham, G. A. (1959). Utilization of labeled substrates by the mycelium and uredospores of flax rust. Can. J. Microbiol. 5, 537--545.

Van Alfen, N. (1969). The identification of a germination factor for basidiospores of *Psilocybe mutans*. M. S. Thesis, Brigham Young University.

Van Sumere, C. F., Van Sumere-De Preter, C., & Ledingham, G. A. (1957). Cell wall splitting enzymes of *Puccinia graminis* var. *tritici*. Can. J. Microbiol. 3, 761--768.

Vargas, J. M., Young, H. C. Jr., & Saari, E. E. (1967). Effect of light and temperature on uredospore germination, infection, and disease development of *Puccinia cynodontis*, and isolation of pathogenic races. Phytopathology 57, 405--409.

Volz, P. A. & Niederpruem, D. J. (1968). Growth, cell division, and branching patterns of *Schizophyllum commune* Fr. single basidiospore. Arch. Mikrobiol. 61, 232--245.

Walkinshaw, C. H. (1968a). Oxygen consumption and germination of *Cronartium fusiforme* aeciospores. Phytopathology 58, 260--261.

Walkinshaw, C. H. (1968b). Dependence of fatty acid chain length of oxygen uptake and decarboxylation by *Cronartium fusiforme* aeciospores. Phytopathology 58, 855--859.

Wasowska, D. (1971). Effect of different substrata and the pH hydrogen-ion concentration. Values on germination of spores of cultivated mushroom (*Agaricus bisporus*) grown in Poland. Polish. Byul Warzywniczy 12, 277--269.

Weete, J. D., Laseter, J. L., Weber, D. J., Hess, W. M., & Stocks, D. L. (1969). Hydrocarbons, fatty acids, and ultrastructure of smut spores. Phytopathology 59, 545--548.

Wessels, J. G. H. (1959). The breakdown of carbohydrate by *Schizophyllum commune*. Fr. The operation of the TCA cycle. Acta. Bot. Neerl. 8, 497--505.

White, G. A. & Ledingham, G. A. (1961). Studies on the cytochrome oxidase and oxidation pathway in uredospores of wheat stem rust. Can. J. Bot. 39, 1131--1148.

Whitney, R. D. (1966). Germination and inoculation tests with basidiospores of *Polyporus tomentosus* (on *Picea glauca*). Can. J. Bot. 44, 1333--1343.

Whitney, R. D. & Bohaychuk, W. P. (1971). Germination of *Polyporus tomentosus* basidiospores on extracts from diseased and healthy trees. Can. J. Bot. 49, 699--703.

Wiese, M. V. & Daly, J. M. (1967). Some effects of pre- and post-germination treatments on germination and differentiation of uredospores of *Puccinia graminis* F. sp. *tritici*. Phytopathology 57, 1211--1215.

Williams, P. G. (1971). A new perspective of the axenic culture of *Puccinia graminis* f. sp. *tritici* from uredospores. Phytopathology 61, 994--1002.

Williams, P. G. & Allen, P. J. (1967). Endogenous respiration of wheat stem rust uredospores. Phytopathology 57, 656--661.

Williams, P. G., Scott, K. J., & Kuhl, J. L. (1966). Vegetative growth of *Puccinia graminis* f. sp. *tritici in vitro*. Phytopathology 56, 1418--1419.

Woodbury, W. & Stahmann, M. A. (1970). Role of surface films in the germination of rust uredospores. Can. J. Bot. 48, 499--511.

Wynn, W. K. (1966). Physiology of uredospores during storage. Contrib. Boyce Thompson Inst. 23, 229--242.

Wynn, W. K. (1968). Gulcan synthesis by germinating bean-D rust spores. Phytopathology 58, 1073.

Wynn, W. K. & Gajdusek, C. (1968). Metabolism of glucomannan-protein during germination of bean rust spores (*Uromyces phaseoli*, *Phaseolus vulgaris* cv. Pinto). Contrib. Boyce Thompson Inst. 24, 123--138.

Wynn, W. K., Staples, R. C., Strouse, B., & Gajdusek, C. (1966).

Physiology of uredospores during storage. Contrib. Boyce Thompson Inst. Plant Rest. 23, 229--242.

Yaniv, Z. & Staples, R. C. (1968). Study of the transfer of amino acids from aminoacyl transfer ribonucleic acid into protein by ribosomes from uredospores of the bean rust fungus. Contrib. Boyce Thompson Inst. 24, 103--108.

Yaniv, Z. & Staples, R. C. (1969). Transfer activity of ribosomes from germinating uredospores. Contrib. Boyce Thompson Inst. 24, 156--163.

Yaniv, Z. & Staples, R. C. (1971). The purification and properities of the aminoacyl RNA binding enzyme from bean rust uredospores. Biochim. Biophys. Acta 232, 717--725.

Zadoks, J. C. & Groenewegen, L. J. M. (1967). On light sensitivity in germinating uredospores of wheat brown rust (*Puccinia recondita*). Neth. J. Plant Pathol. 73, 83--102.

Zogg, H. & Schwinn, F. J. (1971). Surface structures of the spores of the Ustilaginales. Trans. Brit. Mycol. Soc. 57, 403--410.

DISCUSSION

Form and Function of Spores in Basidiomycetes

Chairman: Samuel R. Rushforth
Brigham Young University
Provo, Utah

The question was asked if vesicles were evident in the frozen-etched preparations of the hyphal tips of germinating basidiomycete spores. Professor Hess replied that that was one area that needed further research. He went on to say that in his impression they certainly are not present in levels as high as in many Phycomycetes but that detailed study was necessary to determine their occurrence.

It was mentioned that in a micrograph of a rust spore shown by Paul Allen some vesicles appeared to be present. Professor Hess agreed and stated that further comparative investigations would be valuable.

The question was asked if it was possible that the regular pattern on the surface of the unknown organelle in *Tilletia* teliospores could possibly be caused by a virus since it has been shown several times that viruses often produce such regular patterns. Professor Hess replied that it was true that viruses often produce similar patterns. However, he stated that in the case of the unknown organelle, he and his co-workers considered it unlikely since the organelle changes when the spore germinates, in ways that virus inclusions would not change.

The question was asked of Professor Weber if he had ever detected cyclopropane in any of the spores he had investigated with the mass spectrometer. Professor Weber replied that he could not recall finding that specific compound, although he had sifted through so much data that he could not be certain. He did say, however, that he has definitely found similar compounds, such as cyclohexylpropane.

CHAPTER 16

SPORES IN ASCOMYCTES, THEIR CONTROLLED DIFFERENTIATION

G. Turian
University of Geneva

INTRODUCTION

The characteristic feature of Ascomycetes is their sexually produced spores born in a sac or ascus and therefore called ascospores. They represent the "perfect stage" and there are sometimes four, usually eight and, in certain cases, more or less of them per ascus. Ascospores are not, however, the only means of ascomycetous reproduction as most species have one or more asexually produced sporal types or conidia representing the "imperfect stage."

There are marked structural and functional differences between both types of spores which find their origin in the respective differentiation sequences. All the sexually formed spores are the product of the more uniform pattern of internal reorganization inside the more or less elongated ascal bag. Most important in this process of free cell formation is the involvement of a system of double membranes continuous with the endoplasmic reticulum and then the invagination cutting out the ascal cytoplasm around postmeiotic nuclei to form ascospores (Delay, 1966; Carroll, 1967; Reeves, 1967). A thick wall is then secreted between the two layers of the membrane; this is part of the building-up of ascospores which confers them increased resistance to noxious factors and contributes to their viability over long dormancy periods (Sussman & Halvorson, 1966; Ingold, 1971).

Comparatively, conidia are not the products of a free cell formation process. They differentiate at the ends of mycelia (Sussman & Halvorson, 1966) within thinner walls (except chlamydospores) and are relatively short-lived. They differ widely in their ontogeny (Kendrick, 1971; Mangenot & Reisinger, this book) and structurophysiological comparisons between them are meaningful only when made between asexual spores produced by similar morphogenetic processes. For instance, microconidia of Sphaeriales or of Helotiales, when described as basipetally budded at the level of phialidic collarettes (Subramanian, 1971) are much closer to the phialoconidia of Aspergillales than to their companion macroconidia described as basifugally budded blastoarthroconidia in *Neurospora* (Turian & Bianchi, 1972; Turian et al., 1973).

Spores from the imperfect and perfect states of Ascomycetes can present such close morphological similarities as the characteristic coloration and appendages of Broomella species (Muller, 1971). Even so, they always remain ontogenically very different, and therefore conidia and ascospores will be considered separately as to their respective chemostructural ontogeny, morphogenic controlling factors and the functional morphology of the mature elements. Only two levels of sporogenic control will be presented, the genetic, providing the competence to express sporal morphology, and the metabolic, acting through the shifting of pathways toward specific biosyntheses of sporal compounds. The third level of the various environmental, controlling factors has been frequently reviewed (for conidiogenesis, Turian, 1974; for yeast ascosporogenesis, Fowell, 1969).

CONIDIA

Chemostructural Ontogeny

The concept of conidium is definite enough to allow its segregation from the general concept of the spore: "A conidium is a specialized, nonmotile, asexual propagule, not formed by cleavage" (Kendrick, 1971). In fact, three out of the five basic methods of spore production defined by Carmichael (1971) can be covered by the term conidia. It then includes all asexual ascomycetous spores produced by extrusion (blastoconidia and phialoconidia), fission (arthroconidia), and fragmentation (aleuries), excluding those formed by cleavage, which are the phycomycetous sporangiospores and zoospores.

Both types of extruded conidia are implicated in the dual aspect of conidiogenesis expressed in many Euascomycetes as microconidia = phialoconidia vs macroconidia = blastoarthroconidia. This duality remains potential in the Aspergillales where the blastic, so-called *Cladosarum* type is of mutational origin (see section on genetic control in conidia); it is constitutive only in higher Euascomycetes as a few Sphaeriales of the genus *Neurospora* and in Helotiales mainly of the genus *Sclerotinia*; in Hypocreales, micro- and macroconidia of the form-genus *Fusarium* are, however, both of the phialidic type.

The blastic type of conidium appears to be the most primitive as it is produced by vegetatively reproducing saccharomycetous yeasts, budding morphogenetic equivalents of blastoconidia. Moreover, blastoconidial chains when they appear in Aspergillales (*Cladosarum* type and artificial mutants of *Aspergillus* spp., see below), represent a degradative stage of the conidial extrusion process normally producing phialoconidia in this order. From this, it could be inferred that the macroconidial, monilioid stage of conidiogenesis only formed by a few species of *Neurospora* (the "classics" *N. crassa, N. sitophila, N. tetrasperma*) which is of the blastic type in its first phase (Turian & Bianchi, 1971), is a degradative state of the microconidial stage present in all *Neurospora* species (and Sphaeriales in general) which is known to correspond to the more common phialidic type of conidiogenesis (Dodge, 1932; Lowry et al., 1967; Oulevey-Matikian & Turian, 1968; Rossier et al., 1973).

Following these evolutionary considerations, we will first consider the blastic process at work in vegetative and asexual reproduction in Hemiascomycetes, then the more complex blasto-arthroconidial formation in Euascomycetes, to reach the phialidic type through the Eurotiales to the higher Euascomycetes (Sphaeriales, Hypocreales, Helotiales). Other specialized types of conidia (annello conidia, poroconidia, etc.) are mostly studied in the Fungi imperfecti (Mangenot & Reisinger, this book, Chapter 17).

Buds and Blastoconidia in Hemiascomycetes

Yeast buds can be considered as blastic elements of vegetative reproduction rather than asexual spores.

In some yeasts, including *Saccharomyces cerevisiae* and *Torulopsis glabrata*, budding apparently involved the extension of the existing wall (Grove et al., 1973). According to Gay & Martin (1971) this process of direct expansion of the parent cell wall operates during primary bud formation, while a somewhat different process is responsible for secondary bud formation. In *Metschnikowia krissii*, bud formation takes place by the extension of a small localized area of the existing parent wall and the bud wall also appears to be continuous with the parent wall (Talens et al., 1973 a, b). In apiculate yeasts like *Saccharomycodes ludwigii*, bud formation involves repeated percurrent proliferation, as shown by fluorescent microscopy (Streiblová et al., 1964); it resembles annelloconidium production rather than blastoconidium production and corresponds to septal or secondary budding (Gay & Martin, 1971).

Higher in the Hemiascomycetous series, *Endomycopsis* species have a mycelial form that reproduces by buds which erupt through the mother cell wall in a manner similar to that described as blastic rupture in the yeast phase of the dimorphic fungus *Phialophora dermatitidis* (Grove et al., 1973). Dolipores (septal pores provided with swollen rims bounded by the continuation of the plasmalemma) were found between budding cells in *E. platypodis* (Kreger-Van Rij & Veenhuis, 1969). In culture, *E. fibuliger* can still form budding yeast cells, but can also form arthroconidia by hyphal fragmentation like *Endomyces* species; its blastoconidia germinate by hyphal tube formation but can also bud to form another blastoconidium (Fig. 98 in Webster, 1970). Budding can even occur directly on the spindle-shaped ascospores of a Spermophthoraceae, *Nematospora coryli* (Kreger-Van Rij, 1969, Fig. EM, p. 59). There are also a number of mycelial forms of Hemiascomycetes which form blastoconidia resembling *Endomycopsis*, but in which asci have not been observed (*Candida albicans*, etc.).

At the molecular level, the budding process of yeast has been partially resolved and it can therefore serve as a model for the blastic type of spore formation. Pioneering this experimental field, Nickerson & Falcone (1956) proposed the participation of protein disulfide reductase in the budding mechanism of *Candida albicans*; this proposal was confirmed and expanded to other yeasts by subsequent workers.

Robson & Stockley (1962) and Pitt (1969) detected by autoradiography and other cytochemical procedures, a preferential accumulation of -SH groups at the budding areas of *C. albicans*, *C. utilis*, and *Eremothecium ashbyi*; Brown & Hough (1966) confirmed the existence of protein disulfide reductase in cell-free extracts of *S. cerevisiae*; they localized the enzyme in the mitochondrial fraction, but Moor (1967) ascribed this to smaller particulate contaminants. According to Kanetsuna et al. (1972), the extracts of the budding yeast form of *Paracoccidioides brasiliensis* have five times more protein disulfide reductase activity than the mycelial form which itself is much richer in --S--S-- bridges than the yeast form. Hyphal branching which is initiated by lateral budding is also accompanied by an increased activity of protein disulfide reductase (Fèvre et al., 1975).

Polysaccharides are major components involved in the macromolecular organization of the budding cells of *S. cerevisiae*. By using fluorescein-conjugated concanavalin A to label mannan, Tkacz & Lampen (1972) showed that the new mannan is deposited almost entirely into the wall surrounding the bud and that the major site at which deposition occurs is the distal tip of the growing bud. Most of the newly synthesized glucan had also been found to be located at the poles of the yeast bud, indicating tip growth (Johnson & Gibson, 1966). Both of these results are at variance with those of Chung et al. (1965) showing that the presumed mannan-protein revealed with fluorescent antibodies was layered as an annular band at the base of the bud. The bud tip deposition has however been recently confirmed by Farkas et al. (1974) who, by means of high-resolution autoradiography, have additionally shown the uniform deposition of the new cell wall material over the whole bud surface. The mode of growth they observed in the yeast differs from that of hyphae (Bartnicki-Garcia & Lippman, 1969; Gooday, 1971) in that, besides the polarized tip growth, the nonpolarized or spherical extension also plays an important role. The ellipsoidal shape of the *S. cerevisiae* cells is thus acquired by a combination of these two modes of cell growth (Farkas et al., 1974): the tip growth which apparently predominates shortly after bud emergence and represents the factor responsible for elongation; the cell wall growth by almost uniform insertion of the new building material over the entire surface of the cell, which occurs during maturation of the bud, responsible for expansion.

In the budding yeast, the growth of the cell wall must occur with a well-balanced equilibrium between the degrading and building processes in order to sustain the internal turgor pressure. In the weakening process of the tip site, enzymes such as protein disulfide reductase (Nickerson & Falcone, 1956; Bartnicki-Garcia & McMurrough, 1971) and glucanase (Johnson, 1968; Matile et al., 1969) are active. There is an accumulation of glucanase and mannan-protein containing vesicles, derived from the endoplasmic reticulum, at the budding sites in *S. cerevisiae* (Matile et al., 1971; Cortat et al., 1972; Sentandreu & Northcote, 1969). The spreading of the action of these enzymes all over the surface of the maturing budded cell should then lead to a uniform, spherical extension of the cell wall (Kanetsuna et al., 1972).

Interestingly, in the transition from a typically budding yeast like *Saccharomyces* to the predominant mycelial growth in the Hemiascomycetous series *Endomycopsis* → *Endomyces* → *Eremascus*, there is a diminution of the cell-wall mannan content and an increase in the proportion of chitin until the cell-wall composition approximates that of typical mycelial Euascomycetes (Bartnicki-Garcia & McMurrough, 1971). In *Saccharomyces* species, chitin and glucan are concentrated in a disklike plug from which the bud has detached (Cabib & Bowers, 1971). A chitin synthesizing system is responsible for initiating septum formation between the mother and daughter cells. The specificity both of timing and localizing this event seems to be due to the conversion of a chitin synthetase zymogen to the active enzyme by an activating factor AF (Cabib & Farkas, 1971). Following bud formation, AF-containing vesicles might fuse with the plasma membrane at the site of septation; according to the Cabib & Farkas' hypothesis, this would result in the activation of the presumably uniformly distributed chitin synthetase and thus lead to the localized chitin synthesis. A major concentration of mannan has also been localized in the bud scar of *S. cerevisiae* (Bauer et al., 1972). It is also of importance to note that yeast bud emergence can occur in the absence of DNA replication (Hartwell, 1971).

Blasto-arthroconidia in Euascomycetes

Ontogenically, blastoconidia correspond to holoblastic conidia produced in acropetal (basifugal) chains; in these chains, "the conidiogenous loci are serial, and cytoplasmic continuity must be maintained between members of the chain to permit the translocation of food material to the developing apical conidium" (Kendrick, 1971).

In the ontogeny of the macroconidia of the *Monilia* state of *Sclerotinia* spp., the pure blastic process involves essentially the first phase of marked enlargement of a recognizable conidium initial before the initial is delimited by a septum: This occurs in a second, typically arthric phase (Willetts & Calonge, 1969). The fact that blastic and arthric processes successively participate to achieve macroconidial morphogenesis shows that it is impossible to draw a sharp line between blastoconidia and fission arthroconidia, because they intergrade (Carmichael, 1971). In our time-sequential study of the so-called macroconidiation of the *Monilia* state of *Neurospora crassa*, we have cut through this gradation difficulty in designating the monilioid macroconidia as blasto-arthroconidia (Turian et al., 1970; Turian & Bianchi, 1971). The blastic elements resulting from basifugal budding on the conidiophore were called proconidia (Turian & Bianchi, 1971, 1972); between these proconidia septation was clearly delayed (Fig. 1B) and it is only after development at the constriction (isthmus), first of a simple, then of a double septum by median fission, that the well-known, normally pink-orange conidia were considered to be differentiated (Fig. 1D and 2A--B).

On conidiogenous hyphae of *N. crassa*, septation first occurs between the proconidia; in well-aerated as well as in microcycling conditions, basipetal septation of the conidiophores secondarily intervenes which separates arthroconidial elements (Turian & Bianchi, 1972; Cortat & Turian, 1974). Similar types of septation have been analyzed by time-lapse photomicrography in the *Monilia* state of *N. sitophila* and *S. laxa* (Hashmi et al., 1972). It appears therefore that the answer to the question of the transition from pure blastic states, most frequently through blastic-arthric intergrades, to pure arthric states of the macroconidiogenous process in the *Monilia* forms of *Sclerotinia* and *Neurospora* species depend on how these conidia are initiated.

When the septa appear at the predestined positions of deep constrictions between the units of the proconidial chain, the products are the most comparable to the blastoconidia described in yeasts (see section on buds and blastoconidia in Hemiascomycetes). When broad septa are formed at the level of the light constrictions of elongating monilioid hyphae, they separate elements which are close to the true arthroconidia lately produced by random septation of the supporting hyphae; most often in such cases, the extremity of the monilioid chain liberates the blasto-arthroconidial intergrades ("dry" ovoid conidia) while the conidia falling apart at the base of the chain are arthroconidia (long, angular conidia).

Looking more closely at the blastic process most commonly occurring at the extremity of aerial, fertile hyphae (Turian & Bianchi, 1972), it has been found to be initiated by the budding of a first proconidium at the enlarged tip of the hypha which has momentarily stopped its elongation (Turian et al., 1973); the wall of the enlarged hyphal tip appears to be thicker, more swollen and less dense, admittedly due to a more delayed setting than in normal, vegetative hyphae (in which setting keeps pace with the plasticization of the elongating tip, according to Robertson's scheme, 1965). In our recent views (Turian et al., 1973) the last part of the enlarged tip to be set, which is visible as a papilla (Fig. in Turian & Bianchi, 1971, 1972), is blown-out under the internal cytoplasmic pressure, thus forming the bud of the first proconidium. The budding process appears very similar to that described in the yeast *Metschnikowia* by Talens et al. (1973b). The slow pace of the setting wall process around the proconidium first favors its enlargement and then leaves a new apical papilla bulging into a second proconidium. A kind of meristematic activity is continually moving forward while the nuclei divide in the basifugal chain and, as in yeast buds, make a delayed colonization of the new proconidium (Fig. 1A). Therefore, from the time of the switch from vegetative to reproductive condition in the hyphal tip, elongation growth changes from continuous to rhythmic; this rhythmic elongation is set by the alternates of plasticization and setting, plasticization allows lateral, widening growth (proconidium), setting of it leads to bulging of the papilla tip, that is, provisory resumption of lengthing growth (new bud). It appears that the length of the proconidial chain (average of 10--12 proconidia) is dependent both on nutrient translocation to the tip and on environmental controls among which the oxygen tension appears critical (see section on Conidia Metabolic control); the change to a semianaerobic atmosphere of a proconidiating chain leads to a resumption of continuous, vegetative hyphal growth at the tip ("dedifferentiation," Turian & Bianchi, 1971).

The arthric process intervenes secondarily, most generally when the proconidial chain elongation is close to being arrested (Turian & Bianchi, 1972). In the well-aerated conditions of transferred mycelia at the surface of an acetate medium (conidiogenous, Turian, 1961), the isthmi between induced proconidia are narrow. Turian et al. (1973) have found that, in the midst of each isthmus, small electron transparent ingrowths on the inner surface of the wall

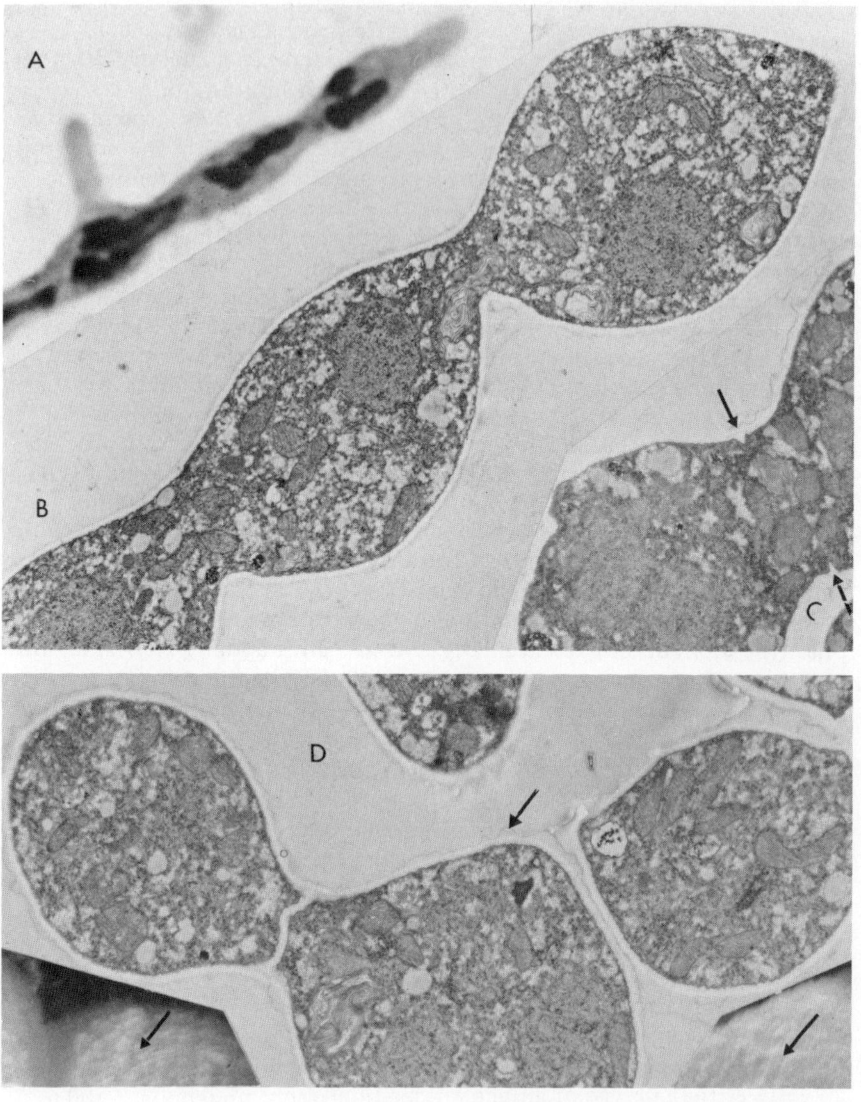

Fig. 1. Blastic (4--B) and early arthric (C--E) phases of macroconidiogenesis in *Neurospora crassa*. Chains of proconidia (blastoconidia) seen by: (A) Optical microscopy. Elongated nuclei (migrating and/or dividing) stained with Giemsa after HCl hydrolysis; anucleated

(Fig. 1C) develop centripetally to form an annulus that becomes coated with denser material before forming a complete septum (Fig. 1D). This simple, interconidial septum then matures into a double septum through the thickening of the primary structure and appearance of a thin, electron transparent median layer or furrow (Fig. 2B). Interestingly, a similar sequence of septal development has been described at the primary bud separation in the yeast *Saccharomycodes ludwigii* (Gay & Martin, 1971). The interconidial septal-plate finally ruptures at the level of the outer wall layer and the broken edges of this layer remain attached as kind of scars to the wall of both individualized conidia (Fig. 2B, C). In aerial, stationary conditions, the successively individualized conidia remain attached in a chain through transient connectives (Turian & Bianchi, 1971, 1972) which must be ruptured (through presumed lysis, deficient in the csp mutants of Selitrennikoff & Nelson, 1973) to liberate conidia. These are now structurally and biochemically (carotenoid pigments, etc.) fully differentiated (see section on mature conidia) and they have to go only through a short period of physiological maturation to acquire full competence for possible germination (average of 10 days).

Phialoconidia of Euascomycetes

This type of asexual spore can be defined as enteroblastic conidia basipetally produced from a fixed conidiogenous locus on a phialide (Kendrick, 1971). The phialide type of conidium development is most frequent in the Eurotiales (essentially *Aspergilli-Penicillia)*, Sphaeriales *(Neurospora, Sordaria,* etc.), and Helotiales

---

proconidial buds (X 3200). (B) Transmission electron microscopy (EM) after glutaraldehyde--osmic acid fixation. Circonvoluted plasmalemma in the presumed budding site (X 11,200). (C) Inception of the interconidial septa (arrows, EM as B (X 11,200). (D) Chain of young conidia separated by single septa (early arthric stage, EM as B--C (X 11,200). Note the peripheral membrane (arrows) contributing to the rough surface pattern of these conidia as shown in the lower corners (arrows) by scanning EM (permanganate fixation, gold--palladium shadowing (X 14,400). (Partially from Turian et al., 1973; scanning micrographs kindly provided by Dr. J.W. Dicker.)

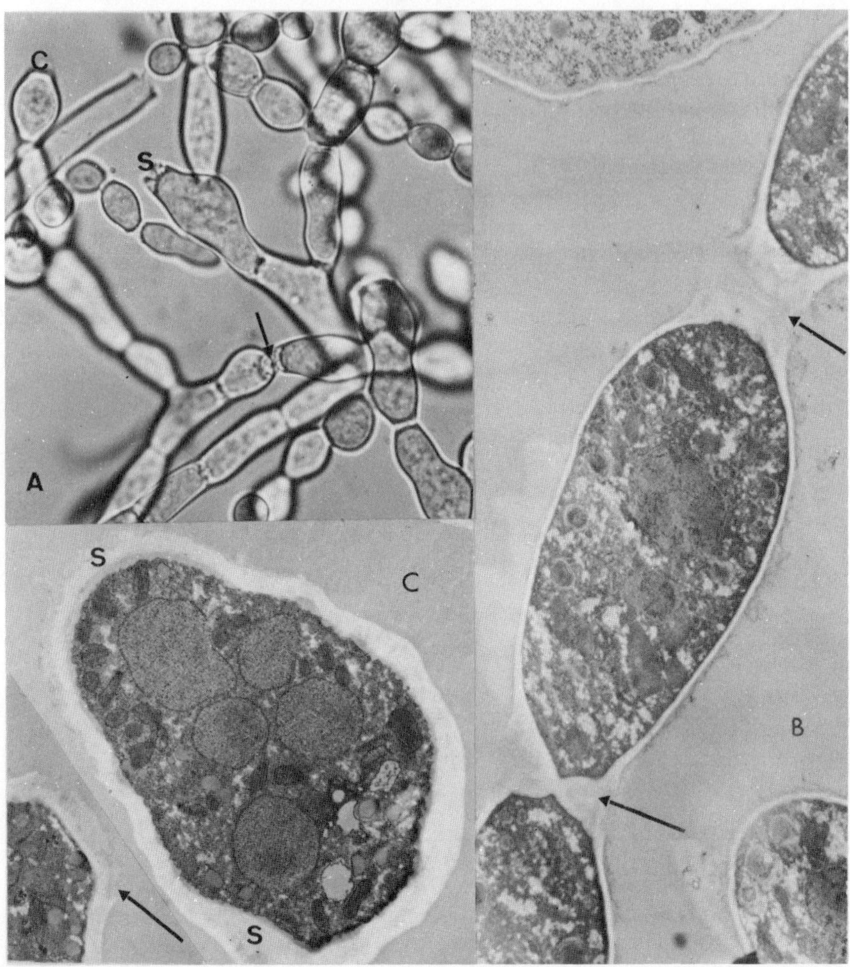

Fig. 2. Arthric phase of macroconidiogenesis in *N. crassa*. Final interconidial disarticulation seen by: (A) optical microscopy (*in vivo* X 1,00). At the level of double septa (arrow) with external wall remains (scars, s) and ruptured interconidial connective (c). (B) Electron microscopy (formaldehyde-glutaraldehyde fixation, osmic acid postfixation (X 11,200). Torn external layer with median, electroluscent discum (arrows). (C) Free, multinucleate conidium

(*Sclerotinia*, etc.). Phialoconidia have not been reported, and are probably lacking, in the Erysiphales (meristem arthroconidia), Pezizales, and Phacidiales while they seem to be rare in the Bitunicatae (Müller, 1971). When phialoconidia predominately exert a sexual function, they are called spermatia as in Laboulbeniales, Sphaeriales, and Helotiales. Thus, in an heterothallic *Sordaria*, where their major role is in spermatization, they have only been considered as spermatia (Fields & Maniotis, 1963). A spermatium-trichogyne relationship has also been found in *S. brevicollis* (Olive & Fantini, 1961) and in species of *Gelasinospora* (Sloan & Wilson, 1958; Goos, 1959). In *Neurospora* species they are called microconidia (Dodge, 1932) as a minor conidiogenous alternative to the macroconidia (blasto-arthroconidia) and can both germinate and spermatize ascogonial trichogynes; in *Sclerotinia* species the microconidia are capable of germination but are also involved in the fertilization of the ascogonial coils borne by overwintered sclerotia (Drayton, 1952).

Conidia of *Aspergilli-Penicillia* are mainly involved in asexual reproduction and their ontogeny corresponds to that of phialoconidia (Subramanian, 1971): they are basipetally budded at the apex of phialides as is dynamically confirmed by time-lapse photomicrography of *P. corylophilum* (Cole & Kendrick, 1969). It has also been confirmed, by transmission electron microscopy, that after each nuclear division, one of the daughter nuclei always remains in the phialide below the delimiting septum bearing a mesosomelike body in *P. claviforme* (Zachariah & Fitz-James, 1967). In *P. claviforme* *P. clavigerum*, and *P. corymbiferum*, the conidium initials are formed by distension of a hollow, pluglike structure of wall material closing the apex of uninucleate phialides (Fletcher, 1971). According to Cole & Kendrick (1969), a layer of wall material is laid down within the still intact phialide apex of *P. corylophilum* before formation of the first conidium protoplast. Sulphydryl groups could be cytochemically detected only in the walls of the phialide tip in *P. notatum* (Pitt, 1969). Later in conidial ontogeny, a newly formed, different looking wall, surrounds the delimited conidium protoplast and conidium chains are finally covered with a thin, electron-opaque surface

---

rich in mitochondria, with polar birth and bud scars (s) more visible in the arrowed left corner (EM, glutaraldehyde-osmic acid fixation (X 11,200). (Unpublished pictures from studies with Dr. M. Cortat & N. Oulevey.)

layer (Fletcher, 1971). Rodlets arranged in characteristic patterns cover wall surfaces on mature conidia of *Penicillia* as well as *Aspergilli* (see sections on mature conidia). An interesting correlation between rodlet orientation and conidiogenesis has been recently described in many Hyphomycetes (Cole, 1973).

The ultrastructure of the conidia-bearing apparatus of *Aspergillus niger* has been investigated first (Tanaka & Yanagita, 1963), followed by that of *A. giganteus* (Trinci et al., 1968) and *A. fumigatus* (Ghiorse & Edwards, 1973). The "apical plug" observed within the apex of the phialide wall in *Penicillia* (see Fletcher, above) has also been seen in *A. giganteus* (Trinci et al., 1968) as wall material distended to surround the emerging conidium. In *A. fumigatus*, successive conidia arise as budlike protuberances of the phialide apical cytoplasm; this contains vesicles which might act as mediators in the synthesis of the wall material distending the cell envelope (Ghiorse & Edwards, 1973). Membranous structures resembling the mesosomes described in phialides of *P. claviforme* (Zachariah & Fitz-James, 1967) were also seen in the peripheral cytoplasm of young conidia and phialides of *A. fumigatus* (Ghiorse & Edwards, 1973). Oppositely, around maturing conidia, the newly synthesized wall described in *Penicillia* (see Fletcher, above) could not be observed in the *Aspergilli* studied.

Deviations in the normal, phialoconidial ontogeny of *Penicillia* and *Aspergilli* have been observed such as the mutational change from basipetal (phialidic-type) to acropetal (blastic-type) cell proliferation (Clutterbuck, 1969; Zachariah & Metitiri, 1970; Vujičić and Mutanjola-Cvetković, 1973; see section on genetic control in conidia).

In *Verticillium albo-atrum*, the development and liberation of the first conidum breaks the phialide tip, leaving a distinct collarette (Buckley et al., 1969). The macroconidia and the microconidia of species of *Fusarium* also blow-out the tip of the phialide. Using the fluorescent antibody staining technique for *F. oxysporum*, Goos and Summers (1964) have shown that part of the blown-out phialide wall usually forms a cap for the first conidium. Moreover, the wall of the successive phialoconidia is apparently formed de novo. The time-lapse study of phialoconidium ontogeny in *Phialophora lagerbergii* has also shown that the wall of the phialide does not seem to contribute to the wall of the conidium which is therefore endogenous (Cole & Kendrick, 1969).

Electron microscope studies of microconidia of *Neurospora crassa*, first carried out on the macroconidialess mutants peach-fluffy grown on sugar-medium (Lowry et al., 1967) and fluffy on acetate-medium (Oulevey-Matikian & Turian, 1968; Fig. 3B) have confirmed their phialoconidial ontogeny (Dodge, 1932). As clearly summarized by Subramanian (1971) "formation of a protuberance from a cell of a microconidiophore presages the development of a microconidium. The cell wall of the conidiogenous cell apparently has several layers and, as the protuberances enlarges, these cell wall layers are ruptured or dissolved following the blowing out of what is apparently a newly formed wall at the site where the protuberance appears and this newly formed wall indeed seems to become the wall of the microconidium." According to Dodge (1932) and Lowry et al. (1967), each conidiogenous cell may produce several conidia; this is indirectly confirmed by the observations of short chains of 2--3 microconidial cells remaining abnormally stuck together out of the phialide collarette in acetate cultures of the fluffy mutant of *N. crassa* (Turian & Viswanath-Reddy, 1971). It is only recently that the main ultrastructural features of the microconidiogenous sequence could be confirmed in the wild-type of *N. crassa* thanks to a chemically directed morphogenesis (Rossier et al., 1973): The iodoacetate-induced microconidia appeared as relatively long uninucleate cells emerging from the middle of a lamellated collar prolongated by a thin collarette (Fig. 3A).

Genetic Control

The sequential transcription or expression of certain genes is needed to change hyphal chemistry as a prerequisite to alter structures to make spores. Compared with sporogenic bacteria, less effort has been devoted in fungi to producing sporulation mutants and few of those obtained can be strictly regarded as such: an important proportion of them have a pleiotropic mutation and are thus metabolic, slow-growing mutants where starvation or toxic accumulation affects both vegetative growth and sporulation. According to Wheeler (1956, cited by Coy & Tuveson, 1964) alteration of a single nuclear gene can switch *Glomerella* from perithecial production to that of "only large mass of conidia," and Coy & Tuveson (1964) have also presented evidence that the production of conidia in *Aspergillus rugulosus* depends upon the activity of a single nuclear gene. In *Aspergillus nidulans*, there is a paucity of mutants specifically and completely blocked in conidium formation, while the number of loci specifically involved in the whole conidiation process has been

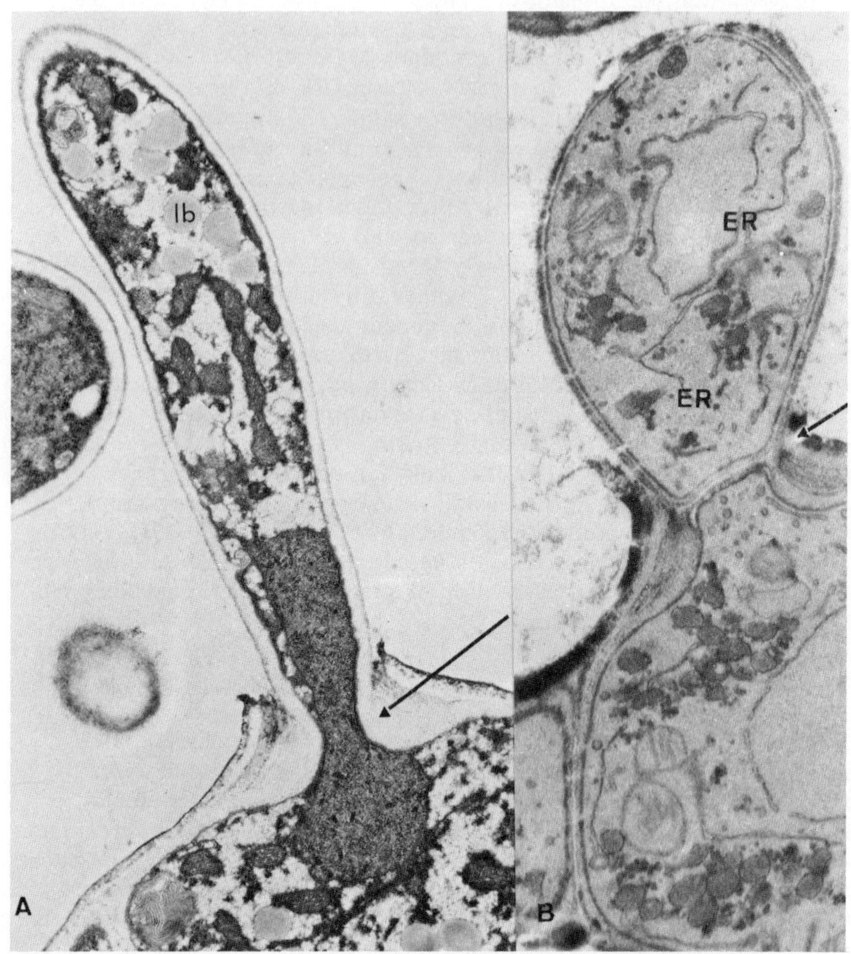

Fig. 3. Electron micrographs of the microconidiogenous apparatus of *Neurospora crassa*. Collars (arrows), each with collarette (external, melanized wall layer) and lamellated cushion (internal wall layer contributing to the conidial wall), surrounding the budding sites of the micro or phialoconidia. These are: (A) Elongated when iodoacetate - induced in the wild type; note the many lipid bodies (lb) preserved and the nucleus entering the conidium (following phialidic mitosis). (Formaldehyde--glutaraldehyde fixation and

estimated to be 45--150 (Martinelli & Clutterbuck, 1971). In 85% of the conidiation mutants the defect was not confined to conidiation (pleiotropic mutations). Mutants were classified into nine developmental stages and into asporogenous (aconidial) and oligosporogenous (oligo conidial) types. The largest group of mutants was blocked before conidiophore formation, and a few only were blocked at the late stages: "Abacus" budding conidia basifugally in a *Monilia*-like way and "bristle" having its site of action at the continuity between the secondary wall of the conidiophore and the single wall of the sterigmata (Oliver, 1972).

The normally basipetal differentiation of true conidia (=phialoconidia) is blocked at nonpermissive conditions in a temperature-sensitive and osmotic-remedial mutant of *Aspergillus aureolatus* (Vujičić & Muntanjola-Cvetković, 1973). The conidial chain elongation becomes apical, blastosporogenic in this *Cladosarum*-like mutant, as in the Clutterbuck's "abaque" mutant of *A. nidulans*. To account for the observed reversion of the young phialide-like element at permissive temperatures, it has been suggested that the defect should be sought in the wall of the mutant (Vujičić & Muntanjola-Cvetković, 1973) rather than in the nuclear differentiation in the phialide, as proposed by Raper & Fennell (1965) to explain the behavior of a similar *Cladosarum* type mutant of *A. niger*. Similarly, in *Penicillium claviforme*, the change from basipetal to acropetal cell proliferation was ascribed to the reversal of a cytoplasmic gradient leading to an altered behavior of daughter nuclei (Zachariah & Metitiri, 1970). In *Penicillium baarnense*, all monogenic mutants concerned with conidiation have shown pleiotropic effects (Parisot, 1972).

Many of the mutants of *A. nidulans* are temperature-sensitive, conditional aconidial strains having their mutations in loci which function essentially only at high temperatures (Martinelli & Clutterbuck, 1971). Mutants defective in their metabolic regulation can also be conditionally aconidial, as shown by the "amycelial" strain of *N. crassa* growing under sterile conditions with vesiculous hyphae on the surface of glucose, restrictive media while recovering the capacity to conidiate on acetate, permissive media (Dicker et al.,

---

osmic acid postfixation, X 16,400, from Rossier et al., 1973.) (B) Rounded from acetate-grown fluffy mutant. Note the paucity in mitochondria contrasting with the development of the endoplasmic reticulum (ER). (Permanganate fixation, X 24,600, from Oulevey-Matikian & Turian, 1968.)

1969; Turian et al., 1971). In *Neurospora*, several induced nuclear-gene mutants, mapping at several loci, affect the type of conidia produced, the density of conidia formed and the presence or absence of conidia (Garnjobst & Tatum, 1967). *Neurospora crassa* can easily mutate to "fluffy" mutants unable to form macroconidia (Lindegren & Lindegren, 1941), while keeping a microconidiogenous ability that is highly increased in the double mutant "peach-fluffy" (Barratt & Garnjobst, 1949). Interestingly, gene-controlled stimulation of production of microconidia by cold treatment has been reported in *Hypomyces solani* f. sp. *cucurbitae* (Wilson & Baker, 1969). Conidial separation defective (csp) mutants have been described in *N. crassa* (Selitrennikoff & Nelson, 1973). Among the three types of morphological mutants induced in *Cochliobolus carbonum* by UV irradiation, one formed abnormally shaped conidia and another was aconidial.

A continuity is apparent between biochemical aconidial mutants and conidiation-specific mutants when we consider that all conidiation genes are in some way concerned with the synthesis of new molecules characteristic for conidia (see also section on mature conidia). We are thus faced with a question of closeness of the relation gene to the spore product. A few genetic loci only could respond to a simple initiating event as starvation causing the transition vegetative growth to sporulation; the transition itself would require the induction of a number of coordinated biosynthetic activities with presumed parallel repression of vegetative pathways leading to the structurally ordinated deposition of the characteristic spore products. Some of the chemical spore products could be more reliable markers than others, some being obligatory such as structural proteins and polysaccharides, especially those associated with spore wall construction, others being facultative, such as pigments which are absent in otherwise conidiating, albino strains of many normally melanizing types of Ascomycetes. White (w) and yellow (y) conidial mutants have been extensively studied in green *Aspergilli* (Fincham & Day, 1971). In the presence of 2-thiouric acid, the wild type of *A. nidulans* produces yellow conidial phenocopies of the y mutant (Darlington & Scazzocchio, 1967). In *A. flavus*, there is an association between pigment accumulation and a free radical electron spin resonance signal produced during conidiation. Mutant colonies, affected in their pigmentation and/or their ability to produce conidia failed to give a free radical signal (Leighton et al., 1971).

It seems however that, from an ecophysiological point of view, melanization as well as other pigmentations such as mixed carotenoid accumulations (see below) can be considered as part of the complete phenotype of many wild type spores. It means that the yellow-spored (also white-spored) strains of *A. nidulans* can be considered as first grade conidiation mutants: they are at least one enzyme---p-diphenol oxidase (laccase)---back from wild type, brownish-green fully differentiated conidia (Clutterbuck, 1972): in mutants like "bristle" that are blocked earlier in their conidial development, the enzyme is not induced. The same considerations can apply to the yellow conidiating mutant Ylo-1 of *N. crassa* described as one gene--one protein--one acidic carotenoid (neurosporaxanthin) back of the normally orange conidiating wild type (Subden & Turian, 1970; Turian & Bianchi, 1972).

The complexity of real genetic control of conidiogenesis will certainly need the study of the regulation of a series of genes and the series of proteins derived from them. A first approach along this line has been made at the level of sequential DNA gene transcription using the selective inhibitor actinomycin D. In *Neurospora crassa*, actinomycin D can interfere with the sequential development of conidiation (Totten & Howe, 1971; Turian & Bianchi, 1972); in synchronously conidiating cultures, DNA-dependent RNA synthesis required for conidiophore formation occurred before 0.5 h and for conidiogenesis before 2 h (Totten & Howe, 1971). There remains however some question concerning the timing of transcription and the extent of RNA systhesis during the period of programmed sporogenic induction; the differential stability of the m-RNAs might explain them as suggested for *Dictyostelium* (Wright, 1973).

Extrachromosomal systems provide an additional genetic control for conidiation in *Aspergilli* (Jinks, 1966) and *Neurospora* (Bertrand et al., 1968).

Metabolic Control

Klebs' generalization that reproduction is initiated by factors which check growth has been echoed in many reviews (Hawker, 1957; Cochrane, 1958; Turian, 1969; Smith & Galbraith, 1971). A definite metabolic shift should thus occur when vegetative growth becomes limited, most generally as a consequence of nutrient exhaustion (mainly of assimilable carbon and/or nitrogen). The problem is then to know how exhaustion of the nutrient supply can cause a metabolic shift to alternate sporogenic pathways. It has been considered (Wright, 1973) that the changes associated with this nutritional deficiency, leading to the depletion of endogenous reserves, act as triggers to the sequential order of events in differentiation. Sporogenesis thus appears most generally as an essentially endogenous, self-sufficient or endotrophic process.

When vegetative growth of fungi stops due to a nutritional deficiency, the intracellular level of primary metabolites is expected to be high and through regulatory processes (Smith & Galbraith, 1971), secondary metabolites are derived from them. Such so-called shunt metabolites can be put at selective advantage for the organisms in being used as specialized products in the formation of spores (Woodruff, 1966). Among them are isoprenoids (carotenoids, steroids), melanin and quinonic pigments which have been found to accumulate in the mature spores (see sections on mature conidia and mature ascospores). Peptide antibiotics are also shunt metabolites which, like sporidesmolides of *Pithomyces chartarum* accumulate in the cultures proportionnally to the number of conidia (Dingley et al., 1962; Di Menna et al., 1970); sporidesmolic acid B produced by sporing felts of *Sporidesmium bakeri* may even contribute to the formation of a water-insoluble coating around conidia (Russel & Brown, 1960).

The nature of the primary metabolic shift initiating the alternate conidiogenic pathways has been investigated in a few filamentous Ascomycetes only (Turian, 1974), mainly with *Neurospora* for the blastic type and with *Aspergillus* species for the phialidic type of conidia.

Conforming to Klebs' principles, growth of *Neurospora* by apical elongation of its hyphae continues so long as enough sugar and nitrogen are available from the medium (Turian, 1970). In subaerial conditions, it stops when translocation of glucose to the hyphal tips becomes limiting and the aerobic stimulus is strong enough to induce a Pasteur effect (Turian, 1969), switching the predominantly alcoholic fermentation of the vegetative hyphae into an oxidative, conidiogenic type of metabolism (Turian, 1973).

The site of the anticonidiogenic effect of glucose appears to be the hyphal tip in which development of the mitochondria is restricted (Turian, 1972) and therefore the oxidative power is catabolically repressed (Turian, 1973). Premature conidiation of *N. crassa* could be obtained by replacing sugar in the nitrate medium by acetate, with parallel induction of the glyoxylate shunt (Turian, 1961); a similar conidiogenic effect could be produced by reducing the glucose level to 0.2% in the repressive ammonium medium, accompanied by a parallel rise of the oxidative activity of the alcohol dehydrogenase and depression of the key enzyme of the glyoxylate cycle, isocitrate lyase (Turian, 1973). Interestingly, the anticonidiogenic effect of high glucose in the ammonium medium (favoring glycolysis, Weiss and Turian, 1966) could be partially lifted by low concentration (10^{-6} M) of cyclic-AMP (Turian & Khandjian, 1973).

The derepressed oxidative activity in the budding, proconidial phase relies mainly on the functioning of the tricarboxylic acid (TCA) cycle as revealed by the higher activities of succinic dehydrogenase and malate dehydrogenase at this stage (Combépine & Turian, 1970) and the anticonidiogenic effect of fluoroacetate (Weiss & Turian, 1966). The activity of the TCA cycle should provide the necessary energy for the syntheses of RNA and protein (Urey, 1971) and, presumably, DNA for the dividing nuclei in the proconidial chains (Turian et al., 1973). The increased activity of the hexose monophosphate shunt (HMP) is apparently confined to the preceding aerial hyphae-conidiophore induction, as shown by the peak of activity of glucose-phosphate dehydrogenase at this stage (Combépine & Turian, 1970; Hochberg & Sargent, 1974), before the progressive appearance of the counteracting activity of NADPase by the end of the blastic phase (Stine, 1969). A progressive slowing down of the TCA cycle against an increased oxidative ADH activity is expected to contribute to the build up of endogenous acetate available for both glyoxylate shunt activity and lipid-carotenoid increased conidial syntheses (see section on mature conidia); this metabolic shift from catabolism to anabolism is critical in opening the gluconeogenic pathway required for the synthesis of, among other things, the structural polysaccharides of the interconidial septa characterizing the arthric phase of *Neurospora* conidiation (see section on the chemostructural ontogeny of conidia; also Turian & Bianchi, 1972).

In *Aspergillus niger*, a dependence of conidia formation on a functionally complete TCA cycle had been reported (Behal & Eakin, 1959) but this has not yet been firmly established (Galbraith & Smith, 1969). At the time of conidiophore differentiation of *Penicillium frequentens*, a transient rise in oxygen consumption might also be due to an activity of the TCA cycle, as suggested from acetate metabolization (Deshpande & Sarje, 1966; Jicinska, 1968; cited from Smith & Galbraith, 1971). Radiorespirometric studies with *Endothia parasitica* (McDowell & De Hertogh, 1968), *A. nidulans* (Carter

& Bull, 1969) and *A. niger* (Galbraith & Smith, 1969) have rather indicated an enhanced contribution of the HMP pathway prior to and during phialoconidiation, with correlated changes found to occur in the levels of glycolytic intermediates (Galbraith & Smith, 1969). These results are in agreement with the changes measured in the levels of glycolytic intermediates during the transition of vegetative mycelia to conidiophore differentiation in *A. niger* (Smith & Anderson, 1973).

That activity of the HMP pathway is conducive to phialoconidiogenesis is also suggested by the selective stimulation of microconidia production in *N. crassa* grown in conditions of iodoacetate-inhibited glycolysis in nitrate medium (Rossier et al., 1973). Under these conditions, macroconidia production is inhibited and cell-free extracts have both high glucose-6-phosphate dehydrogenase and tyrosinase activities (Rossier & Turian, unpublished data 1974); this last enzyme activity is undoubtedly causal to the differential melanism of *Neurospora* microconidia (see section on mature conidia).

Measurements of enzyme activities were also indicative of an active HMP, pentose phosphate pathway in the phialoconidia of *Verticillium albo-atrum* (Throneberry, 1967), in agreement with radiorespirometric data (Brandt & Wang, 1960). Some departure of the TCA cycle from normal has been suspected in these conidia: succinic dehydrogenase, succinic-cytochrome *c* reductase and succinoxidase showed negligible activity in subcellular extracts contrarily to malic dehydrogenase (Throneberry, 1967); succinyl-CoA synthetase, oxoglutarate dehydrogenase and pyruvate dehydrogenase were also found to have low activity (Tokunaga et al., 1970). These results also indicate a disturbance in the cytochrome electron transport system; they thus corroborate current knowledge that an alternate, cyano-insensitive and antimycin A-resistant pathway (Lambowitz & Slayman, 1971; Sherald & Sisler, 1972) is active during conidiation of many filamentous fungi as previously suggested by the paradoxal stimulation of *Neurospora* macroconidiation by KCN (Oulevey-Matikian & Turian, 1968), the conidiogenic response to nitrate of a cytochrome-deficient mutant, and the excretion of flavins in thermo-stimulated conidiating cultures (Ojha & Turian, 1968).

Mature Conidia

At this terminal level, the question is, in what ways does the finished, differentiated product, the conidium, now differ in its chemostructural and functional features from the vegetative hypha? Progressive answers to this question will come from both a better knowledge of the basic ultrastructure of conidia and from refined analyses of their specific chemical constituants.

Following Hawker's review (1965), Sassen et al. (1967) have been among the first to increase our knowledge on the fine structure of conidia in studying those of *Penicillium megasporum* both by chemical fixation (glutaraldehyde--acrolein--osmium procedure) and freeze-etching techniques. Various wall layers could be seen and rodlets presumably composed of cutinlike material were arranged in characteristic patterns on the outer spiked surfaces of conidia (Hess et al., 1968). These structures and the outer wall layer have also been described as electron dense and suggested to consist of fats in the conidial wall of *P. levitum* (McKeen et al., 1966).

In *P. expansum*, the main surface component of conidia is a protein which has an amino acid composition different from that of the total wall protein; the rodlet layer appears to be an integral part of the conidial wall. A surface layer of polyphosphates is present only on mature conidia from cultures grown on a high phosphate medium (Fisher & Richmond, 1970). Related to the apparent richness of *Penicillium* conidia in cytoplasmic lipids is the fact that, when activated by specific compounds, they show an enhanced ability to oxidize octanoid acid (Lawrence & Bailey, 1970).

In resting conidia of *P. megasporum*, the cytoplasm contains many vesicular structures (Sassen et al., 1967) functionally replacing the golgi apparatus, which is absent, and the plasma membrane is externally covered with particles. In the *Aspergilli* resting conidia have also been studied in *A. oryzae* (Tanaka, 1966), *A. niger* (Tsukahara, 1968), and *A. nidulans* (Weisberg & Turian, 1971; Florance et al., 1972). A few invaginations of cytoplasmic membranes were observed in conidia of *A. oryzae* (Tanaka, 1966), and more developed ones stacked into membranosomes have been described in pregerminating conidia of *A. nidulans* (Weisberg & Turian, 1971). The numerous small infoldings of the plasma membrane visible on aldehyde-fixed thin sections of conidia of *A. fumigatus* appeared as long deep groves in frozen-etched preparations (Ghiorse & Edwards, 1973). The freeze-etching technique has also revealed that wall surfaces of mature conidia of *Aspergilli* are covered by rodlets arranged in characteristic patterns (Fig. 4). These rodlets appear to be embedded under an external layer usually lost during the process of freeze-etching (Hess & Stocks, 1969).

In the blastic types, the wall of the dormant conidium of *Botrytis cinerea* (=*Sclerotinia fuckeliana*) is composed of an outer electron-dense layer and an inner electron-translucent layer (Gull & Trinci, 1971); otherwise, the cytoplasm is relatively deficient in both endoplasmic reticulum and mitochondria (Hawker & Hendy, 1963) and contains polarly distributed storage bodies (Buckley et al., 1966).

The mature macroconidium of *Neurospora* is surrounded with a double-layered wall covered with a thin hydrophobic membranous layer sloughed-off in EM preparations (Fig. 1D) presumed to be phospholipidic (Weiss & Turian, 1966; Turian, 1966a). This has been confirmed by chemical analyses showing the presence of phospholipids, fatty acids and hydrocarbons in the surface extracts of *N. crassa* conidia (Fisher et al., 1972). The presence of these surface lipids is however not sufficient to account for the water-repellent properties of all conidial surfaces; according to Fisher et al. (1972), the physical conformation such as the surface rodlet pattern of *Penicillium* (Hess et al., 1968) and *Aspergillus* (Hess & Stocks, 1969) appears to be sufficient to account for hydrophobicity (*P. expansum* has no surface lipids, see below). The liberation scars of the free conidia are remnants of the external wall layer (Fig. 2C) and the main intraconidial structures are: swollen mitochondria, an average of 2--3 nuclei, vacuoles, and sparse elements of endoplasmic reticulum (Weiss, 1965; Weiss & Turian, 1966; Namboodiri, 1966; Turian et al., 1973). Interestingly, the swollen, conidial mitochondria contain only very low succinic dehydrogenase activity (Zalokar, 1959) which presents a different isozymic pattern than that of young hyphae (Ojha et al., 1973); their cytochrome oxidase activity is also relatively low (Owens, 1955b; Ojha & Turian, 1968).

The mature uninucleate microconidia of *Neurospora* are relatively poor in mitochondria (Fig. 3) while their endoplasmic reticulum is more developed than that in macroconidia (Lowry et al., 1967; Oulevey-Matikian & Turian, 1968; Rossier et al., 1973). Similar ultrastructural features have been described in microconidia of Sclerotiniaceae (Willetts & Calonge, 1969).

Chemical composition of the conidial walls is expected to reflect the ultrastructural changes at differentiation. While conidial walls of *A. oryzae* differed only slightly (Horikoshi & Iida, 1964), those of *P. chrysogenum* were rich in galactose compared to the hyphal walls containing glucosamine (Rizza & Kornfeld, 1969). *Neurospora* conidia have a relative ratio of cell wall polymers different from that of the mycelial wall. The higher level of the protein--poly-

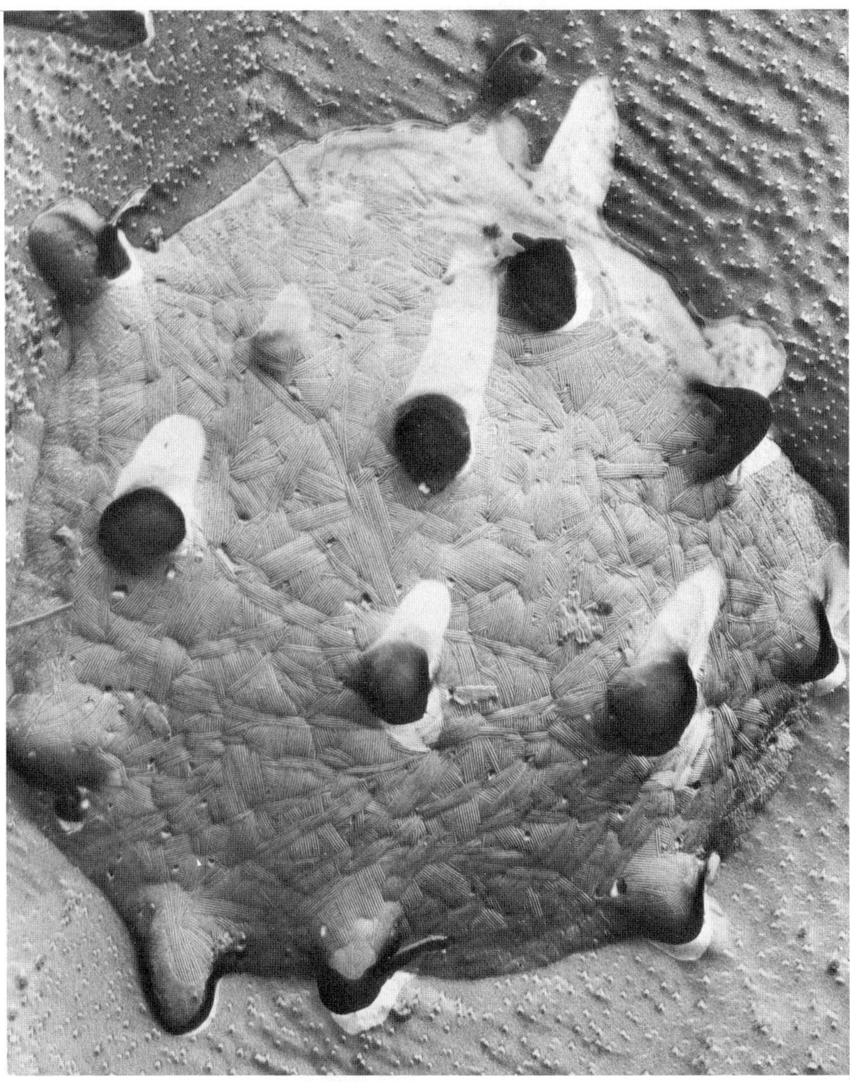

Fig. 4. Surface fracture of a frozen-etched conidium of *Aspergillus aculeatus*. Echinulations (shadowed) and pattern of "rodlets" (X 25,600). (Through the courtesy of Prof. W.M. Hess, Brigham Young University, from Hess & Stocks, Mycologia, 1969, 61, 560--571.)

saccharidic fraction was due to the increased polysaccharide content (Mahadevan & Mahadkar, 1970) which itself includes up to five times more galactosamine than the hyphae in the wild type (Coniordos & Turian, 1973) but not in the "band" mutant of *N. crassa* (Schmit & Brody, 1975). This alcali-soluble fraction densely impregnates the loose glucan-chitinous microfibrillar network constituting the external wall layer of the conidia (Fig. 5; Coniordos & Turian, 1973). Lipids have been extracted (Fisher et al., 1972) from conidial walls of *N. crassa* (9% of their dry weight), *Botrytis fabae* (5%) and *P. expansum* (1%). Surface lipids were present on conidia of *N. crassa* (see below, EM studies), *B. fabae*, and *Alternaria tenuis*, but were absent from conidia of *P. expansum*, *V. albo-atrum*, *Nectria galligena*, and *Erysiphe* spp. (Fisher et al., 1972). Conidial walls of the *Penicillia-Aspergilli* are moreover impregnated with quinonic pigments (Birkinshaw, 1965) and melanic compounds (bound to chitin; Bull, 1970), which are generally absent from the hyphal walls, and are genetically controlled (see section on genetic control in conidia) products of the secondary metabolic pathways (see section on conidial metabolic control) initiated at conidiogenesis. Melanin (extractable with the chitinous fraction) is also deposited in the walls of other phialoconidia such as the microconidia of *Neurospora* (Turian et al., 1967) and, presumably, in those of *Verticillium* (Buckley et al., 1969).

Internal lipids from conidia of *Penicillium roqueforti* (Lawrence, 1967) and *Aspergillus niger* (Gunasekaran et al., 1972) have been extracted and identified. In *A. niger*, the polar lipid content was higher than the neutral lipid fraction and lipid bodies have been revealed on freeze-etched replicas of dormant conidia (Gunasekaran et al., 1972).

Ribosomes have been isolated from free conidia of *N. crassa* and found not to be different than those from hyphae (Henney & Storck, 1963). Those from conidia of *A. oryzae* contain 44% protein and 56% RNA, have a sedimentation coefficient of 80S, and contain a latent ribonuclease (Kimura et al., 1965) which may be associated with a protein inhibitor (Horikoshi, 1971). Absence of polysomes had been reported in ungerminated conidia of *N. crassa* (Henney & Storck, 1964) and *A. oryzae* (Horikoshi et al., 1965). A potential store of prepackaged mRNA has however been detected in conidia of *Botryodiplodia theobromae* (Brambl & Van Etten, 1970) and macroconidia of *Fusarium solani* (Cochrane et al., 1971). The facts that there is conserved mRNA in the free conidia (Bhagwat & Mahadevan, 1970) and that new RNA synthesis is therefore not absolutely necessary for conidial germination (Inoue & Ishikawa, 1970) are in the line of the recent finding of 3% of the ribosomes occurring as polysomes in unhydrated conidia of *N. crassa* (Mirkes, 1974).

In the microconidia, those which are unable to germinate (spermatia of the Sclerotiniaceae) were found to be very low in cytoplasmic RNA concentrations, while those which germinate (*Neurospora*) had high values of RNA (Griffiths, 1959).

Proteins are among the most specific indicators of the differentiated state. New bands of proteins have been electrophoretically detected in conidia compared to nonconidiating mycelia of *P. griseofulvum* (Bent, 1967). The time course levels of several proteins temporally associated with conidial development were assayed in *Microsporum gypseum* (Leighton & Stock, 1970). A protein band has been detected in extracts of free conidia of *N. crassa* analysed by electrophoresis on acrylamide gels (R_f 0.21) which was consistently absent from all mycelial extracts (Combépine & Turian, 1970). The conidial extracts of *V. albo-atrum* contained at least three proteins not detected in extracts from mycelium (Pelletier & Hall, 1971). A glycoprotein with invertase activity is detectable with immunofluorescent staining in greater amounts in the walls of conidiated than in vegetative hyphae of *N. crassa* (Chang & Trevithick, 1972).

The chemostructural organization of conidia is such that it confers them an increased resistance to the deleterious action of environmental factors and thereby contributes to their longevity. Heritable factors influence the resistance to radiation, as reported by Sussman & Halvorson (1966): "Microconidia of *N. crassa* were selected by Goodman (1958) with an LD_{90} for ultraviolet light of 184,000 ergs/cm^2 as compared with 115,000 for the parent organism. Macroconidia of the same strain were shown to have an LD_{90} of 506,000, confirming the observations of Norman (1951) that these multinucleate spores are more resistant than are the microconidia." Moreover, mature conidia of *N. crassa* are less sensitive to ultraviolet light than are germinating ones (Reissig, 1960, in Sussman & Halvorson, 1966), as confirmed with conidia of *A. niger* (Maruyama & Hayashi, 1963). Photoreactivation has been demonstrated for a variety of fungus spores including the conidia of *Penicillia* and *Neurospora* (Sussman & Halvorson, 1966). That the pigments in the conidial walls increase the resistance to radiation has been shown in several cases. Occasional pigmented spores were responsible for the "tailing" in survival curves of *Glomerella cingulata* (Markert, 1953). Conidia of a white-spored strain of *Cochliobolus sativus* were less resistant

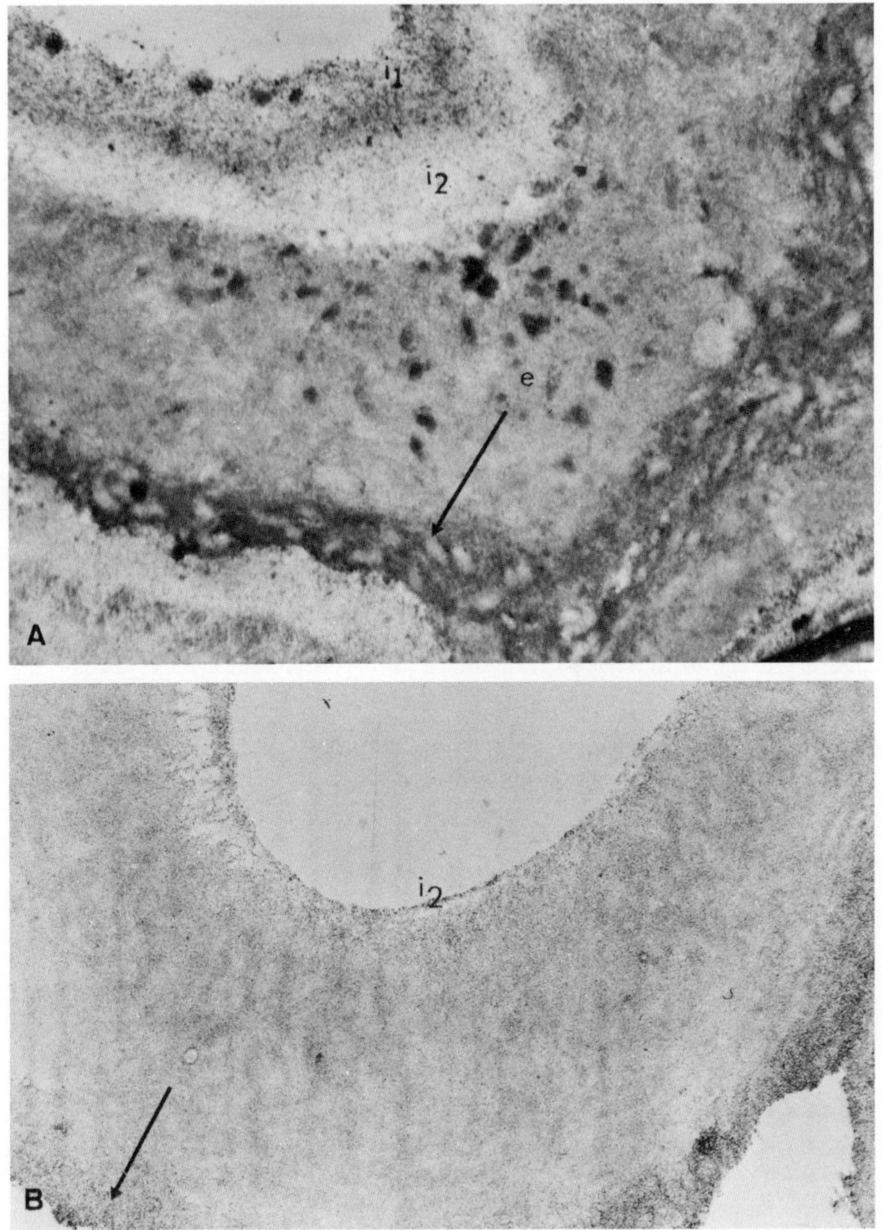
i1
i2
e
A
i2
B

to ultraviolet light than those of a dark-spored strain (Tinline et al., 1960). This may have ecological implications, as shown by the selective survival of the dark-walled Dematiaceae (comparatively to the light pigmented Moniliaceae) in soils irradiated by atomic bomb testing in Nevada (Durrell & Shields, 1960) or exposed to gamma radiation (Johnson & Osborne, 1964).

Conidial resistance and viability are not only influenced by strain differences, but also by the age of conidia (older, presumably more pigmented ones of *A. melleus* are more resistant to UV light according to Dimond & Duggar, cited by Sussman & Halvorson, 1966) and by their cytoplasmic condition (high inviability of the conidia of *A. glaucus* displaying "vegetative death" factor of Jinks (Sussman & Halvorson, 1966).

Among the effects of environmental factors, low temperature (liquid nitrogen; Wellman, 1967) and high humidity have been shown to maintain viability of *Neurospora crassa* conidia (Klingmüller, 1963). These spores retain a viability of greater than 50% when stored for 23 days under conditions of high humidity and constant temperature. In such ageing conidia, the activity of NADP-specific glutamic acid dehydrogenase decreased to a negligible value within 15 days (Tuveson et al., 1967).

Dormant conidia of *A. oryzae* have weak respiratory activity and their shortened longevity at high temperature, high humidity (contrarily to *Neurospora* conidia) and anaerobiosis appeared to be partially dependent on the consumption of mannitol, an

---

Fig. 5. Transverse sections thorugh the walls of free conidia of *N. crassa*, cleared with a detergent--alcohol procedure before their glutaraldehyde--permanganate fixation for EM (X 64,500).

(A) Control section showing duplicate internal layer (i^1--i^2) and external layer (e) densified in the peripheral alveolar network (arrow).

(B) Section through a conidial wall subjected to an additional NaOH 2 N extraction removing the glycoprotein fraction associated with a galactosamine--polyphosphate complex in the peripheral network. Note the clearing of this zone (arrow) and the disappearance of the internal layer (remains of i^2). (From studies with Dr. N. Coniordos, see Coniordos & Turian, 1973.)

intracellular storage substance (Tokoro & Yanagita, 1966). In *A. clavatus*, the storage function of mannitol has been connected with conidiation (Corina & Munday, 1971). Trehalose is also an endogenous reserve in conidia of *Myrothecium verrucaria* (Mandels et al., 1965) and of *A. oryzae* (Horikoshi & Iida, 1966); with trehalase, it is present at high levels in newly formed conidia of *N. crassa* (Hanks & Sussman, 1969). As longevity of conidia is known to be also dependent upon the medium in which they were formed (positive influence of cellulose for conidial viability in *A. luchuensis*, Darby & Mandels, 1955), it might thus well be nutritionally controlled through the type and amount of endogenous reserves accumulated throughout conidiogenesis into the mature conidia.

Contrarily to trehalose, lipids do not appear to be a source of energy at germination but, as in dormant ascospores (see section on mature ascospores of Euascomycetes), are rather slowly utilized for the maintenance metabolism of ageing macroconidia in *N. crassa* (Bianchi & Turian, 1967). Such conidia are richer in lipids, especially phospholipids, than hyphae (Owens et al., 1958). In conidia of *Verticillium albo-atrum*, endogenous respiration decreased with starvation time with a concomitant decrease in RQ also indicating a transition to lipid utilization (Throneberry, 1970). A phospholipidic component, choline sulfate, can accumulate in the conidia of *A. niger* (Takebe, 1960). In this species, esterase activation revealed that lipids can function as a source of carbon and energy during conidiation (Lloyd et al., 1972).

ASCOSPORES

They are "spores produced as a result of meiosis and enclosed within a specialized cover" (Sussman & Halvorson, 1966). Following nuclear divisions, ascospore boundaries become delineated around the clear areas enclosing the nuclei, by a process of "cell free formation" which involves the plasmalemma in a quite different way than at conidiogenesis.

Chemostructural Ontogeny

Hemiascomycetes

During yeast budding, the new plasmalemma is produced by growth in the area of the preexisting membrane, a process in which a revitalization of endoplasmic reticulum or golgi-derived membrane material has been suggested (see section on buds and blastoconidia in Hemiascomycetes). The protoplasts of the mother cell and daughter cell are finally separated by the infolding plasmalemma which produces a closing aperture. In the formation of the ascospore, new plasmalemma is produced late in the process of spore delimitation. The first electron micrographs of thin sections of presporulating *Saccharomyces cerevisiae* (Hashimoto et al., 1958, 1960; Marquardt, 1963; Osumi et al., 1966) have shown that each nucleus produced by the meiotic divisions is surrounded by elements of endoplasmic reticulum which enclose only a part of the cytoplasm available, while a considerable amount of it stays outside in the ascus. Consequently, the ascospores appeared encapsulated by two membranes derived from endoplasmic reticulum, both of which are thought to be involved in the production of the spore cell wall. The inner one alone is involved in the formation of the new plasmalemma of the ascospore (Marquardt, 1963). Lynn & Magee (1970) also observed nuclear enclosure by two unit-membranes but suggested that only one of them was derived from endoplasmic reticulum of the mother cell. Of the double, delimiting membranes, the inner one was confirmed as forming the ascospore plasma membrane, while the outer membrane presumably gives rise to the proteinaceous spore membrane (Briley et al., 1970). Recently, Illingworth et al. (1973) and Beckett et al. (1973) have shown that the principal changes in fine structure which occur between 18 and 24 h in developing asci of *S. cerevisiae* include the appearance of numerous electron-transparent vesicles which become enclosed within the spore-delimiting membranes surrounding the lobes of the meiotic nucleus. This essentially supports not only the findings of Lynn & Magee (1970) with *S. cerevisiae* but also those obtained using the "hat-shaped" ascospores of *Hansenula* yeasts.

In *Hansenula anomala*, Bandoni et al. (1967) had shown that the cytoplasm of the multinucleate (?) ascus is cleaved by a system of coalescing membranes of the endoplasmic reticulum. A double membrane system surrounding each nucleus and a corresponding portion of cytoplasm would then assume the characteristic shape of the mature ascospore wall (Fig. 15 in Bartnicki-Garcia and McMurrough, 1971); this wall is cast in the space between the membranes serving, so to speak, as a mold. In *H. wingei*, Black & Gorman (1971) have shown that the onset of meiosis in preascus cells is indicated by proliferation of the endoplasmic reticulum located between the vacuole and the nucleus. As in *Saccharomyces* (Moens, 1971; see below), the nuclear membrane of *Hansenula* remains intact throughout meiosis. It is by lobulation that the nuclear mass segments into four meiotic products which, enveloped in a double track of endoplasmic membranes, form the "spore initials." Maturation of the ascospores then proceeds with separation of the membranous tracks toward the central nuclear body. Finally, mucopolysaccharide, characteristic of the ascus wall, is deposited in the intercisternal space and the "spore initials" take on the typical hat-shaped profile of mature ascospores (Black & Gorman, 1971).

As described by the above authors, the process by which double membranes enclose the meiotic products finally surrounded by ascospore walls formed by insertion of material between these membranes, would be closely analogous to the process of ascospore formation in Euascomycetes (see next section). Such an analogy has, however, not been confirmed by other detailed ultrastructural studies of *S. cerevisiae* ascosporogenesis. Moens (1971) and Moens & Rapport (1971) have described a new structure associated with the developing spore wall, termed the "prospore wall"; it consists of an electron-transparent region between two dense lines and is observed in association with the cytoplasmic side of the meiotic spindle plaques. These spindle plaques with associated microtubules duplicate after the first meiotic division and finally migrate to face each other for the second meiotic division; the nuclear envelope apparently remains intact ("uninuclear meiosis"), and when the parent nucleus forms four lobes, each lobe is partially enclosed by a "prospore wall." In Moens' view, the transparent region of this lamellated structure expands during ascospore development to form the spore wall. Membranous vesicles pressed to the meiosis II plaques and corresponding to the "prospore wall" have also been seen on sections of spheroplasted *S. cerevisiae* asci fixed with improved techniques (Peterson et al., 1972). A confirmation of the reality of the "prospore wall" structure has been provided by Guth et al. (1972) who investigated meiosis and sporulation in *S. cerevisiae* with the freeze-etching technique. They observed, at the second meiotic division, four bilamellar structures located near the four buds from the parent nucleus which they termed "forespore membranes." Their recognition of these bilamellated structures as membranes extending around the ascospores should contribute to a reduction of the gap between proponents of the double membrane and of the "prospore wall" schemes of yeast ascospore wall development.

The gap between the two main schemes of yeast ascosporogenesis appears to be practically closed by the recent observation of Yoo et al. (1973) of a "forespore membrane" emerging at the beginning of meiosis II during asynchronous ascospore formation in *Schizosaccharomyces pombe*. After enclosure of the nucleus, this membrane is resolved into an inner component which becomes the spore plasmalemma and an outer component which becomes a limiting membrane of the spore wall. Here also, it is in the space between these two membranes that spore wall materials deposit.

In Hemiascomycetes other than the Saccharomycetaceae, it has been observed that *Taphrina deformans* ascospore delimitation in the ascus occurs through the invagination of the plasmalemma at specific sites adjacent to nuclei; moreover, the continuous ascus plasmalemma and delimiting membranes stain preferentially with a phosphotungstic acid--chromic acid mixture (Syrop & Beckett, 1972). In *Nematospora coryli*, a partial dissociation of the nuclear membrane has been seen preceding the appearance of membranes delimiting the ascospores (Plurad, 1972).

Euascomycetes

In the ascus initial the two nuclei of opposite mating type fuse and the fusion nucleus undergoes meiosis to form four haploid daughter nuclei (Olive, 1965). In contrast to budding yeasts, where generally no further division occurs, the four nuclei then undergo a mitotic division so that eight haploid nuclei are present in the elongating ascus when ascospore delimitation starts. A pioneer in this ascal cytology, Harper (1897) and his followers in optical microscopy (summary and refer. in Bellemère, 1969) believed that an astral-ray centrosome complex which developed around each nucleus was involved in the delimitation process. The existence of such structures as nuclear beaks, centrosomes, and astral rays was confirmed with the electron microscope but, even in asci where astral rays are conspicuous (*Pustularia cupularis*), no direct role could be ascribed to them in the individualization of the ascospores (Schrantz, 1966, 1967). Rather, spore delimitation membranes were found to be derived from a peripheral double membrane sac which initially encloses all the nuclei in the ascus; invagination of this sac or ascal vesicle around each nucleus then produces ascospore initials (Carroll, 1967; Delay, 1966; Reeves, 1967; Schrantz, 1970).

Delay (1966) with *Ascobolus immersus* and Bracker (1967) with *Ascodesmis nigricans* noted a structural resemblance between the spore-delimiting membranes and the ascus plasmalemma. This and not the endoplasmic reticulum or the nuclear envelope (Moore, 1965) was therefore vizualized as the source of the delimiting membranes (Bracker, 1967), as confirmed in *Hypoxylon rubiginosum* (Greenhalgh & Griffiths, 1970). A microtubule complex has been found at the invagination point of the plasmalemma in the ascus of *Xylosphaera polymorpha* (Beckett & Crafford, 1970).

Further microscope studies of ascosporogenesis in a variety of Euascomycetes have suggested that major features of spore cleavage and wall formation may be common to all (Bracker, 1967; Carroll, 1969; Oso, 1969). The process of delimitation by "free cell formation" involves invaginations of saclike double membrane system of uncertain origin, but with characteristics of endoplasmic reticulum, located at the periphery of the ascus (Carroll, 1969). These invaginations delimit uninucleate portions of cytoplasm along with its organelles; the excluded cytoplasm becomes the epiplasm surrounding the ascospores. The spore wall develops, as in yeasts between the two single unit components of the double membrane, the inner component becomes the plasmalemma of the spore and the outer, becomes the investing membrane.

The terminology proposed by Le Gal (1947) in her optical microscopic studies of Discomycetous ascospore wall structure, and used by Lowry & Sussman (1958) for *Neurospora*, has been extended at the ultrastructural level by Furtado & Olive (1970) who showed that in *Sordaria fimicola* "wall material deposited between the two membranes of each spore initial during the final mitotic division that makes the ascospore binucleate constitutes the endospore. Later, the investing membrane detaches from the endospore and expands in an undulate fashion as additional wall material is produced, the inner part of which condenses into the epispore, an electron-dense layer that contains a finely granular material and in which spore pigment is located. The outer portion of the secondary wall material becomes the perispore, a thicker, less dense, fibrillar layer. At maturity the perispore is undulate in outline and bounded externally by the investing membrane. This layer corresponds to the gelatinous ascospore sheath seen under the light microscope, the expanded sheath being a characteristic feature of the genus *Sordaria*." The ribbed perispore of the species of *Neurospora* treated with sodium hypochlorite has a very similar appearance under the light microscope (Lowry & Sussman, 1958), emphasizing the probable homology of this layer in the two genera. In *Ascobolus stercorarius*, the perispore is like a mucilaginous sac and, during its growth, a laminated epispore develops between it and the endospore (Wells, 1972). Rather than Le Gal's (1947) descriptive terminology of the relative positions of the various ascospore wall layers, Mainwaring (1972) adopted the sequential terminology of Moore (1963), Carroll (1969), and Beckett et al. (1968) naming the wall layers of *S. fimicola* according to the order in which they are laid down. It is the electron transparent primary wall layer which is first laid down between the outer spore investing and inner spore plasma membranes. The site of origin of the wall material is unknown and neither lomasomes nor endoplasmic reticulum cisternae were observed during wall deposition. The secondary wall material appears outside of the primary wall and its partially fibrillar material might be of epiplasmic origin. The tertiary wall appears as an electron dense layer laid down along the interface between the primary and secondary walls. It contains the characteristic dark pigment of *Sordaria* ascospores. The overall thickness of the ascospore wall increases during maturation, the secondary layer becomes fibrous and gelatinous, the tertiary wall becomes multilayered.

At the germ pore of *Sordaria* ascospores, only the primary and tertiary walls appear thinner (Mainwaring, 1972), while the germ pore passes through all three layers in *Podospora anserina* (Beckett et al., 1968). Some poric structures of *Podospora* were found, by scanning electron microscopy, to be hidden under a sporal shield, others being formed in round, concentric and often unfinished depressions (Melendez-Howell, 1972). In *Neurospora*, the same scanning technique did not reveal any pore before activation (Sullivan et al., 1972; see also section on mature Euascomycete ascospores).

A general pattern of sequential wall deposition in developing ascospores has progressively emerged from all the electron microscope studies presented above, but the site of origin of the material of the primary, secondary and tertiary wall layers remains unclear. Lomasomes have been involved in many cases of primary wall deposition, in others endoplasmic reticulum cisternae (E.R.c.) have been described as having a similar function while in a few cases both of these structures were apparently lacking (Table 1).

In *Ascobolus*, the tertiary wall pigment first appears in the epiplasm and subsequently migrates through the secondary wall to its site of deposition (Delay, 1966; Carroll, 1969; also section on mature ascospores in Euascomycetes). In *Sordaria*, the site of origin of the pigment is unknown and, during its deposition in the wall, no decrease in the lipid content of the spore has been observed (Mainwaring, 1972); this suggests that the incorporation of lipid into the ascospore wall as reported in *Penicillium* (Wilsenach & Kessel, 1965) and various marine Ascomycetes (Kirk, 1966) does not occur in *S. fimicola*.

Genetic Control

Hemiascomycetes

In yeasts, budding and ascosporogenesis are mutually exclusive processes as shown in *Saccharomyces* grown for long periods near the maximum temperature for budding. As first observed by Hansen (1899--1902, as cited by Fowell, 1969) they gradually lose the capacity to form ascospores. Mutations also led to the more vigorous growth of the asporogenous types obtained in the budding *Zygosaccharomyces mandshuricus* (Olenov, 1936) as well as the fission yeast *Schizosaccharomyces pombe* (Leupold, 1950).

As a rule, in *Saccharomyces* species the event of ascosporogenesis overlaps with meiosis. Three strains only are known to produce asci containing two diploid spores, and are therefore formed independently of reductional nuclear division (Grewal & Miller, 1972). Four-spored *S. cerevisiae* conditional mutants have to be used to study impairements in meiosis and sporulation (Tingle et al., 1973) and genetic analysis of temperature-sensitive sporulation specific mutants was carried out at the permissive temperature of 25 ^{0}C (Esposito & Esposito, 1969). By use of adequate selection systems, mutants blocked in the whole sporogenic sequence (vegetative to sporogenic I--V) have been isolated (descriptive Table 3 in Tingle et al., 1973) and it was calculated that 48 $\pm$ 27 loci are specifically involved in meiotic and sporulation function of *S. cerevisiae* (Esposito et al., 1972). None of these Spo-mutants have been found to be defective in protein or RNA synthesis at the restrictive temperature of 34 ^{0}C which blocked the formation of asci. Spo-1 and Spo-2 do not increase their DNA to the parental level during sporulation (Esposito & Esposito, 1969). Similarly, homozygous, incompatible α/α or a/a combinations of mating alleles block the sporogenic sequence of the wild type at the stage I, with parallel inhibition of premeiotic DNA replication (Roth & Lusnak, 1970; Tingle et al., 1973). In vegetative mutants

Table 1 Ascospore Wall Layers in Euascomycetes

Species	Primary involvement of		Secondary positioning		Tertiary migration of
	Lomasomes	E.R.c.	External	Internal (e.t.)[a]	pigment
Ascobolus immersus[1]	−	+	+		+
Ascobolus stercorarius[2]	−	?	+		+
Ascodesmis sphaerospora[3]	−	+	+		
Ascodesmis nigricans[4]	−	−			
Hypoxylon fragiforme[5]	−	−			
Penicillium vermiculatum[6]	+	−			
Podospora anserina[7]	+	−	+		
Pustularia cupularis[8]	+	−			
Pyronema domesticum[9]	−	+		+[b]	
Saccobolus kerverni[10]	+	−	+		+
Sordaria fimicola[11,12]	−	−	+		+

[a]e.t. = electron transparent.

[b]As in yeasts (Bandoni et al., 1967; Lynn & Magee, 1970).

[1]Delay, 1966
[2]Wells, 1972
[3]Moore, 1963
[4]Bracker & Williams, 1966
[5]Greenhalgh & Evans, 1968
[6]Wilsenach & Kessell, 1965
[7]Beckett et al., 1968
[8]Schrantz, 1966
[9]Reeves, 1967
[10]Carroll, 1969
[11]Furtado & Olive, 1970
[12]Mainwaring, 1972

such as the respiratory deficient "petites," diploids homozygous for this character do not form asci while diploids heterozygous for the wild type and "petite" characters are able to sporulate (Ephrussi, 1953).

Although full respiratory activity is required to initiate yeast sporulation, no mitochondrial information needs to be expressed during the process (Küenzi et al., 1974). To demonstrate this, the unique features of mutagenesis with ethidium bromide (EthBr), which makes it possible to generate cells without mtDNA but with an active respiratory system, have been taken advantage of. Under such conditions, sporulation of the EthBr treated cells occurs (primarily two-spored asci) without detectable mtDNA replication but the necessary respiratory activity develops. A functional mitochondrial genetic system is only needed during the period of respiratory adaptation at the time of transfer to the sporulation medium. Because they are respiratory-deficient, established, "petite" mutants that result from alterations in mtDNA cannot sporulate, a finding that is in agreement with Ephrussi's previous data (see above). Sporulation is also blocked when the yeast is grown in the presence of the antibiotic erythromycin which prevents mitochondrial protein synthesis and, consequently, the morphogenic increase in respiratory activity (Puglisi & Zennaro, 1971).

Conditions will no doubt eventually be found for anaerobic sporulation of *Saccharomyces* wild-type cells and even established "petites," as can be forecasted from the finding that *Schizosaccharomyces japonicus* can sporulate anaerobically if provided with sterols (Bulder, 1971).

While most of the Spo-mutants of *S. cerevisiae* are concerned with defects in meiosis, the majority of the 300 asporogenic strains isolated from *Schizosaccharomyces pombe* are disturbed in their ascosporal maturation and wall synthesis (Bresch et al., 1968). These stage V mutants have been detected by the iodine vapour technique (Leupold, 1955) which permits the differentiation between wild-type and metabolically disturbed mutants with abnormal walls.

The importance of the endogenous metabolism in ascospore differentiation has been emphasized by the finding that mutation resulting in low protease activity affected sporogenesis rather than vegetative growth (Chen & Miller, 1968). In other cases, mutations have pleiotropic effects, affecting growth rates and sporulation as recently shown in a glycerol-negative, respiratory-deficient mutant of *Hansenula wingei* (Crandall, 1973).

Euascomycetes

Prominent among previous work with filamentous Ascomycetes is Esser & Straub's (1956, 1958; & Fincham & Day, 1971) finding of a complete series of mutants of *Sordaria macrospora* blocked at the successive steps of sexual morphogenesis normally leading to the formation of eight blackish ascospores. The genes s, min, and pa, were shown to be implicated in the differentiation of the ascospores and the genes ire and lu in their maturation. Similarly, in *Neurospora crassa*, mutations have been produced that affect maturation of ascospores, the best known being the pleiotropic lysine-5 mutant which produces colorless ascospores (Stadler, 1956).

There is genetic control of the number of ascospores differentiated in the asci. The typical ascus of the pseudohomothallic *N. tetrasperma* contains four bisexual ascospores, in contrast to the eight unisexual spores normally found in the ascus of *N. crassa* or *N. sitophila*. This difference in spore number is demonstrably related to the orientation of meiotic and mitotic spindles in the developing ascus (Dodge, 1927, Fincham & Day, 1971). In *N. tetrasperma*, the dominant lethal E routinely produced 8-spored asci but its effects were partially reversible by the growth medium (Dodge, 1939). In the presence of a wild type allele, the lethal one-spored mutant Gsp-1 of *N. tetrasperma* could cause a metabolic deficiency (Novak & Srb, 1971). This one-sporedness is distinct from the thick-walled, indurated asci produced by the dominant gene I (Dodge & Seaver, 1938) or after treatment with chloral hydrate (Zickler, 1931). As stated by Novak & Srb (1971) "just as a biochemical pathway can be reconstructed through a series of specific auxotrophic mutants, a series of (physiological) mutants such as I, Gsp-1, and E may permit one to learn how an ascus containing one, four, or eight ascospores is built."

In *Neurospora*, the spindle behavior in its interactions with cell wall is significant in determining the positions reached by the ascospores. A cell-wall mutant as peak-2 in *N. crassa* shows a nonlinear arrangement of the eight spores in widened ascus: "the observed morphogenetic differences are closely related to differences in nuclear spindle orientation, which in turn are correlated with the geometry of the cell" (Pincheira & Srb, 1969). Biscuit, a "sorbose-type colonial" showing a deviant cell-wall composition (De Terra & Tatum, 1963) and conditional absence of conidia on multiple dichotomously branched hyphae (Dicker et al., 1969), is an allele of peak-2 showing similarly altered asci (Srb & Basl, 1969). Asci homozygous for peak-2 in *N. tetrasperma* also showed a balloon shape, and irregularities in spindle orientation led to the internal differentiation of five to eight ascospores (Pincheira & Srb, 1969). Round spores (with two germ pores) were recently produced by a new, dominant mutant (Rsp-1) of *N. tetrasperma* (Novak & Srb, 1973; Srb et al. 1973).

The size of the ascospores is also under genetic control and a polygenic system has been analysed in *N. crassa* (Lee & Pateman, 1961). Single giant ascospores (average size 20.8 x 69.4 μ) were delimited in the asci of the recessive lethal mutant Gsp-1 spontaneously arisen in *N. tetrasperma* (Novak & Srb, 1971).

Ascospore color mutants have been obtained and used for genetic analysis in *Bombardia lunata* (Zickler, 1934), in *Ascobolus stercorarius* (Bistis & Olive, 1954) and *A. immersus* (Delay, 1966), in *Sordaria fimicola* (Olive, 1956; Heslot, 1958) and in *Podospora anserina* (Esser, 1968).

Wild-type ascospores of *Sordaria fimicola* during maturation pass through the following series of colors: hyaline, yellow, greenish-yellow, deep green, and finally dark gray-brown. According to Olive (1956), all pigments in these and in the spores of three mutant strains (hy, y, g) obtained with the use of ultraviolet irradiation are located in the cell wall, while the cytoplasm is consistently hyaline. The hyaline spored mutant (hy) produces ascospores that are at first hyaline, then subhyaline with a faint yellowish tint. The germination of these spores is drastically reduced. The yellow-spored mutant (y) has bright yellow ascospores that become somewhat duller and darker with age. No green pigments have been observed during spore maturation in the mutant. The gene y appears to have little or no effect upon spore size, but the germination of yellow spores is greatly reduced. The gray-spored mutant (g) produces ascospores that are bluish gray, then gray in color. They are conspicuously lighter than mature wild-type spores. Yellow and green pigments which keep the normal size and full germination ability have not been observed during maturation of these spores.

A tan ascospore color mutant of *S. fimicola* has been used as marker to locate the gene of a so-called blackberry mutant (Pollock & Johnson, 1972). In *N. crassa,* there are nutritional requirements for ascospore color development (Threlkeld, 1965) and also a new white ascospore mutant has been obtained (Phillips & Srb, 1967).

Metabolic Control

Hemiascomycetes

In yeasts, the ability to sporulate is in part governed by the shifting metabolism of a starving cell (Miller & Hoffman-Ostenhof,

1964; Fowell, 1969; Haber & Halvorson, 1972). When enough glucose is provided in the medium for the vegetative cells of *Saccharomyces cerevisiae*, they remain in their log phase growth as actively budding, diploid cells, unable to oxidize acetate and to sporulate. As the concentration of glucose naturally decreases or is induced by starvation, the TCA cycle enzymes are derepressed and the cells in their late growth phase oxidize their ethanol and acetate before starting their early sporogenesis. Such a release of glucose inhibition of sporulation can also be achieved by exogenous cyclic AMP (Tsuboi et al., 1972). As for the trigger action of acetate in sporulation, Croes (1967) has suggested that it acts by speeding up the energy-supplying processes in the cell, resulting in an insufficiency of the glyoxylate cycle. The fact that isocitrate lyase activity increased markedly in a sporogenic strain but not in an asporogenic strain points, however, to an involvement of the glyoxylate pathway in the sporulation process (Miyake et al., 1971).

The sporogenic shift from fermentation to oxidative respiration appears energetically essential to insure sporal syntheses among which storage carbohydrates like a glycogen--glucan fraction and lipids (Tingle et al., 1973). Two periods of lipid synthesis have been detected (stages I--III and V, Esposito et al., 1969). During the 48 h of incubation to complete ascospore formation in *S. cerevisiae*, Illingworth et al. (1973) found a fourfold increase in the lipid content of the cells. This increase has been attributed mainly to an increased synthesis of sterol esters and triacylglycerols and to a lesser extent of phospholipids. According to Henry & Halvorson (1973), phospholipid but not neutral lipid synthesis ceased virtually when the mature asci appeared in *S. cerevisiae*. That some of these lipids, as well as glycogen are also consumed during the morphogenic process has been shown by their decrease when endogenous respiration reaches a maximum paralleling the meiotic divisions (Vezinhet et al., 1971). The carbohydrate reserves are presumed to support carbon requirements for syntheses of ascospore polysaccharides (Tingle et al., 1973).

A period of active proteolysis precedes ascospore differentiation (Chen & Miller, 1968). Extensive protein and RNA turnover and synthesis then occur throughout sporulation (Tingle et al., 1973). Two maxima for protein synthesis occur, one before asci appear and a second during ascospore differentiation (Esposito et al., 1969). An inhibitor of cytoplasmic protein synthesis, cycloheximide, arrests yeast sporulation before the stage of mature ascospores (Esposito et al., 1969), and there is evidence that mitochondrial protein synthesis is also required for ascosporogenesis (see section on genetic control of Hemiascomycetes ascopores).

Euascomycetes

The metabolic study of ascosporogenesis in filamentous Ascomycetes has been hampered by the technical difficulties of obtention of homogenous and synchronously produced material from any of the several stages of the developmental span leading from the initial ascogonial coil through the dicaryotic hyphae and the ascus initials to the maturing ascospores (Turian, 1966b, 1969).

The last stages of this sequence could best be physiologically controlled in *Sordaria fimicola* by the concentration of biotin in the synthetic medium (Barnett & Lilly, 1947). As the biotin concentration was decreased, the percentage of abortive or abnormal asci was greatly increased. Perithecia could, however, still be formed in conditions of severe biotin deficiency (0.1 μg/liter) which totally prevented ascospore differentiation.

In relationship with the known involvement of biotin in carbohydrate metabolism, it has been shown that a complexing agent of *o*-dihydroxy-groups, boric acid, produced similar abortion of ascospores in *Sordaria* grown on optimal concentrations of biotin (Turian, 1955). Boric acid also increased the proportion of asci containing ascospores abnormal in size, number, and pigmentation in *N. tetrasperma* (Liebich & Turian, unpublished results, and Fig. 6).

According to Nelson et al. (1967), sterols are required at each of the principal stages of sexual morphogenesis in *Cochliobolus carbonum*. Ascospore formation, which could be consistently blocked by specific inhibitors of steroid biosynthesis, was restored in crosses treated simultaneously with squalene or sterols such as β-sitosterol, cholesterol, and ergosterol (Nelson et al., 1967). The apparently greater requirement for sterols at ascosporogenesis might be related to the intraascal extension of the membranous system delimiting the spores (see section on ascospore chemostructural ontogeny). Factors such as the carbon source and the concentration of nitrogen can also control the size and septation of ascospores, as shown in a *Leptosphaerulina* (Müller, 1966).

Melanin formation or tyrosinase activity have often been associated with reproductive morphogenesis in fungi (see Turian, 1966b, 1969; Smith and Galbraith, 1972). In the Hypocreaceae, for production of hyaline to slightly pigmented ascospores, tyrosinase has been suggested to be a limiting factor in the initiation of perithecial formation (Wilson & Baker, 1969).

Even in the black-spored *Sordariaceae*, ascosporogenesis can be structurally completed without melanin deposition in the spore walls. In an *albospora* mutant of *Podospora anserina*, there was no difference in the spectrum of phenol oxydase from that of the wild-type, indicating that, rather than melanin, it is the enzyme which is related to sexual morphogenesis (Esser, 1968). The situation is, however, more complex in *Neurospora*, where tyrosinase activity could be induced in tyrosinaseless, female-sterile mutants, but without parallel recovery of fertility (Horowitz et al., 1961) and in wild-type *N. crassa* where tyrosinase was derepressed and fruit body formation was prevented by cycloheximide or actinomycin D (Viswanath-Reddy & Turian, 1972). Therefore, it appears that,

---

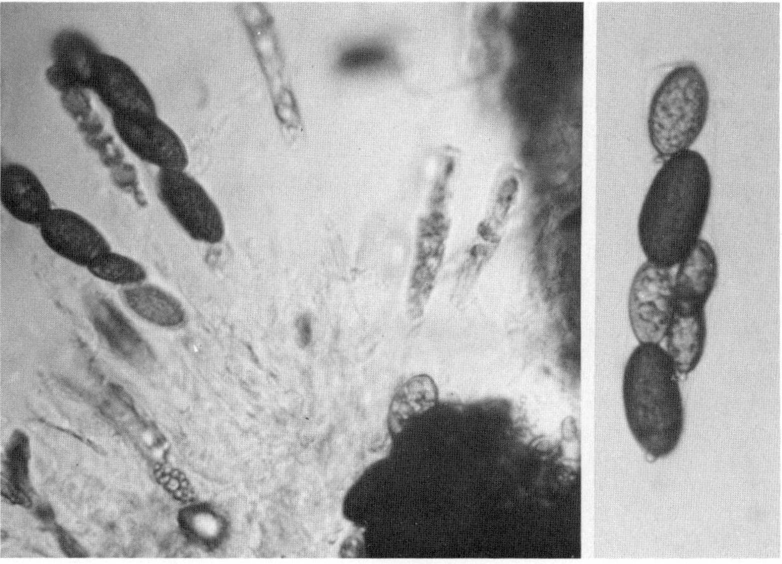

Fig. 6. Aborted and abnormal asci (6-spored on the right) from perithecia of *Neurospora tetrasperma* developed in the presence of boric acid (M/150) (*in vivo*, x 400 and x 540). (From unpublished studies with Dr. B. Liebich, University of Geneva.)

even though blackness of ascospores in *Sordariaceae* is the natural sign of their full differentiation and environmental fitness (see section on mature Euascomycetes ascospores), it is the metabolic shift (pentose--shikimic pathway) leading to this pigmentation which is morphogenetically the most significant.

Concerning the metabolic shifts related to the ascosporogenic developmental span, it has been shown that, while the stage of ascogonium formation relies mainly on fermentative metabolism (high CO^2/O^2 respiratory ratio, genesis at low, 0.5% O_2, inhibition by antiglycolytic agents), the stages of proroperithecial and perithecial formation depend on different types of oxidative activities (Turian & Viswanath-Reddy, 1971). The ascogonium → proroperithecium transformation was blocked by fluoroacetate (10^{-2} M) or citrate (5.10^{-2} M), while the proroperithecia remained sterile (no asci) in the presence of malonate or succinate (10^{-1} M) (Viswanath-Reddy & Turian, unpublished data). Interestingly, quinacrine (mepacrine) which inhibits the cytochrome reductase respiratory pathway can completely block the sexual morphogenesis of *Neurospora* without interfering with the asexual, macroconidiogenic process depending on an alternate oxidative pathway (Turian & Viswanath-Reddy, 1971).

Ascogonial differentiation in *N. tetrasperma* is reversibly prevented at 37 ^{0}C and, in step-down cultures to 25 ^{0}C, both the synchronous induction of ascogonia and the production of a new protein were inhibited by actinomycin D or cycloheximide (Viswanath-Reddy & Turian, 1972). This derepressed protein could be the same as that isolated and characterized from developing fruiting bodies of several species of *Neurospora* (Bowman-Nasrallah & Srb, 1973).

Mature Ascospores

Hemiascomycetes

Variations in the shape of the mature ascospores provides a good criterion for differentiation in yeast systematics (Kreger-Van Rij, 1969). Round, oval, reniform (*Fabospora* spp.), cylindrical, sickle-shaped (*Endomycopsis selenospora*), and needle-shaped (*Nematospora* spp.) spores occur. Among the typical features revealed by light microscopy, there is the ledge, or brim, in the middle or on one side of round or oval spores which gives them a saturn (*Hansenula* spp.), or hat (*Hansenula* spp.), or helmet-shaped

(*Pichia* spp.) appearance. The outer surface of the spore wall can be warty (*Debaryomyces* spp.) or spiny (*Nadsonia* spp.) in contrast to the smoothness of that of *Saccharomyces cerevisiae* as confirmed with the scanning electron microscope (Rousseau et al., 1972).

Ascospore walls are thicker and stronger than those of vegetative cells. In the most easily interpretable sections, two layers of different electron density are visible, a dark outer layer and a lighter inner layer (Besson, 1966). Kreger-Van Rij (1969) has demonstrated the variations in the thickness, and the possible zonation of these layers. According to her observations on the germination of *Hansenula* spores, a light inner zone becomes the wall of the vegetative cell and a darker zone, including the ledge, vanishes. The smooth surface ascospore of *S. cerevisiae* becomes irregular at germination. Along with a loss in light absorbance and refractility, this change may reflect internal chemical modification of their wall (Rousseau et al., 1972).

The ascospore wall possesses a superficial lipid layer which is absent from the wall of the vegetative cell (Fowell, 1969, p. 334), and masses of fat deposits have been seen beside the maturing spore walls at early stage V (Esposito et al., 1969). This richness in surface lipids (also inside, see below) may account for the fact that yeast ascospores are hydrophobic and acid fast. The ability of ascospore walls to stain deeply with gallocyanin has been attributed by Mundkur (1961) to the presence of free amino groups in the large acetylglucosamine fraction of the chitinlike mucopolysaccharides. The fact that these walls are more resistant to snail enzyme extract than ascal walls (Johnston & Mortimer, 1959), is further evidence of their differential composition.

The carbohydrate content of ascospores is higher than that of vegetative cells; this increase is mainly due to glucan, mannan, and trehalose, but not to glycogen (Fowell, 1969). Spores are also rich in lipids, as is shown cytologically (Fowell, 1969; fat globules at the late stage V, Esposito et al., 1969). In contrast, both the protein, amino acid content (except proline), RNA, and the free ribonucleotide content are lower in ascospores than in vegetative or early sporulating cells (Fowell, 1969). In short, the ascospore is in effect a new cell whose components including protein, RNA, polysaccharides are newly synthesized (following switched turnover, see section on Euascomycete metabolic control). That at least some of its new proteins are unique to spores is suggested by the appearance of an immunologically distinct antigen in the ascospore walls of *S. cerevisiae* (Snider & Miller, 1966). There may also be enzymes (or multiple enzymes) unique to sporulating yeasts; cytochrome oxidase is present in relatively high concentrations during the oxidative, sporulating phase, but can remain in negligible amounts in differentiated ascospores as was recently found in germinable "petites" of *S. cerevisiae* produced by brief ethidium bromide mutagenesis (Tingle et al., 1974).

All this chemostructural organization of the yeast ascospore confers it with an increased resistance to the toxic effects of alcohol but, in contrast to the bacterial endospores containing Ca-dipicolinate, only a slightly increased thermoresistance is observed (comparative tables for this and other differential properties of yeast and bacterial spores, in Miller & Hoffmann-Ostenhof, 1964; and Turian, 1969).

Euascomycetes

The generic name of *Neurospora* is derived from the characteristic longitudinal ribs on the surface of its black or greenish-black, mature ascospores (Shear & Dodge, 1927). The ribbed layer is the perispore (Lowry & Sussman, 1958) and corresponds to the "pellicule membrannaire" of Le Gal (1947). Its ribs, elaborated from "rib forming bodies" (Lindegren & Scott, 1937), have been more clearly defined on scanning electron micrographs of ascospores of *N. crassa* (Sullivan et al., 1972; Seale, 1973; Austin et al., 1974; Hohl & Streit, 1975). Interestingly, the ribbed or striated pattern developed on the giant Gsp-1 ascospores of *N. crassa* (see section of Euascomycetes ascospores genetic control) is partially disturbed (Novak & Srb, 1971). Characteristic pits, instead of ribs, develop in the ascospore walls of the related genus *Gelasinospora* (Sung & Alexopoulos, 1954). The episporium has been found laminated in the ascospores of *Ascobolus* (Delay, 1966) and radially striated in the bicellular ascospores of the Lichen *Physcia aipolia* in which the intracellular septum is in continuity with the episporial layer (Rudolph & Giesy, 1966). In *Aspergilli*, the surface structure of their richly ornamented ascospores has been depicted by scanning electron microscopy (Locci, 1972).
(Rudolph & Giesy, 1966).

In addition to the endo-, epi-, and perisporium seen by the light microscope to compose the ascospore wall of *Neurospora* (see section on chemostructural ontogeny in Euascomycetes ascospores), two additional layers were observed with the electron microscope in the walls of dormant spores (Lowry & Sussman, 1968): a relatively electron-transparent one covering ridges of the perisporium and filling the spaces between them; a second, composed of very electron-dense material laying between the episporium and the endosporium; this dense layer contains the melanin and is therefore responsible for the black color of the spores; associated with the endosporium it might also be responsible for the relative impermeability of dormant ascospores, for it is the only layer that appears to be ruptured upon germination. For this later event, there are germ pores, which appear as polar holes through all except the inner layer (Lowry & Sussman, 1968) but become visible on scanning micrographs only after activation of the ascospores (Sullivan et al., 1972). As for the outermost perispore, it can be swollen or removed with a hypochlorite solution without impairing germination and appears to consist of uronic acid residues (Sussman, 1966). This perisporium is imperforate at the level of the germ pore of the dormant ascospore in the several species of *Neurospora* recently studied (Austin et al., 1974; Hohl & Streit, 1975; Fig. 7).

In *Hypoxylon fragiforme*, constrictions in the dense endospore (primary wall layer) were considered to be lines of weakness through which the ascosporal germ tube could emerge (Greenhalgh & Evans, 1968). At the germ pore of *Sordaria* ascospores, only the primary and tertiary walls appear thinner (Mainwaring, 1972), whereas the germ pore passes through all three wall layers in *Podospora anserina* (Beckett et al., 1968). Some poric structures of *Podospora* were found by scanning microscopy to be hidden under a sporal shield, while others were formed in round, concentric, and often unfinished depressions (Melendez-Howell, 1972).

The epispore of *Ascobolus immersus* consists of at least 15 electron-dense bands (Delay, 1966), while the epispore of a mature ascospore of *A. viridulus* has five electron-dense layers (Oso, 1969). In *A. immersus*, granules of blackening pigment are deposited within the perisporal sac and become especially abundant adjacent to the epispore. Since these granules are lacking in the colorless ascospores of a mutant, Delay (1966) considers them as responsible for the color of the wild-type ascospores. In mature ascospores of *A. stercorarius*, only scattered clusters of pigment granules remain in the outer region of the perisporal sac and such spores are colorless under the light microscope after removal of their perisporal sac (Wells, 1972). These findings therefore confirm Chadefaud's (1942) belief that the pigment of the *Ascobolus* ascospores is deposited not in the spore wall proper but in the perisporal sac. This dark violet pigment first appears as a colorless precursor in the vacuolar system of the ascus (Chadefaud, 1942; Malençon, 1962).

In the sporoplasm of *A. stercorarius*, mitochondria change from a tubular to a predominantly spherical shape and, in the more mature ascospores, they are accompanied by two unidentified structures (Wells, 1972). In dormant ascospores of *Neurospora*, endoplasmic reticulum occurs relatively infrequently beside a few large mitochondria. After activation to germination, a whorl of concentric membranes develops in apparent continuity with the endoplasmic reticulum (Lowry & Sussman, 1968).

Like those of yeasts (section on mature Hemiascomycete ascospores), ascospores of *Neurospora* are rich in lipids (26.6%), four-fifths of the total being in the form of esters (Sussman, 1966) while in macroconidia, more than two-thirds are phospholipids (Owens et al., 1958). In contrast to conidia (section on mature conidia), ascospores are not rich in mitochondria and endoplasmic reticulum, and, rather, need "fuel lipids" (presumably esters) for their maintenance metabolism (Lingappa and Sussman, 1959a). They are richer than conidia in the soluble carbohydrate trehalose, the bulk of which is sequestered from the rest of the ascosporal cell and from trehalase up until germination, when it starts to be used fermentatively (Sussman, 1961).

According to Goddard's (1939) terminology (Sussman, 1966), the activated stage is the first to follow dormancy among the different physiological stages of the ascospore. While dormant ascospore extracts of *N. tetrasperma* contained lactate, glycolate, pyruvate, citrate, and malate, those of activated cells contained fumarate and increased amounts of α-ketoglutarate (Sussman et al., 1956). The appearance of these organic acids is indicative of the induction of an oxidative type of metabolism probably different from that functioning (from lipids) in dormant ascospores; before that, a glycolytic activity produces acetaldehyde and ethanol in higher quantities (first, fermentative phase of activation) than the trace amounts present in dormant ascospores (Sussman et al., 1956).

That the enzymes of glycolysis and the TCA cycle are already present in the dormant ascospore has been suggested by the fact that citrate, malate, succinate, and fumarate became labeled without a noticeable lag after the introduction of C_{14}-glucose (Budd et al., 1966). Radioactivity is also found in a hexitol, possibly mannitol, in trehalose, and in a probable glycopeptidic precursor of insoluble spore material.

Of comparative interest is the fact that ascospores accumulate glycolate and some lactate while conidia accumulate glyoxylate (Sussman et al., 1956) and, *N. sitophila*, conidia contain no lactate but succinate (Owens, 1955a); moreover, glycolic acid oxidase could not be detected in dormant ascospores of *N. tetrasperma* (Cheng, 1954). Cytochrome c in the mitochondria was found to be much lower in dormant ascospores than in germinating ones (Holton, 1960).

Ascospores of *Neurospora* have an innate heat-resistance (Lingappa & Sussman, 1959b) and the diversity of enzymes contained in them, among them trehalase and invertase (Hill & Sussman, 1964), should in someway be protected against heat. Protective substance(s) present in ascospore dialysate contribute efficiently toward enhancing

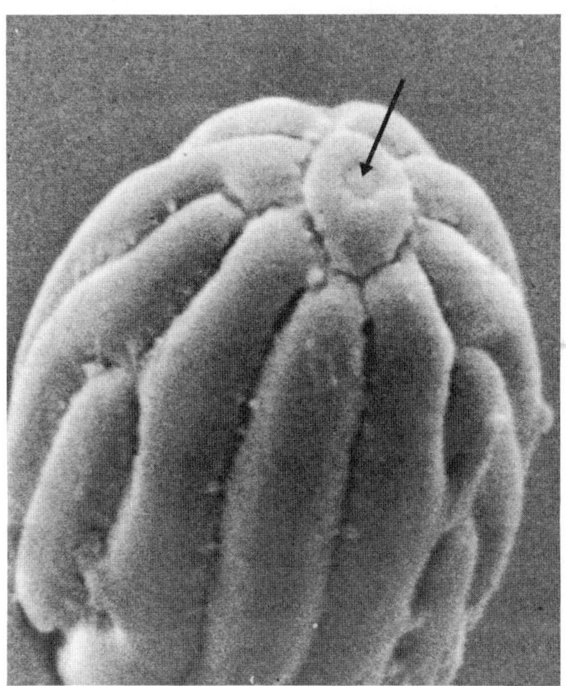

Fig. 7. Scanning electron micrography of a dormant ascospore from *Neurospora dodgei*. Note the wide ribs and the polar, occluded germination pore (arrow) (x 3,200). (Through the courtesy of Prof. H. R. Hohl, University of Zürich.)

the durability of trehalase while being less active toward invertase (Yu et al., 1967). It has been recently shown that trehalases from the ascospores of *N. tetrasperma* and the mycelium of *N. crassa* are very similar if not identical (Hecker & Sussman, 1973a). In the ascospores, trehalase is protected from heat inactivation at 65 ^{0}C by the large amounts of trehalose present (14% of the dry weight; Lingappa & Sussman, 1959b). This disaccharide is intracellularly distributed in two pools, one of which is extractable without breaking the spores, the other only after the spores are disintegrated (Budd et al., 1966). Its protective ability appears however to be coincidental as trehalase has been found to be located within the ascospore wall (Hecker & Sussman, 1973b), as are other glycoproteinic hydrolases such as invertase (Metzenberg, 1963). It is by the elegant immunofluorescent labeling technique that trehalase was shown to reside within the protein and carbohydrate matrix of the endosporium (Hecker & Sussman, 1973b). This association with the cell wall protects the enzyme against the heating which is necessary to activate germination.

Permeability of dormant ascospores of *Neurospora* is far more restricted than that of conidia but they can still incorporate exogenous nutrients such as labeled glucose (Budd et al., 1966); toxic acids such as hydrofluoric or fluoroacetic acids can also penetrate and inhibit the respiration of dormant ascospores at sensitive cytoplasmic sites (Sussman & Halvorson, 1966). On the other hand, the ability of ascospores to bind cations such as Ag_{2+}, Cu_{2+}, or $UO^2{}_{2+}$ on the surface, without permitting their entrance to the cell, has been considered as unique to this stage in the life cycle of *Neurospora*; consequently, a basic antibiotic peptide such as polymyxin B, combines with the wall surface of dormant ascospores, without influencing their respiration (Sussman & Halvorson, 1966). In *Sordaria*, both dark and light ascospores are permeable to water but only light-colored spores are permeable to sodium chloride; the darker, more mature spores have the highest, however variable, osmotic potential (Milburn, 1970).

ACKNOWLEDGMENTS

Thanks are due to Mrs. N. Oulevey and Dr. M. Cortat for their cooperation in the ultrastructural studies, and to Dr. W. Timberlake for reading the manuscript.

REFERENCES

Austin, W.L., Frederick, L. & Roth, I.L. (1974). Scanning electron microscope studies on ascospores of homothallic species of *Neurospora*. Mycologia 66, 130--138.

Bandoni, R.J., Bisalputra, A.A. & Bisalputra, T. (1967). Ascospore development in *Hansenula anomala*. Can. J. Bot. 45, 361--366.

Barnett, H.L. & Lilly, V.G. (1947). The effects of biotin upon the formation and development of perithecia, asci and ascospores by *Sordaria fimicola* Ces. and De Not. Am. J. Bot. 34, 196--204.

Barratt, R.W. & Garnjobst, L. (1949). Genetics of a colonial microconidial strain of *Neurospora crassa*. Genetics 34, 351--369.

Bartnicki-Garcia, S. & Lippman, E. (1969). Fungal morphogenesis: Cell wall construction in *Mucor rouxii*. Science 165, 302--304.

Bartnicki-Garcia, S. & McMurrough, I. (1971). Biochemistry of morphogenesis in yeasts. In The Yeasts, Vol. 2 (ed. A.H. Rose & J.S. Harrison), pp. 441--491. New York and London: Academic Press.

Bauer, H., Horisberger, M., Bush, D.A. & Sigarlakie, E. (1972). Mannan as a major component of the bud scars of *Saccharomyces cerevisiae*. Arch. Mikrobiol. 85, 202--208.

Beckett, A. & Crawford, R.M. (1970). Nuclear behaviour and ascospore delimitation in *Xylosphaera polymorpha*. J. Gen. Microbiol. 63, 269--280.

Beckett, A., Barton, R. & Wilson, I.M. (1968). Fine structure of the wall and appendage formation in ascospores of *Podospora anserina*. J. Gen. Microbiol. 53, 89--94.

Beckett, A., Illingworth, R.F. & Rose, A.H. (1973). Ascospore wall development in *Saccharomyces cerevisiae*. J. Bacteriol. 113, 1054--1057.

Behal, F.J. & Eakin, R.E. (1959). Metabolic changes accompanying the inhibition of spore formation in *Aspergillus niger*. Arch. Biochem. Biophys. 82, 448--454.

Bellemère, A. (1969). Cytologie des Discomycètes (Travaux récents). Rev. Mycol. 34, 122--186.

Bent, K.J. (1967). Electrophoresis of proteins of three *Penicillium* species on acrylamide gels. J. Gen. Microbiol. 49, 195--200.

Bertrand, H., McDougall, K.J. & Pittenger, T.H. (1968). Somatic cell variation during uninterrupted growth of *Neurospora crassa* in continuous growth tubes. J. Gen. Microbiol. 50, 337--350.

Besson, M. (1966). Les membranes des ascospores de levures au microscope électronique. Bull. Soc. Mycol. Fr. 82, 489--503.

Bhagwat, A.S. & Mahadevan, P.R. (1970). Conserved mRNA from the conidia of *Neurospora crassa*. Mol. Gen. Genet. 109, 142--151.

Bianchi, D.E. & Turian, G. (1967). Lipid content of conidia of *Neurospora crassa*. Nat. (Lond.) 214, 1344--1345.

Birkinshaw, J.H. (1965). Chemical constituents of the fungal cell. 2. Special chemical products. In The Fungi, Vol. I (ed. G.C. Ainsworth & A.S. Sussman), pp. 179--228. New York and London: Academic Press.

Bistis, G. & Olive, L.S. (1954). Ascomycete spore mutants and their use in genetic studies. Science 120, 105--106.

Black, S.H. & Gorman, C. (1971). The cytology of *Hansenula*. III. Nuclear segregation and envelopment during ascosporogenesis in *Hansenula wingei*. Arch. Mikrobiol. 79, 231--248.

Bowman Nasrallah, J. & Srb, A.M. (1973). Genetically related protein variants specifically associated with fruiting body maturation in *Neurospora*. Proc. Nat. Acad. Sci. USA 70, 1891--1893.

Bracker, C.E. (1967). Ultrastructure of fungi. Annu. Rev. Phytopathol. 5, 343--374.

Bracker, C.E. & Williams, C.M. (1966). Comparative ultra-structure of developing sporangia and asci in fungi. In Electron microscopy, Proc. 6th Intern. Congr. Electron Microscopy, Kyoto, II, pp. 307--308. Tokyo: Manizen Co.

Brambl, R.M. & Van Etten, J.L. (1970). Protein synthesis during fungal spore germination. V. Evidence that the ungerminated conidiospores of *Botryodiplodia theobromae* contain messenger ribonucleic acid. Arch. Biochem. Biophys. 137, 442--452.

Brandt, W.H. & Wang, C.H. (1960). Catabolism of C_{14}-labelled glucose, gluconate and acetate in *Verticillium albo-atrum*. Am. J. Bot. 47, 50--53.

Bresch, C., Miller, G. & Egel, R. (1968). Genes involved in meiosis and sporulation of yeast. Mol. Gen. Genet. 102, 301--306.

Briley, M.S., Illingworth, R.F., Rose, A.H. & Fisher, D.J. (1970). Evidence for a surface protein layer on the *Saccharomyces cerevisiae* ascospore. J. Bacteriol. 104, 588--589.

Brown, C.M. & Hough, J.S. (1966). Protein-disulfide reduction in yeast. Nat. (Lond.) 211, 201.

Buckley, P.M., Sjaholm, V.E. & Sommer, N.F. (1966). Electron microscopy of *Botrytis cinerea* conidia. J. Bacteriol. 91, 2307--2044.

Buckley, P.M., Wyllie, T.D., & DeVay, J.E. (1969). Fine structure of conidia and conidium formation in *Verticillium albo-atrum* and *V. nigrescens*. Mycologia 61, 240--250.

Budd, K., Sussman, A.S. & Eilers, F.I. (1966). Glucose-C_{14} metabolism of dormant and activated ascospores of *Neurospora*. J. Bacteriol. 90, 551--561.

Bulder, C.J.E.A. (1971). Anaerobic growth, ergosterol content and sensitivity to a polyene antibiotic, of the yeast *Schizosaccharomyces japonicus*. Antonie van Leeuwenhoek, J. Microbiol. Serol. 37, 353-358.

Bull, A.T. (1970). Chemical composition of wild-type and mutant *Aspergillus nidulans* cell walls. The nature of polysaccharide and melanin constituents. J. Gen. Microbiol. 63, 75-94.

Cabib, E. & Bowers, B. (1971). Chitin and yeast budding. Localization of chitin in yeast bud scars. J. Biol. Chem. 246, 152--159.

Cabib, E. & Farkas, V. (1971). The control of morphogenesis: an enzymatic mechanism for the initiation of septum formation in yeast. Proc. Nat. Acad. Sci. USA 68, 2052--2056.

Carmichael, J.W. (1971). Blastospores, aleuriospores, chlamydospores. In Taxonomy of Fungi imperfecti (ed. B. Kendrick), pp. 50--70. Toronto: University Toronto Press.

Carroll, G.C. (1967). The ultrastructure of ascospore delimitation in *Saccobolus kerverni*. J. Cell Biol. 33, 218--224.

Carroll, G.C. (1969). A study of the fine structure of ascosporogenesis in *Saccobolus kerverni*. Arch. Mikrobiol. 66, 321--339.

Carter, B.L.A. & Bull, A.T. (1969). Studies of fungal growth and intermediary carbon metabolism under steady and non-steady state conditions. Biotechnol. Bioengineering 11, 785--804.

Chadefaud, M. (1942). Etudes d'asques III: le mēcanisme de la pigmentation des ascospores chez les *Ascobolus*. Bull. Soc. Bot. Fr. 89, 58--61.

Chang, P.L.Y. & Trevithick, J.R. (1972). Distribution of wall-bound invertase during the asexual life-cycle of *Neurospora crassa*. J. Gen. Microbiol. 70, 23-29.

Chen, A.W. & Miller, J.J. (1968). Proteolytic activity of intact yeast cells during sporulation. Can. J. Microbiol. 14, 957--963.

Cheng, S.G. (1954). The terminal respiratory enzymes in *Neurospora tetrasperma*. Plant Physiol. 29, 458--467.

Chung, K.L., Hawirko, R.Z. & Isaac, P.K. (1965). Cell wall replication in *Saccharomyces cerevisiae*. Can. J. Microbiol. 11, 953--957.

Clutterbuck, A.J. (1969). A mutational analysis of conidial development in *Aspergillus nidulans*. Genetics 63, 317--327.

Clutterbuck, A.J. (1972). Absence of laccase from yellow-spored mutants of *Aspergillus nidulans*. J. Gen. Microbiol. 70, 423-435.

Cochrane, V.W. (1958). Physiology of Fungi. 524 pp. New York and London: John Wiley & Sons.

Cochrane, J.C., Rado, T.A. & Cochrane, V.W. (1971). Synthesis of macromolecules and polyribosome formation in early stages of spore germination in *Fusarium solani*. J. Gen Microbiol. 65, 45--55.

Cole, G.T. (1973). A correlation between rodlet orientation and conidiogenesis in Hyphomycetes. Can. J. Bot. 51, 2413--2422.

Cole, G.T. & Kendrick, W.B. (1969). Conidium ontogeny in Hyphomycetes. The phialides of *Phialophora, Penicillium*, and *Ceratocystis*. Can. J. Bot. 47, 779--789.

Combépine, G. & Turian, G. (1970). Activités de quelques enzymes associés à la conidiogenèse du *Neurospora crassa*. Arch. Mikrobiol. 72, 36--47.

Coniordos, N. & Turian, G. (1973). Recherches sur la différenciation conidienne de *Neurospora crassa*. IV. Modifications chimiostructurales de la paroi chez le type sauvage et chez deux mutants aconidiens. Ann. Microbiol. (Inst. Pasteur) 124 A, 5--28.

Corina, D.L. & Munday, K.A. (1971). Studies on polyol function in *Aspergillus clavatus*: a role for mannitol and ribitol. J. Gen. Microbiol. 69, 221--227.

Cortat, M. & Turian, G. (1974). Conidiation of *Neurospora crassa* in submerged culture without mycelial phase. Arch. Mikrobiol. 95, 305--309.

Cortat, M., Matile, P. & Wiemken, A. (1972). Isolation of glucanase-containing vesicles from budding yeast. Arch. Mikrobiol. 82, 189--205.

Coy, D.O. & Tuveson, R.W. (1964). Genetic control of conidiation in *Aspergillus rugulosus*. Am. J. Bot. 51, 290--293.

Crandall, M. (1973). A respiratory-deficient mutant in the obligately aerobic yeast *Hansenula wingei*. J. Gen. Microbiol. 75, 377--381.

Croes, A.F. (1967). Induction of meiosis in yeast. II. Metabolic factors leading to meiosis. Planta 76, 227--237.

Darby, R.T. & Mandels, G.R. (1955). Effects of sporulation medium and age on fungus spore physiology. Plant. Physiol. 30, 360--366.

Darlington, A.J. & Scazzocchio, C. (1967). Use of analogues and the substrate-sensitivity of mutants in analysis of purine uptake and breakdown in *Aspergillus nidulans*. J. Bacteriol. 93, 937--940.

Delay, C. (1966). Etude de l'infrastructure de l'asque d'*Ascobolus immersus* Pers. pendant la maturation des spores. Ann. Sci. Nat. Bot., 12ème Série, 7, 36 --420.

De Terra, N. & Tatum, E.L. (1963). A relationship between cell wall structure and colonial growth in *Neurospora*. Am. J. Bot. 50, 669-677.

Dicker, J.W., Oulevey, N. & Turian, G. (1969). Amino acid induction of conidiation and morphological alteration in wild type and morphological mutants of *Neurospora crassa*. Arch. Mikrobiol. 65, 241--257.

Di Menna, M.E.,Campbell, J. & Mortimer, P.H. (1970). Sporidesmin production and sporulation in *Pithomyces chartarum*. J. Gen. Microbiol. 61, 87--96.

Dingley, J.M., Done, J., Taylor, A. & Russel, D.W. (1962). The production of sporidesmin and sporidesmolides by wild isolates of *Pithomyces chartarum* in surface and in submerged culture. J. Gen. Microbiol. 29, 127--135.

Dodge, B.O. (1932). The non-sexual and the sexual functions of microconidia of *Neurospora*. Bull. Torrey Bot. Club 59, 347--360.

Dodge, B.O. (1939). A new dominant lethal in *Neurospora*: the E locus in *N. tetrasperma*. J. Hered. 30, 467--474.

Dodge, B.O. & Seaver, B. (1938). The combined effects of the dominant and the recessive lethals for ascus abortion in *Neurospora*. Am. J. Bot. 25, 156--166.

Drayton, F.L. (1952). *Stromatinia narcissi*, a new sexually dimorphic Discomýcete. Mycologia 44, 119--140.

Durrell, L.W. & Shields, L.M. (1960). Fungi isolated in culture from soils of the Nevada test site. Mycologia 52, 636--641.

Ephrussi, B. (1953). Nucleo-cytoplasmic relations in Micro-organisms. 127 pp. Oxford: Clarendon Press.

Esposito, M.S. & Esposito, R.E. (1969). The genetic control of sporulation in *Saccharomyces*. I. The isolation of temperature-sensitive sporulation-deficient mutants. Genetics 61, 79--89.

Esposito, M.S., Esposito, R.E., Arnaud, M. & Halvorson, H.O. (1969). Acetate utilization and macromolecular synthesis during sporulation of yeast. J. Bacteriol. 100, 180--186.

Esposito, R.E., Frink, N., Bernstein, P. & Esposito, M.S. (1972). The genetic control of sporulation in *Saccharomyces*. II. Dominance and complementation of mutants of meiosis and spore formation. Mol. Gen. Genet. 114, 241--248.

Esser, K. (1968). Phenol oxidases and morphogenesis in *Podospora anserina*. Genetics 60, 281--288.

Farkaš, V., Kovařik, J., Košinová, A. & Bauer, Š. (1974). Autoradiographic study of mannan incorporation into the growing cell walls of *Saccharomyces cerevisiae*. J. Bacteriol. 117, 265--269.

Fèvre, M., Larpent, J.-P. & Turian, G. (1975). Bourgeonnement et croissance hyphale fongiques: homologies structurales et fonctionnelles. Physiol. Vég., 13, 23--38.

Fields, W.G. & Maniotis, J. (1963). Some cultural and genetic aspects of a new heterothallic *Sordaria*. Am. J. Bot. 50, 80--85.

Fincham, J.R.S. & Day, P.R. (1971). Fungal Genetics. 402 pp., 3rd., Oxford and Edinburgh: Blackwell Scientific Publications.

Fisher, D.J. & Richmond, D.V. (1970). The electrophoretic properties and some surface components of *Penicillium* conidia. J. Gen. Microbiol. 64, 205--214.

Fisher, D.J., Holloway, P.J. & Richmond, D.V. (1972). Fatty acid and hydrocarbon constituents of the surface and wall lipids of some fungal spores. J. Gen. Microbiol. 72, 71--78.

Fletcher, J. (1971). Conidium ontogeny in *Penicillium*. J. Gen. Microbiol. 67, 207--214.

Florance, E.R., Denison, W.C. & Allen, T.C., Jr. (1972). Ultrastructure of dormant and germinating conidia of *Aspergillus nidulans*. Mycologia 64, 115--123.

Fowell, R.R. (1969). Sporulation and hybridization of yeasts. In The Yeasts, Vol. 1 (ed. A.H. Rose & J.S. Harrison), pp. 303--383. New York and London: Academic Press.

Furtado, J.S. & Olive, L.S. (1970). Ultrastructure of ascospore development in *Sordaria fimicola*. J. Elisha Mitchell, Sci. Soc. 86, 131--138.

Galbraith, J.C. & Smith, J.E. (1969). Changes in activity of certain enzymes of the tricarboxylic acid cycle and the glyoxylate cycle during the initiation of conidiation of *Aspergillus niger*. Can. J. Microbiol. 15, 1207--1212.

Garnjobst, L. & Tatum E.L. (1967). A survey of new morphological mutants in *Neurospora crassa*. Genetics 57, 579--604.

Gay, J.L. & Martin, M. (1971). An electron microscopic study of bud development in *Saccharomycodes ludwigii* and *Saccharomyces cerevisiae*. Arch. Mikrobiol. 78, 145--157.

Ghiorse, W.C. & Edwards, M.R. (1973). Ultrastructure of *Aspergillus fumigatus* conidia development and maturation. Protoplasma 76, 49--59.

Gooday, G.W. (1971). An autoradiographic study of hyphal growth of some Fungi. J. Gen. Microbiol. 67, 125--133.

Goos, R.D. (1959). Spermatium--trichogyne relationship in *Gelasinospora calospora* var. *autosteira*. Mycologia 51, 416--428.

Goos, R.D. & Summers, D.F. (1964). Use of fluorescent antibody techniques in the observations of morphogenesis of fungi. Mycologia 56, 701--707.

Greenhalgh, G.N. & Evans, L.V. (1968). The developing ascospore wall of *Hypoxylon fragiforme*. J. Roy. Microsc. Soc. 88, 545--556.

Greenhalgh, G.N. & Griffiths, H.B. (1970). The ascus vesicle. Trans. Brit. Mycol. Soc. 54, 489--492.

Grewal, N.S. & Miller, J.J. (1972). Formation of asci with two diploid spores by diploid cells of *Saccharomyces*. Can. J. Microbiol. 18, 1897--1905.

Griffiths, E.S. (1959). Cytological and cytochemical aspects of the differentiation of the microconidia of *Gloetinia temulenta*. Trans. Br. Mycol. Soc. 42, 123--124.

Grove, S.N., Oujezdsky, K.B. & Szaniszlo, P.J. (1973). Budding in the dimorphic fungus *Phialophora dermatitidis*. J. Bacteriol. 115, 323-329.

Gull, K. & Trinci, A.P.J. (1971). Fine structure of spore germination in *Botrytis cinerea*. J. Gen. Microbiol. 68, 207--220.

Gunasekaran, M.W., Hess, M. & Weber, D.J. (1972). The fatty acid composition of conidia of *Aspergillus niger* V. Tiegh. Can. J. Microbiol. 18, 1575--1576.

Guth, E., Hashimoto, T. & Conti, S.F. (1972). Morphogenesis of ascospores in *Saccharomyces cerevisiae*. J. Bacteriol. 109, 869--880.

Haber, J.E. & Halvorson, H.O. (1972). Cell cycle dependency of sporulation in *Saccharomyces cerevisiae*. J. Bacteriol. 109, 1027--1033.

Hanks, D.L. & Sussman, A.S. (1969). The relation between growth, conidiation and trehalase activity in *Neurospora crassa*. Am. J. Bot. 56, 1152--1159.

Hartwell, L.H. (1971). Genetic control of the cell division cycle in yeast. IV. Genes controlling bud emergence and cytokinesis. Exp. Cell Res. 69, 265--276.

Hashimoto, T., Conti, S.F., & Naylor, H. (1958). Fine structure of microorganisms. III. Electron microscopy of resting and germinating ascospores of *Saccharomyces cerevisiae*. J. Bacteriol. 76, 406--416.

Hashimoto, T., Gerhardt, P., Conti, S.F. & Naylor, H. (1960). Studies on the fine structure of microorganisms. V. Morphogenesis of nuclear and membrane structures during ascospore formation in yeast. J. Biophys, Biochem. Cytol. 7, 305--317.

Hashmi, M.H., Morgan-Jones, G. & Kendrick, B. (1972). Conidium ontogeny in hyphomycetes. *Monilia* state of *Neurospora sitophila* and *Sclerotinia laxa*. Can. J. Bot. 50., 2419--2421.

Hawker, L.E. (1957). Physiology of Fungi. 360 pp. London: University of London Press.

Hawker, L.E. (1965). Fine structure of Fungi as revealed by electron microscopy. Biol. Rev. 40, 52--92.

Hawker, L.E. & Hendy, R.J. (1963). An electron microscope study of germination of conidia of *Botrytis cinerea*. J. Gen. Microbiol. 33, 43--46.

Hecker, L.I. & Sussman, A.S. (1973a). Activity and heat stability of trehalase from the mycelium and ascospores of *Neurospora*. J. Bacteriol. 115, 582--591.

Hecker, L.I. & Sussman, A.S. (1973b). Localization of trehalase in the ascospores of *Neurospora*: Relation to ascospore dormancy and germination. J. Bacteriol. 115, 592--599.

Henney, H. & Storck, R. (1963). Nucleotide composition of RNA from *Neurospora crassa*. J. Bacteriol. 85, 822--826.

Henney, H. & Storck, R. (1964). Polyribosomes and morphology in *Neurospora crassa*. Proc. Nat. Acad. Sci. USA 51, 1050--1055.

Henry, S.A. & Halvorson, H.O. (1973). Lipid synthesis during sporulation of *Saccharomyces cerevisiae*. J. Bacteriol. 114, 1158--1163.

Heslot, H. (1958). Contribution à l'étude cytogénétique et génétique des Sordariacées. Rev. Cytol. Biol. Vég. 19, Suppl, 2, 1--235.

Hess, W.M. & Stocks, D.L. (1969). Surface characteristics of *Aspergillus* conidia. Mycologia 61, 560--571.

Hess, W.M., Sassen, M.M.A. & Remsen, C.C. (1968). Surface characteristics of *Penicillium* conidia. Mycologia 60, 290--303.

Hill, E.P. & Sussman, A.S. (1964). Development of trehalase and invertase activity in *Neurospora*. J. Bacteriol. 88, 1556--1566.

Hochberg, M. L. & Sargent, M. L. (1974). Rhythms of enzyme activity associated with circadian conidiation in *Neurospora crassa*. J. Bacteriol. 120, 1164--1175.

Hohl, H.R. & Streit, W. (1975). Ultrastructure of ascus and ascospore maturation in *Neurospora lineolata*. Mycologia, In press.

Holton, R.W. (1960). Studies on pyruvate metabolism and cytochrome system in *Neurospora tetrasperma*. Plant Physiol. 35, 757--766.

Horikoshi, K. (1971). Studies on the conidia of *Aspergillus oryzae*. XI. Latent ribonuclease in the conidia of *Aspergillus oryzae*. Biochim. Biophys. Acta. 240, 532--540.

Horikoshi, K. & Iida, S. (1964). Studies of the spore coats of fungi. I. Isolation and composition of the spore coats of *Aspergillus oryzae*. Biochim. Biophys. Acta. 83, 197--203.

Horikoshi, K., Ohtaka, Y. & Ikeda, Y. (1965). Ribosomes in dormant and germinating conidia of *Aspergillus oryzae*. Agr. Biol. Chem. 29, 724--727.

Horowitz, N.H., Fling, M., MacLeod, H., & Watanabe, Y. (1961). Structural and regulative genes controlling tyrosinase synthesis in *Neurospora*. Cold Spring Harbor Sympos. Quant. Biol. 26, 233--238.

Illingworth, R.F., Rose, A.H. & Beckett, A. (1973). Changes in the lipid composition and fine structure of *Saccharomyces cerevisiae* during ascus formation. J. Bacteriol. 113, 373--386.

Ingold, C.T. (1971). Fungal Spores. Oxford: Clarendon Press.

Inoue, H. & Ishikawa, T. (1970). Macromolecule synthesis and germination of conidia in temperature-sensitive mutants of *Neurospora crassa*. Jap. J. Genet. 45, 357--369.

Jinks, J.L. (1966). Extranuclear inheritance. In The Fungi (ed. G.C. Ainsworth & A.S. Sussman), Vol. II, pp. 619--660. New York and London: Academic Press.

Johnson, B.F. (1968). Lysis of yeast cell walls induced by 2-deoxyglucose at their sites of glucan synthesis. J. Bacteriol. 95, 1169--1172.

Johnson, B.F. & Osborne, T.S. (1964). Survival of fungi in soil exposed to gamma radiation. Can. J. Bot. 42, 105--113.

Johnson, B.F. & Gibson, E.J. (1966). Autoradiographic analysis of regional cell wall growth of yeasts. III. *Saccharomyces cerevisiae*. Exp. Cell Res. 41, 580--591.

Johnston, J.R. & Mortimer, R.K. (1959). Use of snail digestive juice in isolation of yeast spore tetrads. J. Bacteriol. 78, 292--295.

Kanetsuna, F., Carbonell, L.M., Azuma, I. & Yamamura, Y. (1972). Biochemical studies on the thermal dimorphism of *Paracoccidioides brasiliensis*. J. Bacteriol. 110, 208--218.

Kendrick, B. (1971). Taxonomy of Fungi Imperfecti. 309 pp., Toronto: University of Toronto Press.

Kimura, K., Ono, T. & Yanagita, T. (1965). Isolation and characterization of ribosomes from dormant and germinating conidia of *Aspergillus oryzae*. J. Biochem. 58, 569--576.

Kirk, P.W., JR. (1966). Morphogenesis and microscopic cytochemistry of marine pyrenomycete ascospores. Beih. Nova Hedwigia 22, 1--127.

Klingmüller, W. (1963). A damaging effect of drying on *Neurospora crassa* conidia and its reversibility. Z. Naturforsch. 18, 55.

Kreger-Van Rij, N.J.W. (1969). Taxonomy and systematics of yeasts. In The Yeasts, Vol. 1 (ed. A.H. Rose & J.S. Harrison), pp. 5--78. New York and London: Academic Press.

Kreger-Van Rij, N.J.W. & Veenhuis, M. (1969). Septal spores in *Endomycopsis platypodis* and *Endomycopsis monospora*. J. Gen. Microbiol. 57, 91--96.

Küenzi, M., Tingle, M.A. & Halvorson, H.O. (1974). Sporulation of *Saccharomyces cerevisiae* in the absence of a functional mitochondrial genome. J. Bacteriol. 117, 80--88.

Lambowitz, A.M. & Slayman, C.W. (1971). Cyanide-resistant respiration in *Neurospora*. J. Bacteriol. 108, 1087--1096.

Lawrence, R.C. (1967). The metabolism of triglycerides by spores of *Penicillium roqueforti*. J. Gen. Microbiol. 46, 65.

Lawrence, R.C. & Bailey, R.W. (1970). Evidence for the role of the citric acid cycle in the activation of spores of *Penicillium roqueforti*. Biochim. Biophys. Acta 208, 77-86.

Lee, B.T.O. & Pateman, J.A. (1961). Studies concerning the inheritance of ascospore length in *Neurospora crassa*. I. Studies on large-spored strains. Autral. J. Biol. Sci. 14, 223--230.

Le Gal, M. (1947). Recherches sur les ornementations sporales des Discomycètes operculés. Ann. Sci. Nat. Sér. XI, Bot. et Biol. 8, 73--286.

Leighton, T.J. & Stock, J.J. (1970). Isolation and preliminary characterization of developmental mutants from *Microsporum gypseum*. J. Bacteriol. 104, 834--838.

Leighton, T.J., Stock, J.J. & Herring, F.G. (1971). Electron spin resonance evaluation of pigment accumulation during asexual sporulation in *Aspergillus flavus*. Biochim. Biophys. Acta 237, 128--131.

Leupold, U. (1950). Die Vererbung von Homothallie und Heterothallie bei *Schizosaccharomyces pombe*. C.R. Lab. Carlsberg, Sér. Physiol. 24, 381--480.

Leupold, U. (1955). Methodisches zur Genetik von *Schizo saccharomyces pombe*. Schweiz. Z. Allgm. Path. Bakt. 18, 1141--1146.

Lindegren, C.C. & Lindegren, G. (1941). X-ray and ultraviolet induced mutations in *Neurospora*. II. Ultraviolet mutations. J. Hered. 32, 435--440.

Lindegren, C.C. & Scott, M.A. (1937). Formation of the ascospore wall in *Neurospora*. Cellule 45, 359--371.

Lingappa, B.T. & Sussman, A.S. (1959a). Endogenous substrates of dormant, activated and germinating ascospores of *Neurospora tetrasperma*. Plant Physiol. 34, 466--472.

Lingappa, Y. & Sussman, A.S. (1959b). Changes in the heat-resistance of ascospores of *Neurospora* upon germination. Am. J. Bot. 46, 671--678.

Lloyd, G.I., Anderson, J.G., Smith, J.E. & Morris, E.O. (1972). Conidiation and esterase synthesis in *Aspergillus niger*. Trans. Br. Mycol. Soc. 59, 63--70.

Locci, R. (1972). Scanning Electron Microscopy of Ascosporic Aspergilli. Supplement to Rivista di Patologia vegetale. Vol. III, ser. IV.

Lowry, R.J. & Sussman, A.S. (1958). Wall structure of ascospores of *Neurospora tetrasperma*. Am. J. Bot. 45, 397--403.

Lowry, R.J. & Sussman, A.S. (1968). Ultrastructural changes during germination of ascospores of *Neurospora tetrasperma*. J. Gen. Microbiol. 51, 403--409.

Lowry, R.J., Durkee, T.L. & Sussman, A.S. (1967). Ultrastructural studies of microconidium formation in *Neurospora crassa*. J. Bacteriol. 94, 1757--1763.

Lynn, R.R. & Magee, P.T. (1970). Development of the spore wall during ascospore formation in *Saccharomyces cerevisiae*. J. Cell Biol. 44, 688--692.

Mahadevan, P.R. & Mahadkar, U.R. (1970). Major constituents of the conidial wall of *Neurospora crassa*. Indian J. Exp. Biol. 8, 207--210.

Mainwaring, H.R. (1972). The fine structure of ascospore wall formation in *Sordaria fimicola*. Arch. Mikrobiol. 81, 126--135.

Malençon, G. (1962). Remarques sur le comportement des spores de l'*Ascobolus furfuraceus*. Bull. Soc. Mycol. Fr. 78, 105--119.

Mandels, G.R., Vitols, R. & Parrish, F.W. (1965). Trehalose as an endogenous reserve in spores of the fungus *Myrothecium verrucaria*. J. Bacteriol. 90, 1589--1598.

Markert, C.L. (1953). Cited by Sussman, A.S. & Halvorson, H.O. (1966), p.81.

Marquardt, H. (1963). Elektroneoptische Untersuchungen über die Ascosporenbildung bei *Saccharomyces cerevisiae* unter cytologischem and cytogenetischem Aspekt. Arch. Mikrobiol. 46, 308--320.

Martinelli, S.D. & Clutterbuck, A.J. (1971). A quantitative survey of conidiation mutants in *Aspergillus nidulans*. J. Gen. Microbiol. 69, 261--268.

Maruyama, Y. & Hayashi, K. (1963). Effect of UV irradiation on nucleic acid synthesis in germinating *Aspergillus niger* conidia. J. Gen. Appl. Microbiol. 9, 425--431.

Matile, Ph., Moor, H. & Robinow, C.F. (1969). Yeast Cytology. In The Yeasts, Vol. 1 (ed. A.H. Rose & J.S. Harrison), pp. 219--302. New York and London: Academic Press.

Matile, Ph., Cortat, M., Wiemken, A. & Frey-Wyssling, A. (1971). Isolation of glucanase-containing particles from budding *Saccharomyces cerevisiae*. Proc. Nat. Acad. Sci. USA 68, 636--640.

McDowell, L.L. & De Hertogh, A.G. (1968). Metabolism of sporulation in filamentous fungi. 1. Glucose and acetate oxidation in sporulating and non-sporulating cultures of *Endothia parasitica*. Can J. Bot. 46, 449--451.

McKeen, W.E., Mitchell, N., Jarvie, W. & Smith R. (1966). Electron microscopy studies of conidial walls of *Sphaerotheca macularis*, *Penicillium levitum* and *Aspergillus niger*. Can. J. Microbiol. 12, 427--428.

Melendez-Howell, L.-M. (1972). Ascospore germinating pore in the genus *Podospora*. Ann. Sci. Nat. Bot. Biol. vég. 13, 129--140.

Metzenberg, R.L. (1963). The localization of b-fructofuranosidase in *Neurospora*. Biochim. Biophys. Acta 77, 455--465.

Milburn, J.A. (1970). Cavitation and osmotic potentials of *Sordaria* ascospores. New Phytol. 69, 133--141.

Miller, J.J. & Hoffman-Ostenhof, O. (1964). Spore formation and germination in *Saccharomyces*. Z. Allg. Mikrobiol. 4, 273--294.

Mirkes, Ph.E. (1974). Polysomes, ribonucleic acid, and protein synthesis during germination of *Neurospora crassa* conidia. J. Bacteriol. 117, 196--202.

Miyake, S., Sando, N. & Sato, S. (1971). Biochemical changes in yeast during sporulation. II. Acetate metabolism. Develop., Growth & Differentiation 12, 285--295.

Moens, P. (1971). Fine structure of ascospore development in the yeast *Saccharomyces cerevisiae*. Can J. Microbiol. 17, 507--510.

Moens, P. & Rapport, E. (1971). Spindles, spindle plaques and meiosis in the yeast *Saccharomyces cerevisiae*. J. Cell Biol. 50, 344--361.

Moor, H. (1967). Endoplasmic reticulum as the initiatior of bud formation in yeast. Arch. Mikrobiol. 57, 135--146.

Moore, R.T. (1963). Fine structure of Mycota. I. Electron microscopy of the Discomycete *Ascodesmis*. Nova Hedwigia 5, 263--278.

Moore, R.T. (1965). The ultrastructure of fungal cells. In the Fungi, Vol. I, (ed. G.C. Ainsworth & A.S. Sussman), pp. 95--118. New York and London: Academic Press.

Müller, E. (1966). Fruchtkörperbildung und Variabilität morphologischer Merkmale bei *Leptosphaerulina australis* McAlpine. Ber. Schweiz. Bot. Ges. 76, 185--229.

Müller, E. (1971). Imperfect-perfect connections in Ascomycetes. In Taxonomy of Fungi imperfecti (ed. B. Kendrick), pp. 184--201. University of Toronto Press.

Mundkur, B. (1961). Electron microscopical studies of frozen-dried yeast. III. Formation of the tetrad in *Saccharomyces*. Exptl. Cell Res. 25, 24--40.

Namboodiri, A.N. (1966). Electron microscopic studies on the conidia and hyphae of *Neurospora crassa*. Caryologia 19, 117--133.

Nelson, R.R., Huisingh, D., & Webster, R.K. (1967). Sexual differentiation in *Cochliobolus carbonum* as influenced by inhibition and repair of steroid biosynthesis. Phytopathology 57, 1081--1085.

Nickerson, W.J. & Falcone, G. (1956). Identification of protein disulfide reductase as a cellular division enzyme in yeasts. Science 124, 722--723.

Novak, D.R. & Srb A.M. (1971). A developmental mutant that affects ascospore delimitation in *Neurospora tetrasperma*. Can J. Gen. Cytol. 13, 1--8.

Novak, D. R. & Srb, A. M. (1973). A dominant round spore mutant in *Neurospora tetrasperma* and its implications for acospore development. Can. J. Gen. Cytol. 15, 685--693.

, M. & Turian, G. (1968). Thermostimulation of conidiation and succinic oxidative metabolism of *Neurospora crassa*. Arch. Mikrobiol. 63, 232--241.

Ojha, M., Combépine, G. & Turian G. (1973). Evolution de l'activité de la succinique déshydrogénase et de son polymorphisme durant les phases précoces du développement du *Neurospora crassa*. C.R. Acad. Sci. Paris 277, 2353--2356.

Olenov, J.M. (1936). On the factors influencing the struggle for life between races of the yeast species *Zygosaccharomyces mandshuricus*. Arch. Mikrobiol. 7, 264--285.

Olive, L.S. (1956). Genetics of *Sordaria fimicola*. I. Ascospore color mutants. Am. J. Bot. 43, 97--107.

Olive, L.S. (1965). Nuclear behavior during meiosis. In The Fungi (ed. G.C. Ainsworth & A.S. Sussman), Vol. I, pp. 143--161. New York and London: Academic Press.

Olive, L.S. & Fantini, A.A. (1961). A new, heterothallic species of *Sordaria*. Am. J. Bot. 48, 124--128.

Oliver, P.T.P. (1972). Conidiophore and spore development in *Aspergillus nidulans*. J. Gen. Microbiol. 73, 45--54.

Oso, B.A. (1969). Electron microscopy of ascus development in *Ascobolus*. Ann. Bot. N.S. 33, 205--209.

Osumi, M., Sando, N. & Miyake, S. (1966). Morphological changes in yeast cell during sporogenesis. Jap. Women's Univ. J. 13, 70--79.

Oulevey-Matikian, N. & Turian, G. (1968). Contrôle métabolique et aspects ultrastructuraux de la conidiation (macro-microconidies) de *Neurospora crassa*. Arch. Mikrobiol. 60, 35--58.

Owens, R.G. (1955a). Metabolism of fungus spores. I. Oxidation and accumulation of organic acids by conidia of *Neurospora sitophila*. Contrib. Boyce Thompson Inst. 18, 125--144.

Owens, R.G. (1955b). Metabolism of fungus spores. II. Cytochrome oxidase, succinoxidase and pyruvate carboxylase systems in homogenates of conidia of *Neurospora sitophila*. Contrib. Boyce Thompson Inst. 18, 145--152.

Owens, R.G., Novotny, H.M. & Michels, M. (1958). Composition of conidia of *Neurospora sitophila*. Contrib. Boyce Thompson Inst. 19, 355--374.

Parisot, D. (1972). Effects pléiotropes de gènes contrôlant les processus de reproduction chez *Penicillium baarnense* (Van Beyma). Ann. Sci. Nat., Bot. Biol. vég. 13, 1--18.

Pelletier, G. & Hall, R. (1971). Relationships among species of *Verticillium*: protein composition of spores and mycelium. Can. J. Bot. 49, 1293--1297.

Peterson, J., Gray, R. & Ris, H. (1972). Meiotic spindle plaques in *Saccharomyces cerevisiae*. J. Cell Biol. 53, 837--841.

Phillips, R.L. & Srb, A.M. (1967). A new white ascospore mutant of *Neurospora crassa*. Can. J. Genet. Cytol. 9, 766--775.

Pincheira, G. & Srb, A.M. (1969). Cytology and genetics of two abnormal ascus mutants of *Neurospora*. Can. J. Genet. Cytol. 11, 281--286.

Pitt, D. (1969). Cytochemical observations on the localization of sulphydryl groups in budding yeast cells and in the phialides of *Penicillium notatum* Westling during conidiation. J. Gen. Microbiol. 59, 257--262.

Plurad, S.B. (1972). Fine structure of ascosporogenesis in *Nematospora coryli* Peglion, a pathogenic yeast. J. Bacteriol. 109, 927--929.

Pollock, R.T. & Johnson, G.A. (1972). The blackberry mutant of *Sordaria fimicola*. Can. J. Genet. Cytol. 14, 943--948.

Puglisi, P.P. & Zennaro, E. (1971). Erythromycin inhibition of sporulation in *Saccharomyces cerevisiae*. Experientia 27, 963--964.

Raper, K.B. & Fennell, D.I. (1965). The genus *Aspergillus*. 686 pp. Baltimore: Williams and Wilkins Co.

Reeves, F. (1967). The fine structure of ascospore formation in *Pyronema domesticum*. Mycologia 59, 1018--1033.

Rizza, V. & Kornfeld, J.M. (1969). Components of conidial and hyphal walls of *Penicillium chrysogenum*. J. Gen. Microbiol. 58, 307--315.

Robertson, N.F. (1965). The mechanism of cellular extension and branching. In The Fungi (ed. G.C. Ainsworth & A.S. Sussman), Vol. I, pp. 613--623. New York and London: Academic Press.

Robson, J.E. & Stockley, H.M. (1962). Sulphydryl metabolism of fungi grown in submerged culture. J. Gen. Microbiol. 28, 57--68.

Rossier, C., Oulevey, N. & Turian, G. (1973). Electron microscopy of selectively stimulated microconidiogenesis in wild type *Neurospora crassa*. Arch. Mikrobiol. 91, 345--353.

Roth, R. & Lusnak, K. (1970). DNA systhesis during yeast sporulation: Genetic control of an early development event. Science 168, 493--494.

Rousseau, P., Halvorson, H.O., Bulla, L.A. & Julian, G.S. (1972). Germination and outgrowth of single spores of *Saccharomyces cerevisiae* viewed by scanning electron and phase-contrast microscopy. J. Bacteriol. 109, 1232--1238.

Rudolph, E.D. & Giesy, R.M. (1966). Electron microscope studies of lichen reproductive structures in *Physcia aipolia*. Mycologia 53, 786--796.

Russel, D.W. & Brown, M.E. (1960). Sporidesmolide acid B, a hydroxyacyldipeptide from *Sporidesmium bakeri*. Biochim. Biophys, Acta 38, 382--383.

Sassen, M.M., Remsen, C.C. & Hess, W.M. (1967). Fine structure of *Penicillium megasporum* conidiospores. Protoplasma 64, 75--88.

Schmit, J. C. & Brody, S. (1975). Developmental control of glucosamine and galactosamine levels during conidiation in *Neurospora crassa*. J. Bacteriol. 122, 1071--1075.

Schrantz, J.-P. (1966). Contribution a l'étude de la formation de la paroi sporale chez *Pustularia cupularis* (L.) Fck. C.R. Acad. Sci. Paris 262, 1212--1215.

Schrantz, J.-P (1967). Présence d'un aster au cours des mitoses de l'asque et de la formation des ascospores chez l'Ascomycète *Pustularia cupularis* (L.) Fck. C.R. Acad. Sci. Paris 264, 1274--1277.

Schrantz, J.-P. (1970). Etude cytologique en microscopie optique et électronique de quelques Ascomycètes. II. La paroi. Rev. Cytol. Biol. Vég. 33, 111--168.

Seale, T. (1973). Life cycle of *Neurospora crassa* viewed by scanning electron microscopy. J. Bacteriol. 113, 1015--1025.

Selitrennikoff, C.P. & Nelson, R.E. (1973). A screening technique for the isolation of macroconidiation mutants. *Neurospora* Newsl. 20, 34.

Sentandreu, R. & Northcote, D.H. (1969). The formation of buds in yeast. J. Gen. Microbiol. 55, 393--398.

Shear, C.L. & Dodge, B.O. (1927). Life-histories and heterothallism of the red bread-mould fungi of the *Monilia sitophila* group. J. Agric. Res. 34, 1019--1042.

Sherald, J.L. & Sisler, H.D. (1972). Selective inhibition of antimycin A-insensitive respiration in *Ustilago maydis* and *Ceratocystis ulmi*. Plant Cell Physiol. 13, 1039--1052.

Sloan, B.J. & Wilson, G.B. (1958). The function of the microspores of *Gelasinospora calospora* var. *autosteira*. Mycologia 50, 111--116.

Smith, J.E. & Anderson, J.G. (1973). Differentiation in the Aspergilli. In Microbial Differentiation (ed. J.M. Ashworth & J.E. Smith), Sympos. Soc. Gen. Microbiol. 23, pp. 295--337. Cambridge, England: Cambridge U. P.

Smith, J.E. & Galbraith, J.C. (1971). Biochemical and physiological aspects of differentiation in the Fungi. Adv. Microb. Physiol. 5, 45--134.

Snider, I.J. & Miller, J.J. (1966). A serological comparison of vegetative cell, ascus walls and the spore coat of *Saccharomyces cerevisiae*. Can. J. Microbiol. 12, 485--488.

Srb, A.M. & Basl, M.(1969). The isolation of mutants affecting ascus development in *Neurospora crassa* and their analysis by a zygote complementation test. Genet. Res. 13, 303--311.

Srb, A. M., Basl, M., Bobst, M. & Leary, J. V. (1973). Mutations in *Neurospora crassa* affecting ascus and ascospore development. The J. Hered. 64, 242--246.

Stadler, D.R. (1956). A map of linkage group VI of *Neurospora crassa*. Genetics 41, 528--543.

Stine, G.J. (1969). Enzyme activities during the asexual cycle of *Neurospora crassa*. III. Nicotinamide adenosine diphosphate glycohydrolase. Can. J. Microbiol. 15, 1249--1254.

Streiblová, E., Beran, K. & Pokorný, V. (1964) Multiple scars, a new type of yeast scar in apiculate yeasts. J. Bacteriol. 88, 1104--1111.

Subden, R.E. & Turian, G. (1970). Analyse électrophorétique des protéines caroténogènes de *Neurospora crassa*. Experientia 26, 935--937.

Subramanian, C.V. (1971). The phialide. In Taxonomy of Fungi Imperfecti (ed. B. Kendrick). pp. 92--119. Toronto, Canada: Toronto U. P.

Sullivan, J.L., Wagner, P.C. & DeBusk, A.G. (1972). Some observations of ascospores of *Neurospora crassa* made with a scanning electron microscope. J. Bacteriol. 111, 825--826.

Sung, H.S. & Alexopoulos C.J. (1954). A cytotaxonomic study of three species of *Gelasinospora*. Cytologia 19, 255--264.

Sussman, A.S. (1961). The role of trehalose in the activation of dormant ascospores of *Neurospora*. Q. Rev. Biol. 36, 109--116.

Sussman, A.S. (1966). Types of dormancy as represented by conidia and ascospores of *Neurospora*. In The Fungus Spore (ed. M.F. Madelin). Colston Papers No. 18, 235--257. London: Butterworths.

Sussman, A.S. & Halvorson H.O. (1966). Spores. Their Dormancy and Germination. 354 pp. New York and London: Harper & Row.

Sussman, A.S., Distler, J.R. & Krakow, J.S. (1956). Metabolic aspects of *Neurospora* activation and germination. Plant Physiol. 31, 126--135.

Syrop, M.J. & Beckett, A. (1972). The origin of ascospore-delimiting membranes in *Taphrina deformans*. Arch. Mikrobiol. 86, 185--191.

Takebe, I. (1960). Choline sulfate as a major soluble sulfur component of conidiospores of *Aspergillus niger*. J. Gen. Appl. Microbiol. 6, 83--89.

Talens, L.T., Miller, M.W. & Miranda, M. (1973a). Electron micrograph study of the asci and ascospores of *Metschnikowia Kamienski*. J. Bacteriol. 115, 316--322.

Talens, L.T., Miranda, M. & Miller, M.W. (1973b). Electron micrography of bud formation in *Metschnikowia krissii*. J. Bacteriol. 114, 413--423.

Tanaka, K. (1966). Change in ultrastructure of *Aspergillus oryzae* conidia during germination. J. Gen. Appl. Microbiol. 12, 239--246.

Tanaka, K. & Yanagita, T. (1963). Electron microscopy of ultra thin sections of *Aspergillus niger*. I. Fine structure of hyphal cells. J. Gen. Appl. Microbiol. 9, 101--118.

Threlkeld, S.F.H. (1965). Pantothenic acid requirement for spore color in *Neurospora crassa*. Can. J. Genet. Cytol. 7, 171--173.

Throneberry, G.O. (1967). Relative activities and characteristics of some oxidative respiratory enzymes from conidia of *Verticillium albo-atrum*. Plant Physiol. 42, 1472--1478.

Throneberry, G.O. (1970). Endogenous and exogenous respiration of conidia of *Verticillium albo-atrum*. Phytopathology 60, 143--147.

Tingle, M., Singh Klar, A.J., Henry, S.A. & Halvorson, H.O. (1973). Ascospore formation in yeast. In Microbial Differentiation (ed. J.M. Ashw rth & J.E. Smith). Sympos. Soc. Gen. Microbiol. 23, 209--243.

Tingle, M., Kuenzi, M.T. & Halvorson, H.O. (1974). Germination of yeast spores lacking mitochondrial deoxyribonucleic acid. J. Bacteriol. 117, 89--93.

Tinline, R.D., Stauffer, J.F. & Dickson, J.G. (1960). *Cochliobolus sativus* 3. Effect of ultraviolet radiation. Can. J. Bot. 38, 275--282.

Tkacz, J.S. & Lampen, J.O. (1972). Surface distribution of invertase on growing *Saccharomyces* cells. J. Bacteriol. 113, 1073--1075.

Tokoro, M. & Yanagita, T. (1966). Physiological and biochemical studies on the longevity of *Aspergillus oryzae* conidia stored under various environmental conditions. J. Gen. Appl. Microbiol. 12, 127--145.

Tokunaga, J., Malca, I., Sims, J.J., Erwin, D.C. & Keen, N.T. (1970). Respiratory enzymes in the spores of *Verticillium albo-atrum*. Phytopathology 59, 1829--1832.

Totten, R.E. & Howe, B., Jr. (1971). Temperature-dependent actinomycin D effect on RNA synthesis during synchronous development in *Neurospora crassa*. Biochem. Genet. 5, 521--532.

Trinci, A.P.J., Peat, A. & Banbury, G.H. (1968). Fine structure of phialide and conidiospore development in *Aspergillus giganteus* "Wehmer." Ann. Bot. 32, 241--249.

Tsuboi, M., Kamisaka, S. & Yanagishima, N. (1972). Effect of cyclic 3',5'-adenosine monophosphate on the sporulation of *Saccharomyces cerevisiae*. Plant Cell Physiol. 13, 585--588.

Tsukahara, T. (1968). Electron microscopy of swelling and germinating conidiospores of *Aspergillus niger*. Sabouraudia 6, 185--191.

Turian, G. (1955). Recherches sur l'action de l'acide borique sur la fructification des *Sordaria*. Phytopathol. Z. 25, 181--189.

Turian, G. (1961). L'acétate et son double effet d'induction isocitratasique et de différenciation conidienne chez les *Neurospora*. C.R. Acad. Sci. Paris 252, 1374--1376.

Turian, G. (1966a). The genesis of macroconidia of *Neurospora*. In The Fungus Spore (ed. M.F. Madelin). Colston Papers No. 18, pp. 61--66. London: Butterworths.

Turian, G. (1966b). Morphogenesis in Ascomycetes. In The Fungi Vol II, (ed. G.C. Ainsworth & A.S. Sussman), pp. 339--385. New York and London: Academic Press.

Turian, G. (1969). Différenciation Fongique. 144 pp. Paris: Masson & Cie.

Turian, G. (1970). Aspects biochimiques de la différenciation fongique (modèle *Neurospora*). Physiol. Vég. 8, 375--386.

Turian, G. (1972). Maintien de l'élongation vegétative par répression catabolique dans l'apex hyphal de *Neurospora crassa*. C.R. Acad. Sci. Paris 275, 1371--1374.

Turian, G. (1973). Induction of conidium formation in *Neurospora* by lifting of catabolite repression. J. Gen. Microbiol. 79, 347--350.

Turian, G. (1974). Sporogenesis in Fungi. Annu. Rev. Phytopathol. 12, 129--137.

Turian, G. & Bianchi, D.E. (1971). Conidiation in *Neurospora crassa*. Arch. Mikrobiol. 77, 262--274.

Turian, G. & Bianchi, D.E. (1972). Conidiation in *Neurospora*. Bot. Rev. 38, 119--154.

Turian, G. & Khandjian, E. (1973). Action biphasique de l'AMP-cyclique sur la conidiogenèse de *Neurospora* et la caroténogenèse en général. Bull. Soc. Bot. Suisse 83, 351--356.

Turian, G. & Viswanath-Reddy, M. (1971). Metabolic and ultrastructural aspects of sexual differentiation in *Allomyces* and *Neurospora*. J. Indian Bot. Soc. Golden Jubilee Vol. 50A, 78--89.

Turian, G., Martin, C. & Bianchi, D.E. (1970). Conidiation par bourgeonnement-fission (blasto-arthrosporogenèse) et réorganisation microfibrillaire de la paroi chez *Neurospora*. C.R. Acad. Sci. Paris 271, 1756--1758.

Turian, G., Oulevey, N. & Coniordos, N. (1971). Recherches sur la différenciation conidienne de *Neurospora crassa*. I. Organisation chimio-structurale de la conidiation conditionnelle d'un mutant amycélien. Ann. Inst. Pasteur 121, 325--335.

Turian, G., Oulevey, N. & Cortat, M. (1973). Recherches sur la différenciation conidienne de *Neurospora crassa*. V. Ultrastructure de la séquence macroconidiogène. Ann. Microbiol. (Inst. Pasteur) 124 A, 443--458.

Turian, G., Oulevey, N. & Tissot, F. (1967). Preliminary studies on pigmentation and ultrastructure of microconidia. *Neurospora* Newsl. 11, 17.

Tuveson, R.W., West. D.J. & Barratt, R.W. (1967). Glutamic acid dehydrogenases in quiescent and germinating conidia of *Neurospora crassa*. J. Gen. Microbiol. 48, 235--248.

Urey, J.C. (1971). Enzyme patterns and protein synthesis during synchronous conidiation in *Neurospora crassa*. Dev. Biol. 26, 17--27.

Vezinhet, F., Arnaud, A. & Galzy, P. (1971). Etude de la respiration de la levure au cours de la sporulation. Can. J. Microbiol. 17, 1179--1184.

Viswanath-Reddy, M. & Turian, G. (1972). Temperature-induced synchronous differentiation of ascogonia in *Neurospora*. Experientia 28, 99--100.

Vujičić R. & Muntanjola-Cvetković, M. (1973). A comparative ultrastructural study on conidium differentiation in the *Cladosarum*-like mutant 22B of *Aspergillus aureolatus*. J. Gen. Microbiol. 79, 45--51.

Webster, J. (1970). Introduction to Fungi. 424 pp. Cambridge, England: Cambridge U. P.

Weisberg, S.H. & Turian, G. (1971). Ultrastructure of *Aspergillus nidulans* conidia and conidial lomasomes. Protoplasma 72, 55--67.

Weiss, B. (1965). An electron microscope and biochemical study of *Neurospora crassa* during development. J. Gen. Microbiol. 38, 85--94.

Weiss, B. & Turian, G. (1966). A study of conidiation in *Neurospora crassa*. J. Gen. Microbiol. 44, 407--418.

Wellman, A.M. (1967). Comparison of viability of strains of *Neurospora* after storage in liquid nitrogen. Can. J. Microbiol. 13, 665--672.

Wells, K. (1972). Light and electron microscopic studies of *Ascobolus stercorarius*. II. Ascus and ascospore ontogeny. University of California Public. in Botany, Vol. 62, pp. 1--93. Berkeley, Los Angeles, London: University of California Press.

Willetts, H.J. & Calonge, F.D. (1969). Spore development in the brown rot fungi *Sclerotinia* spp.). New Phytol. 68, 123--131.

Wilsenach, R. & Kessel, M. (1965). The role of lomasomes in wall formation in *Penicillium vermiculatum*. J. Gen. Microbiol. 40, 401--404.

Wilson, D.M. & Baker, R. (1969). Physiology of production of perithecia and microconidia in *Hypomyces solani* f. sp. *cucurbitae*. Trans. Br. Mycol. Soc. 53, 229--236.

Woodruff, H.E. (1966). The physiology of antibiotic production. The role of producing organisms. Sympos. Soc. Gen. Microbiol. 16, 22--46.

Wright, B.E. (1973). Critical Variables in Differentiation. Concepts of Modern Biology Series (ed. W.D. McElroy & C.P. Swanson), 109 pp. Englewoods Cliffs, New Jersey: Prentice-Hall, Inc.

Yoo, B.Y., Calleja, G.B. & Johnson, B.F. (1973). Ultrastructural changes of the fission yeast (*Schizosaccharomyces pombe*) during ascospore formation. Arch. Mikrobiol. 91, 1--10.

Yu, S.-A., Sussman, A.S. & Wooley, S. (1967). Mechanisms of protection of trehalase against heat inactivation in *Neurospora*. J. Bacteriol. 94, 1306--1312.

Zachariah, K. & Fitz-James, P.C. (1967). The structure of phialides in *Penicillium claviforme*. Can. J. Microbiol. 13, 249--256.

Zachariah, K. & Metitiri, P. O. (1970). The effect of mutation on cell proliferation and nuclear behavior in *Penicillium claviforme* Bainier. Protoplasma 69, 331--339.

Zalokar, M. (1959). Enzyme activity and cell differentiation in *Neurospora*. Am. J. Bot. 46, 555--559.

Zickler, H. (1931). Uber künstliche Erzeugung von Miktohaplonten bei Askomyzeten. Biol. Zentralbl. 51, 540--546.

Zickler, H. (1934). Genetische Untersuchungen an einem heterothallischen Askomyzeten (*Bombardia lunata*, nov. spec.). Planta 22, 573--616.

DISCUSSION

Spores in Ascomycetes, Their Controlled Differentiation

Chairman: Robert M. Brambl
University of Minnesota
St. Paul, Minnesota

The discussion of this paper was concerned with *Neurospora* pigmentation and the general observation that light seems to be required for completion of normal conidiogenesis in this fungus. Professor Turian was asked at what point in his proposed scheme of conidial formation the light becomes necessary and whether he could describe the mechanism of the light reception.

Professor Turian replied that these are indeed interesting questions, but that he could not provide complete answers. He said that at least some *Neurospora* conidia are produced in complete darkness, and---to his knowledge---there is no absolute requirement for light during conidiogenesis. However, carotenoid production in *Neurospora* apparently does not occur in the dark, and it is thought that light is an absolute requirement for production of this pigment during conidial formation. Carotenoid production is probably not essential for conidiogenesis; in cultures containing diphenylamine, an inhibitor of carotenoid biosynthesis, a reduced number of white conidia are produced and albino (carotenoid-less) mutants are also conidiated.

Little is known about the mechanism of light reception or about the stage at which the light intervenes. The light may be received by the hyphal tip, but very little evidence is available to support this idea. In 1968, with Dr. M. Ojha (Arch. Mikrobiol. 63, 232--241), Professor Turian had observed that preparations of *N. crassa* mitochondria always were contaminated with an orange carotenoid pigment. Recently (1974), workers in Germany (Hallermayer, G. & W. Neupert, Hoppe Seyler's Z. Physiol. Chem. 355, 279--288) have reported a detailed examination of this observation and have shown that the pigment, neurosporaxanthin (an acidic oxypolyene), is a normal component of one of the mitochondrial membranes. But at this point, we do not understand the possible relationship (if one exists) between mitochondrial-bound pigment and the light receptor mechanism. Nor do we understand the significance of the light sensitivity of hyphal tips as related to the observation that mitochondria are concentrated in the tip of conidiogenous hyphae only.

The discussion about mitochondrial activity in the hyphal tips of fungi concluded with the observation that the bioenergetics of tip growth and mitochondrial respiratory activities would be extremely interesting and perhaps fruitful areas for future investigations.

CHAPTER 17

FORM AND FUNCTION OF CONIDIA AS RELATED TO THEIR DEVELOPMENT

F. Mangenot and O. Reisinger
University of Nancy

INTRODUCTION

The vegetative cycle of imperfect fungi demonstrates at least some of the phenomena shown in Figure 1.

By loss of the mycelial stage, the vegetative cycle is reduced to the single process of multiplication by budding and fragmentation, in the case of yeasts. In Mycelia sterilia the loss of a differenciated stage of asexual multiplication simplifies its cycle. The arrangement of the conidiogenous apparatuses in a special organ (e.g., pycnidia) adds a supplementary stage to the asexual cycle but does not significantly modify this developmental process. Moreover, the ontogenic processes and the characteristics of conidia in Sphaeropsidales or Melanconiales are similar to those observed in Hyphomycetes (Brewer

& Boerema, 1965; Morgan & Jones et al., 1972; Sutton & Sandhu, 1969).

Although the vegetative cycle may be highly complex, with the exception of Mycelia sterilia, the function of the conidia in dissemination and conservation of the species is maintained.

According to the definition from the Kananaskis Conference (Kendrick, 1971), the conidium is "a specialized, nonmotile, asexual propagule, usually caducous, not developing by cytoplasmic cleavage or free-cell formation."

Aspects of conidium morphogenesis are highly diversified in the Deuteromycetes leading to variation of morphological characteristics. Among these, are color, shape, and septation used in Saccardo's pioneering taxonomic system. Others have been used by authors in later proposals for a taxonomic hierarchy of this group of fungi (e.g., ornamentation, conidium structure, nature of the hilum, etc.). Many of the features of conidia have been considered genetically stable and, in theses cases, have been used in the definition of different taxonomic entities.

In the modern experimental approach to fungal systematics (Hughes, 1953; Subramanian, 1962; Tubaki, 1963; Barron, 1968; Kendrick, 1971), conidium morphology is no longer a primary criterion for separation of major taxonomic entities.

Such studies are oriented particularly toward examinations of conidiogenesis and important results have arisen utilizing the various techniques of light and electron microscopy which have been especially useful in definitions of the origin of the conidium wall.

In a work devoted to the evolution and the role of the wall during the asexual cycle (Reisinger, 1972), we have separately analyzed the formation of the sporogenic element (conidiogenous cell or conidiophore) and the formation of conidia. With respect to the latter, we have distinguished three successive stages:

1. Primordium initiation
2. Conidium maturation
3. Conidium liberation (Mangenot & Reisinger, 1973).

After having examined the mechanisms of thallic growth (Reisinger & Mangenot, 1973) we compared these to the developmental characteristics of primordium initiation, which we now believe may be utilized for the formulation of distinctions between the different kinds of conidiogenesis.

Accepting as much as possible, the major principles outlined in the results of the Kananaskis Conferences (Kendrick, 1971) we have redefined the formation of the thalloconidium and phialoconidum. In

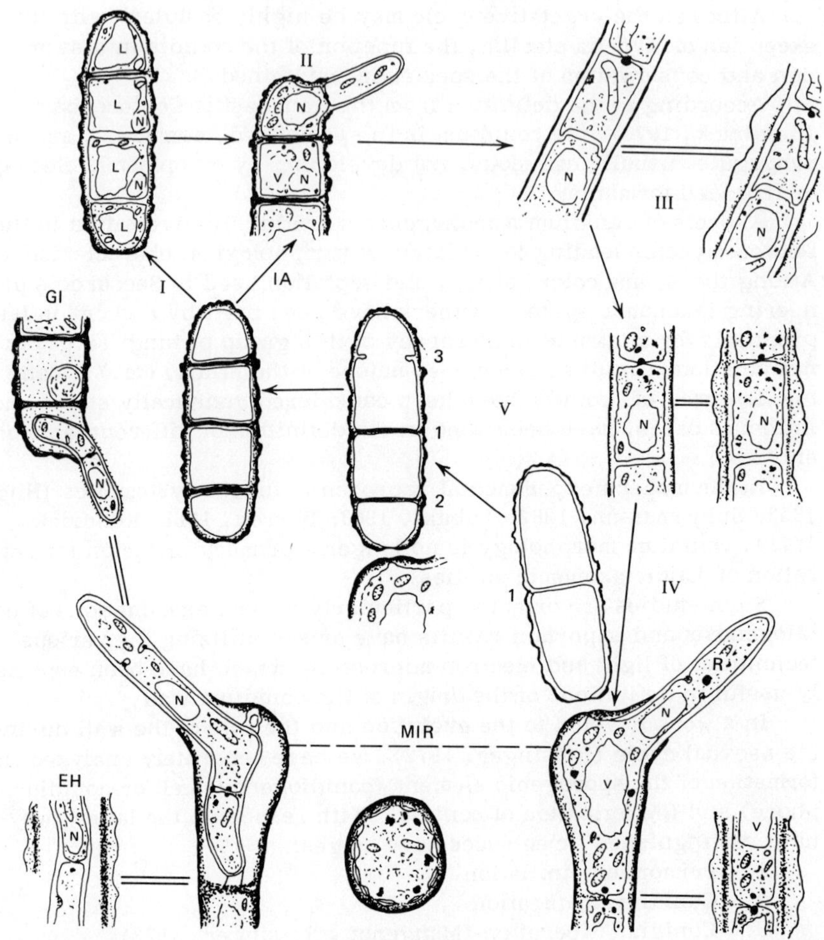

Fig. 1

Schematic representation of important stages of the vegetative cycle and asexual reproduction in *Dendryphiella vinosa*. Arrows outline the fundamental cycles, while dashed and double lines indicate unusual developmental features. (I) Liberation and dispersal of propagules rich in lipid reserve. This is a resting stage, and a new substrate and microhabitat are necessary for germination to occur. (IA) Liberation of propagules which are poor in lipid. In this case, germination is the first process, the primordium is elaborated by mechanisms which were also observed in the development of hyphae and conidiophores, while, in the second process the ontogenetic mechanism has not been observed during the evolution of the vegetative apparatus.

Subsequently we have regrouped (Oláh & Reisinger, 1973) under the term phialoconidium, some annelloconidia whose primordium initiation seems to be clearly enteroblastic (Hammill, 1974).

The primordial initiation stage is characterized by mechanisms delimiting the new cytoplasmic region, which will soon become transformed into a conidium initial (primordium).

Most of the morphological features of the conidium, such as surface structure, definitive shape, and internal anatomy, become differentiated later by phenomena characterizing the stage of conidium maturation.

The developmental processes which occur during the stage of conidium liberation, are generally involved in differentiation of the hilum. However these processes may also contribute to aspects of conidium surface morphology (e.g., *Stephanosporium cerealis*).

FORM

To describe the form of Deuteromycetes conidia, some purely morphological characters can be utilized: size, outline, septation, appendages, and so on.

Culture conditions modify some of these characters such as conidium size (Joly, 1964 for *Alternaria spp.*). Some others seem more stable and they can be used in the definition of conidium type. Such is the case of septum number which permits us to distinguish:

---

possible on the same substrate. (II) Cycle to the mycelial stage by germination. (III) Development of the vegetative apparatus. (IV) Differentiation of the fertile element, conidiogenesis, and sequential proliferation (R) of the conidiophore. (V) Conidium maturation (1, 2, 3, order of appearance of the septal walls: holosporous type). (MIR) Cycle to mycelial stage by formation of an endohyphae within the conidiophore, or vegetative hyphae (EH). Internal germination of a conidium which appears to occur by a process similar to endohyphal formation (MIR, EH). (After Reisinger, 1972; (N) nucleus; (L) lipid; (V) vacuole.)

1. Ameroconidia
2. Didymoconidia
3. Phragmoconidia
4. Dictyoconidia

Fungi with both amero- and phragmoconidia *(Phoma)* or with phragmo- and dictyoconidia *(Embellisia)* in different proportions demonstrate the progressive and interspecifique variability of septation character.

Appendages of diverse forms and origins (Madelin, 1966) can also characterize conidia of some species. In some cases, conidia with tubular appendages have already been investigated with TEM (Griffiths & Swart, 1973, 1974; Swart & Griffiths, 1974b), but ultrastructural basic principles concerning cell-like appendages (e.g., *Acrophragmis* and *Arachnophora*) are actually absent.

Mechanisms Occuring During Conidium Maturation

The developing, or mature, conidium has been studied for several years by different techniques using electron microscopy. The numerous references concerning this subject are gathered in Table 1. Although as complete as possible, this bibliography does not mention the references for works on yeast.

In spite of the large number of studies on the Deuteromycetes, it still remains difficult to summarize data obtained from investigations of the conidium wall. In fact several authors use methods allowing only a surface study (replicas, direct observations with SEM or TEM), while others work especially on the protoplasmic organization and neglect the wall structure. Therefore the number of wall layers is listed in Table 1, after an interpretation of the mcirographs. Furthermore, the interpretations of the parietal subunit (layers, spaces, zones, etc.) are not derived by means of a universal definition and rather depend on personal opinion and on the quality of the techniques used.

Therefore the mechanisms of conidium maturation and the analysis of the mature spore wall shall be described from examples which represent important morphological characteristics.

Characteristics of the Spore Surface

One of the features of the spore which is apparently most stable is its surface structure. This has been examined with the light

Table 1: Species Examined with Electron Microscope

Species	References	SEM or TEM	(Number of layers) or surface structure
Acremoniella sp.	Jones (1967)	SEM	smooth
A. velata.	Jones (1968a)	TEM	(2)
Alternaria brassicicola	Campbell (1968)	TEM	(2)
A. dauci solani	Schwinn (1968)	SEM	ornamented
A. tenuis	Schwinn (1968)	SEM	ornamented
Alternaria spp.	Slifkin (1971)	SEM + TEM	ornamented (3)
Arthroderma quadrifidum	Visset & Vermeil (1973)	SEM	ornamented
A. benhaminae	Visset & Vermeil (1973)	SEM	ornamented
Ascochyta spp.	Brewer & Boerema (1965)	TEM	(2)
Aspergillus aureolatus (*Cladosarium*-like mutant)	Vujičić & Muntanjola-Cvetković (1973)	TEM	(3)
A. carneus	Pore et al. (1969)	TEM	(2)
A. fumigatus	Ghiorse & Edwards (1973)	TEM	(3)
A. fumigatus	Campbell (1971)	TEM	(2)
A. giganteus	Trinci et al. (1968)	TEM	(2)
A. nidulans echinulatus	Fujii & Hess (1969)	TEM	ornamented
A. nidulans	Border & Trinci (1970)	TEM	(3)
	Weisberg & Turian (1971)	TEM	(3)
	Florance et al. (1972)	TEM	(3)
	Oliver (1972)	TEM	(3)
A. niger	Tanaka & Yanagita (1963)	TEM	(2)
	Tsukahara et al. (1966)	TEM	(2+1 zone)
	McKeen et al. (1966)	TEM	(3)
	Jones (1968b)	SEM	ornamented
	Tsukahara (1968)	TEM	(2)
	Sleytr et al. (1969)	TEM	(2)
	Tsukahara (1970)	TEM	(2)

Table 1 continued

Species	References	SEM or TEM	(Number of layers) or surface structure
	Konishi (1971)	TEM	(2)
A. parasiticus	Grove (1972)	TEM	(2)
Aspergillus spp.	Iizuka (1955)	TEM	ornamented
	Dedic & Koch (1957)	TEM	ornamented
	Suguiyama (1967)	TEM	ornamented
	Hess & Stocks (1969)	TEM	ornamented
	Ellis et al. (1970)	SEM	ornamented
	Lazar (1972)	?	?
	Locci & Quaroni (1972)	SEM	ornamented
	Locci (1972)	SEM	ornamented
	Tokunga et al. (1973)	SEM + TEM	ornamented (3)
A. tamarii	Roquebert (1973)	TEM	(3)
Aureobasidium pullulans	Reisinger et al. (1974)	TEM	(2)
Beauveria bassiana	Reisinger & Oláh (1974)	SEM + TEM	smooth (2)
Botryodiplodia theobromae	Ekundayo & Haskins (1969)	TEM	(2)
Botrytis cinerea	Hawker & Hendy (1963)	TEM	(2)
	Buckley et al. (1966)	TEM	(2)
	Gull & Trinci (1971)	TEM	(2)
	Bessis (1972)	SEM	ornamented
B. fabae	Richmond & Pring (1971)	TEM	(2)
Chalara sp.	Reisinger (1972)	TEM	(2)
Chloridium chlamydosporis	Hammill (1971b, 1972e)	TEM	(2)
Cladosporium resinae	Sheridan & Troughton (1973)	SEM	ornamented
Cladosporium spp.	Minnoura (1964, 1966)	TEM	ornamented
Coccosporium maculiforme	Durrell (1964)	TEM	(2)

Table 1 continued

Cochliobolus miyabeanus	Matsui et al. (1962)	TEM	(3)
C. sativus	Old & Robertson (1969)	SEM + TEM	smooth (2)
	Mills (1970)	SEM	smooth
	Old & Robertson (1970)	TEM	(4)
Colletotrichum atramentarium	Griffiths & Campbell (1972)	TEM	(1)
Conioscypha lignicola	Shearer & Motta (1973)	SEM + TEM	smooth (2)
C. varia	Shearer & Motta (1973)	SEM + TEM	smooth (2)
Coniothyrium minitans	Jones & Johnson (1970)		
Coniothyrium spp.	Punithalingam & Jones (1970)	TEM	smooth or ornamented
Conoplea mangenotii	Reisinger (1972)	SEM + TEM	smooth (4)
Coprinus lagopus (oidia)	Heintz & Niederpruem (1971)	TEM	(2)
Cryptosporiopsis sp.	Sutton & Sandhu (1969)	TEM	(2)
Dendryphiella vinosa	Reisinger & Guedenet (1968)	TEM	(3)
	Reisinger & Mangenot (1969)	SEM + TEM	ornamented (3)
Doratomyces nanus	Hammill (1972b)	TEM	(2)
D. purpureofuscus	Kiffer et al. (1971)	TEM	(3)
Drechslera sorokiniana	Cole (1973b)	SEM + TEM	ornamented (3)
Epicoccum nigrum	Brushaber & Haskins (1973)	TEM	(3)
	Griffiths (1973a)	SEM + TEM	ornamented (2)
Epidermophyton floccosum	Ito et al. (1970)	SEM	ornamented
Erysiphe graminis	Plumb & Turner (1972)	SEM	ornamented
E. graminis hordei	Kunoh & Akai (1967)	TEM	smooth

Table 1 continued

Species	References	SEM or TEM	(Number of layers) or surface structure
Fusarium culmorum	Acha et al. (1966)	TEM	(?)
	Marchant (1966a, b)	TEM	(3+1 zone)
F. oxysporum (chlamydosp.)	Stevenson & Becker (1972)	TEM	(3)
	Griffiths (1973b)	TEM	(2)
F. oxysporum cucumerinum	Akai et al. (1969)	TEM	(2)
F. oxysporum elaeides	Van Gool et al. (1970)	TEM	(2)
F. solani cucurbitae (chlamydosp.)	Old & Schippers (1974)	TEM	(3)
F. sulphureum	Schneider & Seaman (1974)	TEM	(2) (chlam. 3)
Geotrichum candidum	Steele & Fraser (1973)	TEM	(3 or 4)
	Hashimoto et al. (1973)	TEM	(2)
	Trinci & Collinge (1974)	TEM	(2)
G. lactis	Duran et al. (1973)	TEM	(?)
Gonatobotryum apiculatum	Cole (1973a)	SEM + TEM	smooth (2)
Helminthosporium maydis	Corlett (1970)	SEM	smooth
	White et al. (1973)	TEM	(2)
H. oryzae	Horino (1964)	TEM	(3)
	Horino & Akai (1965)	TEM	(3)
H. solani	Reisinger (1972)	SEM + TEM	smooth (3)
H. spiciferum	Reisinger (1970b)	SEM + TEM	smooth (4)
Histoplasma capsulatum	Garrison & Lane (1973)	SEM	ornamented
Humicola grisea	Griffiths (1974)	TEM	(Aleuriosp. 1) (chlamyd. 2)

Table 1 continued

H. nigricans	Durrell (1964)	TEM	(2)
Humicola spp.	Bertoldi et al. (1974)	TEM	(multilayered)
Marssonina salicicola	Nival (1972)	TEM	(2)
Melanconium apiocarpum	Sutton & Sandhu (1969)	TEM	(2)
M. bicolor	Sutton & Sandhu (1969)	TEM	(2)
Metarrhizium anisopliae	Zacharuk (1970)	TEM	(2+1)
	Hammill (1972a)	TEM	(2)
Microsporum gypseum	Akin & Michaels (1972)	SEM + TEM	ornamented (3)
M. racemosum	Smith & Sandler (1971)	SEM	ornamented
Monilinia fructicola	Baker & Smith (1970)	TEM	(2)
Monochaetia lutea	Swart & Griffiths (1974a)	TEM	(2+ cuticle)
Monotosporella (Acrogenospora) sphaerocephala	Hammill (1972c)	TEM	(2)
	Reisinger (1972)	SEM + TEM	smooth (2)
Nannizia incurvata	Visset & Vermeil (1973)	SEM	ornamented
N. fulva	Visset & Vermeil (1973)	SEM	ornamented
Nectria haematococca	Hanlin (1971)	SEM	smooth
Neurospora crassa	Nambodiri (1966)	TEM	(2)
	Lowry et al. (1967)	TEM	(2)
	Turian & Bianchi (1972)	TEM	(2)
Pachybasium sp.	Hashioka et al. (1966) (1966)	TEM	(2)
Paecilomyces berlinensis	Oláh (1972)	TEM	(2)
Penicillium chrysogenum	McCoy et al. (1971)	TEM	(?)
P. digitatum	Fujii & Hess (1969)	TEM	ornamented
	Zaki et al. (1973)	TEM	(2)

Table 1 continued

Species	References	SEM or TEM	(Number of layers) or surface structure
P. herquei	Hess et al. (1966)	TEM	ornamented
P. levitum	McKeen et al. (1966)	TEM	(2)
P. megasporum	Sassen et al. (1967)	TEM	(3+1)
P. notatum	Martin et al. (1973)	TEM	(4)
Penicillium sp.	Cole & Ramirez-Mitchell (1974)	SEM + TEM	ornamented
Penicillium spp.	Dedic & Koch (1957)	TEM	ornamented
	Hess et al. (1968)	TEM	ornamented
	Fletcher (1971a, b)	TEM	(2)
Pestalotiopsis funerea	Griffiths & Swart (1974)	TEM	(1)
P. triseta	Griffiths & Swart (1974)	TEM	(1)
Peziza ostracoderma	Hughes & Bisalputra (1970)	TEM	(1)
Phialophora richardsiae	Reisinger (1972)	SEM + TEM	smooth (2)
	Olah & Reisinger (1974a, b)	SEM + TEM	smooth (2)
Phoma fumosa	Sutton & Sandhu (1969)	TEM	(2)
Phoma spp.	Brewer & Boerema (1965)	TEM	(1)
Pithomyces chartarum	Bertaud et al. (1963)	TEM	ornamented
	Dickson (1963)	TEM	(3)
Polyschema congolensis	Reisinger & Kiffer (1974)	SEM + TEM	ornamented (3)
Poronia punctata	Stiers et al. (1973)	SEM + TEM	ornamented (2)
Psathyrella coprophila	Jurand & Kemp (1972)	TEM	(2)
Pyricularia oryzae	Horino & Akai (1965)	TEM	(2)
Riessia semiophora	Goos & Tubaki (1973)	TEM	(1?)

Table 1 continued

Sclerotinia sclerotiorum	Calonge (1970)	TEM	(2)
Scopulariopsis brevicaulis	Hammill (1971a)	TEM	(2)
	Cole & Aldrich (1971)	TEM	(2)
	Balimgart (1973)	SEM	ornamented
S. koningii	Hammill (1971a)	TEM	(2)
Seimatosporium spp.	Griffiths & Swart (1973)	Tem	cf. text
	Swart & Griffiths (1974b)	TEM	cf. text
Sphaerotheca fuliginea	Dekhuijzen & Scheer (1969)	TEM	(2)
	Locci & Bisiach (1972)	SEM	smooth
S. macularis	McKeen et al. (1966)	TEM	(2)
S. pannosa	Akai et al. (1966)	TEM	(2)
Spilocea pomi	Corlett (1970)	SEM	smooth
	Hammill (1973a)	TEM	(2)
Stachybotrys atra	Campbell (1972)	SEM + TEM	ornamented (2)
Steganosporium pyriforme	Hammill (1972d)	TEM	(2)
Stemphylium botryosum	Carroll (1972)	TEM	young conidia
S. botryosum	Corlett (1973)	SEM	ornamented
S. illicis	Durrell (1964)	TEM	(3)
S. sarciniforme	Durrell (1964)	TEM	(1)
Stilbothamnium nudipes	Roquebert & Abadie (1973)	TEM	(2)
	Roquebert (1973)	TEM	(2)
Stephanosporium cerealis	Reisinger (1972)	SEM + TEM	ornamented (3)
Termitaria snyderi	Khan & Aldrich (1973)	TEM	(2)
Thielavia sepedonium	Durrell (1960)	TEM	(2)
Thielaviopsis basicola	Delvecchio et al. (1969)	TEM	(1 conidia 2 chlamyd.)
	Christias & Baker (1970)	SEM + TEM	smooth (2)

Table 1 continued

Species	References	SEM or TEM	(Number of layers) or surface structure
	Tsao & Tsao (1970)	TEM	(2 x 2)
T. paradoxa	Ichinoe & Kurata (1964)	TEM	(?)
Torula sp.	Durrell (1964)	TEM	(2)
Trichocladium asperum	Reisinger (1972)	SEM + TEM	ornamented (2)
Trichoderma saturnisporum	Hammill (1974)	TEM	(2)
T. viride	Hasioka et al. (1966)	TEM	(2)
	Jones (1971)	TEM	(2)
Trichoderma spp.	Walsh (1970)	SEM	ornamented
Trichophyton erinacei	Smith & Sandler (1971)	SEM	ornamented
T. megnini	Ito et al. (1970)	SEM	ornamented
T. rubrum	Ito et al. (1970)	SEM	ornamented
Tritirachium roseum	Hammill (1973b)	TEM	(2)
Ulocladium atrum	Carroll & Carroll (1974)	TEM	(3?)
Verticillium albo-atrum	Buckley et al. (1969)	TEM	(2)
V. nigrescens	Buckley et al. (1969)	TEM	(2)
Xylaria spp.	Greenhalgh (1967)	SEM	smooth
Wardomyces pulvinata	Reisinger (1970a)	SEM + TEM	smooth (3)
W. simplex	Sugiyama et al. (1968)	TEM	smooth

microscope which has differenciated between smooth and rough surfaces. With the aid of references from the taxonomic literature and especially from descriptions of new species, it is possible to distinguish between the following different kinds of ornamented conidia:

1. Echinulate
2. Reticulate
3. Alveolate
4. Spinulose
5. Verrucose.

The intermediate types are usually designated by descriptive words referring to the size of the observed ornamentation.

Results obtained from SEM

The introduction of scanning electron microscopy (SEM) for the examination of the surfaces of mature conidia (Table 1) has confirmed, at first, the existence of smooth and ornamented types. It has also provided a means to describe precisely the form and distribution of the ornamentations of conidia among certain species as well as the description of new structures such as the conidia of *Nannizia incurvata* (Visset, 1972). Actually it is not possible to propose a precise nomenclature to describe ornamentation or arrangement of homologous structures on the surfaces of conidia of different species.

Knowledge of the distribution and of forms of ornamentations seems to be of great developmental and taxonomic interest. Visset (1972) showed that the three conidial forms of *Microsporum gypseum* can be differentiated by their surface structures at an early developmental stage. Punithalingham & Jones (1970) described the ornamentations in *Coniothyrium spp*. in the same way.

In a study of *Aspergillus* Locci & Quaroni (1972) delimited the species in five different categories according to the topography of conidial surfaces, as follows:

1. Smooth to delicately roughened
2. Irregularly roughened to rugose
3. Verrucose
4. Spinulose to spiny
5. Reticulate

This classification of ornamentations, or that proposed by Hawksworth & Wells (1973) for the terminal hairs in the genus *Chaetomium* could also be applied, with some modifications, to rough spores of Deuteromycetes.

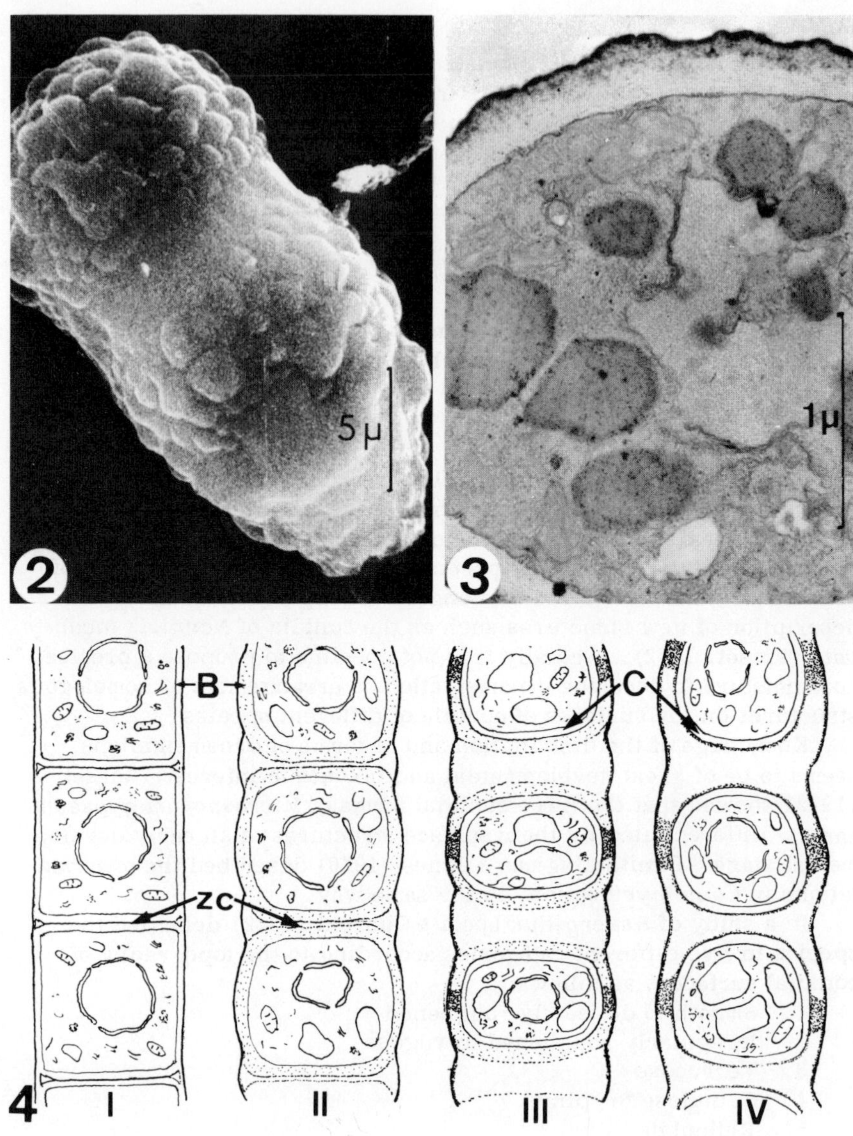

Fig. 2-4

(2) *Trichocladium asperum*. Conidium surface. Ornamentations are absent on a level with septa. (3) *Doratomyces purpureofuscus*. Rudimentary ornamentations and wall melanization. (4) Schematic

The distribution of the ornamentations on the surfaces of conidia of imperfect fungi appear, in most cases, as uniform and seem to depend on the species. Local variations sometimes modify this uniformity. The perihilar zone is always devoid of ornamentations in tretoconidia, as is the surface in contact between two adjacent conidia of a chain. Echinulations are also absent, or very rudimental in *Dendryphiella vinosa, Polyschema congolensis* and in *Trichocladium asperum* (Fig. 2) at the level of the conidial septa. This observation is suggestive that, the wall surface ornamentations are associated with adjacent cytoplasmic activities. These results also suggest that septal wall formation takes place before conidia become echinulate.

A heterogenous aspect of the conidial surface can also be due to the presence of ornamentations of different types. The SEM has shown, for example, a progressive transition from smooth walls to warts with *Polyschema congolensis* (Reisinger & Kiffer, 1974).

Results obtained from TEM

The examination of whole spores in the transmission electron microscope (TEM) allows observations of surfaces and the arrangement of conidia in intact chains (Table 1). The resolution obtained with this technique is very limited and the technique is no longer used since the SEM has become available.

Freeze etching and replica techniques give high quality images of conidial surfaces but the results are mainly concerned with the arrangement of external fibrils or rodlets. Freeze-etch fractures of the wall are particularly well suited for these analyses of wall structure and ornamentation but, so far, only a very small number of species have been examined.

The study of thin sections, now extensively used, allows accurate investigations of the coat as well as ornamentations. Furthermore, this technique permits application of some electron microscope

---

representation of conidium formation, maturation, and liberation in *Stephanosporium cerealis*. (After Reisinger, 1972.) (I) Fertile hyphae septation by euseptation process. (II) Thickening of layer B (B) and beginning of lysis on a level with cleavage zone (zc). (III) Appearance of layer C (C) and melanization on a level with a ring. (IV) Propagule individualization by lysis of apigmented parts of layer B.

cytochemistry, but for analyses of the spatial arrangement of the ornamentations serial sections are necessary, and are very difficult to obtain from spores with thick walls.

Smooth conidia

The smooth conidia described with the light microscope or with the SEM can be delimited into two categories with the TEM.

1. Conidia devoid of an external mucilaginous and hydrated zone;
2. Conidia covered with a mucilaginous compound.

The first case is illustrated by *Chalara sp*. and *Thielaviopsis basicola* (Table 1). In young conidia of *Chalara sp*., there exist some membranous formations which adhere to the wall surface, and are particularly abundant between adjacent conidia where they appear electron dense. This product, when examined with the light microscope, shows a pectic nature. It seems to contain, in all cases, phosphorylated sugars and probably represents some residues of conidial development. In mature conidia, these membranous formations are replaced by a thin, external, electron-dense film. An analogous film also covers the smooth conidia in species of *Helminthosporium (Cochliobolus, Drechslera)*, *Wardomyces*, and so on. (Table 1). In *Beauveria bassiana*, this external film is of a more polysaccharide and proteinaceous nature than the adjacent underlying coat (Reisinger & Oláh, 1974).

In the case of *Fusarium oxysporum* (Van Gool et al, 1970) the smooth surface seems to be due to the presence of a mucilaginous coat. However, the mucilaginous layer may sometimes mask an underlying ornamented surface. The ornamentations remain very visible on *Stachybotrys atra* conidia which protrude through a mucilaginous layer.

Ornamented conidia and formation of ornamentations

Dufour (1888) studied the morphological variations of *Trichocladium asperum* and demonstrated the absence of ornamentations when conidia were grown in some liquid media. Similar observations have been made for conidia which form in agar. When they are examined by light microscopy they appear to be surrounded by an optic halo. Such propagules examined in the TEM are surrounded by a diffuse mucilaginous zone containing fibrils and melanin granules

(Reisinger, 1972), as demonstrated by aleurioconidia of *Humicola nigricans* (Durrell, 1964) and *Humicola grisea* (Griffiths, 1974). With aerial conidia, an external film exists which is irregularly raised from the underlying coat and is identical to the coat mentioned above for smooth conidia. The ornamentation products accumulate in these spaces between the film and underlying coat.

The dispersion of fibrils and melanin granules is due to the irregular and extensive separation of the superficial part of the wall and is well known among hyphae of dematiaceous fungi. Fibrils and melanin granules are present in different reproductive and vegetative cells (Fig. 1) and can also be observed in droplets formed on wall surfaces of cultivated cells of some species. In these regions, the external film is no longer present.

It is suggested that the appearance of ornamentation (echinulation or wart type) depends on the genetic constitution of the species, but the morphological expression of this ornamentation is closely dependent on hygrometric conditions which allow the formation of the external film. The latter is referred to as layer A and possesses the same cytochemical characteristics as the mucilaginous zones of the immersed cells of *Phialophora richardsiae* and *Beauveria bassiana* (Oláh & Reisinger, 1974b, Reisinger & Oláh, 1974). This film may represent a small quantity of mucilaginous products excreted during the conidial maturation and subsequently condensed by rapid dehydration.

On conidia and conidiophores of *Dendryphiella vinosa*, echinulations appear by protrusion of the still unpigmented wall. They are outwardly limited by layer A. Whether this layer remains intact or ruptures during echinulation is not important. During conidial maturation, the substances contained within the echinulations rapidly give rise to melanins and appear similar to the pigmented zone of the wall.

The process of ornamentation is identical in *Polyschema congolensis*, but the same conidial surface may be smooth, echinulated and verrucose. Comparative structural studies of the two types of ornamentations in this species have not revealed noticeable differences. However, in the verrucose type, the space between layer A and the wall is greater than in echinulations and the substances enclosed by the former remain diffuse.

Another example of the formation of echinulations is demonstrated by some species of *Aspergillus* and *Penicillium* (Fletcher, 1971a; Roquebert, 1974). In the thin walls of young conidia of *Aspergillus tamarii* small papilla appear which are covered with a thin external film (layer A). In mature conidia of this species a new internal and striated layer appears. These papilla enclosed a residual material (Roquebert, 1974).

Among other species the ornamentation is more delicate than is demonstrated by echinulate or verrucose surfaces, as illustrated by *Doratomyces purpureofuscus* (Fig. 3). The conidia have a layer firmly attached to the underlying coat, but the surface seems irregularly undulated, or even folded. This surface topography and the more intensive melanization under the convex parts, characterize this particular aspect of these conidia with rudimentary echinulation.

Other types of ornamented conidia (reticulated, alveolated, etc.) have not been examined so far with TEM.

Among the kinds of conidia with unusual surface structure, we can cite the oidia of *Psathyrella coprophila* and the arthroconidia of *Stephanosporium cerealis*. The oidia of the former are surrounded by a mucilaginous capsule in which long filaments appear which are similar to bacterial pili (Jurand & Kemp, 1972).

The propagules of *S. cerealis* on the other hand, are ornamented with a raised and richly pigmented band. This partiuclar surface structure results from enzymic lysis of the primary wall (layer B) of the primordium as shown in Fig. 4.

Conidial anatomy

During maturation of the conidia, the initial wall of the primordium is able to transform by two phenomena, sometimes successive, as follows:

1. Thickening by apposition;
2. Synthesis of a new inner layer.

In addition to these processes in the phaeoconidia, melaninlike substances are also deposited in the wall.

The anatomy of the mature conidia has been frequently described from light microscopic investigations. For example, Nicot (1958) recognized in the conidium wall of *Dendryphiella arenaria* three wall layers which occur from outside to inside in the following sequence:

1. Cuticule
2. Epispore
3. Endospore.

This wall anatomy is comparable to that of *Trichocladium roseum* conidia (Nicot & Leduc, 1957) and is probably found with minor differences in many Deuteromycetes.

Chadefaud (1969), compared conidial walls to the walls of ascospores, and described the following layers:

1. Vagina;
2. Under-vagina;
3. Locula.

Undefined regions which depend on the species under examination are also present but they are always located between the under--vagina and the locula.

In the phragmoconidia of *Helminthosporium* and related genera, Luttrell (1963) distinguished the euseptate and distoseptate conidia as follows: "Euseptate conidia are surrounded by a single wall and have true septa formed as inward extension of the lateral wall," "Distoseptate conidia have a common outer wall enclosing more or less spherical cells, each of which is surrounded by an individual wall."

The sequential construction of the wall in *Conoplea mangenotii* studied with the light microscope and different staining techniques begins by the appearance and thickening of the epispore. At the exterior of this layer the cuticule arises and the subsequent inner layer, the endospore is elaborated (Reisinger, 1967).

The taxonomic bibliography furnishes many examples of basic principles of wall structure but the resolution power of the light microscope is limited and the interpretations are often contradictory.

Basic principles of wall structure based on electron-microscopy studies

The results obtained with electron microscopy (Table 1) and light microscopy indicate a great heterogeneity in the interpretation of the subunits of the wall. Therefore it is necessary to try to define these units.

In the Deuteromycetes so far examined, there exist three essential wall layers which it is possible to characterize by ultrastructure and developmental features, which are demonstrated during the asexual cycle:

1. One outer layer (designated by the letter A);
2. One medial layer, fundamental in all conidia examined (designated by the letter B);
3. One inner layer, which exists only in some conidia (designated by the letter C).

Layer A. This is of secondary importance because it is derived from either the fundamental layer, or from the mucilaginous substances which are probably excreted by the cell. Layer A may be absent during some stages of the asexual cycle. Also in liquid media this layer is replaced by a mucilaginous zone.

In thin sections, it appears amorphous and electron dense. Layer A envelopes most of the dry conidia as a thin pellicle. In the case of mature cells the delimitation of layer A is often difficult to determine, and this problem is compounded by the juxtaposition of the underlying layer. Therefore, it is difficult to be sure that this layer corresponds, in any case, to the cuticule described by light microscopy. In cytochemical studies of aerial cells of *Beauveria bassiana*, in which layer A appears as a continuous pellicle (Reisinger & Oláh, 1974), except at the growing apex, this wall layer appears to be strongly proteinaceous and stains positively for polysaccharides.

However, the studies done with freeze-etching reveal certain fibrils or "rodlets" with regular arrangement on the surfaces of the conidia at the level of layer A (Hess et al., 1968; Hess & Stocks, 1969, Ghiorse & Edwards, 1973; Cole & Ramirez-Mitchell, 1974). This arrangement of rodlets apparently depends on the developmental stage examined (Cole, 1973c). Rodlets (Fig. 5) observed with freeze-etching are not generally well preserved by the thin-sectioning technique (Cole, 1973a). According to Florance et al. (1972) and Ghiorse & Edwards (1973), the striate structure of the outer layer of *Aspergillus* conidia may be due to cross-sectioned rodlets.

Layer A represents an essential unit of the wall of mature, aerial cells and is rarely absent. The outer mucilaginous zone covering some conidia (e.g., *Fusarium*) is, perhaps, the equivalent of the hydrated layer A. It would be possible to experimentally verify this hypothesis by comparing thin sections of walls of *Fusarium* conidia which occur in mucilaginous mass to those isolated and maintained for some time in a dry atmosphere.

Layer B. The thickness of this layer is variable according to the cell or species examined. In thin sections it generally appears striated. It represents the fundamental wall layer and constitutes the first envelope of all cells of the Deuteromycetes. In the case of some conidia its thickness increases considerably by apposition of new wall material. This process of maturation by thickening is generally associated with the presence of lomasomelike structures (Fig. 3).

In dematiaceous fungi, layer B is differentiated into two zones by melanization (Campbell, 1968; Reisinger & Guedenet, 1968; Griffiths, 1974): one zone remains hyaline (B_1) while the other (B_2) becomes pigmented.

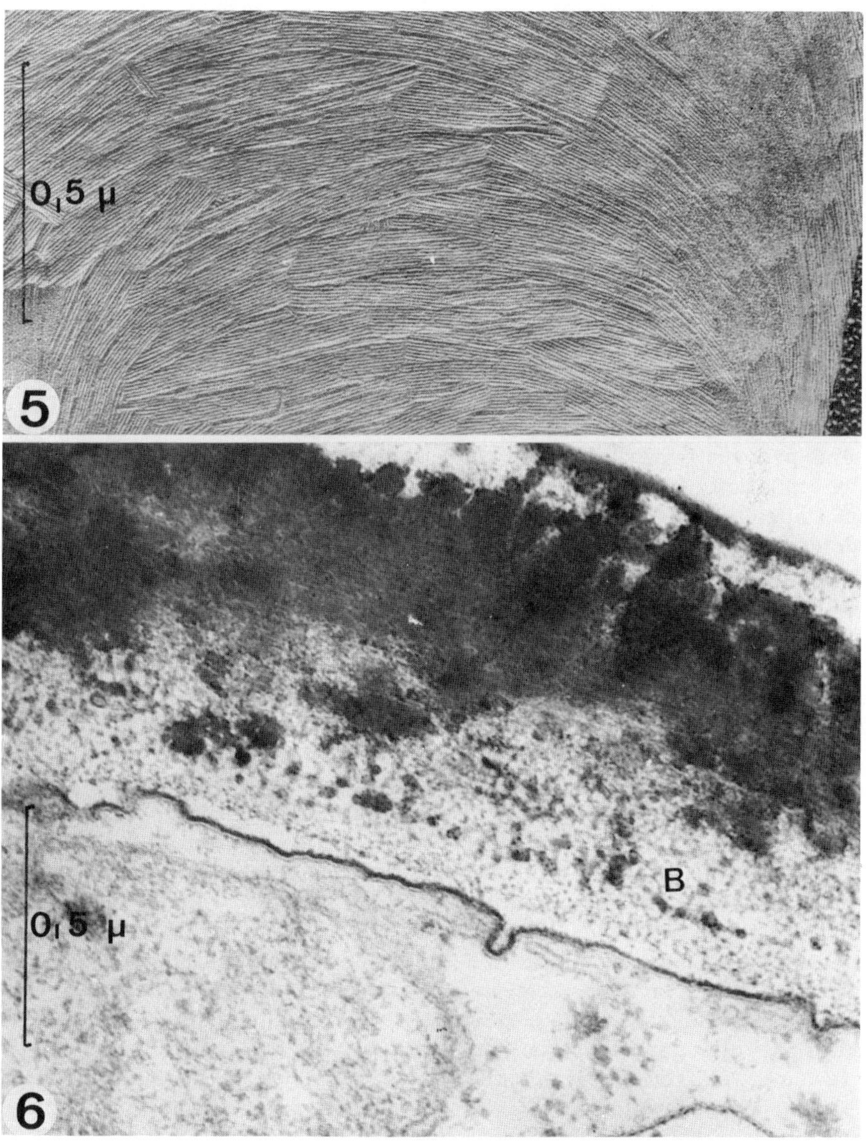

Fig. 5-6
(5) *Gonatobotryum apiculatum*. Conidium surface. (Courtesy of Dr. G. T. Cole.) (6) *Dendryphiella vinosa*. Layer B melanization.

It is possible to destroy zone B_2 by action of a solution of $KMnO_4$ followed by treatment with an organic acid (e.g., acetic acid) while zone B_1 remains (Reisinger & Guedenet, 1968). Moreover, zone B_2 is very rigid and inelastic during the vegetative activity of the cells which it envelopes (e.g., proliferation, germination, etc.). On the other hand, zone B_1 remains elastic and, in some species it forms the new outer wall of the proliferating cell. In spite of the structural and chemical differentiation of these two zones, we do not think that zones B_1 and B_2 can be considered as separate wall layers, but rather subdivisions of a single layer ($B=B_1 + B_2$).

Such a concept permits comparative analyses of wall layers of hyalo- and phaeosporous cells which will probably generate a better interpretation of conidial anatomy in the "albino" mutants.

Figure 6 demonstrates one stage in the melanization of layer B. This process begins at the exterior and progresses inward probably by oxidation of melanin precursors. It is characterized by the appearance of areas in which micrigranules of 7--12 A^o seem to condense and contribute to the electron density of zone B_2. It is now considered that there is no differentiation of the plasmalemma associated with melanization.

While confusion may still exist between differentiation of the endospore and zone B_1, layer B probably corresponds to the epispore.

Layer C. Layer C exists only in some conidia. It is striated, never melanized, and resists the combined action of $KMnO_4$ and acetic acid. It is sometimes separated from layer B by a very thin and generally more electron dense intermediary zone. Actually, it is very difficult to determine the origin of this zone between layers B and C but it may result from a superficial transformation of C or a local modification of B. It is also possible that this zone represents some secreted substances like those of layer A.

In the absence of this intermediary zone it is sometimes difficult to differentiate layer C particularly in phaeospores. However, our research on identification of proteinaceous substances in fungal walls (Oláh & Reisinger, 1974a, b; Reisinger & Oláh, 1974) could furnish a technique to overcome this problem. In fact, our cytochemical studies of propagules of different groups of fungi have shown that a strong and uniform reaction occurs in layer B, without differentiation of zones B_1 and B_2, while the reaction in layer C remains weak. In layer B_1, melanic precursors would react in the same way as melanin itself in layer B_2.

Layer C probably corresponds to the endospore.

Elaboration and structure of septal walls of mature conidia

Brewer & Boerema (1965) showed that septal walls of *Phoma spp.*, which only sometimes occur, develop by invagination of the lateral wall (layer B). This kind of process then leads to the elaboration of euseptate conidia (Luttrell, (1963).

In the case of *Ascochyta pisi*, each cell is surrounded by a new wall layer (layer C) inside the primary envelope (layer B). Therefore the final arrangement of wall layers corresponds to distoseptate conidia (Luttrell, 1963).

During conidium maturation in *Dendryphiella vinosa* the development of the wall is more complex. Layer B invaginates to form a hyphal-like septal wall (process of euseptation). Therefore each conidium cell is surrounded by a new wall layer (layer C) by the same mechanism as the distoseptation process (Reisinger & Guedenet, 1968; Reisinger, 1972).

These three kinds of septation lead to intraconidial septa, whose structure can be defined by the following formulae: (cf. Fig. 7)

1. B', zc, B' (euseptate process);
2. C, B', zc, B', C (double process);
3. C, zc, C (distoseptate process).

Therefore the letter B' is used to designate the thin septal layer which has arisen from layer B.

The "cleavage zone," or splitting zone (zc) corresponds to the thin, electron-transparent medium pellicle, which was particularly well analyzed by Cole & Aldrich (1971).

Except in the case of some separating septal walls, most intraconidial septa demonstrate a symmetry with respect to the cleavage zone.

Formation of phragmosporous septa, which have been examined, can be classified into one of these three kinds of septation defined above. Swart & Griffiths (1974b) showed that the septa of medial cells of conida of *Seimatosporium fusisporum* are structurally different from those of the end cells. In fact,the medial, septal walls seem to result from the double process referred to above (C, B', zc, B', C) in the end cell layer C is absent and the septal wall is formed only by the thin layer B' (locus euseptate). The propagules of this species conform to the developmental concept of the double process and represent an interesting mechanism of wall differentiation, from an evolutionary point of view.

In the case of the distoseptation process, we must examine propagules with four septal wall layers (cf. Table 1). In these conidia a particularly thick zone exists between layers B and C. This zone is more or less deeply inserted into the wall as demonstrated in some *Helminthosporium spp.*, and is structurally different from neighboring layers B and C (Fig. 8). Moreover, this zone swells considerably in water (Fig. 9), and probably acts as a lubricant during the crushing of conidia at which time layer B is ruptured and the conidial cells,

---

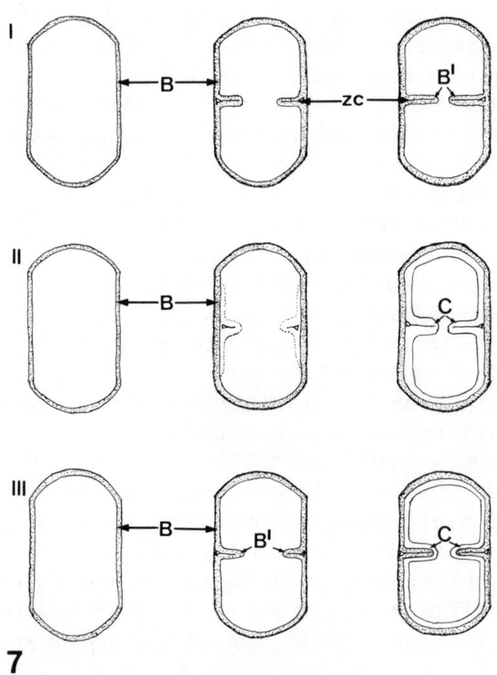

Fig. 7

Schematic representation of the different septation kinds in conidia (After Reisinger, 1972). (I) Euseptation process (septal formula: B', zc, B'). (II) Distoseptation process (septal formula: C, zc, C). (III) Double process (septal formula: C, B', zc, B', C).

surrounded by layer C, are released (Luttrell, 1963, Fig. 27). In mature conidia of *H. spiciferum*, this insertion zone appears as a thin pellicle and when conidia are crushed between glass slides the conidia are ruptured but the cells within the conidia are not released. In this case, the insertion zone has all the characteristics of a mucilaginous zone (broadly fibrillar, swelling in water, becoming thin during the dessication). It is suggested that this zone is equivalent to layer A. Since we do not know the origin of this zone at present, it is temporarily designated by the letter Cm. The septal formula of such conidia is then: C, Cm, zc, Cm, C.

Structure of amerosporus walls.

Studies of several amerospores (cf. Table 1) have shown that the walls of these unicellular conidia may also possess the different wall layers described above. However the number of wall subunits is variable according to the investigator. For example, some authors ignore layer A and others refer to two zones in layer B (B_1 and B_2).

In spite of these difficulties, in a general review of the literature on conidial wall structure, it is possible to recognize two fundamental concepts of wall layers corresponding to our formulae, A,B or A,B,C. Because layer A is always present, these two variations differ only by the presence of the absence of layer C. Therefore the formulae can be designated by the terms haplothecate and diplothecate respectively. Modifications of these two fundamental schemes in some species with diplothecate conidia arise from the presence of a zone of different structure between layers B and C. In order to clearly identify this zone, it will be necessary to define its composition and to know if it is mucilaginous, or whether it arises from layers B or C.

In conclusion, the terms haplothecate or diplothecate can also be used to designate the structure of walls in septate conidia.

The euseptate propagules are haplothecate and the distoseptate conidia, or those demonstrating the process of double septation, are diplothecate.

Shape evolution

Conidial primoridia, initiated by expansion of a sporogenous apex, are transformed into young conidia by the two following mechanisms which are very distinct in some cases:

1. Spherical growth;
2. Cylindrical growth (Turian et al., 1970).

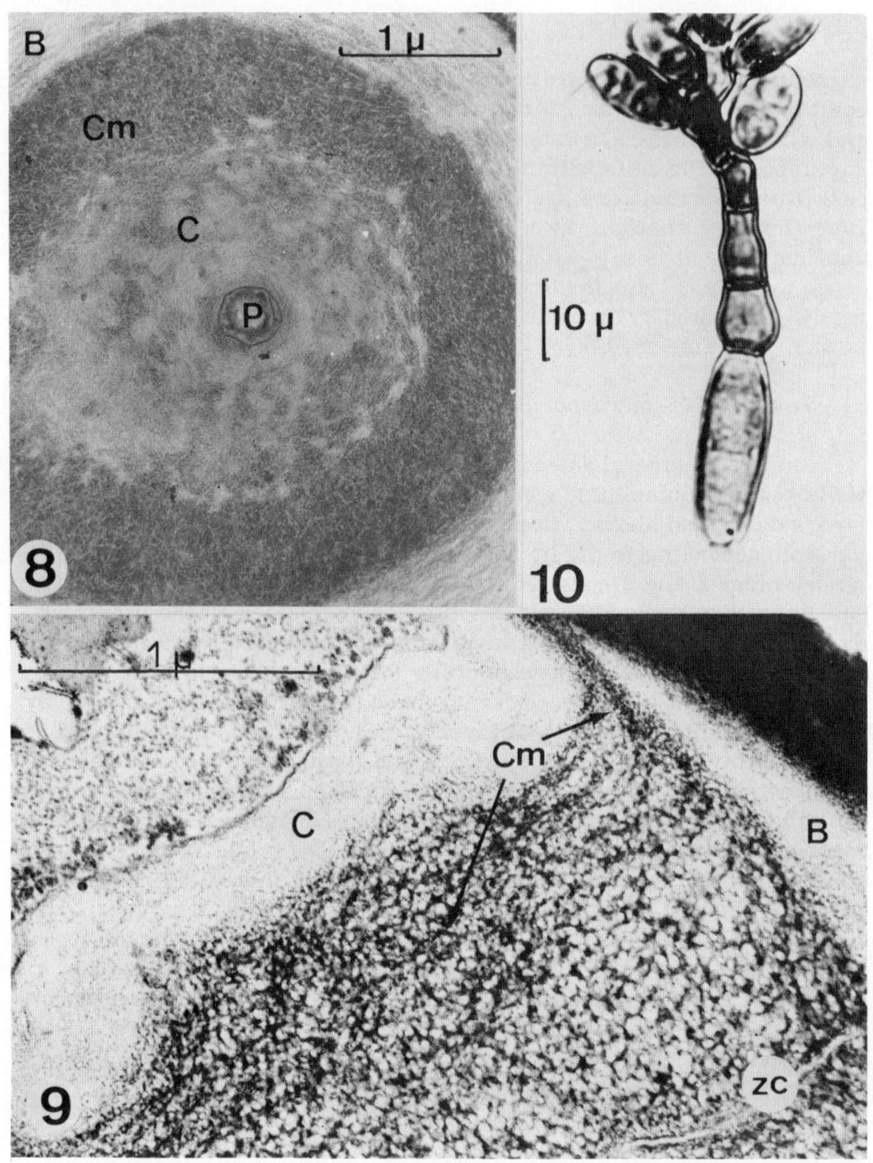

Fig. 8-10

(8) *Helminthosporium spiciferum*. Transversal section on a level with the septum of a nonhydrated conidium. (Lateral wall: B, septum: Cm, C, septal perforation: P) (9) *Helminthosporium*

Definitive shape of most of the ameroconidia is due to spherical growth: An increaseof primoridium volume precedes later wall differentiation (phialoconidia *pro parte*, annelloconidia *pro parte*, radulaspores, blastospores).

Arthroconidia and endogenous phialoconidia generally result from cylindrical growth followed by separation. In these cases propagules are usually truncated at both extremities, except when this truncated shape is modified by later lysis (Ex. *Stephanosporium cerealis*).

In phragmoconidia, the primoridia are first spherical then lengthen by cylindrical growth. In phragmoconidia two shape evolution modalities exist and are designated by the following terms (Luttrell, 1963):

1. Holosporous maturation; "conidium maturation in which the conidium approaches its definitive size and shape before delimiting cells and maturing as a whole." The order of appearance of septa is variable with the species (e.g., *Helminthosporium sorokinianum* and *Dendryphiella spp.*). In holosporous maturation cylindrical growth stops quickly and propagules generally have a rounded top.

2. Acrosporous maturation: "conidium maturation in which cells are delimited and mature in sequence from base to apex as the tip of the conidium continues to expend." In this latter case, cylindrical conidial growth is theoretically indefinite and propagules are generally fusiform. This maturation type appears to be restricted to the murogenous conidia as defined by Luttrell (1963). Consequently it appears in aleuria and in annelloconida *pro parte* (Ex. *Sporidesmium, Clasterosporium*). Acrosporous maturation seems to exclude conidia chain arrangement while holosporous propagules can form acropetal chains.

Acrosporous maturation ultrastructural study and special conidial forms during conidial maturation (e.g., *Dictyosporium*) would help to understand how nutritive substances are transportation to the terminal cell.

Preformed Structure for Germination

Based upon the emergence of germ tubes three kinds of conidia

---

spiciferum. Septum of a hydrated conidium. (Lateral wall: B, septum: C,Cm,zc.) (10) *Helminthosporium spiciferum*. Iterative germination of a conidium.

are distinguished:

1. Conidia with anatomically undefined polarization. In this case, germ tubes emerge from the conidia through the lateral wall, in an area whose structure is the same as that of neighboring sites. Physiological polarization determines the emergence point.

2. Conidia having a zone of weaker resistance by which the germ tubes rarely emerge, as in *Helminthosporium spiciferum*. Conidial apices of this species have a perfectly built germ spore, but the germ tube accidentally emerges at this point. Physiological polarization overcomes anatomical polarization.

3. Conidia with anatomically defined polarization. These conidia have a preformed structure for germination (germ pores or split) which determines the place of germ tube emergence. Anatomical polarization corresponds to physiological polarization.

Ultrastructural studies of the split or of the germ pore have not demonstrated important differences between both structures in which layer B is not melanized or is absent (Reisinger, 1970a, b, 1972; Griffiths, 1974). Splits and germ pores are constituted by a kind of a pigmented plug. This plug, directly against layer A is similar to layer C in diplothecate conidia. The structure of splits and germ pores appear to be similar in ascospores, basidiospores, and conidia.

Dry Conidia and Wet Conidia

Distinction between dry and wet conidia, now rarely used in taxonomy, is very important in the analysis of propagule dispersal.

Investigations with TEM reveal a perfectly formed layer A in solitary dry conidia and in dry catenate conidia. On the other hand, the surfaces of wet conidia are often contaminated by mucilaginous substances of variable origin.

Characters Related to Spore Liberation Mechanism and Conidiogenesis Types

Young conidia are separated from parental cells by a septum but with intercalary chlamydospores and arthroconidia two septa limit propagules.

Later, the separative septum may develop a modification in radulaspores (as *Beauvaria* or *Tritirachium*, Table 1) by transforming its lower part in a spicule. Light microscopy investigations demonstrate that, in some species, conidium separation is executed by rupture of the disjunctive cell lateral wall which is devoid of organized protoplasm. An example of this can be seen with basal propagules of *Dendryphiella vinosa* examined with TEM and SEM.

Besides these particular cases which necessitate other ultrastructural observations, the separative septum in most species is either symmetrical (formula B', zc, B') or asymmetrical. This latter structure appears by thickening of conidial layer B or by elaboration of layer C in conidia whose maturation is realized on sporogenous cells. A formula for such septa is B', zc, B', C from mother-cell to conidium.

The cleavage zone or the structure which replaces it, are important in propagule release. Depending upon the phenomenon intervening in the cleavage zone (Hughes 1971) three main kinds of septa can be defined as follows:

1. Schizogenous;
2. Potentially schizogenous;
3. Aschizogenous.

Septa with schizogenous characters permit spontaneous separation of propagules by lysis which appears in the cleavage zone level. In *Stephanosporium cerealis*, conidiogenous hyphae is fragmented in young conidia by a double septation process (Fig. 4). Lysis begins early on both sides of the cleavage zone and spreads quickly on the lateral wall layer B. Except for a raised and melanized band, fundamental layer B is strongly altered around the conidium. Individualized propagules are connected only by layer A which sometimes adheres to the raised and melanized band and which is generally very resistant to autolytic enzymes. This endogenous enzymatic release usually leads to a hilum which is difficult to recognize and is not very characteristic.

Mechanical liberation is bound to the existence of separative septa which are potentially schizogenous or aschizogenous. In separative potentially aschizogenous septa, the cleavage zone represents a place of less resistance but conidial liberation demands the intervention of a mechanical force. This force can be the pressure of the next primordium (annelloconidia) or the result of different phenomena such as hygroscopic movements (mechanisms and examples see Meredith, 1963; Kenneth, 1964; Webster, 1966; Ingold, 1967, 1971; Reisinger, 1970b) or the consequence of external, physical, or biological factors.

In the case of aschizogenous septa, the cleavage zone is not a privileged rupture zone. Therefore propagule liberation frequently happens by lateral wall rupture of mother-cells (aleuriospores *pro parte*). Consequently the conidial hilum is surrounded by a thin fringe of wall fragments.

Thus nonseceding propagules are released either by mechanical force or by exogenous enzyme intervention. Indeed chlamydospores or acropetal-arthro-aleuria of *Thielaviopsis basicola* are separated, in unicellular elements, by a chitinase effect or by action of soil micro-oorganisms (Tsao & Bricker, 1970). Mycophage microarthropode digestive juice can also have a selective effect on fungal organs. Some chlamydospores and some aleuria (Reisinger, 1972) with thick and melanized walls go through digestive systems without alteration of the germinative power, while hyphae and mother-cells are perfectly digested. Therefore, such propagules are individualized and released in consequence of enzymatic action of external origin.

Conidia with a Typical Hilum

"The practical usefulness of distinctions between types of conidial origin would be increased if the character of the hilum and conidial scars could serve as an index to conidial origin" (Luttrell, 1963). Hilum structure is very often subordinated to the liberation mechanism of propagules. In *Phialophora richardsiae*, *Doratomyces purpureofuscus*, or *Scopulariopsis brevicaulis*, the hilum is constituted by the upper layer of a separative septum and is sometimes surrounded by a thin fringe of layer A fragments. The hilum collarette has the same composition in the radulaspore of the *Beauveria* type, that hilum described above, while it is constituted by layer B fragments in conidia and has a disjonctor cell similar to *Wardomyces* aleuries. Therefore, TEM studies do not permit descriptions of a typical hilum in most ontogenic modalities. However, hilum structure and conidial scar structure can furnish better precision on propagule origin. Nevertheless the morphology of the hilum is a very important character in the recognition of porospores especially in species where primordia appear on poorly pigmented conidiosphore walls or in growth conditions which reduce melanization (Reisinger, 1972).

In spite of some differences from one species to another, the hilum of tretoconidia can be described as a central clear pore surrounded by a thickened strongly melanized and sometimes protuberant area (Leonard & Suggs, 1974).

Conidia in Chains

According to the taxon under examination, conidia are either solitary, arranged in chains, or collapsed to form mucous heads.

In the case of catenate conidia acropetal or basipetal chains can be distinguished. During basipetal development conidia appear successively from a sporogenous locus, and such chains are never ramified (*Aspergillus*, *Penicillium*, *Thielaviopsis*, etc.). In acropetal development the sporogenous activity is transmitted from the mother-cell to successive conidia and formation of ramified chains is possible (*Cladosporium*, *Dendryphiella*, etc.). Conidial maturation in a basipetal chain is equally basipetal while in acropetal chain it is usually acropetal, occasionally basipetal as in *Periconia* species (Mason & Ellis, 1953).

Chain formations and wall relations around catenate conidia were recently analyzed by Subramanian (1972) who gave many examples.

Assumptions Concerning the Relationships Between Sporangiospores and Conidia

Phylogenic relations between conidia and sporangiospores were the object of many speculations. Thus in Mucorales, it is possible to establish an evolutionary series based upon the diminution of the number of formed sporangiospores in a sporangium (Benjamin, 1966) which range to *Cunninghamella* or *Chaetochladium* conidia which are monosporous sporangia. Therefore, an Ascomycete or Deuteromycete conidium would represent an evolutionary stage from a common ancestor.

Hawker and Hendy's hypothesis (1963) concerning a similar rapproachment between conidia and sporangiospores is based on the comparative ultrastructural studies of resting and germinating spores. However this hypothesis is much too general and seems to be based on a misinterpretation of germ tube wall continuity with a preexisting spore layer (Border & Trinci, 1970, Gull & Trinci, 1971).

In various types of septal wall elaboration it is possible to find mechanisms which recall those of sporangiospore formation (Luttrell, 1963). In distoseptate conidia each locus is surrounded by a new wall, and propagule germ pores, which are often nonfunctional as in *Helminthospoporium spiciferum*, could even be compared to sporangial papilles. Loci which can be mechanically or enzymatically separated from the primary wall (layer B) are equivalent to sporangiospores.

In this latter context, distoseptate propagules, retain the sporangial division mechanism in an unlysable cellular envelope. On the other hand this propagule may have acquired an ascomycete like septal perforation, and intercellular connections would be precarious and easy to suppress.

Conidia with double septation (euseptation followed by distoseptation) are difficult to integrate in this scheme. However, if it is admitted that *Stephanosporium cerealis* fertile hyphae are conidiocyste, it is even possible to again find a primary envelope lysis. Therefore organs of this type may have acquired a hyphal septation character and have preserved sporangial features for liberation of disseminative units.

Finally, euseptate propagules may be fragments of hyphae unrelated to sporangia. These propagules represent more evolved asexual multiplication organs if it is admitted that evolution was in the direction of Phycomycetes to Ascomycetes.

However it is necessary to remain cautious in considering the value of such phyllogenetic considerations when they concern asexual forms. As a matter of fact, it is known that similar maturation or separation mechanisms occur in very different groups of microorganisms (e.g., schizogenous septa in Bacteria and arthroconidia).

In imperfect fungi the production of propagules in large numbers and the efficiency of dispersal make possible the establishment of mutants in favorable ecological conditions. Morphological or physiological forms can exist without necessarily being reflected in sexual reproduction. Indeed it would be necessary to admit that, in Deuteromycetes, morphological evolution had, in most cases, masked biological relationships (e.g., *Spilocea, Fusicladium, Pollaccia*). Similarity of the development mechanism in two different asexual cycles does not necessarily implicate a biological relationship. However, it is possible to define an evolutionary series for one character or one given mechanism.

FUNCTIONS

Initiation of Vegetative Cycle.

The most obvious and the most universal conidium function is the initiation of the vegetative cycle. This latter process is realized by germination. Germ tube number and place of emergence are criteria sometimes used for species (Drechsler, 1923; Luttrell, 1963; Rapilly, 1964) or even for genus description (Shoemaker, 1959). The mode of germination of conidia is commonly regarded as a genetically stable character as is suggested by their taxonomic use.

In septate conidia, a prominent distinction of diverse germination types can be based on germ tube number or position. Polar (germ tubes arise from only one or two end cells) and amphigenous (germ tubes arise from anyone cell) germination conidia are distinguished.

Germ tube emergence through the hilum is the definition of germination of a conidia with percurrent germination and tube germinative orientation compared with the main conidial axis which permits differentiation of semi-axial and lateral germ tube types (Luttrell, 1963).

Though germination processes described above were only analyzed in *Helminthosporium* and related genera, their extension to different groups of Deuteromycetes may be possible.

However, it must be pointed out that dictyospore germination is usually amphigenous and that uni- or bipolar germination rate depends upon the medium. Reddy & Bilgrami (1969) have shown that uni- and bipolar germination coexists in some *Helminthosporium* species. With *Dendryphiella vinosa* conidia, Reisinger (1968) has experimentally decreased bipolar germination rate (from 80% to 18%) by reducing the available nutrient supply.

On the other hand, amphigenous germination does not appear to be dependent upon medium conditions. Rapilly (1964) suggested that the septa are obstructed amphigenous conidia. While in polar germination propagules, septum perforations stay open. That hypothesis could explain bipolar germination variation rates by closing septum perforations. Thus the two separated conidial parts would individually germinate. Closure of septum perforations would intervene during small molecule absorption which would modify the osmotic pressure between end and intercalary cells. However ultrastructural verification of this hypothesis is difficult. It is not known whether cytoplasmic continuity between loci can be broken by a difference in osmotic pressure due to end cell wall premeability. This phenomenon can also intervene during the fixation procedure. It is probably for this reason that micrographs reveal opened or closed septal perforations in propagules that undergo polar germination.

Evidence for Hyphal Potential in Conidia

For a long time Dueteromycete conidia, were considered as a modified fragment of the vegetative apparatus. Thus conidia are able to produce the same organs as hyphae (conidiophore, conidium and chlamydospore).

Iterative Germination

Iterative germination describes a particular kind of germination in which the conidia, for diverse reasons, directly form fertile conidiophores. In *Helminthosporium spiciferum* when conidia were incubated on media with reduced nutrient content the bipolar germination rate progressively diminished to become stable at about 20%. Another effect of reduced nutrient content was that conidiophores which appeared after three or five days on young hyphae, were inserted more and more closely to the conidium. In double-distilled water some conidiophores arise directly on propagules. Germ tube numbers which are entirely transformed in conidiophores are the most important when conidia germinate in an atmosphere saturated with water vapor. (Fig. 10). In this case, layer Cm swells, probably by accumulation of water necessary for metabolic initiation.

Iterative germination in *H. spiciferum* takes place in the absence of an external food supply to the conidia but the conidia appear to need water. Formed conidia (8--16) are viable and morphologically comparable with those obtained from a whole vegetative cycle (Reisinger, 1972). However in these new propagules pigmentation is less intense. A similar decrease in pigmentation has also been pointed out for *Aspergillus niger* conidia (Anderson & Smith, 1972).

Iterative germination has recently been designated as "Microcycle conidiation" by Anderson & Smith (1971) and formation of fertile conidiophores has been obtained by thermic variation. In *Penicillium digitatum* a similar reduction of the life cycle depends upon the presence of some amino acids (Zeidler & Margalith, 1973).

Besides these experimental studies in vitro, phenomena similar to iterative germination are described form some observations. Drechsler (1923) pointed out that *Helminthosporium catenarium* can produce short conidial chains and he thinks that the appearance of these alternaria like forms is facilitiated by moisture. However, with *Helminthosporium spiciferum*, humidity is not the single cause for the production of short chains. Absence of an external food supply also seems to be important. Before the analogy between iterative germination of *H. spiciferum* conidia and the appearance of alternaria like forms in *H. catenarium* can be accepted, it will be necessary to demonstrate that successive propagules of the latter species are differentiated from nutrient reserve of the parent conidia.

Apical parts of conidia of diverse *Alternaria spp.* also form fertile conidiophores. Their formation could be likened to iterative germination if it is possible to demonstrate that transport of nutrient substances between conidiophores and conidias stops or decreases after appearance of the first propagules. A similar phenomenon is observed for species of *Annellophora* in which the end conidial cell functions as an annellide.

Germ tube transformation of the conidial apparatus is described for *Geotrichum candidum* by Park & Robinson (1969) where the fungus was grown on Cellophane.

Boosalis (1962) has noticed a phenomenon comparable to iterative germination (precocious sporulation) for *Helminthosporium sativum* spores in a nonsterile soil amended with potato-dextrose broth.

Subsequently it appears that diverse terms such as iterative germination, microcycle conidiation and precocious sporulation all designate a particular morphological expression of germination. According to its strict definition, iterative germination excludes the mycelial phase of the vegetative cycle and reduces it to sporogenous apparatus formation.

Production of Secondary Conidia

Some vegetative fungal structures can produce conidia directly from a hyphal cell which is not morphologically differentiated. Hyphal cell aptitude to from propagules also exists in conidia. This is the case in *Septoria tritici* under near-ultraviolet irradiation (Jones & Lee, 1974) or in species with a yeast growth phase in liquid medium (e.g., *Aureobasidium pullulans*, *Phialophora dermatitidis*, etc.). Several of these latter species have been investigated with TEM (cf. Carbonell, 1969; Lane et al., 1969; Oujezdsky et al., 1973; etc.).

In *Aureobasidium pullulans* the conidiogenesis process remains the same for the two phases: yeasts and mold. Secondary conidia appear from hyphal cells or from propagules which function as annellides (Reisinger et al., 1974).

Production of such secondary conidia is different from iterative germination because conidiophores do not need to be present. Moreover, with the yeastlike stage, the food is directly taken in from the surrounding medium while germination energy comes from the conidial reserves.

The Occasional Transformation into Chlamydospores

In some conditions of environment, nutrients present in conidia can be conserved by morphological differentiation of conidial cell. In this case, one or several loci are transformed to form chlamydospores (Old & Schippers, 1974; Schneider & Seaman, 1974). In *Fusarium sulphureum*, during this process apical and basal cells lose their cytoplasm. Conidial chlamydospores contain dense cytoplasm and are surrounded by a thick wall. Thus, they structurally resemble hyphal chlamydospores (Schneider & Seaman, 1974).

In *Fusarium solani, cucurbitae*, conidial chlamydospores survive a longer time in nonsterile soils than macroconidia (Old & Schippers, 1973).

Dispersal

In most of the Deuteromycetes, conidial germination is inhibited by parental mycelium. Conidia are able to recreate another vegetative cycle if it is carried away on a substrate which is uncolonized by hyphae which have inhibitive action. This substrate change is realized in nature by the different dispersion means. The latter have been particularly well investigated for parasitic fungi and many examples were analyzed by Ingold (1971).

In Deuteromycetes the following kinds of dispersal are recognized:

1. Mechanical dispersal;
2. Trophical dispersal.

Mechanical Dispersal

Mechanical dispersal is due to air or water movements which can carry liberated propagules variable distances. Dispersal by air is usually very effective for dry conidia which have an active liberation mechanism (cf. Meredith, 1963; Ingold, 1971; etc.). On the other hand mechanical dispersal is limited in aleuria and in chlamydospores which are nonseceding propagules.

Mechanical dispersal can also occur for conidia which adhere to animal bodies. So mycophage or detriticole microarthropodes usually live in a habitat characterized by plenty of fungi. The sporogenous apparatus in an aerial position makes the contact easier between conidia and arthropode teguments. Small propagules or mucilaginous surface propagules easily adhere to the different parts of the animal (Fig. 12) and are carried during periodic arthropode migration (Vannier, 1970). When depositied on a petri dish, these arthropodes contaminate the sterile medium. The species which appear are usually those which fructified in microhabitat where the animal lived before (Reisinger, 1972). However this method of dispersion is only effective with conidia which are easily separated from sporogenous organs (propagules in chain, propagule with a schizogenous or potentially schizogenous septum and conidia collapsed in a mucous head).

Trophical Dispersal

With trophical dispersal, propagules pass through the digestive system of an animal. Their ability to germinate in animal excretions is retained. This kind of dispersal can be occasional or frequent and is subordinated to the food preference of the animal.

Thus accidental dispersal of different viable spores of Deuteromycetes by *Musca domestica* excretions is possible if propagule size does not exceed 50 µm (Bolton & Hansen, 1970). This is very interesting because almost all Deuteromycetes in natural conditions form viable propagules smaller than 50 µm.

Besides these accidental samplings of fungal propagules, some microfauna species are regularly fed with fungal cells. This is true for mycophage ladybirds (cf. Turian, 1971). A similar relationship also exists between *Oppia nova* (Acarina, Oribate) and *Oidiodendron echinulatum* (Cancela da Fonseca & Kiffer, 1969). *O. echinulatum* fructifications are used as acarian food.

Investigations of intestinal contents of mycophage or detriticole animals (Forsslund, 1938; Schuster, 1956; Wallwork, 1958) or excretions (Agrell, 1940; Schaller, 1950; Poole, 1959; Singh, 1964a,b; McMillan & Halley, 1971) reveal the presence of spore fragments or more or less intact cells which are sometimes able to germinate (Ghilarov in Dommergues & Mangenot, 1970).

According to Wallwork (1958) in the case of some soil acarina the factors which affect food preferences are:

1. Particle size;
2. Structure of the mouth parts of mites;
3. Nature of the digestive system;
4. Stage of chemical decay of food;
5. Moisture content of food.

Among these factors, the first and second also affect the survival of propagules consumed by microarthropodes with biting mouth parts.

Studies of ultrastructural modifications of dematiaceous conidia ingested by diverse acarina (Reisinger, 1972) show that a preliminary crushing of fungal cell walls by the animal mouth parts is necessary

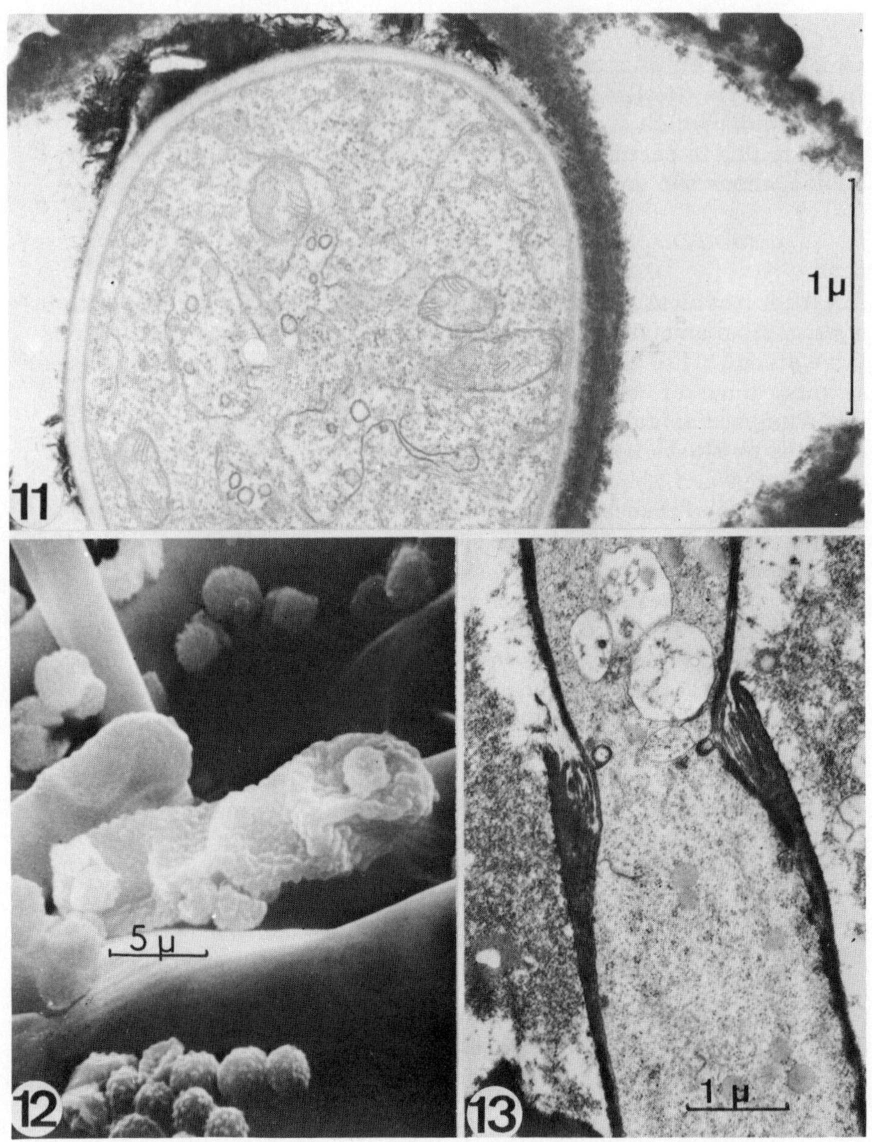

Fig. 11-13

(11) *Dendryphiella vinosa.* Intact cell of a conidium which went through digestive system of an acarina. (12) *Tyrophagus putresentiae* (Acarina). Fungal propagules adhering to the acarina body.

for alteration of protoplasmic cell contents during passage through digestive systems of animals. So, small intact ingested conidia go through the digestive tract without damage and are found entire in feces. On the other hand large conidia are broken. With phragmoconidia some cells protected by lateral walls and transversal septa also survive passage through the digestive tract (Fig. 11).

The germination potential of these propagules depends on physical and biological factors of the microhabitat which surround the dejection. Reisinger (1972) has shown in an experimental model study of the relationship between *Tyrophagus putrescentiae* (Acarina) and *Dendryphiella vinosa* that dessication resistance of fungal propagules contained in the acarina dejections decreases under influence of the bacterial population. As a matter of fact these excrements represent a substrate which is quickly colonized by acarina intestinal microorganisms or by microorganisms of external origin. After having exhausted easily assimilable substances, some bacteria can effect a partial wall lysis and destroy protoplasm. This bacterial destruction of propagules is realized in a way similar to that described for conidia placed in other conditions (cf. Reisinger & Kilbertus, 1973; Wong & Old, 1974; Reisinger & Kilbertus, 1974).

To conclude, diverse arthropode or diptera larva dejections contain viable propagules, subsequently able to form a colony on agar media. However, the germinative power of these propagules is quickly deteriorated by action of bacterial flora. Trophical dispersion appears to be a means of propagation but the efficacy depends upon the following factors:

1. Size and nature of mouth parts compared to conidial size. This report determines the degree of crushing that occurs.
2. Spore wall resistance to the animal digestive juices.
3. Degree of conidium wall melanization, nature, and thickness. These characters determine the longevity of fungal cell life with regard to microbial enzymatic action (cf. Bloomfield & Alexander, 1967; Kuo & Alexander, 1967; Hurst & Wagner, 1969; Webley & Jones, 1971; etc.).
4. Enzymatic ability of the bacterial flora to degrade propagules contained in dejections.
5. Presence of substances required for propagule germination.

---

(13) *Puccinia poarum*. Spermatium formation. (Courtesy of Dr. J. Orcival).

Preservation

Floral inventories carried out in sites where harmful ecological conditions exist (UV, λ, or atomic radiations, frequent dehydrations) generally show a high proportion of Deuteromycetes with melanized walls.

Experiments with species of varying pigmentation appear to confirm this resistance due to melanic wall incrustation (Durrell, 1964; Seiji & Itakura, 1967; Bondar et al., 1971; etc.).

Pigmented fungal cells withstand the attack of different enzymes better than nonpigmented cells (Bloomfield & Alexander, 1967; Kuo & Alexander, 1967; Hurst & Wagner, 1969; Bull, 1970; Chu & Alexander, 1972; etc.).

Though melanized wall propagules can also be destroyed by bacteria (Wang & Old, 1974); Reisinger & Kilbertus, 1974), the studies cited suggest that the degree of propagule pigmentation is one of the main preservation factors.

Chemical constitution of conidial walls, wall thickness, and especially the presence of different layers are also factors which influence preservation efficacy.

Thus in diplothecate conidia such as those of *Helminthosporium spiciferum* (possessing layers A, B, Cm, and C) some bacterial penetration (*Pseudomonas spp.*) is stopped by layer C. This layer surrounds the locus which preserves its germination power for a more or less long period of time (Reisinger, 1972). This observation suggests that some bacteria do not possess the enzymatic equipment necessary for lysis of layer C. Subsequently this layer must have a chemical nature different from other layers. At any rate, it appears that the presence of layer C in special conditions increases the propagule longevity.

Conidial size also plays an important role as a preservation factor especially for trophical dispersal.

Sexual Function

The sexual function of conidia has been known for a long time (Langeron & Van Breuseghem, 1952; Chadefaud, 1960, Gregory, 1966; Kendrick, 1971; Gain, 1972).

Conidia implicated in a sexual cycle are usually small (microconidia) and do not germinate on common media. In Ascomycetes, they arise from phialides. The distinction between phialoconidia and spermatia in Ascomycetes, based on the ability to germinate, is not an available criterion. Phialospores of Mollisioideae are not always able to germinate but this perhaps depends upon the medium used (Le Gal & Mangenot, 1960).

In studies with spermogonium genesis in *Puccinia poarum* (Orcival, 1968) or in *Puccinia sorghi* (Rijkenberg & Truter, 1974) with TEM it has been shown that spermaties of Uredinales were formed as annelloconidia (Fig. 13). Comparisons between phialoconidia and annelloconidia (Oláh & Reisinger, 1973) have shown that annelloconidia, which include, the upper part of separative septa in their walls, do not fundamentally differ from phialoconidia.

Thus, phialoconidiogenesis as defined by Mangenot & Reisinger (1973) appears to be the more frequent mechanism of formation in most fungal sexual propagules.

REFERENCES

Ache, I. G., Aguirre, M. J. R., Uruburu, F. & Villanueva, J. R. (1966). The fine structure of the *Fusarium culmorum* conidium. Trans. Br. Mycol. Soc. 49, 695--702.

Agrell, I. (1940). Die Arthropoden Fauna von Madiera nach den Ergebnissen der Reise von Prof. Dr. O. Lunbdlad Juli--August 1935. XVIII. Collembola. Ark. Zool. 31, 1--7.

Akai, S., Fukotomi, M. & Kobayashi, N. (1969). Fine structure of conidia of *Fusarium oxysporum* f. *cucumerinum*. Ann. Phytopath. Soc. Japan, 35, 351--353.

Akai, S., Fukutomi, M. & Kunoh H. (1966). An observation on fine structure of conidia of *Sphaerotheca pannosa* (Wallr.) Lev. attacking leaves of roses. Mycopathol. Mycol. Appl. 29, 212--216.

Akin, D. E. Michaels, G. E. (1972). *Microsporum gypseum* macroconidial development revealed by transmission and scanning electron microscopy. Sabouraudia 10, 52--55.

Anderson, J. G. & Smith, J. E. (1971). The production of conidiophores and conidia by newly germinated conidia of *Aspergillus niger* (Microcycle conidiation). J. Gen. Microbiol. 69, 185--197.

Anderson, J. G. & Smith, J. E. (1972). The effects of elevated temperatures on spore swelling and germination in *Aspergillus niger*. Can. J. Microbiol. 18, 289--297.

Baker, J. E. & Smith, W. Z., Jr. (1970). Heat-induced ultrastructural changes in germinating spores of *Rhizopus stolonifer* and

Monilinia fructicola. Phytopathology 60, 869--874.

Balimgart, G. (1973). Rasterelektronenmikroskopiche Untersuchungen an *Scopulariopsis brevicaulis* (Sacc.) Bain. (Fungi Imperfecti). Osterr. Bot. Z. 122, 121--124.

Barron, G. L. (1968). The Genera of Hyphomycetes from Soil. 364 pp. Baltimore: William & Wilkins Co.

Benjamin, R. K. (1966). The merosporangium. Mycologia 58, 1--42.

Bertaud, W. S., Morice, I. M., Russell, D. W. & Taylor, A (1963). The spore surface in *Pithomyces chartarum*. J. Gen. Microbiol. 32, 385--395.

Bessis, R. (1972). Etude au microscope electronique à balayage des rapports entre l'hôte et le parasite dans le cas de la Pourriture grise. C.R. Acad. Sci. Paris, D. 274, 2991--2994.

Bloomfield, B. J. & Alexander, M. (1967). Melanins and resistance of fungi to lysis. J. Bacteriol. 93, 1276--1280.

Bolton, H. T. & Hansen, E. J. (1970). Ability of the House Fly, *Musca domestica*, to ingest and transmit viable spores of selected fungi. Ann. Ent. Soc. U.S.A. 63, 98--100.

Bondar, A. I., Zhdanova, N. N., & Pokhodenko , V. D. (1971). (In russian) Izvest. Akad. Nawk. SSSR, Ser. Biol. 1, 80--85.

Boosalis, M. G. (1962). Precocious sporulation and longevity of conidia of *Helminthosporium sativum* in soil. Phytopathology, 52, 1172--1177.

Border, D. J. & Trinci, A. P. J. (1970). Fine structure of the germination of *Aspergillus nidulans* conidia. Trans. Br. Mycol. Soc. 54, 143--152.

Brewer, J. G. & Boerema, G. H. (1965). Electron microscope observations on the development of pycniospores in *Phoma* and *Ascochyta spp*. Proc. Koninkl. Nederl. Akad. Wetensch. Ser. C., 68, 86--97.

Brushaber, J. A. & Haskins, R. H. (1973). Cell wall structures of *Epicoccum nigrum* (Hyphomycetes). Can. J. Bot. 51, 1071--1073.

Buckley, P. M., Sjaholm, V. E. & Sommer, N. F. (1966). Electron microscopy of *Botrytis cinerea* conidia. J. Bacteriol. 91, 2037--2044.

Buckley, P. M., Wyllie, T. D. & Devay, J.E. (1969). Fine structure of conidia and conidium formation in *Verticillium albo-atrum* and *V. nigrescens*. Mycologia, 61, 240--250.

Bull, A. T. (1970). Inhibition of polysaccharase by melanin. Enzyme inhibition in relation to mycolisis. Arch. Biochem. Biophys. 137, 345--356.

Cain, R. F. (1972). Evolution of the fungi. Mycologia, 64, 1--14.

Calonge, F. D. (1970). Notes on the ultrastructure of the microconidium and stroma in *Sclerotinia sclerotiorum*. Arch, Mikrobiol. 71, 191--195.

Campbell, C. K. (1971). Fine structure and physiology of conidial germination in *Aspergillus fumigatus*. Trans. Br. Mycol. Soc. 57, 393--402.

Campbell, R. (1968). An electron microscope study of spore structure and development in *Alternaria brassicicola*. J. Gen. Microbiol. 54, 381--392.

Campbell, R. (1972). Ultrastructure of conidium ontogeny in the deuteromycete fungus *Stachybotrys atra* New. Phytol. 71, 1143--1149.

Cancela da Fonseca, J. P. & Kiffer, E. (1969). A propos des rapports entre certaines espèces d'Acariens Oribates et certaines espèces de champignons existant dans un même sol. C.R.Acad.Sci. Paris, D, 269, 386--389.

Carbonell, L. M. (1969). Ultrastructure of dimorphic transformation in *Paracoccidioides brasiliensis*. J. Bacteriol. 100, 1076--1082.

Carroll, F. E. (1972). A fine-structural study of conidium initiation in *Stemphylium botryosum Wallroth*. J. Cell. Sci. 11, 43--47.

Carroll, F. E. & Carroll, G. C. (1974). The fine structure of conidium initiation in *Ulocladium atrum*. Can. J. Bot. 52, 443--446.

Chadefaud, M. (1960). Traité de Botanique Systématique. Tome I. 1018 pp. Paris: Mason & Co.

Chadefaud, M. (1969). Une interprétation de la paroi des ascospores septées, notamment celles des *Aglospora* et des *Pleospora*. Bull. Soc. Mycol. Fr. 85, 145--157.

Christias, C. & Baker, K. F. (1970). Ultrastructure and cleavage of chlamydospore chains of *Thielaviopsis basicola*. Can. J. Bot. 48, 2305--2308.

Chu, S. B. & Alexander, M. (1972). Resistance and susceptibility of fungal spores to lysis. Trans. Br. Mycol. Soc. 58, 489--497.

Cole, G. T. (1973a). Ultrastructural aspects of conidiogenesis in *Gonatobotryum apiculatum*. Can. J. Bot. 51, 1677--1684.

Cole, G. T. (1973b). Ultrastructure of conidiogenesis in *Drechslera sorokiniana*. Can. J. Bot. 51, 629--638.

Cole. G. T. (1973c). Correlation between rodlet orientation and conidiogenesis in Hyphomycetes. Can. J. Bot. 51, 2413--2422.

Cole, G. T. & Aldrich, H. C. (1971). Ultrastructure of conidiogenesis in *Scopulariopsis brevicaulis*. Can. J. Bot. 49, 745--755.

Cole, G. T. & Ramirez-Mitchell, R. (1974). Comparative scanning electron microscopy of *Penicillium* conidia subjected to critical point drying, freeze drying and freeze-etching. In Scanning Electron Microscopy, Part II. Proceedings of the Workshop on Scanning Electron Microscopy, Chicago. 367--374.

Corlett, M. (1970). Studying fungi with the scanning electron microscope. Greenhouse-Gerden-Grass. 9, 1--5.

Corlett, M. (1973). Surface structure of the conidium and conidiophore of *Stemphylium botryosum*. Can. J. Microbiol. 19, 392--393.

De Bertoldi, M. Mariotti, E. & Filippi, C. (1974). The fine strucutre of some strains of *Humicola* Traaen. Can. J. Microbiol. 20, 237--239.

Dedic, G. A. & Koch, O. G. (1957). Lucht-und Elektronenmikroskopiche Studien an Konidien von Schimmelpilzen. Mikroskopie, 12, 217--227.

Delvecchio, V. G., Corbaz, R. & Turian, G. (1969). An ultrastructural study of the hyphae, endoconidia and chlamydospores of *Thielaviopsis basicola*. J. Gen. Microbiol. 58, 23--27.

Dekhuijzen, H. M. & Van der Scheer, Ch. (1969). The ultrastructure of powdery mildew, *Sphaerotheca fuliginea* isolated from cucumber leaves. Neth. J. Path. 75, 169--177.

Dickson, M. R. (1963). A study of fine structure of the spore of *Pithomyces chartarum*. N.Z. J. Bot. 1, 381--388.

Dommergues, Y. & Mangenot, F. (1970). Ecologie Microbienne du Sol. 796 pp. Paris: Masson & Co.

Drechsler, C. (1923). Some graminicolous species of *Helminthosporium* I. J. Agr. Res. 24, 641--739.

Dufour, L. (1888). Observations sur le developpement et la fructification du *Trichocladium asperum* Harz. Bull. Soc. Bot. Fr. 10, 139--144.

Duran, A., Uruburu, F. & Villanueva, J.R. (1973). Morphogenetic and nutritional studies of *Geotrichum lactis*. Arch. Mikrobiol. 88, 245--256.

Durrell, L. W. (1960). Fine structure of *Thielavia sepedonium*. Mycologia 52, 963--965.

Durrell, L. W. (1964). The composition and structure of walls of dark fungus spores. Mycopathol. Mycol. Appl. 23, 339--345.

Ekundayo, J. A. & Haskins, R. H. (1969). Pycnidium production by *Botryodiplodia theobromae*. II. Development of the pycnidium and fine structure of the maturing pycniospore. Can. J. Bot. 47, 1423--1424.

Ellis, J. J., Bulla, L. A., St. Julian, G. & Hesseltine, C. W. (1970).

Scanning electron microscopy of fungal and bacterial spores. 3rd. Ann. Scanning Electr. Micro. Symp. Chicago 1970, Proc., pp. 145--152.

Fletcher, J. (1971a). Conidium ontogeny in *Penicillium*. J. Gen. Microbiol. 67, 207--214.

Fletcher, J. (1971b). Fine structural changes during germination of conidia of *Penicillium griseofulvum* Dierckx. Ann. Bot. 35, 441--449.

Florance, E. R., Denison, W. C. & Allen, J. T. C. (1972). Ultrastructure of dormant and germinating conidia of *Aspergillus nidulans*. Mycologia 64, 115--123.

Forsslund, K. R. (1938). Uber die Ernährungsverhalthisse der Hormilben und ihre Bedeutung für die Prozesse in Waldboden. Verh. VII. Internat. Kongr. Ent. 3, 1950--1957.

Fujii, R. K. & Hess, W. M. (1969). Surface characteristics of conidia from monosporous cultures of *Penicillium digitatum* and *Aspergillus nidulans* var. *echinulatus*. Can. J. Microbiol. 15, 1472--1473.

Garrison, R. G. & Lane, J. W. (1973). Scanning-beam electron microscopy of the conidia of the brown and albino filamentous varieties of *Histoplasma capsulatum*. Mycopathol. Mycol. Appl. 49, 185--191.

Ghiorse, W. C. & Edwards, M. R. (1973). Ultrastructure of *Aspergillus fumigatus* conidia development and maturation. Protoplasma 76, 49--59.

Goos, R. D. & Tubaki, K. (1973). Conidium ontogeny in *Riessia semiophora*. Can. J. Bot. 51, 1439--1442.

Greenhalgh, G. N. (1967). A note on the conidial scar in the Xylariaceae. N. Phytol. 66, 65--66.

Gregory, P. H. (1966). The fungus spore: What it is and what it does. In The Fungus Spore (ed. M. F. Madelin) pp. 1--13 London: Butterworths.

Griffiths, D. A. (1973a). The fine structure of conidial development in *Epicoccum nigrum*. J. Microsc. 17, 55--64.

Griffiths, D. A. (1973b). Fine structure of the chlamydospore wall in *Fusarium oxysporum*. Trans. Br. Mycol. Soc. 61, 1--6.

Griffiths, D. A. (1974). Development and structure of the aleuriospores of *Humicola grisea* Traaen. Can. J. Microbiol. 20, 55--58.

Griffiths, D. A. & Campbell, W. P. (1972). Fine structure of conidial development in *Colletotrichum atramentarium*. Trans. Br. Mycol. Soc. 59, 483--489.

Griffiths, D. A. & Swart, H. J. (1973). Fine structure of conidia in two species of *Seimatosporium*. Trans. Br. Mycol. Soc. 60, 123--131.

Griffiths, D. A. & Swart, H. J. (1974). Conidial structure in two species of *Pestalotiopsis*. Trans. Br. Mycol. Soc. 62, 295--304.

Grove, S. N. (1972). Apical vesicles in germinating conidia of *Aspergillus parasiticus*. Mycologia. 64, 638--641.

Gull, K. & Trinci, A. P. J. (1971). Fine structure of spore germination in *Botrytis cinerea*. J. Gen. Microbiol. 68, 207--220.

Hammill, T. M. (1971a). Fine structure of annellophores. I. *Scopulariopsis brevicaulis* and *S. koningii*. Am. J. Bot. 58, 88--97.

Hammill, T. M. (1971b). Fine structure of conidiation in *Chloridium chlamydosporis* (Fungi Imperfecti: Hyphomycetes). The ASB Bulletin 18, 36.

Hammill, T. M. (1972a). Electron microscopy of phialoconidiogenesis in *Metarrhizium anisopliae*. Am. J. Bot. 59, 317--326.

Hammill, T. M. (1972b). Fine strucutre of annellophores. II. *Doratomyces nanus*. Trans. Br. Mycol. Soc. 59, 249--253.

Hammill, T. M. (1972c). Fine structure of annellophores. III. *Monotosporella sphaerocephala*. Can. J. Bot. 50, 581--585.

Hammill, T. M. (1972d). Fine structure of annellophores. V. *Steganosoporium pyriforme*. Mycologia 64, 654--657.

Hammill, T. M. (1972e). Electron microscopy of conidiogenesis in *Chloridium chlamydosporis*. Mycologia. 64, 1054--1056.

Hammill, T. M. (1973a). Fine structure of annellophores. IV. *Spilocea pomi*. Trans. Br. Mycol. Soc. 60, 65--68.

Hammill, T. M. (1973b). Fine structure of conidiogenesis in the holoblastic, sympodial *Tritirachium roseum*. Can. J. Bot. 51, 2033--2038.

Hammill, T. M. (1974). Electron microscopy of phialides and conidiogenesis in *Trichoderma saturnisporum*. Am. J. Bot. 61, 15--24.

Hanlin, R. T. (1971). Morphology of *Nectria haematococca*. Am. J. Bot. 58, 105--116.

Hasimoto, T., Morgan, J. & Conti, S. F. (1973). Morphogenesis and ultrastructure of *Geotrichum candidum* septa. J. Bacteriol. 116, 447--455.

Hashioka, Y., Horino, O. & Kamei, T. (1966). Electronmicrographs of *Trichoderma* and *Pachybasium*. Rep. Tottori Mycol. Inst. 5, 18--24.

Hawker, L. E. & Hendy, R. J. (1963). Electron microscope study of germination of conidia of *Botrytis cinerea*. J. Gen. Microbiol. 33, 43--46.

Hawksworth, D. L. & Wells, H. (1973). Ornamentation on the terminal hairs in *Chaetomium* Kunze ex. Fr. and some related genera. Mycol. Pap. Kew. n. 134, 24 pp.

Heintz, C. E. & Niederpruem, D. J. (1971). Ultrastructure of quiescent and germinated basidiospores and oidia of *Coprinus lagopus*. Mycologia 63, 745--766.

Hess, W. M., Sassen, M. M. A. & Remsen, C. C. (1966). Surface structures of frozen-etched *Penicillium* conidiospores. Naturwissften 53, 1--2.

Hess, W. M., Sassen, M. M. A. & Remsen, C. C. (1968). Surface characteristics of *Penicillium* conidia. Mycologia 60, 290--303.

Hess, W. M. & Stocks, D. L. (1969). Surface characteristics of *Aspergillus* conidia. Mycologia 61, 560-571.

Horino, O. (1964). Electron microscopic structure of the conidia of *Helminthosporium oryzae*. Acta Scholae Medic. Gifu, 12, 230--240.

Horino, O. & Akai, S. (1965). Comparison of the electron micrographs of conidia of *Helminthosporium oryzae* Ito et Kurib and *Piricularia oryzae* Cavara. Trans. Mycol. Soc. Japan. 6, 42--46.

Hughes, S. J. (1953). Conidiophores, conidia and classification. Can. J. Bot. 31, 577--659.

Hughes, S. J. (1971). On conidia of fungi and gemmae of algae, bryophytes, and pteridophytes. Can. J. Microbiol. 49, 1319--1339.

Hughes, G. C. & Bisalputra, A. A. (1970). Ultrastructure of hyphomycetes. Conidium ontogeny in *Peziza ostracoderma*. Can. J. Bot. 48, 361--366.

Hurst, H. M. & Wagner, G. H. (1969). Decomposition of $_{14}$C-labelled cell wall and cytoplasmic fractions and from haline and melanic fungi. Soil Sci. Soc. Amer. Proc. 33, 707--711.

Ichinoe, M. & Kurata, H. (1964). Fine structure of endoconidia of *Thielaviopsis basicola*. Trans. Mycol. Soc. Japan, 5, 9--15.

Iizuka, H. (1955). The electron microscopic investigation on classification of conidia of the genus *Aspergilli*. J. Gen. Microbiol. 1, 1--17.

Ingold, C. T. (1967). Liberation mechanisms of fungi. In Airborne microbes (ed. P. H. Gregory, J. L. Monteith), Society for General Microbiology, Symp. No. 17. Cambridge University Press.

Ingold, C. T. (1971). Fungal Spores: Their Liberation and Dispersal, 302 pp. Oxford: Calendon Press.

Ito, Y., Nozawa, Y., Suzuki, M. & Setoguti, T. (1970). Surface structure of dermatophyte as seen by the scanning electron microscope. Sabouraudia, 7, 270--272.

Joly, P. (1964). Le genre *Alternaria*. Encyclopedia mycologique XXXIII. 250 pp. Paris: Lechevalier.

Jones, D. (1967). Examination of mycological specimens in the scanning electron microscope. Trans. Br. Mycol. Soc. 50, 690--691.

Jones, D. (1968a). An electron microscope study of the fine structure of *Acremoniella velata*. Trans. Br. Mycol. Soc. 51, 515--518.

Jones, D. (1968b). Surface features of fungal spores as revealed in a scanning electron microscope. Trans. Br. Mycol. Soc. 51, 608--610.

Jones, D. (1971). Frozen-etched spores of *Trichoderma viride*. Trans. Br. Mycol. Soc. 57, 348--351.

Jones, D. & Johnson, R. P. C. (1970). Ultrastructure of frozen fractured and etched pycnidiospores of *Coniothyrium minitans*. Trans. Br. Mycol. Soc. 55, 83--87.

Jones, G. & Lee, H. P. (1974). Production of secondary conidia by *Septoria tritici* in culture. Trans. Br. Mycol. Soc. 62, 212--213.

Jurand, M. K. & Kemp. R. F. O. (1972). Surface ultrastructure of oidia in the basidiomycete *Psathyrella coprophila*. J. Gen. Microbiol. 72, 575--579.

Kendrick, W. B. (1971). Taxonomy of Fungi Imperfecti, 309 pp. Toronto: Univ. of Toronto Press.

Kenneth, R. (1964). Conidial release in some *Helminthosporia*. Nat. (Lond.) 202, 1025--1026.

Khan, S. R. & Aldrich, H. C. (1973). Conidiogenesis in *Termitaria synderi* (Fungi Imperfecti). Can. J. Bot. 51, 2307--2314.

Kiffer, E., Mangenot, F. & Reisinger, O. (1971). Morphologie ultrastructurale et critères taxinomiques chez les Deuteromycetes. IV. *Doratomyces purpureofuscus* (Fres.) Morton et Smith. Rev. Ecol. Biol. Sol. 8, 397--407.

Konishi, Y. (1971). Fine structure of *Aspergillus niger*. An electron microscope study. Jap. J. Dermatol. 81, 277--291.

Kunoh, H. & Akai, S. (1967). An electron microscopic observation of the surface structure of conidia of *Erysiphe graminis hordei*. Trans. Mycol. Soc. Jap. 8, 77--79.

Kuo, M. J. & Alexander, M. (1967). Inhibition of the lysis of fungi by melanins. J. Bacteriol.94, 624--629.

Lane, J. W. R., Garrision, R. G. & Field, M. F. (1969). Ultrastructural studies on the yeastlike and mycelial phases of *Sporotrichum schenkii*. J. Bacteriol. 100, 1010--1019.

Langeron, M. & Van Breuseghen, R. (1952). Precis de Mycologie, 703 pp. Paris: Masson & Co.

Lazar, V. (1972). Contribution to the electron microscope study of the genus *Aspergillus*. Rev. Roumain Biol. Bot. 17, 357--361.

Le Gal, M. & Mangenot, F. (1960). Contribution a l'étude des Mollisioidées IV. (3° série). Rev. Mycol. 26, 263--331.

Leonard, K. J. & Suggs, E. G. (1974). *Setosphaeria prolata*, the ascigerous state of *Exserohilum prolatum*. Mycologia. 66, 281--297.

Locci, R. (1972). Scanning electron microscopy of ascosporic *Aspergilli*. Riv. Patol. Veg. Suppl. Vol. 8.

Locci. R. & Bisiach, H. (1972). Studies on powdery mildews. I. Scanning electron microscopy of *Sphaerotheca fuliginea*. Riv. Patol. Veg. 8, 239--249.

Locci, R. & Quaroni, S. (1972). Investigations on fungal features by scanning electron microscopy. V. Further studies on ascosporic *Aspergillus* species. Riv. Patol. Veg. 4, 253--320.

Lowry, R. J., Durkee, T. L. & Sussman, A. S. (1967). Ultrastructural studies of microconidium formation in *Neurospora crassa*. J. Bacteriol. 94, 1757--1763.

Luttrell, E. S. (1963). Taxonomic criteria in *Helminthosporium*. Mycologia 55, 643--674.

Madelin, M. F. (1966). The genesis of spores of higher fungi. In The Fungus Spore (ed. M. F. Madelin), pp. 15--36. London: Butterworths.

Mangenot, F. & Reisinger, O. (1973). L'initiation du primordium chez les thalloconidies et les phialoconides. In Proceedings of the International Symposium on the Taxonomy of Fungi. Madras (in press) (ed. B. S. Mehrotra) Cramer, Lehre.

Marchant, R. (1966a). Wall structure and spore germination in *Fusarium culmorum*. Ann. Bot. 30, 821--830.

Marchant, R. (1966b). Fine structure and spore germination in *Fusarium culmorum*. Ann. Bot. 30, 441--445.

Martin, J. F., Uruburu, F. & Villanueva, J. R. (1973). Ultrastructural changes in the conidia of *Penicillium notatum* during germination. Can. J. Microbiol. 19, 797--801.

Mason, E. W. & Ellis, M. B. (1953). British species of *Periconia*. Mycol. Pap. 56, Kew. 127 pp.

Matsui, S., Nozu, M., Kikumoto, T. & Matsura, M. (1962). Electron microscopy of conidial cell wall of *Cochliobolus miyabeanus*. Phytopathology 52, 717--718.

McCoy, E. C., Girard, A. E. & Kornfeld, J. M. (1971). Fine structure of resting and germinating *Penicillium chrysogenum* conidiospores. Protoplasma 73, 443--456.

McKeen, W. E., Mitchell, N., Jarvie, W. M. & Smith, R. (1966). Electronmicroscopy studies of conidial walls of *Sphaerotheca macularis*, *Penicillium levitum* and *Aspergillus niger*. Can. J. Microbiol. 72, 427--428.

McMillan, J. H. & Healey, I. N. (1971). A quantitative technique for the analysis of the gut contents of Collembola. Rev. Ecol. Biol. Sol., 8, 295--300.

Meredith, D. S. (1963). Violent spore release in some Fungi Imperfecti. Ann. Bot. 27, 39--47.

Meyer, F. H. (1970). Abbau von Pilzmycel im Boden. Z. Pflernähr. Bodenk. 127. 193--199.

Mills, J. T. (1970). Morphology of conidia of *Cochliobolus sativus* from untreated and fungicide-treated barley seed. Can. J. Bot. 48, 541--546.

Minoura, K. (1964). Taxonomic studies of the *Cladosporia*. III. Morphological properties (Part I). J. Ferment. Techn. 12, 723--738.

Minoura, K. (1966). Taxonomic studies on Cladosporia. IV. Morphological properties (Part II). J. Ferment. Techn. 4, 137--149.

Morgan-Jones, G., Nag Raj, T. R. & Kendrick, W. B. (1972). Icones genera Coelomycetarum I. Biology Series 3, 42 pp. Univ. of Waterloo .

Nambodiri, A. N. (1966). Electron micropscopic studies on the conidia and hyphae of *Neurospora crassa*. Caryologica 19, 117--133.

Nicot, J. (1958). Une moisissure du littoral Atlantique, *Dendryphiella arenaria* n.sp. Rev. Mycol. 23, 87--99.

Nicot, J. & Leduc, A. (1957). Mise en évidence d'un mucilage dans la paroi des spores de *Trichothecium roseum* Link ex. Fr. C.R. Acad. Sc. Paris, D, 244, 1403--1405.

Nival, M. (1972). Étude du *Marssonina salicicola* en microscopie électronique. Ann. Univ. A.R.E.R.S. 10, 57--66.

Oláh, G. M. (1972). L'ontongenie sporale et l'ultrastructure de la paroi chez *Paecilomyces berlinensis*. Can. J. Microbiol. 18, 1471--1475.

Oláh, G. M. & Reisinger, O. (1973). Etude ultrastructurale et comparative de la genèse des phialoconides subendogènes et des annelloconides. In Proceedings of the International Symposium on the Taxonomy of Fungi. Madras. (ed. B. S. Mehrotra) Cramer, Lehre (in press).

Oláh, G. M. & Reisinger, O. (1974a). L'ontongénie des téguments de la paroi sporale en relation avec le sterigmate et la gouttelette hilaire chez quelques Agarics melanosporés. C. R. Acae Sci. Paris D, 278, 2755--2758.

Oláh, G. M. & Reisinger, O. (1974b). Étude ultrastructurale et cytochimique de l appareil sporifère chez *Phialophora richardsiae* (Nannf.) Conant. Can. J. Bot. 52, 2473--2480.

Old, K. M. & Robertson, W. M. (1969). Examination of conidia of *Cochliobolus sativus* recovered from natural soil using transmission and scanning electron microscopy. Trans. Br. Mycol. Soc. 53, 217--221.

Old, K. M. & Robertson, W. M. (1970). Effects of lytic enzymes and natural soil on the fine structure of conidia of *Cochliobolus sativus*. Trans. Br. Mycol. Soc. 54, 343--350.

Old, K. M. & Schippers, B. (1973). Electron microscopical studies of chlamydospores of *Fusarium solani* f. *cucurbitae* formed in natural soil. Soil. Biol. Biochem. 5, 613-620.

Oliver, P. T. P. (1972). Conidiophore and spore development in *Aspergillus nidulans*. J. Gen. Microbiol. 73, 45--54.

Orcival, J. (1968). Aspect infrastructuraux des spermogonies et de la formation des spermaties chez *Puccinia poarum* Niels. J. Microsc. 7, 48.

Oujezdsky, K. B., Grover, S. N., & Szaniszlo, P. J. (1973). Morphological and structural changes during the yeast-to-mold conversion of *Phialophora dermatitidis*. J. Bacteriol. 113, 468--477.

Park, D. & Robinson, P. M. (1969). Sporulation in *Geotrichum candidum*. Trans. Br. Mycol. Soc. 51, 213--222.

Plumb, R. T. & Turner, R. H. (1972). Scanning electron microscopy of *Erysiphe graminis*. Trans. Br. Mycol. Soc. 59, 149--178.

Poole, T. B. (1959). Studies on the food of Collembola in a Douglas fir plantation. Proc. Zool. Soc. London, 132, 71--82.

Pore, R. S., Pyle, C., Larsh, H. W. & Skvarla. J. J. (1969). *Aspergillus carneus* aleuriospore cell wall ultrastructure. Mycologia 61, 418--422.

Punithalingham, E. & Jones, D. (1970). Spore surface ornamentation in *Coniothyrium* species. Trans. Br. Mycol. Soc. 55, 154--156.

Rapilly, F. (1964). Valeur taxinomique de l'appareil sporifère du genre *Helminthosporium* Link. Ann. Epiphyties, 15, 257--268.

Reddy, S. M. & Bilgrami, K. S. (1969). Studies on some species of *Helminthosporium*. Ind. Phytopathol. 22, 429--437.

Reisinger, O. (1967). Sur *Conoplea mangenotii* sp. nov. isóle à partir de branches mortes de *Rhus cotinus*. Rev. Mycol. 31, 329--340.

Reisinger, O. (1968). Remarques sur les genres *Dendryphiella* et *Dendryphion*. Bull. Soc. Mycol. Fr. 84, 27--51.

Reisinger, O. (1970a). Morphologie ultrastructurale et critères taxinomiques chez les Deuteromycetes. III. Étude au microscope électronique (à balayage et à transmission) de *Wardomyces pulvinata* (Marchal) Dickinson. Acta Phytophathol. 5, 221--230.

Reisinger, O. (1970b). Étude aux microscopes électroniques à balayage et à transmission de la paroi sporale et de son rôle dans la dispersion active des conidies chez *Helminthosporium spiciferum* (Bain.) Nicot. C.R. Acad. Sci. Paris D, 170, 3031--3032.

Reisinger, O. (1972). Contribituon à l'étude ultrastructurale de l'appareil sporifère chez quelques Hyphomycètes à paroi mélanisee. Genèse, modificatons et décomposition. Thèse Sciences Nat. Univ. Nancy, France, 2 vol. 360 pp.

Reisinger, O. & Guedenet, J. C. (1968). Morphologie ultrastructurale et critères taxinomiques chez les Deuteromycètes. I. Les parois sporales chez *Dendryphiella vinosa* (Berk. et Curt.) Reisinger. Bull. Soc. Mycol. Fr. 84, 19--26.

Reisinger, O. & Kiffer, E. (1974). Contribition to the fungal microflora of the Congo. IV. *Polyschema congolensis* sp. nov. taxonomy and ultrastructure. Trans. Br. Mycol. Soc. 62, 289--294.

Reisinger, O. & Kilbertus, G. (1973). Biodegradation et humification. III. Libération des granules. Modèle experimental en présence des bactéries. Conclusions générales. Soil. Biol. Biochem. 5, 187--192.

Reisinger, O. & Kilbertus, G. (1974). Biodegradation et humification. IV. Microorganismes intervenant dans la décomposition des cellules d'*Aureobasidium pullulans* (de Bary) Arnaud. Can. J. Microbiol. 20, 299--306.

Reisinger, O., Kilbertus, G. & Olàh, G. M. (1974). Etudes ultrastructurales du developpement des conidies d'une souche de *Aureobasidium pullulans* Bull. Acad. Soc. Lorr. Sci. 13, 103--111.

Reisinger, O. & Mangenot, F. (1969). Analyses morphologiques au microscope électronique à balayage et étude de l'ontogénie sporale chez *Dendryphiella vinosa* (Berk. et Curt.) Reisinger. C.R. Acad. Sci. Paris D, 269, 1843--1845.

Reisinger, O.& Mangenot, F. (1973). Etude ultrastructurale du développement chez quelques dématiés. Recherche de critères taxonomiques essentiels. In Proceedings of the International Symposium on the Taxonomy of Fungi. Madras. In Press. (ed. B. S. Mehrotra) Cramer, Lehre.

Reisinger, O. & Olàh, G. M. (1974). Étude ultrastructurale et cytochimique de la conidiogenèse chez *Beauveria bassiana*. Can. J. Microbiol. 20, 1387--1392.

Richmond, D. V. & Pring, R. J. (1971). Fine structure of *Botrytis fabea* Sardina conidia. Ann. Bot. 35, 175--182.

Rijkenberg, F. H. J. & Truter, S. J. (1974). The ultrastructure of sporogenesis in the pycnial stage of *Puccinia sorghi*. Mycologia 66, 319--326.

Roquebert, M. F. (1973). Modalités de la sporogenèse chez deux phialidés *Aspergillus tamarii* Kita et *Stilbothamnium nudipes* Haum. Rev. Mycol. 38, 3--8.

Roquebert, M. F. & Abadie, M. (1973). Etude ultrastructurale de la sporogenèse chez un micromycete *Stilbothamnium nudipes* Haum. C.R. Acad. Sci. Paris D, 276, 2883--2885.

Sassen, M. M. A., Remsen, C. C. & Hess, W. M. (1967). Fine structure of *Penicillium megasporum* conidiospores. Protoplasma, 64, 75--88.

Schaller, F. (1950). Biologische Beobachtungen an Humus bildenden Bodentieren, insbesondere an Collembolan. Zool. Jehrb. Abt. Syst. Oakol. 78, 506--525.

Schneider, E. F. & Seaman, W. L. (1974). Development of conidial chlamydospores of *Fusarium sulphureum* in distilled water. Can. J. Microbiol. 20, 247--254.

Schuster, R. (1956). Der Anteil der Oribatiden an den zersetzungsvorgängen im Boden. Z. Morph. Ökol. Tiere, Bd. 45, 1--33.

Schwinn, F. J. (1968). Die Darstellung von Pilzsporen im Raster-Elektronenmikroskop. Phytopathol. Z. 64, 375--379.

Seiji, M. & Itakura, H. (1967). Enzyme inactivation by ultraviolet irradiation and protective effect of melanin *in vitro*. J. Invest. Dermatol. 47, 507--511.

Shearer, C. A. & Motta, J. J. (1973). Ultrastructure and conidiogenesis in *Conioscypha* (Hyphomycetes). Can. J. Bot. 51, 1747--1751.

Sheridan, J. E. & Troughton, J. H. (1973). Conidiophores and conidia of the kerosene fungus *Cladosporium resinae* in the light and scanning electron microscopes. N. Z. J. Bot. 11, 145--152.

Shoemaker, R. A. (1959). Nomenclature of *Dreshslera* and Bipolaris grass parasites segregated from *Helminthosporium*. Can. J. Bot. 37, 879--887.

Singh, S. B. (1964a). Periodicity in the feeding habit of *Tomocerus longicornis* (Müller) Collembola. Ind. J. Entomol. 24, 123--124.

Singh, S. B. (1964b). A preliminary observation on the food and feeding habits of some Collembola. Entomologist, 97, 153--154.

Sleytr, V., Adam, H. & Klaushofer, H. (1969). Die Feinstruktur der Konidien von *Aspergillus niger* v. Tiegh. dargestellt mit Hilfe der Gefrierättechnik. Mikroskopie 25, 320--333.

Slifkin, M. K. (1971). Conidial wall structure and morphology of *Alternaria spp*. J. Ellisha Mitchell Sci. Soc. 87, 231--236.

Smith, J. M. & Sandler, W. J. V. (1971). The surface structure of saprophytic and parasitic dermatophyte spores. Mycopathol. Mycol. Appl. 43, 153--159.

Steele, S. D. & Fraser, T. W. (1973). Ultrastructural changes during germination of *Geotrichum candidum* arthrospores. Can. J. Microbiol. 19, 1031--1034.

Stevenson, I. L. & Becker, S. A. E. W. (1972). The fine structure and development of chlamydospores of *Fusarium oxysporum*. Can. J. Microbiol. 18, 997--1002.

Stiers, D. L., Rogers, J. D. & Russel, D. W. (1973). Conidial state of *Poronia punctata* Can. J. Bot. 51, 481--484.

Subramanian, C. V. (1962). The classification of the Hyphomycetes. Bull. Bot. Surv. Ind. 4, 249--259.

Subramanian, C. V. (1972). Conidial chains, their nature and significance in the taxonomy of Hyphomycetes. Curr. Sci. 41, 43--49.

Suguiyama, J. (1967). Mycoflora in core samples from stratigraphic drillings in Middle Japan. II. The genus *Aspergillus*. J. Fac. Sci. Tokyo, 9, 377--405.

Sugiyama, J, Kawasaki, Y. & Kurata, H. (1968). *Wardomyces simplex* a new Hyphomycete from milled rice. Bot. Mag. Tokyo, 81, 243--250.

Sutton, B. C. & Sandhu, D. K. (1969). Electron microscopy of conidium development in *Cryptosporiopsis sp., Phoma fumosa, Melanconium bicolor,* and *M. apiocarpum*. Can. J. Bot. 47, 745--749.

Swart, H. J. & Griffiths, D. A. (1974a). Australian leaf-inhabiting fungi. IV. Two Coelomycetes on *Acacia pycnantha*. Trans. Br. Mycol. Soc. 62, 151--161.

Swart, H. J. & Griffiths, D. A. (1974b). Australian leaf-inhabiting fungi. V. Two species of *Seimatosporium* on *Eucalyptus*. Trans. Br. Mycol. Soc. 62, 359--366.

Tanaka, K. & Yanagita, T. (1963). Electron microscopy on ultrathin sections of *Aspergillus niger*. II. Fine structure of conidia bear ing apparatus. J. Gen. Appl. Microbiol. 9, 189--203.

Tokunaga, M., Tokunaga, J. I. & Harada, K. (1973). Scanning and transmission electron microscopy of sterigmate and conidiophore formation in *Aspergillus* group. J. Electron Microsc. Japan, 22, 27--38.

Trinci, A. P. J. & Collinge, A. J. (1974). Spore formation in nitrogen and carbon starved cultures of *Geotrichum candidum* and *Mucor racemosus*. Trans. Br. Mycol. Soc. 62, 351--358.

Trinci, A. P. J., Peat, A. & Banbury, G. H. (1968). Fine structure of phialide and conidiospore development in *Aspergillus giganteus*. Ann. Bot. 32, 241--249.

Tsao, P. H. & Bricker, J. L. (1970). Acropetal development in the "chain" formation of chlamydospores of *Thielaviopsis basicola*. Mycologia 62, 960--966.

Tsao, P. W. & Tsao, P. H. (1970). Electron microscopic observations on the spore wall and "operculum" formation in chlamydospores of *Thielaviopsis basicola*. Phytopathology 60, 613--616.

Tsukahara, T. (1968). Electron microscopy of swelling and germinating conidiospores of *Aspergillus niger*. Sabouraudia 6, 185--191.

Tsukahara, T. (1970). Electron microscopy of conidiospore formation in *Aspergillus niger*. Sabouraudia 8, 93--97.

Tsukahara, T., Yamada, M. & Itakagi, K. (1966). Micromorphology of conidiospores of *Aspergillus niger* by electron microscopy. Jap. J. Microbiol. 10, 93--107.

Tubaki, K. (1963). Taxonomic study of Hyphomycetes. Annu. Rep. Inst. Ferment., Osaka, 1, 25--54.

Turian, G. (1971). *Thea 22-punctata* et autres Coccinelles micromycetophages. Nature du pigment élytral jaune. Bull. Soc. Entomol. Suisse, 44, 277--280.

Turian, G. & Bianchi, E. D. (1972). Conidiation in *Neurospora*. Bot. Rev. 38, 119--154.

Turian, G., Martin, C. & Bianchi, D. (1970). Conidiation par bourgeonnement-fission (blastro-arthrosporogenèse) et réorganisation microfibrillaire de la paroi chez *Neurospora*. C.R. Acad. Sci. Paris D, 271, 1756--1758.

Van Gool, A., Meyer, J. & Lambert, R. (1970). The fine structure of frozen etched *Fusarium* conidiospores. J. Micrrosc. 9, 653--660.

Vannier, G. (1970). Réaction des microarthropodes aux variations de l'état hydrique du sol. 319 pp. Paris: Ed. C.N.R.S.

Visset, M. F. (1972). Les formes conidiennes du complex *Microsporum gypseum* observées en microscopie électronique à balayage. Sabouraudia 10, 191--192.

Visset, M. F. & Vermeil, C. (1973). Contribution à la connaissance des fungi keratinophiles de l'Ouest de la France. V. Étude au microscope électronique a balayage. Mycopathol. Mycol. Appl. 49, 89--100.

Vujičić, R. & Muntanjola-Cvetkovič, M. (1973). A comparative ultra-structureal study on conidium differenciation in Cladosarium-like mutant 22B of *Aspergillus aureolatus*. J. Gen. Microbiol. 79, 45--51.

Wallwork, J.A. (1958). Notes on the feeding behavior of some forest soil Acarina. Oikos, 9, 160--271.

Walsh, J. H. (1970). A rapid method for the inspection of phialospore surfaces of *Trichoderma* by electron microscopy. Trans. Br. Mycol. Soc. 55, 491--493.

Webley, D. M. & Jones, D. (1971). Biological transformation of microbial residus in soil. pp. 446--485, (ed. McLarer, D. & Skujins, J.) In Soil Biochemistry, Vol. 2, New York: Marcell Dekker.

Webster, J. (1966). Spore projection in *Epicoccum* and *Arthrinium*. Trans. Br. Mycol. Soc. 49, 339--343.

Weisberg, S. H. & Turian, G. (1971). Ultrastructure of *Aspergillus nidulans* conidia and conidial lomasomes. Protoplasma 72, 55--67.

White, J. A., Calvert, O. H. & Brown, M. F. (1973). Ultrastructure of the conidia of *Helminthosporium maydis*. Can. J. Bot. 51, 2006--2008.

Wong, J. N. F. & Old, K.M. (1974). Electron microscopical studies of the colonization of conidia of *Cochliobolus sativus* by soil microorganisms. Soil Biol. Biochem. 6, 89--96.

Zacharuk, R. Y. (1970). Fine structure of the fungus *Metarrhizium anisopliae* infecting three species of larval Elateridae (Coleoptera). I. Dormant and germinating conidia. J. Invert. Pathol. 15, 63--80.

Zaki, A. I., Eckert, J. W. & Endo, R. M. (1973). The ultrastructure of germinating conidia of *Penicillium digitatum* inhibed by sec-butylamine. Pest. Biochem. Physiol. 3, 7-13.

Zeidler, G. & Margalith, P. (1973). Modification of the sporulation cycle in *Penicillium digitatum* (Sacc.). Can. J. Microbiol. 19, 481--483.

DISCUSSION

Form and Function of Conidia As Related To Their Development

Chairman: G. M. Murray
University of Wisconsin
Madison, Wisconsin

The source and use of the term "iterative germination" was questioned. Dr. Reisinger said that the terms "microcyclic conidiation" and "precocious sporulation" have been used to describe this phenomenon.

Dr. Oláh commented that he and his co-workers had noticed a remarkable similarity in the formation of the various wall layers of basidiospores and conidia examined in their laboratory. A question was raised as to whether there is a similar homologous relationship between wall formation in conidia and in ascospores. However, it appears that ascospore walls generally are formed centrifugally while basidiospore and conidium walls are formed centripetally. Dr. Reisinger observed that ascospores are endogenous spores while conidia and basidiospores are exogenous.

Dr. Frederick asked whether wall layer formation in ascospores could be highly variable. In some, deposition of wall layers is centrifugal, in others, centripetal, and in a few, the last layer may form in between the first two layers formed. He has found in *Neurospora* that the wall layers are centripetal in formation, with the first layer forming outside and subsequent layers within this layer.

D

Workshop

CHAPTER 18 FOSSIL FUNGAL SPORES

W. C. Elsik
Exxon Company

INTRODUCTION

M. F. Madelin (personal communication, 1974) has suggested . . . that it might prove a little restrictive to call the Discussion Topic "Taxonomy of Fossil Fungal Spores": if it were just "Fossil Fungal Spores" it might embrace biological aspects and permit speculation about the sorts of environmental conditions prevailing at the time the spores lived, and about phylogeny.

This view in fact sums up the practical aspect of all paleontological studies---to further elucidate the geologic history of earth through evidences of paleoecology, paleogeography, and phylogeny = evolution or stratigraphic succession. A wealth of applicable data is available through study of fossil fungal spores. Prior to details and generalizations of the significance of these data, however, an easily applied taxonomy of the fungal material is required.

The primary objectives of this paper are:

1. to emphasize the morphologic variety of fossil fungal spores; and
2. to generate a discussion of taxonomic treatment of fossil fungal spores as dispersed entities in fossil sediments.

Emphasis is placed on fossil because of the almost ubiquitous occurrence of modern fungi and the current strong belief among paleontologists that fungal spores are not worthy of study because of the probability of recent contamination.

The occurrence of an abundance and variety of dispersed fungal material in Cenozoic rocks is an established fact and requires elucidation only in detail. Fungal material of the type discussed here also occurs in nonmarine Cretaceous rocks, particularly coals or other sediments rich in carbonaceous material. Current schemes of classification are inadequate for the bulk of this dispersed fungal material.

What are fossil fungal spores? By definition, they are ANY fungal spore, whether *in situ* or dispersed, in sedimentary rocks. *In situ* spores pose less of a taxonomic problem as parent and sometimes host material is available for study; the spores then are classified in existing systems of nomenclature, whether natural or artificial.

For practical purposes in palynology, the term "fossil fungal spores" or "fossil fungal material" implies dispersed spores, microscopic sporangia, hypha or fragmented mycelia, and morphologically related material, some of which may be algal in origin. In the absence of definitive or exclusive fungal spore morphology a further requirement in the recognition of dispersed fungal spores is the presence of the pigment melanin. Dispersed fungal spores are common constituents of some palynologic preparations. Palynology is the study of all microfossils present in insoluble organic residues. The residues are obtained from the enclosing rocks through chemical and mechanical isolation.

FOSSIL FUNGAL SPORES

BIOLOGIC SIGNIFICANCE

Paleoecology

The genus and even some species of *Hypoxylon* is distributed worldwide. Some living species are restricted to the North Temperate Zone or occur outside that area only as varieties. Some species are restricted to local geographic regions. The tropics apparently support more varieties. These conclusions regarding *Hypoxylon* (Miller, 1961) will have validity in the fossil record once it is demonstrated that species diversity is reflected in spore morphology.

The relative abundance of the form genus *Exesisporites* during the Pliocene (Elsik, 1969) is apparently due to a warming climate. Fluctuations before and afterward therefore support interpretations of a cyclic climate. Other stratigraphically and geographically restricted fungal spores are known but likewise have not been identified to a modern counterpart.

Schopf (1946) noted in a study of coal from the Coos Bay Field, Oregon, that ". . . most evident microfossils are the remains of fungi; sclerotia occur in particular abundance . . ." What is the significance of abundant fungal material in coals? Perhaps delineation of specific types of fungi and fungal remains will be needed for precise interpretations. However, Cohen (1973) notes that fungal remains (presclerotinites) are more abundant in modern peats derived from environments with trees (*Cyrilla* and *Taxodium* tree islands and swamps in Georgia as contrasted to herb dominated environments (open marshes, island fringes, or glades).

The distribution of fungal spores in diverse modern marine sediments has been investigated by Müller (1959), Traverse & Ginsburg (1966) and Cross et al. (1966). Unfortunately fungal spores were not differentiated but treated as a group, even so Müller (1959) demonstrated the virtual restriction to the onshore Orinoco delta or very nearshore area of the bulk of fungal material. An inverse relationship, with fungi more abundant than microplankton very near shore and increasing microplankton offshore, has also been noted by Sparks (1967) and Frederiksen (1969). It must be noted therefore, that paleoecologic studies of fossil fungal spore distribution are just beginning.

Phylogeny

The phylogeny of fungal spore types is readily apparent from a study of detailed stratigraphic occurrences. Detailed stratigraphic occurrences for the most part have not been recorded due to general scepticism of valid fossil occurrence plus the fact that an easily applied taxonomy has not been available. Varma & Rawat (1963) were the first to establish the stratigraphic utility in the Tertiary of diporate fungal spores, although they were not certain of the identity of the material they were describing.

Elsik (1969) illustrated the relative abundance of the form genus *Exesisporites* and spores of *Hypoxylon* type in the Neogene of the offshore Louisiana area; following (Elsik, 1970b) with the correlation in several wells from the same area of the Pliocene interval based upon the relative abundance of *Exesisporites* and alder pollen (*Alnus*). The abstract to the latter paper (Elsik, 1970b) is reproduced here:

> Fossil fungal spores are rare to abundant in Cenozoic sediments. Many morphologic types are established by the late Mesozoic and others appear at various intervals of the Cenozoic. One stratigraphically important group, the diporate aseptate to triseptate fungal spores, have generally been misinterpreted as diporate angiosperm pollen. *Fusiformisporites* and related longitudinally ribbed forms appear to be restricted to the Cenozoic. The species exhibit short ranges. Distinctive offshoots of the *Fusiformisporites* line are 1) diporate, aseptate spores and 2) diporate, monoseptate spores, both possessing prominent longitudinal ribs. Additional study is required to elucidate in detail the evolution of this group.
>
> Fungal spores of the extant *Hypoxylon* type and the form genus *Exesisporites* are ubiquitous, especially in Neogene sediments. Specimens of both types, whether from Venezuela, the Gulf Coast, or Alaska, are identical. Relative frequency diagrams of *Exesisporites* from offshore Louisiana are useful for both local and regional correlation.

The compilation of detailed occurrence records of fossil species is in progress. Tentative identifications are possible, that is, *Fusiformisporites* spores are produced by at least extant *Cookeina*. Detailed occurrence data will allow interpretation of phylogenetic relationships of, for example, longitudinally striate and ribbed spores such as *Striatetracellaeites*, *Fusiformisporites*, *Striadiporites*, *Striasporonites*, and the verrucate *Verrusporonites*. The sequential appearance in time of morphologically related spores is already apparent.

Morphology

Wolf (1966a & b; 1967a,b,c,d; 1968a & b; 1969a,b,c,d; 1970) Wolf & Cavaliere (1966) and Wolf & Nease (1970) discussed the fungal spore floras of African lake sediments and other areas, stressing

1. The universal occurrence of many kinds of fungal spores,
2. The favorable conditions for preservation of fungal spores in sediments,
3. The enormous difficulty in identifying fossil fungal spores,
4. That fungal spores document vegational changes, and
5. That analyses of fungal spores can be used to supplement pollen analyses.

The series of line drawings in those papers serve to illustrate the morphologic diversity of Pleistocene to Recent fungal spores.

Spore morphology is just as variable in older sediments (Figs. 1--3). Many of the modern types were already established by the beginning of the Cenozoic. The fossil spores do have two characteristics which are at variance with extant material:

1. The fossilization/laboratory process has obviously removed some wall layers. Light microscopy reveals generally one and, at most, two layers present in the fossil spore. As already indicated, it appears the melanin bearing layer remains.
2. The lack of accessory wall layers is also reflected in the apparent inability to stain fossil spores with safranin-O.

Illustrations of actual specimens are part of a more comprehensive paper in preparation. These examples (Figs. 1--3) illustrate the great variety of fossil fungal material available for study and demonstrate the need for an applicable taxonomy. The dispersed spore is reduced to the following morphologic aspects:

1. Shape and symmetry.
2. Absence or presence and number of apertures.
3. Absence or presence and number of septae.
4. Wall layering and ornament.

TAXONOMY

Is the morphology of an individual spore diagnostic for the complete organism? Are the Fungi Imperfecti of loose enough definition to include isolated spores? M. F. Madelin (personal communication,

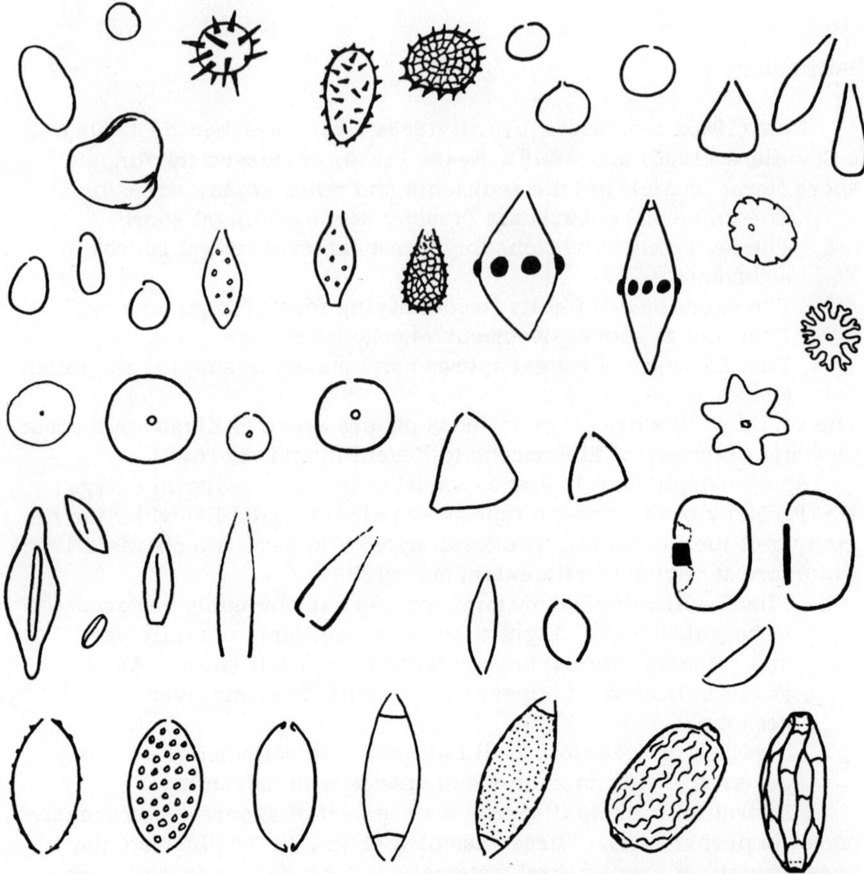

Figs. 1--3 Selected morphologic types of fossil fungal spores and related material. Not to scale. Size ranges from 2 to 4 μm for the smaller inaperturate spores to +600 μm for the largest multiseptate and peltate material. The drawings are loosely arranged in order of increasing complexity.

1974) is of the opinion that

> . . . the morphology of fungal spores is probably not diagnostic for the complete organism, at least at the level of our present knowledge . . . even experienced aerobiologists often have to lump isolated spores together in rather imprecise groups.

A partial analogy can be drawn from the state of knowledge of imperfect Ascomycetes -- even in a group as well studied as the ascomycetous

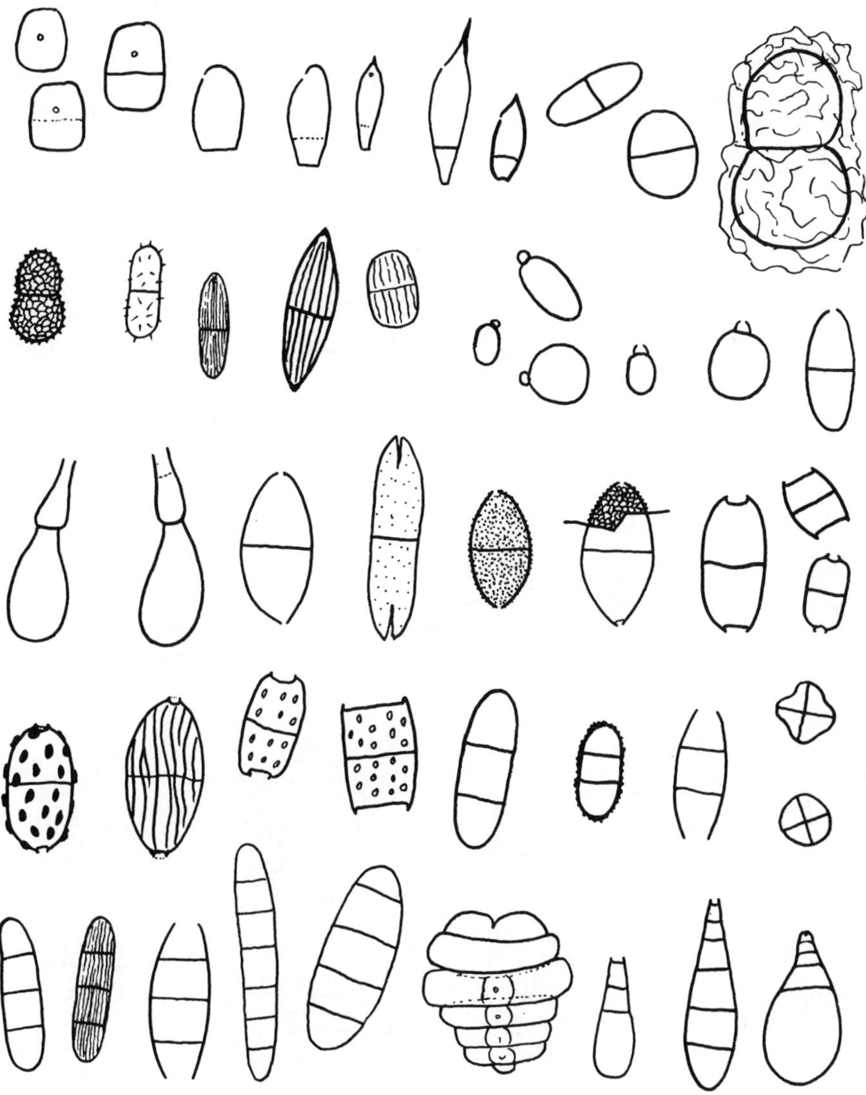

Fungi Imperfecti ontogenetic relationships to the perfect states have only in recent times become of more than isolated possibility (Hughes, 1971a).

More widespread use of the scanning electron microscope is indicated for the study of spore surface features; renewed importance of spore surface morphology in taxonomy is a relatively recent result of

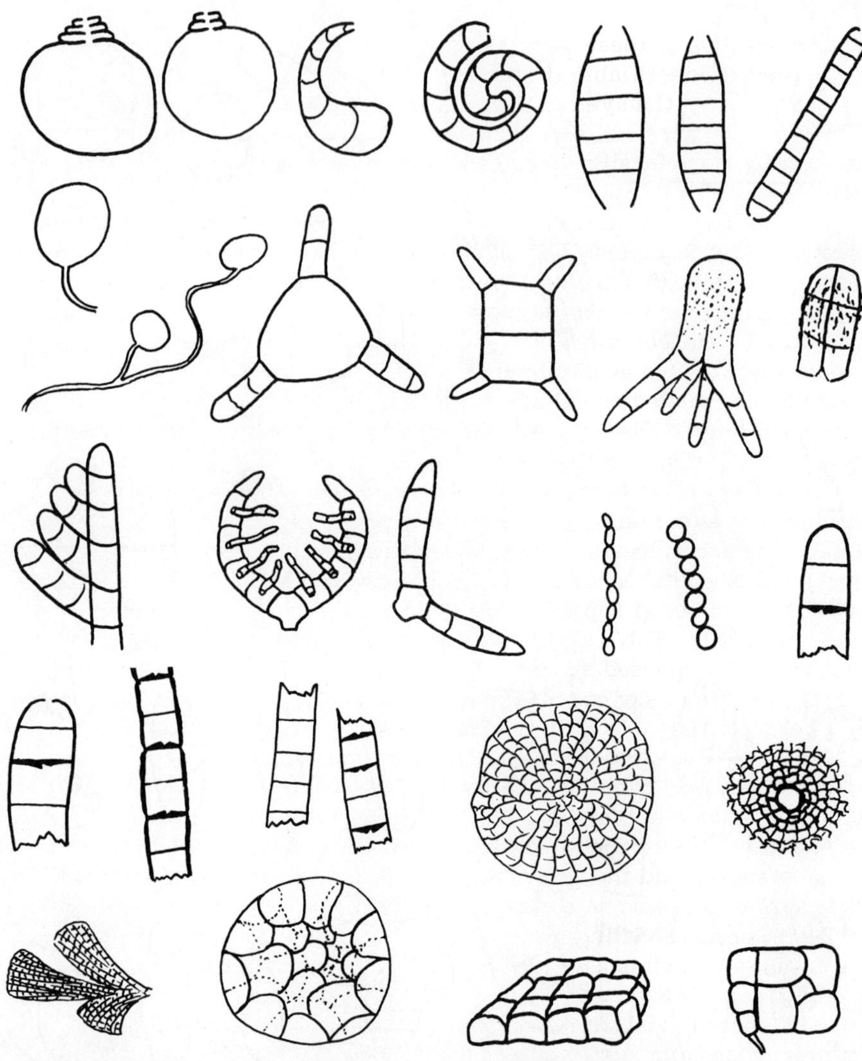

SEM studies (H. R. Hohl, personal communication, 1974; Pegler & Young, 1971). Spores of some fungi, especially conidia and ascospores,

> . . . commonly possess such distinctive and stable morphologic features that identification should be possible, in many instances, from isolated spores. Fossilization should result in excellent

> preservation of these features . . . Fossilized spores . . . should be readily identifiable to genus and in large measure to species . . . (Lafayette Frederick, personal communication, 1974).

Assuredly more fossil material will be moved to the natural system with time.

Fossil fungal material is presently assignable in part to a natural system, Phycomycetes, Ascomycetes, or Basidiomycetes, where diagnostic morphologies are demonstrated. Some fossil material apparently are assignable to the Fungi Imperfecti, where spores or isolated fruiting structures (conidia, pycnidia, or other sporangia) are of exclusive morphology. Also available under the Fungi Imperfecti is Mycelia Sterilia for isolated mycelial structures. Or Fungi Fossiles is available for fossil fungi. Ideally most of the Fungi Fossiles could be classified as Fungi Imperfecti as they are usually recovered *in situ* or *in toto*.

Although all imperfect states belong to the Fungi Imperfecti, relatively little attention has been given to other than ascomycetous fungi, and in the ascomycetous Fungi Imperfecti conidium ontogeny currently has a prime significance in the circumscription of form genera (Hughes, 1971a). The Fungi Imperfecti have been profusely categorized by Hughes (1952, 1953, 1971b), Gilman (1957), and Barnett (1955, 1960). But what of dispersed spores? Madelin (personal communication, 1974) has this to say concerning their taxonomic treatment:

> I personally doubt if the present groupings of the Fungi Imperfecti would stretch to accommodate taxa for isolated spores, but one might well consider erecting a form-order for these alongside the Mycelia Sterilia. The form-class artificiality would do no harm! It all depends on the size of the problem posed by isolated spores and the price one is prepared to pay (in taxonomic terms) to solve it. A conservative approach would be to erect large, accommodating genera in the Fungi Fossiles which Meschinelli writes about in Saccardo's Sylloge Fungorum Vol. 10 (1892), p. 741.

Dispersed fossil fungal spores for some time have been placed in a form-systematic system above the generic level with spores of higher plants, including pollen, by van der Hammen (1954a & b, 1956), Potonie (1960), Varma & Rawat (1963) and Reuda-Gaxiola (1969). Reuda-Gaxiola (1969), however, does take fossil material of algal and fungal origin and classify them separately under the term Sporonites, elevating Sporonites to suprageneric level.

The term *Sporonites* is of questionable standing because of usage:

It was not formally proposed but both fungal and higher plant spores were assigned to it (Jansonius, personal communication, 1974). Other than *Sporonites* (Jansonius, personal communication, 1974), form genera for fossil fungal spores were not widely used by palynologists until the appearance of van der Hammen's papers (1954a,b, 1956); see also Clarke (1965); Elsik (1968, 1970a); Sheffy & Dilcher (1971); Elsik & Dilcher (1974); Elsik & Jansonius (1974).

Granted a certain degree of artificiality, the erection of a Fungi Sporae Dispersae is more naturally a part of Fungi Imperfecti than Meschinelli's (1892, 1902) Fungi Fossiles. Hence the following supra classification is proposed:

Class Fungi Imperfecti (form class)
- Existing form orders
- Order Fungi Sporae Dispersae, new order
- Order Mycelia Sterilia.

Most dispersed fungal material would be classified as either Fungi Sporae Dispersae and/or Mycelia Sterilia.

Upon what morphologic features should the greatest significance be placed for identity? The two most constant features are cell number (septation) and presence or absence of apertures. Both are variable to some extent; some fossil species exhibit a variable septum, that is, *Echmasporites*, *Tetraploa*, or *Cardinalsporonites*, but these are exceptions to the general condition. Subsidiary features are spore size, shape, symmetry, and surface texture.

Both cell number and apertural condition appear to be equally important morphologic features in the fossil material. Accordingly, the following classification is proposed:

Fungi Sporae Dispersae*
- Sporae Monocellae†
- Sporae Monodicellae†
- Sporae Dicellae†
- Sporae Tricellae†
- Sporae Tetracellae†
- Sporae Multicellae†
- Sporae Cellae Indeterminatae†

Mycelia Sterilia
- Cellae†
- Hyphae†
- Peltae†
- Indeterminae†

*New form order.
†New form family.

The additional characters of aperture, shape and surface ornament are used to define genera within the proposed families; those characters are also useful in keying to form genera.

SUMMARY

Fossil fungal spores and morphologically related material are common in Cenozoic sediments. Genera and species are definable on morphologic character including overall shape or symmetry, apertures, septation, and ornament. A suprageneric classification is possible on character of apertures and septation.

Sclerotia, hypha from broken mycelia and certain other cellular structures are accommodated by the Order Mycelia Sterilia of the Class Fungi Imperfecti. A new order, Fungi Sporae Dispersae, is proposed for all dispersed fungal spores.

Detailed occurrence records allow interpretation of phylogenetic trends of fungal form genera. For example, *Fusiformisporites* appears in probable mid-Paleocene sediments and persists to the Recent in extant forms, that is, *Cookeina*. Species of *Fusiformisporites* develop in an orderly procession through the Cenozoic; *Striatetracellaeites*, possibly *Striasporonites* and less likely *Verrusporonites* evolved from this line of longitudinally ribbed spores.

REFERENCES

Barnett, H. L. (1955). Illustrated Genera of Imperfect Fungi, 218 pp. Minneapolis: Burgess.

Barnett, H. L. (1960). Illustrated Genera of Imperfect Fungi. 2nd ed., 218 pp. Minneapolis: Burgess.

Clarke, R. T. (1965). Fungal spores from Vermejo Formation coal beds (Upper Cretaceous) of central Colorado. Mountain Geol. 2, 85--93.

Cohen, A. D. (1973). Petrology of some recent peat sediments from the Okefenokee swamp-marsh complex of southern Georgia. Geology 1, 25.

Cross, A. T., Thompson, G. G., & Zaitzeff, J. B. (1966). Source and distribution of palynomorphs in bottom sediments, southern part of Gulf of California. Marine Geol. 4, 467--524.

Elsik, W. C. (1968). Palynology of a Paleocene Rockdale lignite,

Milam County, Texas. I. Morphology and taxonomy. Pollen Spores 10, 265--313.

Elsik, W. C. (1969). Late Neogene palynomorph diagrams, northern Gulf of Mexico. Trans. Gulf Coast Assoc. Geol. Soc. 19, 509--528.

Elsik, W. C. (1970a). Palynology of a Paleocene Rockdale lignite, Milam County, Texas. III. Errata and taxonomic revisions. Pollen Spores 12, 99--101.

Elsik, W. C. (1970b). Fungal spores in stratigraphy. Geol. Soc. America, Abstracts with Programs 4, 283.

Elsik, W. C. & Dilcher, D. L. (1974). Palynology and age of clays exposed in Lawrence Clay Pit, Henry County, Tennessee. Palaeontographica B 146 , 65--87.

Elsik, W. C. & Jansonius, J. (1974). New genera of Paleogene fungal spores. Can. J. Bot. 52, 953--958.

Frederiksen, N. O. (1969). Stratigraphy and palynology of the Jackson Stage (Upper Eocene) and adjacent strata of Mississippi and western Alabama. Unpublished Ph. D. dissertation, University of Wisconsin, 356 pp.

Gilman, J. C. (1957). A Manual of Soil Fungi. 2nd ed., 450 pp. Ames: Iowa State College Press.

Hughes, S. J. (1952). Fungi from the Gold Coast. I. Mycol. Papers, Commonwealth Mycol. Inst. 48, 1--91.

Hughes, S. J. (1953). Conidiophores, conidia and classification. Can. J. Bot. 31, 577--659.

Hughes, S. J. (1971a). Phycomycetes, Basidiomycetes, and Ascomycetes as Fungi Imperfecti. In Taxonomy of Fungi Imperfecti (ed. B. Kendrick), pp. 7--36. Toronto and Buffalo: University Toronto Press.

Hughes, S. J. (1971b). Annellophores. In Taxonomy of Fungi Imperfecti (ed. B. Kendrick), pp. 132--140. Toronto and Buffalo: University Toronto Press.

Meschinelli, A. (1892). Sylloge fungorum fossilium. In Saccardo, P. A., Sylloge Fungorum, Vol. 10, pp. 741--809. Paravii.

Meschinelli, A. (1902). Fungorum Fossilum Omnium Iconographia. 144 pp. Vicetiae: J. Galla.

Müller, J. (1959). Palynology of Recent Orinoco delta and shelf sediments. Micropaleontology 5, 1--32.

Pegler, D. N. & Young, T. W. K. (1971). Basidiospore Morphology in the Agaricales. 210 pp. Lehre: J. Cramer.

Potonie, R. (1960). Synopsis der Gattungen der Sporae dispersae

III. Beih. Geol. Jb. 39, 1--189.

Rueda-Gaxiola, J. (1969). Una nueva classifacacion morfologico-sistematica para polenasporas fosiles. Nomenclatura y parataxonomia. Inst. Mexicano Petroleo Pub. 69. AG/048, 166 pp.

Schopf, J. M. (1946). Botanical aspects of coal petrology: coal from the Coos Bay field in southwestern Oregon. Am. J. Bot. 33, 833.

Sheffy, M. V. & Dilcher, D. L. (1971). Morphology and taxonomy of fungal spores. Palaeontographica B 133, 34--51.

Sparks, D. M. (1967). Microfloral zonation and correlation of some Lower Tertiary rocks in southwest Washington and northwest Oregon. Unpublished Ph. D. thesis, Michigan State University, 164 pp.

Traverse, A. & Ginsburg, R. N. (1966). Palynology of the surface sediments of Great Bahama Bank, as related to water movement and sedimentation. Marine Geol. 4, 417--459.

Van der Hammen, T. (1954a). El desarrollo de la flora Colombiana en los periodos geologicos. I. Maestrichtiano hasta Terciario mas inferior. Bol. Geol. 2(1), 49--106.

Van der Hammen, T. (1954b). Principios para la nomenclatura palinologica sistematica. Bol. Beol. 2(2), 3--24.

Van der Hammen, T. (1956). A palynological systematic nomenclature. Bol. Geol. 4(2-3), 63--101.

Varma, C. P. & Rawat, M. S. (1963). A note on some diporate grains recovered from Tertiary horizons of India and their potential marker value. Grana Palynologica 4, 130--139.

Wolf, F. A. (1966a). Fungus spores in East African lake sediments. Bull. Torrey Bot. Club 93, 104--113.

Wolf, F. A. (1966b). Fungus spores in East African lake sediments. II. J. Elisha Mitchell Sci. Soc. 82, 57--61.

Wolf, F. A. (1967a). Fungus spores in East African lake sediments. IV. Bull. Torrey Bot. Club 94, 31--34.

Wolf, F. A. (1967b). Fungus spores in East African lake sediments. V. Mycologia 59, 397--404.

Wolf, F. A.(1967c). Fungus spores in East African lake sediments. VI. J. Elisha Mitchell Sci. Soc. 83, 113--117.

Wolf, F. A. (1967d). Fungus spores in East African lake sediments. VII. Bull. Torrey Bot. Club 94, 480--486.

Wolf, F. A. (1968a). Fungus spores in Lake Singletary sediments. J. Elisha Mitchell Sci. Soc. 84, 227--232.

Wolf, F. A. (1968b). Disintegration of fungus spores. J. Elisha Mitchell Sci. Soc. 84, 328--332.

Wolf, F. A. (1969a). Non-petrified fungi in late Pleistocene sediment from eastern North Carolina. J. Elisha Mitchell Sci. Soc. 85, 41--44.

Wolf, F. A. (1969b). A rust and an alga in Eocene sediment from western Kentucky. J. Elisha Mitchell Sci. Soc. 85, 57--58.

Wolf, F. A. (1969c). Naming fossil fungi. Assoc. Southeastern Biol. 16, 106--108.

Wolf, F. A. (1969d). Disintegration of fungal spores. II. J. Elisha Mitchell Sci. Soc. 85, 92--94.

Wolf, F. A. (1970). Reflections regarding nonpetrified fossils. Assoc. Southeastern Biol. 17, 92--94.

Wolf, F. A. & Cavaliere, S. S. (1966). Fungus spores in East African lake sediments. III. J. Elisha Mitchell Sci. Soc. 82, 149--154.

Wolf, F. A. & Nease, F. R. (1970). Nonpetrified fossils from the site of a phosphate mine in eastern North Carolina. J. Elisha Mitchell Sci. Soc. 86, 44--48.

E

Milestones and Predictions of Future Trends in Fungal Spore Research

A. S. Sussman (University of Michigan, USA)

First, I should like to echo the thanks to Dr. Hawker, whose feelings I know are shared by the rest of us who could not speak for the group. As for the symposium itself, among the very valuable things I learned was the complexity of systems we are working with, and I have come around to the point of view that perhaps it is naive to expect in the fungi--that even a dichotomy, like that between memnospores and the xenospores, or between constitutive and exogenous dormancy, is enough to really distinguish between the spore types. I would like to propose that, perhaps, there ought to be more integration between the physiological and morphological in such discriminations. For example, among constitutively dormant spores here already are ways of making a further dichotomy, that is to say, one between spores that swell or do not swell. Furthermore, some of the discrepancies in the literature that were alluded to by Dr. Smith and others in terms of mitochondrial changes might be resolved by a thorough study of further dichotomies among the different spore types. It would be very useful in the future, for a paper to be given six years hence in which these dichotomies and a detailed typification of spores would be the subject of a review.

L. E. Hawker (University of Bristol, United Kingdom)

I would like to see in the next meeting--some consideration given to some of the events immediately leading up to spore formation. I found it extremely difficult to prepare my paper without reference to sporogenesis. I tried very rigorously to exclude developmental factors, but all my life I have been particularly interested in the development of spores. A lot of people are working on that now all over the world, so I think if we could perhaps have one or two papers on that side, particularly the actual swelling, maturation, elaboration, and ornamentation of the spore itself, not going right back to all the physiology that leads up to it but just some of these end points, it would be valuable. Another point is really elaborating what Dr. Sussman has said, and that is that I do feel, and I have always been laying this down to my own students, that the electron microscope should be used wherever possible to support a story, to support physiology, and to support developmental studies. That aspect could be emphasized even more in a future symposium.

G. Turian (University of Geneva, Switzerland)

I agree with Dr. Hawker as I could not dissociate the mature ascomycetous spores from their differentiation factors in the preparation of my chapter. In the future, more could be devoted to the processes of sporogenesis and their triggering or switching agents.

W. M. Hess (Brigham Young University, USA)

Let me add a comment to that. When we were attempting to put a program together we intended to include sporulation as a major topic and it probably would have been better if we would have, but the advice of the people with whom we communicated and corresponded, was that we should not attempt to include the gross aspects of sporulation, but that sporulation ought to be a separate symposium. That is the reason why we did not have it as a major item for the symposium. I agree with the comments, and it was obvious by the discussion that Darrell and I presented, that it is almost impossible to discuss some aspects without discussing differentiation.

D. Gottlieb (University of Illinois, USA)

I don't know why my good friends suddenly turn timid today. I would say that the next session should be devoted entirely, if not mainly, to sporogenesis including biochemical and electron microscopy studies. The whole symposium, by then, could well deserve to be entitled "Sporogenesis."

W. M. Hess (Brigham Young University, USA)

I assume that you are suggesting that sporogenesis should be the topic without adding restrictions or that you may also discuss dormancy and germination if necessary.

D. Gottlieb (University of Illinois, USA)

Such relationships could be discussed, but mainly as they affect or are affected by sporogenesis.

C. Y. Yang (Asian Vegetable and Development Center, Taiwan)

I agree with Dr. Gottlieb. In the next symposium there should be one session on synchronization of the spore, because it is important to connect the ultrastructural studies and physiology and biochemistry of the spore.

E. C. Cantino (Michigan State University, USA)

My crystal ball tells me that one of the waves of the future in mycology is going to be inter- and intraorganelle relationships, irrespective of the stage of development of a fungus, hence it seems to me that one area that should be dealt with is subcellular "ecology" in the fungi.

W. M. Hess (Brigham Young University, USA)

I would like to add that I think that the sophistication of electron microscopy correlated with the topic just mentioned along with the histochemical techniques that are being developed and the sophisticated biochemistry, need to be all blended together much more than they have been blended in the past. I think that this is now becoming possible. Doug Maxwell mentioned that for use of the high voltage electron microscope in Wisconsin that all you need to do, if you have a worthy project, is submit this to a committee and they may even approve travel to Wisconsin to do high voltage microscopy. You may examine thick sections or use the tilt specimen stage with the phase and dark field adaptations. Some of the possibilities are phenomonal and there is tremendous potential to correlate biochemical and histochemical data. I predict that in the next few years the sophistication of electron microscopy and biochemistry will receive more attention and if not, they should.

S. Bartnicki-Garcia (University of California, USA)

I could not help but notice that more than half of the micrographs shown at this symposium, indeed some of the most beautiful micrographs, for instance, those shown by W. M. Hess, were of surface structures. Yet it is really appalling how little we know about the biochemistry of surface structures. I would like to see more people start working on the biochemistry of surface structures. I think this is one of the most urgent and promising areas of spore research for the immediate future.

D. J. Weber (Brigham Young University, USA)

I would like to change the tenor a little bit here. As I think about this group, and meditate about their interests, I wonder about support for fungal spore research. Some of you have been very successful in maintaining support. I imagine Cantino, that you have had support for many years? I know Dr. Gottlieb has continually been able to maintain financial support for his work. Others have found that they have to bootleg their support. For example, some become involved in some plant disease problems and after they get their funds, they do the things that they really want to with fungal spores on the side. I would like to have any comments or advice about obtaining funds for spore research. Have any of you had good experience with private sources or industry? I feel it is important to obtain greater funding for spore research. I don't think any of us are short of ideas, but we are often short of the funds to do them.

J. Smith (Strathclyde University, Scotland)

I would like to make the point that I think we tend to be a little too pure in our research and we tend not to look at many of the applications our work could have to industrial problems. In Britain, several industrial concerns are quite amenable to supporting this type of work, and indeed part of my own work is now being supported by ICI. Thus, I think we should look to the applied side for research support. Applied science used to be rather a naughty word. But I think nowdays when funds are short, we have got to look elsewhere. I feel that many of us could contribute very usefully to certain industrial problems by way of the very studies that we are currently doing in developmental science. Basically, fermentation technology, the making of antibiotics, and other secondary metabolites, is really a problem of the control of the development of the organism. It is developmental mycology.

D. Evenleigh (Rutgers University, USA)

I would simply like to echo two of the former speakers. I believe that the comments on applied research by Dr. Smith are especially pertinent to funding. It amazes me that there are only two representatives from such an applied field of medical mycology. Spores have also been used directly in industrial processes, for instance, Vezina's group in Canada with steroid transformations, and Alex Zeigler's group at the Northern Regional Laboratory in Peoria, who use spores directly to form columns of enzymes. We have had no discussion of such applied approaches.

A. S. Sussman (University of Michigan, USA)

I hate to be contentious on this matter, and I don't really mean to denigrate the approach which would lead one to accept the applied as a legitimate basis for grantsmanship. On the other hand, it has been my experience that when we confront the grant situation, if we compete on an equal basis with those in other fields where the application is surely secondary and far down the road, such as in molecular biology, we are using what would be a tactic that is essential. And it is essential not only for the protection of the approaches we would like to follow, which I think are often basic, but also it establishes the field as one which is appropriate for pursuit in parallel with other fields. If we take too defeatist of a view and convert to problems which may not be central to our interest, again with a nod in the direction of those who do this quite appropriately and well, I think we lose a sense of direction, purpose, and cohesion. I would simply like to balance what has been said, with which I am in agreement up to a point, by the assertion that good research will be funded, recognizing the difficulties, when important questions are asked and meaningful experiments designed.

J. S. Lovett (Purdue University, USA)

My comment is a procedural one. I have really enjoyed myself here, but I think I would like to make a suggestion for future conferences and the way they are organized in terms of distributing the time devoted to sessions. I am sure there have been other people who have gone to Gordon Conferences; the way they are usually run is that they have an intense session all morning, then have all afternoons off and start another session in the evening. I have a very short attention span and I find that that break makes a big difference in my being alert particularly after a nice lunch. I think that this is a very satisfactory format, and I would be interested to know whether other people feel the same way. I don't mean this as a criticism of this meeting in any sense, because it has been an excellent one.

W. M. Hess (Brigham Young University, USA)

Let me just mention that we realize that we have tried to cram a lot into a relatively short period of time. When we conducted our surveys, it seemed obvious that people did not want to spend more than part of the week, or at best a week. It has been an intensive program, although I think we have only been here one evening.

J. S. Lovett (Purdue University, USA)

What I am really saying is that I think you could cover just as much territory, but instead of having people do other things in the evening, let them do these in the afternoon. I am not suggesting that you shorten your program, but rather do in the evening what you do in the afternoon and vice versa. You could have just as many topics and not need to shorten the program at all.

W. M. Hess (Brigham Young University, USA)

I think that will be very useful for the people who plan symposia in the future.

D. J. Weber (Brigham Young University, USA)

I would like to agree with you, Jim. We considered the alternative you suggested, but we had a little trouble deciding what people would do in the afternoon. Hiking is relaxing, but we finally went against the approach because we thought that the Sundance Theater and similar activities were more appealing. But I think you have a good idea.

H. C. Aldrich (University of Florida, USA)

Jim, your comment about the Gordon Conferences moves me to wonder whether the particular thing that we are doing here might be a possible topic for a future Gordon Conference. You have dealt with that more than I have. Could you comment on that perhaps?

J. S. Lovett (Purdue University, USA)

There are probably others who can comment on it better than I can. I have attended a couple of the conferences in developmental biology that are now held every other year. My impression was that they have a very restricted number of slots into which they can fit such meetings, but it certainly would be worth investigating. I believe that the developmental biologists only manage to have conferences every other year because the cell biologist group agreed to meet every other year to provide an opening. I do think it is worthwhile doing this because it is a possibility and there may be money there as well, since they are financed.

E. C. Cantino (Michigan State University, USA)

I think it is out of the question. The possibility of getting the Gordon Conference to sponsor something like this is almost nil in my judgment. They will pay the expenses of the speakers, but they are not likely to partially subsidize the expenses of 100 people or even 50 of them so as to effectively reduce their costs. In my experience, not very many people from universities show up as visitors at a Gordon Conference (except invitees).

J. S. Lovett (Purdue University, USA)

I think the difficulty of arranging it is strictly a matter of whether they have the time open, or the facilities, and the willingness to make more room. The total number of people involved is about the same, and I would guess that the financing is about the same. I presume that most of the people here have paid their own costs with the exception of the speakers. So that, in that sense, is very similar and I don't recall that the Gordon Conferences are very expensive.

E. C. Cantino (Michigan State University, USA)

That is where you are wrong, I think. There is a lot of difference between the registration (and other) costs of coming here to Provo as an uninvited speaker to participate in the spore conference and of attending a Gordon Conference. Am I not right?

J. S. Lovett (Purdue University, USA)

No, it was like $75 or $125 for the whole meeting. That's what I paid last time and I wasn't an invited speaker.

G. Turian (University of Geneva, Switzerland)

In reviewing the field of yeast ascosporogenesis, I was struck by its analogies with that of bacterial endospore formation (in fact, a research group such as Halverson's has been interested in both fields). Moreover, the yeast ascospores are among the only fungal spores considered at the Spore Conference and their outcoming Spores Volumes (ASM) are devoted mainly to bacterial sporulation. Some connecting link should be envisaged in the future between such activities and ours, without forgetting the closely related studies on algal spores. I would like to see some comparative perspectives to the sporal field.

T. Hashimoto (Loyola University, USA)

Probably as some of you know, we have a spore conference, which is not restricted in any way to the bacterial spore. It will be open to any spores that we work on. We have a conference every three years and we will have a meeting (Spore VI) in East Lansing, Michigan in October of this year. Dr. Gerhardt will be sponsoring this conference on behalf of the American Society for Microbiology.

Here we discuss, I think, essentially the same topics, dormancy, activation, germination, mechanisms of germination, sporogenesis and so forth, as we have discussed here. I really think it is very important for all of us to understand what is happening on both sides with regard to the nature of dormancy and breaking of dormancy. I really appreciated the different views expressed during this symposium because I was primarily concerned in the bacterial spores initially. So, those who are interested in participating in the Spore VI, although the date is already past for the closing, should direct any inquiry to Dr. Gerhardt at Michigan State University.

A. S. Sussman (University of Michigan, USA)

While it is true that the spore conferences are open, they are open only nominally and as you will remember from the Fifth Spore Conference, there were only a couple of papers out of some 35 or 40 which were devoted to the fungi. I happen to believe that the amount of crossover since the first two has been negligible. Nevertheless, the point made that we should be dealing with comparative sporology is a very useful one. And I would admit that bacteriologists are far ahead of us in some respects and we can learn a lot from them. I would surely hope that the next spore conference might bring invited bacteriologists to discuss the key issues that they have raised. I would also propose that we lost a great opportunity to extend the results of fungal spore work by not inviting some Algologists who may be dealing with eucaryotic spores that, not only may be parallel, but indeed may be precursors in the evolutionary sense, of fungal systems.

J. S. Lovett (Purdue University, USA)

Philosophically, I agree with the idea that we have to be aware of what is going on in other developing systems such as bacteriology and bacteriological spore differentiation, but I really feel that a group like this will lose a great deal by starting to spread out to include all kinds of different organisms. Again one can use the Gordon Conference as an example. I think that focusing in on one particular kind of problem is more beneficial than the potential generalizations when it comes to this kind of a meeting.

L. E. Hawker (University of Bristol, United Kingdom)

It is all very well for us to put down the kind of things we are interested in at the moment, but this next congress is going to be in eight years time and we have to leave it as open as possible so that the people organizing it can judge what is of importance. Some entirely new aspect may crop up between then and now, and at this congress we should not lay down a firm rule that the next one has got to be about this, that, or the other. It has to be left so that we can take advantage of whatever new lines develop.

D. A. Cotter (Southeastern Massachusettes University, USA)

I was wondering if Al Sussman's comments might also be applied to the field or protozoology. A few selected protozoologists might have insight into eucaryotic dormancy and germination which could be communicated to mycologists in a future meeting.

E. C. Cantino (Michigan State University, USA)

It seems to me this is exactly the point Jim Lovett was making in disagreeing with Al Sussman's suggestion. We will have everybody in the act, and I think thereby lose what we are striving to obtain at this symposium on the fungal spore.

A. S. Sussman (University of Michigan, USA)

The basic question is what are we striving to obtain? If we are striving to make generalizations about developmental systems through the study of spore germination, which is one of the endeavors I think we all would subscribe to, then it seems to me it is well to illuminate the systems we study by understanding what is going on in key systems elsewhere. Of course, we should not dissipate our time and efforts too widely, but invitations to a few key individuals might provide the kind of illumination of these significant issues that would help us in further studies. I am surely not advocating that half of our talks should be distributed so widely as to cut into the emphases and concentrates that we should apply in these meetings. Far from it. I think that what I have been talking about and what I think Dave Cotter and the others have meant, is that surely we need to stay au courant enough so that we learn about the approaches taken by those who deal with other creatures. So I am not advocating destroying the cohesiveness of the group by the intrusion of too many outsiders. But neither should we be xenophobic. We should really consider the possibility of getting some few outsiders in, subject of course to the restrictions voiced by Lilian Hawker.

D. J. Weber (Brigham Young University, USA)

Let me twist the discussion a little bit. As we reviewed the basidiomycetes it is obvious that the biochemists spend most of their time on a few organisms and those who do biochemical work understand why, mainly because you need a backlog of information in order to try to make sense of what is going on. Whereas the ultrastructural people tend to be very broad and are able to look at the whole random area. So the question came to my mind, should there be encouragement to have ultrastructural people concentrate more on, say, the organisms the biochemists are working on or should it be the reverse encouragement to get the biochemist to be broader?

H. C. Aldrich (University of Florida, USA)

I think certainly I have been dealing with that problem just as much as anybody in the case of the myxomycetes, because in some of the earlier work that I did, I dealt with things other than *Physarum polycephalum*, and there is really only one slime mold as far as the biochemists are concerned and that is *Physarum polycephalum*. There exists a second in addition, as far as the geneticists are concerned, and that is *Didymium iridis*. I have come around to the idea that now I am much better off doing my work on *Didymium iridis* and *Physarum polycephalum* simply because once you have done the descriptive ultrastructure, you then start to worry about how to make the organism behave and how to manipulate the organism. You are far far ahead of your own work if you have the data that has been developed by the biochemical people in terms of manipulation of the organism and availability of different genetic strains, should you want to try the effects of crosses of different genomes on the ultrastructure and so forth. This, I think, will give you an idea at least of the advantages. On the other hand, I think if we restrict ourselves too much we may lose some of the variability, but there are many advantages, in fact, in working with the things for which the biochemical and genetic data are freely available.

D. Tyrrell (Sault St. Marie, Canada)

The logic in this kind of view of course is undeniable, but I think there is an opposite point of view in that it does tend to make things relatively inward looking. People tend to work on a very small number of organisms for a variety of reasons. It is nice to work on a spore that germinates readily and is readily obtained, but as I say, it does tend to make things a little inward looking and one then gets into the position of trying to make generalities on the basis of work done on a very small number of spores. I think this gets back to the point Dr. Sussman made at the beginning of this discussion. Lots of spores that come under the heading of constitutive dormancy are still waiting to be looked at. No one knows really whether these are going to conform to the type of things which have been developed with spores which germinate readily in a matter of a few hours. The ones I work with, for instance, will germinate in 24-48 hours after placing on a suitable medium. There are many many spores that come in this category, and I think a lot more work needs to be done. I grant you it is extremely difficult and extremely tedious. This is one of the problems I have been wrestling with for a number of years, and is one area which has to be looked at to find out whether the generalities developed on the basis of conidia which germinate readily can be extended to these other spores.

A. Ostrofsky (Oregon State University, USA)

I agree with Dr. Tyrrell. Perhaps we should note that he is interested in the Entomophthorales, and I am interested in the lichens. I hope that the desire to see a greater diversity of organisms studied is not just wishful thinking.

G. Turian (University of Geneva, Switzerland)

In the background of ultrastructural and biochemical researches there is a third, fundamental dimension, which should be more densely grasped: the genetical, concerned with the fate of informational molecules during the programming of spore formation as well as the factors controlling the expression of the sporogenetic information. Much should be expected from future closer studies of conditional morphological mutants (conditionally asporic) as initiated with those already available from yeasts, *Aspergillus nidulans* and *Neurospora crassa*.

H. C. Aldrich (University of Florida, USA)

I wonder if I might explore one thing that has just occurred to me. This discussion about working with particular experimental organisms brings to mind my very fine experience with the *Physarum* research group that is, in fact, international. We have meetings every two years and exchange information through a *Physarum* newsletter. I would be interested in knowing from the people that are here how many of you deal with another small kind of research group dealing with your particular organism. For example, I am sure Dr. Cotter is familiar with the Cellular Slime Mold Newsletter. Dave, is there also a meeting every so often?

D. A. Cotter (Southeastern Massachusettes University, USA)

There will be a Northeast area conference of cellular slime mold research workers some time this fall of 1974.

H. A. Aldrich (University of Florida, USA)

The *Physarum* thing has been I think, quite successful. I know there have also been *neurospora* meetings from time to time. The *Physarum* people have now settled on meeting informally every year at the Cell Biology Meetings and then every two years at a separate meeting with formal contributed papers. Otherwise it is a very similar meeting to this.

T. Hashimoto (Loyola University, USA)

That reminds me. There is actually a "spore newsletter" being published by the Australian group. (William Murrell, CSIRO, Adelaide, Australia) It comes out every two months or so and includes all the fungal spores. So this is a very good source for knowing what is going on in all areas of spore research, and sometimes they carry even abstracts of papers published elsewhere. They make their own intensive survey of the literature covering all aspects of the spore, sporulation, dormancy, and germination. This might help you to catch up with the literature in any aspect of spore work.

J. S. Lovett (Purdue University, USA)

To get really specialized, we have been having meetings for the last six or eight years about every other year for people interested in *Blastocladiella*. We could call it international because we always have someone from Canada, but ours has been geographically limited only because of the cost. We try and arrange to have a short, informal meeting where people from laboratories in the middlewest can drive and bring all their graduate students. We have found them very successful and very useful. At our last one we had 29 people, though I suspect you may not believe that there are 29 people working on *Blastocladiella*. While I am here can I make a philosophical argument instead of standing up again? Although I recognize the problem of studying enough different fungi, I would like to suggest that what we ought to do is to do what we are doing exceedingly well so that we can get enough other people sufficiently excited that they will start studying other organisms in depth. I think the explosion of progress in molecular biology, based almost entirely on *Escherichia coli*, is an example of what you can achieve when you really exploit a few organisms in depth. Mycologists need to have this approach as well as encouraging people to look at other organisms. With bacteria, people are now looking at more diverse organisms and are finding that some of the generalizations aren't so general. But they are stimulated to do this by the first kind of concentrated activity.

D. Tyrrell (Sault St. Marie, Canada)

I think the comments that Jim has just made beautifully illustrate the point I was trying to make earlier. The in-depth studies of the *Blastocladiella* group, and various other groups, are providing the same sort of examples as the people who worked on *Escherichia coli* have done. I think as more and more spores are studied then different things will come to light. One of two things will happen; more and more different systems will come to light or the principles based on work on *Neurospora* and *Blastocladiella* will be put onto much more sound footing.

J. Smith (Strathclyde University, Scotland)

There does appear to be many specialized meetings carried out throughout the world. One of the great problems is that we never hear about all of them. They are not published. Would it not be possible to find some type of medium whereby these special conferences such as those held by the *Blastocladiella* group could be more widely advertised and some means found for getting some abstracts to the press so that those of us in other parts of the world, or even other parts of the country, can at least know what is happening in these meetings.

W. M. Hess (Brigham Young University, USA)

I might mention that the ASM (American Society of Microbiology) publishes a newsletter which contains a running list of symposia. Perhaps people who are in charge of these small groups could send information to ASM or similar societies and request to have them listed.

H. C. Aldrich (University of Florida, USA)

The Mycological Newsletter might be another medium except that it is published only twice a year and the ASM newsletter is monthly. Science and Bioscience also list meetings and conferences.

D. J. Weber (Brigham Young University, USA)

I think that is a good idea. I believe, though if I got the concept right, the value of these informal meetings is that you talk about the experiments that are almost working except for some flaws. Then you talk about data that has only been run once and you get excited about what you hope will happen. You wouldn't dare put this out in published form and yet it stimulates a small group. Am I right? Is that what you do in your meetings?

J. S. Lovett (Purdue University, USA)

Often one thing you learn is that you don't need to do something because somebody else already tried it and it didn't work! But seriously, it is of great value to see the work that is in process in other labs which may not be published for a long time.

D. J. Weber (Brigham Young University, USA)

But I think in terms of your point. Summary concepts could be made which could carry the implication of the discussion on the experiments, and if you want more details, write to someone.

R. W. Lichtwardt (University of Kansas, USA)

We are talking about two things, though. One is finding out about these meetings in advance. The other is finding out what happens after the fact.

D. J. Weber (Brigham Young University, USA)

In advance would be no problem.

D. Eveleigh (Rutgers University, USA)

This raises another point going back to trying to utilize the Gordon Conference. I think one of the values of this conference will be that it will be published, whereas in the Gordon Conferences, though also emphasizing the concept of the other small "get-together" conferences where ideas are advanced and uncompleted work is unveiled, formal publication is not allowed under their present format.

G. M. Olah (University Laval, Canada)

The domain of fungal spores is very vast and cannot be entirely covered in a one week session. We hope that the results, the thoughts and points of view that have been exposed here will be useful for the general purpose of serving Mycology. We still hope that in future conferences we will see more particular applications of the suggestions.

INDEX